MW00450542

PACIFIC SALMON & THEIR ECOSYSTEMS

A service of ITP

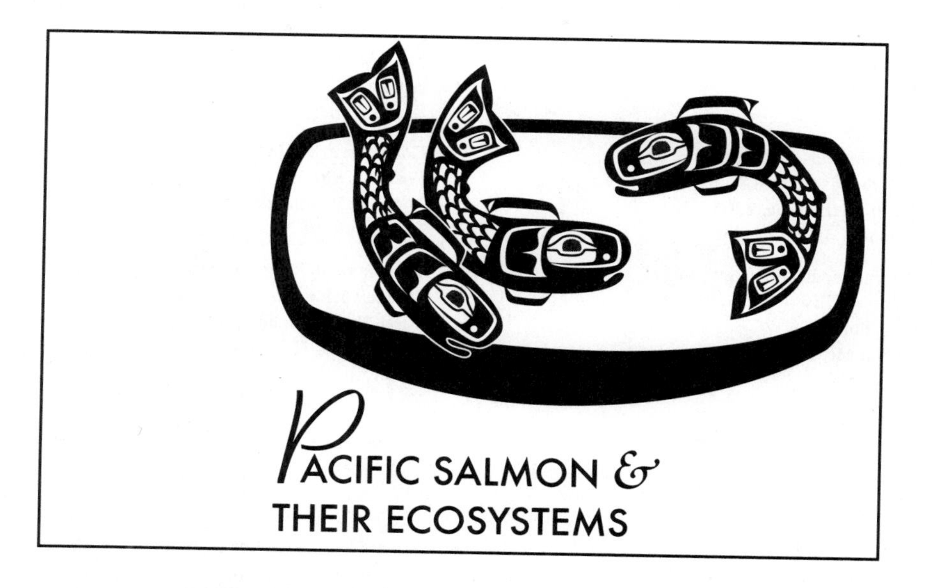

PACIFIC SALMON & THEIR ECOSYSTEMS

STATUS AND FUTURE OPTIONS

Deanna J. Stouder, Peter A. Bisson, Robert J. Naiman, Editors

Marcus G. Duke, Associate Editor

CHAPMAN & HALL

I(T)P® International Thomson Publishing

New York • Albany • Bonn • Boston • Cincinnati • Detroit • London • Madrid • Melbourne
Mexico City • Pacific Grove • Paris • San Francisco • Singapore • Tokyo • Toronto • Washington

Cover Design & Illustration: Cathy Schwartz

Printed in the United States of America

For more information, contact:

Chapman & Hall
115 Fifth Avenue
New York, NY 10003

Chapman & Hall
2-6 Boundary Row
London SE1 8HN
England

Thomas Nelson Australia
102 Dodds Street
South Melbourne, 3205
Victoria, Australia

Chapman & Hall GmbH
Postfach 100 263
D-69442 Weinheim
Germany

International Thomson Editores
Campos Eliseos 385, Piso 7
Col. Polanco
11560 Mexico D.F.
Mexico

International Thomson Publishing-Japan
Hirakawacho-cho Kyowa Building, 3F
1-2-1 Hirakawacho-cho
Chiyoda-ku, 102 Tokyo
Japan

International Thomson Publishing Asia
221 Henderson Road #05-10
Henderson Building
Singapore 0315

2 3 4 5 6 7 8 9 10 XXX 01 00 99 97

Library of Congress Cataloging-in-Publication Data

Pacific salmon & their ecosystems : status and future options / Deanna J. Stouder, Peter A. Bisson, Robert J. Naiman, editors.
p. cm.
"The symposium on Pacific salmon was held in Seattle, Washington, January 10-12, 1994"—Pref.
Includes bibliographical references and index.
ISBN 0-412-98691-4
1. Pacific Salmon. 2. Fishery conservation—Northwest, Pacific. 3. Pacific salmon fisheries—Management. 4. Pacific salmon—Habitat. I. Stouder, Deanna J. II. Bisson, Peter A. III. Naiman, Robert, J.
SH348.P33 1996
333.95'6—dc20 96-38439
CIP

British Library Cataloguing in Publication Data available

To order this or any other Chapman & Hall book, please contact **International Thomson Publishing, 7625 Empire Drive, Florence, KY 41042.** Phone: (606) 525-6600 or 1-800-842-3636.
Fax: (606) 525-7778, e-mail: order@chaphall.com.

For a complete listing of Chapman & Hall's titles, send your requests to
Chapman & Hall, Dept. BC, 115 Fifth Avenue, New York, NY 10003.

Contents

Introduction to a Complex Problem

Status of Pacific Northwest Salmonids

Analyzing Trends and Variability

Regional Trends

Technological Solutions: Cost-Effective Restoration

Institutional Solutions: Effective Long-Term Planning and Management

Where Do We Go From Here?

Contributors

Dana Atagi
BC Ministry of Environment, Lands and Parks
Skeena Region, Fisheries Branch
3726 Alfred Ave., PO Box 5000
Smithers, BC V0J 2N0
604-847-7290
Fax: 604-847-7591
Email: datagi@smithers.env.gov.bc.ca

David A. Bella
Department of Engineering
Oregon State University
Apperson Hall 202
Corvallis, OR 97331-2302
541-737-3500

Robert Beschta
Department of Forest Engineering
Peavy Hall
Oregon State University
Corvallis, OR 97331
541-737-4292
Fax: 541-737-4316
Email: beschtar@ccmail.orst.edu

Robert E. Bilby
Weyerhaeuser Company
Tacoma, WA 98477
206-924-6557
Fax: 206-924-6970
Email: bilbyb@wdni.com

Peter A. Bisson
USDA Forest Service
Olympia Forest Science Lab
3625 93rd Ave SW
Olympia, WA 98502
360-753-7671
Fax: 360-956-2346
Email: bissonp@olywa.net

Daniel L. Bottom
Oregon Department of Fish and Wildlife
Research and Development Section
850 SW 15th St
Corvallis, OR 97333
541-737-7631

Joseph L. Ebersole
Oak Creek Laboratory of Biology
Department of Fisheries and Wildlife
104 Nash Hall
Oregon State University
Corvallis, OR 97331
541-737-4531
Fax: 541-737-3590
Email: ebersolej@ucs.orst.edu

Robert C. Francis
University of Washington
School of Fisheries
Box 357980
Seattle, WA 98195-7980
206-543-7345
Fax: 206-685-7471
Email: rfrancis@pisces.fish.washington.edu

Kurt L. Fresh
Washington Department of Fish and Wildlife
Fish Management Program
600 Capitol Way, M/S 43149
Olympia, WA 98504-3149
360-902-2756
Fax: 360-902-2980
Email: freshkl@dfw.wa.gov

Christopher A. Frissell
Flathead Lake Biological Station
The University of Montana
311 Bio Station Lane
Polson, MT 59860-9659
406-982-3301
Fax: 406-982-3201
Email: frissell@selway.umt.edu

Charles F. Gauvin
Trout Unlimited
1500 Wilson Boulevard, Suite 310
Arlington, VA 22209-2310
703-522-0200

Robert G. Gibbons
Washington Department of Fish and Wildlife
600 Capitol Way North
Olympia, WA 98501-1091
360-902-2329
Fax: 360-902-2980

Gary R. Graves
Northwest Indian Fisheries Commission
6730 Martin Way East
Olympia, WA 98506
360-438-1180
Fax: 360-753-8659
Email: ggraves@nwifc.wa.gov

Stanley Gregory
Department of Fisheries and Wildlife
Oregon State University
Corvallis, OR 97331-3803
541-737-1951
Fax: 541-737-3590
Email: gregorst@ccmail.orst.edu

Robert E. Gresswell
Pacific Northwest Research Station
USDA Forest Service
3200 Jefferson Way
Corvallis, OR 97331
541-750-7410
Fax: 541-750-73229
Email: gresswer@ccmail.orst.edu

Peter F. Hassemer
Idaho Department of Fish and Game
1414 East Locust Lane
Nampa, ID 83686
208-465-8404
Fax: 208-465-8434
Email: PHASSEME@IDFG.STATE.ID.US

Robert M. Hughes
Dynamac
200 SW 35th St
Corvallis, OR 97333
541-754-4516
Fax: 541-754-4716
Email: hughesb@mail.cor.epa.gov

Mark Jennings
National Biological Service
California Science Center
Piedras Blancas Field Station
PO Box 70
San Simeon, CA 93452-0070
805-927-3893
Fax: 805-927-3308
Email: Mark_Jennings@nbs.gov

Thom H. Johnson
Washington Department of Fish and Wildlife
283236 Highway 101
Port Townsend, WA 98368
360-765-3979
Fax: 360-765-4455

Anne Kapuscinski
Department of Fisheries and Wildlife
200 Hodson Hall
University of Minnesota
St. Paul, MN 55108
612-624-3019
Fax: 612-625-5299

Sharon W. Kiefer
Idaho Department of Fish and Game
PO Box 25
Boise, ID 83707
208-334-3791
Fax: 208-334-2114
Email: skiefer@idfg.state.id.us

James F. Kitchell
Center for Limnology
680 N Park St
University of Wisconsin
Madison, WI 53706
608-262-9512
Email: kitchell@macc.wisc.edu

Kathryn Kostow
Oregon Department of Fish and Wildlife
2501 SW First Ave., PO Box 59
Portland, Oregon 97207
503-872-5252, ext. 5414
Fax: (503) 827-5632
Temporary affiliation (1996–97):
National Marine Fisheries Service
525 NE Oregon St, Suite 500
Portland, OR 97232
503-231-2007
Fax: 503-231-2318
Email: Kathryn_Kostow@ccgate.ssp.nmfs.gov

Robert T. Lackey
Environmental Protection Agency
200 SW 35th
Corvallis, Oregon 97333
541-754-4601
Fax: 541-754-4614
Email: lackey.robert@epamail.epa.gov

Kai N. Lee
Center for Environmental Studies
Williams College
PO Box 632, Kellogg House
Williamstown, MA 01267
413-597-2358
Fax: 413-597-3489
Email: Kai.N.Lee@Williams.edu

Robert Lee
College of Forest Resources
University of Washington
Box 352100
Seattle, WA 98195
206-685-0879
Fax: 206-685-0790
Email: boblee@u.washington.edu

Jim Lichatowich
182 Dory Road
Sequim, WA 98382
360-683-0748

Rich Lincoln
Washington Department of Fish and Wildlife
600 Capitol Way North
Olympia, WA 98501-1091
360-902-2325
Fax: 360-902-2980
Email: lincorhl@dfw.wa.gov

William J. Liss
Oak Creek Laboratory of Biology
Department of Fisheries and Wildlife
104 Nash Hall
Oregon State University
Corvallis, OR 97331
541-737-4631
Fax: 541-737-3590
Email: liss@ccmail.orst.edu

Dennis R. McEwan
California Department of Fish and Game
Inland Fisheries Division
1416 Ninth St.
Sacramento, CA 95814
916-653-9642
Fax: 916-654-8099
Email: 102264.3050@compuserve.com

J.D. McPhail
Department of Zoology
University of British Columbia
6270 University Blvd.
Vancouver, BC V6T 2A9
604-822-3326
Fax: 604-822-2416

Terry J. Mills
California Department of Fish and Game
Inland Fisheries Division
1416 Ninth St.
Sacramento, CA 95814
916-653-9642
Fax: 916-654-8099
Email: 76544.551@compuserve.com

Phillip R. Mundy
Fisheries and Aquatic Sciences
1015 Sher Lane
Lake Oswego, OR 97034-1744
503-699-9856
Fax: 503-636-6335
Email: mundy@teleport.com

Robert J. Naiman
School of Fisheries
University of Washington
Box 357980
Seattle, WA 98195-7980
206-685-2025
Fax: 206-685-7471
Email: naiman@u.washington.edu

Richard K. Nawa
Siskiyou Regional Education Project
P.O Box 220
Cave Junction, OR 97523
541-474-7890
Fax: 541-592-2653

Willa Nehlsen
Nehlsen & Associates
2100 SE Hemlock Ave.
Portland, OR 97214
503-235-5628
Email: nehlsenw@aol.com

Jay W. Nicholas
Oregon Department of Fish and Wildlife
28655 Highway 34
Corvallis, OR 97333
541-737-7621
Fax: 541-737-2456

T.G. Northcote
Elderstrand
10193 Giant's Head Rd.
RR2 S77B C10
Summerland, BC V0H 1ZZ0
604-494-8463
Fax: 604-494-8444

William G. Pearcy
College of Oceanic and Atmospheric Sciences
Oregon State University
Oceanography Adm. Bldg. 104
Corvallis, OR 97331-5503
541-737-2601
Fax: 541-737-2064
Email: wpearcy@OCE.ORST.EDU

Charles E. Petrosky
Idaho Department of Fish and Game
PO Box 25
Boise, ID 83707
208-334-3791
Fax: 208-334-2114
Email: cpetrosk@idfg.state.id.us

Gordon H. Reeves
USDA Forest Service
Pacific Northwest Research Station
Forestry Sciences Laboratory
3200 Jefferson Way
Corvallis, Oregon 97331
541-750-7315
Fax: 541-750-7329
Email: reevesg@fsl.orst.edu

Henry A. Regier
Department of Zoology
University of Toronto
Toronto, Canada M5S 1A4
705-924-2763

Reginald R. Reisenbichler
National Biological Service
Northwest Biological Science Center
6505 NE 65th St
Seattle, WA 98115
206-526-6282, ext. 334
Fax: 206-526-6654
Email: Reg_Reisenbichler@nbs.gov

Larry Rutter
Northwest Indian Fisheries Commission
6730 Martin Way E.
Olympia, WA 98516
360-438-1180
Fax: 360-753-8659
Email: lrutter@nwifc.wa.gov

James R. Sedell
USDA Forest Service
Pacific Northwest Research Station
Forestry Sciences Laboratory
3200 Jefferson Way
Corvallis, Oregon 97331
541-750-7315
Fax: 541-750-7329

Curt Smitch
Assistant Regional Director
US Fish and Wildlife Service
3773 MartinWay E.
Bldg. C, Suite 101
Olympia, WA 98501
360-534-9330
Fax: 360-534-9331
Email: curt_smitch@mail.fws.gov

Courtland L. Smith
Department of Anthropology
Oregon State University
Waldo Hall 238
Corvallis, OR 97331-6403
541-737-4515
Fax: 541-737-3650
Email: smithc@cla.orst.edu

Brent S. Steel
Department of Political Science
Washington State University
1812 E. McLoughlin Blvd.
Vancouver, WA 98663-3597
360-546-9731

Deanna J. Stouder
School of Fisheries
Box 357980
University of Washington
Seattle, WA 98195
Current affiliation:
Ohio Cooperative Fish & Wildlife Research Unit
Department of Zoology
The Ohio State University
1735 Neil Ave.
Columbus, OH 43210
614-688-3443
Fax: 614-292-0181
stouder.1@osu.edu

Carl Walters
Department of Zoology
University of British Columbia
Vancouver, B.C. V6T 1Z4
604-822-6320
Fax: 604-822-8934

Cindy Deacon Williams
USDA Forest Service
14th & Independence Ave. SW
Washington, DC 20250
Current affiliation:
Pacific Rivers Council
PO Box 925
Eagle, ID 83616
208-939-8697
Fax: 208-939-4086

Jack Williams
Office of the Director
Bureau of Land Management
1849 C Street NW
Washington, DC 20240
Current affiliation:
Bureau of Land Management
Intermountain Research Station
316 E. Myrtle
Boise, ID 83702
208-364-4376
Fax: 208-364-4346
Email: jwilliam@cyberhighway.net

Alex C. Wertheimer
National Marine Fisheries Service
Auke Bay Laboratory
11305 Glacier Highway
Juneau, AK 99801
907-789-6040
Fax: 907-789-6094
Email: awertheimer@abl.afsc.noaa.gov

Preface

The symposium "Pacific Salmon and Their Ecosystems: Status and Future Options" and this book resulted from initial efforts in 1992 by Robert J. Naiman and Deanna J. Stouder to examine the problem of declining Pacific salmon (*Oncorhynchus* spp.). Our primary goal was to determine informational gaps. As we explored different scientific sources, state, provincial, and federal agencies, as well as non-profit and fishing organizations, we found that the information existed but was not being communicated across institutional and organizational boundaries. At this juncture, we decided to create a steering committee and plan a symposium to bring together researchers, managers, and resource users.

The steering committee consisted of members from state and federal agencies, non-profit organizations, and private industry (see Acknowledgments for names and affiliations). In February 1993, we met at the University of Washington in Seattle to begin planning the symposium. The steering committee spent the next four months developing the conceptual framework for the symposium and the subsequent book. Our objectives were to accomplish the following: (1) assess changes in anadromous Pacific Northwest salmonid populations, (2) examine factors responsible for those changes, and (3) identify options available to society to restore Pacific salmon in the Northwest.

The symposium on Pacific Salmon was held in Seattle, Washington, January 10–12, 1994. Four hundred and thirty-five people listened to oral presentations and examined more than forty posters over two and a half days. We made a deliberate attempt to draw in speakers and attendees from outside the Pacific Northwest.

Most of the papers in this book are up-to-date versions of those talks presented at the symposium. Although authors address their own area of expertise, all look beyond the boundaries of their particular disciplines. The underlying theme of this book is to link information on the status of salmonids to the complex of management, political, social, and biological factors influencing salmon populations. There exists an urgent need to integrate information from all of these arenas to effectively restore the fish and their ecosystems. We hope that this book will enable readers to examine the problem of declining salmonids in a new and different way as well as use it as a tool to evaluate other imperiled natural resources.

Reviewers

Each paper in this book was reviewed by at least three individuals. We realize that the quality of this book is the direct result of their efforts. We would like to acknowledge the following people for their thoughtful and helpful reviews:

F.W. Allendorf, M.B. Bain, M. Black, E.L. Brannon, D.V. Buchanan, A. Byrne, J. Cederholm, D.W. Chapman, C. Coutant, R. Edwards, D.M. Eggers, D.C. Erman, D. Fluharty, M. Fraidenburg, B. Freeman, K.L. Fresh, M. Grayum, S. Hare, B. Harvey, M. Healey, G.R. Heath, J.E. Hightower, D. Huppert, G. Isbell, M. Jennings, M.R. Jennings, K Koski, G.A. Lamberti, M. Landolt, P.W. Lawson, S.A. Levin, G. Lucchetti, C. Luecke, D. Ludwig, L. MacDonald, J. Magnuson, G.K. Meffe, E. Melvin, M. Mesa, D.R. Montgomery, P.B. Moyle, K. Nelson, J. Nielsen, G.J. Niemi, J. Norris, E.P. Odum, A. Olson, S. Petit, T. Pietsch, E. Pinkerton, E.P. Pister, W.S. Platts, D. Policansky, T. Quinn, K. Rawson, B.E. Riddell, G.T. Ruggerone, C. Schreck, S. Schroder, M. Shannon, A. Shedlock, S.M. Sogard, R.A. Stein, F.J. Swanson, C. Warren, M. White, R.J. White, and R. Wissmar.

Acknowledgments

The symposium "Pacific Salmon and Their Ecosystems: Status and Future Options" and the preparation and publication of this book were possible with the generous financial support of a number of organizations and institutions. We thank the following groups for their support: American Fisheries Society, Bonneville Power Administration, Columbia River Inter-Tribal Fish Commission, Environmental Monitoring and Assessment Program of the US Environmental Protection Agency, King County Surface Water Management, National Office of Trout Unlimited, Northwest Indian Fisheries Commission, Northwest Power Planning Council, Northwest Region of the National Marine Fisheries Service, Olympic National Park, Pacific Northwest Region of the USDA Forest Service, Pacific Rivers Council, Region 10 of the US Environmental Protection Agency, Simpson Timber Company, Washington Department of Fisheries, and Washington Department of Wildlife. In addition, we also thank the following colleges and programs at the University of Washington: Center for Streamside Studies, College of Ocean and Fishery Sciences, College of Forest Resources, and Washington Sea Grant Program. Their support partially funded Deanna J. Stouder and paid for the planning, preparation, printing, and mailing of materials for the symposium, and preparation and publication of this book. During the final preparation stages, additional support for Deanna J. Stouder was provided by the Division of Cooperative Research within the National Biological Service. Weyerhaeuser Company and the Northwest Research Station of the USDA Forest Service supported the work efforts of Peter A. Bisson throughout this project.

The support, dedication, and hard work of numerous individuals made the symposium and book possible. We greatly appreciate the help of all concerned. In particular, we thank the steering committee for sharing their ideas, time, and expertise in the development and planning of the symposium. The steering committee included the three book editors and Kurt L. Fresh (Washington Department of Fisheries and Wildlife), Michael Grayum (Northwest Indian Fisheries Commission), Robert Hughes (ManTech Environmental Technology, Inc.), Robert T. Lackey (Environmental Research Laboratory, US Environmental Protection Agency), James Lichatowich (Mobrand Biometrics, Inc.), Gino Lucchetti (King County Surface Water Management), Willa Nehlsen (Pacific Rivers Council), Thomas P. Quinn (School of Fisheries, University of Washington), and Gordon Reeves (Forestry Sciences Laboratory, USDA Forest Service). We are also indebted to Louie S. Echols (Director, Washington Sea Grant), G. Ross Heath (Dean, College of Ocean and Fishery Sciences), Marsha L. Landolt (Director, School of Fisheries), and David B. Thorud (Dean, College of Forest Resources) for their encouragement and support.

This book reflects the enthusiastic and tireless efforts of our superb associate editor, Marcus G. Duke. His technical expertise enhanced the quality of the book; we appreciate his thoroughness, candor, and humor during the entire process. Marcus Duke also provided design, page layout, and typsetting services.We thank Theodore W. Pietsch for providing expertise regarding taxonomic nomenclature. Abby Simpson is greatly appreciated for her assistance in proofing final galleys of the manuscripts. Additional thanks go to Cathy Schwartz for assisting in proofing the final galleys and for designing and illustrating the wonderful logo used for the book cover.

PACIFIC SALMON & THEIR ECOSYSTEMS

Where Are We? Resources at the Brink

Deanna J. Stouder, Peter A. Bisson, and Robert J. Naiman

The population dynamics of anadromous fishes reflect the influences of biological, social, economic, and political factors. As a result, the management, restoration, and conservation of these organisms is unusually complex and challenging. In addition, a poorly developed informational network between researchers, managers, and user groups makes maintaining the vitality of these fishes difficult, especially at this critical juncture in the fate of salmon populations. Linkages need to be established among those who conduct basic research on anadromous fishes and their ecosystems, those who manage resources for recreation and commercial interests, and those who ultimately make legal decisions on the future of natural resources.

Along the Pacific coast of western North America, salmonids (*Oncorhynchus* spp.) are extremely important resources for several reasons. First, they are an important food source, not only for the region, but worldwide. Second, because salmonids migrate through thousands of kilometers, moving from streams and rivers through estuaries to the ocean and back, they provide a valuable indication of environmental conditions in those habitats. Finally, there is a strong cultural bond associated with salmonids among the peoples of this region (Schoonmaker and von Hagen 1996). Thus, these fishes provide a complicated natural resource with which to synthesize information on the complex of management, political, social, and biological factors influencing the maintenance of aquatic biodiversity.

The choices available require difficult decisions. Many salmonid stocks have been driven to or near the point of extinction, and there is little time left to make effective decisions before the options disappear. If salmonid populations are to remain healthy, then human life styles will have to change. In other words, continued land development (habitat destruction), inexpensive power (dams), inexpensive water (diversions), and extensive fisheries (unlimited harvest) are not compatible with sustained populations of salmonids and unlimited numbers of humans. A broad-based view of important factors influencing salmonids is needed. Integrating information and the desire to work across scientific and social boundaries may help select suitable options. The interaction and cooperation of separate disciplines remain critical to future management and restoration.

Salmonids in the Pacific Northwest have been widely studied. Much is known about their life history (see Groot and Margolis 1991) as well as their genetics and the importance of freshwater and estuarine habitat quality (Meehan 1991, Naiman et al. 1992, Nielsen and Lisle 1994, Naiman and Anderson 1996, National Research Council 1996) and ocean productivity (Pearcy 1992, Lawson 1993). Using salmonids as an example, this book provides a case study that integrates basic and applied sciences. In general, this case study stresses the importance of working across many disciplines and spatiotemporal scales for salmonids as well as humans. In the following text, we present a brief overview of each section of the book.

Introduction to a Complex Problem

This section introduces the complexity of the problem, discusses the regional importance of salmonids, and provides background information on salmonid biology, conservation, and management. Smitch (1996) describes the importance of salmon in the Pacific Northwest and how this further extends to a national concern. Many organizations (e.g., academic, state, federal, nonprofit, tribal) consider salmon vital to the health of the region. Regier (1996) then discusses the bases upon which conservation and management of salmonids have been set. Maintaining old traditions as foundations from which to address current issues only exacerbates the crisis for salmonids. McPhail (1996) presents the geologic history of this region, showing how it has influenced the evolutionary biology of salmonids. Because of the young geologic history of the region, as well as long periods of isolation resulting from glaciation, salmonids have uniquely responded to a suite of environmental conditions. Using the foundation provided in the introduction, the following sections address specifics associated with the regional status of salmonids.

Status of Pacific Northwest Salmonids

In this section, the authors describe the data available as well as the information needed to interpret changes in population strength, the status of salmonid populations in different regions, and factors contributing to salmonid declines. In an earlier review, Nehlsen et al. (1991) outlined the decline of salmonid populations; ~47% of the native-spawning populations in California, Oregon, Idaho, and Washington were at high risk of extinction. More recent studies have identified healthy stocks in the same region (e.g., Huntington et al. 1996) that provide opportunities for the rehabilitation and maintenance of stocks. Nehlsen (1996) sets the stage by providing an updated perspective describing stocks at risk throughout the Pacific Northwest.

Analyzing Trends: Data and Variability

Data collected by resource managers vary temporally as well as in quality and consistency. This variability creates complexity and inconsistency in evaluating changes in long-term population patterns. Nicholas (1996) presents an analysis of the types of information that scientists and managers have available to them. Characteristically, some of the available data are collected as verbal and written observations while others are collected with specific goals in mind (e.g., escapement, survival). Thus, not only are the data limited but their uses beyond original goals also are restricted. Given the limitations managers have on the quantity and quality of data, Walters (1996) describes the types of data that would be most useful. Managers and those charged with species conservation need specific kinds of data to address an array of urgent questions such as "How does one determine at what point population numbers have been impacted so that restoration is necessary, critical, or beyond hope?" Lichatowich (1996) continues the evaluation of data to present a provocative discussion on how baselines for comparing stocks today with those in the past influence the determination of the magnitude of changing population sizes. Largely dependent upon the group or organization, the baseline may be historical (>100 years) or within the past 20 years. Thus, interpretation with respect to the magnitude of population change will depend on the time scale evaluated. When we evaluate long-term population fluctuations, incorporating several data types over long temporal scales becomes critical in targeting resources for management, conservation, and rehabilitation.

Regional Trends

For Alaska, California, Idaho, Oregon, Washington, and British Columbia authors from each region briefly describe the respective assessment programs and how each defines a "population" (e.g., wild, natural, hatchery populations). The authors also discuss the tools and methods used, the kinds of available data, and the individual definitions of "baseline." The authors then examine temporal and spatial trends in chinook (*O. tshawytscha*), coho (*O. kisutch*), pink (*O. gorbuscha*), chum (*O. keta*), and sockeye (*O. nerka*) salmon, and steelhead (*O. mykiss*). Overall, it is vividly apparent that stocks in California (Mills et al. 1996) and Idaho (Hassemer et al. 1996) are most threatened, largely resulting from habitat loss (e.g., water diversion, urbanization) and habitat inaccessibility (e.g., migratory corridors blocked by dams). Salmon and steelhead in coastal regions of Washington (Johnson et al. 1996) and Oregon (Kostow 1996) are doing considerably better than stocks in the interior regions of these states (see also Huntington et al. 1996). Salmonid populations in Alaska (Wertheimer 1996) and British Columbia (Northcote and Atagi 1996) appear more stable and productive although the population numbers within some stocks are decreasing. Much of the relative success of the Alaska and British Columbia populations may be attributed to productive oceanic current regimes and the presence of intact freshwater habitats. Each regional perspective also provides suggestions for the future of salmonid stocks that incorporate responsive management regimes, habitat rehabilitation, and habitat conservation.

Factors Contributing to Stock Declines

The factors contributing to stock declines are diverse but also related. Factors include both direct (e.g., habitat degradation, harvest, predation) as well as indirect influences (e.g., oceanic cycles, hydrologic alterations). In total, these factors impact salmonid populations cumulatively, with the strength of interaction varying in a complicated manner. In other words, impacts are rarely additive and linear. One cannot remove one factor and see a direct and corresponding decrease in impact. Initially, the factors contributing to stock declines showed minimal impact. However, as the number, magnitude, and duration of these factors increased, the ecosystem and its salmonid residents could no longer successfully accommodate the accompanying environmental changes. As salmon and steelhead populations declined, populations became isolated, and artificial propagation of salmonid stocks began. In cases where few individuals contribute to the gene pool, there is a strong potential for decreased heterozygosity and changes in life-history traits (Reisenbichler 1996). As populations become more homozygous, they may become more susceptible to disease, may lose their ability to compete successfully against conspecifics and heterospecifics, and may fail to accommodate minor alterations in their habitats. Fresh (1996) examines the role of biological interactions, such as competition and predation, in the decline of salmonids. There is both direct and indirect evidence that biological interactions have become more important owing to anthropogenic environmental changes. Some of these changes include the introduction of nonnative species to habitats previously dominated by salmonids. Salmon and steelhead now encounter a different array of habitats and co-inhabitants thereby shifting the outcome of biological interactions.

Salmonid habitat has been altered by urbanization, channelization, timber practices, and a suite of other factors. Gregory and Bisson (1996) present information on how these changes influence the complex life history of these fishes. Nineteenth century human colonization and habitat development in the Pacific Northwest occurred largely along waterways; current urbanization follows a similar template. As land adjacent to streams, rivers, estuaries, and the ocean is developed, the natural inputs of sediment, woody debris, and nutrients are modified, and much of the water is diverted for other needs. Collectively, these alterations have severe impacts on salmonids.

Salmonids traverse habitats from freshwater to the ocean throughout their life, thus encountering different fisheries (commercial, recreational, and tribal) as well as different management regimes (state or provincial, national, and international). The Pacific Northwest has faced extremes in salmonid harvest; fishers of all varieties have experienced bountiful and non-existent harvests. In concert with salmonid population declines and concerns about estimating potential return, Mundy (1996) describes the difficulty in balancing preservation of the fish and the fishers.

Finally, salmonids encounter a major influence beyond human intervention—ocean productivity. Pearcy (1996) discusses the oceanography and productivity (primary and secondary) of the subarctic Pacific Ocean. Contrasting the fluctuations between the California and Alaskan currents and the associated El Niño Southern Oscillation events facilitates the association of decadal trends with salmon survival. This element of variability cannot be predicted precisely but must be anticipated for the evaluation of stock fluctuations and implementation in management policies.

Salmon Policies and Politics

The vitality of salmonid stocks is intimately related to policies and politics that influence management. Management decisions (e.g., harvest, habitat, water, and hatcheries) are often made independently whereas the fishes are influenced by all these factors throughout the various life-history stages. If salmonid management is to become more successful, it will require coordination with policy and politics.

Rutter (1996) contrasts the processes by which federal, state, and provincial harvest decisions are made. In addition, he includes information on the role of science and politics in the decision-making process and outlines the consequences of the current situation. Salmon harvest decisions include myriad users (recreational, aboriginal, and commercial) across state, provincial, and national borders (United States [US] and Canada), which creates a complex and tedious process. Similarly, Sedell et al. (1996) discuss the fractured nature of habitat management and describe an attempt by the US federal agencies to develop watershed-based management plans. Often terrestrial (e.g., owls) and aquatic organisms (e.g., fish and invertebrates) have overlapping requirements, yet management decisions are made independently. New approaches to resource management (e.g., Forest Ecosystem Management Assessment Team [FEMAT]) attempt to incorporate the full array of biodiversity into future plans.

Complications also arise when one considers how water-management and water-quality decisions are made. Gauvin (1996) provides insights into the processes of water management—how management has proceeded in the past and how it can be improved. In the past, water has been managed with little or no consideration for aquatic organisms or the importance of natural inputs (e.g., woody debris, nutrients) to the overall integrity and productivity of aquatic systems (Naiman et al. 1995a, b).

The absence of a paper on hatchery policy reflects the delicate and controversial nature of this issue. Several potential authors attempted to write papers but were either prevented from or unsuccessful in providing accurate information because of the regional and political pressures to continue using hatcheries as a means to replace lost stocks. Hatchery policies currently remain a topic of considerable debate and concern (Hilborn 1992, Meffe 1992, Stickney 1994). Some of the concern is based on salmonid biology (e.g., genetic diversity) while other aspects are based on the belief that stocks can be rehabilitated through artificial technology rather than through complex and integrative processes incorporating fish ecology with anthropogenic impacts (e.g., habitat loss, urbanization). Hatcheries must be able to provide an adequate mechanism to facilitate recovery, enabling the rehabilitation of stocks. However, it will require a coherent policy facilitated by professionally responsible personnel. This includes shifting hatchery policy from production and augmentation to a policy of the hatchery as a repository for genetic information.

Policy and management decisions are currently made in a fragmented manner. Viewing decisions as separate is dangerous because, in reality, each is intimately connected. What are the consequences of a lack of management coordination? Frissell et al. (1996) examine how to measure the performance of decisions. If the goals are to sustain healthy wild populations, then critical ecological and cultural measures need to be included.

Technological Solutions: Cost-effective Restoration

Given the fragmented nature of salmon policy and a need for integrated management, can technology and science provide us with additional tools for cost-effective restoration? Bisson et al. (1996) propose an ecological framework for setting habitat goals based on the range of conditions generated by natural disturbances. Goals center on watershed management, which includes riparian buffers, presence of large woody debris in streams and rivers, and adequate spawning and rearing habitat. By incorporating a larger-scale view of habitat restoration, the authors suggest including natural spatial and temporal change (e.g., seasonal fluctuations, climatic variation). Beschta (1996) specifically addresses the importance of riparian restoration as a critical step for improving freshwater habitat for salmonids in the Columbia River basin. While waiting for many of the habitat-related problems to be resolved, artificial propagation programs help conserve at-risk stocks. Kapuscinski (1996) evaluates how hatchery programs have functioned, whether they have been successful, and what the future holds.

As resource managers grapple with variability and its inherent uncertainty, methods are needed to assess successes and failures. Fundamental technical problems exist when managing renewable resources with "incomplete information." This somehow implies that the problems would be solved if there was "complete information." Francis (1996) stresses that resources are dynamic; chance and change are becoming regarded as fundamental aspects of the natural world. The future of salmon assessment and management includes explicitly recognizing and dealing with uncertainty. Lackey (1996) further evaluates ecological risk assessment in protecting anadromous salmonid stocks. He suggests that it may be essential to isolate the scientific basis for decision making from the policy-making arena to eliminate real or perceived biases. Using a team approach, scientists examine the risks associated with stocks of interest (i.e., those at or near extinction). Not only can risks be identified, but options available to change activities or to reduce threats to particular stocks can be made available.

In conclusion, Williams and Williams (1996) describe the development of an ecosystem-based management strategy (FEMAT 1993) that was designed to restore and maintain natural production of anadromous salmonids on public lands. This approach consists of sets of goals and objectives that provide guidelines to protect and restore watersheds. It also includes a key component, monitoring, which has been previously lacking in many restoration plans.

Institutional Solutions: Effective Long-Term Planning and Management

The fifth part of this book addresses socioeconomic factors and institutional solutions to planning and management problems. Hughes (1996) introduces this section with an overview of the philosophical basis of values (i.e., contrasting economic and ecological views), development of organizational systems, the integration of information at different scales, and using different values. The fundamental basis for institutional change is the need for resource users and managers to understand past motivations, goals, and agreements. Bottom (1996) presents a thorough review of the ideas inherent in salmonid conservation. A diverse group of people have been involved (e.g., managers, scientists, policy makers) with different motives (e.g., fish as "crops," exploitation). Understanding where salmonid management has been will allow a better evaluation of where future management and conservation should go.

Societal values, organizational systems, and institutional constraints present opposing forces to be reconciled and integrated in order to improve resource condition. Smith and Steel (1996) present the view that where our values lie—in an economic or ecological realm—will influence how resource decisions are made. Little concern is placed on ecosystem or biological preservation when resources are valued primarily for their economic value. Ecological value, however, is difficult to quantify. The problem is further complicated when one considers organizational systems and the burden of proof. Bella's (1996) premise is that organizations incorporate operational failure into the system. These systems are not structured to succeed or to resolve problems; otherwise the organizations might no longer be necessary. This is especially so in the case of providing evidence of harm. The burden of proof currently lies in proving damage after the fact rather than before an alteration or change is made to the system. The situation appears bleak for salmonids if this remains the case.

The science of salmonid ecology requires a perspective that includes the bottom of a stream (or the ocean) as well as the top of a ridge and all the physical, biological, social, and economic components in between. How do we integrate salmon, land and water stewardship, and human values? Robert Lee (1996) proposes a solution that integrates organisms and environment into society without altering our existing life-style quality. Philosophically, society should be able to accomplish this; however, reality includes cross-disciplinary awareness, interaction, and compromise.

Where Do We Go from Here?

This book's organization steps from the basic biology of salmonids to an understanding of the population trends and factors (physical, biological, political, and social) influencing them, and incorporates an attempt to understand options available for restoration and long-term survival. The last two chapters present views for the future of salmonids in the Pacific Northwest. Kitchell (1996) examines the salmon resource from the viewpoint of an "outsider," one who has been actively involved in fishery resources in other regions. He suggests accomplishment of the following: accept humans as components of ecosystems, work with allies in the political arena, and conduct small, potentially successful restoration projects. Success in any of these areas will translate into support and encouragement by the public. Kai Lee (1996), presents three similar principles for sustaining salmon from his own experience in meandering along the challenging path of salmonid restoration, rehabilitation, and conservation. He maintains that sustaining salmon will involve cooperation when conflict is inevitable, incorporation of appropriate spatial and temporal scales into action plans, and learning by experimentation (adaptive management).

In summary, if one may judge by salmonids as an example of resources at the brink, we as a society have been poor stewards. If these resources are an indication of our stewardship, natural resources in the US are at risk. However, if we use their example as a call to consider the complexity of the problem, the critical need for integration across organizations and disciplines, and a willingness to adapt and compromise as resource users, then genuine progress is possible. We hope resource management and conservation will change in such a manner that the health and biodiversity of this planet will be sustained for future generations.

Literature Cited

Bella, D.A. 1996. Organizational systems and the burden of proof, p. 617-638. *In* Stouder, D.J., P.A. Bisson, and R.J. Naiman (eds.), Pacific Salmon and Their Ecosystems: Status and Future Options. Chapman and Hall, New York.

Beschta, R.L. 1996. Restoration of riparian/aquatic systems for improved fisheries habitat in the Upper Columbia Basin, p. 475-491. *In* Stouder, D.J., P.A. Bisson, and R.J. Naiman (eds.), Pacific Salmon and Their Ecosystems: Status and Future Options. Chapman and Hall, New York.

Bisson, P.A., G.H. Reeves, R.E. Bilby, and R.J. Naiman. 1996. Watershed management and Pacific salmon: desired future conditions, p. 447-474. *In* Stouder, D.J., P.A. Bisson, and R.J. Naiman (eds.), Pacific Salmon and Their Ecosystems: Status and Future Options. Chapman and Hall, New York.

Bottom, D.L. 1996. To till the water: a history of ideas in fisheries conservation, p. 569-597. *In* Stouder, D.J., P.A. Bisson, and R.J. Naiman (eds.), Pacific Salmon and Their Ecosystems: Status and Future Options. Chapman and Hall, New York.

Forest Ecosystem Management Assessment Team. 1993. Forest ecosystem assessment: an ecological, economic, and social assessment. USDA Forest Service. Portland, Oregon.

Francis, R.C. 1996. Managing resources with incomplete information: making the best of a bad situation, p. 513-524. *In* Stouder, D.J., P.A. Bisson, and R.J. Naiman (eds.), Pacific Salmon and Their Ecosystems: Status and Future Options. Chapman and Hall, New York.

Fresh, K.L. 1996. The role of competition and predation in the decline of Pacific salmon and steelhead, p. 245-275. *In* Stouder, D.J., P.A. Bisson, and R.J. Naiman (eds.), Pacific Salmon and Their Ecosystems: Status and Future Options. Chapman and Hall, New York.

Frissell, C. A., W. J. Liss, R. E. Gresswell, R. K. Nawa, and J. L. Ebersole. 1996. A resource in crisis: changing the measure of salmon management, p. 411-444. *In* Stouder, D.J., P.A. Bisson, and R.J. Naiman (eds.), Pacific Salmon and Their Ecosystems: Status and Future Options. Chapman and Hall, New York.

Gauvin, C.F. 1996. Water management and water quality decision-making in the range of Pacific salmon, p. 389-409. *In* Stouder, D.J., P.A. Bisson, and R.J. Naiman (eds.), Pacific Salmon and Their Ecosystems: Status and Future Options. Chapman and Hall, New York.

Gregory, S.V. and P.A. Bisson. 1996. Degradation and loss of anadromous salmonid habitat in the Pacific Northwest, p. 277-314. *In* Stouder, D.J., P.A. Bisson, and R.J. Naiman (eds.), Pacific Salmon and Their Ecosystems: Status and Future Options. Chapman and Hall, New York.

Groot, C. and L. Margolis (eds.). 1991. Pacific Salmon Life Histories. University of British Columbia Press, Vancouver, British Columbia.

Hassemer, P.F., S.W. Kiefer, and C.E. Petrosky. 1996. Idaho's salmon: can we count every last one? p. 113-125. *In* Stouder, D.J., P.A. Bisson, and R.J. Naiman (eds.), Pacific Salmon and Their Ecosystems: Status and Future Options. Chapman and Hall, New York.

Hilborn, R. 1992. Hatcheries and the future of salmon in the Northwest. Fisheries 17(1): 5-8.

Hughes, R. 1996. Do we need institutional change? p. 559-568. *In* Stouder, D.J., P.A. Bisson, and R.J. Naiman (eds.), Pacific Salmon and Their Ecosystems: Status and Future Options. Chapman and Hall, New York.

Huntington, C., W. Nehlsen, and J. Bowers. 1996. A survey of healthy native stocks of anadromous salmonids in the Pacific Northwest and California. Fisheries 21(3): 6-14.

Johnson, T.H., R. Lincoln, G.R. Graves, and R.G. Gibbons. 1996. Current status and future of wild salmon and steelhead in Washington state, p. 127-144. *In* Stouder, D.J., P.A. Bisson, and R.J. Naiman (eds.), Pacific Salmon and Their Ecosystems: Status and Future Options. Chapman and Hall, New York.

Kapuscinski, A.R. 1996. Rehabilitation of Pacific salmon in their ecosystems: what can artificial propagation contribute? p. 493-512. *In* Stouder, D.J., P.A. Bisson, and R.J. Naiman (eds.), Pacific Salmon and Their Ecosystems: Status and Future Options. Chapman and Hall, New York.

Kitchell, J.F. 1996. Where do we go from here? An outsider's view, p. 657-663. *In* Stouder, D.J., P.A. Bisson, and R.J. Naiman (eds.), Pacific Salmon and Their Ecosystems: Status and Future Options. Chapman and Hall, New York.

Kostow, K. 1996. The status of salmon and steelhead in Oregon, p.145-178. *In* Stouder, D.J., P.A. Bisson, and R.J. Naiman (eds.), Pacific Salmon and Their Ecosystems: Status and Future Options. Chapman and Hall, New York.

Lackey, R.T. 1996. Is ecological risk assessment useful for resolving complex ecological problems? p. 252-540. *In* Stouder, D.J., P.A. Bisson, and R.J. Naiman (eds.), Pacific Salmon and Their Ecosystems: Status and Future Options. Chapman and Hall, New York.

Lawson, P.W. 1993. Cycles in ocean productivity, trends in habitat quality, and the restoration of salmon runs in Oregon. Fisheries 18(8): 6-10.

Lee, K.N. 1996. Sustaining salmon: three principles, p. 665-675. *In* Stouder, D.J., P.A. Bisson, and R.J. Naiman (eds.), Pacific Salmon and Their Ecosystems: Status and Future Options. Chapman and Hall, New York.

Lee, R.G. 1996. Salmon, stewardship, and human values: the challenge of integration, p. 639-654. *In* Stouder, D.J., P.A. Bisson, and R.J. Naiman (eds.), Pacific Salmon and Their Ecosystems: Status and Future Options. Chapman and Hall, New York.

Lichatowich, J. 1996. Components of management baselines, p. 69-87. *In* Stouder, D.J., P.A. Bisson, and R.J. Naiman (eds.), Pacific Salmon and Their Ecosystems: Status and Future Options. Chapman and Hall, New York.

McPhail, J.D. 1996. The origin and speciation of *Oncorhynchus* revisited, p. 29-38. *In* Stouder, D.J., P.A. Bisson, and R.J. Naiman (eds.), Pacific Salmon and Their Ecosystems: Status and Future Options. Chapman and Hall, New York.

Meehan, W.R. (ed.). 1991. Influences of forest and rangeland management on salmonid fishes and their habitats. American Fisheries Society Special Publication 19. Bethesda, Maryland.

Meffe, G.K. 1992. Techno-arrogance and halfway technologies: salmon hatcheries on the Pacific coast of North America. Conservation Biology 6: 350-354.

Mills, T.J., D. McEwan, and M.R. Jennings. 1996. California salmon and steelhead: beyond the crossroads, p. 91-111. *In* Stouder, D.J., P.A. Bisson, and R.J. Naiman (eds.), Pacific Salmon and Their Ecosystems: Status and Future Options. Chapman and Hall, New York.

Mundy, P.R. 1996. The role of harvest management in the future of Pacific Salmon populations: shaping human behavior to enable the persistence of salmon, p. 315-329. *In* Stouder, D.J., P.A. Bisson, and R.J. Naiman (eds.), Pacific Salmon and Their Ecosystems: Status and Future Options. Chapman and Hall, New York.

Naiman, R.J. and E.C. Anderson. 1996. Ecological implications of latitudinal variation in watershed size, streamflow, and water temperature in coastal temperate rain forest of North America. *In* Schoonmaker, P. and B. von Hagen (eds.), The Rain Forests of Home: An Exploration of People and Place. Island Press, Washington, DC.

Naiman, R. J., T. J. Beechie, L. E. Benda, D. R. Berg, P. A. Bisson, L. H. MacDonald, M. D. O'Connor, P. L. Olson, and E. A. Steel. 1992. Fundamental elements of ecologically healthy watersheds in the Pacific Northwest coastal ecoregion, p. 127-188. *In* Naiman, R.J. (ed.). Watershed Management: Balancing Sustainability and Environmental Change. Springer-Verlag, New York.

Naiman, R.J., J.J. Magnuson, D.M. McKnight, and J.A. Stanford (eds.). 1995a. The Freshwater Imperative: A Research Agenda. Island Press, Washington, DC.

Naiman, R.J., J.J. Magnuson, D.M. McKnight, J.A. Stanford, and J.R. Karr. 1995b. Freshwater Ecosystems and Management: A National Initiative. Science 270: 584-585.

National Research Council. 1996. Upstream: Salmon and Society in the Pacific Northwest. National Academy Press, Washington, DC.

Nehlsen, W.J. 1996. Pacific salmon status and trends—a coastwide perspective, p. 41-50. *In* Stouder, D.J., P.A. Bisson, and R.J. Naiman (eds.), Pacific Salmon and Their Ecosystems: Status and Future Options. Chapman and Hall, New York.

Nehlsen, W., J.E. Williams, and J.A. Lichatowich. 1991. Pacific salmon at the crossroads: stocks at risk from California, Oregon, Idaho, and Washington. Fisheries 16(2): 4-21.

Nicholas, J.W. 1996. On the nature of data and their role in salmon conservation, p. 53-60. *In* Stouder, D.J., P.A. Bisson, and R.J. Naiman (eds.), Pacific Salmon and Their Ecosystems: Status and Future Options. Chapman and Hall, New York.

Nielsen, J.L. and T.E. Lisle. 1994. Thermally stratified pools and their use by steelhead in northern California streams. Transactions of the American Fisheries Society 123: 613-626.

Northcote, T.G. and D.Y. Atagi. 1996. Pacific salmon abundance trends in the Fraser River watershed compared with other British Columbia systems, p. 199-219. *In* Stouder, D.J., P.A. Bisson, and R.J. Naiman (eds.), Pacific Salmon and Their Ecosystems: Status and Future Options. Chapman and Hall, New York.

Pearcy, W.G. 1992. Ocean Ecology of North Pacific Salmonids. University of Washington Press, Seattle, Washington.

Pearcy, W.G. 1996. Salmon production in changing ocean domains, p. 331-352. *In* Stouder, D.J., P.A. Bisson, and R.J. Naiman (eds.), Pacific Salmon and Their Ecosystems: Status and Future Options. Chapman and Hall, New York.

Regier, H.A. 1996. Old traditions that led to abuses of salmon and their ecosystems, p. 17-28. *In* Stouder, D.J., P.A. Bisson, and R.J. Naiman (eds.), Pacific Salmon and Their Ecosystems: Status and Future Options. Chapman and Hall, New York.

Reisenbichler, R.R. 1996. Genetic factors contributing to declines of anadromous salmonids in the Pacific Northwest, p. 223-244. *In* Stouder, D.J., P.A. Bisson, and R.J. Naiman (eds.), Pacific Salmon and Their Ecosystems: Status and Future Options. Chapman and Hall, New York.

Rutter, L.G. 1996. Salmon fisheries in the Pacific Northwest: how are harvest decisions made? p. 355-374. *In* Stouder, D.J., P.A. Bisson, and R.J. Naiman (eds.), Pacific Salmon and Their Ecosystems: Status and Future Options. Chapman and Hall, New York.

Schoonmaker, P. and B. von Hagen (eds.). 1996. The Rain Forests of Home: An Exploration of People and Place. Island Press, Washington, DC.

Sedell, J.R., G.H. Reeves, and P.A. Bisson. 1996. Habitat policy for salmon in the Pacific Northwest, p. 375-387. *In* Stouder, D.J., P.A. Bisson, and R.J. Naiman (eds.), Pacific Salmon and Their Ecosystems: Status and Future Options. Chapman and Hall, New York.

Smitch, C. 1996. Introduction to a complex problem, p. 13-16. *In* Stouder, D.J., P.A. Bisson, and R.J. Naiman (eds.), Pacific Salmon and Their Ecosystems: Status and Future Options. Chapman and Hall, New York.

Smith, C.L. and B.S. Steel. 1996. Values in the valuing of salmon, p. 599-616. *In* Stouder, D.J., P.A. Bisson, and R.J. Naiman (eds.), Pacific Salmon and Their Ecosystems: Status and Future Options. Chapman and Hall, New York.

Stickney, R.R. 1994. Use of hatchery fish in enhancement programs. Fisheries 19(5): 6-13.

Walters, C. 1996. Information requirements for salmon management, p. 61-68. *In* Stouder, D.J., P.A. Bisson, and R.J. Naiman (eds.), Pacific Salmon and Their Ecosystems: Status and Future Options. Chapman and Hall, New York.

Wertheimer, A.C. 1996. Status of Alaska salmon, p. 179-197. *In* Stouder, D.J., P.A. Bisson, and R.J. Naiman (eds.), Pacific Salmon and Their Ecosystems: Status and Future Options. Chapman and Hall, New York.

Williams, J.E. and C.D. Williams. 1996. An ecosystem-based approach to management of salmon and steelhead habitat, p. 541-556. *In* Stouder, D.J., P.A. Bisson, and R.J. Naiman (eds.), Pacific Salmon and Their Ecosystems: Status and Future Options. Chapman and Hall, New York.

Introduction to a Complex Problem

Introduction to a Complex Problem

Curt Smitch

The historical, esthetic, and economic importance of salmon (*Oncorhynchus* spp.) to the Pacific Northwest is well appreciated. However, other aspects of salmon are equally important and possibly are becoming more important as the 21st century approaches. These aspects are the central roles that salmon play in maintaining the biological and cultural integrity of the region. Salmon are biologically important because they are a dominant component of the fish fauna of most Pacific Northwest water bodies. Their feeding habits, reproductive activities, and life-history strategies strongly influence the structure and dynamics of aquatic communities through predation, habitat modification, and nutrient movements. In turn, their role as prey for other vertebrates is central in maintaining biologically diverse and productive communities in the aquatic and associated riparian ecosystems (Naiman et al. 1992). Culturally, they are a symbol of the region's unique richness of resources and traditions that is recognized by indigenous and non-indigenous peoples alike (Schoonmaker and von Hagen 1996).

Despite the general recognition of the regional importance of salmon, we as a society have allowed a large percentage of the salmon populations to decline to dangerously low numbers (Nehlsen et al. 1991). In several cases, specific salmon populations that once numbered in the thousands are now federally protected by the Endangered Species Act (ESA) because of the danger of extinction. For example, in November 1991 the National Marine Fisheries Service (NMFS) listed the Snake River sockeye salmon (*O. nerka*) as endangered and 5 months later (April 1992) listed Snake River spring, summer, and fall chinook (*O. tshawytscha*) as threatened. Fortunately, economic consequences cannot be used to determine if a species should be listed: only biological criteria count in the evaluation of a species' condition.

The Snake River sockeye and chinook salmon populations are perhaps extreme examples of a complex situation but they do illustrate how severe the situation must become before a population can come under federal ESA protection. What is not readily apparent is the large number of populations that are approaching a similarly dire situation. What were the conditions of the Snake River sockeye and chinook salmon runs when they were listed? In 1991, four sockeye salmon returned to spawn in Idaho; in 1992, one returned. In 1993, eight sockeye salmon returned (Fish Passage Center, Portland, Oregon, unpubl. data). Fall chinook did better than sockeye, but not compared with historical run sizes. For example, as recently as 1962, 30,000 adult fall chinook salmon migrated past Ice Harbor Dam on the Snake River on their way to spawn. Today, the Snake River fall chinook population hovers around 600 fish (Northwest Power Planning Council 1992). One does not have to think very long about the condition of the Snake River sockeye and chinook salmon before the obvious question arises. What about the status of the other salmon populations in the Columbia and Snake River basins, which share similar biological needs and habitat requirements? In 1993, 36 other Columbia and Snake River salmon stocks were categorized as being at a high risk of extinction, 14 more were identified as being in moderate risk of extinction, and 26 other stocks were listed as of special concern (Wood 1993). However, the situation gets worse as one continues to assess the biological condition of the Pacific salmon. The Forest Ecosystem Management Assessment Team (FEMAT) report stated that an estimated 316 stocks of anadromous salmonid stocks have been identified as at risk within the range of the northern spotted owl (FEMAT 1993). In a separate action, NMFS, obviously concerned about the growing evidence pointing to the declining health of both coho salmon (*O. kisutch*) and steelhead (*O. mykiss*) stocks, announced a coastwide stock status review of both species. As a result, both species are expected to be proposed for a coastwide listing as threatened under the ESA. At the time of writing, recent indications were that coho would be proposed in March or April 1996 and steelhead in June or July 1996 (W. Stelle, Region Director, NMFS, Seattle, Washington, pers. comm.).

Whether one focuses on the Snake River, the range of the northern spotted owl, or the entire Pacific coastal ecoregion, one comes to the same conclusion. Something is wrong with the salmon's habitat (McIntosh et al. 1994, Wissmar et al. 1994). As the FEMAT (1993) report emphasized, the decline of salmonid stocks is indicative of a historical and continuing trend of aquatic resource degradation. Although several factors are responsible for declines of anadromous salmonid populations, habitat modification and loss are major determinants of their current status.

Habitat is crucial to maintaining salmon in the Pacific Northwest. Salmon need healthy aquatic ecosystems. They need, among other things, adequate amounts of cool, clean freshwater, fully functioning riparian corridors, and wetlands. In short, salmon need the same resources and habitat that humans utilize, occupy, and hold as essential to the very existence of economic communities.

Protecting the salmon will cause society to confront, as never before, the biological nature of human communities. This will not happen easily. The general public has had little reason to deliberate about concepts such as biodiversity and ecosystems. Wilson (1992) said it very well: "Every country has three forms of wealth: material, cultural, and biological. The first two we understand well because they are the substance of our everyday lives. The essence of the biological diversity problem is that biological wealth is taken much less seriously." However, 314 salmon stocks at risk, or one returning sockeye, must be taken seriously, for the salmon is a central species in a complex and important ecosystem that supports many other forms of life, including our own (Naiman 1992).

The decline of salmon populations is telling us that the entire aquatic ecosystem is ill. In discussing why society must support ecosystem protection, Wilson (1992) notes that "whole ecosystems are the targets of choice because even the most charismatic species are but the representatives of thousands of lesser-known species that live with them, and are also threatened." Yet, when an indicator species (like salmon) is endangered, the conflict is not between economic development and the environment, but between the long-term sustainability of the ecosystem and the misguided resource management practices that brought both the species and the ecosystem to the brink of disaster (Reid 1992).

Until now, the public has not responded to this line of argument. Even with the conflict, and the community impacts resulting from the listing of the salmon and the northern spotted owl, there is still not general public concern about the loss of biodiversity or the possible failure of entire ecosystems. A sense of urgency about extinction may exist in the minds of many scientists, but the concern is not widely shared in the political community or among the general public. Furthermore, in the absence of a direct impact on life styles, it is unreasonable to expect people to be concerned with endangerment or extinction of obscure plants or inconspicuous animals (Tobin 1990).

Salmon have already directly impacted life styles and the quality of human life, and they are certainly not inconspicuous. Indeed, it would be hard to select a more charismatic genus for the Pacific Northwest. However, can the salmon cause the regional society to take environmental issues seriously? No one can predict at this time whether the regional society will confront itself. The extent that we as a regional society engage the problems of the region's aquatic ecosystems will be the extent to which challenges are made to current economic and political philosophies. In other words, to confront regional environmental issues, in this case the salmon, will be to confront the forces of the status quo. Preservation of biological diversity will require not only a wholesale change in the way we treat our wilderness and wildlife, but also the way we live and do business (Cooperrider 1991).

In a democratic society, people attempt to change the world through the political process. As a strong defender of the status quo, it is a process of appeasement where there are minor shifts in policy, marginal adjustments in ongoing programs, moderate improvements in laws and regulations, and rhetoric offered in lieu of genuine change (Gore 1992). It is a process that confronts policy makers with the dilemma of developing and implementing policies without adequate scientific data. Thus, does one do nothing when faced with a scarcity of scientific data, or does one act and possibly make things worse? Political systems place the burden of proof on those who want to change the status quo. Those favoring change must produce persuasive and occasionally even incontrovertible evidence that a problem exists. Circumstantial evidence or informed judgment is rarely satisfactory, especially when change threatens the well-being of those who profit from the status quo (Tobin 1990).

It has been my experience that people's values and attitudes determine to a considerable extent how they confront problems like endangered and threatened salmon. For example, if a person happens to believe that the salmon problem is real and serious, then that person is likely to support the ESA, the protection of ecosystems, and biodiversity. If a person does not believe the salmon problem is real and immediate, then he or she is likely to believe that the ESA is being misused, and that ecosystem management is just one more intrusion of government regulations into the private sector and personal lives.

People who believe environmental problems like those associated with salmon are real tend to assert that we humans are part of the biological world and we must act accordingly. Those who do not believe that environmental problems are real tend to argue that technology will resolve environmental problems like the salmon and the spotted owl. Unfortunately, technical solutions to complex environmental problems have a poor record of success (Meffe 1992, Naiman 1992, Ludwig et al. 1993).

These are examples of the range of values and attitudes that people bring to the political process. Just imagining some of the debates can make one slightly cynical or at least pessimistic about changing the status quo. But it is also true that people's values and attitudes are learned, social and environmental literacy are improving, and these are reasons why I remain optimistic.

Synthesis plays an important role in providing a forum for discourse and debate about complex and difficult issues. They help society begin to frame the issues they must come to act upon.

I believe that there is no issue more pressing in the Pacific Northwest—indeed in our society—than the issue of accepting social and ecological interconnectedness. The salmon issue may be one of the best ways—if not the best way—to effectively develop a regional perspective for an integrated socio-environmental system.

Literature Cited

Cooperrider, A. 1991. Reintegrating humans and nature, p. 140-148. *In* E. Hudson (ed.), Landscape Linkages and Biodiversity. Defenders of Wildlife. Island Press, Washington, DC.

Forest Ecosystem Management Assessment Team. 1993. Forest ecosystem management: an ecological, economic, and social assessment. Report of the Forest Ecosystem Management Assessment Team. US Government Printing Office, Washington, DC.

Gore, A. 1992. Earth in the Balance: Ecology and the Human Spirit. Houghton Mifflin Company, New York.

Ludwig, D., R. Hilborn, and C. Walters. 1993. Uncertainty, resource exploitation and conservation: lessons from history. Science 260: 17, 36.

McIntosh, B.A., J.R. Sedell, J.E. Smith, R.C. Wissmar, S.E. Clarke, G.H. Reeves, and L.A. Brown. 1994. Historical changes in fish habitat for select river basins of eastern Oregon and Washington. Northwest Science 68: 268-285.

Meffe, G.K. 1992. Techno-arrogance and halfway technologies: salmon hatcheries on the Pacific coast of North America. Conservation Biology 6: 350-354.

Naiman, R.J. (ed.). 1992. Watershed Management: Balancing Sustainability and Environmental Change. Springer-Verlag, New York.

Naiman, R.J., T.J. Beechie, L.E. Benda, D.R. Berg., P.A. Bisson, L.H. MacDonald, M.D. O'Connor, P.L. Olson, and E.A. Steel. 1992. Fundamental elements of ecologically healthy watersheds in the Pacific Northwest coastal ecoregion, p. 127-188. *In* R.J. Naiman (ed.), Watershed Management: Balancing Sustainability and Environmental Change. Springer-Verlag, New York.

Nehlsen, W., J.E. Williams, and J.A. Lichatowich. 1991. Pacific salmon at the crossroads: stocks at risk from California, Oregon, Idaho and Washington. Fisheries 16(2): 4-21.

Northwest Power Planning Council 1992. Strategy for Salmon, Volume I. Portland, Oregon.

Reid, W.V. 1992. The United States needs a national biodiversity policy, p. 4. *In* World Resources Institute, Issues, and Ideas. World Resources Institute, New York and Washington, DC.

Schoonmaker, P., and B. von Hagen, eds. 1996. Environment and People of the Coastal Temperate Rain Forest. Island Press, Washington, DC.

Tobin, R. 1990. The Expendable Future: US Politics and the Protection of Biological Diversity. Duke University Press, Durham, North Carolina.

Wilson, E.O. 1992. The Diversity of Life. The Belknap Press of Harvard University Press, Cambridge, Massachusetts.

Wissmar, R.C., J.E. Smith, B.A. McIntosh, H.W. Li, G.H. Reeves, and J.R. Sedell. 1994. A history of resource use and disturbance in riverine basins of eastern Oregon and Washington. Northwest Science 68: 233-267.

Wood, C.A. 1993. Implementation and evaluation of the water budget. Fisheries 18: 6-16.

Old Traditions that Led to Abuses of Salmon and Their Ecosystems

Henry A. Regier

Abstract

The salmonid family includes species in three subfamilies: the salmonines (trout, salmon, and char), the thymallines (grayling), and coregonines (whitefish, cisco, and felchen). Complex associations of salmonid taxa evolved in thousands of aquatic ecosystems located between ~35°N and 70°N latitude. On the basis of historical data, many of these associations, with perhaps ten thousand or more identifiably separate stocks within the set of salmonid species, can be classified into several major regional super-complexes. One of these super-complexes uses the tributaries of the North Pacific Basin and another uses the freshwater streams and lakes of the Laurentian Great Lakes Basin. By 1994 only part of one super-complex still exhibits a measure of natural integrity, that of the northern part of the North Pacific. The paper focuses on hemispheric-level cultural phenomena, especially as related to "industrial progress," that are deemed responsible for this great salmonid blight. It also traces some beginnings of reform that may yet reverse the blight's northward movement, especially in the Pacific Basin if climate warming does not occur rapidly. The science and techniques that were created to conserve valued stocks may actually, inadvertently, have contributed to their devastation because it offered too little too late. A hemispheric Wild Salmonid Watch is urged to protect salmon but also monitor the reform of the cultural blight that has harmed all of nature and humans as a result.

Introduction

An introductory paper within a far-ranging book on Pacific salmon should perhaps expose tacit controversies. I decry continued submission to some lingering traditions of the past that should now be consigned to history. What if our civilization really has crossed a historical divide a generation ago? What then are the new traditions that are now emerging and evolving? What old traditions should finally be retired? A societal consensus may be forming, relevant to salmon and other issues, but as yet the consensus may be more tacit than explicit. My aim here is to clarify elements of the new ways and to contrast them with counterpart elements of the older ways.

This paper was not created to stand alone, separate from other papers in this book. Various authors have sketched some new traditions now emerging and perhaps some redeeming features of some old traditions.

To me, on the Atlantic side of the Northern Hemisphere, what has been happening to salmonids on the Pacific side looks something like a replay of what happened starting decades earlier with respect to sea-run salmonids and freshwater salmonids of Atlantic watersheds. Events in Arctic watersheds in turn may be lagging events in Pacific watersheds.

The history of salmonids may be used to indicate the origin, deepening, and spreading of a great cultural blight on nature. This blight started in Northern Europe about the mid-19th century and then spread in all directions, with secondary centers of infestation in Eastern North America and Northeastern Asia. Within the salmonid latitudes the Northern Pacific Basin may yet be protected from severe degradation due to this blight.

It is appropriate to search for causes common to the history of the whole hemispheric array of salmonid complexes. For this general historical sketch, I include salmon and trout (*Oncorhynchus* and *Salmo*) and char (*Salvelinus*) under the salmonines; grayling (*Thymallus*) under the thymallines; and whitefish, cisco, chub, felchen, maraene, etc. (*Coregonus*), under the coregonines.

Salmonid Extinctions and the Beginnings of Reform: A Historical Context

Across our hemisphere, a prolonged human use of salmon and their environment seldom has been sustainable: the usual story is one of unsustainable abuse. The loss of many stocks of salmonids and the great simplification of the salmonid complex, in different times and in different places, have similarities. These cases include the following:

- many anadromous stocks of Atlantic salmon (*Salmo salar*) spawning in streams that flow into the North Atlantic Ocean, the Baltic Sea, and the Gulf of St. Lawrence, where they interact with other less migratory salmonids within relatively simple associations;
- a set of similar complexes in numerous deep mountain lakes of Europe, such as in the Swiss Alps;
- a set of similar complexes in the Laurentian Great Lakes (Canada and the US) and other large lakes and their connecting rivers and tributary streams at the southern edge of the Laurentian Shield;
- a set of similar complexes in deep lakes and high tributaries of the Rocky Mountain system of North America; and
- the super-complex of Pacific salmon and other salmonids that use the stream and lake systems proximate to the North Pacific Ocean for some or all of their life history stages.

By the 1850s, the Atlantic salmon in the Rhine River had already declined, as had brown trout (*Salmo trutta*) in various streams of Europe; some coregonines and chars of the deep Alpine lakes seemed threatened. Fish hatcheries were re-invented by the French and hailed as equivalent in importance to the construction of the Eiffel Tower and the French discovery of the planet Neptune. The invention of fishways around or over dams also dates to those times.

The use of hatcheries and fishways presumably extended the existence of many stocks or mitigated the losses of natural stocks with hatchery stocks. But these techniques usually did not prevent the eventual extinction of threatened salmonid stocks wherever industrial European civilization flourished. Stocks with key habitats in industrialized lowlands toward the southern edge of the salmonid range were generally the first to disappear and then the extinction gradually proceeded upslope and northwards. Presumably, the salmonids of lowlands at the southern edge of their range already were stressed, especially in warm, dry years, by high temperatures and low water. Eventually the degraded, dammed rivers became intolerable or insurmountable even in cool, wet years. Also many lakes became too enriched and polluted for salmonids.

Massively abusive industrial practices were common in urbanized watersheds on both sides of the North Atlantic starting in the mid-19th century and peaking in the mid-20th century. The industry that served militarism was especially abusive of nature; environmental concerns about pollution and other abuses were regularly relaxed for the war effort, far from the theater of war itself. Once relaxed, controls of the abuses were difficult to re-institute, and full recovery from the degradation caused by them was often impossible.

Industrial practices on land and in fresh waters were not solely responsible for extinction of numerous salmonid stocks. Fisheries also had become industrialized and commercialized. Most fishermen and their governmental regulators perceived the ocean as an inexhaustible cornucopia of resources, just as owners of industries perceived the atmosphere's "assimilative capacity" for smoke and waste gases to be unlimited or the "assimilative capacity" of lakes, rivers, and pits in the landscape to be unlimited for liquid and solid wastes. The myth of the unboundedness of nature, as a source of resources and concurrently as a sink for wastes, was implicit in the mindset of western industrial civilization and is still far from being fully discredited, especially around the Pacific rim. The onus is still on the reformer to demonstrate scientifically, case by case, that nature is in essence not cornucopian. And the only kind of science acceptable for such demonstration is logically incapable of providing the requisite proof (see the following).

Historians may decide that the 1930s and 1940s represent the depths of degradation of "Western Culture." A reform process then got underway, almost imperceptibly. In the 1960s a more focused reform broke through in a series of ideological mini-revolutions in all northern geographic regions of western culture including the Socialist Bloc behind the Iron Curtain. As a part of this reform with respect to fisheries, the 1960s led to international regimes on fisheries the world over, though these regimes were still quite tentative as to the nature and depth of the reform that was needed. The difficulties within the International Whaling Commission at that time illustrate the conflict with respect to this reform.

In the late 1960s and early 1970s, international and national reforms emerged with respect to environmental and resource issues of importance to salmonids and many other natural phenomena. Americans know the story well as it relates to US national legislation and programs. Comparable reforms were underway generally in many countries of the Northern Hemisphere, but perhaps mostly only as rhetoric in the Socialist Bloc. Internationally, both intergovernmental and nongovernmental programs began building and proliferating. The United Nations Conference on the Human Environment at Stockholm in 1972 was the first international forum that addressed environmental reform in a relatively comprehensive way.

The massive inertia of conventional industrialism was still continuing to cause extinctions of salmonids and other taxa in the 1970s. The 1980s brought a threat that the balance would be tipped against reform and toward the interests that gloried in the ways of the 1940s (e.g., wars, relaxation of environmental commitments, and all). In the 1990s, reform may again be underway, shakily, at least in our part of the world.

If what is happening to salmonids now is taken as an integrative measure of whether environmental and cognate reforms are winning over conventional industrialism, then the outcome is still in doubt. For example, the rich natural complex of salmonids in the North Pacific still is accorded no special consideration as such and stocks are still suffering extinction. Elsewhere, in what were highly degraded watersheds of the North Atlantic, a few stocks are recovering and some new ones are being created with difficulty. There seems to be more concern to rehabilitate or restore salmonids toward the southern parts of their ranges than to preserve extant stocks toward the northern parts of their ranges through appropriate management of the uses of the salmon and their habitats. Watersheds of the North Pacific are part of a frontier in which conventional industrialism is still dominant politically. The onus is on the pessimists to demonstrate empirically any ecological limits within this perceived cornucopia.

Abuse of nature was inherent in the politics of conventional industrialism and its attendant commercialism. In the 19th century, such abuse was forgiven; industrialism was perceived to serve the progressive goals of nation building. In the 20th century, the abuses were condoned if industrialism served the offensive or defensive needs of national or commercial empires. However, conventional industrialism did not have freedom everywhere all the time. Perhaps the era of excessive industrial freedom ended effectively about 25 years ago. Or perhaps the North American Free Trade Agreement will give new life to old-fashioned commercial industrialism. The imperatives of trade wars may debase ecosystemic qualities in ways similar to the imperatives of military wars.

Direct connections between extinction of salmonid stocks and the devastation of nature caused by industrialism may seem tenuous. Thus, some further analysis may be appropriate.

The Conservation Movement—A Retrospective Glance

The conservation movement that proliferated throughout northern countries of western culture starting over a century ago was ignored in the preceding section. Our environment, and especially salmonids, might have suffered even more harsh abuse had it not been for the conservation movement. But leaders of that movement generally supported the same values as did the relevant industrialists and did not effectively challenge the cornucopian myth. "Multiple" use was interpreted to justify an additional use regardless of how many uses were already being accommodated within an ecosystem. Altogether, the movement successfully blunted some of the abuses of industrialism but ultimately failed to protect fully any natural feature.

The scientific literature contains relatively little clear documentation of precisely what caused the extinction of particular stocks. As with human deaths, extinction may have been the outcome of a number of adverse influences acting together. Adverse influences may often have interacted multiplicatively or even with positive feedback rather than additively; they seldom canceled out each other's effects.

As the conservation movement reacted with some success regarding local consequences of abuse, industrialism reduced its local impacts in part by spreading its abuses more regionally. Higher smokestacks were built; wastes were carted farther before they were dumped; strip mines were opened farther from cities while some nearby were filled with garbage and planted over. When eventually regional abuses came under attack by conservationists, industrialists already were abusing the global ocean and atmosphere. Radionuclides, persistent pesticides, and hazardous contaminants carried everywhere by water and air were largely invisible except to sophisticated scientists, most of whom were biased towards conventional industrialism. As global abuses came to be addressed, industrial and military interests already had begun to litter interplanetary space with junk.

A similar story can be told about the exploitation of forests and fish. In general "too little, too late" may be a fair judgment on the old conservation movement, into which I was baptized about 1953.

A century ago, the conservation movement was driven partly by fears of widespread species extinctions. The poignant stories of the fur seals, bison, passenger pigeons, and some Atlantic salmon runs were well known. The movement resulted in a major political compromise between progressive utilitarians and romantic preservationists; the compromise involved zoning. Most land and water could and should be used, more or less sustainably, for multiple purposes, in the progressive service of nation building. Some of the area, especially terrestrial parts of little interest to progressive utilitarians, should be preserved in a wild state, more or less. In the US, these two policies are often associated with Gifford Pinchot (the industrial utilitarian) and John Muir (the nature romantic).

Apparently, it was realized that simplistic landscape zoning of the "Muir type" could not suffice to protect migratory creatures, such as salmon and waterfowl. Instead, these species were protected by outlawing efficient methods of capturing them and by outlawing capture at particular times and in locales in which the species were particularly vulnerable. The vulnerability of individual salmon to capture by humans is directly proportional to the salmon's preoccupation with its reproductive responsibilities. Given that the main policy of conserving salmon was to limit capture method efficiency (see following), it is consistent with the Muir-Pinchot compromise that salmon should be protected especially at spawning time by devising a stock-specific time-space zone of protection.

The policy of protecting spawning fish, or permitting escapement of enough fish to their spawning streams to permit some reproduction, was not invented by the conservationists. It was standard practice in parts of northern Europe, from traditional origins centuries earlier. But from the perspective of reactive conservationists, this seemed to be a particularly appropriate policy for salmon from a hundred years ago until the 1960s. Perhaps this "zoning approach" became too prominent in the set of efforts to "manage salmon." Too much may have been expected from it, so that other necessary efforts were ignored or underutilized. It would be instructive to explore how salmon fisheries could be managed if major emphasis were focused on life history stages other than the reproductive stage.

Fisheries, and Markets and Property Rights

Fishing for salmon started as artisanal fisheries and eventually evolved into food fisheries and sport fisheries. Food fisheries became vertically integrated into capture, processing, and sales combinations, generally catering to the more affluent in society (e.g., as in the restaurant trade). Sport fisheries have come to be served by an industry that makes many things for anglers and a service sector that facilitates pleasant experiences of all sorts related to this kind of recreation. Financial accountants, or sometimes economists posing as accountants, have helped to "commodify" salmon as things so that their use can be allocated or distributed through the market.

Industry and commerce evolved together with respect to each of the food and the recreational fisheries.

Management of food fisheries was driven by an industrial model related to the formal food market. Most food fishermen fish for dollars, not for fish. This is also the case with professional sport fisheries, as in fishing competitions in which the total mass of fish caught by a contestant determines the dollar winnings. The conventional market measures value rather grossly in monetary terms. It prefers homogeneous quality, and high and constant quantity. It pays a premium for added value through canning, smoking, or transformation into delicacies (or for comparable "things" with some recreational fisheries); thus, the fish themselves are only the initial raw resource in which differences are more a problem than an opportunity. The bioeconomic model of sustainable yield, created in the 1960s, relates to this perception of economic reality and political priority.

The bioeconomic model brought supply and demand measures together, conceptually and in the abstract. Neither resource rent nor costs for administering the fishery were charged directly to fishermen. This financial advantage of free use was offset by the consequences of open access (e.g., excess harvesting and processing capacity, which led to inefficient use of capital and labor). Efficient methods of harvesting fish were not permitted under "effort management," and were not feasible with open access in which an open season was very short; thus, the direct costs of capture were maintained by regulators at an artificially high level. Throughout these developments, the bioeconomic model could offer advice on optimum sustainable yield toward which this quite ungainly system of management should aim. Though ostensibly an equitable regime with respect to opportunities among the fishermen, the old regime progressively led to overfishing and impoverishment of fishermen. Some form of governmental subsidy often followed, especially when it could be argued that some environmental "act of God" was implicated in a particular stock failure. The subsidy usually exacerbated the problem further.

People who make a living in and through the marketplace benefit especially from private property rights of various kinds. (Where these rights are not formally legitimate, cunning entrepreneurs often have ways of coping through arrogation of illegitimate rights, such as in monopsonies and monopolies.) Assigning private property rights to things like migratory salmon is institutionally difficult. Thus, an intricate regulatory system evolved in which rights to access and harvest fish were left mostly open, in practice if not by intent. The information and enforcement services necessary for intricate conventional regulation are impracticable and border on the ineffectual.

The story of recreational fisheries should, perhaps, be different from that of industrial food fisheries. But it has similarities including relatively free use, relatively open access, enforced inefficiency in methods of capture, intricate regulations, and practically impossible demands for information. The catch of the majority of individual anglers is ostensibly limited sharply in many jurisdictions by specifying inefficient capture methods. Such limits can induce the expert, successful angler to take more trips, and incur the costs related to the increased number of trips. Such expenses benefit the regional service economy. Highly prized species are used as loss leaders within a region's commercialized recreational service sector. In this sort of a setting, efforts to induce anglers to internalize the relevant ethic of low-kill or no-kill of trophy fish have been only moderately successful. The different pieces of the overall policy do not really fit together.

I emphasize the role of the market, the regional/national/international market, in this paper. With respect to the fishery issues, however, the "planning systems" of centralized bureaucratic socialist or welfare economies play a similar role to that of the "market system." My intent is not to indict these systems as necessarily wrong; they are not "wrong". Rather I emphasize that nothing of any real importance should be left to the dynamics of the vertically integrated market or the hierarchically centralized bureaucracy. Neither of these institutional mechanisms was created to arbitrate on issues of important cultural value. As institutions, both approach full efficiency asymptotically as the value of the things allocated approaches zero. Each is highly subversive if not kept carefully in its place as a servant to the citizenry. And both mechanisms are highly vulnerable to subversion by bright entrepreneurs or political opportunists.

As with forestry, agriculture, and other cultural activities related to eco-reality, a new approach has been developing within "fisheries" over the past quarter century. The old approach sketched above had developed a relatively coherent political, conceptual, and methodological mindset by the late 1950s. By the late 1960s, this 1950s "synthesis" was suspected to be, at best, insufficient. Many political, practical, and scientific difficulties of the past 25 years can be perceived and understood in the context of the cultural process within which this new mindset has been superseding the old.

An Alternative to the Cornucopian Myth

What acts to limit the presence and abundance of salmon stocks has been difficult to ascertain in particular instances. Two presuppositions have prevailed until recently. First, environmental phenomena closely associated with the reproductive process determined subsequent presence and abundance. Second, hatcheries could mitigate any difficulties. The latter notion fit in nicely with multipurpose progressive utilitarianism; the homogeneous quality and predictable quantities of harvestable resources promised by hatchery optimists suited the cash-oriented food and recreational markets. From a professional perspective, operating a hatchery seems to be a more comfortable, sedentary occupation with predictable products than "managing" stocks and habitats scattered over millions of hectares of difficult terrain and dangerous waters.

The notion that large waters (lakes, rivers, seas, oceans) were effectively limitless with respect to human interests began to be queried by perceptive individuals about 1870, but was not fully rejected by those using the old fisheries approach. Ecological phenomena of the ocean were perceived to hardly limit salmon abundance, especially when compared with phenomena related to reproduction in fresh waters. Thus, the population dynamics approach, which supplied the bio-component in the bioeconomic model, focused largely on the freshwater phase and treated the oceanic phase effectively as relatively constant or varying moderately only in a random way (i.e., without any discernible spatial or temporal trends). Because the ocean was perceived to possess limitless resources for salmon, one could ignore the possibility that artificially augmenting one stock might lead to reducing another through negative interactions between them in the ocean. This is one of the vestiges of the cornucopian myth.

While fisheries interests acted as though the oceans and large lakes provided a limitless supply of things needed by salmon, other industrial interests acted as though these waters had limitless capabilities for accepting and inactivating wastes.

For the Laurentian Great Lakes, the approximate limits of both the maximum sustainable yield of fish resources in aggregate, and the maximum sustainable assimilation of urban and agricultural wastes were inferred by the early 1970s. Lake Erie could supply about 10 kg ha^{-1} yr^{-1}

of fish regardless of the assortment of species harvested. Similarly, it could assimilate about 3 kg $ha^{-1} yr^{-1}$ of phosphates, regardless of the source, and maintain a mesotrophic-oligotrophic state. Little interest, however, has been focused on inferring such generalizations for enclosed seas to parts of the ocean.

Perhaps the combined catch of all fish species in the North Pacific has reached "maximum sustainable yield." Beyond such a point an attempted increase in overall catch would presumably lead to the suppression of K-species (i.e., climax species) with slower turnover rates and to a relative expansion of r-species (i.e., pioneer species) with more rapid turnover rates. Perhaps the salmonids that reach large sizes may tend toward K status while those that do not become large may tend toward r status.

A shift in species dominance, from K to r, might be facilitated in nearshore waters by eutrophication from enrichment of river water flowing into estuaries. A phenomenon might emerge that superficially resembles upwelling of enriched deeper water nearshore, except for the difference in temperature, toxic contaminants, and clay turbidity of the culturally altered waters. Thus, certain r-species, similar to those in upwelling areas (e.g., clupeids) might thrive in culturally enriched nearshore waters. Whether salmonids that thrive near natural upwelling areas would also thrive in the culturally enriched waters has apparently not been examined critically.

The myth of the ocean as cornucopia should be replaced by a working hypothesis that the maximum sustainable yield (MSY) of all species combined in any region of the ocean has already been reached or surpassed. Where the MSY has not yet been surpassed, the practical options now relate to improving the mix of species in the catch rather than expanding the total catch of the mix. Attempts to manage the harvest and augment particular species with hatcheries or improved spawning habitats, then, relate to altering the mix of species in the catch. Subsequently, harvests of some species will need to be reduced in a compensatory way. In the Laurentian Great Lakes, harvests of small clupeids and osmerids have been reduced because these species serve as prey for the highly preferred salmonines.

Management should explicitly operate on the presumed correctness of the above working hypothesis. It would amount to a belated acceptance of the "precautionary principle" urged in the United Nations Agenda 21. The onus should fall on antagonists of this proposed working hypothesis to make their case that the marine fish resources of interest to them are not yet fully tapped.

Perhaps this line of argument should be taken one step further. The working hypothesis, or new conventional wisdom, might be that in aggregate the fishery resources of any region of the world's oceans are now effectively overfished or otherwise abused beyond an optimal level. What is needed is overall rehabilitation with an initial reduction of the aggregate levels of harvest. This is now the working convention in the Laurentian Great Lakes and elsewhere in the North Atlantic Basin. The people of the North Pacific must finally and fully reject a frontier mentality and cornucopianism in all its manifestations.

Changing Emphases in Fisheries Science

The notions of an ecologically limitless ocean but limited fresh waters led to a management focus on "escapement" of migrating adults to their natal spawning grounds and to hatcheries.

Conceptually such management is highly simplistic and risky. Many salmonid stocks were driven to extinction because of cultural and ecological factors that were ignored or by synergistic interactions between effects of fishing and other cultural stresses.

Early fisheries science relied on Newton's paradigm (mass, force, time, space). The bioeconomic model was an application of this paradigm. It was not entirely inappropriate because fish mass was of primary interest to industrial fisheries and many anglers.

In retrospect, however, the science once deemed to be relevant and sufficient for market-related management is inappropriate or insufficient even there. Because market-related regulators attached little value to differences between stocks (or even between some species), and because hatcheries promised to produce raw resources that were often preferred over the wild fish, the science relevant to evolution, biodiversity and eco-integrity had little to offer the regulators. Progressive utilitarians, dominant in government, private enterprises, and even in most charitable foundations, helped fund science relevant to the process of harvesting relatively undifferentiated mass. Naturalists and some scientists served as a cultural remnant of ethical sophistication and appropriate science.

In the 1960s, an attempt to focus on energy flow in food webs attracted much interest in ecology and some in fisheries. The types of energy involved were all first degraded into heat for purposes of measurement. Reducing everything to heat meant that this approach was not much of an advance, if any, over using mass itself. That energy-accounting approach provided some interesting numbers about the apparent inefficiency of trophic networks, but not much insight into the nature of ecosystems as self-organizing, open systems. An exergy accounting approach, in which the quality of energy is of interest, may yet lead to important insights.

Information, as a form of emergent reality in its own right, has been a focus of some attention. What is the "code" by which a self-organizing thing operates? To what extent is the code a copy of a prior code, say through processes of genetic reproduction or through learning of some kind? To what extent is a code itself open to modification, through endogenous and exogenous processes? Through short-term or long-term evolution, living things operate partially on "order from order" and partially on "order from disorder."

When more comes to be understood about information systems, as open and self-organizing, it may be that more of what concerns us about living phenomena, like the salmonid complex or the dynamics of a co-managed fishery as interactive husbandry, may become intelligible. Natural information in the sense of intricate and appropriate awareness, comprehension, and know-how may come to be valued much more than at present. The individual salmon, the stock, the species, and the entire taxonomic complex may each hold some relevant information in self-organizing nested systems. This may also be the case with the expert salmon fishers at different nested levels of self-organization.

With this kind of mindset, more taxonomists and systematicists may again take an interest in salmonids. The salmonid complex has been neglected, perhaps because salmonid evolution was largely unintelligible within the dominant linear-causal, reductionist mindset. It's time for the new ecologists to interact with ethologists and evolutionary biologists using a new mindset.

In pursuit of sustainable use, perhaps we eventually will seek to conserve several kinds of information as crucially important features of reality (i.e., salmon information, habitat information, fishermen information). Then the need for information about information—these have two different though complementary connotations!—will become more apparent.

Shifting Values

We are now some 25 years into a political reform movement to transcend progressive utilitarianism and the old approach to salmon in particular. Gradually the myths, priorities and processes of what is obsolete or secondary in importance are receding into the background and new myths, priorities, and processes are emerging and taking their place. The struggle related to salmon, the spotted owl, and old growth forests in the Pacific Northwest is a leading example of this reform movement operating at a regional level. Comparable struggles are underway in the Laurentian Great Lakes, Chesapeake and Baltic basins. Compatible political initiatives at national and international levels will be influenced by the reforms in the Pacific Northwest, and vice versa. Concurrently, native peoples are reasserting their traditions and local communities are empowering themselves and reforming at these levels.

Salmonids have, for centuries, been perceived to be "special." Ecologically and culturally their role is something like that of ducks and geese. Salmon play a variety of roles in each culture, and the role of salmon as raw resource material for a commercialized food industry has been dominant for relatively short periods (e.g., in the first seven decades of the 20th century for the North Pacific).

Six value orientations have "competed" for salmonids within Northern cultures. The first three values listed here are typical of frontier regions or frontier mentalities and may be losing a struggle for dominance to the advantage of the other three. This is now reflected more strongly in North Atlantic than North Pacific watersheds. The six are as follows:

- industrialized fisheries to supply raw material en masse for the commercial food market;
- commercialized recreational angling in which the individual fish captured is the means by which to score points in a competition with financial awards;
- common-property domestic food harvesting opportunistically by unorganized people, acting as hunter-gatherer-fishers;
- organized artisanal fisheries for ceremonial, subsistence, and trade purposes, with the spirit of the salmonid as a primary focus of concern;
- recreational angling in which capture of a salmonid (perhaps with its subsequent release) may trigger a deep instinctual experience or at least may be one of a number of motives to enjoy nature directly; and
- commitment by "let it be" ethical naturalists to interfere as little as possible in the lives of salmonids and other creatures, with any interactions limited to discreet observation in relatively natural settings.

Interests that relate to the latter three value orientations have been forming tentative alliances. Progress toward achieving an effective politics with respect to this combined mindset could be fostered if scholars and researchers would focus on this as a timely challenge and opportunity. The new ecoscience relevant to these three is beginning to perceive that natural and cultural information is as "real" as mass and energy.

Perhaps some of the code words within the emerging mindset are equivalent to Aldo Leopold's land ethic and his criteria of stability, beauty, and integrity. Recall that his meaning of "land" closely resembles what we now term ecosystem, with no clear boundary to be drawn between culture and nature. Current code words include sustainability, biodiversity, and ecosystem integrity, with connotations that have a rough one-to-one correspondence with Leopold's stability, beauty, and integrity.

In the North Pacific, the notion persists that salmon should serve as a raw resource for the commercial food industry. Simultaneously, the harvesting and marketing is changing to focus more on individual fish of particular species rather than on a relatively undifferentiated mass of salmon flesh. The old science and technique created to assist in managing the exploitive use of salmon is not sufficient for current needs in the new commercial fishery and may play only a minor role with respect to managing the use of salmon by interests with the other value orientations.

A Wild Salmonid Watch

A hemispheric Wild Salmonid Watch was proposed in the late 1970s by an international network of salmonid researchers. It was endorsed formally by the International Union for the Conservation of Nature and Natural Resources (IUCN) in the early 1980s. But the 1980s proved not to be an auspicious period for such an initiative. Now, with strong international interest currently evident in biodiversity, sustainability, the precautionary principle, and migratory fish stocks, the time may have come to formally organize a Wild Salmonid Watch.

Fisheries agencies that have experienced great difficulty extricating themselves from the policy traps of old traditions might welcome the opportunity to demonstrate some leadership in new traditions. Indigenous peoples, angler associations, naturalists, and academics should join as stakeholders and co-leaders of a Wild Salmonid Watch.

The Wild Salmonid Watch may be viewed as an international strategy to promote preservation, conservation, and rehabilitation of wild salmonid stocks and to combat the many harmful human stresses on salmonids and their habitats. Salmonids are good integrative indicators of various human influences on coldwater ecosystems. Thus, a regional element of a Wild Salmonid Watch should be a central part of efforts to rehabilitate regional ecosystems. Salmonids are already being used for this purpose in a formal international agreement for the Laurentian Great Lakes that has as its goal the recovery and maintenance of ecosystem integrity.

A Wild Salmonid Watch could be particularly helpful in global efforts under United Nations auspices with respect to sustainability and biodiversity. It could be made part of a core global monitoring service focusing on climate change; feasibility studies for this application have already been done. A key precondition is that people interested in salmonids, as a hemispheric network, get actively involved in global issues.

Postscript

Much about the unhelpful traditions that I have tried to expose is rather dated, but still active in various ways. Such criticisms about the old fisheries have all been made before now in one way or another. Similar criticisms have been made about old forestry, old agriculture, old wildlife management, old parks management. . . .

The time has come for the permanent retirement of some of these traditions and the further energetic development of some newer traditions. And of course this is what is now happening, as the papers of this book demonstrate. Further, recent personal interactions on these issues with fisheries people of the Northwest show that commitment to a thorough reform is more a rule than an exception.

What if we really have crossed a great cultural divide? Will we find a way to live with Pacific salmon within the new realities?

Acknowledgments and Caveat

The contents of this essay owe something to my participation on a panel of experts convened by Daniel Botkin for the state of Oregon, with financial help from the state of California. The panel was charged to assign the relative importance of forest practices to the decline in anadromous salmon and to recommend how new forest practices can assist in the recovery of salmon. The scope of the present essay extends beyond the Botkin panel's responsibilities. Also the essay may reflect personal idiosyncrasies that may not be shared by other panel members. Hence it should not be taken to represent any consensus within the Botkin panel.

I participated as liaison for the Botkin panel in early meetings of the US National Research Council's Committee on Pacific Salmon, chaired by John Magnuson. Whether the contents of my essay here happen to be consistent with those of the NRC Committee is unknown to me.

L. Talbot's international work on policies concerning sustainable use of renewable resources has provided insights. J. Kay has been teaching me about exergy. J. Magnuson and two anonymous reviewers helped with the editing of the present paper.

The Origin and Speciation of *Oncorhynchus* Revisited

J.D. McPhail

Abstract

Thirty-seven years have passed since Ferris Neave published his classic paper on the origin and speciation of the salmonid genus, *Oncorhynchus*. Since then, new data on fossils, chromosomes, molecules, and morphology have accumulated, and new analytical techniques for quantifying phylogenetic data have been developed. These new data are reviewed, summarized, and used to reexamine the evolutionary history of the genus. Apparently, *Salmo* and *Oncorhynchus* diverged sometime in the early Miocene (~20 million years ago [mya]) and most of the early speciation (lineage splitting) in the genus occurred during the Miocene. By the late Miocene, or early Pliocene (~6 mya), members of the chum (*O. keta*), pink (*O. gorbuscha*) and sockeye (*O. nerka*) salmon lineages were present in Idaho and Oregon. Geographically, most of the living members of the earliest divergences (the Pacific trout) are concentrated in western North America near the southern margin of the distribution of *Oncorhynchus*. Later divergences (the Pacific salmon) probably occurred in the Pacific Northwest and in Asia. Although processes that produced lineage splitting in the past can never be identified with certainty, one can examine genetic divergence in modern species. Thus, I use the sockeye-kokanee salmon divergence as an example of lineage splitting. I conclude that speciation in modern *Oncorhynchus* usually involves geographic isolation, followed by local adaptation, genetic divergence and, where divergent forms come into secondary contact, competitive interactions among forms. There is no reason to suppose that past divergences were driven by different processes. The evolutionary history of the genus suggests that most local adaptation is ephemeral and that clusters of populations (metapopulations) may be appropriate management units.

Introduction

> I have gathered a posie of other men's flowers and nothing but the thread that binds them is my own (Montaigne).

Thirty-seven years ago Ferris Neave published a paper entitled "The origin and speciation of *Oncorhynchus*" (Neave 1958). This classic article summarized what was known at that time about the evolutionary history, and mode of evolution, of Pacific salmon (*Oncorhynchus* spp.).

Much has changed in the intervening years. *Oncorhynchus* now is not the genus Neave knew; it has been expanded to include Pacific trout as well as Pacific salmon (Smith and Stearley 1989). New techniques using chromosome morphology, DNA sequencing, and phylogenetic analyses have clarified relationships within both the family Salmonidae and the genus *Oncorhynchus* (Wilson et al. 1985, Thomas et al. 1986, Hartley 1987, Thomas and Beckenbach 1989, Sanford 1990, Shedlock et al. 1992, Stearley 1992, Murata et al. 1993, Stearley and Smith 1993). In addition, since 1958, discoveries of fossil salmonids (Cavender and Miller 1972; Kimmel 1975; Smith and Miller 1985; Smith 1975, 1981; Wilson 1977) have pushed the origin of *Oncorhynchus* farther back in time than originally postulated by Neave.

My task is to revisit the question of the origin and evolution of *Oncorhynchus* and, by reviewing the recent literature, provide a brief summary of the evolutionary history of the genus. For any group of organisms, a reliable phylogeny is fundamental to understanding not only the pattern of their evolution but also the processes that have driven their divergences (Brooks and McLennan 1992). For salmonoid fishes, three recent studies provide the necessary phylogenies (Sanford 1990, Stearley 1992, and Stearley and Smith 1993). Although these phylogenies differ in some details, for our purposes they are complementary: Sanford's (1990) phylogeny emphasizes families, subfamilies and genera, while Stearley's (1992) and Stearley and Smith's (1993) phylogenies emphasize relationships among *Oncorhynchus* species.

Origin of *Oncorhynchus*

The living salmonoids consist of three lineages: the whitefish (Coregoninae), the graylings (Thymallinae), and the char, trout, and salmon (Salmoninae). Most recent authors (e.g., Kendall and Behnke 1984, Smith and Stearley 1989, Wilson and Williams 1992, Stearley and Smith 1993) consider the Salmoninae and the Thymallinae to be sister taxa that together form the sister group of the Coregoninae. These authors place all three of the lineages in the family Salmonidae. In contrast, at least one author (Sanford 1990) places the whitefish in a separate family, the Coregonidae. However, regardless of the details of interrelationships among the three lineages most recent studies accept that these lineages share a suite of derived traits and that, together, they form a monophyletic group (Kendall and Behnke 1984, Smith and Stearley 1989, Sanford 1990, Wilson and Williams 1992, Stearley and Smith 1993).

One of the derived traits shared by salmonoids is a tetraploid genome. This implies that salmonoids evolved after a polyploid event that doubled the chromosome number in some common ancestor. Allendorf and Thorgaard (1984) presented a thorough review of the karyological and molecular evidence for tetraploidy in the salmonoids. One consequence of tetraploidy is gene duplication. Allendorf and Thorgaard (1984) suggested that the existence of some undifferentiated duplicate loci and some multivalents at meiosis, as well as tetrasomic inheritance at some loci, all indicate that the proto-salmonoid karyotype doubled through autoploidy (a polyploid event within a single genome). Thus, initially, the two sets of homologous chromosomes probably associated randomly at meiosis. However, over time, a combination of mutation and selection produced divergence between many of the duplicate loci. Eventually, enough differences accumulated to prevent some of what had originally been homologous chromosomes from pairing randomly. Apparently, in salmonoids, this diploidization of the duplicate genome is ~50% complete (Allendorf 1978). Using the degree of divergence between duplicate LDH loci, Lim et al. (1975) and Lim and Bailey (1977) estimated that the tetraploid event in salmonoid evolution occurred about 100 million years ago (mya). Allendorf and Thorgaard (1984) cautioned that this estimate is based on dubious assumptions and suggested that the tetraploid event could have occurred anywhere over a range of 25–100 mya. Recently, Devlin (1993) used sequence differences in the duplicated growth hormone gene in three species of salmonids—Atlantic salmon (*Salmo salar*), rainbow trout (*O. mykiss*) and sockeye salmon (*O. nerka*)—to estimate that ~30 million years had lapsed since the evolution of disomic inheritance of the two genes. Since divergence between duplicate loci can only evolve after some of the original four sets of homologous chromosomes no longer assort randomly, the tetraploid event must have occurred some time before 30 mya. For convenience, Devlin (1993) used 50 million years in his calculations, but the fossil and phylogenetic evidence (see ensuing text) suggests that the tetraploid event occurred even earlier.

The first clear evidence of the family Salmonidae is a middle Eocene fossil, *Eosalmo driftwoodensis*. This fossil was first described from the Driftwood Creek site in central British Columbia (BC) (Wilson 1977) and later found at several sites in southern BC and northern Washington (Wilson and Williams 1992). For a number of reasons, *E. driftwoodensis* represents an important stage in salmonid evolution. First, it shares some but not all of the derived characters that define the salmonine lineage. This is strong evidence that the whitefish and grayling lineages had diverged from the salmonine lineage (char, trout and salmon) before the evolution of *Eosalmo* (Wilson and Williams 1992). Stearley (1992) suggests *Eosalmo* is an "archaic trout." He includes among the archaic trout the extant Eurasian salmonines (*Brachymystax, Acantholingua, Salmothymus*, and *Platysalmo*). These fish are often ignored by North American authors but are important to understanding the evolutionary history of the salmonine lineage (Norden 1961). The presence of *Eosalmo* in British Columbia in the Eocene (Wilson and Williams 1992) places salmonine fish in the Pacific Northwest ~40–50 mya. Since the whitefish and grayling lineages share a tetraploid genome with salmonines, and these lineages diverged before the evolution of *Eosalmo*, the tetraploid event that produced the salmonoid fishes must have occurred >50 mya.

After *Eosalmo* there is a gap in the salmonine fossil record until the late Miocene (~7 mya). Smith and Miller (1985) reported a trout-like fossil, which they provisionally assigned to the extant Eurasian genus *Hucho*, from the Miocene Clarkia Lake beds in Idaho. Later strata (Miocene-Pliocene boundary, and Pliocene) from Idaho and Oregon provide an array of salmonine fossils (Cavender and Miller 1972, Kimmel 1975, Smith 1975) including several species assigned to *Oncorhynchus* (one of which was a non-anadromous planktivore remarkably like kokanee) and an extinct genus, *Paleolox*. Stearley's (1992) phylogeny suggests *Paleolox* is the sister taxon of the chars, *Salvelinus*. Later, however, Stearley and Smith (1993) placed *P. larsoni* in the genus *Salvelinus*. Together, these fossils indicate that most of the salmonine diversity, at least at the generic level, had evolved before the beginning of the Pliocene (about 5-6 mya). Thus, the fossil record clearly establishes that not only was *Oncorhynchus* present in the Pacific drainages of western North America at this time, but also that the rainbow, cutthroat, and Pacific salmon lineages had diverged by the beginning of the Pliocene (Stearley and Smith 1993).

Consequently, the split between *Oncorhynchus* and *Salmo* must have occurred sometime well before the Pliocene. All known *Oncorhynchus* species, extinct and extant, are associated with North Pacific drainages (except for a few western Arctic populations of Pacific salmon), and all *Salmo* species are associated with Atlantic drainages (including the Mediterranean, Black,

Caspian and Aral seas). This distribution pattern suggests that the divergence between *Salmo* and *Oncorhynchus* occurred after the formation of the North Atlantic Ocean and its peripheral seas, but before a northern connection opened between the Atlantic and Pacific oceans. The North Atlantic Ocean started to form in the late Mesozoic, and sea-floor spreading continued into the late Eocene. Throughout this time (~50 million years), a northern land connection between Eurasia and North America separated the two oceans. This isolation lasted until the mid-Pliocene (~4 mya) when the sudden appearance of fossil mollusks of Pacific origin in Icelandic strata indicates the opening of a connection between the two oceans (Einarsson et al. 1967). Because species resembling the modern sockeye, pink (*O. gorbuscha*), and chum (*O. keta*) salmon clades of *Oncorhynchus* were present in Idaho and Oregon at least 6 mya (Smith 1992), the origin of the genus must predate the opening of this northern connection. Unfortunately, the absence of fossil material from between the Eocene and Miocene periods leaves the question of *Oncorhynchus*' time of origin unanswered.

However, how long the two genera have been separate can be estimated indirectly. This method involves gene sequence differences and the use of a molecular clock. It is a method fraught with assumptions and must be used with caution; however, Devlin (1993) has compared sequence data for the growth hormone genes in rainbow trout, sockeye salmon and Atlantic salmon. His paper is a model of caution, clearly spelling out both the assumptions and sources of error in his estimates of divergence times. Using shared and derived sequence differences, he suggested that *Salmo* and *Oncorhynchus* diverged ~20 mya (i.e., early Miocene) at a time when rainbow trout and sockeye salmon still shared a common gene pool.

Speciation in *Oncorhynchus*

Once gene exchange between populations is severed, genetic divergence is inevitable. Even in the unlikely case of identical selection regimes, given enough time, mutations at neutral loci will produce genetic divergence. However, normally, geographically isolated populations are subject to different selection regimes and this accelerates the process of divergence. Thus, if populations are isolated long enough, their gene pools will diverge and eventually give rise to separate species. Speciation (the division of a single species into two or more species) is a particular case in the general process of lineage splitting (cladogenesis). Although cladogenesis in *Oncorhynchus* has proceeded for millions of years, we can only study the process directly in living animals. Thus, past divergences can be inferred through phylogenetic analyses, but insights into the process of divergence are obtained through observation of extant populations.

Stearley (1992) analyzed relationships in most of the living, and fossil, members of *Oncorhynchus* and recognized 13 extant species; however, the specific status of some forms like the "redband" trout (*O. mykiss* sspp.), the Apache (*O. apache*) and Gila trout (*O. gilae*), the golden trout (*O. aguabonita*), and various members of the Asian *O. masou* clade is uncertain (Hikita 1962, Kato 1978, Nakano et al. 1990, Behnke 1992, Stearley and Smith 1993, Utter et al. 1993). Nevertheless, considerable speciation clearly has occurred since the origin of the genus. Although the cladograms in Stearley (1992, Fig. 1c) and Stearley and Smith (1993, Fig. 12) are provisional, they suggest that the Mexican golden trout (*O. chrysogaster*) represents a basal split in the *Oncorhynchus* lineage, followed closely by the divergence of the cutthroat lineage and later by the divergence of the Apache and Gila trout lineage. Interestingly, their data suggest that the closest sister group to the rainbow lineage (including golden and redband trout) is not the cutthroat trout but the Pacific salmon lineage. In contrast, mtDNA sequence data support the traditional view that rainbow and cutthroat trout are sister species (Shedlock et al. 1992). Among Pacific salmon, the cherry salmon (*O. masou*) usually is considered the earliest divergence (Hikita 1962, Stearley 1992, Stearley and Smith 1993); however, the mtDNA data suggest that the coho (*O. kisutch*) and chinook (*O. tshawytscha*) salmon lineage diverged before the cherry salmon (Shedlock et al. 1992). There is general agreement that the chum, pink, and sockeye salmon lineages diverged in sequence after the other species.

The date of the earliest divergence (Mexican golden trout) is unknown; however, if *Salmo* and *Oncorhynchus* diverged ~20 mya (Devlin 1993) and pink and sockeye salmon (the most recent divergence) ~6 mya (Smith 1992), most of the speciation in *Oncorhynchus* must have occurred during the Miocene. However, this does not exclude more recent divergences. Indeed, the splits between Apache and Gila trout, and between golden, redband, and rainbow trout, probably are of Pleistocene origin. The geographic distributions of members of the early divergences in *Oncorhynchus* are concentrated near the southern margin of the North American range of the genus. They include most of the extant Pacific trout and an extinct species, *O. australis*, described by Cavender and Miller (1982) from Pleistocene deposits near Lake Chapala, Jalisco, Mexico. This geographic pattern suggests that most of the speciation in Pacific trout occurred near the southern edge of the North American range of *Oncorhynchus*.

The distribution of fossil and living Pacific salmon presents a more complex picture. The fossil material argues for a center of speciation in the Pacific northwest of North America; however, if the masou clade is the earliest divergence of a Pacific salmon (Hikita 1962, Stearley 1992, Stearley and Smith 1993), this lineage is restricted to Asia. Interestingly, the pre-Pleistocene fossil material from North America represents the later divergences (i.e., the chum, pink, and sockeye salmon lineages; Smith 1992) but not the coho and chinook salmon lineages; however, a chinook salmon otolith is recorded from the Pleistocene of northern California (Fitch 1970). Perhaps the chum, pink, and sockeye salmon lineages originated in the Pacific Northwest while the coho and chinook originated elsewhere. The molecular evidence suggests that coho and chinook salmon are sister species. Shedlock et al. (1992) analyzed sequence data from the mtDNA control region (D-loop) and indicated that coho and chinook salmon are sister species. In addition, using SINEs (short interspersed elements), Murata et al. (1993) also concluded that coho and chinook salmon are sister species, and Du et al. (1993) indicated that coho and chinook share a male-specific growth hormone pseudogene that, so far, has not been found in other Pacific salmon or in Pacific trout (R. H. Devlin, Dep. Fisheries and Oceans, West Vancouver Laboratory, Vancouver, BC, pers. comm.). Apparently, this pseudogene evolved in the coho-chinook salmon lineage after that lineage diverged from the other Pacific salmon.

Regardless of when or where the various lineages diverged, to understand speciation in *Oncorhynchus* we need to understand what drives divergence. Basically, speciation involves the splitting of a single genome into two or more independent genetic units, and to understand this process we shift from phylogenetics to evolutionary ecology and from fossils to living fish. In nature, selection constantly "fine tunes" organisms to their local environments and, if undisturbed by gene flow from adjacent populations, this microevolution leads to local adaptation and eventually to genetic divergence. One striking feature of microevolution in salmonines is the repeated independent evolution of similar phenotypes (behavioral and morphological) under similar ecological conditions (Kelso et al. 1981, Healey 1983, Leider et al. 1984, Foote et al. 1989, Swain and Holtby 1989, Verspoor and Cole 1989, Taylor 1990, Taylor et al. 1996). This pervasive parallel evolution implies that ecological factors drive microevolution in *Oncorhynchus*; perhaps the best example of parallel evolution in the genus is the kokanee salmon. I use this nonmigratory form of sockeye salmon to argue that microevolution can produce speciation in *Oncorhynchus*.

Allozyme (Foote et al. 1989), mitochondrial and minisatellite DNA (Taylor et al. 1996) data from both North America and Asia indicate that kokanee salmon have evolved repeatedly from anadromous sockeye. In some cases the kokanee salmon are landlocked (isolated from anadromous sockeye by a physical barrier), but in other cases sockeye and kokanee intermingle at spawning time. The two forms have different life histories and are subject to different selection pressures; sockeye salmon undergo long, often arduous migrations and spend years at sea whereas kokanee are permanent residents of freshwater. Experimental studies (Wood and Foote 1990, Taylor and Foote 1991, Foote et al. 1992) have demonstrated behavioral, developmental, and ecological differences between sockeye and kokanee salmon and, for a number of life history traits, that sockeye-kokanee hybrids are inferior to the parental forms. Together, these studies provide compelling evidence of adaptive divergence between the two life history types; however, is it sufficient to prevent interbreeding? Obviously, this question cannot be answered when sockeye and kokanee salmon are allopatric; however, where breeding sockeye and kokanee intermingle, we can test for reproductive isolation.

Like many salmonines, sockeye display size assortative mating (Hanson and Smith 1967, Foote and Larkin 1988) and, normally, anadromous sockeye are larger than kokanee. Consequently, interbreeding between the two forms is limited. However, reproductive isolation usually is incomplete and kokanee salmon males can act as sneakers in sockeye spawnings (Foote and Larkin 1988). Yet, at some sites (e.g., in the Babine, Takla, and Shuswap river systems, BC), even in the face of persistent gene flow, the two life history forms maintain genetic differences (Foote et al. 1989, Taylor et al. 1996). This situation suggests substantial selection against hybrids. Thus, genetically, at least some sympatric sockeye and kokanee salmon populations have diverged to the point that a balance between selection and gene flow maintains the two life history forms as separate, but linked, entities.

However, in one remarkable case a divergence involving kokanee salmon appears to have reached the level of biological species. Biological species, as opposed to morphological species, are defined by the capacity to maintain themselves in sympatry as independent genetic entities. This species concept is vexing to taxonomists because it can be applied only in sympatry and, morphologically, the species often are difficult to define. These limitations restrict the applicability of the concept but, where it can be applied, the biological species concept provides the least arbitrary of the many definitions of species. In Lake Kronotskiy, Kamchatka, two kinds of kokanee salmon occur: one feeds primarily on benthic organisms and spawns in July and early August; the other feeds on plankton and spawns from mid-August into September (Kurenkov 1978). They also differ in average gill raker number (a trophic trait?). Thus, these two kokanee salmon display the principal attributes of biological species: they are reproductively isolated, and show strong resource partitioning.

Lake Kronotskiy was formed when a lava flow blocked the river and access to the sea (~10,000–15,000 BP). Apparently, this barrier resulted in landlocked spring and summer races of anadromous sockeye salmon, which then gave rise to the two kokanee species (Kurenkov 1978). Thus, the initial separation into seasonal races predates geographic isolation, but once contact with anadromous sockeye salmon was eliminated, divergence between the two kokanee stocks was rapid. This suggests that competitive interactions, both reproductive and ecological, between diverging forms can accelerate the process of speciation. Indeed, in freshwater fish where reproductive isolation and resource partitioning have evolved rapidly (i.e., post-glacially), interactions between diverging forms appear to have accelerated divergence (Schluter and McPhail 1992, 1993).

In summary, experimental and field observations on sockeye salmon indicate that speciation in *Oncorhynchus* usually involves geographic isolation, followed by local adaptation and, in cases where divergent forms come into secondary contact, competitive interactions between forms. Although we can never be certain, the process of speciation in *Oncorhynchus* likely was the same in the past.

Management Implications

Aside from the academic interest, are there management implications in the evolutionary history of *Oncorhynchus*? I think so. Genetic divergence is the inevitable outcome of geographic isolation, and the propensity of Pacific salmon and trout to home to their natal streams ensures some degree of isolation among most populations; homing reduces gene flow and allows selection to fine tune populations to local conditions. On the basis of the adaptive nature of most local divergence in Pacific salmon and trout (Taylor 1991) and the prevalence of parallel evolution in the genus, natural selection clearly is the dominant force driving interpopulation divergence in *Oncorhynchus*.

A contentious issue in salmonid management is how to conserve population diversity but at the same time supplement or enhance exploited populations (Meffe 1992). Theoretically, with modern biochemical and statistical techniques it is possible to show that most *Oncorhynchus* populations are genetically different from other populations. In the Pacific Northwest, this means that there are hundreds, perhaps thousands, of distinct populations. Managing all of these populations as separate entities is a formidable if not impossible task. Clearly, some management unit other than the population is desirable, and Utter et al. (1993) argued that populations can be grouped logically through their evolutionary history. This is similar to Williams' (1992) argument that the process of cladogenesis is essentially the same regardless of the level of divergence. It starts with populations and, if genetic isolation is maintained, continues through all taxonomic levels. Although the process is continuous, the probability that genetic isolation among divergent clades will break down is different at different taxonomic levels. Thus, divergence among populations can be reversed by gene flow, but at the species level (and above) gene exchange is less likely to produce genetic fusion. This is because genetic differences among species usually are greater than genetic differences among populations, and the gene pools of species often are more integrated (co-adapted). Consequently, recombinant genotypes associated with interspecific hybridization usually are less fit (relative to parental genotypes) than recombinant genotypes associated with interpopulation hybrids. This suggests that most interpopulation divergence is ephemeral, and that the majority of divergent populations will either go extinct or be swamped by gene flow from adjacent populations.

Given that cladogenesis is a continuous process, there will be divergences that lie somewhere between populations and recognized taxa like subspecies and species. These are geographically contiguous clusters of populations (metapopulations) that share a relatively recent common evolutionary history. Typically, populations within such clusters are not genetically identical. They often differ in life histories, behaviors, and other ecologically significant traits that respond rapidly to local selection, but they usually share distinctive biochemical or morphological characters (Utter et al. 1993). Normally, these metapopulations have a geographic distribution and there is some gene flow among the constituent populations. Thus, the genetic makeup of individual populations within the assemblage probably represents a balance between local selection and gene flow among the populations. The widespread occurrence of such clusters of populations in *Oncorhynchus* (Utter et al. 1993, Taylor et al. 1996) suggests that it may be possible to manage metapopulations as single units. The main problem with this approach lies in defining the metapopulations. Their evolutionary history and the geological history of the regions they occupy, as well as physical barriers to gene flow within regions, define metapopulations, not political boundaries or administratively convenient management units. As an aid in determining metapopulation boundaries, managers may profit from exploring the biogeographic literature on native species in their region. After all, salmonids in most regions share a recent geological history with the rest of the fauna; thus, some concordance is expected between the pattern of genetic structuring in salmonid and nonsalmonid species.

Acknowledgments

R.H. Devlin and E.B. Taylor guided me through the molecular literature, and C.J. Foote provided insights into the evolution of kokanee salmon. I am grateful for their help and, of course, any misinterpretations are my own.

Literature Cited

Allendorf, F.W. 1978. Protein polymorphism and the rate of loss of duplicate gene expression. Nature 272: 76-79.

Allendorf, F.W. and G.H. Thorgaard. 1984. Tetraploidy and the evolution of salmon fishes, p. 1-53. *In* B. Turner (ed.), Evolutionary Genetics of Fishes. Plenum Press, New York.

Behnke, R.J. 1992. Native trout of western North America. American Fisheries Society, Monograph 6. American Fisheries Society, Bethesda, Maryland.

Brooks, D.R. and D.A. McLennan. 1992. Historical ecology as a research program, p. 76-113. *In* R. L. Mayden (ed.), Systematics, Historical Ecology, and North American Freshwater Fishes. Stanford University Press, Stanford, California.

Cavender, T.M. and R.R. Miller. 1972. *Smilidonichthys rastrosus* a new Pliocene salmonid fish. Museum of Natural History, University of Oregon, Bulletin 18.

Cavender, T.M. and R.R. Miller. 1982. *Salmo australis*, a new species of fossil salmonid from southwestern Mexico. Contributions from the Museum of Paleontology, University of Michigan 26: 1-17.

Devlin, R.H. 1993. Sequence of sockeye salmon type 1 and 2 growth hormone genes and the relationship of rainbow trout with Atlantic and Pacific salmon. Canadian Journal of Fisheries and Aquatic Sciences 50: 1738-1748.

Du, S.J., R.H. Devlin, and C.L. Hew. 1993. Genomic structure of growth hormone genes in chinook salmon (*Oncorhynchus tshawytscha*): presence of two functional genes, GH-I and GH-II, and a male-specific pseudogene, GH-Ψ. DNA and Cell Biology 12: 739-751.

Einarsson, T., D.M. Hopkins, and R.R. Doell. 1967. The stratigraphy of Tjornes, northern Iceland, and the history of the Bering land bridge, p. 312-325. *In* D. M. Hopkins (ed.), The Bering Land Bridge. Stanford University Press, Stanford, California.

Fitch, J.E. 1970. Fish remains, mostly otoliths and teeth, from the Palos Verdes Sand (late Pleistocene) of California. Contributions to Science, Los Angeles County Museum 199.

Foote, C.J. and P.A. Larkin. 1988. The role of mate choice in the assortative mating of anadromous and non-anadromous sockeye salmon (*Oncorhynchus nerka*). Behaviour 106: 43-62.

Foote, C.J., C.C. Wood, and R.E. Withler. 1989. Biochemical genetic comparison between sockeye salmon and kokanee, the anadromous and non-anadromous forms of *Oncorhynchus nerka*. Canadian Journal of Fisheries and Aquatic Sciences 46: 149-158.

Foote, C.J., C.C. Wood, W.C. Clarke, and J. Blackburn. 1992. Circannual cycle of seawater adaptability in *Oncorhynchus nerka*: genetic difference between sympatric sockeye salmon and kokanee. Canadian Journal of Fisheries and Aquatic Sciences 49: 99-109.

Hanson, S.J. and H.D. Smith. 1967. Mate selection in a population of sockeye salmon (*Oncorhynchus nerka*) of mixed age groups. Journal of the Fisheries Research Board of Canada 24: 1955-1977.

Hartley, S.E. 1987. The chromosomes of salmonid fishes. Biological Reviews 62: 197-214.

Healey, M.C. 1983. Coastwide distribution and ocean migration patterns of stream- and ocean-type chinook salmon (*Oncorhynchus tshawytscha*). Canadian Field-Naturalist 97: 427-433.

Hikita, T. 1962. Ecological and morphological studies of the genus *Oncorhynchus* (Salmonidae) with particular consideration on phylogeny. Scientific Reports of the Hokkaido Salmon Hatchery, No. 17.

Kato, F. 1978. Morphological and ecological studies on two forms of *Oncorhynchus rhodurus* found in Lake Biwa and adjoining inlets. Japanese Journal of Ichthyology 25: 197-204.

Kelso, B.W., T.G. Northcote, and C.F. Wehrhahn. 1981. Genetic and environmental aspects of the response to water current by rainbow trout (*Salmo gairdneri*) originating from inlet and outlet streams of two lakes. Canadian Journal of Zoology 59: 2177-2185.

Kendall, A.W. and R.J. Behnke. 1984. Salmonidae: development and relationships, p. 142-149. *In* Moser, H. G. (ed.), Ontogeny and Systematics of Fishes. American Society of Ichthyologists and Herpetologists, Special Publication 1. Allen Press, Lawrence, Kansas.

Kimmel, P.G. 1975. Fishes of the Miocene-Pliocene Deer Butte formation, southeast Oregon. Museum of Paleontology, University of Michigan, Papers on Paleontology, no. 14: 69-87.

Kurenkov, S.I. 1978. Two reproductively isolated groups of Kokanee salmon, *Oncorhynchus nerka kennerlyi*, from Lake Kronotskiy. Journal of Ichthyology 18: 526-533.

Lieder, S.A., M.W. Chilcote, and J.J. Loch. 1984. Spawning characteristics of sympatric populations of steelhead trout (*Salmo gairdneri*): evidence for partial reproductive isolation. Canadian Journal of Fisheries and Aquatic Sciences 41: 1454-1467.

Lim, S. T., and G. S. Bailey. 1977. Gene duplication in salmonid fish: evidence for duplicated but catalytically equivalent A(4) lactate dehydrogenases. Biochemical Genetics 15: 707-721.

Lim, S.T., R.M. Kay, and G.S. Bailey. 1975. Lactate dehydrogenase isoenzymes of salmonid fish: evidence for unique and rapid functional divergence of duplicated H_4 lactate dehydrogenases. Journal of Biological Chemistry 250: 1790-1800.

Meffe, G. 1992. Techno-arrogance and halfway technologies: salmon hatcheries on the Pacific coast of North America. Conservation Biology 6: 350-354.

Murata, S., Takasaki, N., Saitoh, H., and N. Okada. 1993. Determination of the phylogenetic relationships among Pacific salmonids by using short interspersed elements (SINEs) as temporal landmarks of evolution. Proceedings of the National Academy of Science, USA 90: 6995-6999.

Nakano, S., T. Kachi, and M. Nagoshi. 1990. Restricted movement of the fluvial form of red-spotted masu salmon, *Oncorhynchus masou rhodurus*, in a mountain stream, central Japan. Japanese Journal of Ichthyology 38: 158-163.

Neave, F. 1958. The origin and speciation of *Oncorhynchus*. Transactions of the Royal Society of Canada, Third Series, Vol. 52: 25-39.

Norden, C.R. 1961. Comparative osteology of representative salmonid fishes, with particular reference to the grayling (*Thymallus arcticus*) and its phylogeny. Journal of the Fisheries Research Board of Canada 18: 679-791.

Sanford, C.P.J. 1990. The phylogenetic relationships of salmonoid fishes. Bulletin British Museum of Natural History (Zoology) 56: 145-153.

Schluter, D. and J.D. McPhail. 1992. Ecological character displacement and speciation in sticklebacks. American Naturalist 140: 85-108.
Schluter, D. and J.D. McPhail. 1993. Character displacement and replicate adaptive radiation. Trends in Evolution and Ecology 8: 197-200.
Shedlock, A.M., J.D. Parker, D.A. Crispin, T.W. Pietsch, and G.C. Burmer. 1992. Evolution of the salmonid mitochondrial control region. Molecular Phylogenetics and Evolution 1: 179-192.
Smith, G.R. 1975. Fishes of the Pliocene Glenns Ferry formation, southwest Idaho. Museum of Paleontology, University of Michigan, Papers on Paleontology, no. 14: 1-68.
Smith, G.R. 1981. Late Cenozoic freshwater fishes of North America. Annual Review of Ecology and Systematics 12: 163-193.
Smith, G.R. 1992. Introgression in fishes: significance for paleontology, cladistics, and evolutionary rates. Systematic Biology 41: 41-57.
Smith, G.R. and R.R. Miller. 1985. Taxonomy of fishes from Miocene Clarkia Lake beds, Idaho, p. 75-83. *In* C.J. Smiley (ed.), Late Cenozoic History of the Pacific Northwest. Pacific Division, American Association for the Advancement of Science, San Francisco, California.
Smith, G.R. and R.F. Stearley. 1989. The classification and scientific names of rainbow and cutthroat trout. Fisheries 14: 4-10.
Stearley, R.F. 1992. Historical ecology of the Salmoninae, p. 622-658. *In* R. L. Mayden (ed.), Systematics, Historical Ecology, and North American Freshwater Fishes. Stanford University Press, Stanford, California.
Stearley, R.F. and G.R. Smith. 1993. Phylogeny of the Pacific trouts and salmons (*Oncorhynchus*) and genera of the family Salmonidae. Transactions of the American Fisheries Society 122: 1-33.
Swain, D.P. and L.B. Holtby. 1989. Differences in morphology and agonistic behaviour in coho salmon (*Oncorhynchus kisutch*) rearing in a lake or its tributary stream. Canadian Journal of Fisheries and Aquatic Sciences 46: 1406-1414.
Taylor, E.B. 1990. Environmental correlates of life history variation in juvenile chinook salmon, *Oncorhynchus tshawytscha* (Walbaum). Journal of Fish Biology 37: 1-17.
Taylor, E.B. 1991. A review of local adaptation in Salmonidae, with particular reference to Pacific and Atlantic salmon. Aquaculture 98: 185-207.
Taylor, E.B. and C.J. Foote. 1991. Critical swimming velocities of juvenile sockeye salmon and kokanee, the anadromous and non-anadromous forms of *Oncorhynchus nerka*. Journal of Fish Biology 38: 407-419.
Taylor, E. B., C. J. Foote, and C. C. Wood. 1996. Molecular evidence for parallel life history evolution within a Pacific salmon (sockeye salmon and kokanee, *Oncorhynchus nerka*). Evolution 50: 401-416.
Thomas, W.K. and A.T. Beckenbach. 1989. Variation in salmonid mitochondrial DNA: evolutionary constraints and mechanisms of substitution. Journal of Molecular Evolution 29: 233-245.
Thomas, W.K., R.E. Withler, and A.T. Beckenbach. 1986. Mitochondrial DNA analysis of Pacific salmonid evolution. Canadian Journal of Zoology 64: 1058-1064.
Utter, F.M., J.E. Seeb, and L.W. Seeb. 1993. Complementary uses of ecological and biochemical genetic data in identifying and conserving salmon populations. Fisheries Research 18: 59-76.
Verspoor, E. and L.J. Cole. 1989. Genetically distinct sympatric populations of resident and anadromous Atlantic salmon, *Salmo salar*. Canadian Journal of Zoology 67: 1453-1461.
Williams, G.C. 1992. Natural selection: domains levels and challenges. Oxford Series in Ecology and Evolution. Oxford.
Wilson, G.M., W.K. Thomas, and A.T. Beckenbach. 1985. Intra- and inter-specific mitochondrial DNA sequence divergence in *Salmo*: rainbow, steelhead, and cutthroat trouts. Canadian Journal of Zoology 63: 2088-2094.
Wilson, M.V.H. 1977. Middle Eocene Freshwater Fish from British Columbia. Royal Ontario Museum, Life Sciences Contribution 11B: 1-61.
Wilson, M.V.H. and R.R.G. Williams. 1992. Phylogenetic, biogeographic, and ecological significance of early fossil records of North American freshwater teleostean fishes, p. 224-244. *In* R. L. Mayden (ed.), Systematics, Historical Ecology, and North American Freshwater Fishes. Stanford University Press, Stanford, California.
Wood, C.C. and C.J. Foote. 1990. Genetic differences in the early development and growth of sympatric sockeye salmon and kokanee (*Oncorhynchus nerka*). Canadian Journal of Fisheries and Aquatic Sciences 47: 2250-2260.

Status of Pacific Northwest Salmonids

Pacific Salmon Status and Trends—A Coastwide Perspective

Willa Nehlsen

Abstract

The diversity of Pacific salmon (*Oncorhynchus* spp.) varies from north to south with the greatest diversity of species found in the intermediate latitudes (northern Oregon to southern Alaska). Pink (*O. gorbuscha*), chum (*O. keta*) and sockeye (*O. nerka*) salmon predominate in the northern part of the range of Pacific salmon (British Columbia and Alaska), while chinook (*O. tshawytscha*), coho (*O. kisutch*) and steelhead (*O. mykiss*) are stronger in the south (Washington, Oregon, Idaho and California). Reviews of the status of Pacific salmon in this book suggest that the health of salmon populations generally deteriorates in a southward direction. This is due partly to greater human-caused damage to freshwater habitats outside Alaska and to ocean productivity cycles creating unfavorable conditions off Washington, Oregon, and California. Degradation of freshwater habitats in the southern part of the range must be reversed before the next low ocean productivity cycle, and productive habitats in the northern part of the range should be protected in anticipation of low ocean productivity.

Introduction

Pacific salmon (*Oncorhynchus* spp.) spawn and rear in rivers at the margins of the North Pacific Ocean from ~32° to 70°N latitude (Meehan and Bjornn 1991) in Asia and North America. The North American spawning distribution of the species considered here (pink, *O. gorbuscha*; chum, *O. keta*; sockeye, *O. nerka*; chinook, *O. tshawytscha*; coho, *O. kisutch*; and steelhead, *O. mykiss*) ranges from southern California to northern Alaska.

Pacific salmon species all are generally anadromous, spending a portion of their lives at sea and returning to freshwater to spawn. They migrate from one ecosystem to another, riverine to estuarine to marine and back again, to complete their life histories. However, the Pacific salmon species vary considerably with regard to how much time they spend in each type of ecosystem and when the migratory movements occur. Juvenile salmon may remain in freshwater for a few days, a few months, or several years. Adult salmon may remain at sea for a period from <1 year to 5 years. Considerable variation also occurs in life-history characteristics within species of Pacific salmon, which is the result of local stock adaptations (see Nehlsen et al. 1991 for discussion). Local stock adaptations are created and maintained by the well-known tendency of Pacific salmon to home to the stream of origin with nearly 100% fidelity.

The diversity of Pacific salmon species varies from north to south, and this can partly be explained by the variation in freshwater life-history requirements among the species (Schalk 1977). The greatest diversity in species is found in the intermediate latitude zone, from ~ 45° to 60° N latitude (northern Oregon to southern Alaska). All salmon species are found in this zone. North of the intermediate zone, extreme variations in temperature reduce the stability of riparian environments, favoring species that spend less time in freshwater (pink and chum salmon) (Schalk 1977); species requiring longer freshwater residence (chinook and coho salmon and steelhead) are not favored in this region. Sockeye salmon, which also are favored in northern latitudes, circumvent the need for stability of riparian environments by rearing in lakes (Jim Lichatowich, Alder Fork Consulting, Sequim, Washington, pers. comm.). South of the intermediate zone, increased freshwater productivity makes freshwater residence a more profitable strategy, favoring chinook and coho (Schalk 1977).

Schalk's (1977) concept is illustrated by the analysis of Fredin (1980), who examined geographic trends in commercial harvest over 6 decades for sockeye, pink, chum, coho, and chinook salmon. The majority of pink and sockeye salmon were caught in Alaska, with fewer caught in British Columbia and Washington, and negligible numbers caught in Oregon and California. Conversely, chinook and coho salmon made up the majority of the catch for Washington, Oregon, and California (Fredin 1980).

Since 1987, four reviews of the status of Pacific salmon and steelhead over large parts of their North American ranges have been published (Konkel and McIntyre 1987, Nehlsen et al. 1991, The Wilderness Society 1993, Huntington et al. 1996). Konkel and McIntyre (1987) analyzed trends in escapement (number of adults returning to spawn) in 886 Pacific salmon populations in Oregon, Washington, Idaho, California, and Alaska. They found statistically significant trends in ~30% of these populations. Coho and chum salmon escapement trends were predominantly decreasing (the ratio of increasing to decreasing was 1 to 3) throughout the five-state area. For chinook, sockeye, and pink salmon, trends were predominantly increasing in Alaska and either lacking or predominantly decreasing in the other states. Konkel and McIntyre (1987) concluded that the status of coho and chum salmon was of greatest concern because declines in these species occurred throughout the study area. However, although chinook, sockeye, and pink salmon escapements showed increasing trends in Alaska, Konkel and McIntyre (1987) considered their status by no means secure because of widespread declines in the other states.

Four years later, Nehlsen et al. (1991) identified 214 native Pacific salmon populations in Oregon, Washington, Idaho, and California whose persistence was in jeopardy: 101 populations at high risk of extinction, 58 at moderate risk of extinction, 54 of special concern, and one (Sacramento River winter chinook) listed as threatened under the Endangered Species Act of 1973 and as endangered under the laws of the state of California. These imperiled populations were generally distributed throughout the four western states, in both inland and coastal basins. All Pacific salmon, steelhead, and sea-run cutthroat trout (*O. clarki*) species were represented. We also identified 106 extinct populations. For the Columbia River Basin, Williams et al. (1992) estimated that of the populations historically present, ~35% were extinct, 40% met the criteria for our list of depleted stocks, and 25% were considered secure. The Forest Ecosystem Management Assessment Team (FEMAT 1993), reviewing Nehlsen et al. (1991) and subsequent assessments by others, listed 314 stocks of salmon and steelhead at risk of extinction in Oregon, Washington, Idaho, and California.

The Wilderness Society (1993) mapped the distribution of at-risk native Pacific salmon populations in California, Oregon, Washington, and Idaho using data from Nehlsen et al. (1991)

and other sources. It concluded that Pacific salmon were either extinct or at risk of extinction across the vast majority of the survey area. They found that in over half of the total freshwater habitat of Pacific salmon, 50–100% of the native salmon species were extinct or at risk of extinction; in over one-third of the total freshwater habitat, all Pacific salmon species were extinct; in only 6% of the freshwater habitat area were no Pacific salmon species known to be declining.

In a study that complements the assessments of at-risk populations, Huntington et al. (1996) surveyed healthy native stocks in California, Oregon, Washington, and Idaho. The survey identified 99 healthy native stocks whose abundance is believed to be at least one-third that of historical levels; 20 of the 99 are believed to be at least two-thirds as abundant as they were historically. The numbers of healthy stocks identified by Huntington et al. (1996) are considerably smaller than the numbers of at-risk stocks identified by Nehlsen et al. (1991) and FEMAT (1993) for the same geographic area.

Coastwide Overview of Pacific Salmon and Steelhead—Current Status And Trends

Papers in this book (Hassemer et al. 1996, Johnson et al. 1996, Kostow 1996, Mills et al. 1996, Northcote and Atagi 1996, and Wertheimer 1996) comprise the first comprehensive (five US states and British Columbia) review of the status of Pacific salmon and steelhead across their North American range. These papers consider the status of salmon and steelhead stocks; a stock is explicitly or implicitly defined as the fish spawning in a particular lake or stream(s) at a particular season, which to a substantial degree do not interbreed with any group spawning in a different place, or in the same place at a different season (Ricker 1972). The authors emphasize native, naturally spawning stocks although Johnson et al. (1996) give equal consideration to the status of other (i.e., non-native and hatchery) stocks as well. The role and importance of native stocks are described by Nehlsen et al. (1991) and Huntington et al. (1996).

The temporal frame for consideration of current status and trends of Pacific salmon is variable but does not include information prior to the early 1950s. The state and province reviews in this book used varying baselines, depending primarily on the availability of data. For example, Northcote and Atagi (1996) explicitly used two baselines, the turn of the century and 1953 (the latter is reflected in Table 1). The earliest baseline used by Johnson et al. (1996) is the late 1960s. The other authors used whatever baseline was available on a case-by-case basis, referring to the historical context where possible, but generally focusing on the period since the 1950s or later. In general, because all stocks outside Alaska have declined since before European settlement, historically based information would show no resolution among stocks today (all would be considered to be much depleted; Table 1). However, Lichatowich (1996) cautions against using too recent a baseline, pointing out that this masks the magnitude of long-term declines and leads to errors in management; rather, he considers that reconstruction of historical stock abundance and habitat status is a necessary part of understanding Pacific salmon status and trends and developing appropriate management responses.

The terms used to describe status and trends vary considerably; therefore, I attempted to provide as much consistency across the North American range as possible for comparative purposes, while conveying as much as possible of the original authors' intent (Table 1). Verbal

Table 1. Recent population status and trends for Pacific salmon species (based on native naturally spawning stocks). Source: Wertheimer (1996), Alaska; Northcote and Atagi (1996), British Columbia; Johnson et al. (1996), Washington; Kostow (1996), Oregon; Hassemer et al. (1996), Idaho; and Mills et al. (1996), California. Other references used are Nehlsen et al. (1991), Cooper and Johnson (1992), and Washington Department of Fisheries et al. (1993).[a] The designations "endangered" or "threatened" refer to current status under the federal Endangered Species Act (ESA) of 1973. "Under review" means that National Marine Fisheries Service is reviewing the status in response to a petition for listing under the ESA or on its own initiative.

	Alaska	British Columbia	Washington[b]	Oregon	Idaho	California
Pink	Increasing	Increasing	Predominantly healthy, but two Puget Sound stocks under review	Not applicable	Not applicable	Extinct
Sockeye	Increasing except in southeast	Increasing	Mixed	Remnant	Endangered	Not applicable
Chum	Decreasing or stable since 1984 record high	Increasing in Fraser River; mixed elsewhere	Predominantly healthy except two stocks in Puget Sound under review, and lower Columbia populations declining or at low levels	At very low levels with scattered extinctions	Not applicable	Extinct
Chinook	Escapement increasing statewide; harvest constrained	Increasing in Fraser River; decreasing elsewhere	Majority of Columbia River and Puget Sound stocks declining or at low levels (five Puget Sound and mid-Columbia River summer stocks under review). Snake River stocks endangered. Majority of coastal stocks healthy	Snake River: declining, endangered; elsewhere: fall run—declining south coast; stable otherwise; spring run—declining or very small except Deschutes, John Day, North Umpqua rivers	Endangered (except Clearwater River)	Fall run—declining; winter run—endangered; spring run—remnant
Coho	Increasing except in southeast	Declining	Predominantly healthy (Puget Sound and coast); under review	Extinct east of Deschutes; very small or extinct elsewhere (except Clackamas); under review	Not applicable	Declining or remnant; under review
Steelhead	Declining	Declining	Known stocks predominantly healthy in Puget Sound and coast (one under review). Predominantly declining or at low levels, or unknown in Columbia. Statewide declining trend. Under review.[c]	Declining; under review	Declining; under review	Declining or remnant; under review

[a]The papers in this volume were the primary sources of information for the table; other sources were relied upon where additional information was needed for completeness.
[b]Descriptors for Washington are based on definitions in the Salmon and Steelhead Stock Inventory (SASSI; Washington Department of Fisheries et al. 1993). Stocks identified therein as "depressed" are referred to here as "declining or at low levels." Stocks identified in the SASSI as "critical" (defined as experiencing production levels that are so low that permanent damage to the stock is likely or has already occurred) are referred to here as "under review" because they have been petitioned for listing under the ESA. The "healthy" designation used in SASSI—"a stock of fish experiencing production levels consistent with its available habitat and within the natural variations in survival for the stock"—is used here.
[c]Washington Department of Fisheries et al. (1993) considered Puget Sound and coast stocks to be healthy, but these stocks have been declining since the mid-1980s (Cooper and Johnson 1992).

descriptions of trends are relied on as much as possible; however, the term "healthy" as used by Johnson et al. (1996) could not be described as a trend in a straightforward way. Johnson et al. (1996) use the term to refer to "a stock of fish experiencing production levels consistent with its available habitat and within the natural variations in survival for the stock." Under this definition, a stock undergoing a declining trend (such as most steelhead stocks in Washington) still is considered to be healthy. Regardless of its merits (see Lichatowich 1996), this definition does create an apparent inconsistency with the descriptions of trends provided for the other areas.

Pink, Sockeye and Chum Salmon

Alaska currently produces ~79% of the total North American commercial salmon harvest (British Columbia produces ~17% and the Pacific Northwest ~4%) (Wertheimer 1996). In 1993, Alaska enjoyed an all-time (since 1878) record commercial harvest of 193 million fish. The bulk of this catch (57%) was provided by pink salmon, followed by sockeye (33%), chum (6%), coho (3%), and chinook (<1%) (H. Savikko, Alaska Dep. Fish and Game, Juneau, Alaska, unpubl. data). On average, 27% of the pink and 7% of the sockeye salmon catch are composed of hatchery fish. Record catches of pink and sockeye salmon also have been accompanied by increasing escapement trends, with the exception of sockeye in the southeast region (Wertheimer 1996).

In the Fraser River drainage (British Columbia), pink salmon escapement has increased from <1 million fish to >3 million in the past decade, and sockeye salmon escapement increased from 1.5 to 2 million. However, these recent increases should be considered in the context of historical average runs of nearly 50 million pink and 160 million sockeye (Northcote and Atagi 1996). Pink and sockeye salmon runs returning to other drainages of the province also have shown recent increases (Northcote and Atagi 1996).

The generally positive status of pink salmon also is demonstrated in Washington, where 9 of 14 Puget Sound pink salmon stocks are considered to be healthy. However, two stocks are considered to be critical (defined as experiencing production levels that are so low that permanent damage to the stock is likely or has already occurred) (Washington Department of Fisheries [WDF] et al. 1993, Johnson et al. 1996). The two critical stocks are under review for listing under the Endangered Species Act (ESA) of 1973. Pink salmon apparently never occurred in Oregon and are extinct in California (Nehlsen et al. 1991, The Wilderness Society 1993).

The status of sockeye salmon begins to deteriorate in the southern and more eastern portions of its range. In Washington, the Columbia River and one coastal stock are considered to be healthy, while another coastal stock is considered to be declining or at low levels and one is of unknown status (WDF et al. 1993, Johnson et al. 1996). In Oregon, only one very small population (Deschutes River) remains of several historical sockeye salmon stocks (Kostow 1996), and of several historical stocks in Idaho, only the Redfish Lake population (listed as endangered under the ESA) remains (Hassemer et al. 1996).

Chum salmon runs are at high levels in Alaska, although a decreasing trend is evident since the 1984 record high commercial harvest of >13 million fish (Wertheimer 1996). In British Columbia, Fraser River runs have increased but trends have been mixed in other areas, resulting in no significant overall trend. However, British Columbia chum salmon runs have decreased only twofold from turn-of-the-century levels compared with much greater decreases for the other species (Northcote and Atagi 1996). Chum salmon runs in Washington are considered to be predominantly healthy with the exception of two critical Puget Sound stocks and lower Columbia River populations that are declining or at low levels (WDF et al. 1993, Johnson et al. 1996). The two critical Puget Sound stocks are under review for listing under the ESA. Chum salmon populations are at very low levels in Oregon and many may be extinct (Kostow 1996). In California, chum salmon are extinct (Nehlsen et al. 1991, Mills et al. 1996).

Chinook and Coho Salmon and Steelhead

Chinook and coho salmon exhibit a similar southward deterioration in status, and also exhibit increasing declines in the eastern portion of their ranges. In Alaska, chinook salmon escapements are increasing. Commercial harvests are at relatively high levels (e.g., 747,000 in the record year of 1993), but not as high as the late 1970s and early 1980s when harvests reached over 800,000. Commercial harvest of chinook salmon is constrained by Pacific Salmon Commission rebuilding efforts (Wertheimer 1996). In British Columbia, recent escapements to the Fraser River are increasing while escapements to other regions are decreasing. The majority of Washington coastal stocks are considered to be healthy, but the majority of Puget Sound stocks are declining or at low levels, and five are considered to be critical (WDF et al. 1993, Johnson et al. 1996). Critical stocks are being reviewed for listing under the ESA.

In the Columbia River Basin, all Snake River chinook stocks have been recently classified as endangered under the ESA. The majority of Washington's Columbia River stocks are considered to be declining or at low levels (WDF et al. 1993, Johnson et al. 1996), and mid-Columbia summer chinook stocks are being reviewed for listing under the ESA. In Oregon's portion of the Columbia River below the Snake River confluence, most fall runs are considered to be stable, but the majority of spring runs are declining or very small (Kostow 1996). On the Oregon coast, fall runs are declining on the south coast but otherwise stable; nearly all spring runs are declining.

In California, fall chinook runs have declined from record highs in 1986 to below the escapement goal for the past 3 years. Spring chinook have been extirpated from a large portion of their range and remaining populations are at low levels. Winter chinook in the Sacramento River are listed as endangered under the ESA (Mills et al. 1996).

Coho salmon runs in Alaska yielded a record commercial harvest of >7 million fish in 1992. Escapement trends are increasing in the state, with the exception of the southeast region (Wertheimer 1996). In British Columbia, recent escapement trends are declining in all but transboundary drainages with total escapement in 1992 reaching the lowest level since the early 1950s. Coho have exhibited the largest decrease (eightfold) from historical levels in the Fraser River of all salmon species (Northcote and Atagi 1996). In Washington, Puget Sound and coastal coho runs predominantly are considered to be healthy (WDF et al. 1993, Johnson et al. 1996).

In Oregon, nearly all coho runs are very small or extinct, and in the Columbia River Basin they are extinct east of the Deschutes River (Kostow 1996). Many coho runs in California have been extirpated, and remaining runs are at very low levels (Mills et al. 1996). Coho populations in California are probably <6% of their previous levels in the 1940s, and there has been a ≥70% decline since the 1960s (Brown et al. 1994). Brown et al. (1994) found that the farther south in California a stream is located, the more likely it is to have lost its coho salmon populations. The status of coho salmon in Oregon, Washington, California, and Idaho is being reviewed for listing under the ESA.

Steelhead runs are declining throughout their range (Cooper and Johnson 1992, and papers in this book). However, as noted previously, Puget Sound and Washington coast runs are considered to be healthy (WDF et al. 1993, Johnson et al. 1996). Populations south of San Francisco Bay in California are at very low levels (Mills et al. 1996) and likely are in the most jeopardy.

Discussion

The status of Pacific salmon and steelhead in North America covers the extremes: excellent in Alaska, where all species except steelhead have experienced record high abundance in the recent decade; good in British Columbia, where only coho salmon and steelhead are declining province-wide; and mixed in Washington, bleak in Oregon, and perhaps desperate in Idaho and California.

The reviews in this book have upheld and reinforced the prognoses of previous assessments and make possible a broader understanding of the coastwide picture. Nevertheless, an assessment that is focused on general trends understates the magnitude of the problem. Such an assessment neglects the extirpation of individual spawning populations, which may not contribute significantly in terms of numbers but does contribute to genetic diversity. The loss of individual spawning populations likely is significant even in British Columbia (Northcote and Atagi 1996) and may be much greater in other areas, such as California (for example, the loss of coho salmon populations described by Brown et al. 1994). Moreover, an assessment that is focused on current trends cannot convey the full magnitude of the historical loss (Lichatowich 1996).

As the reviews in this book attest, the factors responsible for the condition of North American Pacific salmon and steelhead include climatic conditions affecting freshwater and marine habitats, human-caused damage to freshwater and estuarine habitats, and fish management. The high abundance of Pacific salmon in Alaska results from its relatively undeveloped habitat base, salmon management policies, the elimination of high-seas driftnet interceptions, enhancement by hatcheries, and favorable environmental conditions (Wertheimer 1996).

Outside Alaska, freshwater habitats have been damaged and destroyed by mining, logging, agriculture, grazing, water development projects, hydropower, urbanization, and other land and water development activities. Natural climatic conditions, such as floods and drought, have had a major impact on freshwater habitats. Factors related to fish management have contributed to declines in native salmon and steelhead runs, including hatchery programs, non-native fish introductions, and overfishing (see individual papers in this book for more detail on causes of declines).

From a coastwide perspective, there is an obvious north-to-south cline in the status of salmon populations, from relatively good in the north to very poor in the south. There also is an east-west cline, reflected by the poorer status of salmon and steelhead populations in eastern Oregon, eastern Washington, and Idaho than in western Oregon and Washington. The survey of healthy native salmon populations by Huntington et al. (1996) reinforces this conclusion. Ninety-eight of the 99 healthy salmon populations identified by Huntington et al. (1996) are found in Oregon or Washington; only one resides in California and none in Idaho.

These geographic trends are due in part to deterioration of freshwater habitats. Human-caused habitat destruction has generally been more severe in southern portions of the range of Pacific salmon, especially California; hydropower development has severely damaged anadromous salmonid populations in Idaho. In addition, the effects of the recent drought have been greater in the southern and eastern portions of the range of salmon.

Ocean productivity cycles are of particular interest from a coastwide perspective. Cooper and Johnson (1992), reviewing the status of steelhead in British Columbia, Washington, and Oregon, concluded that cyclic changes in ocean productivity and competition for food in the marine environment played a significant role in the coastwide decline in steelhead populations. For salmon species, ocean productivity cycles appear to play a major role in the north-to-south cline in population status. Beamish and Bouillon (1993) found that commercial catches of pink, sockeye, and chum salmon by all nations correlated with climatic conditions over the North Pacific Ocean. Catches of these northern species were increasing until 1989 (the last year of record in Beamish and Bouillon's 1993 study) and continue to be high. Conversely, Lawson (1993) and Lichatowich (1996) summarize literature indicating that Oregon coastal coho salmon populations are in a period of low ocean productivity. Francis and Sibley (1991) observed a reciprocal relationship between pink salmon catch in the Gulf of Alaska and coho salmon catch in the Washington-Oregon-California region. That is, when pink salmon harvests are high, coho salmon harvests are low, and vice versa. The current high productivity of northern species (pink, chum, and sockeye) and low productivity of southern species (coho, chinook) help explain the north-to-south trend in status of Pacific salmon described here.

Lawson (1993) suggested that the imperiled status of coho salmon in Oregon is due to a low ocean productivity phase superimposed on a continuing declining trend in freshwater habitat status. He predicted that coho salmon populations will show a resurgence as ocean productivity cycles reverse but recommended that this not be cause for complacency. Rather, long-term restoration of freshwater habitat should be initiated now while public concern is high and early success can be anticipated. The alternative, allowing freshwater habitat to continue to degrade, may mean large-scale extinction of coho salmon in the next low ocean-productivity phase (Lawson 1993).

The same recommendation holds true for salmon and steelhead throughout California, Oregon, Washington, and Idaho. Failure to reverse freshwater habitat destruction could mean that the foreseeable future will see Pacific salmon all but extinct south of British Columbia. At the same time, Alaska and British Columbia, which enjoy relatively robust salmon populations, should prevent further freshwater habitat degradation; the ocean conditions that favor them today will not last.

Acknowledgments

I am grateful for the constructive suggestions of three anonymous reviewers, and, as always, for inspiring discussions with J. Lichatowich.

Literature Cited

Beamish, R.J. and D.R. Bouillon. 1993. Pacific salmon production trends in relation to climate. Canadian Journal of Fisheries and Aquatic Sciences 50: 1002-1016.

Brown. L.R., P.B. Moyle, and R.M Yoshiyama. 1994. Historical decline and current status of coho salmon in California. North American Journal of Fisheries Management 14(2):237-261.

Cooper, R. and T.H. Johnson. 1992. Trends in Steelhead (*Oncorhynchus mykiss*) abundance in Washington and along the Pacific Coast of North America. Report No. 92-20, Washington Department of Wildlife Fisheries Management Division. Olympia, Washington.

Forest Ecosystem Management Assessment Team. 1993. Forest ecosystem management: an ecological, economic, and social assessment. US Forest Service, National Marine Fisheries Service, Bureau of Land Management, US Fish and Wildlife Service, National Park Service, US Environmental Protection Agency. Portland, Oregon, and Washington, DC.

Francis, R.C. and T.H. Sibley. 1991. Climate change and fisheries: what are the real issues? Northwest Environmental Journal 7:295-307.

Fredin, R.A. 1980. Trends in North Pacific salmon fisheries, p 59-119. *In* W.J. McNeil and D.C. Himsworth (ed.), Salmonid Ecosystems of the North Pacific Ocean. Oregon State University, Corvallis, Oregon.

Hassemer, P.F., S.W. Kiefer, and C.E. Petrosky. 1996. Idaho's salmon: can we count every last one?, p. 113-125. *In* D.J. Stouder, P.A. Bisson, and R.J. Naiman (eds.), Pacific Salmon and Their Ecosystems: Status and Future Options. Chapman and Hall, New York.

Huntington, C., W. Nehlsen, and J. Bowers. 1996. A survey of healthy native stocks of anadromous salmonids in the Pacific Northwest and California. Fisheries 21(3): 6-14.

Johnson, T.H., R. Lincoln, G.R. Graves, and R.G. Gibbons. 1996. Status and wild salmon and steelhead stocks in Washington State, p. 127-144. *In* D.J. Stouder, P.A. Bisson, and R.J. Naiman (eds.), Pacific Salmon and Their Ecosystems: Status and Future Options. Chapman and Hall, New York.

Konkel, G.W. and J.D. McIntyre. 1987. Trends in spawning populations of Pacific anadromous salmonids. Technical Report 9, US Fish and Wildlife Service Fish and Wildlife. Washington, DC.

Kostow, K. 1996. The status of salmon and steelhead in Oregon, p. 145-178. *In* D.J. Stouder, P.A. Bisson, and R.J. Naiman (eds.), Pacific Salmon and Their Ecosystems: Status and Future Options. Chapman and Hall, New York.

Lawson, P.W. 1993. Cycles in ocean productivity, trends in habitat quality, and the restoration of salmon runs in Oregon. Fisheries 18: 6-10.

Lichatowich, J. 1996. Evaluating salmon management institutions: the importance of performance measures, temporal scales, and production cycles, p. 69-87. *In* D.J. Stouder, P.A. Bisson, and R.J. Naiman (eds.), Pacific Salmon and Their Ecosystems: Status and Future Options. Chapman and Hall, New York.

Meehan, W.R. and T.C. Bjornn. 1991. Salmonid distributions and life histories, p. 47-82. *In* W.R. Meehan (ed.), Influences of Forest and Rangeland Management on Salmonid Fishes and Their Habitats. American Fisheries Society Special Publication 19, Bethesda, Maryland.

Mills, T.J., D.R. McEwan, and M.R. Jennings. 1996. California salmon and steelhead: beyond the crossroads, p. 91-111. *In* D.J. Stouder, P.A. Bisson, and R.J. Naiman (eds.), Pacific Salmon and Their Ecosystems: Status and Future Options. Chapman and Hall, New York.

Nehlsen, W., J.E. Williams, and J. A. Lichatowich. 1991. Pacific salmon at the crossroads: stocks at risk from California, Oregon, Idaho and Washington. Fisheries 16(2):4-21.

Northcote, T.G. and D.Y. Atagi. 1996. Pacific salmon abundance trends in the Fraser River watershed compared with other British Columbia systems, p. 199-219. *In* D.J. Stouder, P.A. Bisson, and R.J. Naiman (eds.), Pacific Salmon and Their Ecosystems: Status and Future Options. Chapman and Hall, New York.

Ricker, W. E. 1972. Hereditary and environmental factors affecting certain salmonid populations, p. 19-160 *in* R.C. Simon and P.A. Larkin (eds.), The Stock Concept in Pacific Salmon. University of British Columbia, Vancouver.

Schalk, R.F. 1977. The structure of an anadromous fish resource, p. 207-249. *In* L.R. Binford (ed.), For Theory Building in Archaeology. Essays on Faunal Remains, Aquatic Resources, Spatial Analysis, and Systemic Modeling. Academic Press, New York.

Washington Department of Fisheries, Washington Department of Wildlife, and Western Washington Treaty Indian Tribes. 1993. 1992 Washington State Salmon and Steelhead Stock Inventory. Olympia, Washington.

The Wilderness Society. 1993. Pacific salmon and federal lands, a regional analysis. The Wilderness Society, Washington, DC.

Wertheimer, A. 1996. Status of Alaska salmon, p. 179-197. *In* D.J. Stouder, P.A. Bisson, and R.J. Naiman (eds.), Pacific Salmon and Their Ecosystems: Status and Future Options. Chapman and Hall, New York.
Williams, J.E., J.A. Lichatowich, and W. Nehlsen. 1992. Declining salmon and steelhead stocks: new endangered species concerns for the West. Endangered Species Update 9(4): 1-8.

Analyzing Trends and Variability

On the Nature of Data and Their Role in Salmon Conservation

Jay W. Nicholas

Abstract

Data provide the foundation of salmon management. As important as data are, the nature of data and their role in salmon conservation are rarely discussed. In this paper, I describe the historical origin, evolution, and application of salmon databases in Oregon, which are representative of data collected from California through British Columbia and Alaska. In an effort to broaden understanding of the nature of data, I discuss two awkward aspects of databases: (1) all populations are not equally represented in databases, and (2) databases contain some incorrect information. In conclusion, I speculate that the decade of the 1990s is a moment of opportunity for salmon conservation, not because of the quantity of data that are available, but primarily because relatively few salmon are left to conserve.

The Historical Perspective Of Data

Pacific salmon (*Oncorhynchus* spp.) have probably been harvested on a subsistence-scale by Native Americans for at least the last 10,000 years (Schalk 1986). Native Americans managed salmon populations and altered ecosystems to varying degrees (McEvoy 1986) but recorded relatively little numerical data regarding salmon. Non-native peoples began making large-scale alterations of landscape features and harvesting natural resources on an industrial scale in the Pacific Northwest during the mid-1800s.

As a result, numerical data began to accumulate, and the acquisition of data on salmon and other natural resources has accelerated through the present day. Data first recorded typically included records of harvests: ounces of gold mined, board-feet of lumber cut, beaver pelts traded, barrels of salmon salted, and cases of salmon canned. The societal preoccupation was one of achieving yields of natural resources, conquering nature, and reaping nature's bounty (McEvoy 1986, Worster 1977, Bottom 1996). Thus, data acquisition reflected the societal preoccupation. Salmon were thought of as a natural resource commodity. Scant historical data were collected regarding life-history characteristics, population characteristics, species distributions within basins, ecological relationships, morphological characteristics, or behavior, because this sort of information was not considered important.

From the late-1800s through the mid-1900s, the emphasis of salmon management in Oregon and other states and regions was twofold: first, to harvest salmon, and second, to support harvest levels by producing hatchery fish (Lichatowich and Nicholas in press). Therefore, historical data on salmon emphasize harvest and hatchery production. Somewhere around the 1960s, Oregon's natural resource management programs began to emphasize ecological principles, evidenced by the adoption of a variety of Oregon Administrative Rules (e.g., Oregon Forest Practices Act, Division of State Lands Fill and Removal Rules).

This change in salmon management reflected an increased emphasis on conservation of wild populations. The shift in emphasis of salmon management generated support for development of broader databases. Although contemporary databases still include considerable information directly related to catch and hatchery production, they also include considerable information on distribution and abundance of wild populations.

Data Resources

The focus of this paper is restricted to data that permit inference of trends in the geographic distribution and abundance of salmon. Throughout this discussion, I refer primarily to data representing salmonids, especially salmon, using Oregon as an example. However, the nature of data about salmon in Oregon represents the nature of data throughout the Pacific Northwest. Data are used by scientists, harvest managers, commercial fishers, recreational anglers, and people who may simply consider salmon as a symbol of the health of natural resources and ecosystems. Data are often thought of as consisting entirely of numbers; however, some historical information about fish is available in other formats (Table 1). I describe several of these types of data below.

NARRATIVE REPORTS: These reports prepared by the Master Fish Warden of Oregon detail his experiences as he traveled by horse and wagon throughout the state. These annual reports provide an irreplaceable narrative source of information about Oregon's fish populations around the turn of the century.

PHOTOGRAPHS: Photographic records of an angler displaying the catch from an afternoon of trout fishing on a particular river contain information about species composition, size range, and, perhaps, abundance. Even a photograph of a barge overloaded with salmon in a coastal estuary may yield information regarding run timing, size at maturity, and species composition. In addition to information about fish, historical photographs may provide information about the previous condition of ecosystems.

ANGLING REPORTS: Around the turn of the century, popular journals published useful information for interpreting the historical record. These reports provide information on species composition, size, and numbers taken.

SURVEYOR'S RECORDS: These records have proven to be rich sources of historical descriptions of habitat quality (P. Benner, United States [US] Department of Agriculture, Forest Service, Pacific Northwest Forest and Range Experiment Station, Corvallis, unpubl. data), as have records of the US Army Corps of Engineers (Sedell and Luchessa 1992).

NATIVE AMERICAN SPIRITUAL VALUES AND LEGENDS: Such cultural resources provide information useful in forming a picture of salmon populations before the region was altered by

Table 1. Many types of data are useful in interpreting differences between historical and contemporary distribution and abundance of salmon in Oregon. Similar types of data are available from other states and regions.

Type of information	Period available	Source of information
Photographs	late 1800s–early 1900s	Various Fish Commission annual reports, Oregon Historical Society, Oregon State Library
Oregon Fish Commission Annual Reports, Oregon Game Commission Annual Reports	1897 through about 1940	Oregon State Library
The Oregon Sportsman, a popular fishing and hunting journal	1913–27	Oregon State Library, Oregon State University Library
Records of the cannery pack at individual canneries in Oregon	1892–1922	Gharrett and Hodges (1950), Mullen (1981a), Oregon Department of Fish and Wildlife (ODFW) unpubl. data
Landings of salmon in gillnet fisheries in specific coastal rivers	1923–61, depending on years that fishery was open in individual river	Gharret and Hodges (1950), Mullen (1981a), ODFW unpubl. data
Landings of salmon at coastal ports from commercial ocean salmon fisheries	1925–present	Johnson (1984), Pacific Fishery Management Council (1994), ODFW unpubl. documents
Estimates of salmon caught by recreational anglers in individual river basins	1953–present	ODFW unpubl. data
Counts of fish at dams	Various, usually starting when dam was constructed	Mullen (1981b), Nicholas and Hankin (1988), ODFW unpubl. data
Counts of fish in spawning areas of coastal rivers, run-size estimates of salmon entering or spawning in coastal river basins	1950–present	Nicholas and Hankin (1988), Cooney and Jacobs (1993), Jacobs and Cooney (1993), various ODFW unpubl. documents and annual reports

nonnative peoples. Legend and spiritual beliefs, for example, imply years of great abundance as well as years of poor salmon runs (e.g., McEvoy 1986).

ESTIMATES OF SALMON HARVEST (i.e., salmon catch data): Harvest estimates may be based on voluntary logbook systems, fish processing plant records, subsamples of fish landings, poundage tax records, and such.

ESTIMATES OF SALMON PASSING DAMS (i.e., counts): These estimates are based on observations of fish passing upstream over the dam during a subsample of the migration period. Video cameras and electronic counting devices have been used in recent years to assist and validate the estimation process.

ESTIMATES OF SALMON IN SPAWNING AREAS: Such estimates are usually based on counts of adults made during a portion of the spawning area and time period.

Awkward Aspects Of Databases

Unequal Representation

Some salmon populations and species are better represented in databases than others. This unequal pattern of representation creates informational distortions with important management implications. As stated previously, more data are available for contemporary populations than for historical populations. In addition, fewer data are available for populations that occur in remote locales, contain relatively few individuals, and are primarily a target of recreational rather than commercial harvest. For example, it is possible to rank Oregon salmonid fish species in general order of their representation in databases, starting with the best represented and ending with the least represented: coho *(O. kisutch)*, chinook *(O. tshawytscha)*, chum *(O. keta)*, steelhead *(O. mykiss)*, and anadromous cutthroat trout *(O. clarki)*.

The range of some salmon species in Oregon today is greatly reduced from historical conditions (Kostow 1996); therefore, many populations are extinct. Because of unequal representation in databases, documented extirpations and reductions in population abundance are probably only a small fraction of the actual degradation of wild salmon populations. Although this implication is crucial to natural resource management decisions, it is rarely emphasized by resource management agencies. Public resource management agencies usually do not wish to appear inflammatory or exaggerate the extent of detrimental change in the environment. Thus, formal pronouncements of extirpations and reductions in population abundance are likely to be confined to cases that can be proved.

Inclusion of Incorrect Information

Although generally reliable, most databases include some incorrect information. Therefore, judgment is an essential aspect of any analysis using these databases. Potential sources of incorrect information in databases include the following.

Species misidentification: This is much less a problem in contemporary databases than in historical databases. In some written accounts of fish in the late 1800s and early 1900s, writers used terms like silversides, salmon-trouts, and similar variations. Thus, the reader must infer identity of species the writer reported, and some uncertainty may remain as to whether the writer correctly identified species.

Deliberate omission, alteration, or fabrication: Many historical catch summaries are based on records of salmon landings that were required for tax purposes. In other words, the more fish or pounds recorded by an individual, the more tax the person paid. Estimates of historical landings have not been adjusted to account for possible deliberate under-reporting.

Inadvertent omission: Data are collected by many people throughout the state. People who compile summary databases from a variety of sources may overlook the existence of relevant data.

Unreported illegal catch (poaching): Annual reports of the Master Fish Warden of Oregon contain numerous references to individuals who netted (or otherwise gaffed, pitch-forked, or speared) salmon illegally. Apparently, a lot of this activity occurred, but there was only one Master Fish Warden. It was difficult to travel by train, wagon, and horseback to distant locations to catch poachers. Estimates of historical catch have not been adjusted to account for unreported illegal catch.

REPORTING ERRORS: People make mistakes. This has always been true, and even with the best of intentions some mistakes occur.

REPETITIVE REPORTING ERRORS: Some published reports contain errors. Later, other authors write new reports based on earlier reports. Sometimes, the new writer discovers and corrects errors made by the original author. However, the original errors often are preserved.

UNKNOWN VARIANCE OF ESTIMATES: As noted earlier, many databases are lists of estimates constructed from sample data. Only rarely are confidence intervals of these estimates noted. Usually, confidence intervals could not be calculated because of inadequacies in the sampling design. It is often difficult to judge how much confidence one should have in comparing values in a table: Is the value 200 really distinguishable from the value 250, or from the value 300, or even from the value 400?

Data and Salmon Conservation

Since 1850, many populations of salmon have become extinct (Hassemer et al. 1996, Johnson et al. 1996, Kostow 1996, Mills et al. 1996, Northcote and Atagi 1996, Wertheimer 1996), resulting in reduction in the geographic distribution of all salmon species. In addition, the abundance of many wild salmonid populations in Oregon has declined sharply (Nickelson et al. 1991). Today, many of the remaining populations of salmon in the Pacific Northwest are thought to be at risk of extinction (Nehlsen et al. 1991). During 1995, the National Marine Fisheries Service recommended listing coho salmon in Oregon as threatened under the federal Endangered Species Act.

Currently available databases have allowed scientists and resource managers to make a variety of important decisions. Data provide a basis for restricting salmon harvest as well as documenting reductions in the geographic distribution of salmon and steelhead. As a result of these data, managers have also concluded that abundance of populations in some local regions or basins has declined. Furthermore, interpretation of these databases provides a conceptual basis for public support of a wide variety of salmon conservation and restoration efforts.

One may ask whether sufficient data are available to manage salmon populations. Although vast quantities of data about salmon are available, opinions expressed by professionals vary considerably regarding the answer to this question. Certainly, answering the question necessitates defining what "salmon management" means. If it means harvesting the greatest possible annual yield from natural resources in the Pacific Northwest without driving salmon populations to the brink of extinction, we probably do not know enough. If it means generally understanding the kinds of actions by humans that will be beneficial or detrimental to salmon conservation efforts, I believe that we do know enough. No scientific formula exists that prescribes a single course of action that is both necessary and sufficient to conserve a natural resource like salmon. I think that scientists and fishery managers know enough to make reasonable recommendations regarding the needs of salmon. It will then be up to society to decide whether to make the adjustments that may be needed.

Data about salmon influence societal beliefs, choices, and expectations. However, these data probably exert less influence on societal beliefs than economic values and underlying societal views of resource utilization. Generally, a belief exists in the ability of technology to improve on nature, in contrast to a belief in conserving natural systems and living in harmony with nature (see Meffe 1992).

When natural resources are abundant, many people stand to gain economically from use of these resources. Advocates of caution and restraint are likely to be ignored, especially when underlying management theories support a belief in sustained harvest, maximum sustainable yield, and an industrial economy. Under these circumstances, sufficient doubt regarding the reliability of data will exist to support perpetuation of economically popular management regimes. Overall, society will refuse to believe data predicting deterioration or collapse of the resource base. People are generally reluctant to heed warnings of a distant or future resource disaster. A culturally based sense of optimism usually allows people to believe that a solution will be found, that people will have access to natural resources at little economic cost or at little risk to the future availability of the resource.

Given what fishery scientists and managers now know about salmon, would people have chosen not to build dams on the Columbia River? Would all irrigation diversions have been built with effective fish exclusion devices? Would all livestock grazing, gravel mining, timber harvest, pollution disposal, gold mining, home building, harvesting in mixed stock fisheries, and development of hatchery programs have been conducted in a manner and at a pace that would have ensured against the extinction or serious reductions of salmon populations? I think not. Consider, for example, the following prediction, made in 1947:

> Construction of potential main stream dams will convert the Columbia and Snake Rivers into a series of lakes. This, together with construction of additional dams on tributaries will create anadromous fish problems of such magnitude that, despite remedial measures, runs to upper waters of the basin may be seriously depleted or even eliminated entirely (US Department of the Interior 1947).

The warning in this statement is clear, but the dams were built anyway. Today, society struggles with economic and scientific debates concerning the costs and methods of restoring salmon populations to the upper Columbia basin. However, no amount of data would have been sufficient to change the course of natural resource management in 1947.

Conceptually, I imagine that two curves describe accumulation of data on and understanding of salmon during the period since 1850. I contrast these two curves to a general decline in salmon abundance during the same time frame (Fig. 1). Data regarding salmon have accumulated, at first slowly, and then rapidly, and are now accumulating at historically high rates. Understanding of salmon populations increased rapidly at first, then slowly. Contemporary increase in fundamental understanding of salmon is relatively small. Acceptance of my general premise suggests that more than an increase in data or understanding of salmon, per se, will be needed to accomplish conservation of salmon populations in the future.

However, I believe that when a resource collapses to the point where traditional economic uses are no longer possible, a moment of opportunity exists when it may be possible to achieve fundamental change in resource management emphasis. A variety of beliefs regarding the merits of resource consumption and conservation simultaneously coexist within society as a whole. Support for consumption tends to be highest when resources are most abundant but erodes to lower levels as the resource declines in abundance. Support for consumption is strongly shaped

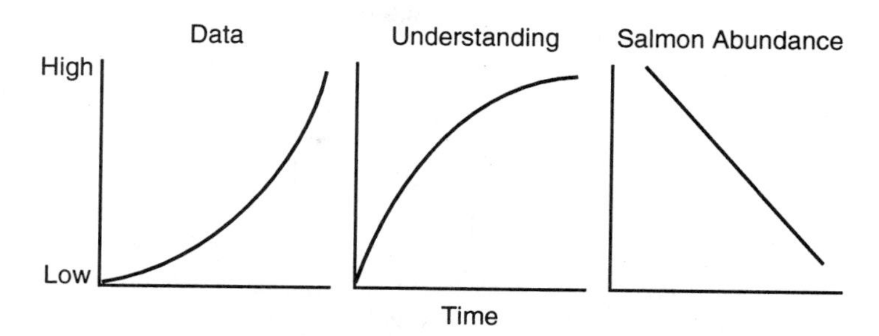

Figure 1. Illustration of conceptual relationships of data, understanding, and salmon abundance. Since the mid-1800s, salmon data have accumulated and are being collected at an increasing rate; understanding of salmon was low at first, increased rapidly, but has leveled off recently; abundance of salmon was highest when data and understanding were low, and has declined as data and understanding increased.

by economic concerns and a cultural belief in the right of humankind to dominate nature. In contrast, support for conservation is lowest when resources are abundant but may increase through time as resources decline in abundance. Support for conservation also reflects cultural and economic concerns, but the role of economics is less significant. When support for resource consumption is very strong, even the most convincing scientific data are not likely to result in effective conservation measures. When a resource declines so substantially that accustomed patterns of use are no longer possible, societal support for consumption and conservation may be at similar levels. The decade of the 1990s is what I refer to as a moment of opportunity to achieve fundamental change in salmon management; it is occurring primarily because there are relatively few salmon left to save, not because of the amount or quality of data that are available on salmon. The reader may agree or disagree with this proposition. Ludwig et al. (1993) made a similar proposition and were challenged by Aron et al. (1993). Scientists and others will continue to debate the question. Society, as a whole, will decide the issue.

Literature Cited

Aron, W., D. Fluharty, D. McCaughran, and J.F. Roos. 1993. Fisheries management. Science 261: 813.

Bottom, D.L. 1996. To till the water—a history of ideas in fisheries conservation, p. 569-597. *In* D.J. Stouder, P.A. Bisson, and R.J. Naiman (eds.), Pacific Salmon and Their Ecosystems: Status and Future Options. Chapman and Hall, New York.

Cooney, C.X. and S.E. Jacobs. 1993. Oregon coastal salmon spawning surveys, 1991. Oregon Department of Fish and Wildlife Information Report (Fish) 93-1. Portland.

Gharrett, J.T. and J.I. Hodges. 1950. Salmon fisheries of the coastal rivers of Oregon south of the Columbia. Oregon Fish Commission Contribution 13. Portland.

Hassemer, P.F., S.W. Kiefer, and C.E. Petrosky. 1996. Idaho's salmon: can we count every last one? p. 113-125. *In* D.J. Stouder, P.A. Bisson, and R.J. Naiman (eds.), Pacific Salmon and Their Ecosystems: Status and Future Options. Chapman and Hall, New York.

Jacobs, S.E. and C.X. Cooney. 1993. Improvement of methods used to estimate the spawning escapement of Oregon coastal natural coho salmon. Annual progress report, Oregon Department of Fish and Wildlife. Corvallis.

Johnson, K. 1984. A history of coho fisheries and management in Oregon through 1982. Oregon Department of Fish and Wildlife Information Reports (Fish) 84-12. Portland.

Johnson, T.H., R. Lincoln, G.R. Graves, and R.G. Gibbons. 1996. Status of wild salmon and steelhead stocks in Washington State, p. 127-144. *In* D.J. Stouder, P.A. Bisson, and R.J. Naiman (eds.), Pacific Salmon and Their Ecosystems: Status and Future Options. Chapman and Hall, New York.

Kostow, K. 1996. The status of salmon and steelhead in Oregon, p. 145-178. *In* D.J. Stouder, P.A. Bisson, and R.J. Naiman (eds.), Pacific Salmon and Their Ecosystems: Status and Future Options. Chapman and Hall, New York.

Lichatowich J.A. and J.W. Nicholas. In press. Oregon's first century of hatchery intervention in salmon production: Evolution of the hatchery program, legacy of a utilitarian philosophy and management recommendations. *In* Proceedings of the Symposium on Biological Interactions of Enhanced and Wild Salmonids, Nanaimo, B.C., June 17-20, 1991. Canada Department of Fisheries and Oceans, Pacific Region.

Ludwig, D., R. Hilborn, and C. Walters. 1993. Uncertainty, resource exploitation, and conservation: lessons from history. Science 260: 17.

McEvoy, A.F. 1986. The Fisherman's Problem: Ecology and Law in the California Fisheries 1850-1980. Cambridge University Press, Cambridge

Meffe, G.K. 1992. Techno-arrogance and halfway technologies: salmon hatcheries on the Pacific coast of North America. Conservation Biology 6: 350-354.

Mills, T.J., D. McEwan, and M.R. Jennings. 1996. California salmon and steelhead: beyond the crossroads, p. 91-111. *In* D.J. Stouder, P.A. Bisson, and R.J. Naiman (eds.), Pacific Salmon and Their Ecosystems: Status and Future Options. Chapman and Hall, New York.

Mullen, R.E. 1981a. Oregon's commercial harvest of coho salmon *Oncorhynchus kisutch* (Walbaum), 1892-1960. Oregon Department of Fish and Wildlife Information Reports (Fish) 81-3. Portland.

Mullen, R.E. 1981b. Estimates of historical abundance of coho salmon *Oncorhynchus kisutch* (Walbaum), in Oregon coastal streams and in the Oregon Production Index Area. Oregon Department of Fish and Wildlife Information Reports (Fish) 81-5. Portland.

Nehlsen, W., J. E. Williams, and J. A. Lichatowich. 1991. Pacific salmon at the crossroads: Stocks at risk from California, Oregon, Idaho, and Washington. Fisheries 16: 4-21.

Nicholas, J.W. and D.G. Hankin. 1988. Chinook salmon populations in Oregon coastal river basins: Description of life histories and assessment of recent trends in run strengths. Oregon Department of Fish and Wildlife, Information Reports (Fish) 88-1, Corvallis.

Nickelson, T.E., J.W. Nicholas, A.M. McGie, R.B. Lindsay, and D.L. Bottom. 1992. Status of Anadromous Salmonids in Oregon coastal basins. Oregon Department of Fish and Wildlife. Portland.

Northcote, T.G. and D.Y. Atagi. 1996. Pacific salmon abundance trends in the Fraser River watershed compared with other British Columbia systems, p. 199-219. *In* D.J. Stouder, P.A. Bisson, and R.J. Naiman (eds.), Pacific Salmon and Their Ecosystems: Status and Future Options. Chapman and Hall, New York.

Pacific Fishery Management Council. 1994. Review of 1993 ocean salmon fisheries. Portland, Oregon.

Schalk, R.F. 1986. Estimating salmon and steelhead usage in the Columbia basin before 1850: the anthropological perspective. Northwest Environmental Journal 2: 2, 1-25.

Sedell, J.R. and K.J. Luchessa. 1982. Using the historical record as an aid to salmonid habitat enhancement. Pages 210-223. *In* N.B. Armantrout (ed.), Acquisition and Utilization of Aquatic Habitat Inventory Information. American Fisheries Society, Bethesda, Maryland.

US Department of the Interior. 1947. The Columbia River. A comprehensive department Report on the development of the water resources of the Columbia River basin for review prior to submission to Congress. Washington, DC.

Wertheimer, A.C. 1996. Status of Alaska salmon, p. 179-197. *In* D.J. Stouder, P.A. Bisson, and R.J. Naiman (eds.), Pacific Salmon and Their Ecosystems: Status and Future Options. Chapman and Hall, New York.

Worster, D. 1977 Nature's Economy: A History of Ecological Ideas. Cambridge University Press, New York.

Information Requirements for Salmon Management

Carl Walters

Abstract

Management of salmon takes place on three very different time scales, with substantially different data needs at each scale. Much mismanagement in the past has come from preoccupation with one of these scales, while using very simplistic data or assumptions about the others. For long-term management, the most fundamental need is to monitor and maintain suitable habitats. For the medium term (each salmon generation), the need will always be for empirical experience of the basic relationship between how many fish spawn and subsequent production of recruits, and how this relationship is linked to freshwater habitat factors subject to management. Knowing the key environmental factors that cause variation in the relationship would be helpful, but the most important need is to establish the average relationship so that managers can set wise escapement goals. For short-term or "inseason" management, where we try to meet both escapement goals defined by long-term studies and complex allocation goals, the most critical need today is for innovative abundance indexing systems to measure interannual variation in spatial distribution and timing of migration through fishing areas.

Introduction

Any experienced salmon manager should be able to fill many pages with lists of data that are potentially useful in making decisions. These decisions create an almost endless list of different information priorities. If management is interested in protecting a particular local stock (e.g., a stock being considered for rare and endangered listing), it must know a host of details about precisely where and when that stock is vulnerable to various mortality agents (i.e., to know the life history of the stock in substantial detail). At an opposite extreme, people concerned with investment planning for regional enhancement or habitat management programs do not need to examine the detailed behavior of each individual stock, and will instead express information needs about broad production/harvest patterns and potential tradeoffs between enhanced and wild populations.

Given this variety of decision situations and information needs, it would be pointless (and probably misleading) to attempt any comprehensive description of data needs. What this paper attempts instead is to show that information needs fall into three categories representing biological dynamics and decision making on different time and space scales; it also attempts to identify a few of the most important uncertainties and needs at each scale. Failure to recognize that decision making does take place on very different scales has led to much confusion among biologists about when and how to use various types of information. This confusion is particularly evident in the review by Fletcher and Deriso (1988), where salmon managers were asked whether they use stock-recruit curves (relating average abundance of new recruits to the abundance of spawners that produced them). The review points out that salmon managers rightly do not use stock-recruit relations in deciding on inseason harvest regulations, but it fails to emphasize that the escapement (spawning abundance) targets at which these regulations are aimed must ultimately come from a stock-recruit assessment of some sort.

A theme that appears repeatedly in the following discussion is the extent to which the ability to meet information needs is limited by management itself. Management activities often actively prevent informative variation in population sizes (i.e., to relate recruits to spawner abundance, one needs to examine how both low and high numbers of spawners perform), cause confounding of environmental effects with the effects of progressive management changes (such as development of hatchery production), and often lead to rigid licensing and allocation programs that prevent development of new fisheries that could be used as sources of key information (i.e., as test fisheries). Indeed, the main conclusion of this paper is not simply that we need more data, but rather that we need to fundamentally rethink the interplay between management and data gathering so as to develop management systems that will enhance both our ability to conserve stocks and our ability to obtain the information that is critical for conservation.

Management Time Scales

Decision making for salmon management falls into three broad categories representing different time scales of habitat and population response. On time scales of decades and longer, the most fundamental concern is with maintaining and restoring suitable habitats for salmon spawning and rearing, and with the potential for replacing these habitats with various engineered facilities. On time scales of one to a few salmon generations (2 years to 2 decades), the essential concern is with maintaining and restoring adequate spawning escapements. Here the definition of "adequate" is obviously a target that is expected to change over time with changes in habitat factors. On time scales of weeks to months within each fishing season, decision making is generally concerned with a mixture of objectives involving not only how to ensure adequate spawning escapement, but also how to allocate the available harvest among competing user groups. Obviously, decision making on this short time scale cannot proceed wisely without some basic input of information from longer-term analyses (despite the pressures of short-term decision making, you still must have a basic escapement goal, or allocation to conservation needs, or whatever you may choose to call your concern with the future).

The "downward" (long-term to short-term) linkages between these scales are well recognized by most managers (habitat factors set stock-recruit limits, stock-recruit relations set inseason goals). But there are equally important and less well understood "upward" linkages as well. The practicalities and limited options of inseason management can place severe bounds on what escapement goals can be achieved, especially in mixed-stock fisheries. Setting escapement goals based on historical stock-recruitment relationships can result in spawning stock levels that are too low to provide adequate tests of habitat capacity (low escapement levels may fail to result in enough intraspecific competition to stimulate dispersal and colonization of potential habitats).

Information for Habitat Maintenance

The key variables that should be measured at decadal time scales are indices of habitat "capacity" or potential (whether the habitat is currently used by salmon or not) and spatial distributions of actual fish use relative to these capacity measures. Changes in habitat capacity can occur rapidly as a result of local disturbances such as logging or dam building, can exhibit roughly cyclic patterns due to large-scale climatic changes such as shifts in the position of the Aleutian low off Alaska (Beamish and Bouillon 1993, Pearcy 1996), or can occur very gradually owing to accumulated small changes associated with urban, agricultural, and forestry developments.

Long-term programs for monitoring habitat capacity and use have historically been concentrated on spawning habitat and adult fish. There is a key need to develop similar, routine monitoring programs for juvenile distributions of at least those species and stocks that have extended periods of freshwater and coastal marine rearing. Much attention in recent monitoring programs has been focused on detecting genetic variation and stock structure with associated differences in life-history characteristics of adult fish. After all, biologists have been taught that careful attention to stock structure is important for production predictions (Ricker 1973), and genetic variation is relatively easy for biologists to think about and measure. But for both adults and juveniles, it will be equally important in the long term to also monitor lack of habitat utilization: where are there no or few fish when there should be far more? Mapping changes in spatial distributions of fish relative to habitats and generating measures of changing habitat utilization such as stock concentration curves (Walters and Cahoon 1985) involve tedious and relatively expensive field work, and the results are not especially interesting to scientists. The crucial importance of better distributional monitoring data become apparent only when it comes time to argue for habitat protection and restoration and changes in harvest policies to allow better habitat seeding (i.e., when biologists pay closer attention to management questions than to questions that are scientifically interesting).

Thanks to experimental and comparative watershed studies along the Pacific Coast (reviews in Salo and Cundy 1987; see also Holtby 1988, Thedinga et al. 1989), there is now a great deal of experience to define requirements for spawning and juvenile fish. Some of this experience has produced surprising results (such as the apparent enhancement in coho salmon [*Oncorhynchus kisutch*] smolt production following some logging operations) that might be used in the future to design more cooperative programs for linking watershed development with enhanced fish production. Now we should critically examine such surprising results, and use this examination to design a new series of experimental projects aimed specifically at testing possible ways to enhance salmon production through more careful watershed development.

Information for Setting Escapement Goals

Setting an appropriate escapement target or average spawning stock goal is probably the most important step in developing effective conservation and sustainable management programs. But working directly with stock-recruitment data for salmon is always frustrating: clear patterns are seldom evident in the data, bad statistical effects often occur as a result of the time-series structure of the data and difficulties in measuring spawner numbers accurately, and great gaps in experience usually exist for both low and high spawning stock sizes. By fitting models to fake data (with known population parameters), we can easily see that existing methods for estimating average recruitment relationships (fitting mean stock-recruitment curves) are surprisingly good at finding the average relationship even in data so noisy that a biologist can see only a shotgun scatter of points in visual examination. Such tests indicate that biologists should not trust their eyes when doing recruitment analysis, but few biologists are willing to trust statistical methods without visual confirmation. Moreover, many biologists have turned away from stock-recruitment analysis entirely in favor of simpler, more comfortable methods such as calculating an average spawning area needed per fish and then deriving the optimum spawning number from total habitat area available divided by this area per fish. Surprisingly, not many biologists seem to realize the fundamental scientific error in supposing that such calculations or models can be substituted for direct stock-recruit analysis (Hilborn and Walters 1992). In principle, such calculations can only produce a prediction of the mean stock-recruit relation, and such predictions must ultimately be tested through direct comparison with observed overall stock-recruit relations (simple calculations are subject to numerous sources of bias, e.g., not including limiting factors at life stages in the calculation, or not accounting for compensatory survival changes in baseline data used to parameterize the calculations).

Many biologists have made the gross error of assuming that stock-recruit relationships can be ignored because spawning stock seldom explains much recruitment variation in a statistical sense. This error represents a fundamental misunderstanding about why stock-recruitment analysis is done in the first place. The analysis is not done to find a good year-to-year predictor of recruitment; far better ways exist to generate short-term forecasts of recruitment based on environmental factors and monitoring of pre-recruit abundances. Rather, the purpose of stock-recruit analysis is to determine the longer-term average effect of spawning stock size on production in order to set long-term stock size goals correctly. Indeed, if there were no such average effect (if spawning numbers really did not explain recruitment variation at all), then there would be no point in having any fisheries regulations aimed at conservation in the first place. Managers could all go home and relax, confident that the fish could take care of themselves in spite of any fishing pressures they might encounter.

For major sockeye (*O. nerka*) and pink (*O. gorbuscha*) salmon stocks along the Pacific Coast, a broad enough range of experience has been accrued with different spawning stock levels to estimate the average recruitment relationship well enough to set general escapement goals. There are interesting debates now about the possible deleterious effects of setting very high spawning stock levels such as in Alaska after oil spills or in Rivers Inlet, British Columbia (BC), as a deliberate management experiment (Walters et al. 1993). But there is little debate about the more important conservation question of the minimum number of spawners needed. Chum (*O. keta*) salmon have generally not been monitored very well, but available data indicate that spawning stock sizes generally need to be increased substantially; only time will tell whether this prediction is correct.

The greatest uncertainties about stock and recruitment are for chinook (*O. tshawytscha*) and coho salmon. Here there are two basic problems. First, little information exists on recruitments produced by very low spawning stocks, except during a few overall declines that might well be due to factors other than inadequate spawning numbers. Second, it has been impossible to untangle the effects of four factors that have been changing simultaneously over the past few decades: (1) decreases in spawning stock sizes, (2) deterioration in freshwater rearing habitats, (3) decreases in marine survival rate that may be due to oceanographic changes (e.g., warm water off the BC coast), and (4) increases in releases of hatchery fish. Off at least the Oregon and southern BC coasts, increases in chinook and coho hatchery releases have not led to increases in total harvestable abundance; instead, marine survival rates of both wild and hatchery stocks have decreased, with average marine production remaining about the same or even declining (Peterman and Routledge 1983, Nickelson 1986, Walters and Riddell 1986, Emlen et al. 1990, Peterman 1991, Walters in press). Perhaps the measurements of marine survival rate are not good, so in fact wild stocks have declined due to overfishing and habitat loss at just about the same rate as hatchery stocks have increased. Or perhaps oceanographic changes just happen to have coincided with the increases in hatchery releases. But the most likely hypothesis is that there is a very limited marine carrying capacity for chinook and coho, and that wild stocks are being driven down by increasing the proportion of this capacity used by hatchery fish (Walters 1993).

Uncertainties about chinook and coho recruitment relationships cannot be resolved just by collecting more data. To determine the minimum number of spawners needed to fully seed freshwater rearing environments, we need to either do some simple experiments involving controlled reductions in spawning numbers on a few stocks, or else wait for fisheries and other factors to produce these low numbers for us in a large number of stocks at once. By probing ahead and risking a few stocks, we may obtain the data needed to argue more effectively for policies that will prevent a similar risk for many stocks (Walters in press). To discriminate between the oceanographic and carrying capacity hypotheses, it will be necessary to do a very major management experiment involving deliberate alternation between high and low hatchery releases to see if marine survival rates can be improved.

Information for Inseason Management

The most complex and difficult management activities, and by far the most demanding data needs, are associated with fisheries regulation within each fishing season. Yet surprisingly few scientific papers describe either how inseason regulation is currently done (Wright 1981, Sprout and Kadowaki 1987, Woody 1987) or how in theory it might be improved (Walters and Buckingham 1975, Fried and Hilborn 1988; see also Rutter 1996). The general scheme that is widely used is as follows: beginning with a preseason forecast of stock strength(s), target escapements, and allocations among gear types, managers first define a preseason fishing plan specifying expected times and places of fishery openings (this is often a complex specification of many openings for many areas, and is derived from a combination of historical experience and explicit accounting models). They then follow this plan for the first few days or weeks of the fishing season, while gathering fishery and test indices of actual run strength. If these indices suggest a substantially larger or smaller run than expected, managers then begin to adaptively revise fishery openings. As the season progresses, they continue updating the run strength estimates and adjusting fishing patterns as more and more data become available. Obviously, this adaptive adjustment process can go awry, leading to inadequate escapements or waste of potential harvest if misleading monitoring signals are gathered along the way.

There are two major types of salmon fisheries: migration or gauntlet fisheries on maturing fish during their return to natal rivers (mainly by net gear), and feeder or pool fisheries on immature, growing fish (mainly by troll and sport gear). Because the fish become highly concentrated and particularly vulnerable to overfishing during the gauntlet fisheries, inseason management monitoring and adjustment processes are focused mainly on such situations. The pool fisheries are rarely adjusted within each fishing season, except in areas of concentrated fishing effort where there are conservation concerns about particular stocks.

Since most pool fisheries involve large areas and a variety of stocks that are mixed through at least part of the fishing season, the key data needs for managing them generally are as follows: indices of overall abundance (such as commercial catch per effort or catch rates in survey test fisheries) that could be used to trigger emergency closures (or extended openings) when extreme index low (or high) values are seen near the start of a season; and stock composition/contribution information, particularly as related to wild vs. hatchery contribution and presence of particular weak (unproductive or depressed because of historical accidents) stocks for which exceptional protective measures might be warranted. But because there is currently no broad consensus in salmon management about when it is worthwhile to forgo harvest of stronger stocks to protect weaker ones, stock composition data have been used inconsistently in inseason decision making (composition data ultimately are used mainly in international negotiations and preseason planning for broad allocations between the US and Canada, and among states).

Gauntlet fisheries obviously could be run much more efficiently and safely if it were possible to develop very accurate preseason forecasts of run strengths. However, such forecasts are unlikely to become available soon; nature seems to have many ways to surprise any biologist who develops a seemingly viable forecasting scheme, such as those based on environmental factor correlations or juvenile abundance indices. In the absence of good forecasts, the most important factor for successful inseason management is the ability to update run size estimates rapidly within the season based on index information (catches, catch rates, index escapement counts, density indices from echo sounding).

The key limiting factor today for updating inseason run size estimates is not in gathering precise index information, but rather in using that information in conjunction with estimates of run timing. The basic problem is as follows. Suppose there is a weak showing of fish in the index data early in a season. Even if the index is very precise, should the manager infer a weak run, or instead that the fish are arriving late? Suppose there is a strong showing early. Should the manager conclude that a strong run is coming, or instead that there is about to be the salmon manager's worst nightmare, a "little run coming early"? These questions emphasize that the inseason salmon manager's worst data problem arises from run-timing anomalies. Runs can arrive in fishing areas as much as 2 wk earlier or later than expected, which would be an extreme variation considering that most single-stock runs only last 4–6 wk.

Thus, gauntlet fisheries managers need to gather better, earlier information on run timing within each fishing season. In a few cases the very first or "outside" commercial fisheries now act as early monitoring locations for run timing (e.g., the BC troll fisheries for Fraser River sockeye salmon provide timing information ≤1 wk before fish reach the large net fisheries in Johnstone Strait and the Strait of Juan de Fuca). Also, a few deliberate test fisheries are conducted for early timing assessment (e.g., the Port Moller test fishery at the entrance to Bristol Bay, Alaska). A genuine high seas test fishery, using troll gear or perhaps even carefully placed drift nets, would provide a useful complement to existing sources of timing data. This high-seas fishery would involve a very limited number of fishing licenses, with precise conditions placed on the fishermen in terms of fishing times, locations, and standardization of gear deployment. The development of such an offshore fishery would in effect create a new allocation structure (with attendant headaches for managers) but could dramatically improve the ability to allocate catches among other fisheries along the gauntlets.

Conclusion

The information requirements outlined above emphasize that the need today is not just more data of all sorts, but rather information collected under carefully controlled experimental conditions and aimed at resolving particular uncertainties that cannot be eliminated by waiting for more experience to accumulate. Further, these experiments are not ones that researchers can undertake by themselves; the required manipulations will need to be designed and conducted through close cooperation among managers, fishermen, and researchers. The necessary experiments involve deliberate tests of habitat capacities and productivities through planned variation in spawning fish numbers, tests of altered hatchery release effects on marine survival of wild fish, and development of offshore fisheries to provide better timing information for inseason management.

Literature Cited

Beamish, R.J. and D. Bouillon. 1993. Pacific salmon production trends in relation to climate. Canadian Journal of Fisheries and Aquatic Sciences 50: 1002-1016.

Emlen, J.M., R.R. Reisenbichler, A.M. McGie, and T.E. Nickelson. 1990. Density-dependence at sea for coho salmon (*Oncorhynchus kisutch*). Canadian Journal of Fisheries and Aquatic Sciences 47: 1765-1772.

Fletcher, R.I. and R.B. Deriso. 1988. Fishing in dangerous waters: remarks on a controversial appeal to spawner-recruit theory for long-term impact assessment. American Fisheries Society Monograph 4: 232-244.

Fried, S.M. and R. Hilborn. 1988. In-season forecasting of Bristol Bay, Alaska sockeye salmon (*Oncorhynchus nerka*) abundance using Bayesian probability theory. Canadian Journal of Fisheries and Aquatic Sciences 45: 850-855.

Hilborn, R. and C. Walters. 1992. Quantitative Fisheries Stock Assessment and Management: Choice, Dynamics, and Uncertainty. Chapman and Hall, New York.

Holtby, L.B. 1988. Effects on logging on stream temperatures in Carnation Creek, British Columbia, and associated impacts on the coho salmon (*Oncorhynchus kisutch*). Canadian Journal of Fisheries and Aquatic Sciences 45: 502-515.

Nickelson, T.E. 1986. Influence of upwelling, ocean temperature, and smolt abundance on marine survival of coho salmon (*Oncorhynchus kisutch*) in the Oregon Production Area. Canadian Journal of Fisheries and Aquatic Sciences 43: 527-535.

Pearcy, W.T. 1996. Salmon production in changing ocean domains, p. 331-352. *In* D.J. Stouder, P.A. Bisson, and R.J. Naiman (eds.), Pacific Salmon and Their Ecosystems: Status and Future Options. Chapman and Hall, New York.

Peterman, R.M. 1991. Density-dependent marine processes in Pacific salmonids: lessons for experimental design of large-scale manipulations of fish stocks, p. 69-77. *In* S. Lockwood (ed.), The Ecology and Management Aspects of Extensive Aquaculture. International Council for the Exploration of The Sea, Marine Science Symposium 192. Copenhagen.

Peterman, R.M. and D.M. Routledge. 1983. Experimental management of Oregon coho salmon: designing for yield of information. Canadian Journal of Fisheries and Aquatic Sciences 40: 1212-1223.

Ricker, W.E. 1973. Two mechanisms that make it impossible to maintain peak-periods yield from stocks of Pacific salmon and other fishes. Journal of the Fisheries Research Board of Canada 30: 1275-1286.

Rutter, LG. 1996. Salmon fisheries in the Pacific Northwest: How are harvest decisions made? p. 355-374. *In* D.J. Stouder, P.A. Bisson, and R.J. Naiman (eds.), Pacific Salmon and Their Ecosystems: Status and Future Options. Chapman and Hall, New York.

Salo, E.A. and T.W. Cundy (eds.). 1987. Streamside Management: Forestry and Fishery Interactions. Center for Streamside Studies, Univ. of Washington, Seattle, Washington.

Sprout, P.E. and R.K. Kadowaki. 1987. Managing the Skeena River sockeye salmon (*Oncorhynchus nerka*) fishery—the process and the problems, p. 385-395. *In* H.D. Smith, L. Margolis, and C.C. Wood (eds.), Sockeye Salmon (*Oncorhynchus nerka*) Population Biology and Future Management. Canadian Special Publication of Fisheries and Aquatic Sciences 96.

Thedinga, J.F., M.L. Murphy, J. Heifetz, KV. Koski, and S.W. Johnson. 1989. Effects of logging on size and age composition of juvenile coho salmon (*Oncorhynchus kisutch*) and density of presmolts in Southeast Alaska streams. Canadian Journal of Fisheries and Aquatic Sciences 46: 1383-1391.

Walters, C.J. In press. Where have all the coho gone? *In* Proceedings, Symposium on Wild and Enhanced Coho Stocks. British Columbia Association of Professional Biologists and the Northwest International Chapter of the American Fisheries Society. Department of Fisheries and Oceans, Vancouver, British Columbia.

Walters, C.J. and S.L. Buckingham. 1975. A control system for intra-season salmon management, p. 105-137. *In* Proceedings, International Institute for Applied Systems Analysis, Laxenburg, Austria, CP-75-2.

Walters, C.J. and P. Cahoon. 1985. Evidence of decreasing spatial diversity in British Columbia salmon stocks. Canadian Journal of Fisheries and Aquatic Sciences 42: 1033-1037.

Walters, C.J., R. Goruk, and D. Radford. 1993. Rivers Inlet sockeye salmon: an experiment in adaptive management. North American Journal of Fisheries Management 13: 253-262.

Walters, C.J. and B. Riddell. 1986. Multiple objectives in salmon management: the chinook sport fishery in the Strait of Georgia, B.C. Northwest Environmental Journal 2: 1-15.

Woody, J.C. 1987. In-season management of Fraser River sockeye salmon (*Oncorhynchus nerka*): meeting multiple objectives, p 367-374. *In* H.D. Smith, L. Margolis, and C.C. Wood (eds.), Sockeye Salmon (*Oncorhynchus nerka*) Population Biology and Future Management. Canadian Special Publication of Fisheries and Aquatic Sciences 96.

Wright, S. 1981. Contemporary Pacific salmon fisheries management. North American Journal of Fisheries Management 1: 29-40.

Evaluating Salmon Management Institutions: The Importance of Performance Measures, Temporal Scales, and Production Cycles

Jim Lichatowich

Abstract

Management of renewable resources from an ecosystem perspective will require institutional changes. Management institutions will have to enhance their ability to learn, use what they learn, and promote more effective evaluation of performance. Development of appropriate management baselines is an important prerequisite to institutional learning and evaluation. Three important components of management baselines are biological performance measures, temporal scale of institutional evaluation, and natural environmental fluctuations in the ecosystems that produce Pacific salmon (*Oncorhynchus* spp.). Performance measures should broaden the traditional focus on harvest and economics to include measures of the condition of important ecological processes. Temporal scale will have to include the entire history of management institutions. Compressing the temporal scale and shifting baselines forward will create inevitable extinction of the resource. Natural cycles in production require flexibility in management programs and the development of realistic expectations.

Introduction

About 130 years ago, exploitation and management of Pacific salmon (*Oncorhynchus* spp.) passed from the exclusive domain of Native Americans to a growing dominance by Euro-Americans. Native Americans continued to participate in the fishery to varying degrees, often with a great deal of controversy (Boxberger 1989), and in recent decades they have been recognized as co-managers. Even though Native Americans have participated in the exploitation and management of Pacific salmon in recent decades, western science and economics have dominated the decision-making process. For salmon, the change was decidedly traumatic: salmon shifted from a gift in a natural economy to a commodity in a global market. Before contact with Euro-Americans, management systems were based on thousands of years of observation embedded in myth and ceremony. After contact, management systems were based on science embedded in politics. The relationship between salmon and Native Americans was guided by short feedback loops between the resource and user. In the industrial economy of Euro-Americans, feedback loops dissolved and control over exploitation shifted to the demands of distant markets and short-term economic efficiencies.

How have the salmon fared as a result of the change? Few would argue against the conclusion that they have not done very well.

Throughout most of the past 130 years, management was primarily concerned with the orderly extraction of commodities through commercial and sport fisheries. That approach to management was consistent with the values held by the community at the time: ecosystems were merely storehouses where commodities were held until needed by the market economy (Worster 1977). The role of ecosystems in the production of those commodities was given little attention (McEvoy 1986). Renewable resource management was primarily concerned with the extraction of free wealth provided by the virgin productivity of Pacific Northwest ecosystems.

The extraction of free wealth, like the frontier, is part of American history. Natural productivity of the Pacific salmon south of British Columbia has declined by ~80% (Anderson 1993). In response to the declines in abundance, salmon management institutions are in the process of shifting emphasis from exploitation to restoration (e.g., Northwest Power Planning Council [NPPC] 1987, 1992; Bowles and Leitzinger 1991; Potter 1992; Washington Department of Fisheries et al. 1993). The change will require a better understanding of the basis for the productivity of Pacific Northwest watersheds. The industrial economy extracted free wealth from Pacific Northwest ecosystems while resource managers remained largely ignorant of the ecological relationships that determined the productivity of those systems (McEvoy 1986). Effective restoration cannot be achieved with the same level of ignorance. Successful restoration will be a test of our understanding of ecological relationships (Bradshaw 1990).

In response to changing societal values, a new resource management paradigm based on an ecosystem perspective is emerging (Franklin 1992, Bottom 1996, Frissell et al. 1996). Adopting an ecosystem perspective will require new management science, institutional change, and a reexamination of old information from new perspectives. Management institutions must enhance their ability to learn (Lee 1993), become more reflexive or better able to use what they learn (Wright 1992), and become better at incorporating the public in the management process (Slocombe 1993; M.A. Shannon and C. Robinson, Inst. for Resources in Society, College of Forest Resources, Univ. Washington, Seattle, unpubl. data). The latter should include a process for legitimizing institutional evaluation and criticism (Wright 1992). The need for institutional learning will be critical over the next few decades while ecosystem science is being transformed into practical management prescriptions. An important part of institutional learning is the development of appropriate baselines.

This paper considers the performance of management institutions from the biological perspective (i.e., the productivity of the Pacific salmon resource). Social performance of management institutions, such as stability of resource-dependent communities, promotion of a stewardship ethic, and public participation in management, although important, is not considered.

Conventional Baselines

The term baseline is often associated with environmental impact assessment. A conventional definition of a baseline is a data set collected and analyzed to define the present state of a biological community and its environment (Green 1979). Usually, a change from the baseline condition is anticipated as a result of human activity. Change is detected by comparative analysis of the baseline with a similar data set collected in the future. The statistical aspects of baselines in fisheries research and management, particularly the power of comparative statistical tests, have been addressed (e.g., Green 1979; Lichatowich and Cramer 1979; Peterman and Bradford 1987; Peterman 1989, 1990; McAllister and Peterman 1992). Thus, they are not reviewed here.

Management Baselines

Institutional learning involves more than well-designed studies and the accumulation of knowledge by individual scientists. It is the ability to incorporate the collective wisdom of resource managers and scientists into agency programs, policies, and guiding philosophies. Since it is the institutions that persist and provide continuity, rather than individuals, it is at the institutional level that relevant learning takes place. However, there are many impediments to institutional learning (Hilborn 1992) including an emphasis on legitimizing established programs and practices and a reluctance to legitimize institutional evaluation (Wright 1992).

Management baselines discussed in this paper are the basis for evaluating institutional learning and performance. They are a synthesis of appropriate data to provide a historical and scientific context for evaluating institutional performance. As defined here, management baseline allows that institutional evaluation is not the product of a series of experiments conducted on a production system in stable equilibrium. When considered at the institutional level, salmon management is one long, unique event embedded in a changing natural-cultural system. Therefore, the evaluation of management performance cannot be treated as a series of repeatable and controlled scientific experiments. Here I describe some components that should be considered in developing baselines for the evaluation of institutional performance, including biological performance measures, temporal scales, and production cycles.

Components of Management Baselines

Biological Performance Measures

An important step in developing any baseline is the selection of things to measure. For management baselines, these measures are reported to the public to apprise them of the status of the resource. Performance measures should, in part, reflect community values regarding the salmon resource. In the past the dominant values were utilitarian (Worster 1977, Bottom 1996). Those values are reflected in current performance measures, which are almost entirely focused on production, harvest, and economics of the commercial fishery (Table 1). The one exception is stock status (e.g., Nehlsen et al. 1991, Nickelson et al. 1992, Washington Department of Fisheries et al. 1993), which was instituted in response to the continuing decline in salmon.

Community values that guide the management of renewable resources are changing. The emphasis on production of commodities, which dominated resource management until recently, is giving way to a realization that other values, symbolized by the terms biodiversity and ecosystem

Table 1. Measures of management performance in current use.

Measures
Catch (sport & commercial)
Angler days
Economic value of the catch
Licenses sold
Pounds of fish released from hatcheries
Number of habitat projects completed
Escapement
Stock status

conservation, need to be incorporated into management (Franklin 1992). In salmon management, this change in values is reflected as an increased emphasis on the conservation of wild stocks.

In salmon management, the concern for wild salmon, for the genetic resources they contain, and for their role as indicators of healthy ecosystems began to receive scientific attention (Calaprice 1969, Ricker 1972, Reisenbichler and McIntyre 1977) and policy recognition (Oregon's wild fish policy adopted by the Fish and Wildlife Commission, May 25, 1978) in the 1970s. Recent inventories and status reports on wild stocks of salmon (Nehlsen et al. 1991, Nickelson et al. 1992, Washington Department of Fisheries et al. 1993) suggest that concern for wild stocks is becoming institutionalized, recognizing that salmon managers have a dual responsibility: (1) to manage for a sustained flow of commodities and the economic benefits derived therein, and (2) to manage the human exploitation of salmon and salmon habitat to ensure continuous production (Lichatowich 1992).

The current list of performance measures adequately determines the performance of commodity production, and they should be retained for that purpose. However, if the performance of management institutions is to reflect changing community values, and if the institutions are to organize their activities around an ecosystem perspective, the list of performance measures must be revised to include those that reflect ecosystem health (Table 2). Habitat complexity and connectivity, the condition of the riparian zone and floodplain, and life-history diversity are consistent with ecosystem health defined as the maintenance of complexity and self-organizing capacity (Norton 1992). Post-release survival of hatchery fish and genetic diversity of mixed hatchery and wild production systems evaluate the effect of technology on resource quality. Recruit/spawner ratios are better indices of production trends than simple catch data.

Bottom (1996) suggests additional areas from which performance measures could be derived. A complete list would include those that also evaluate social performance of management institutions.

Temporal Scales

Two consistent and contradictory themes have persisted throughout the past 100 years of Pacific salmon management in the Pacific Northwest: pessimism due to the continuous decline in salmon production (e.g. Hume 1893, McGuire 1894, Rich 1942, Tait 1961, Nehlsen et al. 1991), and optimistic predictions regarding future production (e.g., Oregon State Board of Fish

Table 2. Indicators of ecosystem health from which measures of management performance might be derived.

Habitat complexity and connectivity
Life-history diversity
Post-release survival of hatchery fish
Recovery following low periods in productivity cycles
Conditions of the riparian zones and floodplains
Genetic diversity of mixed hatchery-wild production system
Long-term recruit/spawner ratios

Commissioners 1887, Hume 1893, Oregon Fish and Game Commission 1919, Bell 1937, Oregon Fish Commission 1968, NPPC 1987). Pacific salmon managers were able to maintain optimistic predictions in the face of continued declines in salmon production, in part, because baselines lacked the appropriate temporal scale. Two examples to illustrate this phenomenon are discussed below.

The Oregon hatchery program was initiated in 1897 and by 1907, 12 hatcheries were releasing 27 million salmon fry. However, the artificial propagation program failed to increase the catch, which discouraged fishery managers (Oregon Department of Fisheries 1908) and led to a series of innovative experiments. With advice and financial assistance from several of the cannery operators, state hatcheries that normally released sac fry began to rear juvenile salmon for several months and release them at a larger size. Coincident with this experiment, the catch increased in 1914, and after 5 successive years of improved catches in the Columbia River (Fig. 1), the Oregon Fish and Game Commission announced the success of their experiments:

> this new method has now passed the experimental stage, and . . . the Columbia River as a salmon producer has 'come back.' By following the present system, and adding to the capacity of our hatcheries, thereby increasing the output of young fish, there is no reason to doubt but that the annual pack can in time be built up to greater numbers than ever before known in the history of the industry. . . . (Oregon Fish and Game Commission 1919).

Subsequent review indicated that the increased catch probably reflected environmental variability and could not be attributed to the new hatchery methodology (Johnson 1984). The original conclusion, drawn from a baseline of compressed temporal scale (Fig. 1), resulted in an overestimate of the new technology's effectiveness.

Misinterpretation resulting from the use of baselines with inadequate temporal scales is not limited to the early decades of this century. A more recent use of a compressed temporal baseline led to an overly optimistic projection of the recovery of chum salmon (*O. keta*) on the north Oregon coast. An experimental hatchery on Whiskey Creek, a tributary to Netarts Bay, was operated to develop new technology to improve the artificial propagation of chum salmon. The hatchery began operation in 1969 and by 1974, based on one good return year (Fig. 2), a news release from Oregon State University (1974) concluded the following:

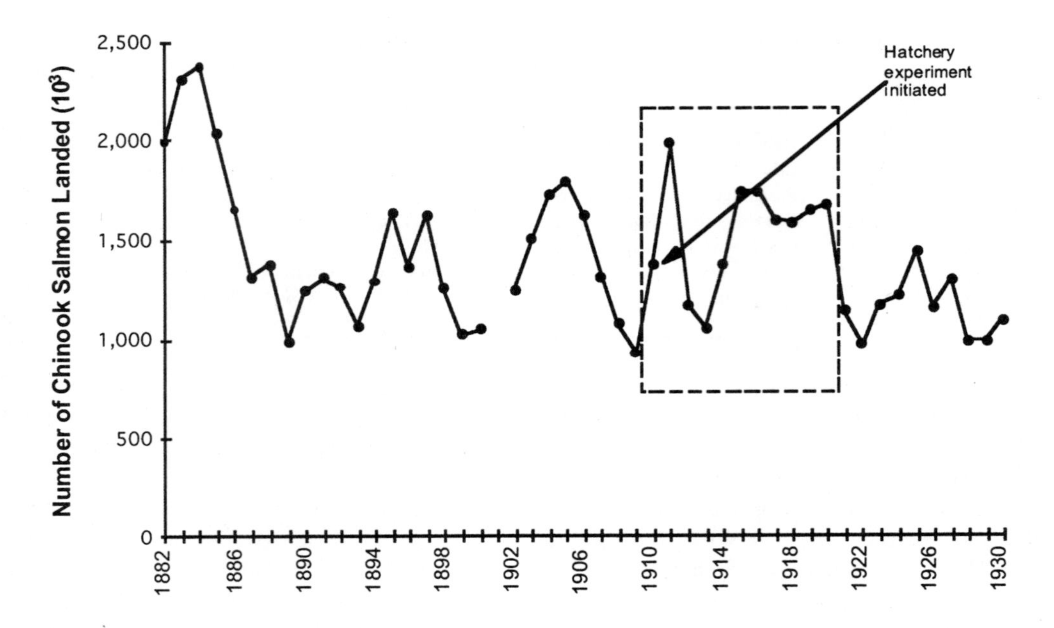

Figure 1. The number of chinook salmon landed in the Columbia River, 1882-1930. Source: Beiningen (1976).

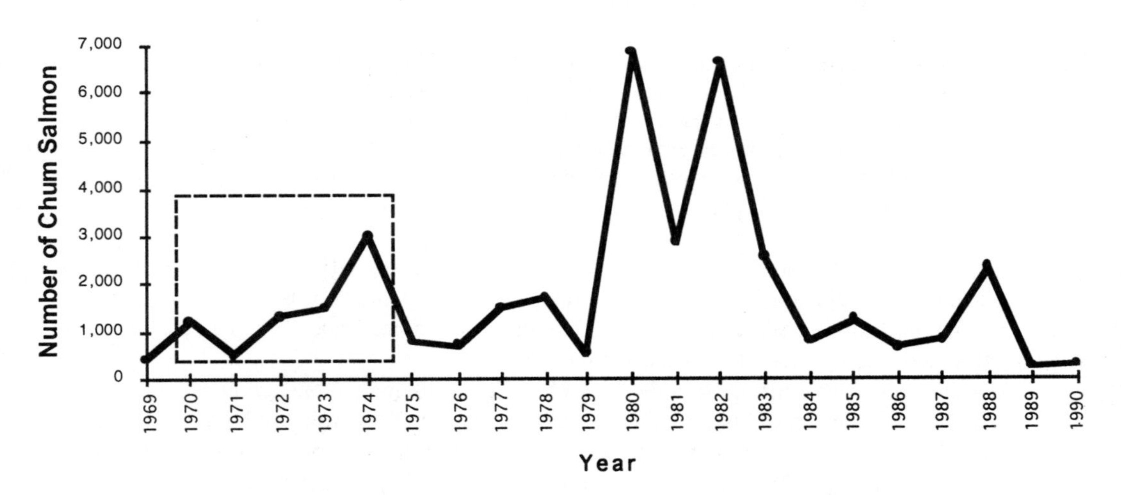

Figure 2. The number of chum salmon returning to the hatchery rack on Whiskey Creek. Source: Nickelson et al. (1992).

> This year over 3,000 salmon returned (to Whiskey Creek). . . . The Oregon legislature made private chum salmon hatcheries legal in 1971. That law and the new hatchery design will lead to a new fishing industry in Oregon. . . . By 1980 chum salmon returns to private hatcheries may reach three million pounds annually.

Whiskey Creek Hatchery failed to achieve that optimistic projection.

On the surface it would appear these two examples simply illustrate poor experimental design, which is probably true. However, these examples need to be considered in their historical context. They are but two of the many overly optimistic projections of the benefits of hatchery technology (Bottom 1996), which demonstrates a persistent failure in institutional learning. Contributing to that failure was the use of baselines of inappropriate temporal scale.

Cobb (1930) noted the lack of evaluation of hatchery programs and the ". . . almost idolatrous faith in the efficacy of artificial propagation. . . ." Although he believed hatchery programs could increase the production of salmon, the blind optimism with which hatchery technology was pursued caused Cobb (1930) to list hatcheries as one of the threats to the fishing industry. The lack of evaluation of hatchery performance or evaluation baselines of inappropriate scale have impeded institutional learning and prevented effective use of hatchery technology (see Hindar et al. 1991, Ryman 1991, Waples 1991, and Reisenbichler 1996 for recent discussions of threats posed by hatcheries).

A second example of compressed temporal scale in management baselines is the current definition of healthy stocks employed by the states of Oregon and Washington. Oregon's definition of a healthy stock is "A stock is healthy if available spawning habitat has generally been fully seeded and abundance trends have remained stable or increased over the last 20 years" (Nickelson et al. 1992). Washington's definition of a healthy stock is "a stock of fish experiencing production levels consistent with its available habitat and within natural variation in survival of the stock" (Washington Department of Fisheries et al. 1993).

By considering only the currently available habitat, Oregon and Washington have compressed the habitat baseline. There is no historical component. Past habitat loss is not taken into account in either definition. The implication is that we are satisfied with the status quo, the current condition of salmon habitat. This has led to stocks being listed as healthy that have exhibited dramatic declines. For example, the Tillamook Bay chum salmon is listed as a healthy stock even though the annual run declined from an average of 92,000 fish in the late 1940s and early 1950s to an average of 17,000 fish in recent years (averages calculated from Table 6 in Nickelson et al. 1992).

Selecting the temporal scale of management baselines presents the individual working for a resource agency with a dilemma. Responsibility for declining economic and recreational opportunities is often placed on individuals currently employed by a management agency. Those individuals are reluctant to be held accountable for losses in production that occurred before they assumed management responsibility. However, good intentions to hold the line at current habitat loss are thwarted when each generation of managers shifts the baseline forward to avoid accountability for past mistakes. Under this scenario institutional learning and adaptive management are impossible and Lee's (1993) hope for an environmentally sustainable economy is greatly diminished.

If each generation is allowed to reset the temporal scale of the baselines used to evaluate management performance, the result will be no evaluation or accountability, a misinformed public, and a spiral of decline for which the outcome will be inevitable extinction of Pacific salmon in much of their range. Although the careers of individuals may encompass a short time, and individuals can only be held accountable for their own performance, institutional performance must be evaluated over its entire history of management.

Developing information on historical habitat and stock status for incorporation into management baselines is often difficult because the data are not always available in the appropriate form and some reconstruction is often necessary. Nicholas (1996) gives a detailed discussion of the sources and uses of historical fisheries data. Sometimes reconstruction has to be inferred from incomplete information or from unconventional sources. Sedell and Luchessa (1981) reconstructed the historic habitat complexity of Pacific Northwest streams from geography texts, Army journals, personal diaries, reports from other scientific disciplines such as forestry and agriculture, and the reports from the United States Secretary of War Chief of Engineers. Developing baselines from historical reconstruction of salmon abundance and habitats is a critical step in salmon restoration (Sedell and Luchessa 1981, Doppelt et al. 1993, Williams 1993). However, managers have been reluctant to undertake reconstruction of historical stock abundance and habitat status (e.g., Nickelson et al. 1992, Washington Department of Fisheries et al. 1993).

When evaluating the benefits and drawbacks of historical reconstruction, we need to remember that ecosystems are the product of their history (Lewis 1969). In an important sense, ecosystems are composed of combined histories: the geological and erosional histories of the land forms and river channels, the evolutionary history of the biota in the watershed, and the cultural history of human economies. These histories influence the trajectory of an ecosystem's development: They determine the system's present state and help establish the range of possibilities for future developmental trajectories.

Salmonid ecosystems in the Pacific Northwest will continue to evolve and change. Understanding the historical roots and incorporating the possibilities for future change into management programs will be a major challenge for resource managers attempting to adopt an ecosystem perspective. Achieving that understanding will require that long-standing myths generated by the mechanistic world view be dispelled (Timmerman 1986, Botkin 1990). In the mechanistic world view, ecosystems operate as machines in stable equilibriums. The machine's operation in the past and future can be inferred from today's performance—there is no meaningful history since the machine's operation does not change (Botkin 1990). In the future, resource managers will have to accept the need to manage under conditions of uncertainty and change (Botkin 1990). Under that scenario, historical reconstruction and incorporation of that information into management baselines are critical.

Fortunately, there is a recent example of the value of historical reconstruction. Sedell and Luchessa (1981) reconstructed the riverine habitats in Oregon and Washington largely based on inferences drawn from non-fisheries literature. Their work led to a hypothesis regarding the role of large woody debris in salmon habitat. Subsequent empirical verification of their hypotheses has had a major impact on habitat management and will be a positive influence on the future developmental trajectories of Pacific Northwest watersheds.

Production Cycles

Management baselines of appropriate temporal scale are the only way to differentiate between the kinds of uncertainty that Hilborn (1992) has labelled noise and state of nature. Noise is the annual variability in production, whereas uncertainty in the state of nature may result from long-term cycles in productivity or a permanent shift in the salmon's ecosystem to a higher or lower productive state. Because our experience with shifting states of nature is limited, their likelihood or nature cannot be quantified (Hilborn 1992, Ludwig et al. 1993).

The following three recent papers have presented evidence for long-term fluctuations in fisheries productivity in the region. First, primary and secondary production and biomasses of pelagic fishes in the California Current fluctuate on a 40- to 60-year oscillation (Ware and Thompson 1991). Second, the abundance of salmon in the North Pacific Ocean corresponds to the long-term fluctuation in the Aleutian low pressure system (Beamish and Bouillon 1993). Third, survival of coho salmon (*O. kisutch*) in the Oregon Production Index (OPI) is determined by the intensity of coastal upwelling (Nickelson 1986), which at least partially explains a 50-year cycle in coho salmon production (Fig. 3).

The existence of long-term fluctuations in productivity has important consequences for the evaluation of salmon management programs and institutions. Management baselines must incorporate knowledge of past fluctuations in productivity so a change in state can be detected and differentiated from annual noise early enough for managers to respond with a program correction (e.g., adjustment in harvest or hatchery programs). When the temporal scale of a management baseline is compressed, the expected performance of a system may be derived from erroneous assumptions. Performance measured as salmon abundance over a short time interval might suggest an ecosystem in a stable equilibrium, whereas the same system, viewed over a longer time scale might reveal a system with cyclic fluctuations in productivity. Interpreting measurements

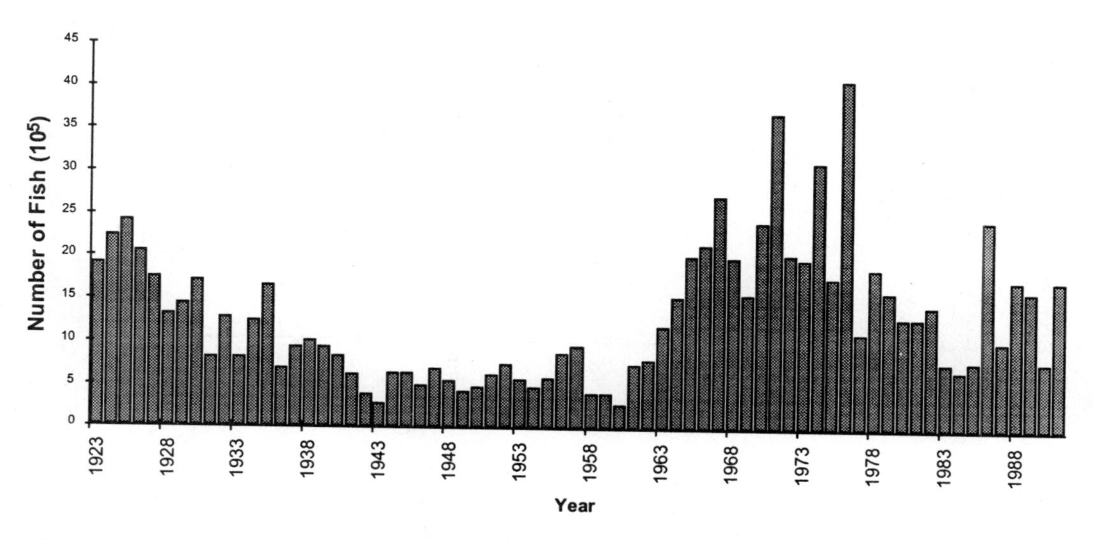

Figure 3. Harvest of coho salmon in the Oregon Production Index (OPI). Sources: 1923–70, Oregon Dep. Fish and Wildlife, unpubl. data; 1970–91, Pacific Fishery Management Council (1992).

of a system's performance depends on the scale at which the system responds and not the scale within which we operate (O'Neill et al. 1986).

A failure to correctly interpret the state of a system can produce surprise—a contrary response to management activity—and lead to erroneous management prescriptions. For example, in a system subjected to natural, long-term fluctuations in productivity, hatchery releases during improving conditions may assist in the rate of increase in abundance, whereas continued releases of large numbers of hatchery-reared fish may not be the appropriate strategy during periods of deteriorating conditions and decreasing survival (Beamish and Bouillon 1993). Baselines that fail to account for fluctuations in productivity may mask the effects of habitat degradation or cause a misinterpretation of the effects of restoration programs (Lawson 1993). A failure to recognize a large shift in productivity during the first half of this century may have led to erroneous management prescriptions.

Historical standing stocks of pelagic fishes (hake, *Merluccius productus*; sardine, *Sardinops sagax*; and anchovy, *Engraulis mordax*) in the California Current were reconstructed from scales contained in core samples taken from anaerobic sediments (Soutar and Isaacs 1974, Smith 1978). Those data show fluctuation in abundance over a 200-year period (Fig. 4). An important feature of that data set is the occurrence of a 200-year peak in standing stocks near the turn of the 19th century followed by a 200-year low in standing stocks in the 1930s and 1940s. The magnitude of the change between 1900 and 1940 was the largest in the 200-year data set. The Oregon harvest of coho salmon parallels the trend in marine standing stocks (Fig. 4).

The fluctuation in the catch of coho salmon and standing stocks of pelagic fishes in the California Current corresponds to indices of climatic conditions in the freshwater habitat of the

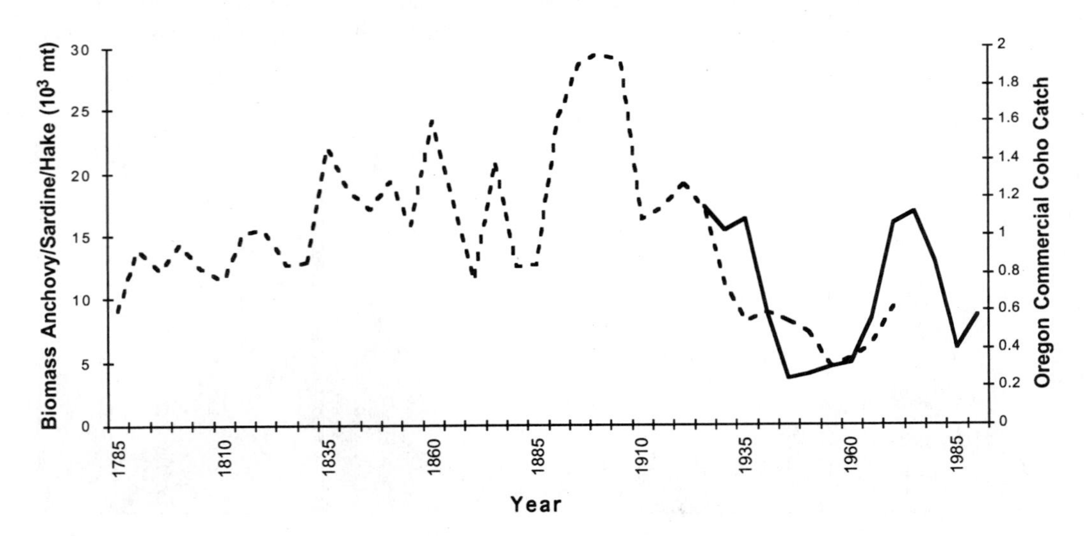

Figure 4. Dashed line: Total biomass of anchovy, sardine, and hake in the California Current in thousands of metric tons. Standing stock inferred from contemporary stock size and scale deposition rates in 18th and 19th centuries. Solid line: Commercial catch of coho salmon in millions of fish. Annual coho salmon harvest averaged by 5-year intervals. Sources: anchovy, sardine, and hake, Smith (1978); coho salmon, Lichatowich (1993).

Columbia River Basin. Climatic conditions inferred from spacing of growth rings on trees from the Columbia and Snake rivers also show cyclic fluctuations (Fig. 5; Fritts 1965). A period of cool, wet weather especially in the Snake River around 1900 was followed by a severe hot, dry period, which lasted through mid-1940s. Large-scale climate change probably influenced salmon production both in freshwater and in the marine environment during 1900–40. Commercial landings of chinook (*O. tshawytscha*), coho, sockeye (*O. nerka*) and chum salmon in the Columbia River and chinook and coho salmon in Oregon coastal streams were in decline between 1920 and 1940 (Figs. 6 and 7). In addition, the catch of chinook and coho salmon in Puget Sound showed significant declines between 1896 and 1934 (Bledsoe et al. 1989).

From 1880 to 1920, standing stocks of pelagic marine fishes in the California Current were at a 200-year high and favorable climatic conditions prevailed in freshwater. Those conditions coincided with the rapid development and early peaks in commercial salmon fisheries in the Pacific Northwest (Table 3). Declining marine standing stocks and a shift to hot, dry climate in the Columbia River Basin coincided with declining commercial harvest of salmon (Figs. 6 and 7). The importance of these observations to the development of management baselines is this: The rapid development of the salmon fishery in the Pacific Northwest during a period of high productivity in the marine environment and favorable climate in freshwater established expectations that could not be maintained in the long term. This conclusion is evident now only because sufficient historical information is available to identify natural fluctuations in production and because biologists

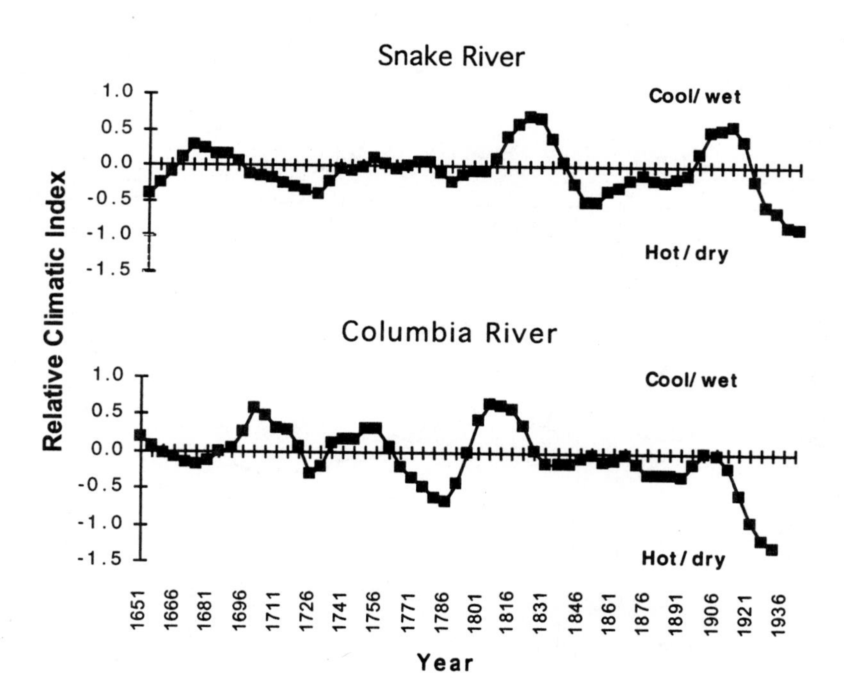

Figure 5. Fluctuation in an index of climate inferred from growth rings of trees in the Columbia Basin. Shown are 5-year moving averages of relative departures from a 270-year mean. Positive departure indicates cool, wet climate and negative departure indicates hot, dry climate. Source: Fritts (1965).

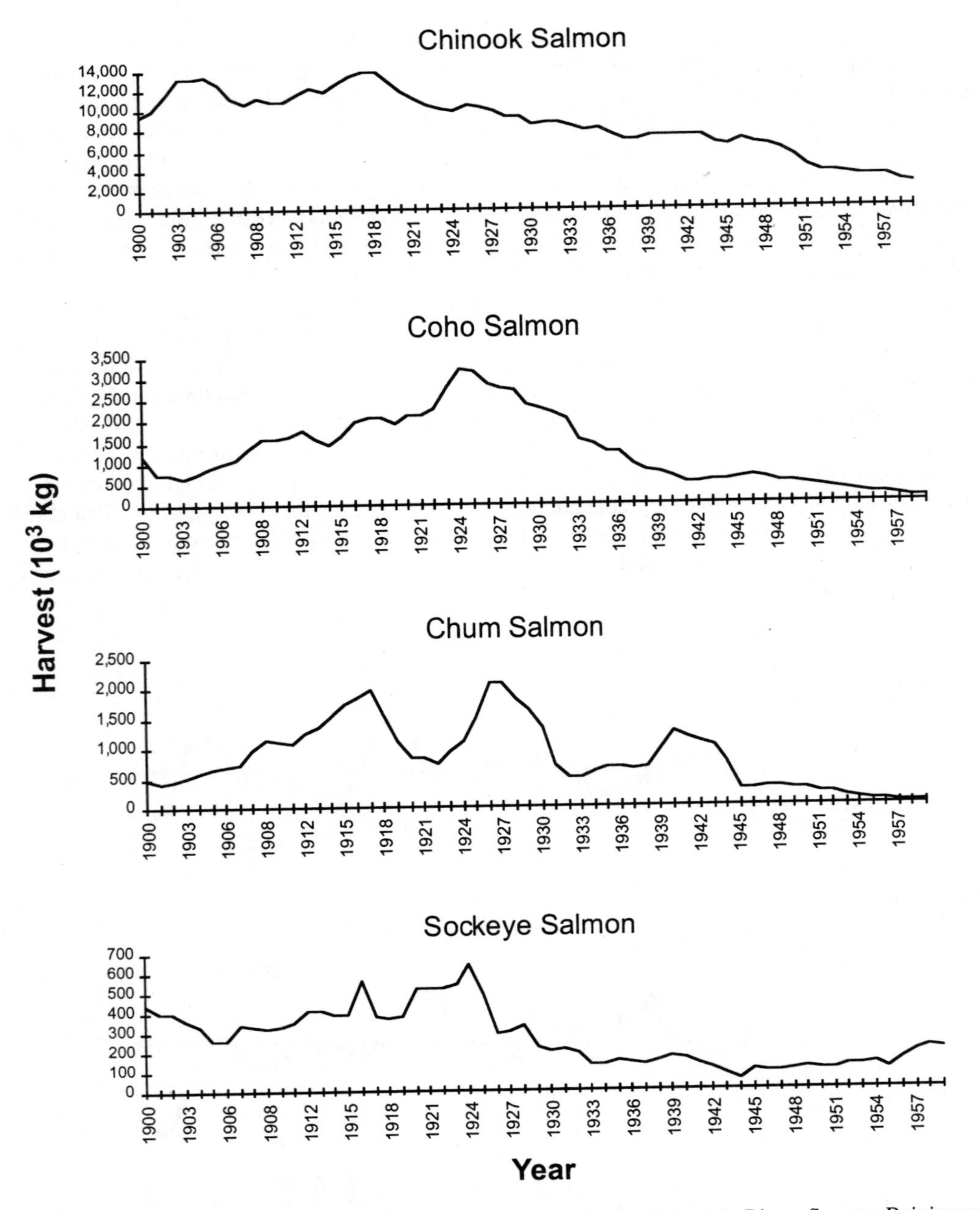

Figure 6. Five-year moving average of commercial harvest in the Columbia River. Source: Beiningen (1976).

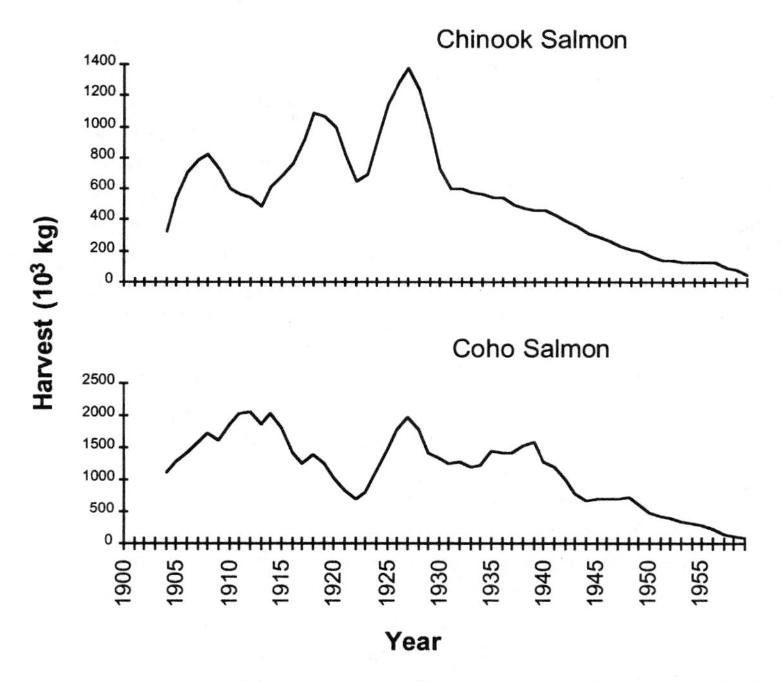

Figure 7. Five-year moving average of the commercial chinook and coho salmon harvest in Oregon coastal rivers. Sources: Mullen (1981); R. Mullen, Oregon Dep. of Fish and Wildlife, unpubl. data.

Table 3. Peak cannery packs of all species of Pacific salmon in the Northwest prior to 1930. Source: Cobb 1930.

Area	Peak year	Harvest (metric tons)
Coastal Oregon rivers	1911	3,002
Coastal Washington rivers	1915	691
Columbia Basin	1895	13,803
Fraser River	1901	21,749
Grays Harbor	1911	1,653
Klamath River	1912	392
Puget Sound	1913	56,248
Sacramento River	1882	4,354
Willapa Harbor	1902	886
Total		102,778

have accepted a different view of ecosystem stability. Until recently, undisturbed ecosystems were believed to be in a stable equilibrium, and after disturbances the system would return to this equilibrium (Timmerman 1986, Botkin 1990). This view was evident in the early Columbia River fishery investigations, which endeavored to return the fishery to stable productivity (Craig and Hacker 1940).

The decline in the Columbia River fishery was attributed to overharvest and habitat destruction (Craig and Hacker 1940). If human disturbance in the form of habitat degradation and overharvest had disturbed the system's stability, it was consistent with the prevailing world view to search for ways to let the system return to its natural equilibrium. The decision to use artificial propagation to circumvent anthropomorphic disturbance (Lichatowich and Nicholas in press) was consistent with the managers' understanding of the causes for the declining salmon production and the prevailing world view. However, emphasizing hatcheries as an alternative to habitat protection and harvest regulation probably intensified production problems. A period of hot, dry climate (1920–40, Fig. 5) may have reduced the total habitat capacity of the Columbia River. The habitat and system capacity was constricted even more by the construction of mainstem dams and continued habitat destruction due to grazing, irrigation, and timber harvest. The situation was further aggravated by the release of hatchery fish exhibiting low survival (Lichatowich and Nicholas in press) into the habitats with shrinking capacity.

The use of hatcheries to circumvent habitat degradation has become a cornerstone of management programs. Hatcheries consume a large share of agency budgets. As mentioned earlier, managers are recognizing the value of naturally reproducing populations of salmon and that management must maximize the effectiveness of both the natural and artificial production systems. Naturally produced and artificially propagated fish occupy different freshwater habitats; artificially propagated fish are protected in the hatchery through much of their freshwater residence. Therefore, wild and artificially propagated salmon could be impacted differently by climate fluctuations. Even in the ocean, where hatchery and wild salmon share a common environment, the survival of juvenile hatchery and wild fish might differ in response to changing patterns of coastal upwelling (Nickelson 1986). In addition, the presence of artificially propagated fish can adversely impact naturally produced salmon (Hindar et al. 1991, Ryman 1991, Waples 1991, Meffe 1992, Reisenbichler 1996). The adverse effects include overharvest in mixed stock fisheries and interbreeding between hatchery and wild fish in natural spawning areas. Production and survival differences and the potential for negative interaction between hatchery and wild fish make it imperative that management baselines separate hatchery and wild components of production.

The ocean harvest of coho salmon between 1923 and 1992 in the OPI illustrates how a failure to differentiate natural and artificial production can mask the true response of wild stocks to natural fluctuations in productivity (Figs. 3 and 8). The decline in production during the 1920s, 1930s, and 1940s discussed earlier continued through the 1950s, followed by recovery through the 1960s and 1970s. Unfortunately, separate estimates of the hatchery and wild components of production were made through only part of the data set. Prior to 1960, the harvest was almost entirely wild. After 1960, with improving diets and disease control the contribution of hatchery fish to the harvest increased. In addition, ocean conditions were improving in the 1960s, which also influenced survival of hatchery fish (Nickelson 1986). Historical records of the total harvest of coho salmon (Fig. 3) suggest a robust recovery in the OPI in response to changing ocean conditions (Oregon Department of Fish and Wildlife 1982). However, when both artificially and naturally produced fish are accounted for, the wild components of the harvest were negligible in comparison to hatchery fish (Fig. 8).

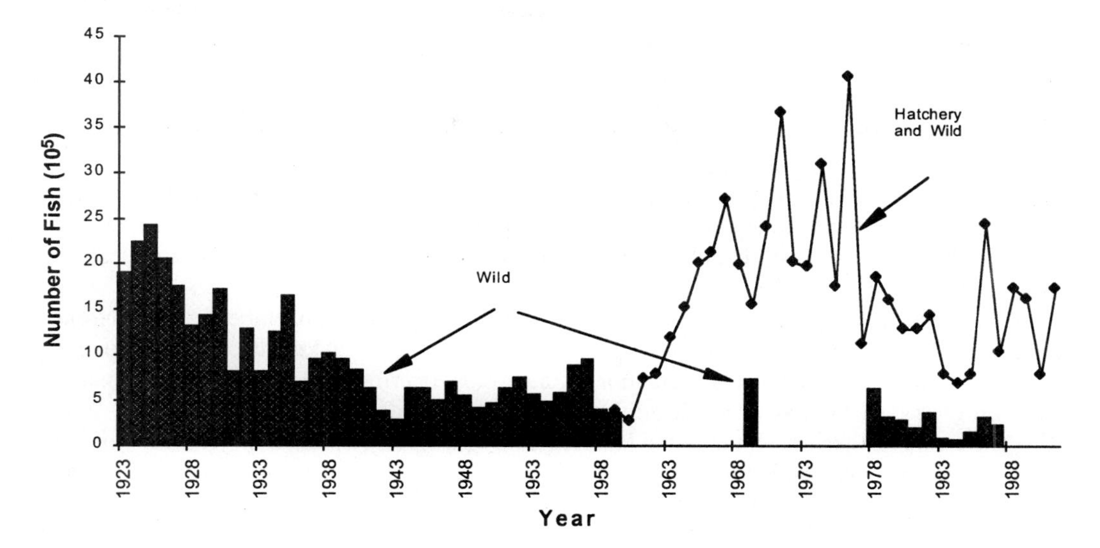

Figure 8. Harvest of coho salmon in the Oregon Production Index partitioned into wild and hatchery fish. Solid bars are catch of wild coho salmon. All coho salmon are assumed to be wild before 1960. Sources: OPI harvest, 1923–70, Oregon Dep. Fish and Wildlife, unpubl. data; 1971–91, Pacific Fishery Management Council (1992). Harvest of wild coho salmon 1959 and 1969, Oregon Dep. Fish and Wildlife (1982); 1978–87, L. Borgerson, Oregon Dep. Fish and Wildlife, pers. comm.

Conclusions

Salmon managers cannot reset the clock. They will never again have the opportunity to manage the exploitation of a virgin salmon resource in unspoiled habitat. Unfortunately, we still have not learned how to achieve an environmentally stable economy (Lee 1993) with what is remaining of the natural productivity of the Pacific Northwest ecosystems. Lee's (1993) pessimism is certainly warranted for those local economies heavily dependent on salmon; they do not appear to be sustainable.

Given the decline in abundance of Pacific salmon over the last 130 years, it seems reasonable to conclude that the performance of salmon management can be improved. A major step in the direction of improved performance is the general recognition that management must shift emphasis from single species to an ecosystem perspective. The need for that shift is an important theme of this book.

The shift to an ecosystem perspective in management of natural resources entails institutional changes as well as the transfer of ecosystem science into management prescriptions. An important institutional change is the ability to learn (adaptive management) and the development of processes for evaluating institutional performance. Selection and design of management baselines can facilitate those institutional changes.

In this paper I described three components of management baselines that are either lacking or need revision: (1) performance measures, (2) temporal scales, and (3) production cycles. Until recently, performance measures reflected the management emphasis on production and harvest.

As freshwater ecosystems degraded, resource managers measured the effects of that degradation indirectly through declining harvest. Resource managers failed to monitor the condition and remained largely ignorant of the ecological process that produced salmon. Existing performance measures need to be expanded to include those that reflect the health of the ecological process important to the production of salmon.

Salmon-producing ecosystems are products of history. The components include geologic and erosional histories, the evolutionary histories of biota, and histories of human economies. To deny the importance of history by continually shifting baselines forward can only lead to extinction spirals.

The ecosystems that produce salmon in the Pacific Northwest are not embedded in environments consistent with expectations of long-term stable production. Natural cycles in productivity must be incorporated into the evaluation of institutional performance. In addition, management programs need to have the flexibility to respond to natural changes in productivity.

The current status of salmon in the Pacific Northwest clearly calls for an evaluation of the effectiveness of management institutions. The suggestions and observations contained in this paper are offered to facilitate that evaluation.

Acknowledgments

R. Naiman, C. Warren, D. Bottom, and an anonymous reviewer offered several suggestions which improved the manuscript.

Literature Cited

Anderson, M. 1993. The living landscape. Volume 2. Pacific Salmon and Federal Lands. The Wilderness Society, Bolle Center for Forest Ecosystem Management, Washington, DC.

Beamish, R.J. and D.R. Bouillon. 1993. Pacific salmon production trends in relation to climate. Canadian Journal of Fisheries and Aquatic Sciences 50: 1002-1016.

Beiningen, K.T. 1976. Fish Runs, Report E. *In* Investigative Reports of Columbia River Fisheries Project. Pacific Northwest Regional Commission, Vancouver, Washington.

Bell, F.T. 1937. Guarding the Columbia's silver horde. Nature Magazine 29: 43-46.

Bledsoe, L.J., D.A. Somerton, and C.M. Lynde. 1989. The Puget Sound runs of salmon: An examination of the changes in run size since 1896, p 50-61. *In* C.D. Levings, L.B. Holtby and M.A. Henderson (eds.), Proceedings of the National Workshop on Effects of Habitat Alteration on Salmonid Stocks. Canadian Special Publication Fisheries and Aquatic Sciences 105. Ottawa, Ontario.

Botkin, D.B. 1990. Discordant Harmonies: A New Ecology for the Twenty-First Century. Oxford University Press, New York, New York.

Bottom, D.L. 1996. To till the water: a history of ideas in fisheries conservation, p. 569-597. *In* D.J. Stouder, P.A. Bisson, and R.J. Naiman (eds.), Pacific Salmon and Their Ecosystems: Status and Future Options. Chapman Hall, New York.

Bowles, E. and E. Leitzinger. 1991. Salmon supplementation studies in Idaho rivers: Experimental design. Idaho Department of Fish and Game, Project 89-098. Boise, Idaho.

Boxberger. D.L. 1989. To Fish in Common: The Ethohistory of Lummi Indian Salmon Fishing. Univ. Nebraska Press, Lincoln, Nebraska.

Bradshaw, A.D. 1990. Restoration: An acid test for ecology, p. 23-29. *In* W.R. Jordan III, M.E. Gilpin, and J.D. Aber (eds.), Restoration Ecology. Cambridge University Press, Cambridge, Massachusetts.
Calaprice, J.R. 1969. Production and genetic factors in managed salmonid populations, p. 377-388. *In* T.G. Northcote (ed.), Symposium on Salmon and Trout in Stream. H.R. MacMillan Lectures in Fisheries, Institute of Fisheries, The University of British Columbia, Vancouver, British Columbia, Canada.
Cobb, J.N. 1930. Pacific salmon fisheries. US Department of Commerce, Bureau of Fisheries Document No. 1092, Washington, DC.
Craig, J.A. and R.L. Hacker. 1940. The history and development of the fisheries of the Columbia River. Bulletin of the Bureau of Fisheries No. 32. Washington, DC.
Doppelt, B., M. Scurlock, C. Frissell, and J. Karr. 1993. Entering the Watershed. Island Press, Covelo, California.
Franklin, J.F. 1992. Scientific basis for new perspectives in forests and streams, p. 25-72. *In* R.J. Naiman (ed.), Watershed Management: Balancing Sustainability and Environmental Change. Springer-Verlag, New York.
Frissell, C.A., W.J. Liss, R.E. Gresswell, R.J. Nawa, and J. Ebersole. 1996. A resource in crisis: changing the measure of salmon management, p. 411-444. *In* D.J. Stouder, P.A. Bisson, and R.J. Naiman (eds.), Pacific Salmon and Their Ecosystems: Status and Future Options. Chapman and Hall, New York.
Fritts, H.C. 1965. Tree-ring evidence for climatic changes in western North America. Monthly Weather Review 93: 965-441.
Green, R.H. 1979. Sampling Design and Statistical Methods for Environmental Biologists. Wiley and Sons, New York, New York.
Hilborn, R. 1992. Can fisheries agencies learn from experience? Fisheries 17: 6-14.
Hindar, K., N. Ryman and F. Utter. 1991. Genetic effects of cultured fish on natural fish populations. Canadian Journal of Fisheries and Aquatic Sciences 48: 945-957.
Hume, R. D. 1893. Salmon of the Pacific Coast. Schmidt Label & Lithographic Co., San Francisco, California.
Johnson, S. 1984. Freshwater environmental problems and coho production in Oregon. Oregon Department of Fish and Wildlife, Information Reports No. 84-11. Portland, Oregon.
Lawson, P.W. 1993. Cycles in ocean productivity, trends in habitat quality, and restoration of salmon runs in Oregon. Fisheries 18(8): 6-10.
Lee, K.N. 1993. Compass and Gyroscope: Integrating Science and Politics for the Environment. Island Press, Covelo, California.
Lewis, J.K. 1969. Range management viewed in the ecosystem framework, p. 97-187. *In* G.M. Van Dyne (ed.), The Ecosystem Concept in Natural Resource Management. Academic Press, New York.
Lichatowich, J.A. 1992. Management for sustainable fisheries: Some social, economic and ethical considerations, p. 11-17. *In* G. Reeves, D. Bottom and M. Brooks (eds.), Ethical Questions for Resource Managers. US Department of Agriculture, Forest Service, Pacific Northwest Research Station, General Technical Report PNW-GTR-288. Portland, Oregon.
Lichatowich, J.A. 1993. Ocean carrying capacity: Recovery issues for threatened and endangered Snake River salmon. Bonneville Power Administration, Technical Report 6 of 11, P.O. Box 3621. Portland, Oregon.
Lichatowich, J. and S. Cramer. 1979. Parameter selection and sample sizes in studies of anadromous salmonids. Oregon Department of Fish and Wildlife Information Report Series, Fisheries Number 80-1. Portland, Oregon.
Lichatowich, J.A. and J.W. Nicholas. In press. Oregon's first century of hatchery intervention in salmon production: Evolution of the hatchery program, legacy of a utilitarian philosophy and management recommendations. International Symposium on Biological Interactions of Enhanced and Wild Salmonids, Nanaimo, British Columbia, Canada, June 17-20, 1991. Canadian Journal of Fisheries and Aquatic Sciences.
Ludwig, D., R. Hilborn, and C. Walters. 1993. Uncertainty, resource exploitation, and conservation lessons from history. Science 260: 17, 36.
McAllister, M.K. and R.M. Peterman. 1992. Experimental design in the management of fisheries: A review. North American Journal of Fisheries Management 12: 1-18.
McEvoy, A. F. 1986. The Fisherman's Problem: Ecology and Law in the California Fisheries 1850-1980. Cambridge University Press, New York, New York.
McGuire, H.D. 1894. Fish and game protector to the Governor. 1st & 2nd Annual Report. Salem, Oregon.

Meffe, G.K. 1992. Techno-arrogance and halfway technologies: Salmon hatcheries on the Pacific coast of North American. Conservation Biology 6: 350-354.

Mullen, R.E. 1981. Oregon's commercial harvest of coho salmon, *Oncorhynchus kisutch* (Walbaum), 1892-1960. Oregon Department of Fish and Wildlife, Information Report Series, Fisheries Number 81-3. Portland, Oregon.

Nehlsen, W., J.E. Williams and J. Lichatowich. 1991. Pacific salmon at the crossroads: Stocks at risk from California, Oregon, Idaho and Washington. Fisheries 16: 4-21.

Nicholas, J.W. 1996. On the nature of data and their role in salmon conservation, p. 53-60. *In* D.J. Stouder, P.A. Bisson, and R.J. Naiman (eds.), Pacific Salmon and Their Ecosystems: Status and Future Options. Chapman and Hall, New York.

Nickelson, T.E. 1986. Influences of upwelling, ocean temperature, and smolt abundance on marine survival of coho salmon (*Oncorhynchus kisutch*) area. Canadian Journal of Fisheries and Aquatic Sciences 43: 527-535.

Nickelson, T.E., J.W. Nicholas, A.M. McGie, R.B. Lindsay, D.L. Bottom, R.J. Kaiser, and S.E. Jacobs. 1992. Status of anadromous salmonids in Oregon coastal basins. Oregon Department of Fish and Wildlife. Portland, Oregon.

Northwest Power Planning Council. 1987. Columbia River Basin Fish and Wildlife Program (amended). Northwest Power Planning Council, Portland, Oregon.

Northwest Power Planning Council. 1992. Strategy for Salmon. Vol ll. Northwest Power Planning Council, Portland, Oregon.

Norton, B.G. 1992. A new paradigm for environmental management, p. 23-41. *In* R. Costanza, B.G. Norton, and B.D. Haskell (eds.), Ecosystem Health. Island Press, Washington, DC.

O'Neill, R.W., D.L. DeAngelis, J.B. Waide, and T.F.H. Allen. 1986. A Hierarchical Concept of Ecosystems. Monographs in Population Biology 23. Princeton University Press, Princeton, New Jersey.

Oregon Department of Fish and Wildlife. 1982. Comprehensive plan for production and management of Oregon's anadromous salmon and trout. Part II. Coho salmon plan. Portland, Oregon.

Oregon Department of Fisheries. 1908. Annual report of the Department of Fisheries of the State of Oregon to the Legislative Assembly, Twenty-Fifth Regular Session, 1909. Portland, Oregon.

Oregon Fish and Game Commission. 1919. Biennial report of the Fish and Game Commission of the State of Oregon, 1919. Portland, Oregon.

Oregon Fish Commission. 1968. Biennial Report. Portland, Oregon.

Oregon State Board of Fish Commissioners. 1887. First annual report of the Oregon Fish Commissioners. Salem, Oregon.

Oregon State University. 1974. Chum salmon hatchery: A new industry for Oregon. News release, Sea Grant College Program. Corvallis, Oregon.

Pacific Fishery Management Council. 1992. Review of the 1991 ocean salmon fisheries. 2000 SW First Avenue, Portland, Oregon.

Peterman, R.M. 1990. Statistical power analysis can improve fisheries research and management. Canadian Journal of Fisheries and Aquatic Sciences 47: 2-15.

Peterman, R.M. 1989. Application of statistical power analysis to the Oregon coho salmon (*Oncorhynchus kisutch*) problem. Canadian Journal of Fisheries and Aquatic Sciences 46: 1183-1187.

Peterman, R.M. and M. Bradford. 1987. Statistical power of trends in fish abundance. Canadian Journal of Fisheries and Aquatic Sciences 44: 1879-1889

Potter, M.S. 1992. Governor's coastal salmonid restoration initiative. Summary Report. Oregon Department of Fish and Wildlife. Portland, Oregon.

Reisenbichler, R. 1996. Genetic factors contributing to declines of anadromous salmonids in the Pacific Northwest, p. 223-244. *In* D.J. Stouder, P.A. Bisson, and R.N. Naiman (eds.), Pacific Salmon and Their Ecosystems: Status and Future Options. Chapman and Hall, New York.

Reisenbichler, R.R. and J.D. McIntyre. 1977. Genetic differences in growth and survival of juvenile hatchery and wild steelhead trout, *Salmo gairdneri*. Journal of Fisheries Research Board of Canada 34: 123-128.

Rich, W.H. 1942. The salmon runs of the Columbia River in 1938. US Department of the Interior, Fish and Wildlife Service, Fishery Bulletin 37. Washington, DC.

Ricker, W.E. 1972. Hereditary and environmental factors affecting certain salmonid populations, p. 19-160. *In* R.C. Simon and A. Larkin (eds.), The Stock Concept in Pacific Salmon. H.R. MacMillan Lectures in Fisheries, The University of British Columbia, Vancouver, British Columbia, Canada.

Ryman, N. 1991. Conservation genetics considerations in fishery management. Journal of Fish Biology 39 (Supplement A): 211-224.

Sedell, J.R. and K.J. Luchessa. 1981. Using the historical record as an aid to salmonid habitat enhancement, p. 210-223. *In* N.B. Armantrout (ed.), Acquisition and Utilization of Aquatic Habitat Inventory Information. American Fisheries Society, Bethesda, Maryland.

Slocombe, D.C. 1993. Implementing ecosystem-based management: Development of theory, practice, and research for planning and managing a region. BioScience 43: 612-622.

Smith, P.E. 1978. Biological effects of ocean variability: Time and space scales of biological response. Rapports et Proces-Verbaux des Reunions Conseil International pour l'Exploration de la Mer 173: 117-127.

Soutar, A. and J.D. Isaacs. 1974. Abundance of pelagic fish during the 19th and 20th Centuries as recorded in anaerobic sediment off the Californias. Fishery Bulletin 72: 257-273.

Tait, H.D. 1961. Pacific salmon rehabilitation: Highlight and recommendations of the 1961 governors' conference on salmon. Conference called by Governor William A. Egan of Alaska, in January 1961. Juneau, Alaska.

Timmerman, P. 1986. Mythology and surprise in the sustainable development of the biosphere, p. 435-452. *In* W.C. Clark and R.E. Munn (eds.), Sustainable Development of the Biosphere. International Institute for Applied Systems Analysis, Laxenburg, Austria.

Waples, R.S. 1991. Genetic interactions between hatchery and wild salmonids: Lessons from the Pacific Northwest. Canadian Journal of Fisheries and Aquatic Sciences 48 (Supplement 1): 124-133.

Ware, D.M. and R.E. Thomson. 1991. Link between long-term variability in upwelling and fish production in the Northeast Pacific Ocean. Canadian Journal of Fisheries and Aquatic Sciences 48: 2296-2306.

Washington Department of Fisheries, Washington Department of Wildlife, and Western Washington Treaty Indian Tribes. 1993. 1992 Washington State salmon and steelhead stock inventory. Washington Department of Fisheries. Olympia, Washington.

Williams, J. E. 1993. Restoring watershed health on federal lands. Trout 34: 18-21.

Worster, D. 1977. Nature's Economy: A History of Ecological Ideas. Cambridge University Press, Cambridge, Massachusetts.

Wright, W. 1992. Wild Knowledge: Science, Language, and Social Life in a Fragile Environment. University of Minnesota Press, St. Paul.

Regional Trends

California Salmon And Steelhead: Beyond The Crossroads

Terry J. Mills, Dennis R. McEwan, and Mark R. Jennings

Abstract

Virtually all California salmon (*Oncorhynchus* spp.) and steelhead (*O. mykiss*) stocks have declined to record or near-record low levels during 1980–95. Escapement of naturally spawning Klamath and Sacramento basin fall-run chinook salmon (*O. tshawytscha*) stocks has fallen consistently below the goals of 35,000 adults (Klamath) and 120,000 adults (Sacramento) established by the Pacific Fishery Management Council. These two stocks constitute the primary management units for ocean harvest regulations in California and southern Oregon. This decline triggered a mandatory review of ocean harvest and inland production conditions in each basin. The Sacramento winter-run chinook salmon, once numbering >100,000 adult spawners, was listed as threatened in 1990 and endangered in 1994 under the Endangered Species Act. The listing occurred as a result of a precipitous decline in abundance (to <200 adult spawners) and significant threats to this stock's continued existence.

Spring-run chinook salmon, historically an abundant component of California's inland fish fauna with >500,000 adult spawners, has been extirpated from the San Joaquin River basin. However, remnant populations of this naturally spawning stock remain within the Klamath, Smith, and Sacramento river basins. Unfortunately, annual counts of 3,000–25,000 spawners in the Sacramento River basin during the past 25 years are largely of hatchery origin. Recent steelhead data from the same region indicate that many stocks are close to extinction, and nearly all steelhead in the Sacramento River are also of hatchery origin. Both spring-run chinook salmon and summer steelhead are considered to be species of special concern by the California Department of Fish and Game because of their limited distributions and sensitivities to degraded habitat conditions. The southern race of winter steelhead south of Point Conception is nearly extinct and remnant populations have been recently recorded in only 9 streams.

Coastal cutthroat trout (*O. clarki*), which are restricted to lowland drainages from the Eel River northward, are greatly depleted. Coho salmon (*O. kisutch*), which once probably numbered close to 1,000,000 fish per year in coastal California streams, have dwindled to ~5,000 natural spawners per year. Chum salmon (*O. keta*), never a significant part of the state's native fish fauna, are currently restricted to <10 spawners in three different streams in the Sacramento River basin and occasionally in the South Fork of the Trinity River. The historically small runs of pink salmon (*O. gorbuscha*) in the Sacramento and Russian rivers are probably now extirpated. Anadromous sockeye salmon (*O. nerka*) are only recorded as strays.

In response to serious declines in salmon and steelhead stocks, numerous legislative and congressional actions have been undertaken and California has embarked on an ambitious plan to restore riparian habitats, improve fish passage, and increase natural production. Additionally, many currently unlisted California salmon and steelhead stocks are potential candidates for protection under the Endangered Species Act. These include coho, chum, spring-run chinook, and San Joaquin fall-run chinook salmon, as well as summer steelhead and the southern race of winter steelhead.

Introduction

California possesses a rich diversity of anadromous salmonid habitats that constitute the southernmost range for many species of trout and salmon (*Oncorhynchus* spp.; Table 1). Together, these fish exhibit a broad spectrum of unique life-history behaviors that allowed their continued existence in suitable aquatic habitats prior to changes brought about by human intervention. For example, the Sacramento winter-run chinook salmon (*O. tshawytscha*) spawns in a large river system during May, June, and July, and its life history is not replicated by another chinook salmon stock along the Pacific Coast (Healy 1991).

Historically, the most significant producer of salmon in California was the Central Valley, which includes the Sacramento and San Joaquin rivers, major tributaries, and the ecologically important Sacramento-San Joaquin Delta estuary (Skinner 1962; Fig. 1). This region once provided 1.8–5 million kg of fish annually for commercial gillnet fishers prior to the turn of the century (Clark 1929) and was the source from which many introductions of various species of trout and salmon were made throughout California and other parts of the United States (US), South America, Europe, and New Zealand (Stone 1897, Hedgpeth 1941). Since 1850, salmon resources of the Central Valley have declined by ~90% owing to overfishing and habitat loss, and are currently supplemented to a large extent by hatchery operations, a process which began in 1872 (Fisher 1994).

Besides the Central Valley, historically significant numbers of California salmon and steelhead (*O. mykiss*) were found in the Smith River, Klamath basin (including the Trinity River), Mad River, Eel River, Russian River, and numerous smaller North Coast streams (Fig. 1). As a result of overfishing, placer mining, and habitat loss, all of these streams have seen a >85% decline in adult salmon and steelhead resources from historical 19th century observations (Moyle et al. 1989).

Because of the drastic declines observed in all anadromous trout and salmon species in California over the past century, many attempts have been made to significantly increase the numbers of these economically valuable fish (see review in Lufkin 1991a). In this paper, we review the current status of each of the major stocks of native anadromous salmonids in the state, with emphasis on trends observed during the past 25 years. Additionally, we provide information on how the California Department of Fish and Game (CDFG) currently manages these fisheries resources.

Stock Management Policies

The state of California has a number of policies regarding the management and restoration of salmon and steelhead stocks. These policies have been developed by the California Fish and

Table 1. Diversity of native anadromous salmonids within selected California streams and rivers (adapted from data presented in Fry [1973] and Moyle [1976]). An "X" indicates the presence of each fish species (or stock) and a "?" indicates uncertainty within the selected river basin or stream.

River basin	Fish species (or stock)									
or stream	Fall-run chinook	Spring-run chinook	Late fall-run chinook	Winter-run chinook	Chum salmon	Coho salmon	Pink salmon	Coastal cutthroat	Winter steelhead	Summer steelhead
Smith River	X	—	—	—	—	X	—	X	X	X
Klamath River	X	X	X	—	—	X	—	X	X	X
Trinity River	X	X	—	—	X	X	—	X	X	X
Mad River	X	—	—	—	—	X	—	X	X	X
Eel River	X	—	—	—	—	X	—	X	X	X
Russian River	X	—	—	—	—	X	X	—	X	—
Sacramento River	X	X	X	X	X	—	X	—	X	—
San Joaquin River	X	—	X	?	—	—	—	—	?	—
North Coast streams	X	—	—	—	—	X	—	—	X	X
South Coast streams	—	—	—	—	—	X	—	—	X	—

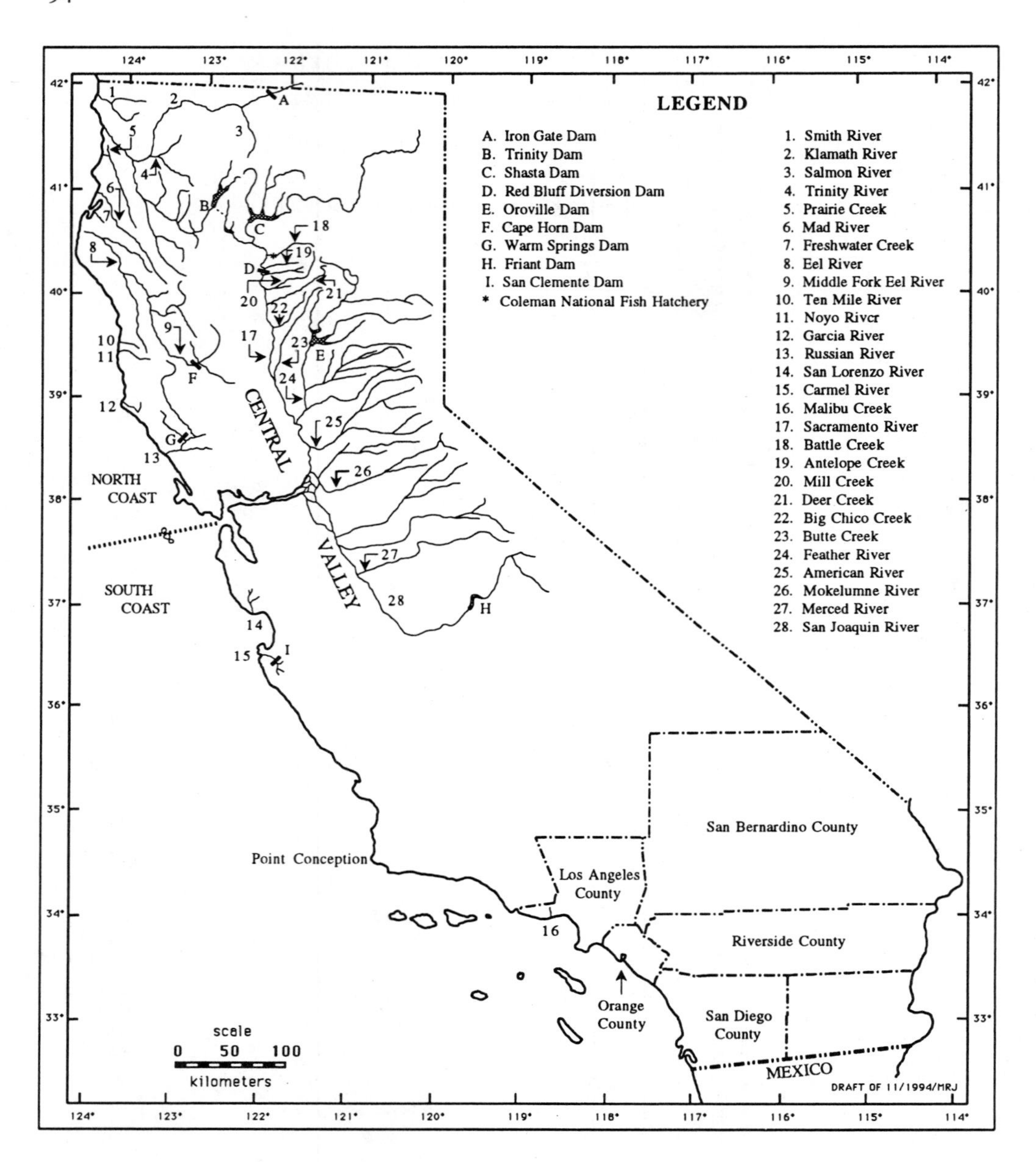

Figure 1. Map of California showing the locations of major rivers, dams, hatcheries, counties, etc., as discussed in the text. Degrees noted indicate 33–42° N and 114–124° W.

Game Commission, through state legislation, and through recent congressional actions such as the passage of the Central Valley Project Improvement Act (Public Law 102-575). Reproducing populations of California salmon and steelhead are generally categorized as either "naturally spawning" or "artificially propagated" (Gould 1994). Naturally spawning is a composite description of fish that reproduce in the wild regardless of parentage. Similarly, artificially spawning fish are any fish that return to a hatchery or other artificial propagation facility (Reynolds et al. 1990).

A key policy of the state of California in managing its salmon and steelhead populations is to maintain adequate breeding stocks and suitable spawning areas, and provide for the natural rearing of young fish to a migratory size. A complementary policy of the CDFG is to maintain the genetic integrity of all identifiable stocks of salmon and steelhead in California (Gould 1994).

California's stock classification and management system has defined the appropriate stock and the role of artificial production for management of each salmon and steelhead stream in California. This classification may be applied to drainages, individual streams, or segments of streams as necessary to protect discrete stocks of salmon or steelhead. The six management classes range from "endemic" to "any stock" (Table 2). Management and restoration efforts are broadly guided by this classification system. Policies relating to artificial production must also be compatible with this classification system.

California Salmon and Steelhead Hatchery System

California has an extensive hatchery system (see McGinnis 1984) that annually produces ~41.7 million chinook salmon, 1.1 million coho salmon (*O. kisutch*), and 5.4 million steelhead (Table 3). The CDFG operates four anadromous salmonid hatcheries in the Central Valley, two in the Klamath basin, one in the Mad River drainage, and one in the Russian River drainage. The US Fish and Wildlife Service operates one anadromous salmonid hatchery on Battle Creek, a tributary to the Sacramento River. Despite continued declines in the state's anadromous salmonid resources, these hatcheries have continued to catch, spawn, raise, and release as many fish as possible, which has resulted in an ever-increasing percentage of salmon and steelhead stocks in California being composed of hatchery origin (e.g., see Moyle et al. 1989, Reynolds et al. 1990, and Brown et al. 1994).

California Stock Assessment Program

The development of trend data for salmon and steelhead populations drives many of the statewide assessment programs. These trend data assist in the evaluation of the effects of timber harvest, water development and dam construction, harvest regulations, and restoration efforts on salmon and steelhead. Methods to annually enumerate spawning stocks include mark-recapture programs using the Petersen, Schaefer, and Jolly-Seber methods, fishway or ladder counts, aerial redd counts, and direct observation.

Table 2. California salmon and steelhead classification and management system (from Reynolds et al. 1990).

Classification	Description
Endemic	Only historically naturally reproducing fish originating from the same stream or tributary
Naturally reproducing from within basin	Naturally reproducing stocks from the drainage of which the stream is part
Hatchery stocks within basin	Stocks that may include hatchery-produced fish from streams within the drainage
Naturally reproducing from out of basin	Naturally reproducing stocks from streams outside the basin of which the stream is part
Hatchery stocks out of basin	Stocks that may include hatchery-produced fish from streams outside the basin
Any stock	Any stock that appears to exhibit characteristics suitable for the stream system

The ocean and inland harvest regulatory process is the basis for intensive fall-run chinook stock assessment within the Klamath basin and the Sacramento Valley. With this assessment information, the Pacific Fishery Management Council (PFMC) is required to evaluate the performance of ocean regulations, seasons, and quotas. Much of the emphasis for studies of other salmon and steelhead stocks is derived from regulatory and administrative requirements to monitor populations and develop trend data. These efforts include stock assessment, collection of population data, and analysis of the fisheries as required by provisions of the Federal Energy Regulatory Commission process for licensing hydropower projects and the State Water Resources Control Board regulatory process, which sets flow and water quality standards for rivers and streams in California. Additionally, anadromous fish populations are routinely monitored as required by state legislation to determine whether current numbers of salmon and steelhead are nearing the stated goal of doubling fish numbers from those counted during 1988 by the year 2000 (Lufkin 1991b). Similar federal legislation requires a doubling of anadromous fish populations over the average numbers present during 1967 through 1991. The annual monitoring of Sacramento River winter-run chinook salmon populations is required by a biological opinion regarding the operation of the state and federal water projects in the Central Valley (National Marine Fisheries Service 1993).

Recent declines in chinook salmon populations have caused CDFG anadromous fisheries research and management efforts to be largely redirected to the more economically important salmon stocks; consequently, steelhead stock assessment and management has been greatly reduced. Because of this and the difficulty in assessing steelhead populations, as most adults migrate on high, turbid flows, the CDFG has only a few reliable population estimates of returning adult steelhead.

Table 3. List of California salmon and steelhead production hatcheries and the average annual production of chinook salmon, coho salmon, and all stocks of steelhead (data from unpublished records of the California Department of Fish and Game and the US Fish and Wildlife Service).

		Average annual production					
Facility[a] and period of record	Location	Fall-run chinook	Late fall-run chinook	Winter-run chinook	Spring-run chinook	Coho salmon	Steelhead
Feather River Hatchery (1968–93)	Feather River	7,433,921	N.P.[b]	N.P.	1,218,837[c]	N.P.	751,162
Nimbus Hatchery (1965–93)	American River	8,809,514	N.P.	N.P.	N.P.	N.P.	766,993
Mokelumne River Hatchery (1965–93)	Mokelumne River	946,166	N.P.	N.P.	N.P.	N.P.	160,993
Merced River Hatchery (1970–93)	Merced River	579,331	N.P.	N.P.	N.P.	N.P.	N.P.
Coleman National Fish Hatchery (1940–93)	Battle Creek[d]	14,940,749[e]	639,221[f]	26,219[g]	N.P.[e]	N.P.	813,896[h]
Trinity River Hatchery (1964–93)	Trinity River	2,547,223	N.P.	N.P.	1,209,051	587,599	792,718
Iron Gate Hatchery (1966–93)	Klamath River	5,823,573	N.P.	N.P.	20,196[i]	121,757	391,885
Mad River Hatchery (1970–93)	Mad River	420,229	N.P.	N.P.	N.P.	232,361	865,304
Warm Springs Hatchery (1980–93)	Dry Creek[j]	189,501	N.P.	N.P.	N.P.	163,361	845,369
Sum of average statewide production		41,690,207	639,221	26,219	2,448,084	1,105,078	5,388,320

[a]All facilities are operated by the California Department of Fish and Game, except that Coleman National Fish Hatchery is operated by the US Fish and Wildlife Service.
[b]N.P. = not produced.
[c]Spring-run chinook salmon propagated at Feather River Hatchery are believed to have interbred with fall-run chinook salmon.
[d]Battle Creek is a tributary of the Sacramento River.
[e]During 1944–51, fall-run chinook salmon numbers are combined with spring-run chinook salmon numbers.
[f]During 1952–81, fall-run chinook salmon and late fall-run chinook salmon numbers were combined.
[g]Data from 1989–93. Small numbers of winter-run chinook salmon were also produced at Keswick during 1963, 1965, 1966, 1967, and 1983 (US Fish and Wildlife Service, unpubl. data).
[h]Steelhead were produced from 1952 to 1993.
[i]Spring-run chinook salmon were produced from 1968 through 1977.
[j]Dry Creek is a tributary of the Russian River.

Population Status and Trends

Water resource development within California's Central Valley has benefited agriculture, power production, municipal water supplies, recreation, and other uses, but it has caused serious harm to fisheries resources and aquatic ecosystems (Moyle 1976, McGinnis 1984). In particular, substantial declines in naturally spawning stocks of chinook salmon and steelhead in the Sacramento and San Joaquin basins over the past several decades are closely linked to water development, particularly water storage and export (McEvoy 1986). However, overharvest (both poaching and ocean gillnetting), drought, pollution, and climatic change have also taken their toll (Lufkin 1991a).

Central Valley anadromous salmonid declines since 1955 are attributed to many causes. Dam construction (especially the closure of Shasta Dam on the Sacramento River in 1944 and Friant Dam on the San Joaquin River in 1949; Fig. 1) and increased water diversions in the late 1950s and early 1960s have played major roles in these declines (Fry 1961, Hallock et al. 1970). During this period, the US Bureau of Reclamation completed the Trinity Division of the Central Valley Project in 1962 and the Red Bluff Diversion Dam (RBDD) in 1966 on the mainstem of the Sacramento River, and the California Department of Water Resources completed Oroville Dam in 1968 on the Feather River (Fig. 1). These state and federal water projects annually export vast quantities of water through the Sacramento-San Joaquin Delta for distribution in central and southern California—all at the expense of native resident and anadromous fisheries (Moyle 1976).

Historical and present-day fisheries problems in the Klamath basin are complex and linked to water development on the Trinity and upper Klamath rivers, timber harvest, road construction, placer mining, illegal fishing, and a variety of other lesser causes. The substantial decline in the Klamath basin chinook population is not a recent phenomenon. Severe declines in the chinook population occurred between the late 1800s and the 1920s (Scofield 1929, Snyder 1931, Moffett and Smith 1950). These early declines were attributed mainly to in-river commercial fishing, placer and hydraulic mining, and local dam construction. The problems of reduced anadromous fish habitat in various parts of the basin and increased harvests of anadromous fish stocks have been studied, but much remains to be identified and corrected. The primary management goal for the Klamath basin is to increase numbers of naturally spawning salmon and steelhead to levels appropriate for the available habitat. These naturally spawning stocks include chinook salmon, coho salmon, and steelhead.

Interspersed through the period of 1850–1995 were floods and droughts in both coastal and interior sections of the state. El Niño oceanic conditions became prevalent in the mid-1980s. The cumulative effects of water development, timber harvest, and increasing urbanization, coupled with natural climatic events, led to significant declines in California's salmon and steelhead resources during the past decade (Moyle et al. 1989, Lufkin 1991a, Brown et al. 1994). The effects of changing climatic conditions on California's salmon stocks during their ocean phase are unknown but presumed to be largely negative (Lufkin 1991b).

Chinook Salmon

Fall-run chinook salmon are presently the most abundant and widespread salmon in California. Adult returns to the Sacramento River basin since 1967 have ranged from >212,000 in

1986 to 62,000 in 1992 (Fig. 2a). The escapement goal of the PFMC was not attained in three consecutive years (1990–92), a situation that has triggered a PFMC habitat subcommittee review of ocean harvest and inland habitat conditions. Klamath fall-run chinook salmon escapement has been estimated annually since 1978, ranging from >113,000 naturally spawning adults in 1986 to ~12,000 in 1992 (Fig. 3a). The year 1993 was the fourth consecutive year that natural escapement was below the escapement baseline of 35,000 adult fish established by the PFMC. San Joaquin fall-run chinook are particularly vulnerable to habitat disturbance and water management operations in the Sacramento-San Joaquin Delta. Their numbers since 1967 have ranged from ~65,000 adults in 1985 to as few as several hundred in recent years (Fig. 3b). Naturally spawning escapements of Sacramento late fall-run chinook salmon since 1967 have ranged from

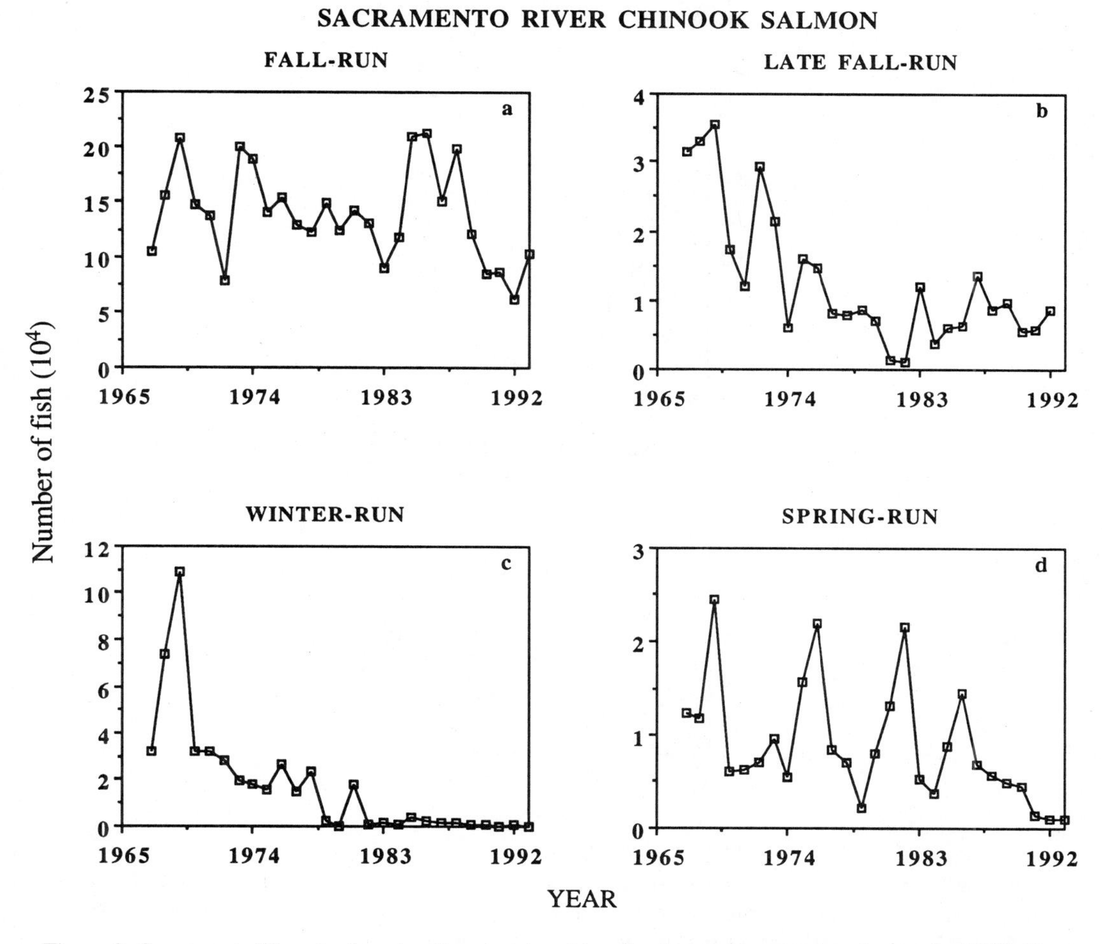

Figure 2. Sacramento River basin naturally spawning chinook salmon escapement estimates for 1967–93. Stocks are divided up into (a) fall-run, (b) late fall-run, (c) winter-run, and (d) spring-run. Data are from estimates published by the Pacific Fishery Management Council (1994).

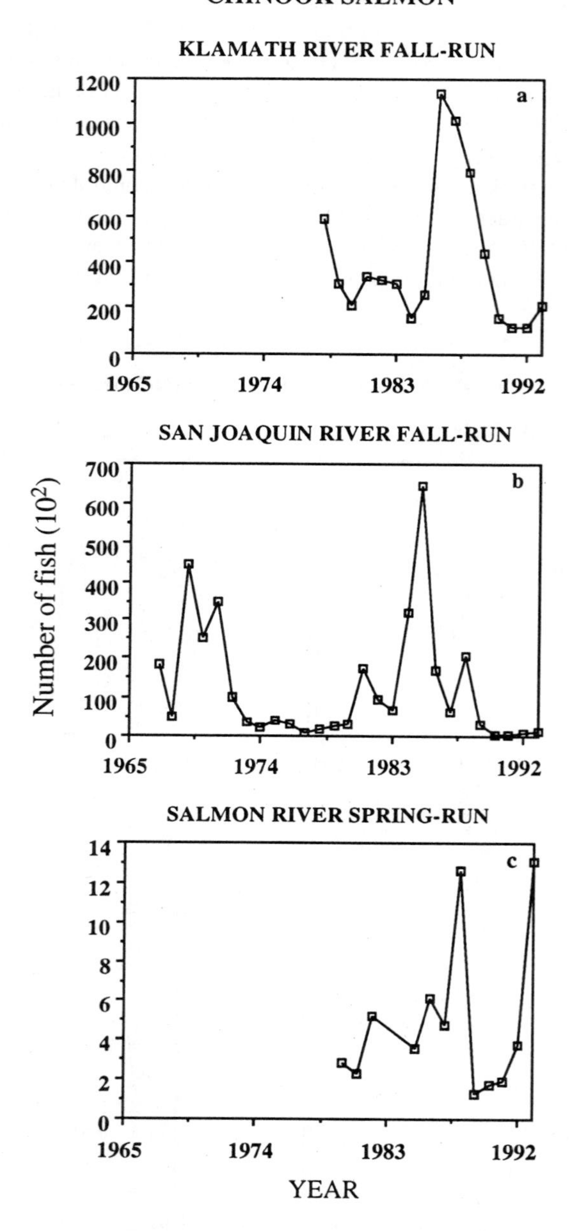

Figure 3. Naturally spawning chinook salmon escapement estimates for 1967–93: Klamath River (a) fall-run, (b) San Joaquin River fall-run, and (c) Salmon River spring-run. Data are from estimates published by the Pacific Fishery Management Council (1994).

36,000 in 1969 to ~1,000 adults in 1982 (Fig. 2b). Although the population declined dramatically between the late 1960s and 1982, it has generally remained between 5,000 and 10,000 spawners in recent years.

Sacramento winter-run chinook salmon have exhibited one of the most significant declines of salmon stocks within California, resulting in protection by the National Marine Fisheries Service (as threatened in 1990 and endangered in 1994) under the Endangered Species Act (ESA) (National Oceanographic and Atmospheric Administration [NOAA] 1990, 1994a). Since 1967, the population has declined from >100,000 in 1969 to <200 in recent years (Fig. 2c). Spawning escapements are expected to remain at low levels during the 1990s even though significant conservation measures have been implemented to permit unobstructed upstream and downstream passage for adults and juveniles; improved spawning, incubation, and fry rearing temperature control; reduced inland harvest by anglers; and curtailment of water for agricultural purposes during the times when juveniles are susceptible to entrainment (Williams and Williams 1991).

Spring-run chinook salmon in the Klamath and Sacramento basins declined in numbers to low levels and, thus, are considered a species of special concern by CDFG because of limited distribution and their sensitivity to degraded habitat conditions. Recent escapements are of particular consequence in the Sacramento basin owing to the questionable genetic identity of mainstem Sacramento River and Feather River stocks. Each of these populations had most of their access to historical spawning areas eliminated by the dam construction and were forced to spawn in new areas proximate to or overlapping with areas used by fall-run chinook (Slater 1963). Recent observations indicate that the spring-run and fall-run chinook salmon stocks in these areas may have interbred or continue to do so (Vogel 1987a, 1987b). Spring-run chinook stocks in the Sacramento River tributary streams such as Mill, Deer, Antelope, Big Chico, Battle, and Butte creeks (Fig. 1) are thought to most closely represent pre-1900 stocks of spring-run chinook (Moyle et al. 1989). Escapements of spring-run chinook into key Sacramento River indicator streams since 1967 have ranged from 24,000 in 1969 to 500 in 1991 (Fig. 2d). The status of spring-run chinook stocks in the Klamath basin is typified by the Salmon River stock. Since 1980, the population has been as high as 1,300 and as low as 175 (Fig. 3c). Although there are no clear trends in the Salmon River spring-run chinook population, its prospect for continued existence is better than for Sacramento spring-run chinook stocks because of the absence of major water development projects in the Salmon River drainage.

Coho Salmon

Coho salmon historically occurred in as many as 582 coastal streams from the California-Oregon border, south to the San Lorenzo River along the northern edge of Monterey Bay (Brown et al. 1994; Fig. 1). They are rarely observed in the Central Valley (Hallock and Fry 1967). Although precise data are lacking (e.g., Brown et al. [1994] cite a personal communication from E. Gerstung of 200,000–500,000 coho spawners during the 1940s), limited evidence indicates that historical populations of coho salmon in California probably numbered close to 1 million fish in the mid-1800s. This estimate is based on the generally accepted figure of 100,000 spawners present in the 1950s and 1960s (California Department of Fish and Game 1965), and the finding that chinook and steelhead stocks during the same time period had declined to 10% of historical observations made during the 1800s. However, there have been further sharp declines during the past decade (from 30,480 natural spawners during 1984–85 to ~13,000 natural spawners during 1987–91), and coho salmon have apparently disappeared from approximately half of the California streams in which they once occurred (Brown et al. 1994). With a 67% projected decline for Oregon and California stocks alone between 1993 and 1994 (National Oceanic and Atmospheric Administration 1994b), 1994 estimates for this species indicate ~5,000 natural spawners in California, a figure that closely matches the prediction of Brown et al. (1994). Such a precipitous decline probably qualifies coho salmon for protection under the ESA in California and contiguous states (Moyle 1994, Moyle and Yoshiyama 1994). Although coho salmon have been relatively well studied during the 1930s and early 1940s at one small stream at the southern edge of its range (Shapovalov and Taft 1954), the best data available for returning spawning adults in the state come from dams on the Klamath, Trinity, and Noyo rivers along the North Coast where numbers have varied from 11 in 1965 to 23,338 in 1987 for all streams combined (see Fig. 4 for individual river counts). The majority of these fish are presently of hatchery origin (Brown et al. 1994). Unfortunately, the winter spawning requirements of coho salmon reduced the opportunity for estimating spawner escapements in many small streams owing to high flows and typically turbid water conditions. Regardless, diverse sources of data regarding the distribution and abundance of juvenile coho indicate that a large number of stocks became extirpated and the abundance of individual populations is very low (many with <100 spawners) throughout their range in the state (Moyle et al. 1989, Brown et al. 1994).

Chum Salmon

Spawning populations of chum salmon (*O. keta*) were historically confined to the Sacramento River and also possibly the Trinity River, although strays have been taken in the Smith, Klamath, Mad, and San Lorenzo rivers and Freshwater Creek (Scofield 1916, Rogers 1974; Fig. 1). Estimates of spawners in the Sacramento River basin have varied from 34 to 210 (68 total adult fish observed) for 1950–58 (Hallock and Fry 1967), while 6 adults total have been caught on the South Fork of the Trinity River for 1985–87 and 1990 combined (California Dep. Fish and Game, unpubl. data). The species is currently restricted to <10 spawners in three different streams in the Sacramento River basin (Moyle 1994) and possibly the South Fork of the Trinity River where one pair of spawners was observed over a redd in 1987 (T. Mills, California Dep. Fish and Game, Sacramento, pers. observ.).

Pink Salmon

Reproducing populations of pink salmon (*O. gorbuscha*) were historically recorded from the Sacramento and Russian rivers where the total number of adult fish probably numbered <50 spawners (Fry 1967, Hallock and Fry 1967). Pink salmon recorded from the Klamath, Garcia, Mad, Ten Mile, and San Lorenzo rivers (Scofield 1916, Snyder 1931, Taft 1938), as well as Prairie Creek (Smedley 1952), were undoubtedly strays (Fry 1973). Today, the species is apparently extinct in California as a result of unsuitable spawning conditions in the Sacramento River, as well as habitat loss in the Russian River from the construction of Warm Springs Dam on Dry Creek in 1982 (Fig. 1) and major pollution events in the lower river (Moyle et al. 1989, Moyle 1994).

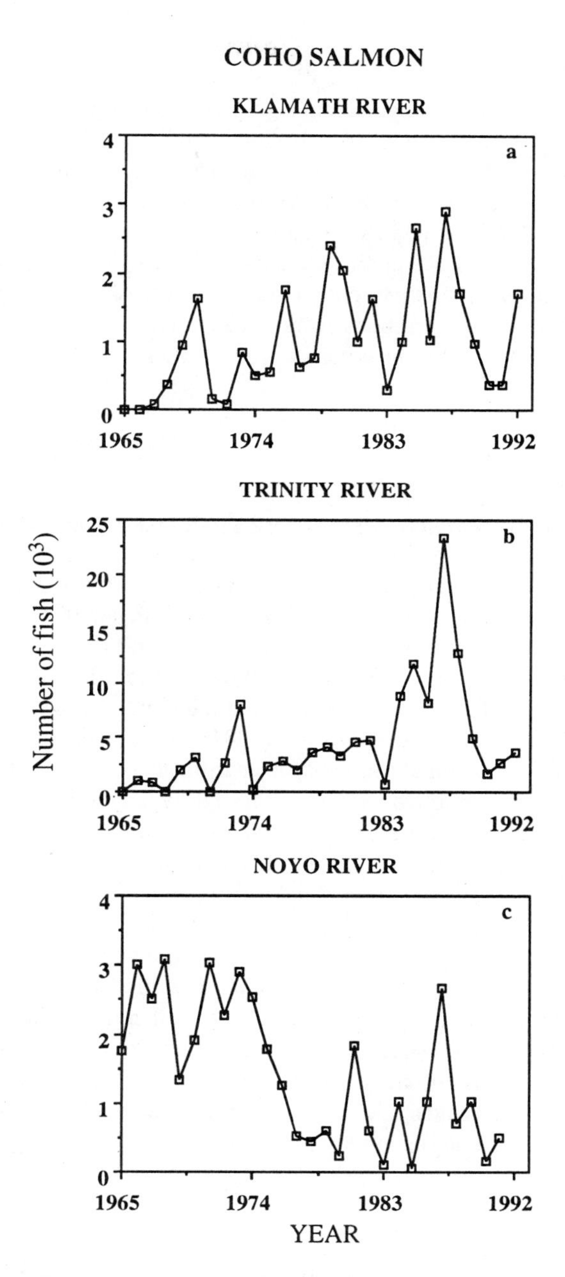

Figure 4. Counts of coho salmon from 1967 to 1993 at (a) Iron Gate Hatchery on the Klamath River, (b) Trinity Hatchery on the Trinity River, and (c) the egg-taking station on the Noyo River. Data are from California Department of Fish and Game (CDFG) hatchery records as published in CDFG Anadromous Fisheries Branch Administrative Reports (1967–85) and CDFG Inland Fisheries Administrative Reports (1986–93).

Sockeye Salmon

Anadromous forms of the sockeye salmon (*O. nerka*) have historically been recorded as strays in the Sacramento River (Rutter 1904, Hallock and Fry 1967, Fry 1973) because of the lack of suitable spawning and rearing habitat (Moyle 1976). A small run may have existed in the Klamath River (see Jordan and Evermann 1896) prior to 1917 when Copco Dam blocked all access to upper Klamath Lake (Moyle 1976). However, only a single adult sockeye salmon has been recorded from the Klamath River since that date (Fry 1973). We do not consider sockeye salmon to be part of California's native ichthyofauna because of the lack of suitable spawning and rearing habitat. The non-anadromous form, frequently referred to as kokanee, has been widely introduced into Central Valley reservoirs by the CDFG since 1951 (Moyle 1976, McGinnis 1984).

Coastal Cutthroat Trout

Coastal cutthroat trout (*O. clarki*) were historically found in North Coast streams from the Oregon-California border south to the Eel River (DeWitt 1954). Anadromous populations are restricted to low-gradient streams of the humid forested regions of coastal California (Gerstung 1981). Although coastal cutthroat trout were once abundant in small streams north of the Mad River during the late 1970s (M. Jennings, National Biological Service, San Simeon, pers. observ.), they have since become greatly depleted because of damage to riparian habitats by timber harvest, road construction, residential development, and other anthropogenic activities (Moyle et al. 1989, Moyle 1994). According to Gerstung (1981), surveys in the Smith River drainage (where the largest state population of this subspecies is located) found 15% of the streams were severely degraded, 29% moderately degraded, 35% slightly degraded, and 21% pristine. These percentages have not improved over the past decade owing to the effects of poor logging practices during the previous 25 years and current timber salvage operations in headwater streams (M. Jennings, National Biological Service, San Simeon, pers. observ.).

Recently, extensive declines occurred in coastal cutthroat populations in Oregon and Washington where stocks were augmented by hatchery operations (Moring 1993). The decline in native Oregon populations has been extensive enough that the National Marine Fisheries Service has proposed listing the Umpqua River stocks as endangered under the ESA (National Oceanic and Atmospheric Administration 1994c). The status of other Oregon and Washington coastal cutthroat trout stocks is currently under review (H. Weeks, Oregon Dep. Fish and Wildlife, Portland, pers. comm.). The decline in these northern populations is mirrored in California where hatchery augmentation has been very minimal.

Steelhead

The distribution of steelhead in the North Coast rivers and tributaries has not changed as drastically as in other areas of California. Major dams that blocked access to historical spawning and rearing areas are mostly far enough inland and at elevations >460 m, so that a considerable amount of habitat is still available downstream (Fig. 1). The major exception to this is on the Trinity River at Trinity Dam where a significant amount of steelhead habitat was badly degraded or eliminated in 1962. The relatively wide distribution of steelhead on the North Coast contrasts sharply with the Central Valley where dams at low elevations blocked access to thousands of kilometers of spawning and rearing habitat and severely reduced steelhead distribution (Reynolds et al. 1990). Steelhead once ranged throughout the tributaries and headwaters of the Sacramento and San Joaquin rivers below about 2,000 m prior to dam construction, water development, and watershed perturbations of the 19th and 20th centuries. Now they are mostly restricted to the lower elevations of the Sacramento River system (Moyle 1976). On the South Coast, water development, urbanization, and consequent localized extinctions have caused a severe reduction in steelhead range (Fry 1973).

Few escapement estimates of naturally spawning steelhead stocks within California have been developed. Light (1987) estimated the 1987 abundance of adult steelhead in California to be 275,000 adults. This is a rough estimate and is probably high, but it provides a general view of the magnitude of steelhead abundance during the past decade. This estimate represents a >50% decline from the 603,000 adult steelhead statewide estimate made 23 years earlier (California Department of Fish and Game 1965).

The present Central Valley annual steelhead run size, based on RBDD counts, hatchery counts, and natural spawning escapement estimates for tributaries, is probably <10,000 adult fish. This results in a substantial decline from the estimated 30,000 steelhead that returned to Central Valley rivers and streams in the early 1960s (California Department of Fish and Game 1965). A striking indicator of the magnitude of the decline of Central Valley steelhead stocks is the trend reflected in the RBDD counts. Steelhead counts at the RBDD have declined from an average annual count of 11,187 adults for 1967–76 to 4,391 adults for 1977–86 and 1,411 adults for 1987–93 (Fig. 5a). The decline of naturally reproducing populations in the Central Valley has been more precipitous than that of hatchery stocks. Hallock et al. (1961) reported that the composition of naturally produced steelhead in the 1950s ranged from 82% to 97% and averaged 88%. At present, only about 10% to 30% of the adults returning to spawn in the upper Sacramento River system are of natural origin (Reynolds et al. 1990). Naturally spawning stocks currently in the Sacramento River system are mostly confined to the upper Sacramento River tributaries. Recent counts at Coleman National Fish Hatchery (CNFH), Feather River Fish Hatchery, and Nimbus Hatchery are also well below the averages for these hatcheries (California Dep. Fish and Game, unpubl. data). Most steelhead ascending the RBDD in recent years likely result from hatchery fish destined to return to CNFH on Battle Creek, a tributary to the Sacramento River.

The upper Eel River at Cape Horn Dam is another area where steelhead trend data are currently being collected. Since 1967, counts of winter steelhead ranged from nearly 2,200 to <100 and declined substantially in recent years, although during 1993 returning adults increased (Fig. 5b). The decline of naturally spawning winter steelhead in this system has been even more precipitous in recent years, from >1,000 adult fish in 1987 to ~50 adult fish in 1993. In addition to winter steelhead, the Eel River also supports one of the largest populations of summer steelhead in California. Since 1972, counts of adult summer steelhead in the Middle Fork of the Eel River ranged from 1,600 to ~400 (Fig. 5c). This population has also declined in recent years owing to drought, timber harvest practices, and poaching (Moyle et al. 1989). These fish, along with all populations of summer steelhead, are now considered a species of special concern by CDFG.

Southern steelhead stocks (those occurring south of San Francisco Bay) were formerly found in coastal drainages as far south as the Santo Domingo River in northern Baja California and were present in all accessible streams and rivers of Los Angeles, Orange, Riverside, San

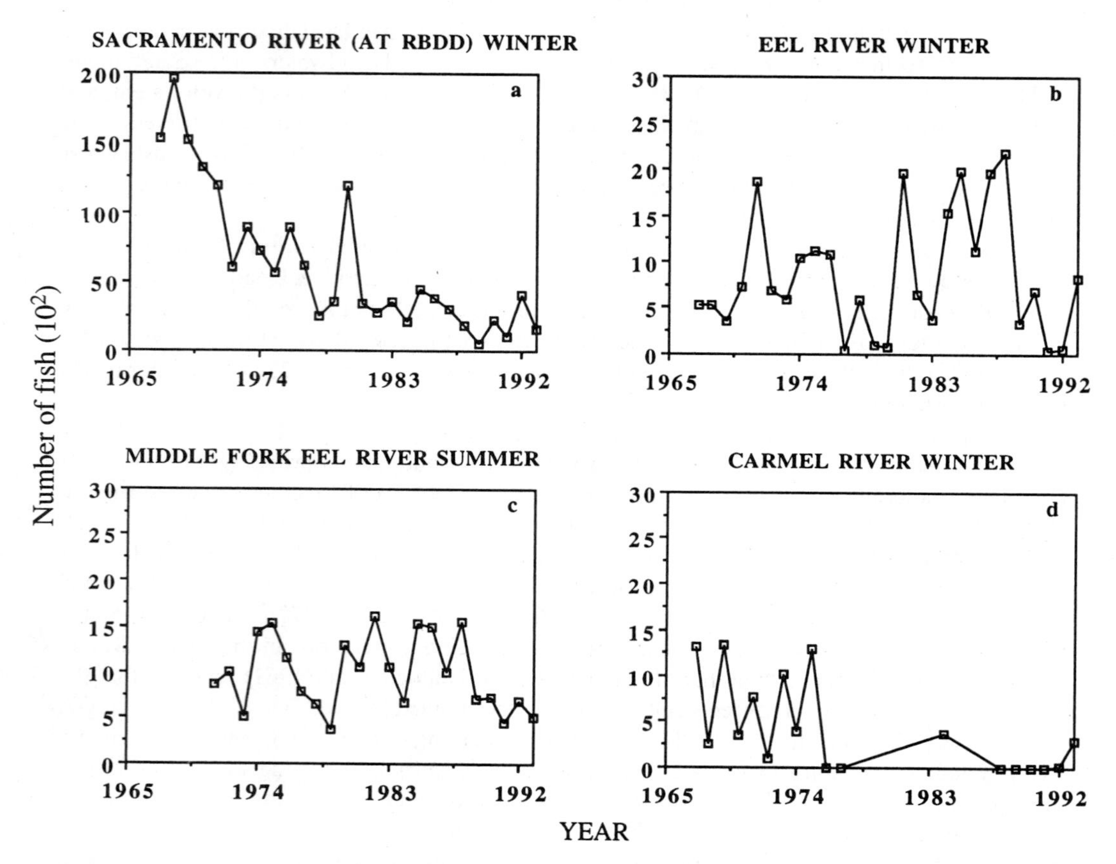

Figure 5. Counts from 1967 to 1993 of (a) winter steelhead at Red Bluff Diversion Dam (RBDD) on the Sacramento River, (b) winter steelhead in the upper Eel River, (c) summer steelhead in the Middle Fork Eel River, and (d) winter steelhead at San Clemente Dam on the Carmel River. No data are available for the Carmel River for the years 1978–83 and 1985–87. Data are from California Department of Fish and Game records.

Bernardino, and San Diego counties (Hubbs 1946; Fig. 1). At present, Malibu Creek in northern Los Angeles County is the southernmost stream containing a known spawning population, although additional isolated remnant populations may result from sporadic spawning successes in other streams farther south (such as the Santa Margarita River; Swift et al. 1993).

Southern steelhead stocks are the most jeopardized of all of California's steelhead populations. Steelhead numbers have declined drastically in nearly all southern streams as a result of habitat loss from water development projects. The California Fish and Wildlife Plan (California Department of Fish and Game 1965) estimated an annual spawning escapement of 59,750 steelhead for coastal areas south of San Francisco Bay. Present estimates of steelhead abundance for this area are unavailable but are undoubtedly quite low given that remnant populations south of Point Conception have been recorded from only 9 streams during recent years (Swift et al. 1993). The precipitous decline of southern steelhead is well described by Titus et al. (in press). On the basis of a review of agency files and other sources of information, they concluded that of 168 streams south of San Francisco Bay known to have historically contained a steelhead population, 24% no longer supported populations, 31% had steelhead populations that were reduced from historical levels, 32% had an unknown status, and only 13% of the streams had populations that had not declined significantly from historical levels.

The Carmel River in Monterey County is the only southern steelhead stream for which we have reliable population trend data (Fig. 5d). The mean number of adults counted at the San Clemente Dam fish ladder from 1964 to 1975 was 821 fish per year. These counts should be considered trend or index counts rather than total fish counts given the methodology used. During the drought year of 1976–77, biologists observed no adults at the fish ladder. During the 3-year period from 1988 to 1990, the river never breached the sand bar at the mouth, making it inaccessible to adult steelhead. Observations included one adult in the ladder in 1991, 14 adults in 1992, and 285 adults in 1993.

Approaches for the Future

Nehlsen et al. (1991) observed that many California salmon and steelhead stocks were at a crossroads in terms of their continued existence. The situation for these populations has not changed, and many stocks have gone beyond the crossroads and now require extreme conservation measures to ensure their continued existence. The statewide decline in salmon and steelhead populations is principally due to losses of migration, spawning, and rearing habitats, as well as increases in juvenile mortality resulting from water management projects and the establishment of an extensive exotic fauna of predatory fishes over the past century (Moyle 1994). Water projects result in myriad effects. They have blocked or impeded migration to spawning grounds; reduced instream flows needed for successful spawning, rearing, and migration; reduced the suitability of the streambed for spawning; and diverted large inflows of freshwater, resulting in higher mortality for migrating smolts and fry. Poor land-use practices including timber harvest and agricultural operations have also contributed to freshwater habitat degradation.

Restoration of salmon and steelhead populations is intimately tied to the establishment of a new management ethic for California's rivers and streams. This ethic should place a much higher priority on the continuance of essential physical, biological, and ecological processes in rivers that are regulated or proposed for development. In other words, rivers need to flow and contain sufficient amounts of water at the proper times to maintain native aquatic biota in viable condition. Without this, aquatic habitats will continue to degrade, species will continue to decline, and there will be continued impasses on water usage and development.

Numerous mitigation measures have been implemented, including the construction and operation of nine hatcheries (Table 3) and several fish passage structures (see Reynolds et al. 1990). However, natural production of salmon and steelhead remains well below the levels that the basins were once capable of producing. Additionally, the extensive use of hatcheries to replace native salmon stocks lost in California over the past century has now been shown to generally be a failure (Meffe 1992, Black 1995).

Habitat restoration will be the key in the management of California salmon and steelhead resources through the remainder of this century and into the next. In 1984, the Trinity River Basin Fish and Wildlife Act (Public Law 98-546) was passed by Congress, requiring implementation of an 11-point Trinity River Basin Management Program and authorizing the expenditure of $57 million over a 10-year period. Many of the expenditures were directed at salmon and steelhead habitat manipulation and restoration. In 1986, the Klamath Basin Fisheries Resources Restoration Act (Public Law 99-552) passed, authorizing the expenditure of $20 million to restore and enhance salmon and steelhead habitat. In 1992, Congress passed the Central Valley Project Improvement Act, which required the Secretary of the Interior to implement a wide variety of structural repairs to improve survival of juvenile and adult salmon and steelhead. A restoration fund was also established by this Act in the United States Treasury and authorized the expenditure of ≤$50 million per year to implement provision of the Act.

A major element in the management and restoration of California's salmon and steelhead populations centers on the ability to accurately assess population trends and status. In recent years, CDFG expended considerable effort to monitor key stocks. These stocks form the basis of intensive ocean sport and commercial fisheries, and represent either depleted stocks or stocks protected under the ESA. Future monitoring of all stocks in the mainstem Sacramento River will require developing and testing more accurate fish-counting methods to account for fish migrating upstream of the RBDD. The juvenile life histories of the various anadromous salmonids are so intertwined that new methods to identify individual fish on a stock-specific basis need to be developed and verified. Maintenance of individual stocks will rely heavily upon the ability to accurately identify and manage stocks on a genetic basis.

Single-species management, while perhaps effective in preventing local extinctions of some stocks, has pitted the needs of listed stocks against unlisted stocks, resulting in an unworkable tool in fisheries management and restoration. An ecosystem approach has become a critical tool in fostering the broad conservation of biological diversity at an environmentally sustainable level. Designing and implementing an ecosystem approach to the problem of dwindling salmon and steelhead abundance requires the full participation of federal, state, local, tribal, public, and private partners, as well as strengthening the current ESA to allow for the protection of "endangered ecosystems" (see Moyle and Yoshiyama 1994). In such a program, restoration and management strategies must be based on the best available science, must be flexible and innovative, and must promote local action and involvement. It remains to be seen if the vast expenditure of capital can indeed "purchase" restored and healthy fish populations without the support and involvement of all stakeholders.

Acknowledgments

We thank M. Black and R.G. Titus for allowing us to use information from their manuscripts accepted for publication, and H.J. Weeks for information on Oregon and Washington anadromous cutthroat stocks. R. Camacho retrieved much of the CDFG salmon and steelhead data used in this report. We also appreciate the helpful comments of three anonymous reviewers which greatly improved the quality of the manuscript.

Literature Cited

Black, M. 1995. Tragic remedies: a century of failed fishery policy on California's Sacramento River. Pacific Historical Review 64: 37-70.

Brown, L.R., P.B. Moyle, and R.M. Yoshiyama. 1994. Historical decline and current status of coho salmon in California. North American Journal of Fisheries Management 14: 237-261.

California Department of Fish and Game. 1965. California Fish and Wildlife Plan. Volume III, Supporting Data; Part B, Inventory (Salmon-Steelhead and Marine Resources). California Department of Fish and Game, Sacramento.

Clark, F.H. 1929. Sacramento-San Joaquin salmon (*Oncorhynchus tshawytscha*) fishery of California. California Department of Fish and Game, Fish Bulletin 17: 1-73.

DeWitt, J.W., Jr. 1954. A survey of the coast cutthroat trout, *Salmo clarki clarki* Richardson, in California. California Fish and Game 40: 329-335.

Fisher, F.W. 1994. Past and present status of Central Valley chinook salmon. Conservation Biology 8: 870-873.

Fry, D.H., Jr. 1961. King salmon spawning stocks of the California Central Valley, 1940-1959. California Fish and Game 47: 55-72.

Fry, D.H., Jr. 1967. A 1955 record of pink salmon, *Oncorhynchus gorbuscha*, spawning in the Russian River. California Fish and Game 53: 210-211.

Fry, D.H., Jr. 1973. Anadromous fishes of California. California Department of Fish and Game, Sacramento.

Gerstung, E.R. 1981. Status and management of the coast cutthroat trout (*Salmo clarki clarki*) in California. Cal-Neva Wildlife Transactions 1981: 25-32.

Gould, J. and B. Gould. 1994. Policies adopted by the California Fish and Game Commission pursuant to section 703 of the Fish and Game Code, p. 570-620. *In* Fish and Game Code of California, 1994. Gould Publications, Inc., Longwood, Florida.

Hallock, R.J. and D.H. Fry, Jr. 1967. Five species of salmon, *Oncorhynchus*, in the Sacramento River, California. California Fish and Game 53: 5-22.

Hallock, R.J., R.T. Elwell, and D.H. Fry, Jr. 1970. Migrations of adult king salmon (*Oncorhynchus tshawytscha*) in the San Joaquin Delta, as demonstrated by the use of sonic tags. California Department of Fish and Game, Fish Bulletin 151: 1-92.

Hallock, R.J., W.F. Van Woert, and L. Shapovalov. 1961. An evaluation of stocking hatchery-reared steelhead rainbow trout (*Salmo gairdnerii gairdnerii*) in the Sacramento River system. California Department of Fish and Game, Fish Bulletin 114: 1-74.

Healy, M.C. 1991. Life history of chinook salmon (*Oncorhynchus tshawytscha*), p. 313-393. *In* C. Groot and L. Margolis (eds.), Pacific Salmon Life Histories. University of British Columbia Press, Victoria.

Hedgpeth, J.W. 1941. Livingston Stone and fish culture in California. California Fish and Game 27: 126-148.

Hubbs, C.L. 1946. Wandering of pink and other salmonid fishes into southern California. California Fish and Game 32: 81-86.

Jordan, D.S. and B.W. Evermann. 1896. The fishes of North and Middle America. A descriptive catalogue of the species of fish-like vertebrates found in the waters of North America, north of the isthmus of Panama. United States National Museum Bulletin (47), part I: 1-1240.

Light, J.T. 1987. Coastwide abundance of North American steelhead trout. Document submitted to the annual meeting of the International North Pacific Fisheries Council 1987. University of Washington School of Fisheries, Fisheries Research Institute FRI-UW-8710. Seattle.

Lufkin, A. 1991a. Historical highlights, p. 6-36. *In* A. Lufkin (ed.), California's Salmon and Steelhead; the Struggle to Restore an Imperiled Resource. University of California Press, Berkeley, California.

Lufkin, A. 1991b. Summary and conclusions, p. 263-268. *In* A. Lufkin (ed.), California's Salmon and Steelhead; the Struggle to Restore an Imperiled Resource. University of California Press, Berkeley, California.

McEvoy, A.F. 1986. The Fisherman's Problem: Ecology and Law in the California Fisheries, 1850-1980. Cambridge University Press, New York.

McGinnis, S.M. 1984. Freshwater fishes of California. California Natural History Guides (49). University of California Press, Berkeley.

Meffe, G.K. 1992. Techno-arrogance and halfway technologies: salmon hatcheries on the Pacific coast of North America. Conservation Biology 6: 350-354.

Moffett, J.W. and S.H. Smith. 1950. Biological investigations of the fishery resources of the Trinity River, California. United States Fish and Wildlife Service, Special Scientific Report, Fisheries (12): 1-71.

Moring, J.R. 1993. Anadromous stocks. p. 553-580. *In* C.C. Kohler and W.A. Hubert (eds.), Inland Fisheries Management in North America. American Fisheries Society, Bethesda, Maryland.

Moyle, P.B. 1976. Inland Fishes of California. University of California Press, Berkeley.

Moyle, P.B. 1994. The decline of anadromous fishes in California. Conservation Biology 8: 869-870.

Moyle, P.B. and R.M. Yoshiyama. 1994. Protection of aquatic biodiversity in California; a five-tiered approach. Fisheries 19: 6-18.

Moyle, P.B., J.E. Williams, and E.D. Wikramanayake. 1989. Fish species of special concern of California. California Department of Fish and Game, Inland Fisheries Division, Final Report under contract number 7337. Rancho Cordova.

National Marine Fisheries Service. 1993. Biological opinion on the operation of the Federal Central Valley Project and California State Water Project. National Marine Fisheries Service, Terminal Island, California. [February 12, 1993].

National Oceanographic and Atmospheric Administration. 1990. Endangered and threatened species; winter-run chinook salmon. Federal Register 55: 46515-46523. [Monday, November 5, 1990].

National Oceanographic and Atmospheric Administration. 1994a. Endangered and threatened species; status of Sacramento River winter-run chinook salmon. Federal Register 59: 440-450. [Tuesday, January 4, 1994].

National Oceanographic and Atmospheric Administration. 1994b. Ocean salmon fisheries off the coasts of Washington, Oregon, and California. Federal Register 59: 22999-23012. [Wednesday, May 4, 1994].

National Oceanographic and Atmospheric Administration. 1994c. Endangered and threatened species; proposed endangered status for North and South Umpqua River cutthroat trout in Oregon. Federal Register 59: 35089-35093. [Friday, July 8, 1994].

Nehlsen, W., J.E. Williams, and J.A. Lichatowich. 1991. Pacific salmon at the crossroads: stocks at risk from California, Oregon, Idaho, and Washington. Fisheries 16: 4-21.

Pacific Fishery Management Council. 1994. Review of the 1993 ocean salmon fisheries. Pacific Fishery Management Council, Portland, Oregon.

Reynolds, F.L., R.L. Reavis, and J. Schuler. 1990. Central Valley salmon and steelhead restoration and enhancement plan. California Department of Fish and Game, Sacramento.

Rogers, D.W. 1974. Chum salmon observations in four North Coast streams. California Fish and Game 60: 148.

Rutter, C. 1904. Natural history of the quinnat salmon. A report on investigations in the Sacramento River, 1896-1901. Bulletin of the United States Fish Commission 22: 65-141.

Scofield, N.B. 1916. The humpback and dog salmon taken in the San Lorenzo River. California Fish and Game 2: 41.

Scofield, N. B. 1929. The status of salmon in California. California Fish and Game 15: 13-18.

Shapovalov, L. and A.C. Taft. 1954. Life histories of the steelhead rainbow trout (*Salmo gairdneri gairdneri*) and silver salmon (*Oncorhynchus kisutch*) with special reference to Waddell Creek, California, and recommendations regarding their management. California Department of Fish and Game, Fish Bulletin (98): 1-375.

Skinner, J.E. 1962. An historical review of the fish and wildlife resources of the San Francisco Bay area. California Department of Fish and Game, Water Projects Branch Report (1): 1-226.

Slater, D.W. 1963. Winter-run chinook salmon in the Sacramento River, California with notes on water temperature requirements at spawning. United States Fish and Wildlife Service, Special Scientific Report, Fisheries (461): 1-9.

Smedley, S.C. 1952. Pink salmon in Prairie Creek. California Fish and Game 38: 275.

Snyder, J.O. 1931. Salmon of the Klamath River, California. California Department of Fish and Game, Fish Bulletin (34): 1-130.

Stone, L. 1897. The artificial propagation of salmon on the Pacific Coast of the United States, with notes on the natural history of the quinnat salmon. Bulletin of the United States Fish Commission 16: 205-238.

Swift, C.C., T.R. Haglund, M. Ruiz, and R.N. Fisher. 1993. The status and distribution of the freshwater fishes of southern California. Bulletin of the Southern California Academy of Sciences 92: 101-167.

Taft, A.C. 1938. Pink salmon in California. California Fish and Game 24: 197-198.

Titus, R.G., D.C. Erman, and W.M. Snider. In press. History and status of steelhead in California coastal drainages south of San Francisco Bay. Hilgardia.

Vogel, D.A. 1987a. Estimation of the 1986 spring chinook salmon run in Deer Creek, California. United States Fish and Wildlife Service Report Number FR1/FAO-87-3. Red Bluff, California.

Vogel, D.A. 1987b. Estimation of the 1986 spring chinook salmon run in Mill Creek, California. United States Fish and Wildlife Service Report Number FR1/FAO-87-12. Red Bluff, California.

Williams, J.E. and C.D. Williams. 1991. The Sacramento River winter chinook salmon: threatened with extinction, p. 105-115. *In* A. Lufkin (ed.), California's Salmon and Steelhead; the Struggle to Restore an Imperiled Resource. University of California Press, Berkeley.

Idaho's Salmon: Can We Count Every Last One?

Peter F. Hassemer, Sharon W. Kiefer and Charles E. Petrosky

Abstract

Since the late 1950s, Idaho Department of Fish and Game (IDFG) biologists have collected standardized information about the status of salmon (*Oncorhynchus* spp.) in Idaho to assist fisheries management. Currently, IDFG uses indices of production such as redd and parr counts for streams, and adult salmon counts at dams and weirs to assess the recent history of our salmon runs. We depict Idaho anadromous salmon trends using indices for sockeye (*O. nerka*); spring, summer, and fall chinook salmon (*O. tshawytscha*); and steelhead (*O. mykiss*). Counts of redds in index areas provide the best historical indicator of trends and status of spring and summer chinook salmon. Numbers of redds have declined sharply through the last 33 years. All of the naturally reproducing Snake River anadromous salmonid populations, excluding chinook salmon in the Clearwater River drainage and Snake River steelhead, have been listed as endangered pursuant to the federal Endangered Species Act. Hydroelectric development of the mainstem Snake and Columbia rivers is widely recognized as the major factor causing the decline of these populations.

Aggregated trend information is useful, but alone does not provide information needed for future management. We must augment historical information with knowledge about brood-year performance of multiple populations. We are using our indexed production information for life-cycle modeling to assess productivity, a key to assessing recovery actions. Increased monitoring and analyses of indicator populations will better describe salmon status and trends to meet management needs for the future.

Introduction

Idaho's anadromous salmonids are unique among salmonids in the contiguous United States. They migrate as far as 1,540 km downstream as smolts to reach the ocean and return upstream as adults to their natal spawning grounds. The freshwater migration distance has not changed during the recent history of the species. However, the character of the journey has been radically altered during the past 60 years. Hydropower development in the lower Snake and Columbia rivers has impeded salmonid migration by converting a once free-flowing river into a system of eight dams and reservoirs (Fig. 1). This migration corridor has remained important for the survival

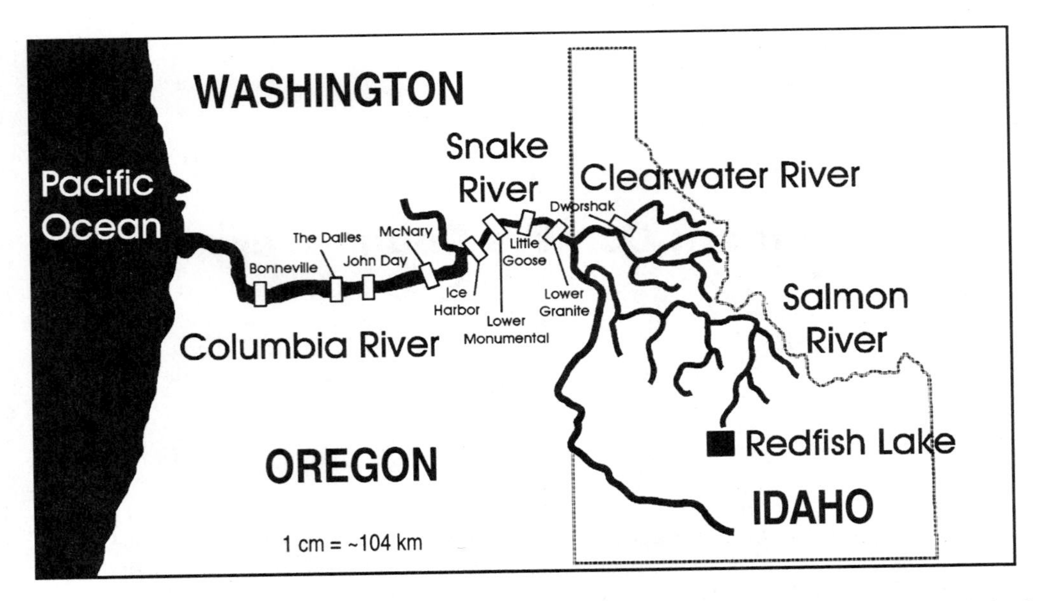

Figure 1. Lower Snake and Columbia rivers anadromous fish migration corridor, illustrating the Federal Columbia River Power System. Redfish Lake, Idaho, is 1,445 km from the Pacific Ocean.

of anadromous salmonids because it has provided the critical link between the rich ocean environment and Idaho's freshwater spawning and rearing habitat.

The amount of freshwater production habitat in Idaho accessible to anadromous salmonids declined during the last 100 years. Dam construction has been the major factor resulting in loss of access to spawning and rearing habitat (Waples et al. 1991a). However, ~62% of Idaho's historical spawning and rearing habitat for spring/summer chinook salmon (*Oncorhynchus tshawytscha*) remains available (Fig. 2); a similar amount of steelhead (*O. mykiss*) habitat remains after dam construction. Approximately 25% of the historical surface area of sockeye salmon (*O. nerka*) nursery lakes in Idaho remains accessible. The greatest loss of production habitat has occurred for Snake River fall chinook salmon, for which 17% of the historical habitat is currently accessible. The construction of the three-dam Hells Canyon complex on the Snake River during the 1960s eliminated access to most fall chinook production habitat (Waples et al. 1991a).

A modest amount of the existing freshwater production habitat for anadromous salmonids remains protected from further impacts or alterations. Approximately 30% of Idaho's stream kilometers inhabited by salmon and steelhead are located within designated wilderness areas or waterways classified as wild and scenic rivers. These designations provide special protection to the streams and stream corridors so that much of the habitat in these areas experiences little or no anthropogenic disturbance. Habitat outside these areas remains affected to varying degrees by land-use activities.

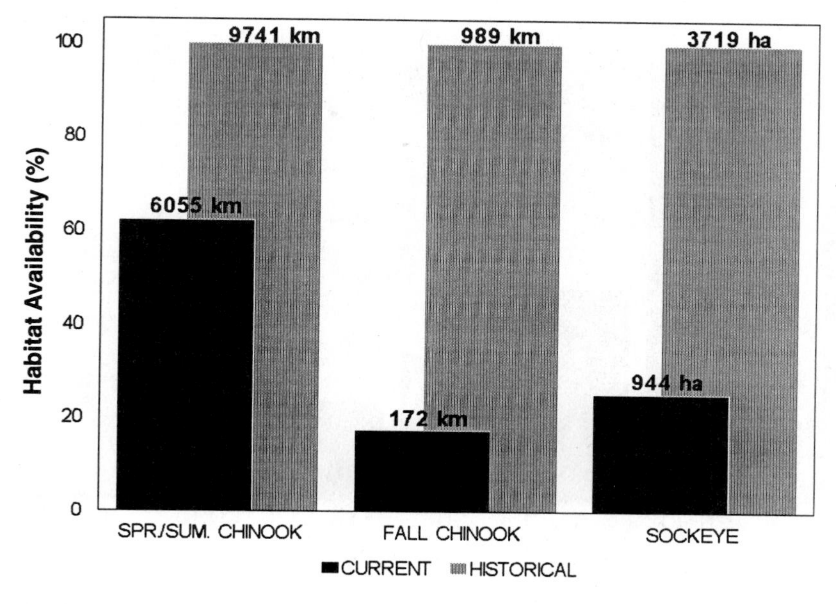

Figure 2. Current and predevelopment habitat availability for spring and summer chinook, fall chinook, and sockeye salmon in Idaho. Habitat availability is denoted as the proportion of remaining river kilometers (km) for chinook salmon and lake area (ha) for sockeye salmon.

Historical Run Sizes

Before the Federal Columbia River Power System in the Snake and Columbia rivers was completed in the mid-1970s, the Snake River produced ~40% of the adult spring chinook salmon, 45% of the adult summer chinook salmon, and 55% of the summer steelhead entering the Columbia River (Mallet 1974). Fish entering the Snake River were destined for tributaries in Idaho, Oregon, and Washington. In this paper we focus on the recent status and trends of naturally produced anadromous salmonids in Idaho. The precipitous decline of these fish is evident. For example, from 1962 to 1971 an average of 148,000 adult anadromous salmonids crossed Ice Harbor Dam into the Snake River each year (Fig. 3). These numbers represented primarily wild fish, for which chinook salmon comprised 51% of the total. In contrast, the average number of adult anadromous salmonids crossing Ice Harbor Dam from 1982 to 1991 decreased to a total of 137,000 fish, but the proportion of chinook salmon declined to 25%. During this period, most of the anadromous salmonids crossing the dam outmigrated from hatcheries. By 1992, Snake River coho salmon (*O. kisutch*) were declared extinct and Snake River sockeye salmon were listed as endangered pursuant to the federal Endangered Species Act (ESA). All three runs of Snake River chinook salmon (spring, summer, and fall) in Idaho were listed as federally threatened in 1992, excluding spring and summer chinook salmon produced in the Clearwater River drainage (Fig. 1). Wild steelhead in the Snake River Basin were petitioned for listing in February 1994.

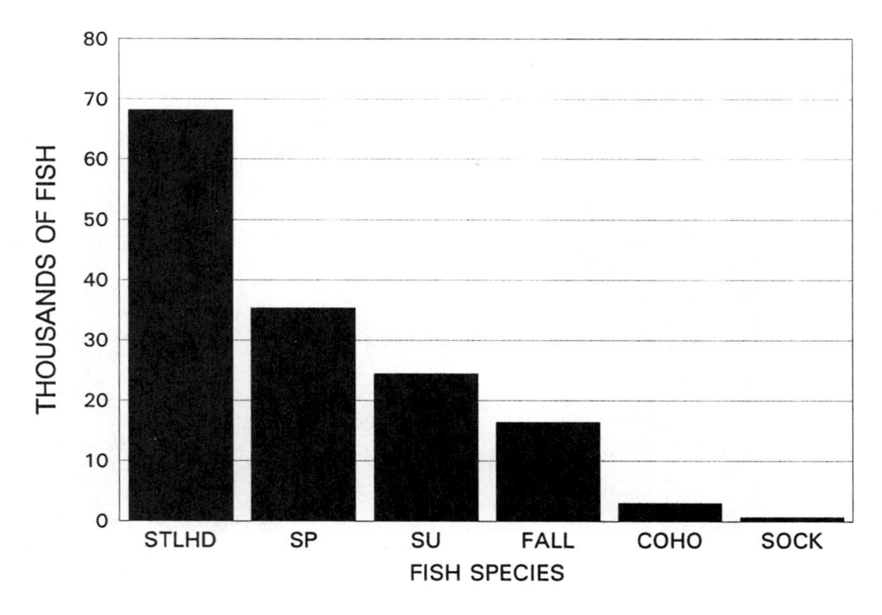

Figure 3. Average counts of steelhead (STLHD), spring chinook (SP), summer chinook (SU), fall chinook (FALL), coho (COHO), and sockeye (SOCK) salmon at Ice Harbor Dam during 1962–71. These counts represent primarily wild fish.

Hydroelectric development in the mainstem Snake and Columbia rivers is widely recognized as the major factor causing the decline in the number of anadromous salmon returning to the Snake River. Loss of stock productivity and recruitment coincided directly with this development (Raymond 1979, 1988; Petrosky and Schaller 1994). In 1993, the National Marine Fisheries Service (NMFS) estimated that juvenile mortality rates ranged from 55% to 92% (depending on species) through the Federal Columbia River Power System (Fig. 4). The NMFS also estimated that adult mortality ranged from 8% to 41%.

Management Program

The basic goal of anadromous salmonid management in Idaho is to provide the maximum amount of fishing opportunity consistent with meeting spawning escapements, preserving genetic resources of naturally sustained populations, and utilizing available natural habitat (Idaho Department of Fish and Game 1992). To achieve this goal, the primary focus areas of IDFG's anadromous management program include production and harvest management, and genetic preservation. During the 1960s and early 1970s, fishery managers implemented harvest management. Idaho sport anglers enjoyed numerous tributary fisheries, primarily on self-sustaining wild stocks (Ortmann and Richards 1964, Reingold 1976). Fisheries were managed based on contemporary stock-recruitment theory and numbers of harvestable fish. As many as 21,000 salmon and 30,000 steelhead were harvested annually during this period (Idaho Department of Fish and Game 1992).

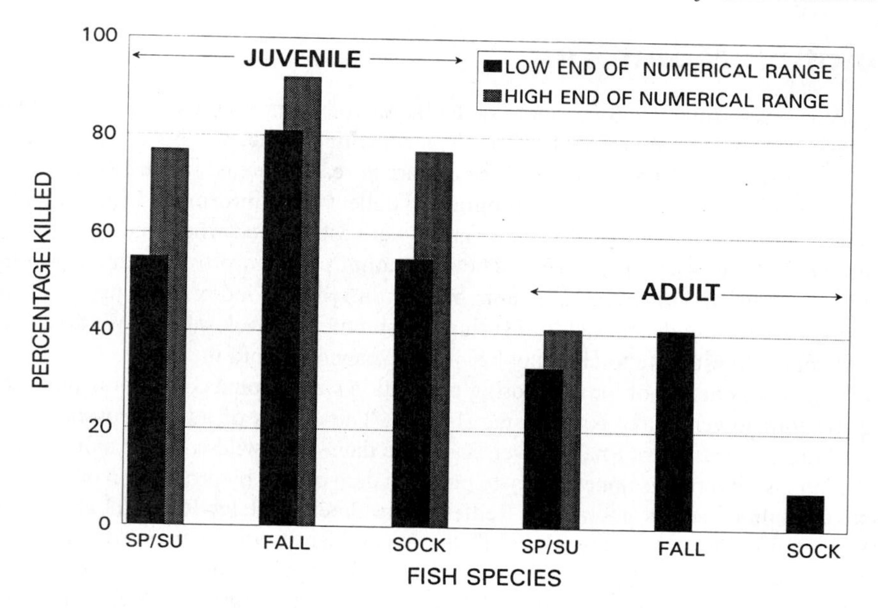

Figure 4. Estimated ranges of juvenile and adult Snake River spring and summer chinook (SP/SU), fall chinook (FALL), and sockeye (SOCK) salmon killed in 1993 by passage through the Federal Columbia River Power System, including associated reservoirs (National Marine Fisheries Service 1993). Numbers for chinook and sockeye salmon were not available because annual variation in the outmigration—NMFS provides estimates of mortality only in percentages, which has been acceptable for the Columbia Basin.

The last general chinook salmon season in Idaho occurred in 1978. Idaho anglers since have had very few and intermittent opportunities for salmon harvest, and only for surplus hatchery returns (Hassemer 1991). Steelhead sport harvest has been restricted to only marked hatchery fish since 1984 (Technical Advisory Committee 1991). In the face of declining salmonid returns, ESA listings, and diminished harvest opportunities, management emphasis shifted to genetic preservation and production management. Until adequate flow and passage conditions improve smolt and adult survival through the Federal Columbia River Power System, the preservation of existing naturally reproducing stocks is the primary objective of IDFG's management program. To meet this objective, our current management focuses on restoring stock productivity while maintaining genetic diversity and integrity of wild populations.

The IDFG monitors three management categories for naturally reproducing populations: wild, natural, and hatchery-influenced. Wild populations are those with no history of hatchery outplants or supplementation activities. Natural populations have some history of hatchery outplantings or interactions, but are currently managed only for natural spawning and production. In spawning areas where returning adults are hatchery-origin fish mixed with adults of natural origin, we categorize the population as hatchery-influenced. Future implementation of supplementation research and management pursuant to ESA listings may require additional management categories.

Population Monitoring

We use several assessment tools to monitor the status of salmon and steelhead populations. These assessment tools include counts of redds in spawning areas, numbers of adults crossing dams in the lower Snake River, and densities of parr in rearing areas. In the late 1950s, IDFG developed a standardized redd counting program to collect trend information for adult chinook salmon populations. As early as 1957, index areas were established where redds were and still are counted each year (Hassemer 1993). The redd count program provides trend information about annual spawning escapement. Counts of salmon redds in index areas provide our best historical indicator of trends and status of spring and summer chinook salmon. We do not use the index redd counts to estimate total chinook salmon escapement into the state.

A second assessment tool for monitoring population size is counts of salmon and steelhead crossing the four lower Snake River dams (Fig. 1). The counts of adults crossing the dams represent returns to the entire Snake River Basin. To date, only wild sockeye salmon have returned to Idaho, so it remains unnecessary to partition dam counts by production origin. We also count sockeye salmon adults at a weir on Redfish Lake Creek (outlet to Redfish Lake), giving us complete enumeration of returning adults. Thus, we can identify loss between the dams and the lake. Chinook salmon counted at the dams are classified as spring, summer, or fall run depending on the time when they cross each dam (US Army Corps of Engineers 1993). Furthermore, a mixture of natural-origin and hatchery-origin chinook salmon return, making further classification necessary. We estimate the number of wild and natural chinook salmon by subtracting hatchery fish returns enumerated at upstream weirs from the dam count. The estimated number of wild and natural fish crossing the Lower Granite dam represents all fish not accounted for at upstream hatchery weirs and other enumeration points. We use 13 weirs for Idaho chinook salmon programs (Leitzinger et al. 1993). In 1996, we will be able to externally distinguish hatchery and naturally produced chinook salmon based on fin clips. Currently, only one Snake River hatchery releases fall chinook salmon in the Snake River Basin. By subtracting the number of hatchery-origin fish, we estimate the number of natural-origin fall chinook salmon crossing Lower Granite Dam (uppermost passable dam). Enumeration of coded-wire tagged fish at Lower Granite Dam enables us to estimate the size of the hatchery origin component.

Fish biologists also count the adult steelhead entering the Snake River Basin at the four lower Snake River dams. Using scale samples collected from steelhead trapped at Lower Granite Dam, we separated the run into wild and hatchery components based on scale analysis (Technical Advisory Committee 1991). Since 1984, the steelhead run crossing Lower Granite Dam has been separated into wild and hatchery components. Using life-history characteristics, we further classify steelhead entering Idaho as either A-run or B-run summer steelhead. In general, B-run steelhead grow larger, mature later, and enter the Columbia River later than A-run steelhead (Idaho Department of Fish and Game 1992). In addition, the B-run steelhead occur only in Idaho whereas A-run steelhead are found throughout the Columbia River Basin (Technical Advisory Committee 1991). Because all hatchery-origin steelhead have their adipose fin removed, we can visibly differentiate natural from hatchery-origin fish based on the presence or absence of this fin. This allows us to provide protection to depleted wild stocks in sport fisheries while addressing adult and juvenile migration survival. Retention of adult steelhead with an adipose fin is prohibited in all Idaho sport fisheries.

Our third major monitoring tool involves an annual index of salmon and steelhead parr densities in all major drainages with anadromous fish (Rich and Petrosky 1994). Aggregate populations are indexed from ~100 streams representing classes of different runs, stocks, habitat conditions, and production strategies. Parr density data complement and integrate a number of our more intensive studies. For example, the Intensive Smolt Monitoring project (Kiefer and Lockhart 1993) estimates juvenile abundance and survival for individual populations. We express the status and trends of various classes in terms of mean density and as a percentage of parr carrying capacity. Field observations and experimental fish stocking, combined with standardized planning data, allow us to estimate carrying capacity (Rich et al. 1992).

All of our standard index programs for monitoring stock status have expanded over the last 2 decades. We now include more intensive research and more sophisticated methods (e.g., genetic monitoring).

Status and Trends

Salmon

The numbers of spring and summer chinook salmon redds have declined sharply over the last 33 years (Fig. 5). The average annual count of redds for 1989–93 declined 79% from the average annual count for 1964–68. Numbers of redds counted prior to 1970 substantially underrepresent total escapement to the state because they do not account for in-state harvest (Hassemer 1993). Sport harvest prior to 1970 ranged from an estimated 39,000 in 1957 to 6,500 chinook salmon in 1968 (Idaho Department of Fish and Game 1992). The redd index in both wild and natural trend

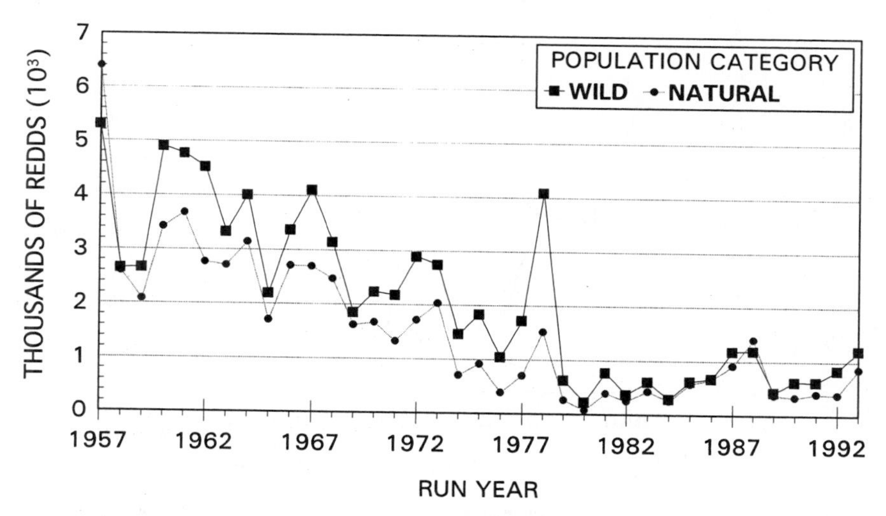

Figure 5. Numbers of spring and summer chinook salmon redds counted in Salmon River drainage wild and natural index areas, 1957–93.

areas indicates primarily wild-origin fish prior to 1970 (Fig. 5), and these trend areas were managed as wild production areas. In response to run declines and mitigation programs, juvenile and adult hatchery chinook were released in some wild trend areas beginning about 1971. This resulted in classification of some of the wild trend areas as natural trend areas, hence the beginning of our current management categories. Beginning in 1980, redd counts in trend areas categorized as hatchery-influenced were substantially affected by hatchery operations. In these trend areas, spring and summer chinook salmon of natural origin were intercepted at hatchery weirs and retained for broodstock. Also, adults of hatchery origin were released upstream from the weir to spawn with returning natural-origin chinook salmon. Index redd counts provide NMFS with a basis for evaluating probabilities of persistence and population stabilization for ESA-listed Snake River salmon.

Similar to the redd count index, the number of wild and natural spring and summer chinook salmon counted at the uppermost Snake River dam declined precipitously over the last 3 decades. The average annual count at the uppermost dam for 1989–93 was 84% less than the average annual count for 1964–68 (Fig. 6). Whereas the redd count index provided a direct accounting of naturally spawning spring and summer chinook salmon, dam counts indirectly account for naturally spawning fish. Concurrence of these two enumeration methods indicates accurate monitoring of the wild and natural fish using each method. Overall, juvenile salmon densities correspond with adult salmon escapements (Fig. 7).

The trend for the wild Snake River fall chinook salmon population is more difficult to interpret. Prior to 1986, wild and hatchery fish counts were combined (Fig. 8). Counts in recent years represent Snake River wild fish only; hatchery strays and Lyons Ferry Hatchery stock have been subtracted. The total annual count at Lower Granite Dam averaged 441 fish from 1979 through 1992, with a record low return of only 78 fish in 1990.

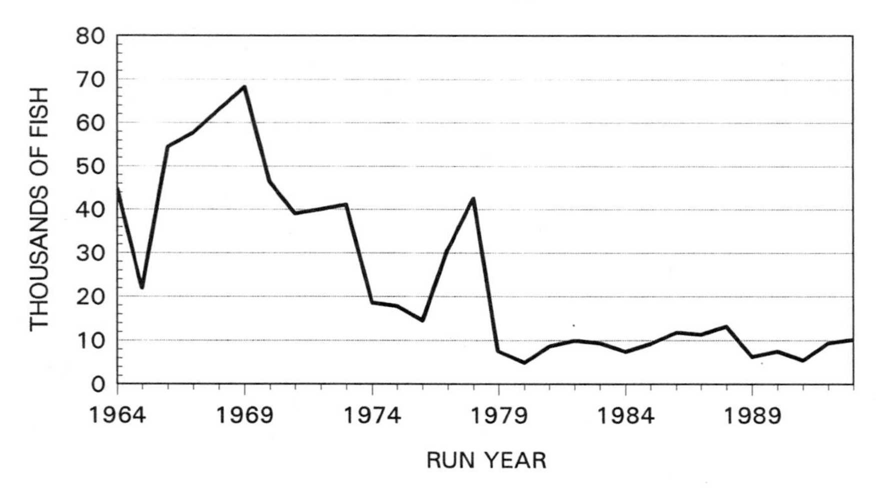

Figure 6. Numbers of wild spring/summer chinook salmon counted at the uppermost dam on the lower Snake River, 1964–93. Uppermost dam is indicated by years: 1964–68, Ice Harbor; 1969, Lower Monumental; 1970–74, Little Goose; and 1975–93, Lower Granite.

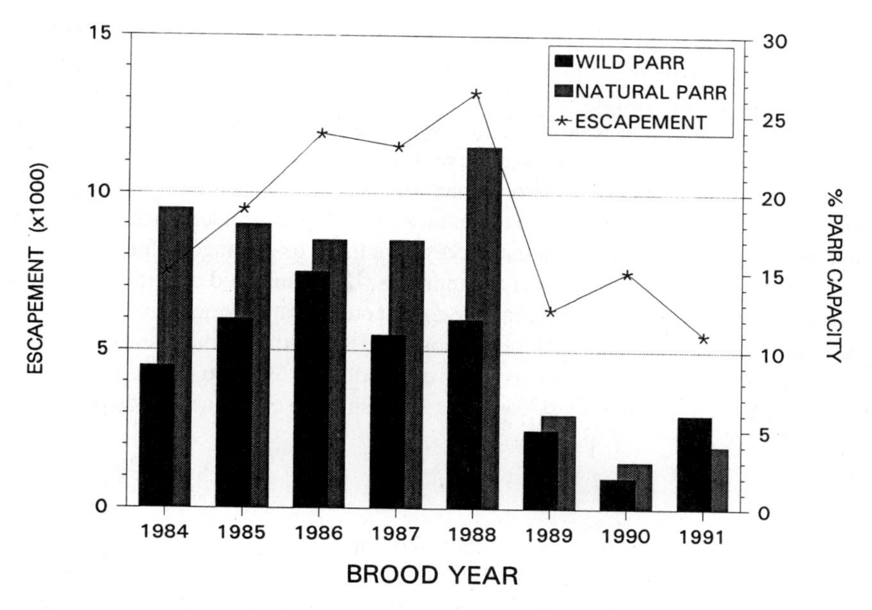

Figure 7. Escapement of adult chinook salmon of wild and natural origin past Lower Granite Dam and resulting parr density indices for brood years 1984–91 wild and natural chinook salmon populations in Idaho (Rich and Petrosky 1994). Histograms represent percent of parr carrying capacity.

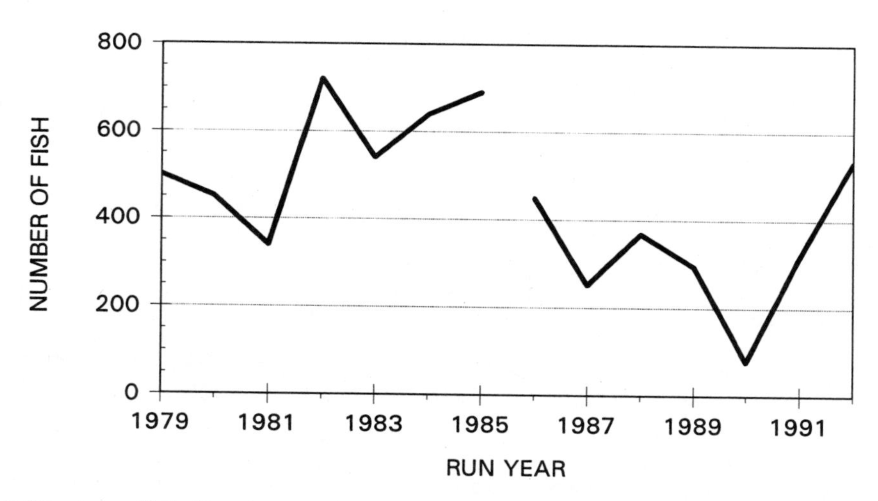

Figure 8. Numbers of fall chinook counted at Lower Granite Dam, 1979–93. Counts for 1979–85 include hatchery strays; 1986–93 counts are estimated numbers of wild fish only.

Sockeye salmon counted at Snake River dams have declined at a similar rate to other species (Fig. 9). The headwaters of the Salmon River represent the only historical sockeye salmon production habitat still accessible in the Snake River Basin. In recent years, the only known sockeye salmon spawning occurred in Redfish Lake (Waples et al. 1991b). Even in the early 1960s, Redfish Lake was the only lake with a sizable return. Since 1990, from 1 to 8 sockeye (never >2 females) returned from the ocean. Research and monitoring efforts identified both anadromous and residual forms of sockeye in Redfish Lake, as well as the resident kokanee (*O. nerka*) salmon population. Both anadromous and residual sockeye are listed as endangered pursuant to the ESA. To preserve the run until migration survival improves, IDFG initiated a captive broodstock program in spring 1991. As part of the program, we collect outmigrating smolts and returning adults at a weir located in the Redfish Lake outlet (Johnson 1993). Returning adults are spawned and their progeny released or reared to maturity as captive broodstock. We also rear a portion of the outmigrating smolts to maturity to release back into Redfish Lake or use as broodstock.

STEELHEAD

Unlike chinook salmon, numbers of wild and natural A-run steelhead counted at the uppermost Snake River dam increased from 1974 through 1986 (Fig. 10). The number counted at Lower Granite Dam then declined precipitously after 1987 and, in 1990, fish reached a record low of 2,500 fish. Wild and natural B-run steelhead counts increased slightly from 1974 through 1985. Since 1985 a trend is lacking; however, in 1991 we counted a record low return of 1,100

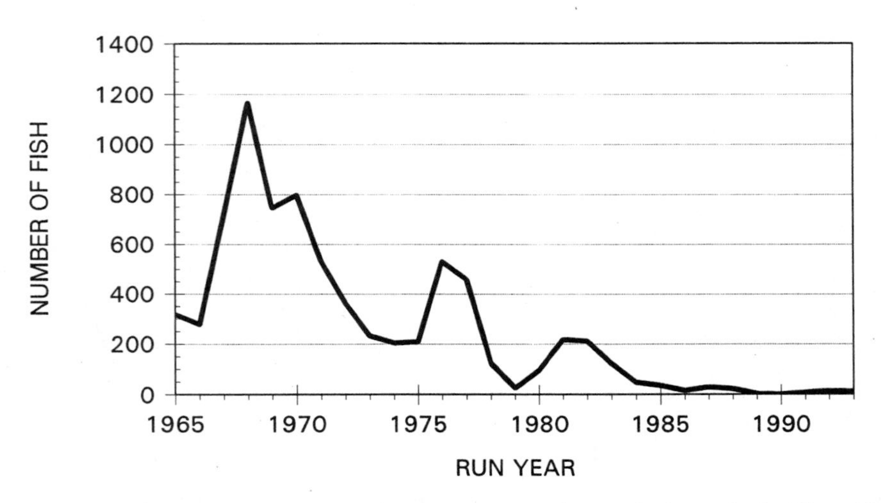

Figure 9. Numbers of sockeye salmon counted at the uppermost dam on the lower Snake River, 1965–93. Uppermost dam is indicated by years: 1964–68, Ice Harbor; 1969, Lower Monumental; 1970–74, Little Goose; and 1975–93, Lower Granite.

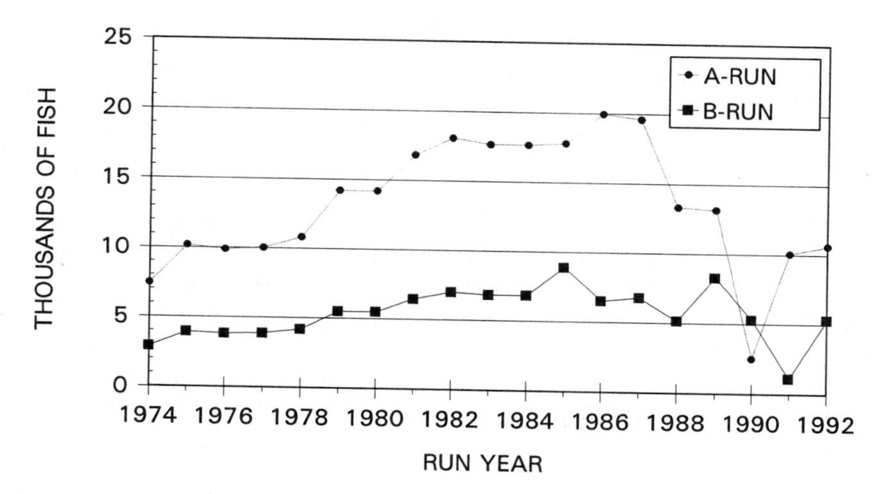

Figure 10. Estimated numbers of wild/natural A-run and B-run summer steelhead crossing the uppermost dam on the lower Snake River, 1974–92. Run year indicated is for the fall of the adult migration year (e.g., 1974 = 1974–75 migration year). Uppermost dam is indicated by years: 1970–74, Little Goose; 1975–93, Lower Granite.

fish at Lower Granite Dam. These numbers are well below IDFG escapement goals for wild and natural summer steelhead at Lower Granite Dam (22,000 A-run and 32,700 B-run fish) (Idaho Department of Fish and Game 1992). Since 1988, A- and B-runs have averaged 45% and 15%, respectively, of IDFG's escapement goals.

The current status of wild and natural steelhead is masked by Snake Basin hatchery programs, which produce substantial numbers of adults in most years. From 1975 through 1979, the total steelhead run at Lower Granite Dam consisted of an average of 41% hatchery fish. As a result of declining numbers of wild and natural fish and slightly increased returns of hatchery fish, the hatchery composition of the run increased to an average of 76% for 1985 through 1989.

Future Status Assessments

Our future assessments of salmon and steelhead status will be directed at intensively monitoring several individual populations. We plan to use a variety of techniques to obtain complementary information. Currently we augment our index redd counts with more intensive counts of redds and compare this with adults passed upstream of weirs to estimate spawning escapement and adult-to-adult survival. In addition to collecting parr density information, we also estimate total parr and smolt production for some streams using snorkeling techniques and juvenile traps (Kiefer and Lockhart 1993, Leitzinger et al. 1993). In the mid-1980s we began using passive integrated transponder (PIT) tags to monitor chinook and sockeye salmon and steelhead migration (Buettner and Brimmer 1993, Kiefer and Lockhart 1993). All of these monitoring tools enable us to collectively assess populations and evaluate the effectiveness of production actions (e.g., supplementation), habitat enhancement, and actions to improve migration survival.

We recently expanded our genetic monitoring of salmon and steelhead. Our focus guards against management actions (e.g., supplementation or other hatchery practices) that may adversely affect the genetic complement of non-target populations (Bowles and Leitzinger 1991).

We continue to annually monitor adult escapements. This is important because status determination relies on measurements of adult-to-adult survival and adult escapement. Nonetheless, we must provide more than just an accounting of fish; population productivity must also be monitored. The productivity of populations remains paramount to evaluating the effectiveness of our management actions and, ultimately, the success of our recovery attempts. For Idaho fish populations, we develop production and survival indices by brood year and smolt migration year. This information is important in determining the extent of successful mainstem passage and survival, which are critical to the persistence and recovery of Snake River salmon populations (Petrosky and Schaller 1994). Linking our information with tribal and federal cooperators will result in an Idaho-coordinated monitoring program that incorporates both our standard indices and intensive research.

The future direction relies on an integration of our monitoring program into system-wide application. A system-wide monitoring plan that incorporates indicator stocks has been proposed within the Columbia Basin by the Northwest Power Planning Council's Fish and Wildlife Program (Northwest Power Planning Council 1992). Implementation of such a plan can provide a valuable framework in which performance of Snake River salmon and steelhead populations can be assessed simultaneously with other Columbia Basin fish populations and perhaps even other West Coast salmon stocks. This level of integration will provide better answers about fish stock status and trends from the population to the regional level. We hope the answers will allow us to determine the factors limiting the productivity and recovery of salmon populations in Idaho and elsewhere in the Pacific Northwest before we count the last fish.

Literature Cited

Bowles, E. and E. Leitzinger. 1991. Salmon supplementation studies in Idaho rivers, experimental design. Idaho Department of Fish and Game Report to US DOE, Bonneville Power Administration, Division of Fish and Wildlife, Project No. 89-098, Contract No. DE-BI79-89BP01466. Portland, Oregon.

Buettner, E.W. and A.F. Brimmer. 1993. Smolt monitoring at the head of Lower Granite Reservoir and Lower Granite Dam. Idaho Department of Fish and Game Annual Report 1992 to US DOE, Bonneville Power Administration, Division of Fish and Wildlife, Project 83-323 B, Contract DE-B179-83BP11631. Portland, Oregon.

Hassemer, P.F. 1991. Little Salmon River spring chinook *Oncorhynchus tshawytscha* sport harvest, 1986 to 1990. Idaho Department of Fish and Game. Boise, Idaho.

Hassemer, P.F. 1993. Salmon spawning ground surveys, 1989-92. Idaho Department of Fish and Game Project F-73-R-15, Pacific Salmon Treaty Program Award No. NA17FP0168-02. Boise, Idaho.

Idaho Department of Fish and Game. 1992. Anadromous fish management plan, 1992-1996. Boise, Idaho.

Johnson, K.A. 1993. Research and recovery of Snake River sockeye salmon. Idaho Department of Fish and Game Annual Report 1991 to US DOE, Bonneville Power Administration, Division of Fish and Wildlife, Project 91-72, Contract Number DE-BI79-91BP21065. Portland, Oregon.

Kiefer, R.B. and J.N. Lockhart. 1993. Idaho habitat and natural production monitoring: Part II. Idaho Department of Fish and Game Annual Report 1991 to US DOE, Bonneville Power Administration, Division of Fish and Wildlife, Project 91-73, Contract DE-BI79-91BP21182. Portland, Oregon.

Leitzinger, E.J., K. Plaster, and E. Bowles. 1993. Idaho supplementation studies. Idaho Department of Fish and Game Annual Report 1991-1992 to US DOE, Bonneville Power Administration, Division of Fish and Wildlife, Project 89-098, Contract DE-BI79-89BP01466. Portland, Oregon.

Mallet, J. 1974. Inventory of salmon and steelhead resources, habitat, use and demands. Idaho Department of Fish and Game Job Performance Report, Project F-58-R-1. Boise, Idaho.

National Marine Fisheries Service. 1993. Biological opinion on 1993 operation of the federal Columbia River power system. Portland, Oregon.

Northwest Power Planning Council. 1992. Strategy for salmon, Volume II. Portland, Oregon.

Petrosky, C.E. and H.A. Schaller. 1994. A comparison of productivities for Snake River and lower Columbia River spring and summer chinook stocks, pages 247-268. *In* Salmon Management in the 21st Century: Recovering Stocks in Decline. Proceedings of the 1992 Northeast Pacific Chinook and Coho Workshop, Idaho Chapter of the American Fisheries Society. Boise, Idaho.

Ortmann, D. and M. Richards. 1964. Chinook salmon and steelhead sport fisheries in the South and Middle Fork drainages of the Salmon River, 1961-1962. Idaho Department of Fish and Game. Boise, Idaho.

Raymond, H.L. 1979. Effects of dams and impoundments on migrations of juvenile chinook salmon and steelhead from the Snake River, 1966 to 1975. Transactions of the American Fisheries Society 108(6):505-529.

Raymond, H.L. 1988. Effects of hydroelectric development and fisheries enhancement on spring and summer chinook and steelhead in the Columbia River Basin. North American Journal of Fisheries Management 8(1): 3.

Reingold, M. 1976. Check station surveillance of major salmon and steelhead fisheries in Idaho (Salmon only). Idaho Department of Fish and Game Job Completion Report, Project F-18-R-22. Boise, Idaho.

Rich, B.A., R.J. Scully, and C.E. Petrosky. 1992. Idaho habitat/natural production monitoring: Part I. Idaho Department of Fish and Game Annual Report 1990 to US DOE, Bonneville Power Administration, Division of Fish and Wildlife, Project 83-7, Contract DE-A179-84BP13381. Portland, Oregon.

Rich, B.A. and C.E. Petrosky. 1994. Idaho habitat and natural production monitoring: Part I. Idaho Department of Fish and Game Annual Report 1992 to US DOE, Bonneville Power Administration, Division of Fish and Wildlife, Project Number 91-73, Contract Number DE-BI79-91BP21182. Portland, Oregon.

Ricker, W.E. 1938. "Residual" and kokanee salmon in Cultus Lake. Journal of the Fisheries Research Board of Canada 4(3): 192-218.

Technical Advisory Committee. 1991. 1991 all species review—Columbia River Fish Management Plan. US versus Oregon Technical Advisory Committee. Copy available from Oregon Department of Fish and Wildlife. Clackamas, Oregon.

US Army Corps of Engineers. 1993. Annual fish passage report, Columbia and Snake rivers for salmon, steelhead and shad. North Pacific Division, US Corps of Engineer Districts, Portland and Walla Walla.

Waples, R.S., R.P. Jones, Jr., B.R. Beckman, and G.A. Swan. 1991a. Status review for Snake River fall chinook salmon. US Department of Commerce, NOAA Technical Memorandum, National Marine Fisheries Service F/NWC-201. Portland, Oregon.

Waples, R.S., O.W. Johnson, R.P. Jones, Jr. 1991b. Status review for Snake River sockeye salmon. US Department of Commerce, NOAA Technical Memorandum, National Marine Fisheries Service F/NWC-195. Portland, Oregon.

Status of Wild Salmon and Steelhead Stocks in Washington State

Thom H. Johnson, Rich Lincoln, Gary R. Graves, and Robert G. Gibbons

Abstract

We describe the status and trends of wild salmon (*Oncorhynchus* spp.) and steelhead (*O. mykiss*) populations in Washington State. For each species, we consider statewide catch trends, and run size trends for the Puget Sound, coastal, and Columbia River regions. The utility of catch data for identifying trends in wild salmonid populations is limited by the omission of escapements and the inclusion of hatchery and non-local fish. Wild salmon and steelhead run size data have greater utility for measuring variations in abundance but typically range from the mid-1960s to the mid-1990s owing to the limited availability of reliable escapement estimates. For most species through the 1970s and 1980s, run sizes increased, followed by a declining trend in the 1990s. Recent population declines are attributed to a combination of factors, including poor survival in the ocean (in part due to elevated coastal water temperatures related to a series of El Niño events), freshwater and estuarine habitat alterations, and fishing harvest.

We review a recent study by the Washington State Departments of Fisheries and Wildlife and the Western Washington Treaty Indian Tribes, which presented an initial wild salmon and steelhead stock status inventory. The Salmon and Steelhead Stock Inventory (SASSI) was the first attempt at a comprehensive, statewide inventory for these species and provided an approach for developing a list of salmon and steelhead stocks and a process for rating their current status. The term "wild stock" as used in SASSI refers to how fish reproduce (i.e., by spawning and rearing in the natural habitat, regardless of parentage), and does not refer to genetic heritage. A total of 294 wild salmon stocks and 141 wild steelhead stocks were identified in Washington State, of which 187 (43%) were rated as healthy, 122 (28%) were rated as depressed, 12 (3%) were rated as critical, 113 (26%) were rated as unknown, and one stock was rated as recently extinct. The number and percentage of stocks in each stock status category varied in different regions and by species.

The SASSI report is the first step in a Wild Stock Restoration Initiative (WSRI) that has an ultimate goal of maintaining healthy wild salmon and steelhead stocks and their habitats in order to support the region's fisheries, economies, and other societal values. Objectives of the WSRI and other future actions include periodically reviewing and updating the salmon and steelhead resource status inventory, reviewing current resource management goals and objectives for hatchery and wild stocks and the region's fisheries, developing and implementing recovery programs for priority stocks and habitats, and monitoring and evaluating programs. The inventory will become a part of the salmon and steelhead management cycle for the state agencies and treaty Indian tribes, guide future data collection programs, and facilitate the measurement of the short-term and long-term success in rehabilitating priority stocks. For the WSRI to succeed as a long-term restoration strategy, improved coordination and funding of programs designed to protect and restore salmon and steelhead stocks will be needed.

Introduction

Throughout history, Washington State has been linked with abundant salmon (*Oncorhynchus* spp.) and steelhead (*O. mykiss*) populations. Native Americans found salmon in such great abundance that large parts of their traditions and culture developed around these fish. In the 1800s, newly arriving settlers introduced methods of canning and transporting salmon, thereby transforming use of the resource from primarily subsistence to major commercialization; as a result, Pacific salmon became a worldwide commodity. More recently, recreational fishing and nonconsumptive uses (e.g., fish viewing) have increased in importance. Today, most state residents would have a difficult time imagining a future for Washington State that did not include salmon and steelhead.

Washington State has numerous habitats and geographic areas that support diverse assemblages of anadromous salmonids. Five salmon species, chinook (*O. tshawytscha)*, chum (*O. keta*), coho (*O. kisutch*), pink (*O. gorbuscha*), and sockeye (*O. nerka*), and steelhead occur in the state. Washington can be divided into three major salmon and steelhead production regions: Puget Sound (which includes the Strait of Juan de Fuca, Hood Canal, and Puget Sound), the Washington coast (which includes the Pacific Coast, Grays Harbor, and Willapa Bay), and the Columbia River (Fig. 1). These regions reflect differences in habitat types, species use, and the nature and locations of traditional fisheries.

The majority of salmon and nearly all wild steelhead runs in Washington are managed to achieve natural spawning escapement goals (i.e., a designated number of spawners). In this paper, the term "wild" refers to how fish reproduce (i.e., by spawning and rearing in natural habitats) and does not refer only to genetic heritage; this terminology is not intended to diminish the importance of native stocks, but rather emphasizes the need to protect a wide range of genetic resources maintained by natural reproduction. Numerically, many stocks are relatively healthy, and naturally reproducing populations still account for >50% of the total salmon and steelhead production in Washington. However, some wild stocks are significantly depleted and most runs experience periodic declines. Annual and long-term resource management planning efforts are commonly undertaken to assess status of stocks and fisheries. Specific harvest, culture, and habitat measures have been recommended and implemented in many areas to maintain or improve the status of wild stocks (Pacific Fishery Management Council 1992).

Despite the relatively healthy condition of many wild salmon and steelhead stocks, factors such as habitat degradation, some poorly designed hatchery programs, and overfishing have caused some wild stocks to decline to historically low levels (Nehlsen et al. 1991, Pacific Fishery Management Council 1992, Washington Department of Fisheries 1992, Washington Department of Fisheries et al. 1993). These factors have caused past and continuing losses of stocks as well as diminished abundance and genetic diversity. In particular, land-use activities (e.g., urban and industrial growth, forest practices, agricultural practices, water diversions, and hydropower)

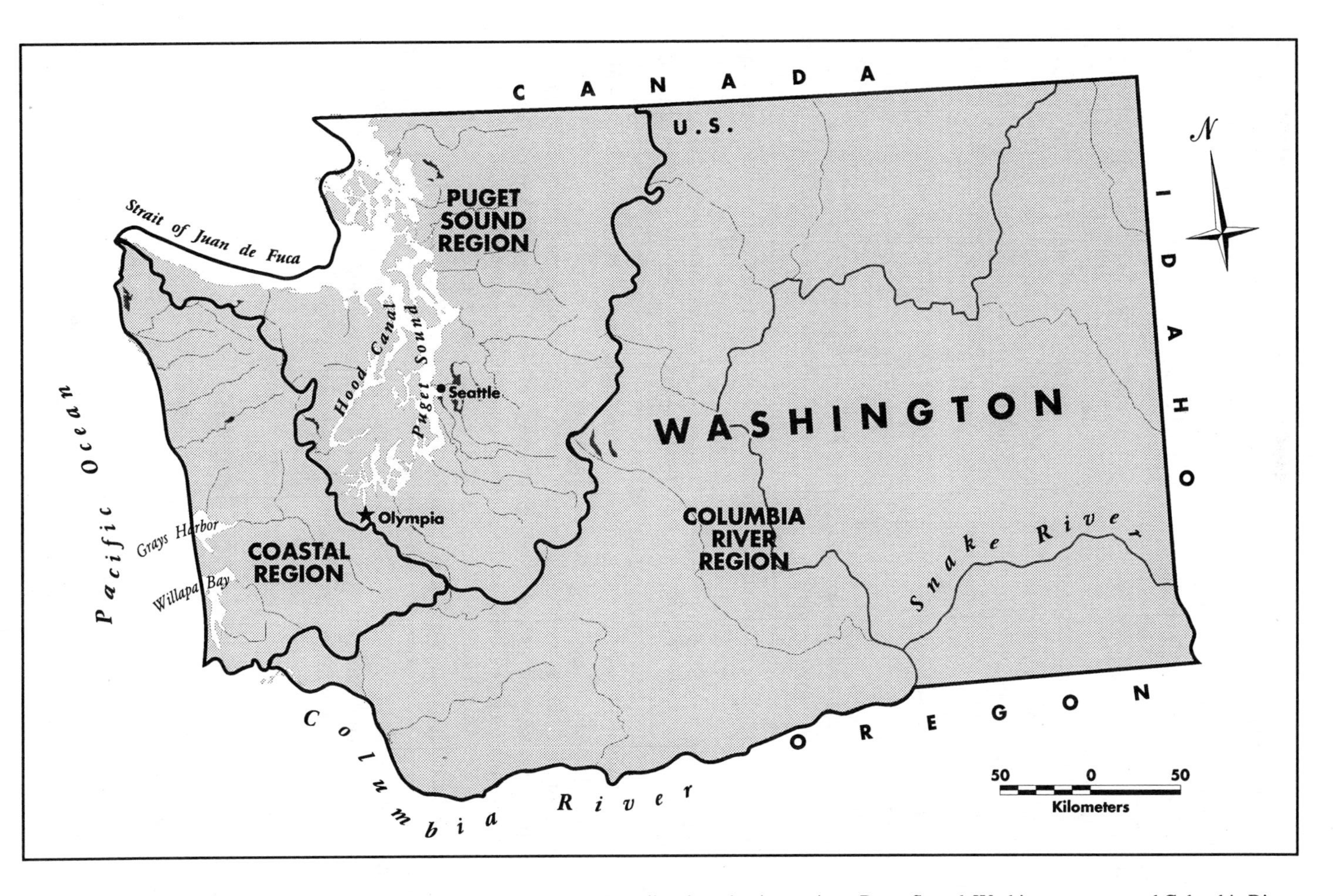

Figure 1. Washington State can be divided into three major salmon and steelhead production regions: Puget Sound, Washington coast, and Columbia River. Source: Washington Sea Grant Program (B. Feist).

have impacted wild salmonid habitat and survival. These factors have reduced Washington's salmon and steelhead production over the years and continue to do so at an escalating rate (Washington Department of Fisheries 1992). Wild fish resources and their habitats must be protected and restored in order to maintain viable and healthy fisheries and to provide for associated ecological, cultural, and aesthetic values.

In this paper, we describe the status and trends for wild salmon and steelhead in Washington State. We consider the use of statewide and regional data to evaluate population trends for the six species of anadromous salmonids. In addition, we review a recent study that identified and rated the status of individual stocks of salmon and steelhead. For each species within a region, there are multiple, reproductively isolated population units, or stocks, that represent important components of overall diversity. Maintenance of genetic diversity requires identifying, evaluating, protecting, and restoring individual stocks.

In 1992, the Washington Department of Fisheries and Washington Department of Wildlife and the Western Washington Treaty Indian Tribes completed a wild salmon and steelhead stock status inventory. The inventory was the first step in a statewide effort to help identify currently available information for each salmon and steelhead spawning population and to guide future restoration planning and implementation. The inventory is focused primarily on the current condition of six of Washington's naturally reproducing anadromous salmonid species and not on the adequacy of current resource management objectives. As part of a statewide Wild Stock Restoration Initiative (WSRI), assessment of management objectives and strategies will be one of many subsequent steps aimed at improving the status of wild salmon and steelhead resources in Washington. The Salmon and Steelhead Stock Inventory (SASSI; Washington Department of Fisheries et al. 1993) is the first step in the WSRI and provides an approach for developing a list of salmon and steelhead stocks and a process for rating their current status based on trends in fish population abundance, spawning escapement, or survival.

The concept of resource inventories is not new; fishery management agencies spend considerable time collecting and assessing resource status data (e.g., spawning escapements, harvests). This information is routinely used for decision making but often is not well documented or visible to others. As a result, an objective of SASSI has been to develop a simple, clear, consistent, and meaningful system of collating and reporting statewide salmon and steelhead resource assessment information while recognizing that the inventory will change over time. The information is meant to provide a look at current stock status and build a foundation for future restoration and inventory efforts. Future updates will accommodate new information and be integrated with developing regional resource information systems.

Statewide and Regional Trends

Statewide Catch

Washington State salmon and steelhead populations have been commercially exploited for >100 years, and long-term catch databases have traditionally been used to examine abundance trends for salmonid species. Statewide catches of salmonids during the 1900s provide some general insights into long-term trends in salmon and steelhead abundance. However, catch data have several weaknesses when used to measure variations in salmonid populations. They omit the numbers of fish allowed to escape to freshwater spawning grounds; escapement can vary substantially as a result of changes in relative run sizes and management strategies. Catch data include salmon originating from outside the state that are caught by fisheries within Washington. For example, sockeye salmon landings in the state are almost exclusively Canadian-origin fish (primarily from the Fraser River), and Columbia River catches include fish originating in Idaho, Oregon, and Canada. Washington salmon are also intercepted by fisheries outside of state waters. Hatchery programs in Washington State contribute large numbers of fish to some fisheries; however, catches can remain relatively stable or even increase while wild populations are declining. Finally, major changes have occurred in the gear types (e.g., purse seines have changed in size, different mesh sizes have been used by gill netters), vessel types (e.g., bigger, more powerful boats), fishing locations, and efficiencies of the state's fisheries over the last 100 years, all of which can affect catch patterns.

An examination of statewide catches from 1913 through 1992 is informative but shows the typical variability associated with such databases (Fig. 2). Record catches for salmon occurred between 1910 and 1920 and in most cases were followed by years of low catches, suggesting that the high catches may have been overharvests. Washington chinook salmon catches show evidence of a long-term decline probably resulting from habitat alterations and harvest pressures. Chum and coho salmon catches are at or near historically high levels; however, returns from hatchery production contribute substantially to fish catches. The catches of pink and sockeye salmon in the state are dominated by Canadian-origin fish and cannot be used to assess the status of Washington pink and sockeye salmon. Statewide steelhead catches fluctuated cyclically from 1962 to 1992, but declined to historically low levels during the early 1990s. Cooper and Johnson (1992) concluded that similarities in overall and year-to-year trends of steelhead abundance in British Columbia, Washington, and Oregon indicated that common factors were partially responsible for this decline. These factors include low productivity in the North Pacific Ocean, competition for food resulting from increases in numbers of hatchery smolts released, and catches of steelhead in high-seas driftnet fisheries. Demonstrating limitations with catch data, however, Bledsoe et al. (1989) attempted a detailed examination of the trends in run sizes of salmon returning to Puget Sound from 1896 through 1975, and found that the inherent variability of catch made it impossible to demonstrate changes in wild run abundance.

Regional Run Size

Total run size, which combines both catch and spawning escapement data for specific runs of fish, is a more informative measure of fluctuations in wild salmonid populations than catch alone. Various run reconstruction approaches are used to assemble escapements and catches (by time and area) into total run size estimates. These run size estimates are limited to the time periods for which reliable escapement estimates are available (i.e., from the mid-1960s to the mid-1990s), which vary by species and region. Run reconstruction methodologies typically estimate the numbers of open-ocean fish returning to inside waters (e.g., individual river systems or management areas in the Puget Sound, coastal, and Columbia River regions). Washington

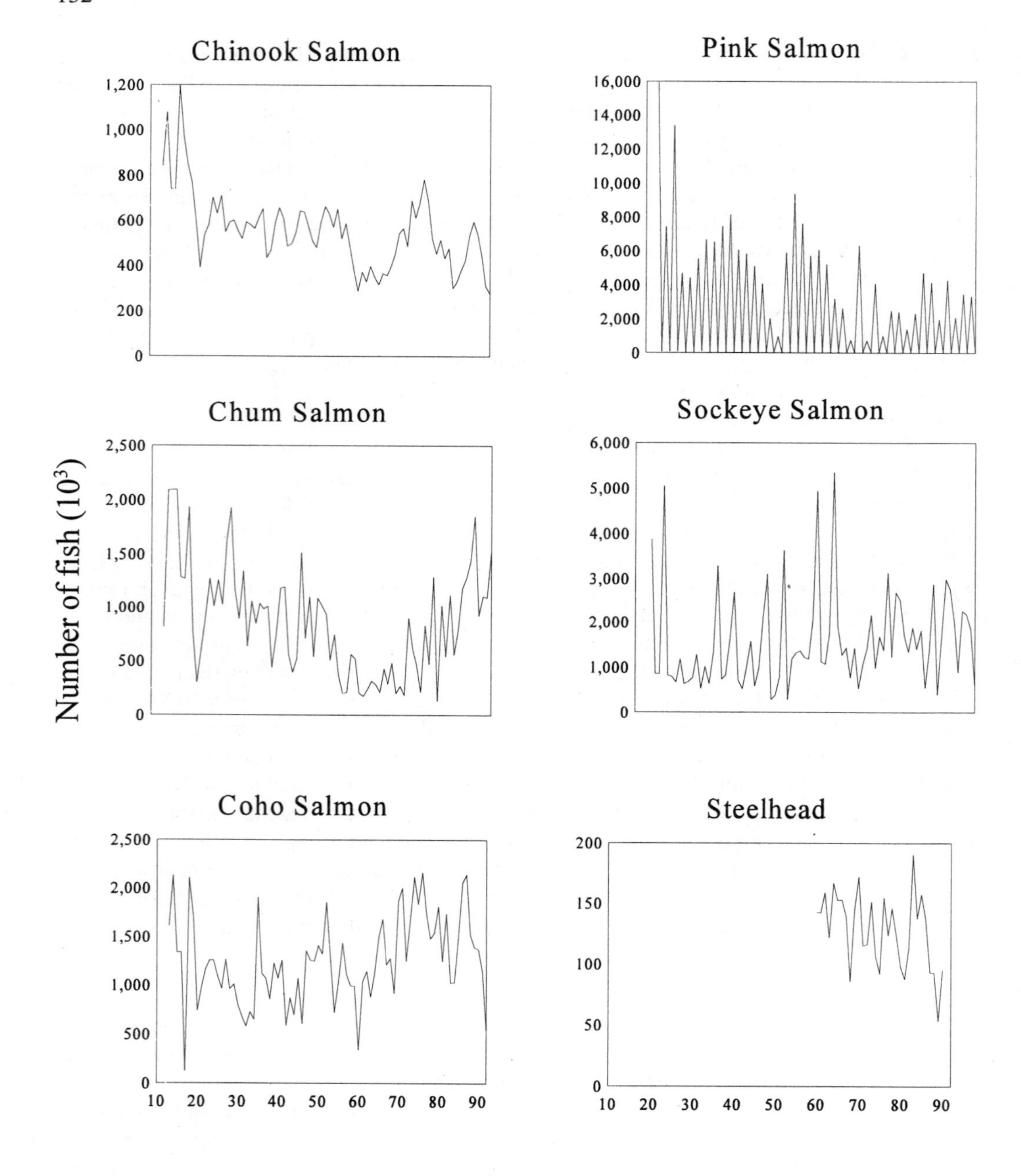

Figure 2. Statewide harvest of salmon and steelhead from 1913 through 1992. Source: Cooper and Johnson (1992); R. Geist, Washington Dep. Fish and Wildlife, Olympia, unpubl. data.

salmon catches in the ocean include fish from other regions, and identifying that portion of ocean catch that represents Washington-origin fish is a complex process. Reductions and closures of ocean fisheries in recent years have transferred a higher proportion of the total numbers of fish to the runs entering Puget Sound, the coastal bays, and Columbia River. These changes in fishing patterns have primarily affected chinook and coho harvests and tend to cause an overestimation of relative run strength when using the run size entering inside waters for examining trends. The following discussions present wild salmon and steelhead abundance trends from run size data for each region.

Puget Sound

The Puget Sound region has abundant runs of all five North American salmon species and steelhead that return to hundreds of individual streams. Maturing fish returning from the ocean to spawn in Puget Sound streams share the straits of Georgia and Juan de Fuca with vast numbers of fish destined for Canadian streams. Fisheries conducted in these mixed stock areas harvest both Canadian and Puget Sound fish, and a number of techniques are used to identify fish by region of origin (e.g., scale characteristics and biochemical stock identification). The Puget Sound salmon-run reconstruction methodology attempts to define Puget Sound wild fish in the mixed stock areas, and includes all Washington-origin fish leaving the ocean (at Cape Flattery) to enter inside waters. The run size numbers generated represent the majority of the salmon returning to inside waters; however, some catch components may be missing for the various species. For example, sport-caught salmon are generally not included because data are currently unavailable to identify the origin of the fish caught. The exception is the run reconstruction for south Puget Sound sockeye salmon, which is a total run size including all ocean, Canadian, and sport catches. Even though most Puget Sound salmon run size databases do not include all catch components, they are useful for examining trends in salmon abundance. Generally, most steelhead harvest occurs in freshwater areas; therefore, we combined wild steelhead run sizes from major individual Puget Sound rivers for this paper.

Wild run size estimates for the various salmon species in Puget Sound are available for ≥27 years (Fig. 3). Differences in life histories and factors like spawning distributions, harvest patterns, and the impacts of environmental variability have caused different patterns of abundance for each species. Run size estimates for wild chinook salmon from 1968 through 1994 show relatively stable returns until a sharp drop in abundance beginning in 1991. This drop resulted from a combination of poor ocean survivals, habitat alterations, and harvest pressures. Chum salmon abundance has increased steadily over the 27-year database and is now approaching historically high levels. The returns of Puget Sound coho salmon showed a steady rate of increase through the 1970s and 1980s, with a steep drop beginning in 1991. As with chinook salmon, poor ocean survival, habitat alterations, and harvest pressures likely contribute to the lower returns of coho salmon. Puget Sound pink salmon return predominantly in odd-numbered years and display a pattern of increasing abundance throughout the 1965–93 database. However, sockeye salmon have declined in the 1990s as a result of a drop in freshwater survival in the dominant south Puget Sound run. The smaller north Puget Sound sockeye salmon run has increased in recent years in response to restoration efforts. Wild steelhead run size increased through the 1970s and mid-1980s but declined to record lows during the 1990s.

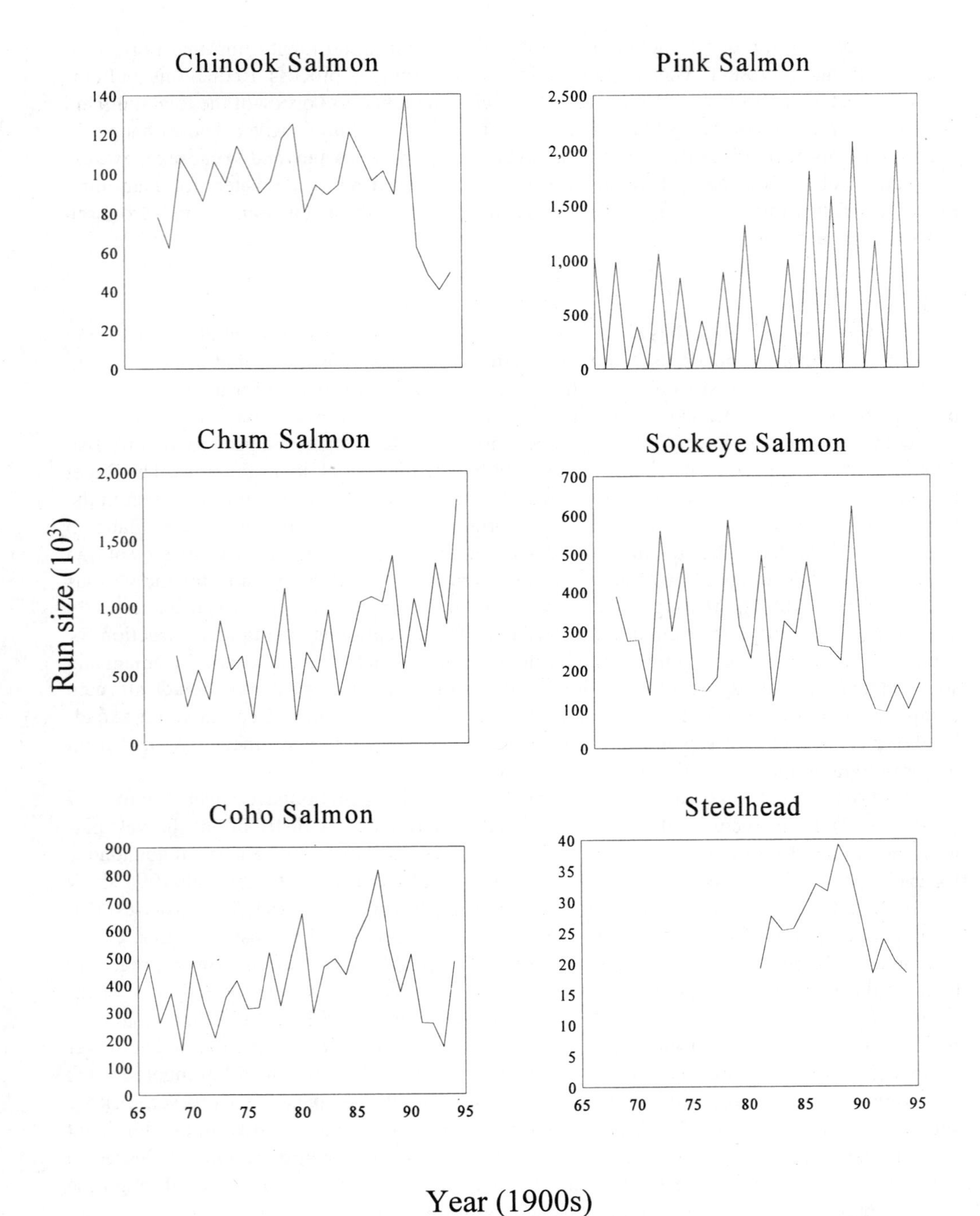

Figure 3. Run size trends for wild salmon and steelhead in the Puget Sound region, 1965–94.

The recent drops in abundance for chinook and coho salmon and steelhead are thought to be in part related to poor ocean survival resulting from high water temperatures associated with a recent series of El Niño events along the Pacific coast of North America (Pearcy 1996). Puget Sound chum and pink salmon populations seem to be unaffected by these El Niño occurrences. Their lack of a negative response to the warm waters along the coasts of Washington and southern Canada may be related to the more northern distribution of these species during ocean migrations.

Coastal

Salmon and steelhead returning to coastal Washington leave the ocean and enter directly into the river mouths of their natal streams, or enter one of two coastal estuaries, Grays Harbor and Willapa Bay. Run reconstruction for coastal salmon separates catches for each species by timing and area for each run and adds the number of fish that escaped to the spawning grounds. Escapement estimates and regional run size data for all species are available for different time periods. As with Puget Sound, coastal run sizes may not include all run components; however, the majority of fish are represented and trends in abundance can be identified. Again, most sport and tribal steelhead harvest occurs in freshwater areas, and we combined wild steelhead run sizes from the major coastal rivers for this discussion.

The run size estimates for the four species of Pacific salmon returning to coastal streams (pink salmon are absent) show similar trends in abundance over the available database (Fig. 4). Chinook and coho salmon run sizes increased from the mid-1960s through the 1970s and dropped in the 1990s, probably as a result of increased coastal water temperatures related to recent El Niño events. Chum salmon increased substantially in abundance during the 1980s and have since displayed a modest reduction in run sizes. Sockeye salmon have been fairly stable throughout the 30 years of run size data. Wild steelhead run sizes increased through the 1970s and mid-1980s but declined during the 1990s, similar to trends in steelhead abundance in Oregon and British Columbia along the Pacific coast of North America (Cooper and Johnson 1992).

Columbia River

The decline of Columbia River salmon and steelhead has been well documented (Washington Department of Fish and Wildlife and Oregon Department of Fish and Wildlife 1994). Prior to significant habitat alteration in the basin, average annual run-size estimates of adult salmon and steelhead had a range of 10–16 x 10^6 fish (mid-to-late 1800s). Beginning in the 1860s, a commercial fishery developed rapidly. The fisheries annually harvested over 18.2 x 10^6 kg of Columbia River salmon and steelhead ten times between 1883 and 1925. Between 1940 and 1980, the minimum number of salmon and steelhead entering the Columbia River fluctuated over a range of ~1–2.4 x 10^6 fish. Recently, fish abundance reached a high of 3.2 x 10^6 fish (1986) and a low of ~0.95 x 10^6 fish (1993; Fig. 5); steelhead generally accounted for ~10–25% of the run. Currently, ~75% of all fish returning to the Columbia River system are produced in hatcheries and other artificial production facilities as compensation for the loss of wild salmonid production (Washington Department of Fish and Wildlife and Oregon Department of Fish and Wildlife 1994), which is primarily due to loss of habitat following construction of hydroelectric dams on the mainstem Columbia River and major tributaries.

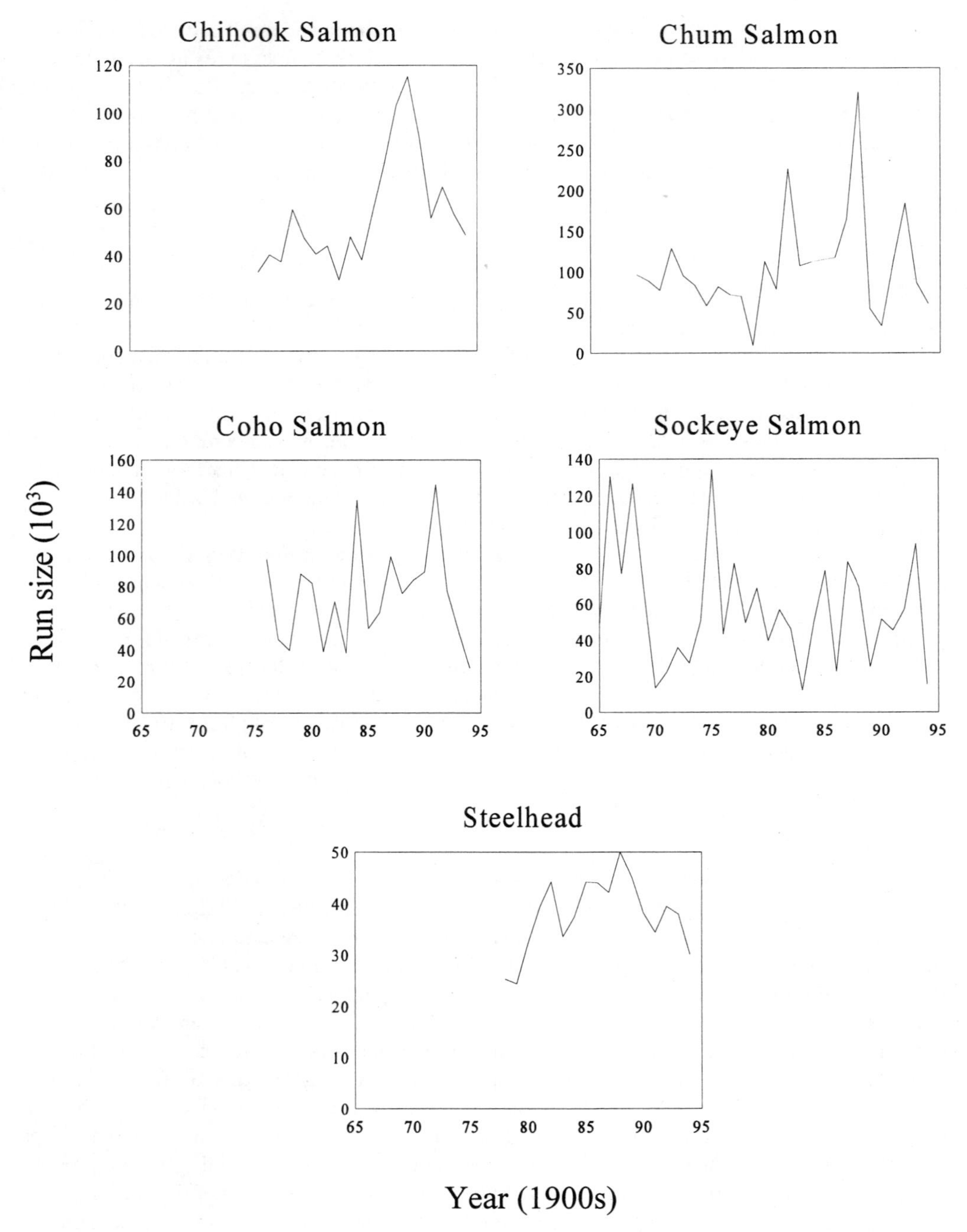

Figure 4. Run size trends for wild salmon and steelhead in the coastal Washington region, 1976–94. Pink salmon are not present during the years covered.

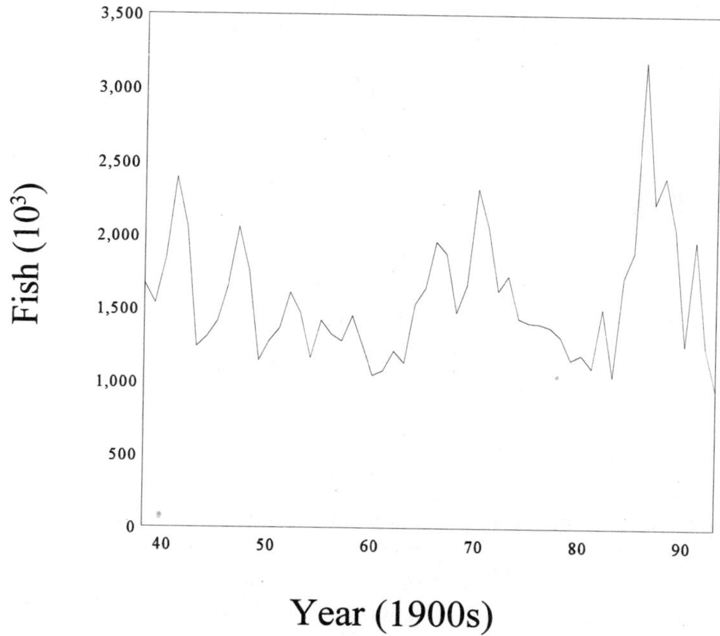

Figure 5. Minimum numbers of salmon and steelhead entering the Columbia River from 1938 through 1993. Source: Washington Department of Fish and Wildlife and Oregon Department of Fish and Wildlife 1994.

The magnitude of declines in Columbia River salmon and steelhead has varied widely for different species and stocks, leading to several threatened and endangered listings under the federal Endangered Species Act (ESA). For example, the Snake River portion of the Columbia River sockeye salmon run was listed as endangered under the ESA in December 1991. In May 1992, the wild portions of the Snake River spring chinook and summer chinook runs were combined into a single species and listed as threatened under the ESA, and the wild Snake River fall chinook run was listed as threatened (Matthews and Waples 1991, Waples et al. 1991). In 1994, the status of both Snake River spring/summer chinook and fall chinook was revised to endangered (Waples 1995).

Columbia River runs are reconstructed by computing the minimum estimates for the numbers of salmon and steelhead entering the mouth of the Columbia River (Fig. 5). These estimates have been generated since 1938 and are calculated separately for each species (Washington Department of Fish and Wildlife and Oregon Department of Fish and Wildlife 1994); however, this approach suffers from many of the same problems discussed previously for catch data.

Total run size data (fish entering the river) have been assembled for Columbia River wild salmon and steelhead originating in Washington State streams (W. Dammers, Washington Department of Fish and Wildlife, Battleground, pers. comm.). These run size estimates are limited to 1980–95 return years because of the availability of escapement estimates for wild fish populations. Wild summer steelhead have been enumerated at Bonneville Dam since 1984. All species have experienced similar trends during this period. Run sizes increased during the 1980s, followed by a steep decline in the 1990s (Fig. 6). As with the coastal stocks, the recent decrease in Columbia River runs has been attributed to unfavorable ocean rearing conditions due to recent El Niño events (Pearcy 1996). Harvests have been greatly reduced, and some fisheries were closed to provide protection to these small returns.

Stock Status

The 1992 Washington State Salmon and Steelhead Stock Inventory (SASSI; Washington Department of Fisheries et al. 1993) identified and evaluated the current status of 435 salmon and steelhead stocks statewide. This study was a major step beyond the more traditional review of abundance trends from catch and run size databases because it examined individual stocks for the first time. The following is a review of this inventory.

Definition and Identification of Stocks

Stocks were chosen because they form the basic building blocks of Northwest salmon and steelhead management and allow assessment of production in larger units (e.g., runs, watersheds) by combining stocks. Stocks are also widely accepted within the scientific community as a basis for evaluating fish populations. The definition of the term "stock" and its application can be problematic because distinctions among different groups of animals are often difficult to identify, and because the term is used for a variety of purposes (Ricker 1972, Waples 1991; see also *US vs. Washington* 1985, No. 9213, Subproceeding 85-2, Order Adopting Puget Sound Management Plan [Western Washington District, October 15, 1985]). However, in salmonid management a stock is generally considered to be a discrete breeding population.

The SASSI report uses stock (see Ricker 1972) to mean fish spawning in a particular lake or stream in a particular season that, to a substantial degree, do not interbreed with any group spawning in a different place, or in the same place during a different season (Washington Department of Fisheries et al. 1993). Thus, SASSI does not imply that this definition should be applied for other uses, or that even smaller units of production are unimportant, or that management of fish or their habitat, or both, should be done on this basis. Where reproductive isolation was shown or presumed to exist, it may or may not have indicated genetic uniqueness from other stocks. The term "wild stock" is used as described previously in this paper. A salmon or steelhead population is considered a distinct stock (i.e., reproductively isolated) if it has one or more of the following characteristics: a distinct spawning distribution, a distinct temporal distribution (including spawning or run timing), or distinct biological characteristics (e.g., genetics, size, age structure). The term "distinct" does not imply complete isolation from other stocks because genetic interchange between some salmonid populations naturally occurs. Thus, more detailed analysis in the future likely will change some stock designations.

Categories and Criteria Used

For each stock, current status was identified in SASSI on the basis of trends in abundance, spawning escapement, or survival. Stocks were placed into one of five stock status categories:

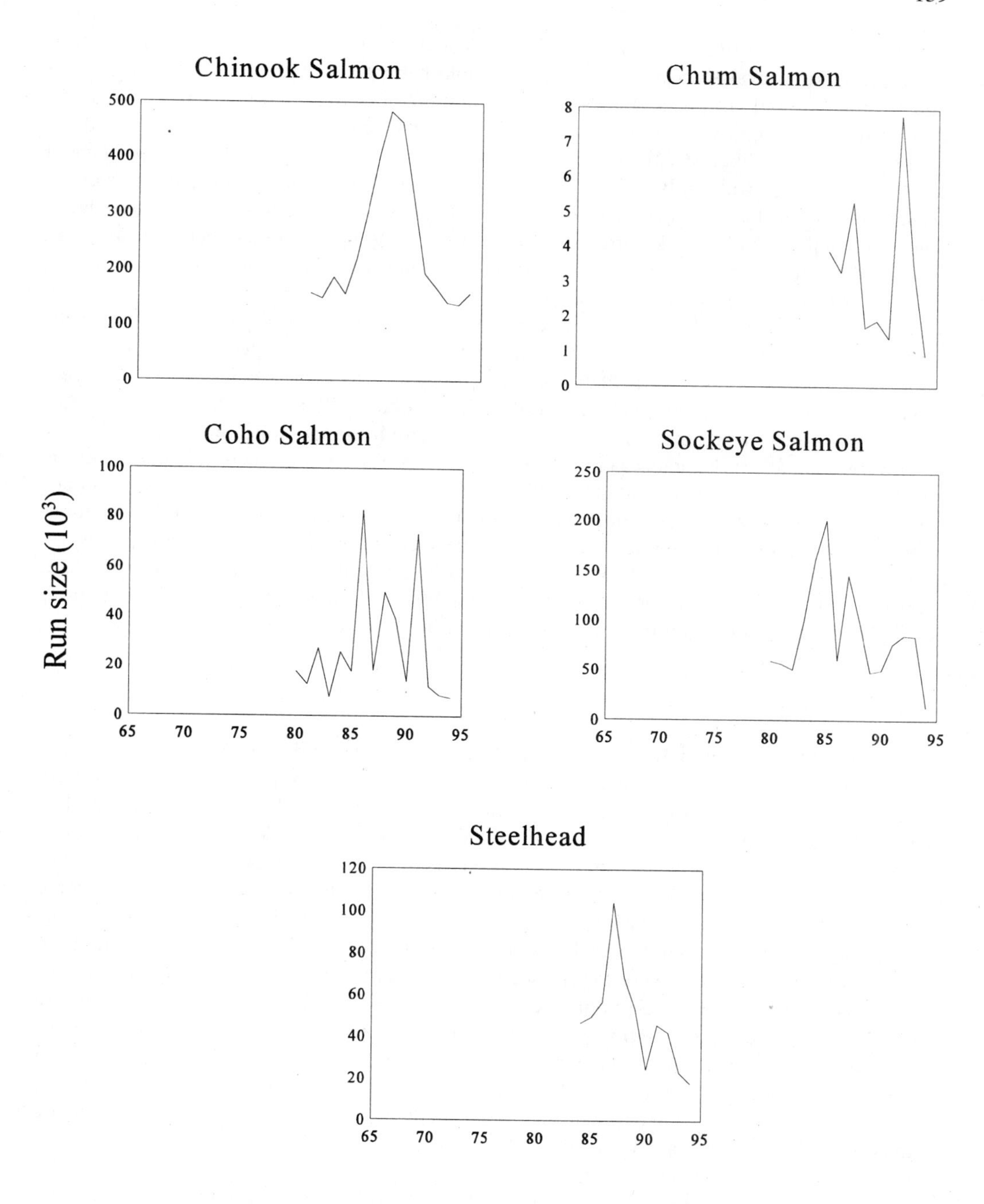

Figure 6. Run size trends for wild salmon and steelhead in the Columbia River region, 1980–94. Pink salmon are not present in this region.

healthy, depressed, critical, extinct, or unknown. A stock was classified as healthy if it experienced production levels consistent with its available habitat and within natural variations in survival for the stock. Depressed stocks had production below expected levels based on available habitat and natural variations in survival rates, but above the level where permanent damage to the stock was likely. Critical stocks experienced production levels so low that permanent damage to the stock was likely or had already occurred. A major concern with critical stocks is the loss of within-stock diversity, which could significantly limit a stock's ability to respond to changing conditions. If a stock currently being tracked in escapement or fishery management databases was found to occur no longer in its original range or as a distinct stock elsewhere, it was rated as extinct. However, past extinctions were not included in this category because it would be impossible to assemble any kind of comprehensive listing of past extinctions since many of these losses occurred prior to the time when inventory programs were initiated. Finally, where trend information was unavailable or could not be used to assess stock status, stocks were rated as unknown. The various stocks rated as unknown likely will be rated in each of the status categories (healthy, depressed, critical, or extinct) as more information becomes available.

Four points about the status ratings bear mentioning. First, some trend information exists for each of the healthy, depressed, and critical stocks. Second, the abundance of these stocks varies considerably (e.g., some stocks have hundreds of thousands of fish while others number only in the hundreds). Third, various types of data are used to assess stock status. Fourth, several criteria are used to identify trends within the depressed and critical categories. For example, a depressed stock may have become recently depressed (short-term severe decline), or may have been depressed but stable (chronically low) or declining (long-term negative trend).

Stock Status

The 1992 SASSI report identified 294 wild salmon and 141 wild steelhead stocks statewide in Washington. Of these stocks, 187 were rated as healthy, 122 were rated as depressed, 12 were rated as critical, 113 were rated as unknown, and one stock was rated as extinct (Table 1). Excluding the unknown stocks, the status of the remaining stocks statewide in terms of percentage was as follows: 58% healthy, 38% depressed, 4% critical (Washington Department of Fisheries et al. 1993). While only 43% of the wild salmon and steelhead stocks in Washington State were rated as healthy, these stocks include many medium to large spawning populations and thus represent a large proportion of the total production of wild salmon and steelhead in Washington. Depressed stocks have not likely experienced permanent damage, but many may need active efforts to return them to more productive levels. Most of the 12 stocks rated as critical are currently the subject of restoration programs. Approximately one-quarter of the wild stocks were rated as unknown, which indicates that there was insufficient information on trends in abundance to rate stock status as either healthy, depressed, or critical. This emphasizes the need to collect spawner escapement or other stock assessment information for unknown stocks. Many of the unknown stocks comprise a historically small number of salmon or steelhead and could be especially vulnerable to any negative impacts.

The percentage of stocks in each status category varied in different regions of the state and by species. The Puget Sound and Washington coastal regions (Fig. 1) have the highest proportion of healthy stocks (Table 1). However, 21% of the stocks in the Puget Sound region were

Table 1. Summary of stock status for wild salmon and steelhead in Washington State. Values indicate the numbers of stocks within each category. Washington Department of Fisheries et al. (1993).

	Healthy	Depressed	Critical	Unknown	Extinct	Total
Puget Sound						
Chinook salmon	10	8	4	7	0	29
Chum salmon	38	1	2	13	1	55
Coho salmon	20	16	1	9	0	46
Pink salmon	9	2	2	2	0	15
Sockeye salmon	0	3	1	0	0	4
Steelhead	16	14	1	29	0	60
Total stocks (%)	93 (44%)	44 (21%)	11 (5%)	60 (29%)	1 (0.5%)	209
Coastal						
Chinook salmon	20	5	0	7	0	32
Chum salmon	9	0	0	5	0	14
Coho salmon	17	0	0	9	0	26
Pink salmon	-	-	-	-	-	-
Sockeye salmon	1	1	0	1	0	3
Steelhead	18	2	0	20	0	40
Total stocks (%)	65 (56%)	8 (7%)	0 (0%)	42 (37%)	0 (0%)	115
Columbia River						
Chinook salmon	24	22	1	0	0	47
Chum salmon	1	2	0	0	0	3
Coho salmon	0	18	0	0	0	18
Pink salmon	-	-	-	-	-	-
Sockeye salmon	2	0	0	0	0	2
Steelhead	2	28	0	11	0	41
Total stocks (%)	29 (26%)	70 (63%)	1 (1%)	11 (10%)	0 (0%)	111
Washington State total						
Chinook salmon	54	35	5	14	0	108
Chum salmon	48	3	2	18	1	72
Coho salmon	37	34	1	18	0	90
Pink salmon	9	2	2	2	0	15
Sockeye salmon	3	4	1	1	0	9
Steelhead	36	44	1	60	0	141
Total stocks (%)	187 (43%)	122 (28%)	12 (3%)	113 (26%)	1 (0%)	435

rated as depressed, and 11 of 12 critical stocks statewide were in the Puget Sound region. In the coastal region, only 7% of the stocks were rated as depressed, and none were rated as critical. The Columbia River region (which includes the lower and upper Columbia River and Snake River) had the greatest proportion of depressed stocks, and only 26% of the stocks were rated as healthy (Table 1).

Future Actions

In Washington, SASSI is envisioned as the first step in an ongoing process that has the ultimate goal of maintaining healthy wild salmon and steelhead stocks and their habitats (Washington Department of Fisheries et al. 1993). Accomplishments toward this goal will help directly support the region's fisheries, economies, and other societal values. Defining and managing future change (e.g., urban growth, land-use activities, fisheries) will be at least as difficult as creating technical solutions. Because habitat, harvest, hatchery, and other species' impacts all contribute to wild stock status, coordinated management of these factors will be needed to provide comprehensive strategies for restoring healthy stocks and fisheries. Recent calls for an ecosystem approach to the ESA (Reeves and Sedell 1992) indicate a need for a system-wide evaluation of watersheds and the various species they support while developing plans to restore depleted wild stocks.

A hierarchy of responses will be needed to establish an ecosystem approach. First, it will be important to complete and maintain a resource status inventory of Washington's wild salmon and steelhead stocks and their habitats. Second, managers must review current resource management goals and objectives for hatchery and wild stocks and the region's fisheries. Third, recovery programs should be developed and implemented for priority stocks and habitats. Finally, success can be evaluated only by maintaining adequate monitoring programs.

The potential for success will be affected by several key factors. One important element is the availability of adequate funding. State and tribal managers face declining fiscal resource bases juxtaposed with the need to improve wild stock status. Budget reductions in other programs such as harvest management, hatchery production, and habitat protection could also result in risks to wild stocks and reduced harvest opportunity. Without significant financial support, fish management entities will be limited in their abilities to address and successfully implement priority wild stock issues. Besides adequate fiscal resources, we must be willing to solve difficult resource management issues and adopt new approaches to complex problems. For instance, the long-term status of fishery resources ultimately will be determined by public support and willingness of land-use regulators to deal effectively with growth management and land- and water-use issues. Resolving conflicts between stock restoration and habitat loss/degradation is central to maintaining healthy wild stocks and fisheries.

Successful wild stock management and restoration will depend upon filling key information gaps and developing new management objectives and approaches. During 1993, state and tribal managers prioritized stock and habitat restoration needs and identified the absence of important information. At a minimum, management actions need to address the following: a critical review of harvest, habitat management, and hatchery production goals and objectives; development of wild stock management and genetics policies and associated guidelines; evaluation of costs and benefits of alternate resource management strategies; and the increased monitoring of wild spawning populations to address critical information gaps and to improve assessment of wild stock status. As an example of how resource management objectives and approaches can be modified, the Washington Department of Fish and Wildlife has generally opted for conservative regulations (such as wild steelhead release, closed areas, and season closures) to protect wild steelhead stocks/runs where run sizes, escapement goals, escapements and status cannot be readily estimated.

The initial step provided by SASSI (Washington Department of Fisheries et al. 1993) was designed to facilitate a process of periodic reviews and updates of salmon and steelhead stock status tailored to particular regions and habitats. Specific strategies will be developed for establishing complementary hatchery stock and habitat inventories. A continually updated inventory will become a part of the salmon and steelhead management cycle for the state agencies and tribes. This inventory will guide future data collection programs and function as a tool to measure the short- and long-term success in rehabilitating priority stocks and in protecting and maintaining healthy stocks and fisheries. A long-term restoration strategy will rely on improved coordination and reliable funding, as well as implementation and evaluation of programs designed to protect and rehabilitate salmon and steelhead habitat.

Acknowledgments

The authors wish to thank R. Brix, W. Dammers, and R. Geist of the Washington Department of Fish and Wildlife for providing and updating various catch and run-size databases for this paper. It would be inappropriate to present a review of the 1992 Salmon and Steelhead Stock Inventory without acknowledging the dedicated efforts of the numerous staff within the Northwest Indian Fisheries Commission, Washington Department of Fish and Wildlife, and the Western Washington Treaty Indian Tribes who contributed as authors and made SASSI possible.

Literature Cited

Bledsoe, L.J., D.A. Somerton, and C.M. Lynde. 1989. The Puget Sound runs of salmon: an examination of the changes in run size since 1896, p. 50-61. *In* C.D. Levings, L.B. Holtby, and M.A. Henderson (eds.), Proceedings of the National Workshop on Effects of Habitat Alteration on Salmonid Stocks. Canadian Special Publication of Fisheries and Aquatic Sciences 105.

Cooper, R. and T.H. Johnson. 1992. Trends in steelhead abundance in Washington and along the Pacific coast of North America. Washington Department of Wildlife, Fisheries Management Division. Report No. 92-20. Olympia.

Matthews, G.M. and R.S. Waples. 1991. Status review for Snake River spring and summer chinook salmon. NOAA National Marine Fisheries Service Technical Memorandum NMFS F/NWC-200. Seattle.

Nehlsen, W., J.E. Williams, and J.A. Lichatowich. 1991. Pacific salmon at the crossroads: stocks at risk from California, Oregon, Idaho, and Washington. Fisheries 16(2): 4-21.

Pacific Fishery Management Council. 1992. Assessment of the status of five stocks of Puget Sound chinook and coho as required under the PFMC definition of overfishing. Summary Report for Pacific Fishery Management Council. Prepared by Puget Sound Stock Review Group. Sept. 1992. Portland, Oregon.

Pearcy, W.G. 1996. Salmon production in changing ocean domains, p. 331-352. *In* D.J. Stouder, P.A. Bisson, and R.J. Naiman (eds.), Pacific Salmon and Their Ecosystems: Status and Future Options. Chapman and Hall, New York.

Reeves, G.H. and J.R. Sedell. 1992. An ecosystem approach to the conservation and management of freshwater habitat for anadromous salmonids in the Pacific Northwest, p. 408-415. *In* R.E. McCabe (ed.), Biological Diversity in Aquatic Management. Transactions of the 57th North American Wildlife and Natural Resources Conference. Wildlife Management Institute, Washington DC.

Ricker, W.E. 1972. Hereditary and environmental factors affecting certain salmonid populations, p. 19-160. *In* R.C. Simon and P.A. Larkin (eds.), The Stock Concept in Pacific Salmon. University of British Columbia, Vancouver.

Waples, R.S. 1991. Pacific salmon, *Oncorhynchus* spp., and the definition of "species" under the Endangered Species Act. Marine Fisheries Review 53(3): 11-22.

Waples, R.S. 1995. Evolutionarily significant units and the conservation of biological diversity under the Endangered Species Act, p. 8-27. *In* J.L. Nielsen (ed.), Evolution and the Aquatic Ecosystem: Defining Unique Units in Population Conservation. American Fisheries Society Symposium No. 17. American Fisheries Society, Bethesda, Maryland.

Waples, R.S., R.P. Jones, Jr., B.R. Beckman, and G.A. Swan. 1991. Status review for Snake River fall chinook salmon. NOAA National Marine Fisheries Service Technical Memorandum NMFS F/NWC-201. Seattle.

Washington Department of Fish and Wildlife and Oregon Department of Fish and Wildlife. 1994. Status report—Columbia River fish runs and fisheries, 1938-93. Joint Columbia River Management Staff, Battleground. Washington/Clackamas, Oregon.

Washington Department of Fisheries. 1992. Salmon 2000—Phase 2: Puget Sound, Washington coast, and integrated planning. Washington Department of Fisheries Technical Report. Olympia.

Washington Department of Fisheries, Washington Department of Wildlife, and Western Washington Treaty Indian Tribes. 1993. 1992 Washington State salmon and steelhead stock inventory. Olympia.

The Status of Salmon and Steelhead in Oregon

Kathryn Kostow

Abstract

Oregon's conservation laws and regulations provide direction to the Oregon Department of Fish and Wildlife for the conservation management of all wild fishes in Oregon. Conservation management includes systematic monitoring and status assessment of species, metapopulations, and local breeding populations, which is followed by management planning and implementation to ensure the conservation of wild fish into the future. Status reviews of the five species of Pacific salmon (*Oncorhynchus* spp.) indicate highly variable trends. The status of chinook salmon (*O. tshawytscha*) populations varies geographically, ranging from good along the mid- to north coast to depressed on the south coast, Columbia River, and Snake River. Coho salmon (*O. kisutch*) populations are depressed to nearly extinct in the Columbia River Basin while coastal populations are small and many are declining. Most coastal and inland steelhead trout (*O. mykiss*) populations are stable or slightly declining apparently in cyclic trends. Both chum (*O. keta*) and sockeye (*O. nerka*) salmon populations are depressed to nearly extinct throughout their range in Oregon. The anthropogenic causes for population declines include loss and degradation of freshwater habitats, historical overharvest, and impacts from interbreeding and competition due to hatchery programs. Natural cyclic trends in ocean productivity and droughts also affect populations.

Overview of Fish Conservation Management in Oregon

Anadromous salmonids (*Oncorhynchus* spp.) have supported the spirit, subsistence, and economy of Oregon's human population since early settlement of the region. The explorers Lewis and Clark documented the extensive Native American culture built around the Columbia River Basin salmon runs (DeVoto 1953). In 1867 European settlers began the exploitation of Columbia River fisheries. By 1890 canneries were operating throughout the lower Columbia River and on every major Oregon coastal bay. Salmon fishing gave rise to a million-dollar industry in Oregon by the 1890s (Hume 1893).

By the turn of the century, people recognized that salmon were not an inexhaustible resource and protection of freshwater habitats and regulation of harvest were needed to sustain the industry (Hume 1893). Throughout the first half of the 20th century, concern was voiced by resource managers about mining impacts, unscreened irrigation diversions, extensive water withdrawals, logging, migration blockages caused by dams, poor water quality, and overharvest (Ellis 1938, Dimick and Merryfield 1945, Moore and Suomela 1947, McKernan et al. 1950, Rivers 1991). By the 1920s, salmon populations in many Oregon rivers had already become extinct (Thompson and Haas 1960, Fulton 1968).

The impacts to Oregon fish populations have continued to the present despite the early recognition of the problems and despite legislation, regulation, and mitigation funding to protect the species. Snake River fall and spring chinook salmon (*O. tshawytscha*) became the first Oregon salmon species listed under the federal Endangered Species Act (ESA) in 1992 (United States [US] Department of Commerce 1992). The National Marine Fisheries Service (NMFS) is now reviewing all other Oregon anadromous salmonid species for potential ESA listings.

The Oregon Department of Fish and Wildlife (ODFW) has substantially increased its emphasis on wild fish conservation management over the last decade. Laws and regulations adopted by the Oregon State Legislature and the ODFW Commission (Oregon Revised Statute [ORS] 496.012; ORS 496.172, 176, and 182; Oregon Administrative Rules [OAR] Division 7, 100, and 500) provide policy direction, standards, and definitions, in addition to requiring systematic monitoring, status assessment, management planning, and implementation. Ultimately, when necessary, these regulations affect state listing and direct recovery for all wild fish species in the state. The laws also give the public a legal basis for holding ODFW accountable for the proper management and conservation of Oregon's wild fish.

In Oregon, wild fish are defined as those produced by natural spawning in indigenous populations. To the best of our knowledge, these populations have an unbroken lineage to pre-European settlement in the state, defined as the year 1800 in OAR 635-07-501(26). This definition permits natural changes in population boundaries including natural expansions and colonizations. Reintroduction of fish following an extinction, using artificial production or transport, creates a "natural" population. Broken lineages constitute extinctions.

Wild fish are managed for conservation purposes at the following three levels: (1) species, (2) metapopulations, and (3) breeding populations. Each taxonomic species contains one or more reproductively isolated gene pools or metapopulations termed "gene conservation groups." These groups allow ODFW to protect the natural genetic variation present across the range of the species. The Oregon Department of Fish and Wildlife's overriding responsibility in fish management is to prevent the serious depletion of fish species including the extinction of any gene conservation group (ORS 496.012, OAR 635-07-522 and 537(2)).

Gene conservation groups contain one or more local breeding populations. Breeding populations may be linked together by gene flow under present conditions or, where population fragmentation has occurred, under recent historical conditions. Populations are expected to be elastic with some local contractions, expansions, mergers, extinctions, and colonizations over time. Monitoring, risk assessment, and many management decisions occur at the population level. For example, hatchery programs are designed and monitored at the level of the affected wild breeding population. Similarly, risks due to habitat alteration, overharvest, or species introductions are evaluated with respect to their influence on the populations. Conservation actions such as a state listing or a recovery program would occur at the larger gene conservation group level.

Every 2 years, ODFW is required to report on both the status of all wild fish species and the progress of conservation policy implementation. The status of each species is determined using assessment standards. These standards address population size, stability, extinction, fragmentation and isolation, losses of genetic variation, and breeding failure and survival. In addition, ODFW biologists evaluate the impacts of hatchery programs and species introductions, and causes of unsatisfactory conditions including overharvest and habitat degradation.

Oregon state fishery biologists are currently developing a systematic, statewide monitoring program to provide information for the biennial status review. Meanwhile, the quality and kind of data available for status assessment varies across fish species and river basins. Previous information on fish and aquatic environments had not been collected for the purpose of conducting conservation status assessments and is often inadequate for this purpose. Long-term population data have been collected from fish counts at traps, weirs, or dams, from spawning ground surveys, or from sampling fishery catches.

Oregon has six anadromous salmonid species, each subdivided into multiple gene conservation groups and populations. These species occupy river basins along the Oregon coast and accessible portions of the Columbia River Basin (Fig. 1). Extensive reviews of these stocks are available (Kostow 1995); however, I focus my discussion on the five representative species: chinook, coho (*O. kisutch*), chum (*O. keta*), and sockeye (*O. nerka*) salmon, and steelhead (*O. mykiss*).

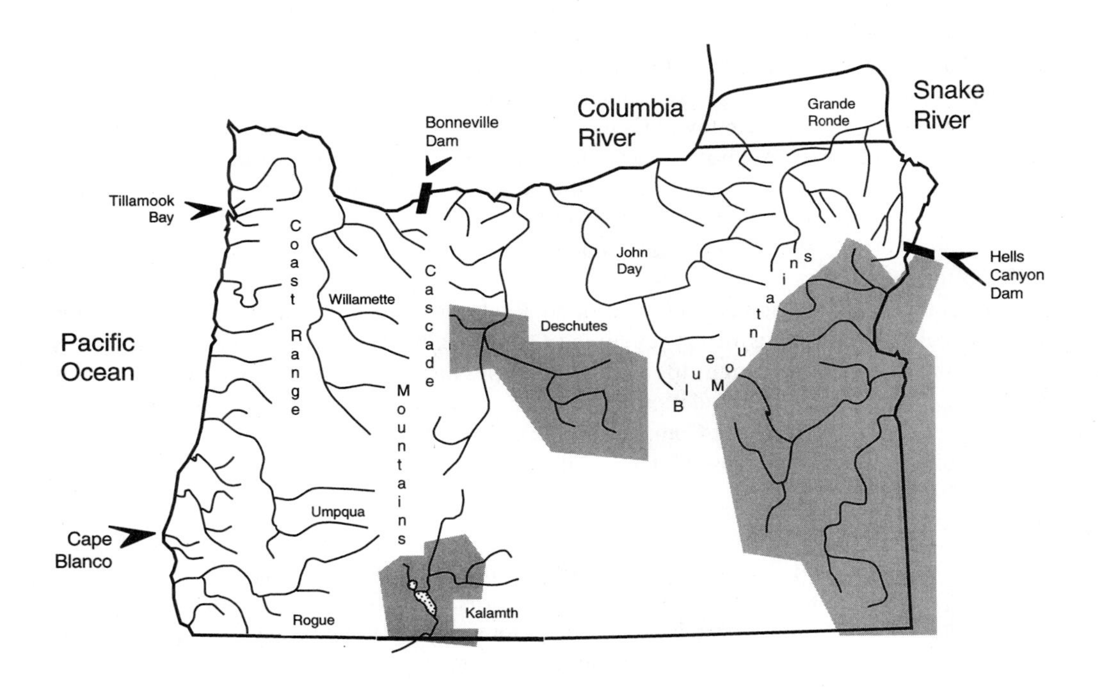

Figure 1. Oregon river basins currently and historically occupied by salmon and steelhead. Populations in shaded basins are extinct.

Status Reviews of Oregon Salmon and Steelhead

In the following sections I present information on long-term changes in salmon and steelhead populations. Within each basin or region I describe overall trends in abundance as well as those factors responsible for these changes. The most important contributions to change include habitat influences, fish released from hatcheries, overharvest in commercial and sports fisheries, and cycles in freshwater and ocean productivity.

Chinook Salmon

Chinook salmon were among the first salmon populations to become extinct in Oregon as a result of preferred harvest (Hume 1893) and extensive early habitat degradation in eastern Oregon (Thompson and Haas 1960). The range of chinook salmon in Oregon historically included all Columbia and Snake river sub-basins and all coastal streams below natural barriers (Fig. 1). Where adequate habitat was available, adult chinook entered rivers throughout the year with peaks during different months that formed separate breeding populations.

The species is completely extinct in the Snake River above the Hells Canyon dam complex, in the Deschutes River above the Pelton/Round Butte dam complex, in the Oregon portion of the upper Klamath basin, above dams in the Sandy, Willamette, Umpqua and Rogue rivers, and in the Umatilla and Walla Walla rivers. Populations have been lost from other basins including an early fall-run population in the lower Grande Ronde River, the fall-run population in the John Day River, and the spring-run population in the upper Hood River. Winter-run populations may have been lost in the Sandy River and in Tillamook Bay. Irrigation diversions, hydroelectric dams, and other habitat problems that decreased flows, caused blockages, and increased summer and fall water temperatures caused many of the early extinctions (Thompson and Haas 1960, Fulton 1968).

Currently ODFW recognizes 100 wild chinook salmon populations in 13 gene conservation groups (Table 1). The status of the remaining Oregon chinook salmon populations is highly varied owing in part to the various life-history strategies and habitat use by this species (Healey 1994). The Snake River fish populations are listed as both state threatened and federally threatened species (US Department of Commerce 1992). The lower Columbia River and south coast fall chinook populations are classified as a state sensitive species.

North to Mid-Coast Basins

Abundance trends of Oregon coastal chinook salmon can be estimated using historical catch records, spawning ground counts for fall-run populations, dam counts in a few locations, as well as incidental observations. With the exception of some dam counts, actual wild spawning escapements cannot be enumerated from the available data (Jacobs and Cooney 1993). Chinook population declines had already become a concern in Oregon by the 1880s (Hume 1893). Coastwide fall and spring chinook salmon production has declined by 30% to 50% from the estimated 300,000–600,000 fish present in 1900 (Nickelson et al. 1992).

Most of the fall-run populations in the three gene conservation groups (i.e., Nehalem, Mid-Oregon Coast, and Coos, Coquille, and Sixes basins) that occupy coastal basins draining the

Table 1. Current and extinct gene conservation groups and populations of Oregon salmon and steelhead.

Species	Gene conservation group	Current number of populations[a]	Number of extinct populations[a]
Chinook	Nehalem	1 sm, 3 f	
	North to mid-coast	7 sp, 20 f	
	Umpqua	2 sp, 3 f	
	Mid- to south coast	2 sp, 7 f	
	South coast/Rogue	1 sp, 10 f	
	Lower Columbia fall	19 f	
	Willamette spring	5 sp	4 sp
	Sandy fall	2 f/w	1 w?
	Mid-Columbia spring	6 sp	3 sp
	Mid-Columbia (Deschutes) fall	1 f	2 f
	Lower Snake fall	3 f	
	Lower Snake spring/summer	6 sp, 2 sm	
	Upper Snake fall		Multiple
	Upper Snake spring/summer		Multiple
Coho	North to mid-coast	56	
	Umpqua	4	
	Mid- to south coast	21	
	South coast/Rogue	14	
	Lower Columbia	28	
	Willamette/Clatskanie	4	
	Mid-Columbia		Multiple
	Snake		Multiple
Coastal steelhead[b]	North to mid-coast/Umpqua	64 w, 2 sm	
	South coast/lower Rogue	10 w	
	Upper Rogue	3 w, 3 sm	
	Lower Columbia	24 w, 1 sm	
	Willamette	9 w	
	Klamath		1 sm
Inland steelhead[b]	Mid-Columbia	4 w, 7 sm	1 sm
	South Fork John Day	1 sm	
	Lower Snake	6 sm	
	Upper Snake		Multiple
Chum	(Not defined) Coast	32	
	(Not defined) Columbia	23	
Sockeye[b]	Deschutes/Suttle Lake	1	
	Grande Ronde/Wallowa Lake		1

[a]sm = summer-run, f = fall-run, w = winter-run, sp = spring-run.
[b]Additional resident populations also present in group.

Coast Range from Cape Blanco to the Columbia River (Fig. 1) appear to have reached their lowest abundance levels between 1930 and 1960. Generally, within these three conservation groups, fall chinook abundance increased along the mid- to north Oregon coast (Fig. 2). While actual fall chinook spawning escapements cannot be measured, current spawning populations vary from 200–500 individuals in the smaller basins to 1,000–2,000 individuals in the larger basins. The spring-run populations in these groups are not well monitored. Incidental observations of current populations indicate that all of the populations are small and have remained depressed since declines in the 1950s.

Chinook salmon were historically very abundant in the Umpqua basin and present year-round based on catch records. Population levels fell to such low numbers by the 1940s that the basin was not included in the fall chinook salmon spawning ground monitoring program established for the rest of the mid- and north Oregon coast. Since 1947, counts of the North Umpqua fall and spring chinook populations have been made at Winchester Dam. The fall chinook population has averaged only 148 fish over the monitoring period. The wild spring chinook population increased to over 5,000 fish by the mid 1960s (Fig. 3A). Incidental observations indicate an increase in both fall-run populations in the South Umpqua (≅2,500 fish) and the Smith River tributary.

Most of Oregon's coastal basins north of Cape Blanco drain the Coast Range and lie completely within the temperate coniferous rain forest belt. These small basins drain into large estuaries and were historically dominated by complex old growth forests, swamps, and marsh lands. Chinook salmon populations have been impacted by many habitat alterations in the basins, which have affected the abundance, stability, and accessibility of mainstem gravel bars used for spawning. In addition, alterations have eliminated large holding pools required by spring chinook salmon and have decreased lower basin habitat complexity, water quality, and estuary productivity required by rearing juveniles. Most of the upland impacts result from past logging practices while lower basins and estuaries have been impacted by agricultural practices and urbanization. Over the last few decades, improvements in fall-run populations have been in part due to improved logging practices, which have decreased impacts to mainstem spawning gravels. The relatively poor status of spring-run populations is due to the continued lack of deep holding pools and decreased flows and water quality during the summer.

The Umpqua River cuts through the Coast Range and drains the Cascade Mountains. The upper basin includes drier grassland and deciduous forest ecosystems as well as conifer forests. Habitat impacts are similar to those in other coastal rivers. Generally, the addition of extensive water withdrawals and development has affected the upper basin, and river channelization and a naturalized smallmouth bass population in the lower mainstem river have affected juvenile salmon survival.

Chinook hatchery programs along the Oregon coast are concentrated in the Tillamook and Coos bay areas, and in the Salmon, Alsea, Coquille, and Elk rivers. The spring chinook hatchery program in Tillamook Bay and the fall chinook programs in the Salmon and Elk rivers may be impacting the wild populations in those basins, based on estimates of the number of hatchery fish straying to wild spawning areas. Hatchery programs in the Umpqua basin have had similar effects. Other populations in this area are not affected by current hatchery programs.

Oregon coastal fall chinook are harvested in the North Pacific Ocean off southeast Alaska and British Columbia and in freshwater recreational fisheries. Since 1982, adult equivalent brood exploitation rates for coastal fall chinook have ranged from 52% to 68% based on coded-wire

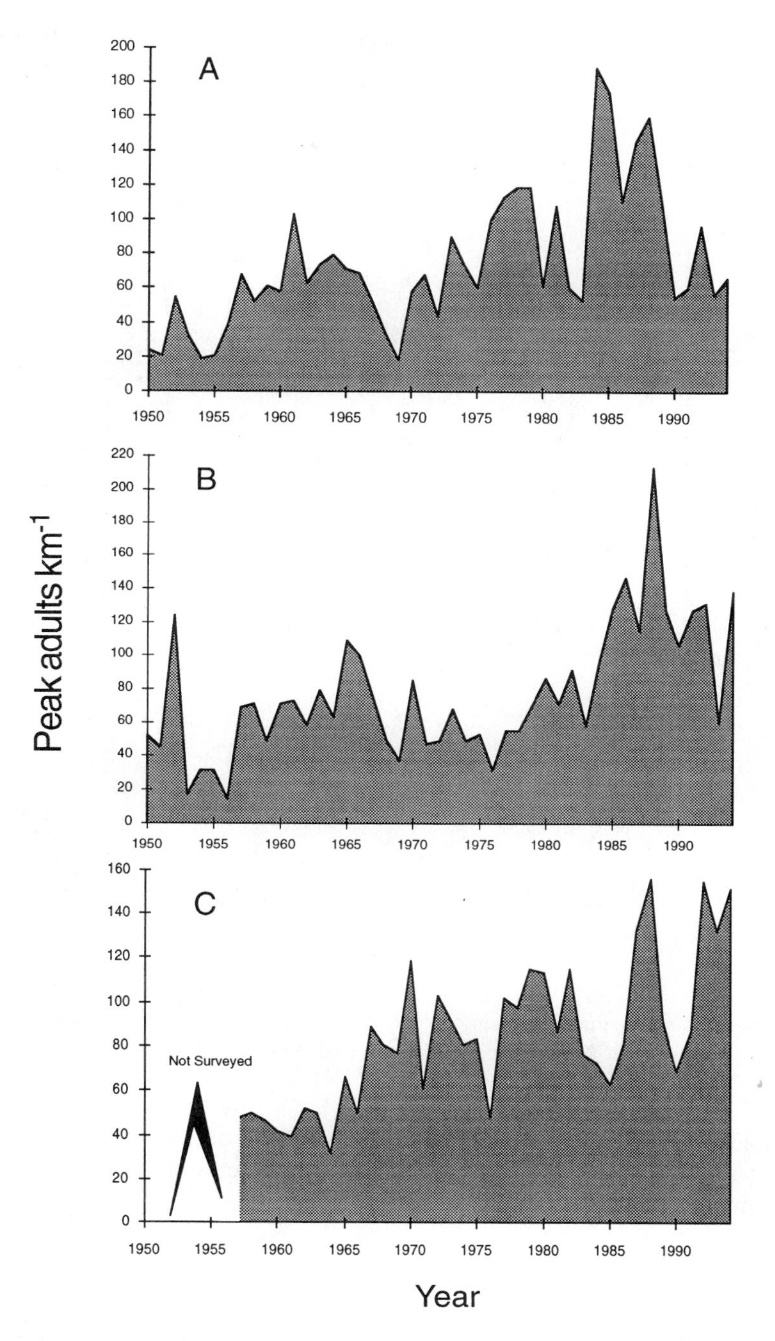

Figure 2. Average peak spawning ground counts of wild fall chinook salmon per kilometer in index survey areas in Oregon coastal basins north of Cape Blanco, 1950–94: (A) Nehalem group, (B) mid-Oregon coast group, and (C) Coos-Coquille group.

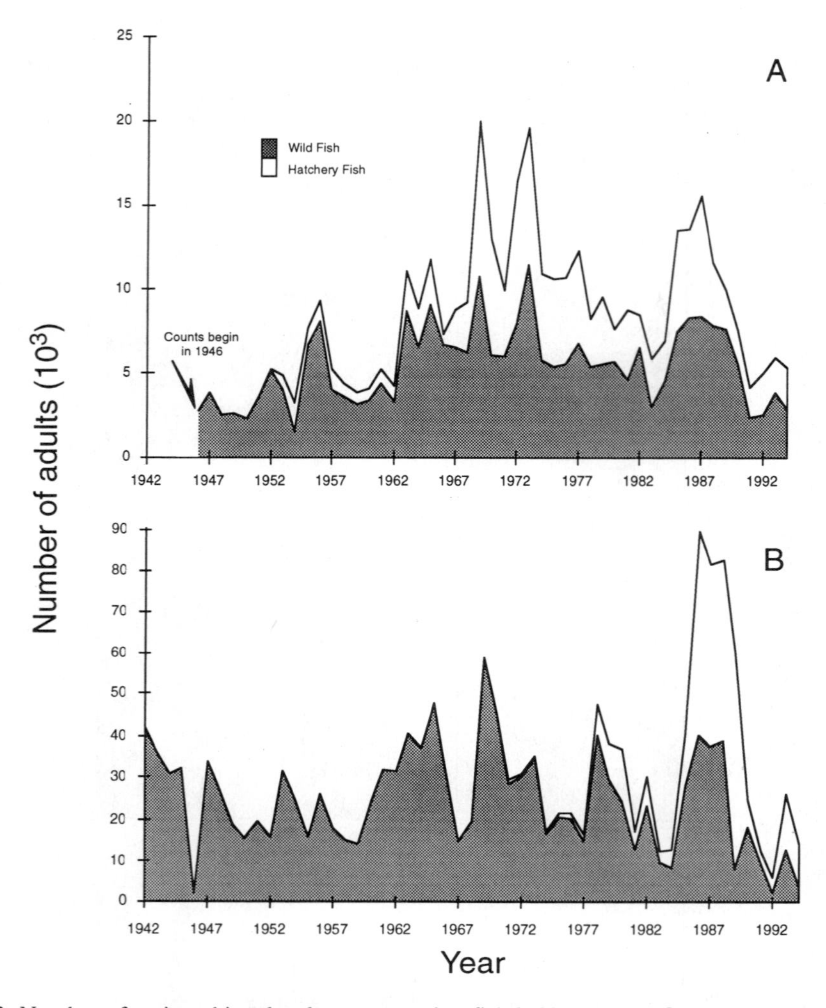

Figure 3. Number of spring chinook salmon counted at fish ladders at two Oregon coastal basin dams, 1942–94: (A) Winchester Dam, North Umpqua River; and (B) Gold Ray Dam, Rogue River.

tag returns from a coastal hatchery indicator stock. The intent of ocean fisheries management over this period has been to increase wild spawning escapements in coastal streams. Harvest rates on the spring chinook populations are thought to be lower.

Rogue River and South Coast Basins

The status of salmon populations in the gene conservation group that includes the Rogue River (Fig. 1) and small south coast basins is highly variable. Some populations in the upper Rogue basin contain >10,000 fish, while some populations in the lower Rogue basin and coastal streams appear to have dropped to <50 fish in the early 1990s.

Historical catch records indicate that chinook runs into the Rogue basin were very large. In-river catches between 1890 and 1920 reached >70,000 fish in some years (Nicholas and Hankin 1989). Since 1942, the spring chinook population in the upper Rogue River has been monitored by counts at Gold Ray Dam (Fig. 3b), varying from 10,000 to 40,000 fish annually. The wild population dropped to <10,000 fish in consecutive years during the early 1990s for the first time in monitoring history, but it is unclear whether this indicates a long-term decline. Since the late 1970s, the upper basin fall chinook populations have been monitored using spawning ground counts. During this time populations appear to have remained stable, with 2,000–4,000 fish each.

In comparison, populations in the lower Rogue and south coast basins have declined in the last two decades. The largest populations in this area are in the Chetco and Winchuck rivers (Fig. 4). These populations have declined over the monitoring period. Populations in some of the other basins appear to contain <100 wild fish each.

The river basins south of Cape Blanco drain the Siskiyou Mountains in the Klamath geological province. The exception is the Rogue River, which cuts through these mountains to the southern Cascade Mountains. The geology of the area is very different from the rest of the Oregon coast. Streams are characterized by steep gradients, high levels of gravels, and very small estuaries. The climate is also different, with a higher proportion of the annual precipitation in the winter, resulting in very flashy flows within the drainage. The upper Rogue River is drier than the coast and includes deciduous forest and grassland as well as coniferous forest ecosystems.

Habitat impacts in the upper Rogue basin began with a gold rush in the 1850s. Hydraulic dredge mining occurred in the larger and intermediate tributaries used by chinook salmon. This resulted in disturbed spawning gravels, serious erosion, and high sedimentation levels. Gold mining in the basin declined after 1900. Irrigated agricultural development occurred early in the upper Rogue basin with the construction of hundreds of small irrigation diversions and several large dams, many of which were originally unladdered and unscreened (Rivers 1991). As a

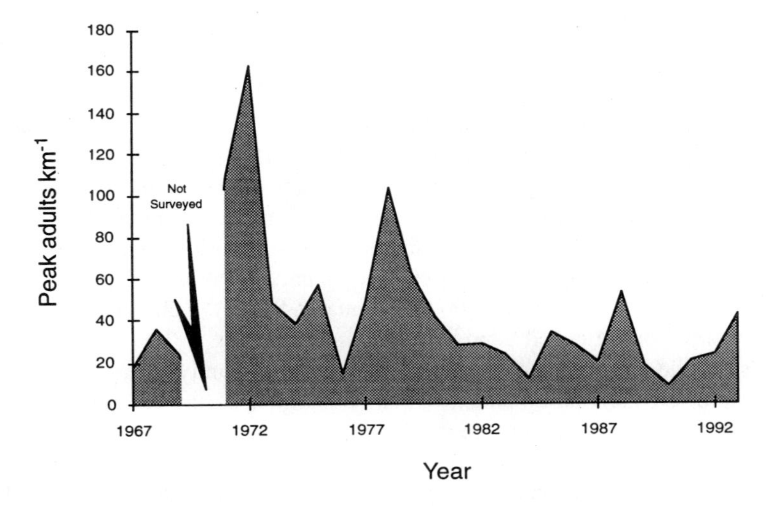

Figure 4. Average peak spawning ground counts of wild fall chinook salmon per kilometer in index survey areas in two south Oregon coastal basins, Winchuck and Chetco rivers, 1967–93.

result of construction of the Lost Creek Dam on the mainstem Rogue River in 1977, ~33% of the spring chinook spawning area in the basin was made inaccessible. Despite improvement for fish passage at most of these facilities, water withdrawals continue to cause summer and fall flow and temperature problems.

Intense logging activity and road construction in the other south coast basins during the 1950s and 1960s seriously destabilized stream banks (Nawa et al. 1989). These activities contributed to high sediment and bed load materials. When combined with flashy flows the results include destabilized gravel bars, decreased number of pools, and increased sedimentation. High bed load movement during winter flows buries chinook eggs and scours redds. The loss of riparian vegetation has caused summer water temperatures to increase. The mouths of some of the smaller basins often become closed by sand bars during low summer flows because of channelization and sedimentation in the rivers. Summer-migrating chinook juveniles become trapped in these basins and are subsequently exposed to water temperatures that may exceed tolerance levels.

The spring chinook salmon hatchery program below Lost Creek Dam in the upper Rogue basin is one of the largest on the Oregon coast. The proportion of hatchery fish on the natural spawning grounds currently averages >50%, which exceeds Oregon standards for the program. Several small fall chinook salmon hatchery programs in coastal basins, implemented with the intent of rehabilitating the wild populations, have actually impacted wild populations by decreasing already critical effective population sizes (see Ryman and Laikre 1991) and increasing straying.

Harvest rates on Oregon chinook salmon populations south of Cape Blanco are unknown. These populations have a different ocean distribution compared with other Oregon chinook. Generally, these fish are located off the coast of Oregon and northern California. Ocean harvest rates are probably similar to those on fish from the Klamath River in northern California, which have declined from 75% to <5% on 4-year-old fish since 1979. Additional recreational harvest occurs in the Rogue River.

Lower Columbia River Basin

Fall chinook salmon were historically very abundant in the lower Columbia River Basin, based on early catch records and observations. Presently, the populations are inadequately monitored because of large numbers of natural-spawning, unmarked hatchery fish that confound populations estimates. Wild fish population sizes are much smaller than those present in the early 1900s and appear to be depressed relative to population sizes observed in the 1950s and 1960s. The abundance of Sandy River fall chinook in the early 1900s has been estimated at ~10,000 fish. Since 1984, Sandy River fish populations have been monitored using spawning ground counts, with estimates of 500 to 2,000 fish annually. Some fall chinook populations may be extinct in the lower Columbia and Sandy river basins.

Abundances of wild spring chinook in the Willamette and Sandy rivers are also unknown owing to large numbers of unmarked hatchery fish. Willamette spring chinook are well monitored via dam counts at several locations (e.g., Fig. 5), indicating the presence 20,000–100,000 fish. However, hatchery traps and coded-wire tag returns that provide estimates of the number of hatchery fish in the basin indicate that the proportion of hatchery fish far exceeds wild fish and that the wild populations are very small.

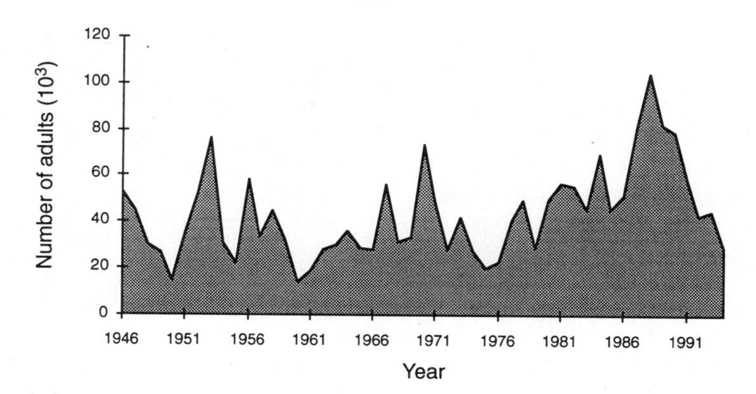

Figure 5. Number of spring chinook salmon, hatchery and wild fish combined, counted at Willamette Falls fish ladder, Willamette River, 1946–94.

Lower Columbia and Willamette river basin tributaries drain the northern Coast Range and Cascade Mountains. Spring chinook salmon were historically present only in the tributaries that drain the Cascades. Spring chinook populations are extinct in several of the Willamette tributaries and distribution is reduced in others owing to blockages caused by six dams. Both spring and fall chinook salmon populations have been impacted by historical logging practices similar to those on the north Oregon coast and by extensive urban development associated with the Portland metropolitan area. Stream channelization, loss of riparian areas, blockages caused by culverts, dams, and diversions, and poor water quality have decreased both spawning success and juvenile survival. The Sandy River fall chinook salmon populations have primarily been impacted by dams on the mainstem and on a major lower basin tributary, which have blocked access to historical habitat. In contrast, spring chinook salmon habitats in the upper Sandy and Clackamas basins are in relatively good condition.

Fall chinook hatchery programs in the lower Columbia River Basin are extensive. In 1992 and 1993, 29–40 million chinook smolts and fry, including fish from locally founded broodstocks and from three introduced stocks, were released from Oregon hatcheries into the lower Columbia River Basin. Additional fish were released from Washington hatcheries. The hatchery programs in the Willamette River released 6–8 million spring chinook smolts in 1992 and 1993. A much smaller number of smolts were also released into the Sandy basin. The hatchery fish are mostly unmarked; however, limited mark-recapture studies indicate that hatchery straying and probable impacts to the wild populations are high. Hatchery fall chinook have not been released into the Sandy basin since 1977; thus, the populations in this basin are probably not affected by current hatchery programs.

The fall chinook hatchery programs in the lower Columbia River Basin were developed to mitigate habitat losses with production. The intention is to provide fish for harvest in the ocean and in the lower Columbia River. Since 1982, ODFW has monitored harvest rates for three different fall chinook hatchery indicator stocks in the lower river. Adult equivalent brood year harvest rates ranged from 48% to 93% over the monitoring period (Pacific Salmon Commission 1993). Because wild populations are harvested along with the hatchery fish, both probably experience similar harvest rates. Harvest rates on the Sandy River fish are unknown, but may be lower.

Willamette spring chinook salmon have a unique ocean distribution compared with other Columbia River Basin spring chinook. These fish are harvested in ocean fisheries off the coast of southeast Alaska and British Columbia. An extensive commercial and recreational harvest also occurs in the lower Columbia and Willamette rivers. Since 1982, estimated adult equivalent brood-year harvest rates on the hatchery stocks have ranged from 66% to 88% (Pacific Salmon Commission 1993). Wild and hatchery fish are harvested together.

Mid-Columbia River and Snake River Basins

Chinook populations in Oregon's Columbia River Basin tributaries between the Cascade Mountains and the confluence of the Snake River are now restricted to the Deschutes and John Day basins. Total abundance in the area has declined as a result of many population extinctions. The remaining populations have been monitored by weir or spawning ground counts. The Deschutes basin includes a population of 500 to 1,000 spring chinook salmon in the Warm Springs River and a second population of ~100–300 below Pelton/Round Butte dams. The Warm Springs population is monitored at a weir (Fig. 6A) and included about 1,500 wild fish until the 1990s when it declined to <1,000 fish. The four John Day spring chinook populations have been stable or increasing (combined in Fig. 6B). The North Fork John Day population had over 1,500 fish in the early 1990s while the others had about 200 to 500 fish each. The mid-Columbia fall chinook gene conservation group includes only one remaining population in the mainstem Deschutes River. The population ranged from ~4,000 to ~8,000 fish from 1977 to 93 and then increased to >15,000 fish in 1994 (Fig. 6C).

The abundances of fall, spring, and summer chinook in Oregon tributaries of the Snake River Basin have decreased because of extensive population extinctions. Remaining populations are restricted to the Imnaha, Grande Ronde, and mainstem Snake rivers below Hells Canyon Dam. Fall chinook salmon abundance has been monitored since the closure of the first lower river dam in 1964 (Fig. 7A). Abundance has declined precipitously during this period. Wild fish run sizes during the 1990s ranged from <100 fish to ~700 fish. Since the early 1950s, Grande Ronde spring and Imnaha summer chinook salmon have been monitored by spawning ground counts. All populations have declined (Fig. 7B, C). Most of the fish in the Imnaha River are concentrated into a single population while the Grande Ronde River fish have become fragmented into six separate populations. The Imnaha River contained 200–400 fish in the early 1990s while most of the Grande Ronde River populations had <100 fish each. All of these populations have continued to decline through the mid-1990s.

Chinook salmon habitats in the mid-Columbia and Snake river basins range from those that have been severely impacted to others in good condition. Chinook salmon are extinct in the most impacted basins such as the Umatilla. In the Deschutes and John Day basins, habitat has been affected by agricultural practices, including grazing and irrigation withdrawals that decrease water flow and quality. A dam complex on the Deschutes River greatly reduced the available habitat in that basin by blocking access to historical spawning and rearing habitat. Mining and logging impacts have led to habitat loss by increasing sedimentation, changing water quality, and altering temperature regimes. Over the past decade, the remaining habitats have generally improved because of extensive riparian fencing projects implemented as mitigation for cattle grazing impacts. Chinook salmon habitat in the Imnaha River is in good condition while some tributaries and the mainstem Grande Ronde River have been affected by river channelization, water withdrawals, grazing, and logging.

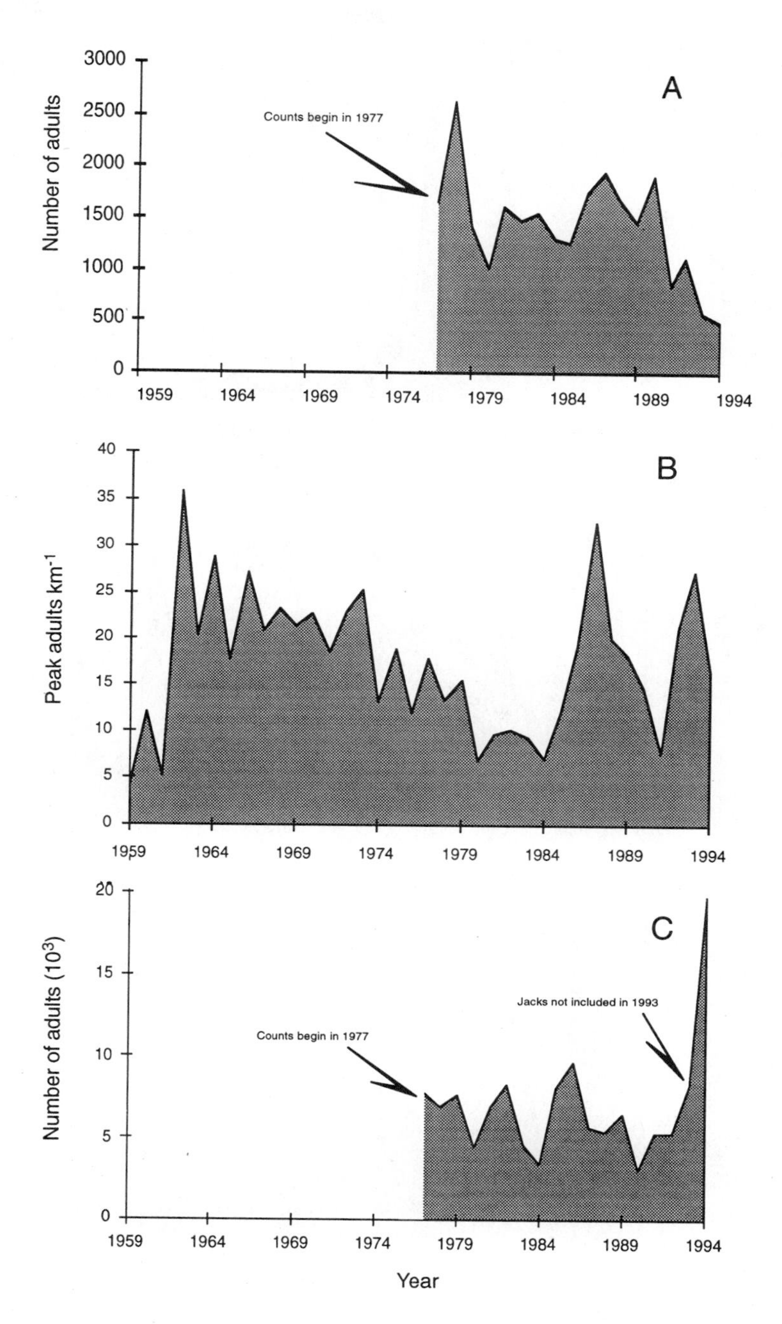

Figure 6. Estimates of wild chinook salmon abundance in mid-Columbia River basins, 1959–94: (A) spring chinook counts at Warm Springs weir, Warm Springs River, Deschutes basin, (B) average peak spawning ground counts of spring chinook per kilometer in the John Day basin, and (C) estimated fall chinook abundance calculated from spawning ground counts in the Deschutes River.

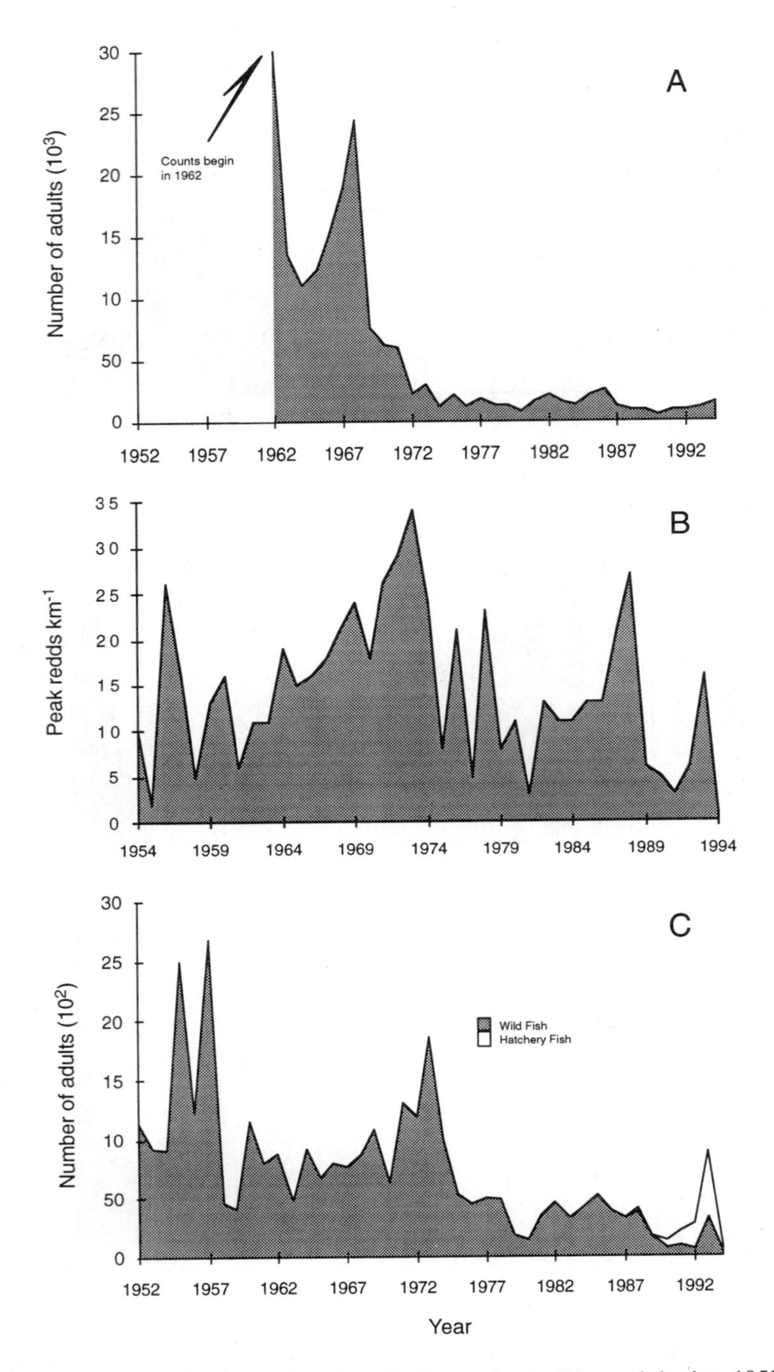

Figure 7. Estimates of chinook salmon abundance in Oregon Snake River sub-basins, 1952–94: (A) fall chinook counted at fish ladders at the uppermost mainstem Snake River Dam, (B) average peak spawning ground counts of spring chinook per kilometer in the Grande Ronde sub-basin, and (C) estimated summer chinook abundance calculated from spawning ground counts in the Imnaha sub-basin.

A major impact affecting all of these populations has been the construction of hydropower dams on the mainstem Columbia and Snake rivers. The Deschutes, John Day, and Snake River populations cross two, three, and eight mainstem dams, respectively. The impacts to spring and summer chinook salmon include physical injury to juveniles when passing through the dams, increased migration time, increased predation in reservoirs, stress, and decreased condition, all of which increase mortality as these fish migrate to salt water. Adults also experience injury and stress during upstream migration. The impacts to fall chinook include the same passage mortalities plus extensive loss of spawning and rearing habitat due to blockage and inundation.

Oregon operates spring and summer chinook salmon hatchery programs in the Deschutes, Imnaha, and Grande Ronde river basins. Although the hatchery programs are small, they potentially impact some wild populations through excessive hatchery strays. The broodstock used in the Grande Ronde River was developed from outside the basin and thus presents a concern. A sorting weir on the Warm Springs River protects most of the wild fish in the Deschutes basin from hatchery strays. The fall chinook salmon reintroduction program in the Umatilla River, below the confluence of the Snake River, produced strays into the Snake River in the late 1980s and early 1990s. All of the Umatilla strays are now marked and can be removed at ladders in the dams if they enter the Snake River.

Since the 1950s most harvest on Oregon's mid-Columbia and Snake River spring chinook salmon occurred in mainstem Columbia River fisheries, while fall chinook have been taken in ocean fisheries. Since the mid-1970s, spring chinook salmon harvest rates have been low, and all harvest rates have decreased substantially since 1990 (Pacific Salmon Commission 1993).

Coho Salmon

The historical range of coho salmon in Oregon included Columbia River tributaries to the Malheur sub-basin in the Snake River and all coastal basins. The species is now extinct east of the Cascade Mountains. Habitat degradation that decreased fall flows and increased water temperatures led to extinctions in mid-Columbia River sub-basins by the early 1960s (Thompson and Haas 1960). Mixed-stock harvest in the ocean and lower Columbia along with the construction of eight mainstem hydropower dams on the Columbia and Snake rivers completed the extinctions in the Snake in the 1980s. A reintroduction program is underway in the Umatilla basin. Coho salmon are listed as a state sensitive species throughout Oregon, and populations on the coast have been proposed for federal threatened status (US Department of Commerce 1995a). Currently, ODFW recognizes 128 coho populations. Six gene conservation groups have been described to date (Table 1).

North to Mid-Coast Basins

Historical abundance trends in the three coho salmon gene conservation groups north of Cape Blanco can be estimated from historical catch data and from spawning ground surveys conducted since 1950. The pre-harvest abundance of coho in these combined groups was estimated at 1.7 million adults in 1900, but at only 70,000 adults in the early 1990s (Fig. 8).

Pre-harvest abundance trend indices, based on peak spawning ground counts in standard index streams conducted since 1950 and estimated harvest rates, are provided for each gene

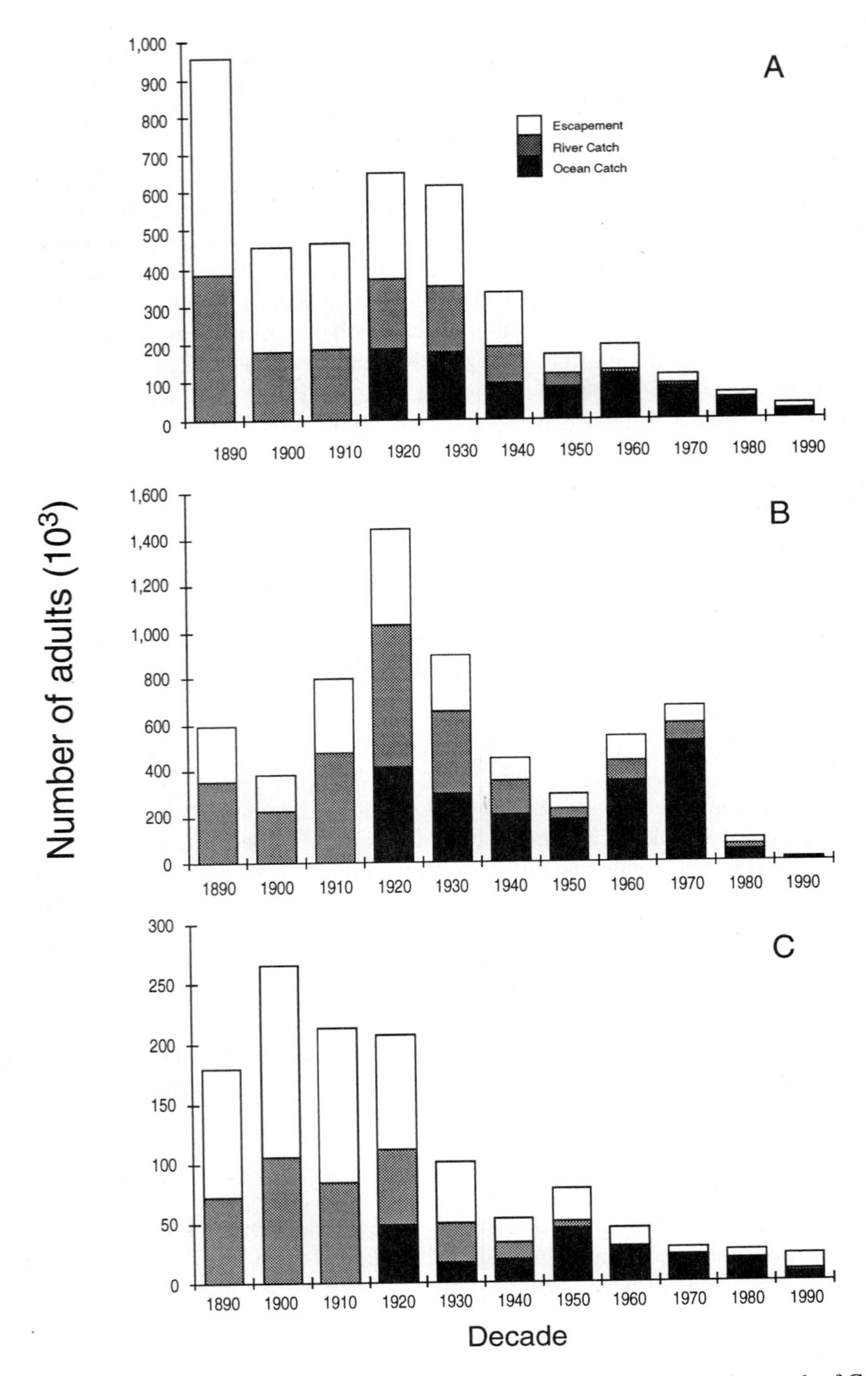

Figure 8. Estimated historical abundance of coho salmon in Oregon coastal basins north of Cape Blanco, including harvested fish, by decade 1890–1990: (A) mid- to north coast basins, (B) Umpqua basin, and (C) Coos, Coquille, and lake basins.

conservation group (Fig. 9). According to these indices, all of the groups declined in abundance in the mid-1970s. The pre-harvest index for the mid- to north coast group continued to decline an average of 7% yr^{-1} between 1978 and 1994 while the Umpqua showed no trend and the Coos and Coquille increased slightly. The decline in productivity is also demonstrated by the trend in the ratio of offspring per spawner in the mid- to north coast groups (Fig. 10). This index also declined by an average of 7% yr^{-1} for the 1975 through 1991 broods, indicating a decrease in survival rates during this period. Again, the other groups showed no trend with this index.

Actual estimates of coho salmon spawning escapements in coastal streams cannot be made from the peak spawning ground counts in the standard index areas because the spawner densities in the index areas are much higher than in other spawning habitats (Jacobs and Cooney 1993). A stratified random sampling of spawning areas was initiated in 1990 to correct for this error. Using this improved sampling design, we estimated the number of coho for several aggregations of populations to range from 450 fish in the Tillamook Bay area in 1990 to 23,340 fish in the Coos Bay area in 1993; however, confidence intervals are wide in all cases (Fig. 11). An additional 3,300 coho were estimated to spawn in three large coastal lake systems located near the mouth of the Umpqua River during this period.

Historically, coho salmon habitat in coastal basins was dominated by old growth coniferous forests that provided shading, controlled flows, stable river banks, and ample large woody debris to river channels. In these forests, activities that have decreased riparian habitat complexity (e.g., loss of deep rearing pools and complex channels) have also decreased rearing capacity for coho juveniles. The major impact to coho in the upland areas of the Coast Range has been due to timber harvest, while lowland area impacts are attributed to agricultural and urban development. The upper Umpqua basin has also been affected by water withdrawals and urban development.

Since the late 1970s, coho abundance levels along the Oregon coast appear to be more affected by low productivity in the ocean than by limiting factors in the freshwater environment. The effect of low ocean productivity on current coho production is indicated by significant declines over the same time period in the survival of hatchery smolts from coastal facilities (Oregon Department of Fish and Wildlife 1995). Hatchery smolts migrate directly to the ocean after release and their survival is not affected by the freshwater environment.

Releases of hatchery coho salmon along the mid- to north Oregon coast began to increase during the mid-1960s. Competitive interactions in both freshwater and marine environments, in addition to interbreeding, are suspected to have impacted wild coho (McGie 1980, Nickelson et al. 1986). Hatchery releases in the 1990s have included primarily acclimated smolts, which decrease competition impacts and, although ~5 million fish are still released annually in this area, scale samples collected from weirs and natural spawning areas indicate few hatchery strays into most wild populations. However, the wild populations in the north fork Nehalem, Trask, Salmon, Siletz and Umpqua river basins may still be affected by strays since hatchery facilities are located adjacent to wild spawning grounds.

Hatchery coho salmon on the Oregon coast are produced for ocean fisheries, which also harvest wild coho. Harvest rates on wild populations have been calculated since the 1950s (Fig. 9A-C). Harvest rates reached a high of 87% in the 1970s but declined following the adoption of a management plan for Oregon coho (Oregon Department of Fish and Wildlife 1982). Harvest rates in the 1990s declined further to only incidental harvests in 1994.

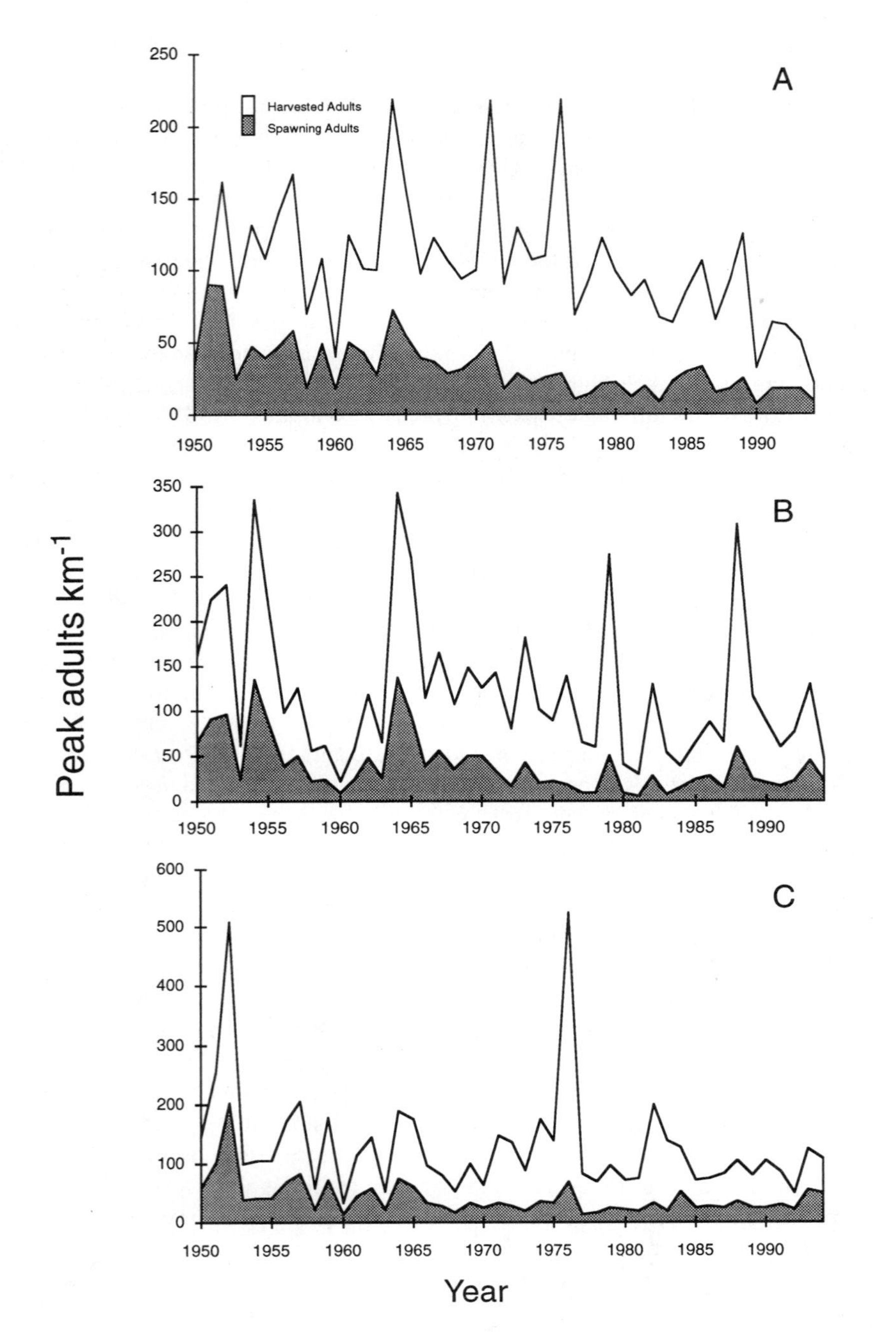

Figure 9. Average peak spawning ground counts of coho salmon per kilometer in index survey areas and estimated total abundance based on harvest rates in Oregon coastal basins north of Cape Blanco, 1950–94: (A) mid- to north coast basins where total abundance declined an average of 7% per year from 1978 to 1994 ($p < 0.005$; linear regression), (B) Umpqua basin where there is no significant trend, and (C) Coos and Coquille basins where spawning adult abundance increased an average of 4% per year from 1978 to 1994 ($p < 0.05$; linear regression).

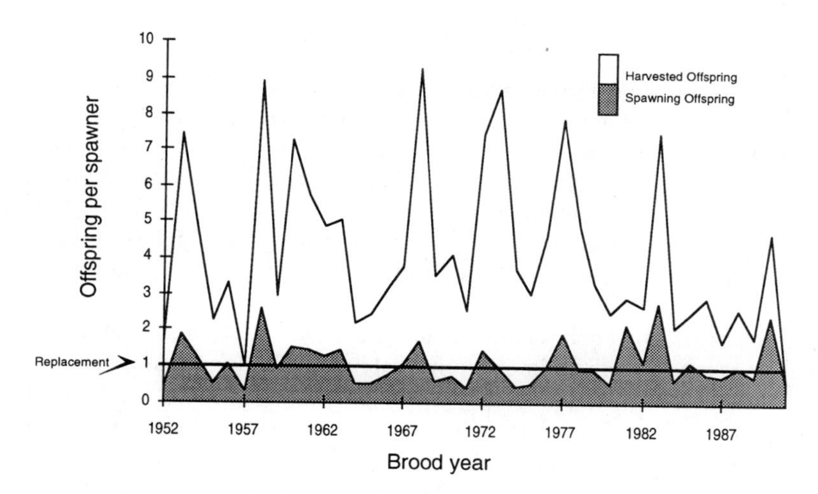

Figure 10. Number of offspring, including both spawning offspring and harvested offspring, produced per spawning adult coho salmon by brood year in mid- to north Oregon coast basins, 1952–91 broods. Total number of offspring per spawner decreased by an average of 7% per year for the 1975 through 1991 broods ($p = 0.07$; linear regression). The Umpqua, Coos and Coquille basins (not shown) showed no trend in the number of offspring produced per spawning adult.

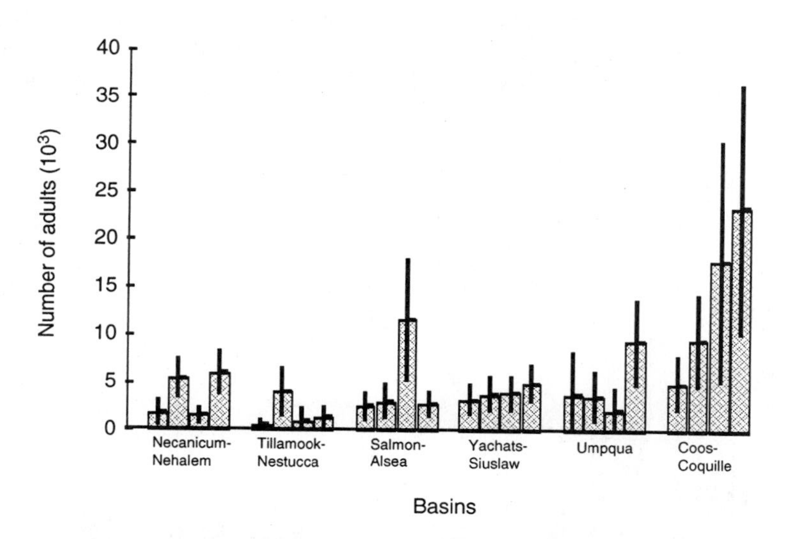

Figure 11. Estimated number of adult coho salmon in major Oregon coastal basins north of Cape Blanco for 1990–93 with 95% confidence intervals. Bars indicate individual years (i.e., 1990, 1991, etc.). Estimates are calculated from spawning ground survey data collected in a random sampling of all available spawning habitat.

Rogue River and South Coast Basins

Coho salmon are rare in the small coastal basins south of Cape Blanco. Most of the coho in this group are in the Rogue River. Abundance information for Rogue River populations is available from historical catch records and seining in the lower river. Pre-harvest abundance in the basin was estimated at 750,000 adults in 1900. By the 1920s, only ~6,000 adults remained (Fig. 12). This abundance level has continued, on the average, into the 1990s. The largest remaining population is in the Illinois River, a tributary of the lower Rogue River. The current monitoring method is not precise, but indicates that 5,000–6,000 adults are currently produced in the Rogue River basin.

Habitat impacts affecting coho salmon in the south coast streams are similar to those elsewhere on the Oregon coast. Coho salmon rearing habitats in the Rogue basin have been affected by extensive water withdrawals, particularly in the Illinois basin, and by urban development in the upper basin. Lost Creek Dam on the mainstem Rogue River blocked access to historical coho habitats in the upper basin. The only Oregon coho hatchery program south of Cape Blanco is located below this dam and does not affect most of the wild coho in this group. Harvest rates on this group during the past decade have been somewhat lower than on other coastal coho.

Lower Columbia River Basin

Coho salmon in the Columbia River Basin have decreased to <5% of the estimate for pre-harvest abundance (600,000 to 1.4 million fish) in the early 1900s (Fig. 13). This decline from historical levels has been due both to complete extinctions of some populations and to declines of the remaining populations. Since the 1950s, abundance trends have been monitored by use of dam counts in the Clackamas basin and by spawning ground counts in several lower Columbia River Basin tributaries. The Clackamas population has shown no trend over the monitoring period, ranging from 500 to >3,000 spawning fish (Fig. 14), although the 1993–94 count of 190

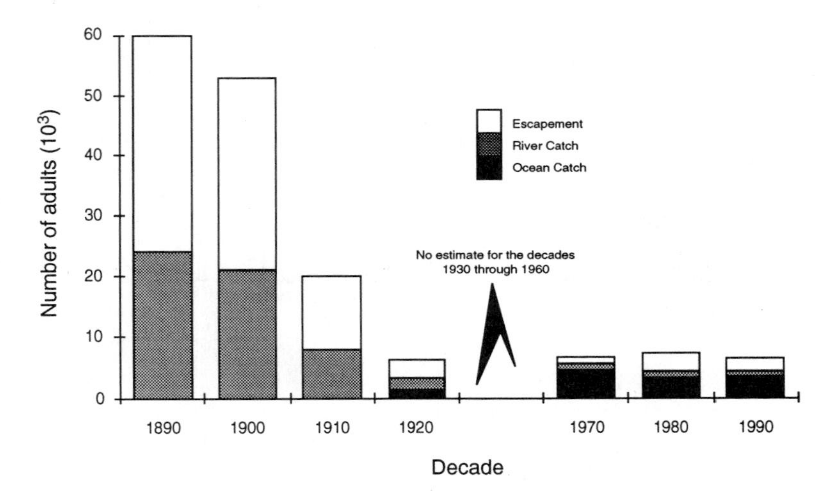

Figure 12. Estimated historical abundance of coho salmon in Oregon coastal streams south of Cape Blanco, including harvested fish, by decade 1900–20 and 1970–90.

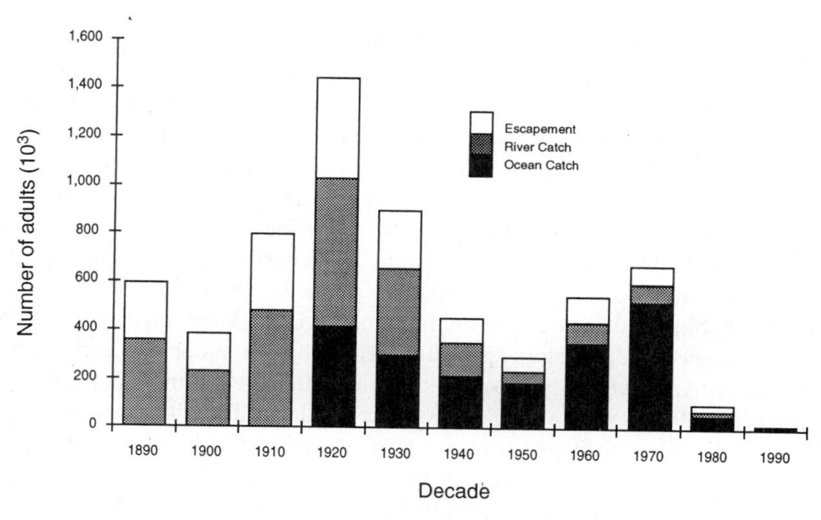

Figure 13. Estimated historical abundance of coho salmon in the Columbia Basin, including harvested fish, by decade 1890–1990.

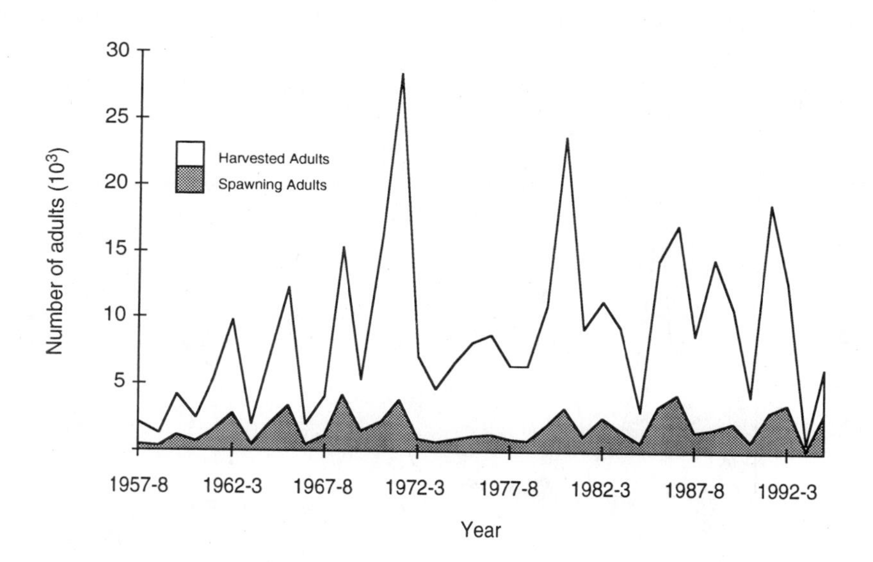

Figure 14. Number of coho salmon, hatchery and wild fish combined, counted at North Fork Dam fish ladder, with number of harvested adults estimated from harvest rates, Clackamas River, Columbia Basin, 1957–94.

fish was a record low number. However, lower Columbia River Basin spawning ground counts indicate a precipitous decline in coho abundance over the monitoring period (Fig. 15). Since 1978, the pre-harvest abundance index, which is calculated from these counts and estimated harvest rates, has declined an average 21% yr^{-1}, while the escapement index has declined an average of 14% yr^{-1}. Current population sizes in the small Columbia tributaries cannot be accurately estimated, but only a few hundred fish are scattered across the 25 tributaries that contain spawning habitat. Coho salmon have also have been counted at dams in the Sandy and Hood rivers in recent years: The Sandy has had runs of 500–1,500 fish over the past decade, although the 1993–94 count was only 220 fish; the Hood River population has included 30–100 fish in the 1990s.

Coho salmon habitat in the lower Columbia River Basin has been impacted by a combination of factors associated with logging, agricultural and urban development, and hatchery construction in the basin. Approximately 100 river km of habitat below Bonneville Dam historically used by the species has been lost as a result of blockages by dams, diversion structures, and hatchery weirs. Decreased water quality, river channelization, and wetland draining and filling have decreased coho rearing capacity in portions of the lower Sandy and Willamette rivers and Multnomah Channel that are associated with the Portland metropolitan area. Agricultural and historical logging activities in the Coast Range have caused the loss of large pools and rearing habitat complexity in upper basins and have increased sedimentation, elevated summer temperatures, and decreased water quality in lower basins (Oregon Department of Fish and Wildlife 1991).

Since 1900, hatchery coho smolts and fry have been released extensively in the lower Columbia River Basin. Historical presmolt releases in wild rearing areas probably impacted wild populations because of competitive interactions (Nickelson et al. 1986). Current releases from

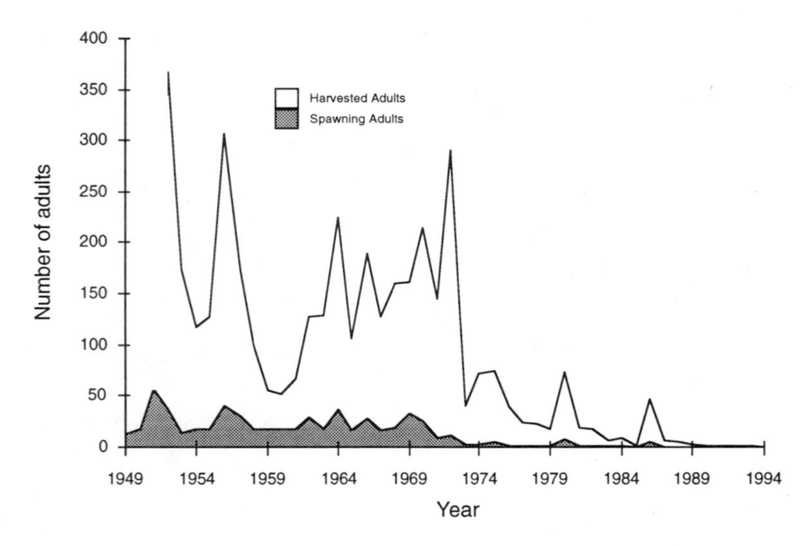

Figure 15. Estimated number of coho salmon adults, including both spawning and harvested adults, based on spawning ground counts and harvest rates in Oregon sub-basins of the Columbia River below Hood River (excluding the Sandy and Clackamas rivers), 1950–94. Total abundance decreased by an average of 21% per year ($p < 0.001$; linear regression) from 1978 through 1994, while the abundance of spawners decreased by an average of 14% per year ($p < 0.005$; linear regression) over the same period.

Oregon facilities are acclimated smolts although release numbers reached ~10.5 million smolts annually in the early 1990s (Kostow 1995). Additional hatchery coho are released from Washington State facilities. These high releases continue to produce excessive hatchery strays into most of the small Columbia River wild populations.

The hatchery coho salmon are produced for the purpose of providing fish for harvesting in the ocean and lower Columbia River. As a consequence, wild fish are harvested with the hatchery fish (see Figs. 14 and 15 for estimates). The gillnet harvest in the lower Columbia River may have also affected run timing in the populations. Harvest rates on Oregon coho salmon have been sharply curtailed in the 1990s: In 1994 only incidental harvests occurred in the ocean and at the mouth of the Columbia while the mainstem Columbia River fishery was severely restricted.

Steelhead Trout

Oregon has two *O. mykiss* subspecies that include anadromous forms: Coastal steelhead/rainbow (*O. m. irideus*) and inland Columbia River Basin steelhead/redband (*O. m. gairdneri*) (see Behnke 1992 for subspecies and range designations). Coastal steelhead populations occupy all major Oregon basins below barriers west of the Cascade Mountains. Coastal *O. mykiss* comprises a wide variety of life-history phenotypes including steelhead with various adult run-times and nonanadromous rainbow trout (Schreck et al. 1986). Populations with different adult life histories are rarely sympatric in Oregon with exceptions in larger basins such as the Rogue, Umpqua, and Willamette rivers. Summer steelhead that may have belonged to the coastal subspecies (Behnke 1992) historically traveled up the Klamath River to the area below Klamath Lake in Oregon. This population is now extinct owing to blockage caused by dam construction. Oregon steelhead populations south of Cape Blanco have been proposed for federal threatened status (US Department of Commerce 1995b). Within this subspecies, ODFW recognizes 119 steelhead populations, 113 of which are winter-run, in five provisional gene conservation groups (Table 1).

Inland Columbia River Basin steelhead occur below barriers in all Oregon sub-basins of the Columbia east of the Cascades, including the Snake River Basin below Hells Canyon Dam. Most Oregon inland steelhead are summer-run with the exception of four winter-run populations along the western boundary of the subspecies (Schreck et al. 1986). All steelhead populations in this subspecies are sympatric with resident redband trout populations, and while the two life histories form different breeding populations due to assortative mating, they do not appear to be completely reproductively isolated from each other (e.g., Currens 1987). Inland steelhead are extinct in the Snake River Basin above the Hells Canyon dam complex and in the Deschutes basin above the Pelton/Round Butte dam complex. Redband trout persist in both areas. Currently, ODFW recognizes 18 inland steelhead populations in three gene conservation groups (Table 1).

North to Mid-Coast

Winter steelhead population trends along the mid- to north Oregon coast have been difficult to monitor because techniques used for other species, such as spawning ground counts, have proven ineffective for these populations. Current steelhead abundance information for most populations is, therefore, limited to catch data. On the basis of this information, all coastal steelhead populations have declined from historical sizes. These populations have also experienced a recent, mild decline, with a cluster of populations between the mouth of the Umpqua and the Nestucca rivers in a more severe decline. The current downward trend coastwide appears to be the result of the current long-term decrease in ocean productivity. Certain populations are further affected by activities in the freshwater environment including habitat degradation, species introductions, and hatchery impacts.

The Umpqua River has one summer-run and four winter-run steelhead populations. Resident rainbow may be sympatric with the steelhead in the upper basin. The North Umpqua winter and summer steelhead populations have been counted at Winchester Dam since 1947 (Fig. 16). These populations have been stable over the monitoring period at about 6,000 winter-run and 2,500 summer-run fish, but the other Umpqua basin populations are thought to be declining. The

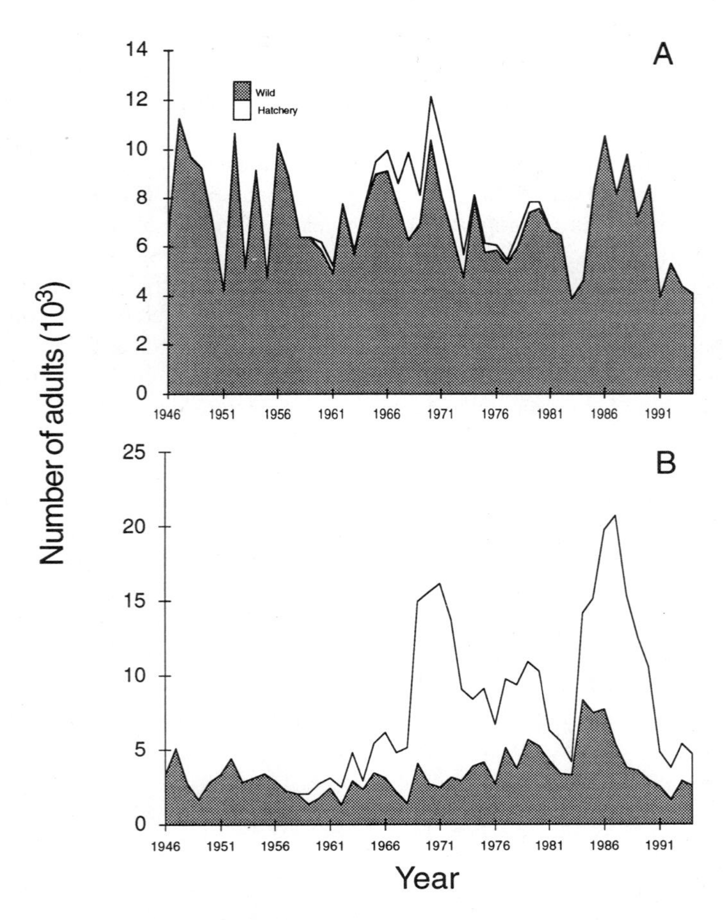

Figure 16. Number of steelhead trout counted at Winchester Dam fish ladder, North Umpqua River, 1946–94: (A) winter-run steelhead, (B) summer-run steelhead.

productivity of most steelhead habitat in the basin has decreased primarily as a result of impacts from logging and road construction.

Steelhead habitats in coastal rivers are primarily in the upper basins of the Coast Range where logging and road construction have been the major land uses. Stream siltation, loss of riparian cover, and loss of habitat complexity have impacted juvenile survival and probably contributed to the long-term, coastwide declines from historical abundances. The only summer steelhead population in a Coast Range basin, located in the Siletz River, is estimated to be ~50 fish. This population declined after a natural falls that effected selective passage of the population and exclusion of all other anadromous fish was breached in 1953 to allow the passage of other species. This breach was closed in 1995 as part of a recovery program for this population.

Hatchery steelhead from a broodstock founded in the Alsea River have been released into many Oregon coastal rivers. Most of these releases use unacclimated smolts from a central rearing facility in the Alsea basin, resulting in high straying of hatchery fish. In recent years, populations in stocked streams have included ≤75% hatchery fish while those in adjacent unstocked streams have had ≤40% hatchery fish. While genetic variation among these steelhead populations has remained high (Reisenbichler et al. 1992), the hatchery programs are thought to be impacting wild populations, particularly along the mid-coast. Since 1990, stocking of Alsea hatchery steelhead has been discontinued or decreased in many streams, and local broodstocks and acclimation sites are being developed. Angling regulations in 35 coastal streams now target hatchery fish and require the release of all wild steelhead.

Rogue River and South Coast Basins

On the basis of catch data, most of the populations in the coastal basins south of Cape Blanco appear to be stable. The exception is the winter-run population in the Illinois River, a tributary of the lower Rogue River, which has declined from an estimated 2,500 fish in the 1970s to ~200 fish in the 1990s. *Oncorhynchus mykiss* populations in the upper Rogue River express 16 different life-history patterns including both summer and winter-run adults, a subadult nonspawning freshwater migration, and resident rainbow trout. One winter-run population and part of the summer-run population are counted at Gold Ray Dam in the upper Rogue River basin (Fig. 17). The summer steelhead are also monitored by seining in the lower river. The summer-run fish declined in the early 1990s to ~25% of the abundance measured in the mid-1980s, but the pattern may be cyclic rather than an indication of a long-term decline. The winter-run populations are also currently low. All populations, however, still contain thousands of wild fish.

In the lower Rogue River basin, irrigation withdrawals have caused serious habitat impacts in the Illinois River area, particularly during recent droughts when many stream reaches used by rearing steelhead have been dewatered and summer water temperatures in remaining reaches have exceeded tolerance levels. Angling regulations require the release of all steelhead in the Illinois basin. The only steelhead hatchery program along this section of the coast is a small release in the Chetco River.

Similar to the Illinois basin, irrigation withdrawals and recent droughts are affecting all steelhead populations in the upper Rogue River basin; channelization and urbanization along several summer steelhead tributaries have particularly affected this population. Two dams in the upper mainstem Rogue and Applegate rivers have blocked access to historical steelhead habitats, although rainbow trout persist in both areas. Hatchery winter and summer steelhead from

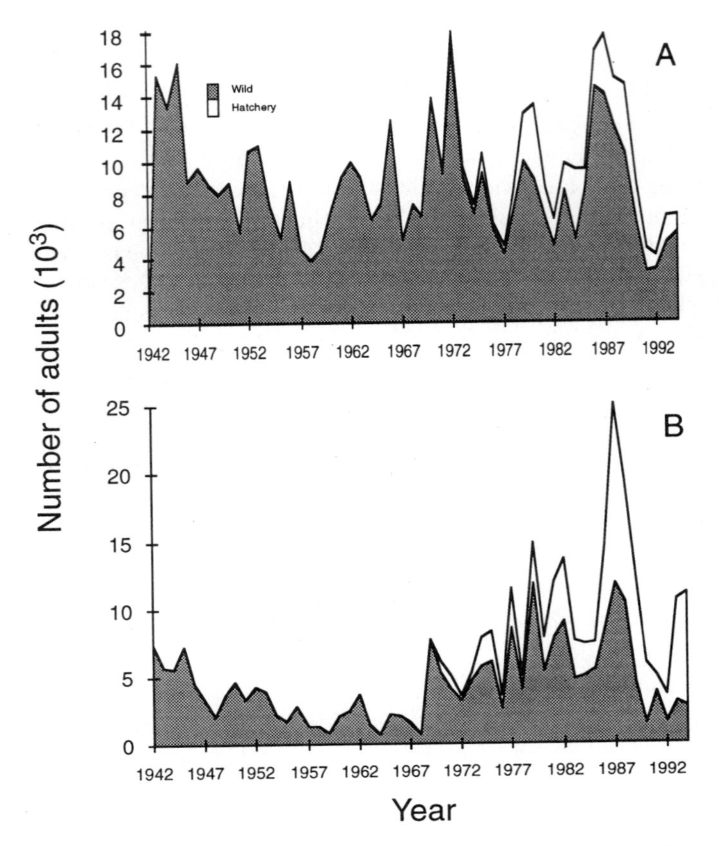

Figure 17. Number of steelhead trout counted at Gold Ray Dam fish ladder, Rogue River, 1942–94: (A) winter-run steelhead, (B) summer-run steelhead.

local broodstocks are released in the upper Rogue basin. Unharvested returning adults are removed at trapping facilities and interactions between hatchery and wild fish are thought to be minor.

Lower Columbia River Basin

Coastal steelhead are present in most lower Columbia River tributaries, with the largest populations occurring in the Willamette, Sandy, and Hood rivers. The distribution in the Willamette River is naturally restricted to lower basin tributaries, primarily those draining the Cascade Mountains, up to the Calapooia River. All populations in the lower Columbia River Basin are winter steelhead, excepting one summer steelhead population in the Hood River basin. Rainbow trout are sympatric with steelhead in the Willamette, Sandy, and Hood rivers.

Steelhead abundance is currently monitored by using dam counts in the Hood, Sandy, and Clackamas rivers and at Willamette Falls. Only the Clackamas and Willamette river counts provide long-term trends (Figs. 18 and 19). The Clackamas River population has shown no trend over the monitoring period, but the Willamette River counts, which are an aggregate of all populations

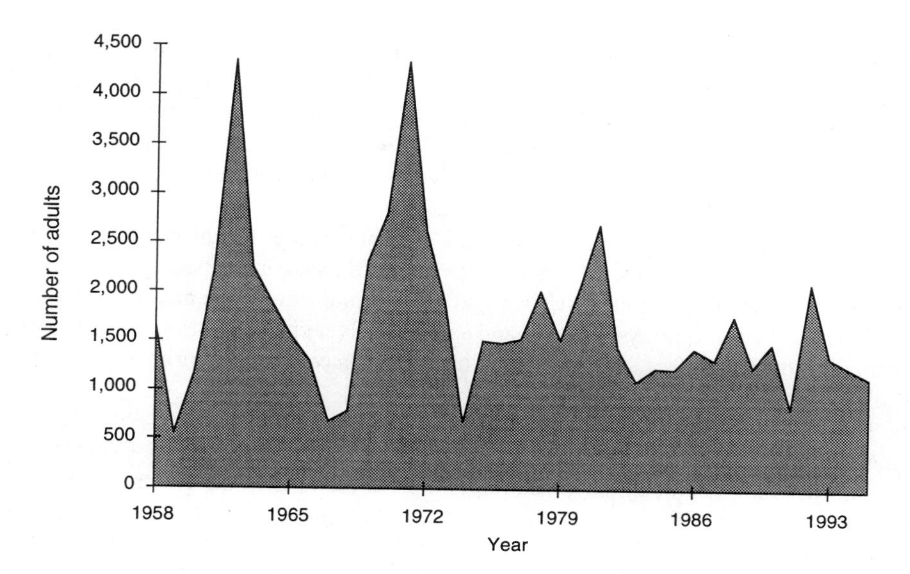

Figure 18. Number of winter steelhead trout, hatchery and wild fish combined, counted at North Fork Dam fish ladder, Clackamas River, Columbia Basin, 1958–95.

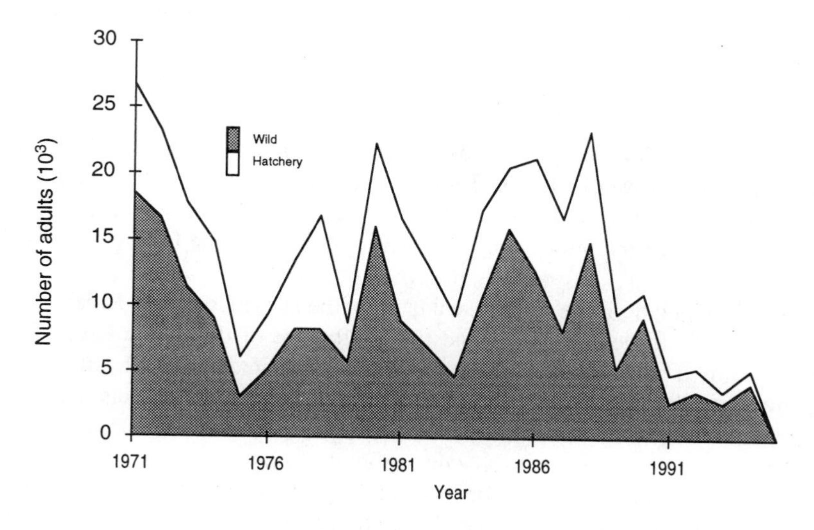

Figure 19. Number of winter steelhead trout counted at Willamette Falls fish ladder, Willamette River, Columbia Basin, 1971–94.

above Willamette Falls, dropped to record lows during the 1990s. The Clackamas River population has ranged from 1,000 to 2,000 fish in recent years, while the Sandy River population has ranged from 2,000 to 4,000 fish and the Hood River populations from 200 to 600 fish each. Between 5,000 and 15,000 wild fish passed Willamette Falls until recently when counts dropped to ~3,000 fish. All of the other lower Columbia River populations are small, primarily because the basins occupied by them are very small.

Coastal steelhead in the lower Columbia and Willamette river basins have been impacted by dams and diversion structures, which have decreased flows, caused passage problems, and blocked access to historical habitats. Urban and agricultural developments in the Willamette Valley have caused water quality problems and extensive river channelization, which decreased the juvenile rearing capacity of the basin. Logging impacts have occurred primarily in the Coast Range sub-basins.

The history of steelhead hatchery programs in the lower Columbia and Willamette river basins includes extensive use of two broodstocks: one from Big Creek near the Columbia River estuary, and one from a summer steelhead population in the Washougal River in Washington. The winter steelhead hatchery fish have been planted throughout the lower Columbia sub-basins, while the summer steelhead were introduced into the Willamette and Sandy river basins (where fish with a summer-run life history are not native) and planted into the Hood River basin. Some of the programs using these stocks have been discontinued or redesigned to use local broodstocks since 1990. All hatchery steelhead are now marked, and programs to monitor hatchery straying into wild populations are in place or under development. Angling regulations in the lower Columbia and Willamette river basins now target the hatchery fish and require the release of all wild steelhead.

Mid-Columbia and Snake River Basins

Monitoring of inland Columbia River Basin steelhead abundance began in 1939 with counts at Bonneville Dam on the lower Columbia River. The abundance of the entire subspecies in the upper Columbia River Basin varied between 50,000 and 150,000 wild fish during the first few decades of counts. In 1993 the number of wild fish in the Bonneville Dam counts declined to 22,600 fish.

Wild steelhead populations are still present in all of the larger Columbia River tributaries in Oregon between the Cascade Mountains and the confluence of the Snake River. Population abundance is monitored using dam or weir counts in the Deschutes, Umatilla, and Walla Walla rivers, and using spawning ground counts in the John Day River. Most data sets only extend into the 1980s, so long-term trends are generally unavailable for these populations. The exception is in the John Day basin where steelhead spawning ground counts extend back to 1959 (Fig. 20). All of the populations appear to have declined since the mid-1980s although the John Day data suggest that the populations cycle over time. The two best enumerated populations in this area are in the Deschutes River, which has ranged from 900 to 9,600 wild fish since 1983, and in the Umatilla River, which has ranged from 700 to 3,200 wild fish since 1980.

The steelhead populations in these rivers pass one to four mainstem Columbia River hydroelectric dams during their migrations to and from the ocean. Both juvenile and adult fish experience passage mortalities and stress during these migrations. Habitat degradation has also lowered the productivity of the sub-basins. The most severe impacts include the complete blockage of historical habitats in the upper Deschutes basin due to dam construction and serious decreases

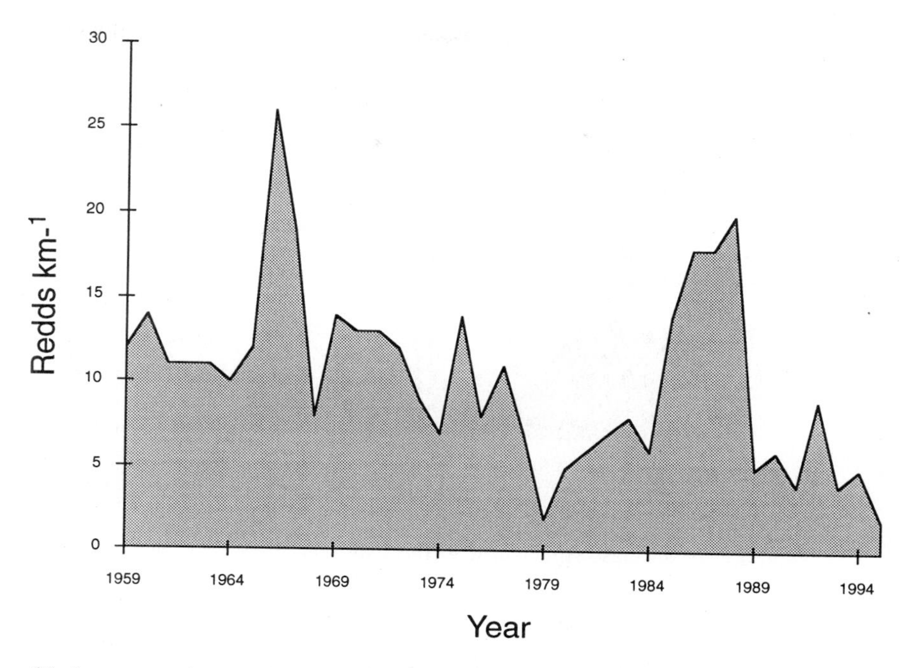

Figure 20. Average peak spawning ground counts of steelhead redds per kilometer, John Day basin, Columbia River Basin, 1959–95.

in stream flows due to irrigation diversions in all basins, but particularly in the Umatilla basin. Cattle grazing has also impacted all basins by removing riparian vegetation and degrading stream channels. Grazing impacts have been recently mitigated by extensive riparian fencing projects in the John Day and Deschutes basins.

Summer steelhead in Oregon's tributaries of the Snake River are now restricted to the Grande Ronde and Imnaha river basins. These populations have been monitored by spawning ground counts. Although long-term trends and current spawning escapements cannot be determined from the data, the populations have declined since the 1980s. The most severe habitat impacts to these populations are caused by eight mainstem Columbia and Snake river hydroelectric dams, which cause both juvenile and adult steelhead mortalities during migration.

Hatchery programs are present in the Deschutes, Umatilla, Grande Ronde, and Imnaha rivers. The number of hatchery adults returning to spawning grounds in the Umatilla basin can be strictly controlled at the counting station in the basin. Hatchery strays are a greater concern in the Deschutes River basin where an excessive number of hatchery fish, including strays from elsewhere in the Columbia basin, have been counted at the monitoring station at Sherars Falls (Fig. 21). Many of the hatchery fish in the other eastern Oregon basins that escape the harvest are now removed at traps. Angling regulations in these populations target hatchery fish and require the release of all wild steelhead.

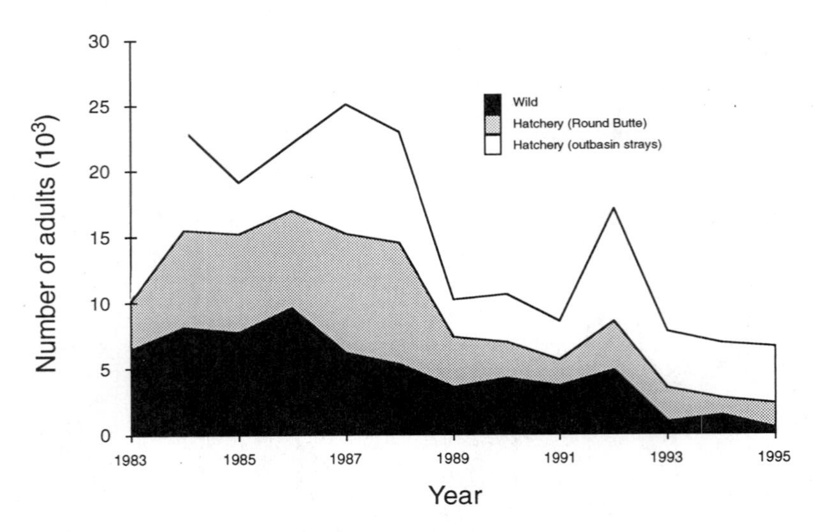

Figure 21. Number of summer steelhead estimated at Sherars Falls fish ladder, Deschutes River, Columbia River Basin, 1983–95.

Chum Salmon

The range of chum salmon in Oregon extends from Cascade Rapids (now Bonneville Dam) on the Columbia River to the Coquille River on the south coast. The species has not yet been declared extinct in any part of its range in Oregon, but it is listed as an Oregon state sensitive species throughout its range. Gene conservation groups have not been described for the species because of insufficient information. Currently, ODFW recognizes 55 chum populations in the state including 32 on the coast and 23 in the lower Columbia River Basin (Table 1).

The largest chum salmon populations on the Oregon coast are in Tillamook and Netarts bays. Since the 1920s, a combination of catch records and spawning ground counts have been used to annually estimate spawner abundance in Tillamook Bay, which currently contains eight populations (Fig. 22). Spawner abundance in the bay varied between 50,000 and 150,000 adults, with occasional higher peaks, until a steep decline in the 1950s, after which abundance declined to between 5,000 and 20,000 adults annually. Since 1953, chum abundance also has been measured at a weir on Whiskey Creek in Netarts Bay where it has varied between 300 and 1,500 fish annually. A population of several hundred fish is also present in the Nehalem basin, north of Tillamook Bay.

Factors that caused the initial decline of chum salmon probably included a combination of freshwater habitat degradation and overharvest. Chum have only a brief freshwater residency but are vulnerable to habitat degradation in lower mainstem rivers where spawning occurs and in bays and estuaries where early juvenile rearing occurs. Adults are unable to pass minor migration barriers that have been constructed in Oregon coastal basins, such as tide gates, road culverts, and gravel berms. Diking, dredging, filling, deforestation, and channelization of coastal estuarine wetlands, along with logging, gravel mining, and road construction along mainstem river reaches, degraded chum habitat along the Oregon coast.

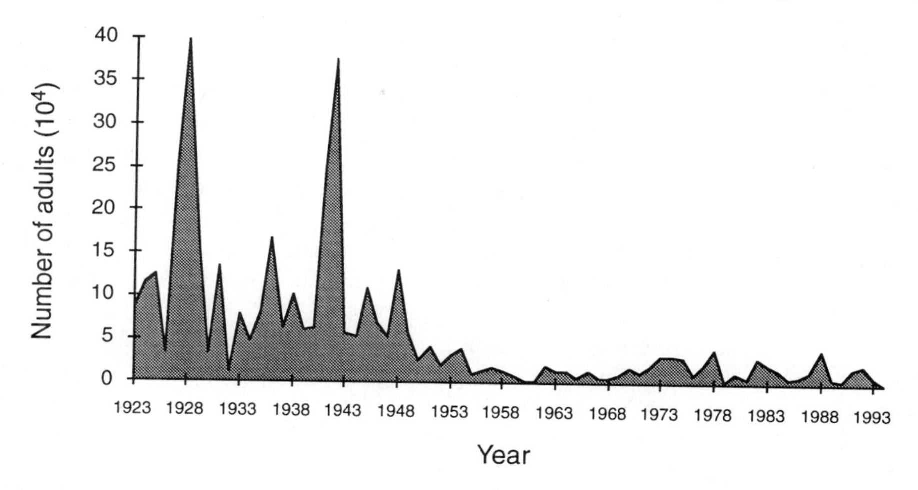

Figure 22. Estimated number of spawning chum salmon in the Tillamook Bay basin, Oregon coast, calculated from historical harvests (1923–47) and from spawning ground counts (1948–94).

The reason why Oregon chum salmon populations have not increased from the lows reached in the 1950s is unclear. Freshwater environments are still very degraded in many areas, particularly in the lower Columbia River, but the remaining habitat appears to be underseeded. Since the late 1970s, the poor ocean productivity conditions that have occurred off the Oregon coast have probably affected the species. Since the 1960s, chum also may be affected by the large releases of hatchery coho in some chum rearing areas given that coho can be a major predator on the species (Hargreaves and LeBrasseur 1986).

Limited harvest reports and anecdotal observations suggest that similar 1950s-era abundance declines occurred in chum salmon populations statewide with possible population extinctions in areas where habitat and harvest impacts have been more severe (as in the lower Columbia) or where the historical populations were naturally small (as south of the Nestucca River). Prior to the 1940s, chum salmon landings of 100,000–600,000 fish were reported on the lower Columbia River (Norman and King 1994). Only scattered, possibly nomadic, adults are still observed along Oregon's side of the Columbia River and many populations may be extinct. The scattered individuals may provide the opportunity for recolonizing the Oregon tributaries if conditions permit.

Historically, chum salmon populations south of the Nestucca River may have been naturally small. For example, harvest records in the Umpqua estuary estimate only 50–3,300 chum in this basin in the 1920s and 1930s. Small populations with probably <300 fish each remain in the Salmon, Alsea, Yaquina, Siletz, and Coos Bay systems. Populations in the Nestucca, Umpqua, and Coquille river basins are extremely small and may be extinct although potential colonizing individuals appear to persist.

Prior to the 1950s, chum salmon were one of the target species taken in gillnet fisheries in Oregon bays and in the lower Columbia River and have also been taken incidentally in river and ocean fisheries for coho salmon (Henry 1953). Targeted commercial harvest on the species was discontinued in 1962 although chum salmon continue to be vulnerable to Columbia River gillnets set for other species (Norman and King 1994).

Sockeye Salmon

The historical distribution of sockeye salmon in Oregon included the Grande Ronde River (Wallowa Lake) in the Snake River Basin and the Deschutes River (Suttle Lake) in the mid-Columbia River Basin. Anadromous sockeye and resident kokanee (*O. nerka*) salmon were historically sympatric in both locations. Spawning populations of sockeye salmon are nearly extinct in the state although wild kokanee are still present in both basins. Sockeye salmon became extinct in the Grande Ronde River in the 1930s following the blockage of the Wallowa Lake outlet by a dam in 1916. The Deschutes River may contain a remnant sockeye population (Table 1).

A few sockeye salmon return annually to the Deschutes River. The outlet of Suttle Lake was blocked in the early 1900s; however, sockeye persisted in the Metolius River below the lake until the mainstem Deschutes River was blocked by the Pelton/Round Butte dam complex in the 1960s. Sockeye salmon currently are only observed when they enter a fish trap at the base of the dams. Seven individuals were captured in 1992, 1 was captured in 1993, 14 were captured in 1994, and 9 were captured in 1995. The sockeye salmon are suspected of spawning below the dam, but reproductive success is uncertain. Alternatively, the returning fish may be produced by the wild kokanee population that remains in the Metolius River above the dams or by hatchery kokanee planted in one of the reservoirs in the dam complex.

Future Direction of Conservation Management in Oregon

It is the intent of ODFW's conservation management program to detect and correct problems affecting the status of wild fish species before populations decline to the point that recovery would be difficult, expensive, and restrictive to other activities. The agency (ODFW) is charged with implementing the following actions: an extensive monitoring program, research on conservation issues, guidelines for about 1,500 hatchery programs, habitat rehabilitation projects, risk assessments for all major management actions, improved management planning, biennial species status reviews, and recovery programs for depressed groups of populations. Improved population monitoring techniques are under development for coastal coho and steelhead, and improved habitat inventories are being implemented statewide. Current population recovery and habitat rehabilitation activities for salmon and steelhead are focused on the Snake River chinook salmon populations, on coho salmon statewide, and on all Rogue River basin and south coast populations. Improvements in steelhead hatchery programs are being researched and implemented on the mid-coast and in the lower Columbia River.

There are still many natural wonders left in Oregon. But the challenges of conserving Oregon's wild fishes into the future are considerable as the state faces a growing human population and increasing demands on all natural resources. Only a committed partnership between resource management agencies and the citizens of Oregon dedicated to achieving this mission will ensure that the wonder will remain for tomorrow's generations.

Acknowledgments

Information for this paper was provided by the following ODFW staff: T. Nickleson, M. Jennings, S. Jacobs, R. Williams, H. Schaller, S. Spangler, and R. Beamesderfer. The following district biologists also contributed data: B. Buckman, G. Stewart, W. Beidler, P. Reimers, T. Unterwegner, D. Loomis, G. MacLeod, M. Everson, J. Newton, E. Claire, D. West, B. Smith, J. Phelps, J. Massey, T. Bailey, J. Zakel, and R. Klumph.

Literature Cited

Behnke, R.J. 1992. Native Trout of Western North America. American Fisheries Society Monograph 6. Bethesda, Maryland.

Currens, K.P. 1987. Genetic differentiation of resident and anadromous rainbow trout (*Salmo gairdneri*) in the Deschutes River basin, Oregon. MS thesis, Department of Fish and Wildlife, Oregon State University. Corvallis, Oregon.

DeVoto, B. (ed.). 1953. The Journals of Lewis and Clark. Houghton Mifflin Co., Boston, Massachusetts.

Dimick, R.E. and F. Merryfield. 1945. The fishes of the Willamette River system in relation to pollution. Oregon State College, Engineering Experiment Station, Bulletin 20. Salem, Oregon.

Ellis, M.N. 1938. Stream mining on the Rogue River, Oregon in its relation to fish and fishing in that stream. Oregon Department of Geology and Mineral Industries, Bulletin No. 10. Salem, Oregon.

Fulton, L.A. 1968. Spawning areas and abundance of chinook salmon (*Oncorhynchus tshawytscha*) in the Columbia River basin—past and present. US Fish and Wildlife Service, Special Scientific Report Fisheries No. 571. Washington, DC.

Hargreaves, N.B. and R.J. LeBrasseur. 1986. Size selectivity of coho (*Oncorhynchus kisutch*) preying on juvenile chum salmon (*O. keta*). Canadian Journal of Fisheries and Aquatic Sciences 43: 581-586.

Healey, M.C. 1994. Variation in the life history characteristics of chinook salmon and its relevance to conservation of the Sacramento winter run of chinook salmon. Conservation Biology 8: 876-877.

Henry, K.S. 1953. Analysis of factors affecting the production of chum salmon (*Oncorhynchus keta*) in Tillamook Bay. Fish Commission of Oregon Contribution No. 18. Portland, Oregon.

Hume, R.D. 1893. Salmon of the Pacific coast. Manuscript available from Oregon Department of Fish and Wildlife. Portland, Oregon.

Jacobs, S.E. and C.X. Cooney. 1993. Improvement of methods used to estimate the spawning escapement of Oregon coastal natural coho salmon. Oregon Department of Fish and Wildlife. Portland, Oregon.

Kostow, K.E. (ed.). 1995. Biennial Report on the Status of Wild Fish in Oregon. Oregon Department of Fish and Wildlife. Portland, Oregon.

McGie, A.M. 1980. Analysis of relationships between hatchery coho salmon transplants and adult escapements in Oregon coastal watersheds. Oregon Department of Fish and Wildlife Information Report Series, Fisheries, No. 80.6. Portland, Oregon.

McKernan, D.L., D.R. Johnson, and J.I. Hodges. 1950. Some factors influencing the trends of salmon populations in Oregon. Transactions of the Fifteenth North American Wildlife Conference. Wildlife Management Institute, Washington, DC.

Moore, M. and A.J. Suomela. 1947. A program of rehabilitation of the Columbia River fisheries. State of Washington Department of Fisheries, Olympia, Washington, State of Oregon Fish Commission, Portland, Oregon, US Fish and Wildlife Service, Portland, Oregon.

Nawa, R.K., C.A. Frissell, and W.J. Liss. 1989. Life history and persistence of anadromous fish stocks in relation to stream habitat and watershed classification. Annual Progress Report. Department of Fisheries and Wildlife, Oregon State University, Corvallis, Oregon.

Nickelson, T.E., J.W. Nicholas, A.M. McGie, R.B. Lindsay, D.L. Bottom, R.J. Kaiser, and S.E. Jacobs. 1992. Status of anadromous salmonids in Oregon coastal basins. Oregon Department of Fish and Wildlife. Corvallis, Oregon.

Nickelson, T.E., M.F. Solazzi, and S.L. Johnson. 1986. Use of hatchery coho salmon (*Oncorhynchus kisutch*) presmolts to rebuild wild populations in Oregon coastal streams. Canadian Journal of Fisheries and Aquatic Sciences 43: 2443-2449.

Nickelson, T.E., J.W. Nicholas, A.M. McGie, R.B. Lindsay, D.L. Bottom, R.J. Kaiser, and S.E. Jacobs. 1992. Status of anadromous salmonids in Oregon coastal basins. Oregon Department of Fish and Wildlife. Corvallis, Oregon.

Nicholas, J.W. and D.G. Hankin. 1989. Chinook salmon populations in Oregon coastal river basins: Description of life histories and assessment of recent trends in run strengths, 2nd ed. Oregon Department of Fish and Wildlife. Corvallis, Oregon,

Norman, G. and S.D. King. 1994. Status report: Columbia River fish runs and fisheries, 1938-93. Washington Department of Fish and Wildlife, Olympia, Washington; Oregon Department of Fish and Wildlife, Portland, Oregon.

Oregon Department of Fish and Wildlife. 1982. Comprehensive plan for production and management of Oregon's anadromous salmon and trout. Part II: Coho salmon plan. Oregon Department of Fish and Wildlife. Portland, Oregon.

Oregon Department of Fish and Wildlife. 1991. Lower Columbia River coho salmon: evaluation of stock status, causes for decline and critical habitat; part 2. Oregon Department of Fish and Wildlife. Portland, Oregon.

Oregon Department of Fish and Wildlife. 1995. Oregon coho salmon biological status assessment. Oregon Department of Fish and Wildlife, Portland, Oregon.

Pacific Salmon Commission. 1993. Joint chinook technical committee annual report. Report TCChinook (94-1). Pacific Salmon Commission. Vancouver, British Columbia.

Reisenbichler, R.R., J.D. McIntyre, M.F. Solazzi, and S.W. Landino. 1992. Genetic variation in steelhead of Oregon and northern California. Transactions of the American Fisheries Society 121: 158-169.

Rivers, C.M. 1991. Rogue River fisheries Vol. 1: History and development of the Rogue River Basin as related to its fisheries prior to 1941. Oregon Department of Fish and Wildlife. Portland, Oregon.

Ryman, N. and L. Laikre. 1991. Effects of supportive breeding on the genetically effective population size. Conservation Biology 5: 325-329.

Schreck , C.B., H.L. Li, R.C. Hjort, C.S. Sharpe. 1986. Stock identification of Columbia River chinook salmon and steelhead trout. Bonneville Power Administration Project 83-451. Portland, Oregon.

Thompson, R.W. and J.B. Haas. 1960. Environmental survey report pertaining to salmon and steelhead in certain rivers of eastern Oregon and the Willamette River and its tributaries. Fish Commission of Oregon. Portland, Oregon.

US Department of Commerce. 1992. Endangered and threatened species: threatened status for Snake River spring/summer chinook salmon, threatened status for Snake River fall chinook salmon. Federal Register 57: 14653-14683.

US Department of Commerce. 1995a. Endangered and threatened species; proposed threatened status for three contiguous ESUs of coho salmon ranging from Oregon through central California. Federal Register 60: 38011-38030.

US Department of Commerce. 1995b. Endangered and threatened species; proposed threatened status for southern Oregon and northern California steelhead. Federal Register 60: 14253-14261.

Status of Alaska Salmon

Alex C. Wertheimer

Abstract

Alaska currently produces ~80% of the salmon (*Oncorhynchus* spp.) harvested in North America. The total Alaska salmon catch has increased dramatically since the 1970s and is now at historically high levels. Commercial catch has averaged 135 million salmon since 1980 and set a new harvest record of 192 million salmon in 1993. Catches of all five species of Pacific salmon in each of the three International North Pacific Fisheries Commission statistical regions for Alaska have increased since the 1970s, and generally are at or are near historically high levels; an exception is the catch of chinook salmon (*O. tshawytscha*) in southeast Alaska. Escapements for all species evaluated had predominantly no trend or increasing trends over time, indicating that the current high harvest levels are reflective of abundance and productivity, and not over-exploitation of the resource such as occurred in the first half of the century. Factors that have influenced the recent high productivity of Alaska salmon include a relatively pristine and undeveloped habitat base, salmon management policies within the state, the elimination of high-seas driftnet fisheries, enhancement by hatcheries, and favorable environmental conditions.

Introduction

Alaska is derived from the Aleut word for "the great land, that which the sea breaks against." The name is appropriate to the geographic scope of the state. Alaska covers >1.5 million km^2, ~17% of the total area of the US. It has 10,700 km of coastline, more than the rest of the US combined, and borders the Gulf of Alaska, North Pacific Ocean, Bering Sea, Chukchi Sea, and Arctic Ocean. Alaska contains 3,000 rivers and has over 3 million lakes >8 ha (Metcalfe 1993).

Many of these rivers and lakes provide habitat for anadromous salmonids (*Oncorhynchus* spp.). Over 14,000 water bodies containing anadromous salmonids have been identified in the state (Alaska Department of Fish and Game 1992). These waters are the spawning and initial rearing habitat for most of the salmon produced in North America. In 1991, Alaska produced 79% of the Pacific salmon harvest compared with 17% from British Columbia and 4% for the Pacific Northwest (Rigby et al. 1991, Canada Department of Fisheries and Oceans 1993, Henry 1993).

This paper is a simple overview of the current status of Pacific salmon in Alaska based on historical comparisons of catch and escapement data. I review harvest records and escapement counts as indicators of the relative magnitude and status of recent returns. I also discuss factors that have contributed to the recent high productivity of salmon in the state and factors that may threaten productivity and diversity of the resource.

Statewide Harvest History

Salmon have served as a primary source of food for native American people in Alaska since well before the colonization of Alaska by European nations. The use of salmon along the coast of southeast and southwest Alaska has been described as the most highly developed aboriginal fishing complex on the continent (Cooley 1961). Annual subsistence harvest before the era of commercial exploitation is approximated to be >12 million salmon (Pennoyer 1988).

The commercial harvest of salmon constituted a minor activity during the period of Russian occupation (1741–1867). Alaska was remote from the major markets for salmon, and drying and salting of fish proved an inadequate means of preservation for large-scale commercial exploitation. The development of a technique for preserving salmon in tin cans was critical in the expansion of the commercial fishing industry in Alaska (Cobb 1911). In 1878, 11 years after the purchase of Alaska by the US, the first salmon canneries opened in southeast Alaska. By 1910, 52 canneries operated in the state (Cobb 1911).

Harvest increased rapidly with the expanding processing capability. In 1878 estimated catch of salmon in Alaska totaled 56,000 fish; by 1900 it was >21 million fish (Rigby et al. 1991). The harvest continued to increase to >100 million fish in 1918, declined sharply to 39 million in

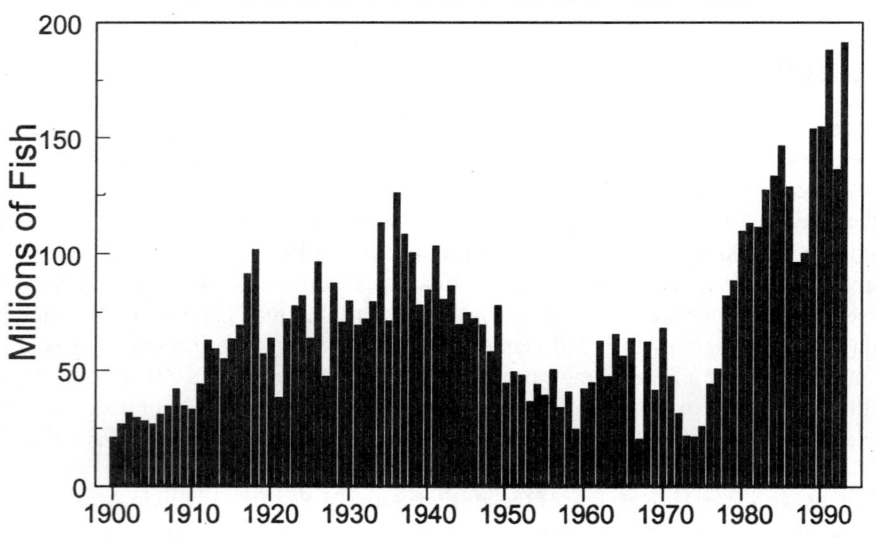

Figure 1. Commercial salmon harvest in Alaska, 1900–93.

1921, then generally increased to a new record level of 126 million in 1936 (Fig. 1). This new record was followed by a long, continuous decline; in 1959, the year of Alaska statehood, the harvest was 25 million salmon, the lowest since 1900 (Fig. 1).

Overfishing was a major factor in the extent and duration of the decline (Royce 1989). Cooley (1961) reviewed the early management of Alaska's salmon fisheries and described it as a "pathetic history of ruinous overexploitation." The federal government managed the resource until statehood. Conservation concerns were recognized early in the commercial era. Congress prohibited the use of dams or barricades in streams to harvest returning fish in 1889 and, in 1896, made fishing in streams <152 m in width illegal. Following the decline in 1921, Congress passed the White Act, which included a declaration of intent that "not less than 50% of the salmon shall be allowed to escape to the spawning beds." Unfortunately, Congress did not provide adequate resources to develop sound management policies or to enforce the established regulations (Cooley 1961, Royce 1989).

The decline of the fishery in the 1940s and 50s precipitated a new era of research, management, and conservation. The federal government pursued international agreements to constrain high-seas fishing, increased research efforts, and began to apply more scientifically based management practices in the 1950s. The University of Washington established a strong research program in Alaska on salmon population dynamics with the support of the salmon industry (Royce 1989). This commitment to science-based management continued in the Alaska Department of Fish and Game (ADFG), which was established following statehood. The state emphasized management at the local district level rather than the centralized approach of the federal era. In 1974, the state also imposed limited entry to control the amount of gear in its salmon fisheries and initiated a large hatchery program to enhance the production from wild stocks.

Conservation measures imposed through the 1960s were followed by a rapid increase in salmon abundance during favorable environmental conditions in the late 1970s. By 1980, commercial harvest again exceeded 100 million fish. The commercial catch has averaged 135 million fish from 1980 to 1993 and, except for 1987 (97 million), has been >100 million annually (Fig. 1). The catch for 1993 totaled 192 million salmon, a new harvest record (H. Savikko, Alaska Dep. Fish and Game, Juneau, pers. comm.). In addition to the commercial harvest, the recreational and subsistence harvest each approximated 1 million salmon annually in recent years (International North Pacific Fisheries Commission 1993, Mills 1993).

Catch and Escapement Trends by Species

Data Sources

Data on catch and escapement of Pacific salmon and anadromous steelhead (*O. mykiss*) in Alaska are compiled by ADFG. I summarized catch data by species statewide and for the three major International North Pacific Fisheries Commission (INPFC) statistical regions for Alaska salmon (Fig. 2). (The INPFC was dissolved and replaced by the North Pacific Anadromous Fish Commission in 1993; the statistical reporting regions remained the same for Alaska salmon.) I obtained data as follows: commercial harvest statewide and by INPFC statistical region from Rigby et al. (1991) and H. Savikko (Alaska Dep. of Fish and Game, Juneau, pers. comm.); recreational

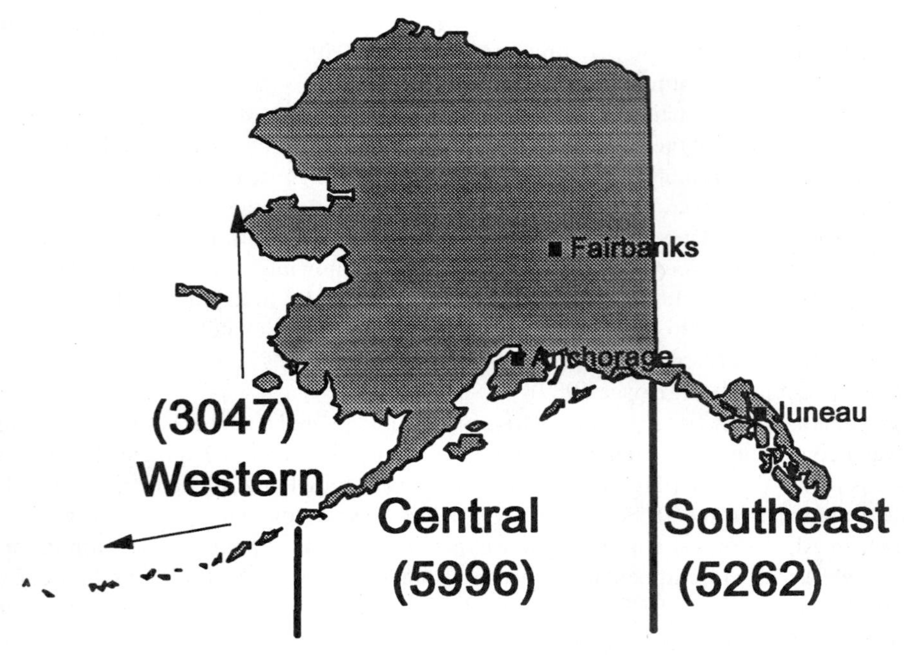

Figure 2. Location of International North Pacific Fisheries Commission Statistical Regions for Pacific salmon in Alaska.

harvest from Mills (1993); and the components of the catch from enhancement operations from M. McNair (Alaska Dep. Fish and Game, Juneau, pers. comm.) and McNair and Holland (1994).

I present recent trends in escapement based on my analysis of ADFG escapement counts of salmon and steelhead populations in southeast Alaska and steelhead in central Alaska, and a prior analysis by Konkle and McIntyre (1987) of population escapement trends for salmon spawning in central and western Alaska. I used the term "population" to refer to the group of spawning adults of a species represented by an escapement count (sensu Konkle and McIntyre 1987). These populations do not necessarily represent a stock as defined by Ricker (1972).

To determine trends in escapement counts from southeast Alaska, I used linear regression (see Konkle and McIntyre 1987). I restricted my analysis to populations with ≥10 years of counts using the same technique (e.g., foot survey, aerial survey, weir), and with ≥3 years of counts pre-1980 and 3 years post-1979. A significant regression slope ($p < 0.05$) indicated a trend in the data. Potential problems exist using linear regression as an analytical technique for escapement time-series data; the data may be nonlinear, and autocorrelation of the observations can cause an overestimate of the statistical significance of a slope (Freund and Littell 1992). It is also important to note that management for escapement goals could obscure increasing trends in escapement time series. Despite these problems, I used linear regression as a simple measure of trends in the data owing to consistency with past analyses.

Another shortfall of the escapement analysis is the use of a dated study (Konkle and McIntyre 1987) for evaluating current status. Unfortunately, I was unable to update the analysis for central and western Alaska for this paper. Because the analysis by Konkle and McIntyre (1987) did extend into the 1980s, when harvest numbers increased dramatically from the previous decades, I felt that their analysis was still useful in evaluating whether these large harvests have coincided with trends in escapement counts.

PINK SALMON

The catch of pink salmon (*O. gorbuscha*) followed the same pattern as the total statewide salmon harvest, declining from high catches in the late 1930s and early 1940s, then increasing to record catches in the 1980s and early 1990s (Fig. 3A). The annual catch since 1980 averages 74 million fish. As hatchery programs initiated in the 1970s developed, hatchery production became a larger component of the harvest, averaging 27% of the annual catch since 1985. A record catch of both hatchery and wild pink salmon occurred in 1991 when 93 million wild fish and 35 million hatchery fish were harvested.

Pink salmon are harvested primarily in the southeast and central regions; relatively little harvest occurs in the western region (Fig. 3B). In southeast Alaska, wild stocks produced most (>98%) of the catch. Catch over the last 10 years (1984–93) averaged 39 million, some 2 million fish greater than the previous high 10-year period (1933–42). In the central region, catch over the last 10 years averaged 42 million fish, far above historical levels (Fig. 3B). Hatcheries produced much of the catch of pink salmon in the central region in recent years; during 1990–93, fish from hatcheries composed 46% of the catch.

Escapement analyses for pink salmon were limited to the central and southeast regions. In both regions, escapements generally showed no trend or were increasing (Table 1). In the central region, 76% of the 134 populations evaluated had no trend in escapement whereas 24% increased significantly (Konkle and McIntyre 1987). Similarly, in the southeast region 66% of the 471 populations exhibited no trend in escapement, 32% increased, and 2% decreased.

SOCKEYE SALMON

Current statewide catches of sockeye salmon (*O. nerka*) are at record highs (Fig. 4A). The 10-year catch (1984–93) averaged 44 million fish; the previous high 10-year period, 1913–1922, averaged 28 million fish. Wild stocks produced most of the catch although production from enhancement operations also increased (Fig. 4A), averaging 6% of the 1990–93 total catch.

Since the early 1970s, catches of sockeye salmon increased sharply in all three statistical regions (Fig. 4B) and are now at record levels in the central and western regions. In the southeast region, catches also increased markedly in recent years, although they were still below historical high levels. Since 1984, catches averaged 2.0 million fish in southeast Alaska annually, compared with 1910–19 when catches averaged 2.6 million fish.

In all regions, escapements predominantly exhibited either the lack of a trend or an increasing one (Table 1). In the western region, slightly more than half (58%) of the populations analyzed had no trend in escapement whereas the remainder increased (Konkle and McIntyre 1987). Most of the populations (62%) in the central region also did not show significant trends, and the remainder increased (Konkle and McIntyre 1987). In the southeast region, the majority (87%)

Pink Salmon

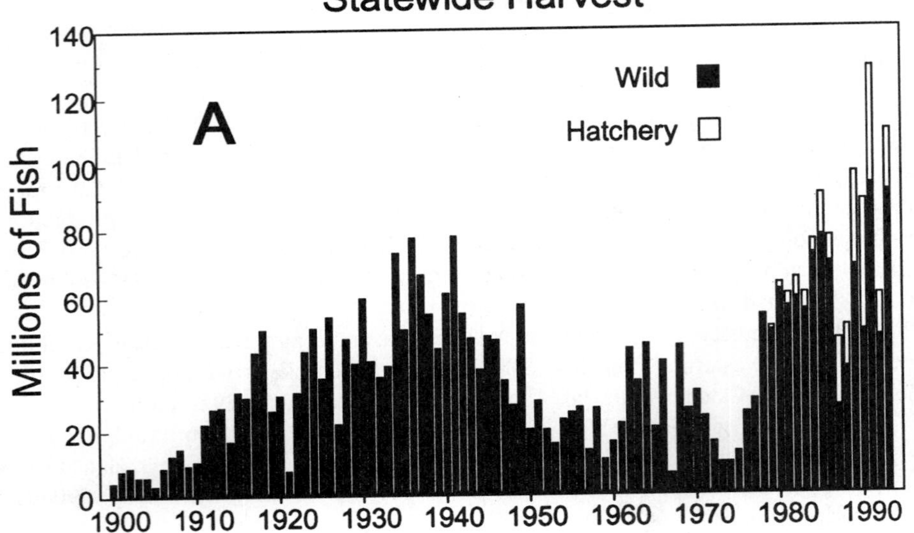

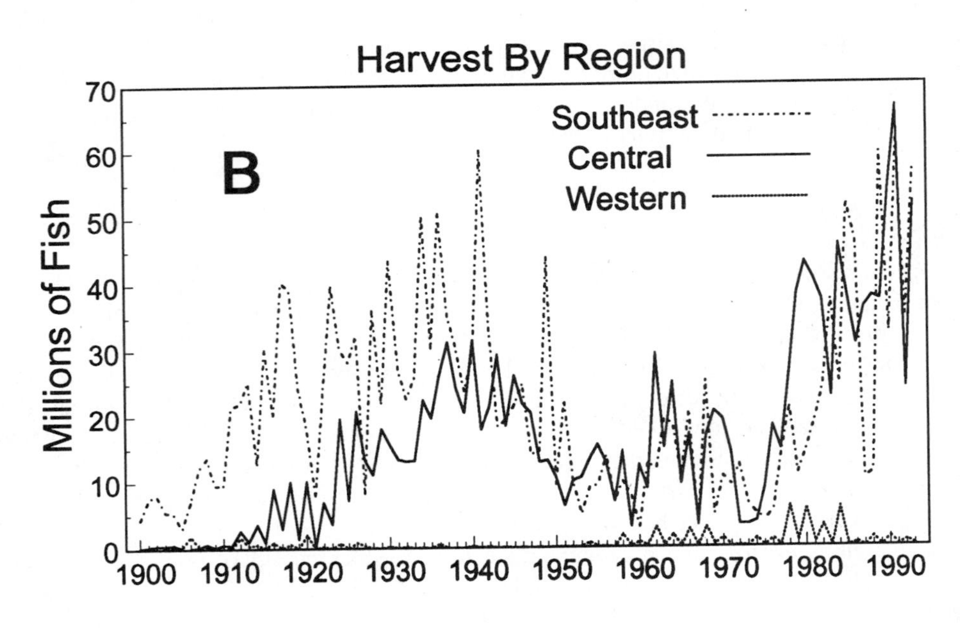

Figure 3. Statewide (A) and regional (B) commercial harvests of pink salmon in Alaska, 1900–93.

Table 1. Trends in escapement counts for Pacific salmon and steelhead in Alaska. The number of populations evaluated was restricted by availability of data, and represents a small subset of the total populations in the state. Escapements were classified as increasing, decreasing, or no trend based on whether the slope of the regression of escapement over time diverged significantly ($p < 0.05$) from zero. See text for data sources.

Species/Region	Years	No. of populations		
		No trend	Decrease	Increase
Pink Salmon				
Southeast	1960–1993	312	9	150
Central	1968–1984	102	0	32
Sockeye Salmon				
Southeast	1960–1993	58	4	5
Central	1968–1984	58	0	35
Western	1968–1984	16	0	12
Chum Salmon				
Southeast	1960–1993	28	5	5
Central	1968–1984	61	11	3
Western	1968–1984	10	2	0
Coho Salmon				
Southeast	1960–1993	33	3	5
Central	1968–1984	4	0	4
Chinook Salmon				
Southeast	1960–1993	18	2	13
Central	1968–1984	20	0	23
Western	1968–1984	24	1	15
Steelhead				
Southeast	1960–1993	1	0	1
Central	1960–1993	1	0	0

of the 67 populations analyzed again had no trend in escapement whereas 7% had an increasing trend and 6% had a decreasing trend.

Chum Salmon

Current statewide catches of chum salmon (*O. keta*) are at record levels (Fig. 5A). Total catch averaged 10.9 million fish for 1984–93; the previous high 10-year period was 8.7 million (1935–44). Hatcheries produced an increasing percentage of the catch (Fig. 5A). From 1990 to 1993, hatchery fish averaged 27% of the catch.

Harvests increased sharply in all regions after the early 1970s (Fig. 5B). The catch of chum salmon in the western region reached record levels in the 1980s, reflecting both recent high productivity and the development of commercial fisheries in the region. However, catch dropped from a high of 5 million fish in 1988 to 1 million fish in 1993, raising concerns for some

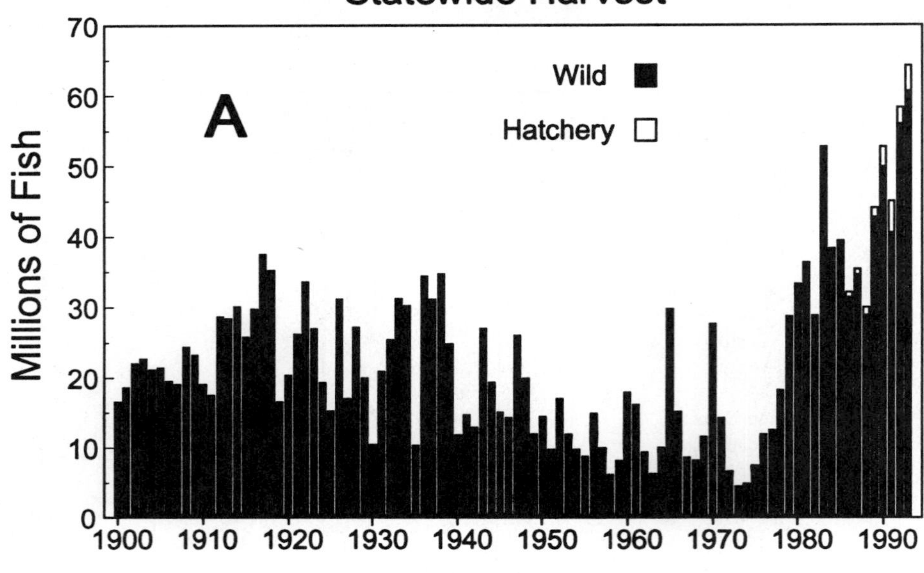

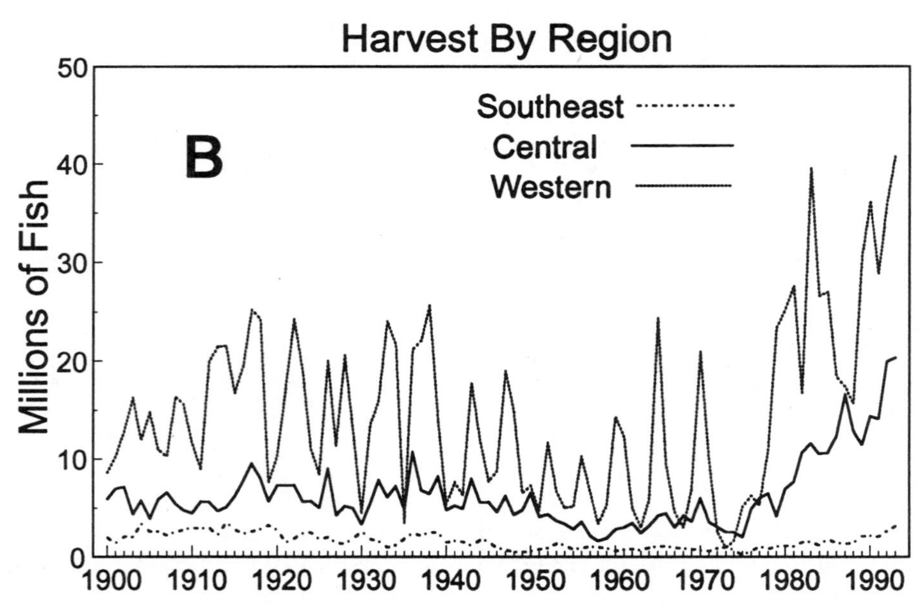

Figure 4. Statewide (A) and regional (B) commercial harvests of sockeye salmon in Alaska, 1900–93.

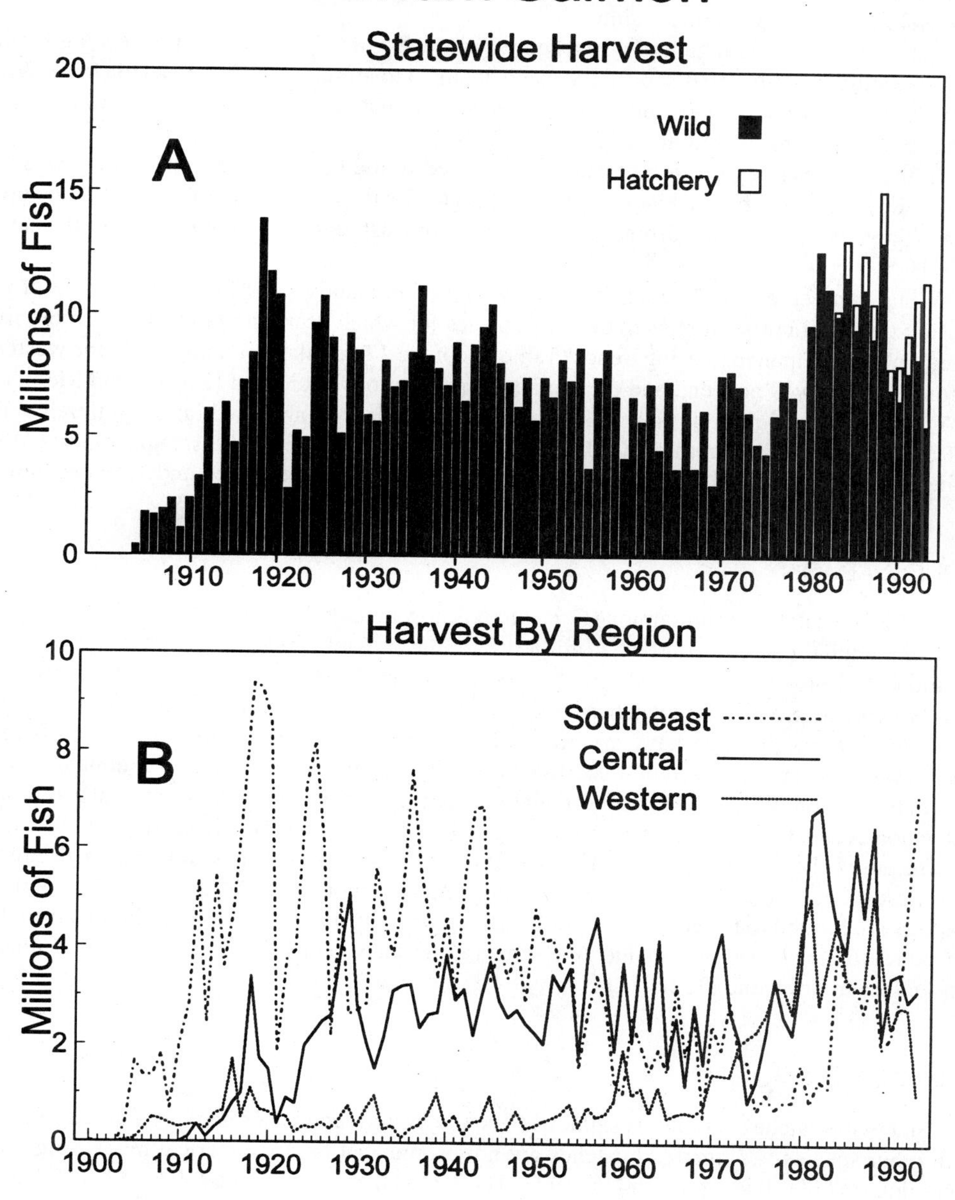

Figure 5. Statewide (A) and regional (B) commercial harvests of chum salmon in Alaska, 1900–93.

populations of western Alaska chum salmon. Wild stocks produced almost all (99%) of the chum salmon harvested in this region.

Catch in the central region also reached record levels in the 1980s, often exceeding 5 million fish annually. After 1988, the annual catch declined to around 3 million fish (Fig. 5B). Wild stocks also produced most of this catch; however, the hatchery component of the harvest is increasing, averaging 19% of the catch since 1990.

In the southeast region, recent catches increased to the highest levels since the 1940s although they were still below historical (1910s) highs for the region (Fig. 5B). Hatcheries produced a large component of current harvest in the southeast region, averaging 41% of the catch since 1990.

While escapements of chum salmon showed predominantly (>80%) either no trend or increasing trends, chum salmon were the only species for which decreasing trends were more frequently observed than increasing trends (Table 1). Of the 12 populations analyzed in the western region, 83% showed no trend and the remaining populations decreased (Konkle and McIntyre 1987). In the central region, most (81%) of the 75 populations analyzed lacked a trend while 15% decreased and 4% increased (Konkle and McIntyre 1987). In the southeast region, most (74%) of the 38 populations analyzed also showed no trend while 13% increased and 13% decreased.

Coho Salmon

Statewide catch of coho salmon *(O. kisutch)* is now at record levels (Fig. 6A). From 1984 to 1993, the annual catch averaged 5.4 million fish; the previous high 10-year average was 2.9 million fish (1940–49). While wild stocks produced most of the catch, an increasing proportion of the catch was produced by hatchery operations (Fig. 6A), averaging 15% for 1990–93.

Catches in recent years have reached record levels in all regions of the state (Fig. 6B). Increases were most pronounced in western Alaska, reflecting the development of commercial fisheries in this region. Wild stocks produced all of the harvest in this region. In the central and southeast regions, catch from Alaska hatcheries averaged 9% and 26%, respectively, of the 1990–93 catch.

Because coho run timing is late and typically coincides with adverse weather and high stream flows, escapement data are sparse, especially in more remote areas of the state. Escapement analyses are limited to the central and southeast regions. Equal numbers of populations from the central region increased or remained the same (Table 1). Most (80%) escapement in southeast Alaska lacked any trend (Table 1).

Chinook Salmon

Statewide catches of chinook salmon (*O. tshawytscha*) remained relatively stable compared with other species. Commercial catches are now somewhat below historical levels (Fig. 7A), averaging 637,000 fish from 1984 to 1993. The record 10-year average (1930–39) is 773,000. Recreational harvest of chinook salmon has increased rapidly in recent years (Mills 1993). Chinook salmon are the only Pacific salmon in Alaska for which the recreational harvest is >10% of the commercial harvest (5% for coho salmon and <1% for sockeye, pink, and chum salmon). Recreational harvest of chinook salmon over 1983–92 averaged 107,000 fish; thus, the total commercial and recreational harvests of chinook were very similar to the historically high

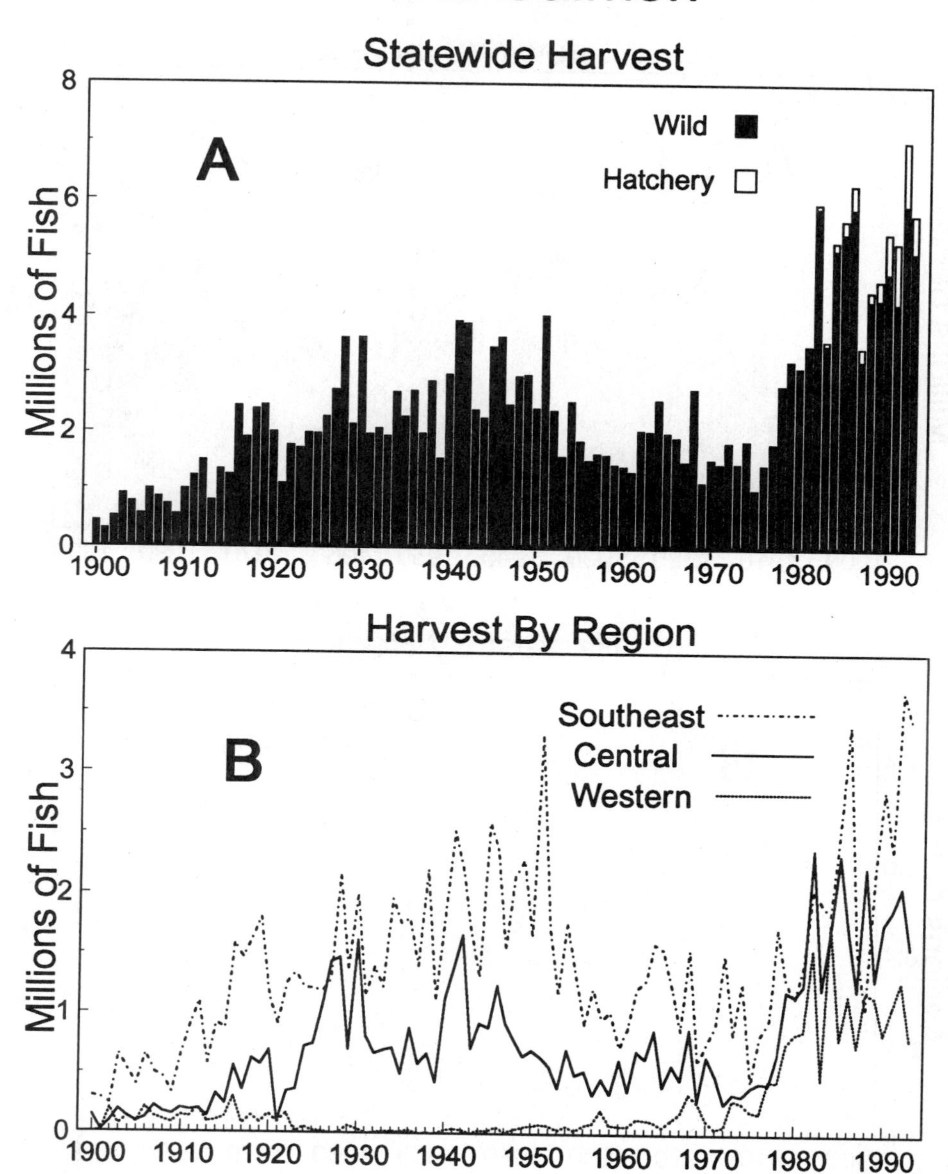

Figure 6. Statewide (A) and regional (B) commercial harvests of coho salmon in Alaska, 1900–93.

Chinook Salmon

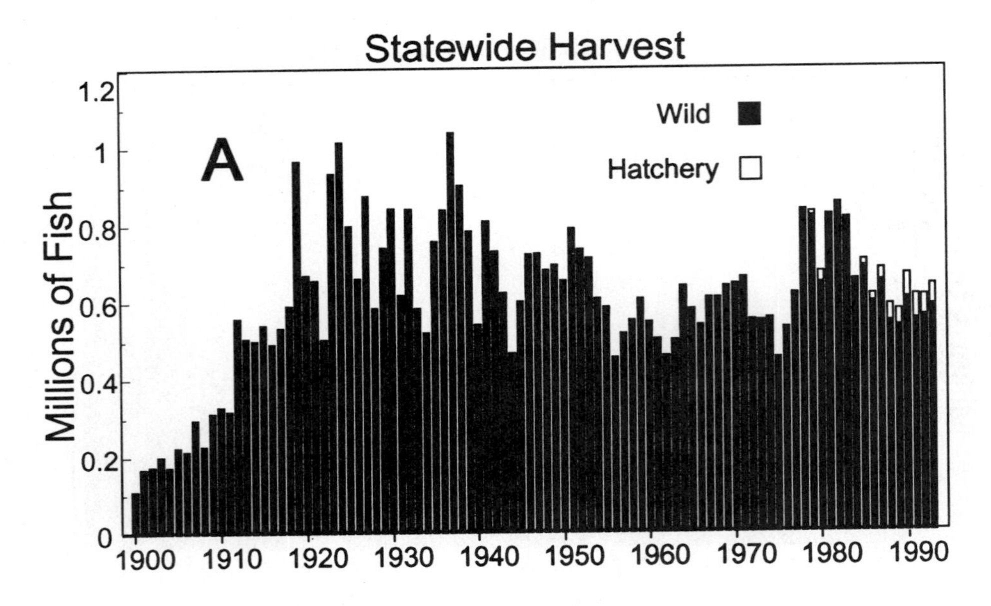

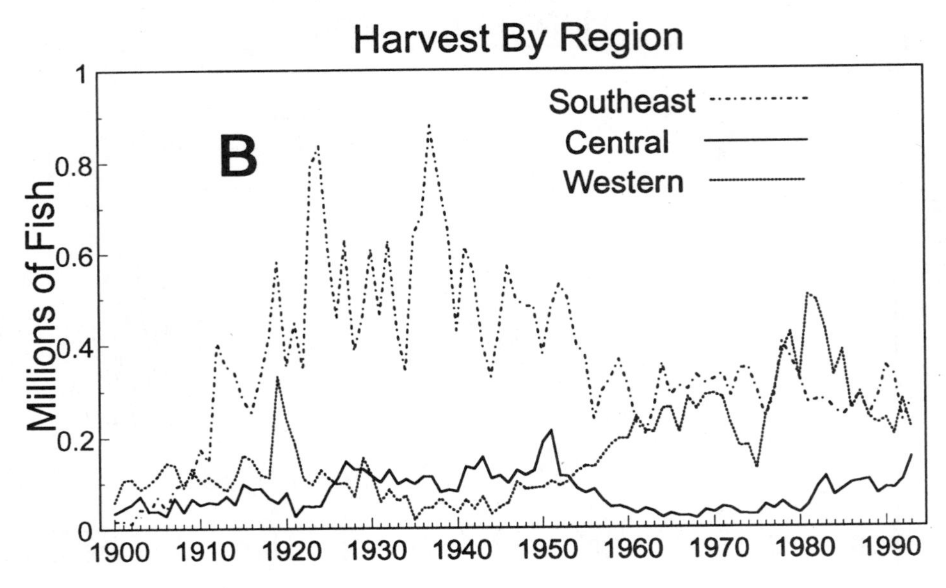

Figure 7. Statewide (A) and regional (B) commercial harvests of chinook salmon in Alaska, 1900–93.

harvest levels. Contributions from Alaska hatchery programs have increased in recent years (Fig. 7A) and averaged 10% of the commercial catch from 1990 to 1992.

The catch patterns over time were very different between the three regions (Fig. 7B). In the western region, catches increased in the 1950s, reflecting more extensive development of commercial fishing for this species in the region. Catch declined in the early 1970s, then increased to historically high levels in the late 1970s and 1980s. The record 10-year (1977–86) catch average for the region reached 381,000. Since that period, catches declined somewhat and averaged 249,000 fish. Foreign high-seas driftnet fishing and bycatch in trawl fisheries may have negatively affected domestic catch of western Alaska chinook (Olsen 1994). Wild stocks produced virtually all the chinook salmon harvested in the western region.

Commercial catches in the central region were highest during 1942–51 (Fig. 7B) when the average catch was 138,000 chinook salmon. Catches declined to 20,000–55,000 fish in the 1960s and 1970s, and then increased in the 1980s, averaging 94,000 from 1984 to 1993. Most of the chinook salmon recreational harvest occurred in the central region, and much of the chinook production in this region is now allocated for recreational harvest. From 1983 to 1992, recreational catch in the central region averaged 81,000 chinook; thus the combined recreational and commercial harvest is at historically high levels. Wild stocks produced almost all (>97%) the fish harvested.

In the southeast region, the recent (1984–93) average catch of 276,000 chinook salmon was less than half of the historical high 10-year (1930–39) average of 607,000. The chinook salmon harvest in southeast Alaska is predominantly an interception fishery on stocks of chinook salmon originating from Canada and the Pacific Northwest that migrate through or that rear in the southeast region. Catch limits set by the Pacific Salmon Commission (PSC), as part of the coastwide chinook rebuilding program established by the US/Canada Salmon Treaty in 1985, directly constrained current harvest levels. The large decline in catch from the historical high levels of the 1930s to 200,000–400,000 fish after 1953 (Fig. 7B) was due in large part to the decline and extirpation of chinook populations in the Columbia River.

Chinook production from Alaska hatcheries is an important component of the catch in southeast Alaska. Most of the Alaskan hatchery production is exempt from the PSC catch limits. Alaskan hatcheries contributed an average of 19% of the commercial chinook catch in the region over 1990–93. Hatchery fish from British Columbia, Oregon, and Washington also made up a large component of the catch in this region, contributing 36% of the commercial harvest in 1992 (K. Crandall, Alaska Dep. Fish and Game, Juneau, pers. comm.).

Chinook salmon escapements generally exhibited no trend or increased across all three regions (Table 1). In the western region, 60% of the populations indicated no trend while 38% increased (Konkle and McIntyre 1987). In the central region, similar percentages of the populations either lacked a trend (47%) or increased (53%) (Konkle and McIntyre 1987). In the southeast region, most (55%) of the populations exhibited no trend while 39% increased and 6% decreased.

STEELHEAD

In Alaska, steelhead (*O. mykiss*) are both highly valued for recreational fishing and harvested incidentally in commercial salmon fisheries. Although steelhead are distributed from southeast Alaska to the Alaska Peninsula, the species is not abundant in the state. The annual recreational harvest of steelhead averaged 5,500 fish for 1983–92 (Mills 1993). Harvest declined since 1989 to 3,149 fish in 1992 (the last year for which harvest data were available), the lowest level since 1979. This decline in harvest has been due, at least in part, to implementation of catch-and-release regulations on some of the major steelhead watersheds.

There are few long-term data sets on steelhead in Alaska, leaving only three watersheds for my analysis (two in southeast and one in central Alaska) (Table 1). Although these long-term data sets did not indicate declines, catch and release regulations were recently implemented because of concern over low escapements of steelhead (Doug Jones, Alaska Dep. Fish and Game, pers. comm.). Escapements in the late 1980s and early 1990s were generally below the long-term average in the three populations. In 1992 and 1993, escapement counts rebounded to historical highs for one of the southeast populations and for the central population, but remained below average for the other southeast population.

Factors Influencing High Productivity

Five major factors contribute to the general high abundance and productivity of Pacific salmon in Alaska:

1. relatively pristine and undeveloped habitats,
2. salmon management policies within the state,
3. the elimination of high-seas driftnet fisheries,
4. enhancement by hatcheries, and
5. favorable environmental conditions.

I discuss each in turn.

Habitat

Historically the remoteness of Alaska and its climate have been the main reasons for its relatively undeveloped habitat. Now >50% of Alaskan lands are designated as National Park, Preserve, Wildlife Refuge, or National Forest Wilderness (Metcalfe 1993). Sustained yield from Alaskan fish and wildlife is a specific mandate in the State Constitution, and anadromous fish habitat receives special protection under the state's Anadromous Stream Act. A number of state and federal agencies are actively involved in regulating development to minimize impacts on anadromous fish habitat.

Anadromous salmonid habitat actually increased in Alaska in recent history as a result of rapid recession of glacial ice sheets since the 18th century. In Glacier Bay, southeast Alaska, four species of Pacific salmon have colonized at least 12 watersheds since glacial recession (Milner and Bailey 1989, Alaska Department of Fish and Game 1992). Geomorphological changes associated with glacial recession also provided new salmon habitat in the Yakutat forelands in the southeast region and in the Kenai Fiords in the central region.

Management

Enlightened management played a critical role in the recovery of Alaskan salmon resources from decades of overfishing. Initial federal management originated as passive regulation under central authority; the state's approach included intensive regulation under local authority, limited entry, and a policy of discouraging and reducing interception fisheries (Eggers 1992). Royce (1989) described the current management approach as "a model fishery management program that is produced, supported, and accepted politically by those that are managed." This program has been specifically designed to provide for predetermined escapement needs within ranges thought to ensure the state's constitutional mandate for sustained yield while avoiding resource overutilization and destruction (Meacham and Clark 1994).

Elimination of High-Seas Driftnets

The elimination of the high-seas driftnet fisheries in the North Pacific resulted from decades of international negotiations. Research under INPFC auspices showed that large numbers of North American salmon were intercepted. During the 1950s, the Japanese high-seas salmon fleet took ≤50% of the harvest of western Alaska sockeye salmon (Fredin and Worlund 1974). Through INPFC, the US and Canada effectively reduced the directed catch of North American salmon and eventually phased out the high-seas salmon fisheries (Harris 1988, Myers et al. 1993). International concern over bycatch of salmon, birds, and marine mammals further led to a United Nations moratorium on all high-seas driftnet fishing.

Hatcheries

Hatchery production significantly enhanced the total catch of salmon in Alaska in recent years. For the recent 5-year period (1989–93), hatcheries produced an average of 20% of the total commercial catch of salmon in the state. Hatcheries now produce large percentages of the harvest of pink salmon in the central region and chum, coho, and chinook salmon in the southeast region.

Favorable Environmental Conditions

Environmental conditions influence both the freshwater and marine survival of anadromous fish. Trends in abundance of Alaska salmon correspond with temperature and climate patterns in the North Pacific Ocean (Beacham and Bouillon 1993, Pearcy 1996). Since 1976, water temperatures in the Northeast Pacific have increased (Royer 1993). Warmer temperatures in the North Pacific are generally associated with higher abundance of salmon in Alaska (Rogers and Ruggerone 1993).

Factors Threatening Productivity and Diversity

Nehlsen et al. (1991) identified factors that have placed a large proportion of salmon populations in the Pacific Northwest at risk of extinction. Alaska is not immune to the same types of resource development and management choices that led to the decline of salmon elsewhere. I discuss four potential threats to the health of Alaskan salmon populations: (1) anthropogenic habitat degradation, (2) overfishing, (3) salmon enhancement programs, and (4) climate and geological processes.

Habitat Degradation

Despite the relatively undeveloped nature of Alaska and the presence of many land-use designations, certain development and resource extraction activities can affect anadromous salmonid habitats. The Alaska Department of Environmental Conservation (1992) has identified over 200 surface water bodies with impaired or suspected impaired water quality. Major sources of damage include urbanization, timber harvest, mining, and oil and gas development. A large number of restoration projects have been attempted in the state in response to both natural and anthropogenic impacts on aquatic habitats (Parry et al. 1993)

Overfishing

Managers of fish resources in and adjacent to Alaska face a constant challenge of achieving the potential harvest from highly productive populations while not overfishing less productive species or populations. For example, western Alaska chum salmon are caught as bycatch in directed sockeye salmon fisheries or in groundfish fisheries. Management policies using constant escapement goals and stock-specific harvests can provide protection to weak populations, but such policies require an expensive, intensive management system (Eggers 1993).

Enhancement Programs

A major factor contributing to the high salmon harvest levels in Alaska is salmon enhancement. These enhancement programs can generate problems as well as benefits. Biological concerns include competition between wild and hatchery fish, displacement on the spawning grounds, and genetic introgression (Hindar et al. 1991). Productive hatcheries also can lead to overfishing of wild stocks; this concern is especially acute in Prince William Sound, where hatchery production now dominates the salmon fisheries (Geiger et al. in press). While the State has established specific policies on fish transport and hatchery permitting in order to minimize such interactions (e.g., Davis and Burkett 1989), enhancement programs should be continually evaluated to ensure that the policies are achieving their goals of increased production while minimizing impacts on wild stock productivity.

Climate

On the basis of the 18.6-year cycle identified by Royer (1993), the Northeast Pacific should be entering a period of cooling temperatures, which could result in lower production of salmon. Climate patterns also influence glacial movement patterns, which can have severe impacts on salmon habitats. The advance of the Lowell Glacier in the mid-18th century exterminated anadromous fish populations in the main branch of the Alsek River (Lindsey 1975). The current advance of the Hubbard Glacier threatens the salmonid populations in the Situk River (Thedinga et al. 1993).

Conclusion

Alaska salmon have reached unprecedented levels of abundance. The numbers of salmon harvested are generally at or approaching record high levels throughout the state. Escapements have been predominantly stable or increasing, indicating that the high harvest levels reflect abundance and productivity of salmon stocks and not the overexploitation of the resource that occurred in the first half of the century.

The excellent current status of the resource does not indicate the absence of problems or concerns for the well-being of salmon in Alaska. As noted previously, the same factors associated with the decline of salmon in the Pacific Northwest exist in Alaska. Escapements for some populations of salmon have declined in recent years. Other populations, although stable, are considered depressed and need to be rebuilt (e.g., some chinook salmon populations in southeast Alaska [Chinook Technical Committee 1993]). A number of anadromous stream systems are in need of habitat restoration (Parry et al. 1993). Ongoing harvest management problems include bycatch of salmon in groundfish fisheries and incidental catch of non-targeted populations in directed salmon fisheries. These issues certainly need to be addressed to ensure continued high productivity; however, such problems in Alaska represent islands of disturbance in a matrix of intact habitat and robust salmon populations. This is a very different situation from that of salmon in the Pacific Northwest, where healthy populations are often isolated in a geography of anthropogenic disturbance.

Resource managers in Alaska will be challenged to maintain the current health of salmon and their habitat in the state as environmental conditions fluctuate and urbanization and development of other resources increase. Managers must evaluate the impact of habitat, harvest, and enhancement policies on productivity and diversity of salmon populations in order to act effectively to maintain these populations. To help meet this need, a cooperative effort involving ADFG and the Alaska Chapter of the American Fisheries Society is now underway to produce a population status inventory compiling currently available information on salmon and steelhead in the state.

Acknowledgments

I thank T. Baker, H. Savikko, P. Rigby, R. Harding, D. Jones, and M. McNair for providing me with their compilations of current and historical data sets. I also appreciate the insightful and critical reviews by T. Baker, P. Rigby, D. Eggers, B. Heard, J.H. Clark, R. Wilbur, B. Van Alen, and two anonymous reviewers. The perspectives presented in this paper are those of the author and do not necessarily reflect the views of either the reviewers or my employer.

Literature Cited

Alaska Department of Environmental Conservation. 1992. Alaska water quality assessment. Alaska Department of Environmental Conservation EPA Section 305(b) Report.

Alaska Department of Fish and Game. 1992. Catalog of waters important for spawning, rearing, or migration of anadromous fishes. Alaska Department of Fish and Game Habitat Division. Juneau, Alaska.

Beacham, R.J. and D.R. Boullion. 1993. Pacific salmon production trends in relation to climate. Canadian Journal of Fisheries and Aquatic Sciences 50: 1002-1016.

Canada Department of Fisheries and Oceans. 1993. Annual summary of British Columbia commercial catch statistics: 1991. Department Fisheries and Oceans Canada. Vancouver, British Columbia.

Chinook Technical Committee. 1993. Joint Chinook Technical Committee 1992 Annual Report. Report TCCHINOOK (93)-2. Pacific Salmon Commission. Vancouver, British Columbia.

Cobb, J.N. 1911. The salmon fisheries of the Pacific coast. Document No. 751. US Department Commerce and Labor Bureau of Fisheries. Washington, DC.

Cooley, R.A. 1961. Decline of the Alaska salmon: a case study in resource conservation policy. PhD dissertation, University of Michigan. Ann Arbor.

Davis, B. and B. Burkett. 1989. Background of the genetic policy of the Alaska Department of Fish and Game. Alaska Department of Fish and Game FRED Report 95.

Eggers, D.M. 1992. The benefits and costs of the management for natural sockeye salmon stocks in Bristol Bay, Alaska. Fisheries Research 14: 159-177.

Eggers, D.M. 1993. Robust harvest policies for Pacific salmon fisheries, p. 85-106. *In* G. Kruse, D.M. Eggers, R.J. Marasco, C. Pautzke, and T.J. Quinn II (eds.), International Symposium on Management Strategies for Exploited Fish Populations. Anchorage, Alaska. Alaska Sea Grant Program Rep. 93-02, University of Alaska, Fairbanks.

Fredin, R.A. and D.D. Worlund. 1974. Catches of sockeye salmon of Bristol Bay origin by the Japanese mothership salmon fishery, 1956-70. International North Pacific Fisheries Commission Bulletin 30: 1-80.

Freund, R.J. and R.C. Littell. 1992. SAS system for regression, 2nd ed. SAS Institute Inc., Cary, North Carolina.

Geiger, H., S. McGee, D.M. Gaudet, L. Peltz, and J.A. Brady. In press. Trends in abundance of hatchery and wild stocks of pink salmon in Cook Inlet, Prince William Sound, and Kodiak, Alaska. *In* B. Riddle (ed.), Proceedings of the International Symposium on Biological Interactions of Enhanced and wild Salmonids. Canadian Special Publication of Fisheries and Aquatic Sciences.

Harris, C.K. 1988. Recent changes in the pattern of catch of North American salmonids by the Japanese high seas salmon fisheries, p.41-65. *In* W.J. McNeil (ed.), Salmon Production, Management, and Allocation. Oregon State University Press, Corvallis, Oregon.

Henry, K.A. 1993. Pacific coast salmon, p.63-68. *In* L. Low (ed.), Status of Living Marine Resources off the Pacific Coast of the United States for 1993. US Department of Commerce, NOAA Technical Memorandum NMFS-AFSC-26. Seattle, Washington.

Hindar, K., N. Ryman, and F. Utter. 1991. Genetic effects of cultured fish on natural fish populations. Canadian Journal of Fisheries and Aquatic Sciences 48: 945-957.

International North Pacific Fisheries Commission. 1993. Statistical Yearbook 1990. International North Pacific Fisheries Commission. Vancouver, British Columbia.

Konkle, G.W. and J.D. McIntyre. 1987. Trends in spawning populations of Pacific anadromous salmonids. US Fish and Wildlife Service Technical Report 9.

Lindsey, C.C. 1975. Proglacial lakes and fish dispersal in southwestern Yukon Territory. Internationale Vereinigung fuer Theoretische und Angewand Limnologie Verhandlungen 19: 2364-2370.

Meacham, C.P. and J.H. Clark. 1994. Pacific salmon management—the view from Alaska. Alaska Fishery Research Bulletin 1(1): 76-80.

Metcalfe, P. M. (ed). 1993. Alaska Blue Book 1993-94. Alaska Department of Education. Juneau, Alaska.

McNair, M. and J.S. Holland. 1994. Alaska fisheries enhancement program 1993 annual report. Alaska Department of Fish and Game, Commercial Fisheries Management and Development Division.

Mills, M.J. 1993. Harvest, catch, and participation in Alaska sport fisheries during 1992. Alaska Department of Fish and Game Fisheries Data Series 93-42.

Milner, A.W. and R.G. Bailey. 1989. Salmonid colonization of new streams in Glacier Bay National Park, Alaska. Aquaculture and Fisheries Management 20: 179-192.

Myers, K.W., C.K. Harris, Y. Ishida, L. Margolis, and M. Ogura. 1993. Review of the Japanese landbased driftnet salmon fishery in the Western North Pacific Ocean and the continent of origin of salmonids in this area. International North Pacific Fisheries Commission Bulletin 52.

Nehlson, W., J.E. Williams, and J.A. Lichatowich. 1991. Pacific salmon at the crossroads: stocks at risk from California, Oregon, Idaho, and Washington. Fisheries 16(2): 4-21.

Olsen, J.C. 1994. Alaska salmon resources, p. 77-89. *In* L. Low, J. C. Olsen, and H. W. Braham (eds.), Status of Living Marine Resources off Alaska, 1993. US Department of Commerce, NOAA Technical Memorandum NMFS-AFSC-27.

Parry, B.L., G.A. Seaman, and C.M. Rozen. 1993. Restoration and enhancement of aquatic habitats in Alaska: project inventory, case study selection, and bibliography. Alaska Department of Fish and Game Habitat and Restoration Division Technical Report 93-8.

Pearcy, W. 1996. Salmon production in ocean domains, p.331-352. *In* D.J. Stouder, P.A. Bisson, and R.J. Naiman (eds.), Pacific Salmon and Their Ecosystems: Status and Future Options. Chapman and Hall, New York.

Pennoyer, S. 1988. Early management of Alaska fisheries. Marine Fisheries Review 50: 194-197.

Ricker, W.E. 1972. Hereditary and environmental factors affecting certain salmonid populations, p. 19-160. *In* R.C. Simon and P.A. Larkin (eds.), The Stock Concept in Pacific Salmon. University of British Columbia, Institute of Animal Resource Ecology. H.R. MacMillan Lectures in Fisheries. Vancouver, British Columbia.

Rigby, P., J. McConnaughey, and H. Savikko. 1991. Alaska commercial salmon fisheries, 1878-1991. Alaska Department of Fish and Game Regional Information Report 5J91-16.

Rogers, D.E. and G.T. Ruggerone. 1993. Factors affecting marine growth of Bristol Bay sockeye salmon. Fisheries Research 18: 89-103.

Royce, W.F. 1989. Managing Alaska's salmon fisheries for a prosperous future. Fisheries 14(2): 8-13.

Royer, T.C. 1993. High latitude oceanic variability associated with the 18.6 year nodal tide. Journal of Geophysical Research 98(3): 4639-4644.

Thedinga, J.F., S.W. Johnson, KV. Koski, J.M. Lorenz, and M.L. Murphy. 1993. Potential effects of flooding from Russell Fiord on salmonids and habitat in the Situk River, Alaska. NMFS Auke Bay Laboratory Processed Report AFSC 93-01.

Pacific Salmon Abundance Trends in the Fraser River Watershed Compared with Other British Columbia Systems

T.G. Northcote and D.Y. Atagi

Abstract

Overall abundance—catch plus escapement—of adult Pacific salmon (*Oncorhynchus* spp.) in the Fraser River basin (about one quarter of the area of British Columbia [BC]) has decreased sharply for sockeye (*O. nerka*), pink (*O. gorbuscha*), chum (*O. keta*), chinook (*O. tshawytscha*), and coho (*O. kisutch*) salmon between estimated levels near the turn of the century and those of recent decades. The extent of decline ranges from a low of about twofold for chum salmon to a high of nearly eightfold for coho salmon. Nevertheless, the Fraser River still remains the world's greatest single river producer of Pacific salmon. Trend analyses for escapements over recent decades show statistically significant increases for all Pacific salmon escapements combined and for pink, chum, chinook, and sockeye individually. Sub-basins of the Fraser watershed support distinctive mixes of the five eastern Pacific salmon species; extent, causes, and management implications of these regional differences in species abundance are explored.

Parts of the remaining three quarters of BC also support sizable stocks of Pacific salmon. The largest other producer is the Skeena River system. Small to moderate stocks occur in the Okanagan River (Columbia drainage), a number of Vancouver Island and Queen Charlotte Island rivers, various coastal drainages of the south-central mainland, the Nass and Stikine river systems, as well as in other transboundary coastal rivers (Alaska panhandle), and in tributaries of the Yukon and Mackenzie rivers. Since the early 1900s, two upper Columbia River chinook stocks have become extinct in BC and another virtually so in the Okanagan River. The latter river continues to support the largest sockeye population for the Columbia River system, but escapements there have not exceeded 40,000 in the last two decades and have dropped to as low as 1,000 in 1978. Long-term decreases have occurred in coho and chinook salmon escapements to Vancouver Island, south-central coast and Skeena-Nass rivers, and in all five eastern Pacific salmon species to the Queen Charlotte Islands. In contrast long-term sockeye escapement to the Skeena-Nass system probably has increased.

In all major BC drainage areas, except the transboundary, recent coho escapements have declined significantly. Sockeye escapements show a significant increase in Vancouver Island, south-central coast, and Skeena-Nass drainage areas. In British Columbia overall, total salmon escapements have significantly increased over the last 4 decades, especially for even- and odd-year pink as well as sockeye salmon. Chum and chinook salmon escapements have shown no significant trend, but coho salmon have declined significantly. Causes for these recent changes in salmon escapement to BC systems are considered briefly.

Introduction

A comprehensive review of the status of British Columbia (BC) anadromous salmonid stocks (*Oncorhynchus* spp.) was initiated in 1993 under the auspices of the American Fisheries Society, North Pacific International Chapter, and a grant from The Pew Charitable Trusts of Philadelphia, USA. The objective in part was to identify and rank stocks of such salmonids at risk and to complete coverage along the eastern Pacific of similar reviews already conducted for California, Oregon, Idaho, and Washington (Nehlsen et al. 1991) and currently underway in Alaska. Unfortunately, results of the BC review are still unavailable. Alternatively, recent reviews have been completed on salmonid stocks (excluding steelhead trout, *O. mykiss*) of the Fraser River, the largest river system within BC, draining over 220,000 km^2 (about one quarter of the province) (Fig. 1), and the largest single river producer of Pacific salmon in the world (Northcote and Larkin 1989, Northcote and Burwash 1991). Therefore, it would be useful to summarize both long-term (around turn of the century to present) and short-term (the recent four decades) abundance trends for each of the five species of Pacific salmon in the Fraser River watershed. Furthermore, this watershed is formed by several sub-basins wherein the great diversity in geology, physiography, climate, and hydrology is also reflected in both the species and stock mix of Pacific salmon and their abundance. Insight into factors controlling such diversity in the sub-basins should be instructive in considering those same factors for the province as a whole.

Other regions (Fig. 1) within BC include the Skeena River watershed, another very large producer of Pacific salmon (second only to the Fraser River), and the Nass River. In addition there are a number of smaller coastal and insular (Vancouver Island, Queen Charlotte Islands) drainages along with parts of the Columbia, Yukon, and Mackenzie river systems, many of which contribute sizable salmon stocks. Again, it would be useful to compare Pacific salmon species distribution and abundance in these systems (Table 1) with those of the Fraser River watershed. Finally, it should be possible to estimate the present overall recorded escapement of salmon to BC systems.

Fraser River Salmon Abundance

Long-Term Trends

Abundance of all five species of eastern Pacific salmon in the Fraser River has declined sharply since the 1950s compared with estimated levels near the turn of the 19th century (Table 2). During the latter, pink salmon (*O. gorbuscha*) abundance in odd-numbered years (virtually none occur in even-numbered years) approached 24 million. Chum (*O. keta*), coho (*O. kisutch*), and chinook (*O. tshawytscha*) salmon numbers probably were close to 1 million annually, and sockeye averaged well over 34 million annually but with large variation between year-to-year "lines" (i.e., stocks) (see Ricker 1950, 1987). The historical abundance of pink and sockeye salmon spawners bound for the Fraser River was enormous with runs up to nearly 50 million pink and ≤160 million sockeye on a big cycle year (Ricker 1950, 1987, 1989). Coho have decreased nearly eightfold in abundance between historical and recent periods (Table 2). Pink, chinook, and sockeye abundance has decreased about fivefold and chum about twofold.

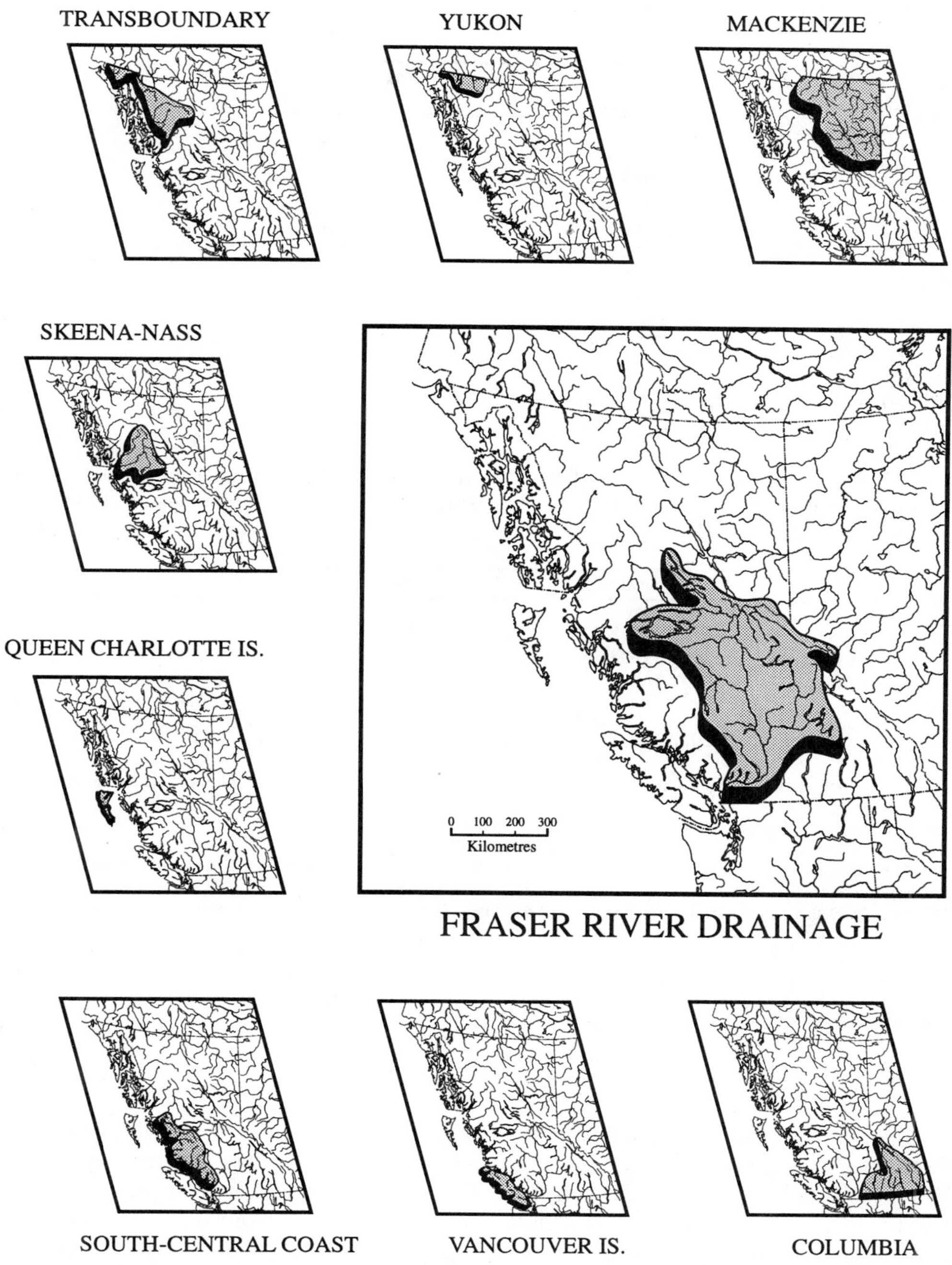

Figure 1. The Fraser River drainage, British Columbia, and the eight other drainage areas that support some stocks of anadromous Pacific salmon. Adapted in part from figures in McPhail and Carveth (1992).

Table 1. Number of streams or rivers in each drainage area where escapements have been estimated from 1953 to 1992.

Drainage area	Number of recorded rivers/streams
Fraser River	347
Columbia River	1[a]
Vancouver Island	392
South-central coast	752
Queen Charlotte Islands	236
Skeena-Nass rivers	226
Transboundary rivers	3[b]
TOTAL	1957

[a]Okanagan River.
[b]System-wide escapement estimates for Alsek, Taku, and Stikine rivers.

Table 2. Average annual abundance (catch plus escapement in millions) of Pacific salmon in historical (late 1800s to early 1900s) and recent (1951 to 1980s) periods for the Fraser River system. Adapted from Northcote and Burwash (1991).

Species	Historical	Recent	Change
Pink	23.85[a]	4.32[b]	5.5-fold decrease
Chum	0.80[c]	0.39[d]	2.0-fold decrease
Coho	1.23[e]	0.16[f]	7.7-fold decrease
Chinook	0.75[e]	0.15[f]	5.0-fold decrease
Sockeye	34.23[g]	6.75[h]	5.1-fold decrease

[a]1907–13 estimated abundance (per 2 years) of adult fish bound mainly for the Fraser River and a few much smaller nearby streams (Ricker 1989).
[b]Average return 1961–79 brood years per 2 years from Beacham (1984a).
[c]1940–50 catch estimates from Palmer (1972) with 75% exploitation rate.
[d]1951–81 escapement data (Canada Department of Fisheries and Oceans, Pacific Biological Station, Nanaimo, BC, unpubl. data) plus catch data from Beacham (1984b).
[e]1920s to early 1930s average annual catch estimates for Sooke traps, Swiftsure Bank troll fishery, Fraser River gillnets (Fraser et al. 1982), with assumed 50% exploitation rate.
[f]1951–80 average return from Fraser et al. (1982).
[g]From Ricker (1987) for 1901–1913, using 100 million for the 1901 "line" and 5 million for the 1902, 1903, and 1904 "lines"; see also Ricker (1950).
[h]Average of early plus late runs, 1970–82 from Starr et al. (1984).

Over three-quarters of the pink salmon historically spawned above the middle mainstem of the Fraser River canyon, as did most large stocks of sockeye. Railway construction in 1913 caused blockages at Hells Gate in this canyon, which exterminated upriver pink stocks and sharply reduced those of sockeye. Overfishing in general and selectively on certain stocks and sizes, along with natural and human-induced changes to spawning habitat, have been implicated as contributory causes for declines in pink and sockeye abundance (Ricker 1987, 1989) and probably were involved in declines for the other three species of Pacific salmon.

SHORT-TERM TRENDS

Between 1951 and 1980, average salmon escapement to the Fraser River remained close to 2.5 million annually (Fig. 2), but then doubled to >5.6 million fish during 1980–90. Indeed, if considered over the last 13 years, including preliminary 1993 data, the average annual escapement would be >7 million salmon. This increase occurred mainly in pink salmon, whose average annual escapements in the earlier decades were <1 million fish but rose to >3 million recently, and to sockeye salmon which showed a smaller increase (from ~1.5 million in earlier decades to >2 million in 1981–92; Fig. 2).

Because of problems with repeated time series data, the usual tests for statistical significance cannot be applied. The method developed by Cox and Stuart (1955) permits a test for

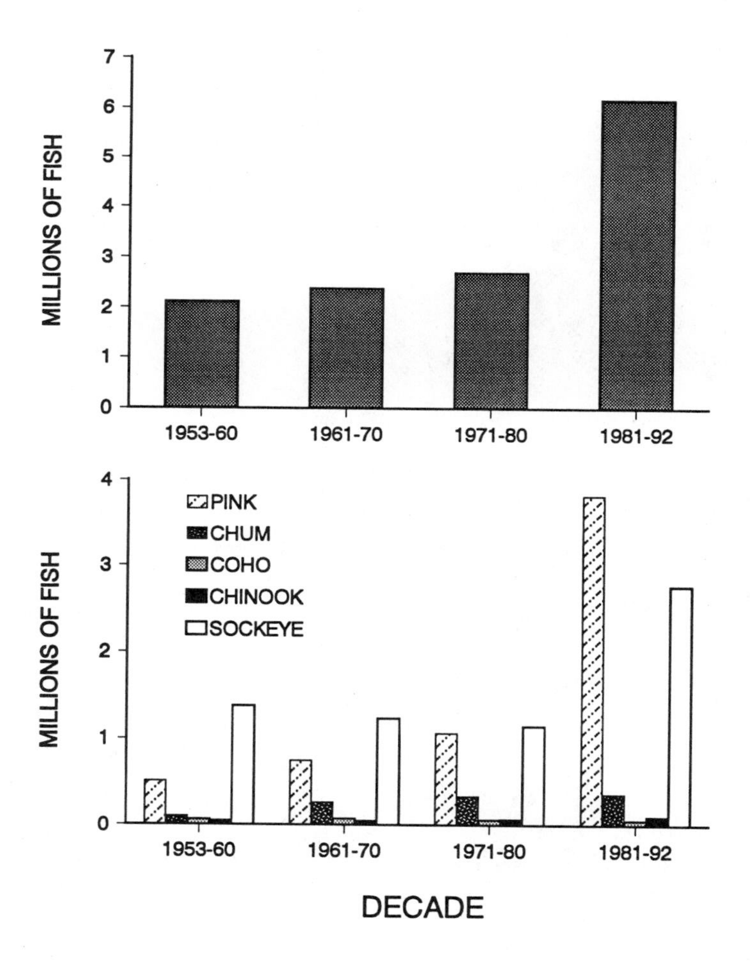

Figure 2. Upper: average annual escapement of Pacific salmon to the Fraser River, 1953–92. Lower: as above for each of the five species. Pink salmon escapements given on an annual average basis (i.e., half of the odd-numbered year escapement). Adapted and updated from Northcote and Burwash (1991).

trends over time without making assumptions of independence or normality for the data (see also Sachs 1978). The occurrence of pink salmon adults only in odd-numbered years is another complication in examining trends for all species together, but this can be adjusted by splitting pink salmon escapements between sequential odd and even years. The increasing trend for total salmon escapement over the four decades is statistically significant ($p < 0.05$; Fig. 3). There are also significant ascending trends ($p < 0.05$) for pink, chum, chinook, and sockeye alone, but not so for coho escapements, which have significantly declined during the same period.

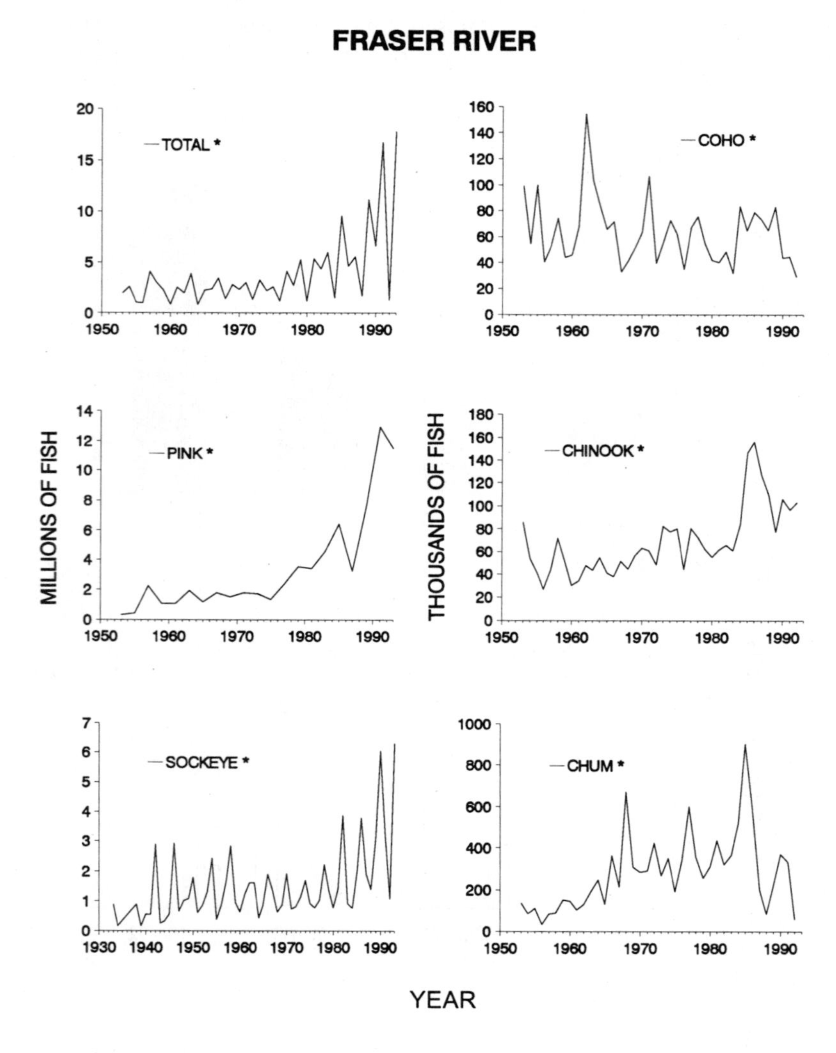

Figure 3. Annual Pacific salmon escapement to the Fraser River, 1953–92. Preliminary 1993 sockeye and pink escapement estimates are also included. Adapted and updated from Northcote and Burwash (1991). * = trend statistically significant at $p \leq 0.05$.

Regional and Sub-Basin Diversity

The 1,375-km mainstem of the Fraser River first sweeps northwestward, following the Rocky Mountain Trench in its headwaters; it then turns south, running through its deep central canyon to where it swings west and flows at low gradient over its last 160 km to the Pacific Ocean (Figs. 1, 4). Its smallest region is formed by the Lower Fraser Valley plus several smaller drainages and one larger drainage, the Lillooet Sub-basin (Fig. 4). By far its largest region is in its middle reaches, containing the mainstem canyon along with one small (Bridge-Seton) and

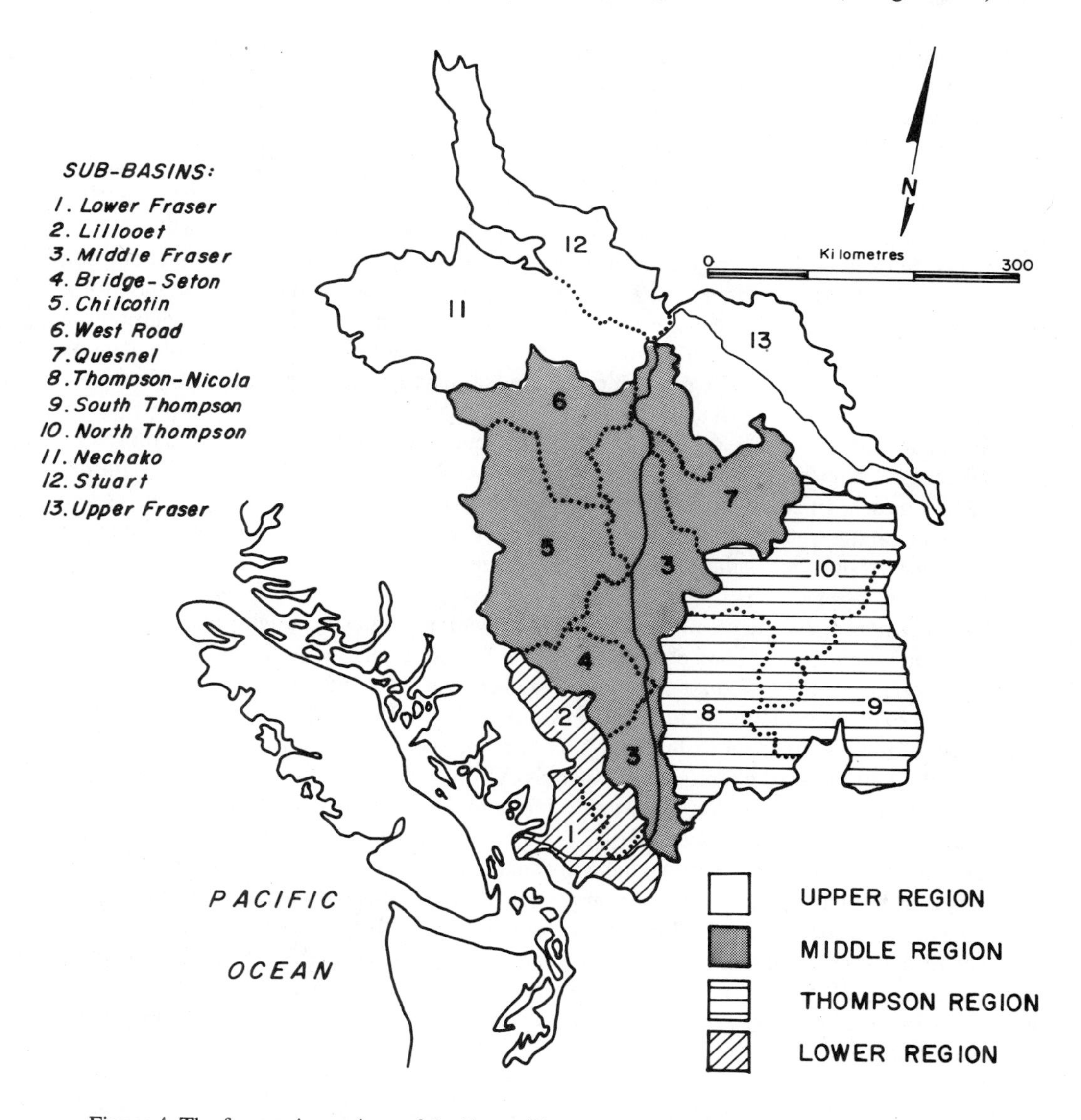

Figure 4. The four major regions of the Fraser River watershed and their respective sub-basins.

three larger sub-basins (Chilcotin, West Road, and Quesnel). The Thompson region forms the second largest area, with three sub-basins of approximately equal size. The Upper Region has three main parts, the Nechako Sub-basin draining from the west and having lost about 44% of its area by the Kemano diversion, the Stuart and Salmon river systems (Stuart Sub-basin) draining from the northwest, and the Upper Fraser Sub-basin arising from the western slopes of the Rocky Mountains.

The four major regions of the Fraser River vary greatly not only in magnitude of salmon escapement but also in species composition (Fig. 5). Almost half (~47%) of the average escapement of salmon to the river during the last four decades (>3 million fish annually, adjusted for adult pink salmon only on odd-numbered years) has been accommodated in its smallest (Lower) region. The largest (Middle) region has accounted for only 19% of the annual average escapement whereas that to the Thompson Region was 25%. The Upper Region, though large in area, supported only 8% of the total escapement.

Pink salmon make up over two thirds of the average annual escapement to the Lower Region and are the second largest component of the Middle and Thompson regions (~30% and 25%, respectively; Fig. 5). Chum salmon contribute appreciably (~18%) only to the Lower Region escapements. Sockeye salmon are the most important species to the Middle (68%), Thompson (70%), and particularly the Upper (96%) region escapements. Coho and chinook salmon form only a small portion of the escapement to all regions. Coho make their highest contribution (both by percentage and numerically) to the Lower Region (3.4%; 47,000 fish) while chinook contribute >4% to the Upper Region. Numerically, chinook are the highest in the Thompson Region, averaging ~27,000 fish annually.

Regional differences in escapement by the five species of salmon become even more distinct when sub-basins within regions are considered (see Northcote and Burwash 1991 for details). Salmon escapement to the Fraser River is highly partitioned between the five species into different parts of the watershed, with some species such as pink salmon dominating in the lower drainage sub-divisions and others such as sockeye salmon dominating in outer and upper extremities of the system.

The mainstem Fraser River also is used as a spawning habitat by at least three species of Pacific salmon—pink, chum, and chinook. Over the 17 odd-numbered years between 1957 and 1989, an average of 1.53 million pink salmon spawned every odd-numbered year in the section of mainstem river between the communities of Chilliwack and Hope. This represents 90% of the total pink salmon escapement to the Lower Fraser Sub-basin, making the mainstem river of paramount importance to the reproductive success of this species. Chum salmon were thought to spawn in a similar portion of the mainstem river as that used by pink salmon. Palmer (1972) indicated that in the 1960s an average of 77,000 chum salmon may have used the mainstem reaches. Recent estimates have shown that only small numbers (average of 8,500 annually) use the mainstem river itself. Chinook salmon spawn in the headwater reaches of the mainstem Fraser River, representing >32% of the salmon escapement to this drainage sub-basin.

The recent apparent increase in average escapement of salmon to the Fraser River seems to have occurred in all four regions, but most clearly in the Lower Region and less so in the Upper Region (Northcote and Burwash 1991). In the Lower Region, chum, pink, and to a small extent sockeye salmon have contributed to the increase. Recorded escapements of pink, sockeye, and chinook salmon have increased over the last 4 decades in the Middle Region. Pink salmon in the 2 more recent decades and sockeye in the most recent decade are the major contributors to the

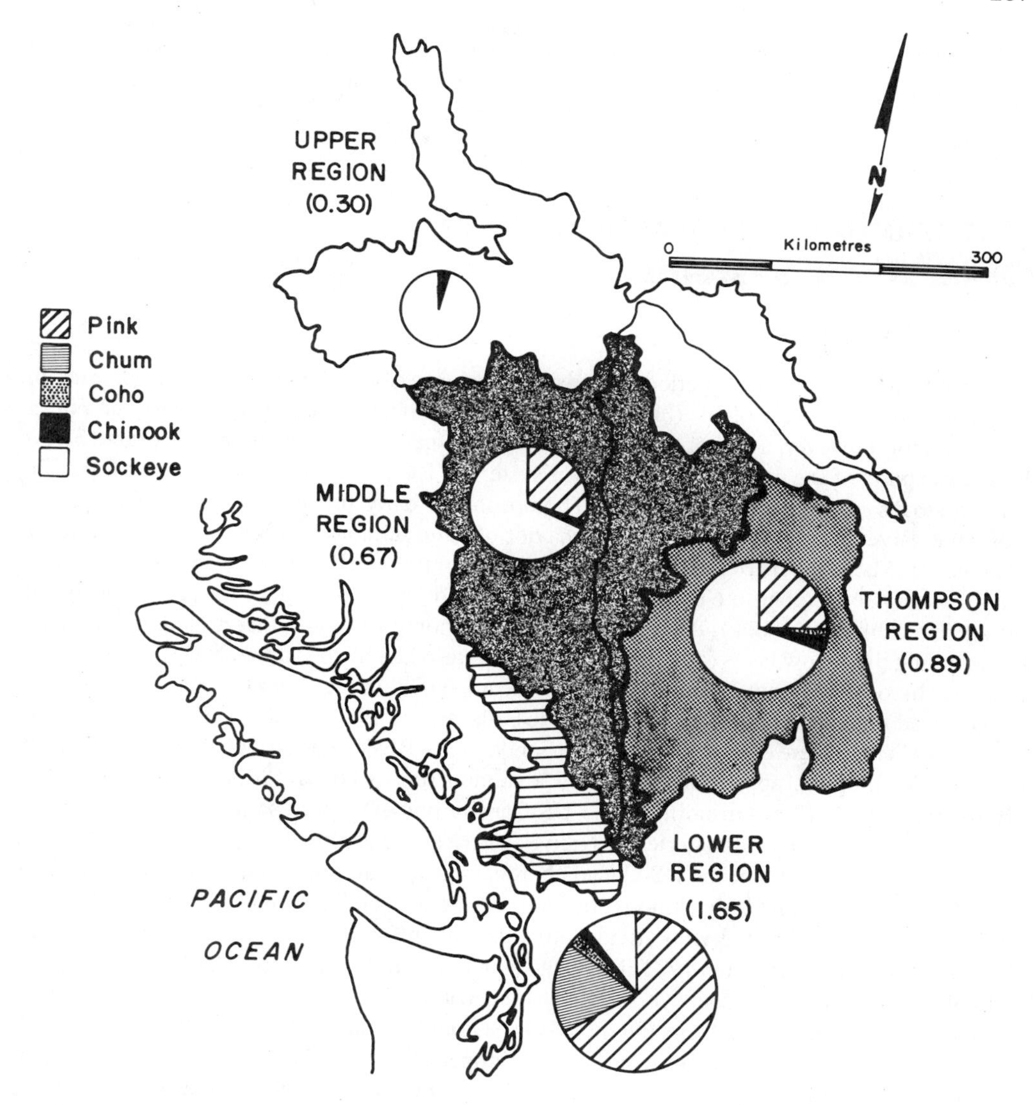

Figure 5. Average annual escapement (millions; 1953–92) of Pacific salmon to the four major regions of the Fraser River watershed. Pie chart areas are proportional to averages given in parentheses, representing an overall average for the whole system of >3.4 million salmon. Adapted and updated from Northcote and Burwash (1991).

overall increased escapements to the Thompson Region. In the Upper Region, sockeye salmon escapements did not increase significantly until the most recent decade, and chinook salmon escapements have nearly doubled in the same period.

Distribution and Abundance in Other River Systems

Long-Term Trends

Between historical and recent periods, trends in abundance of Pacific salmon in the other major drainage systems or areas of BC (Fig. 1) cannot be determined as clearly as those for the Fraser River system; however, some observations indicate the general pattern of change (Table 3). At least three parts of the upper Columbia River system in BC probably supported sizable Pacific salmon stocks in historical times. David Thompson, the early fur-trade explorer of the upper Columbia River, recorded the spawning behavior of large salmon in a tributary to Windermere Lake near the headwaters of the Columbia River system in BC. His party consumed the salmon for ~3 wk in late October to November 1807 (Hopwood 1971). Several of the salmon weighed >9 kg (most likely chinook). Also in late October, Thompson noted that a Native American family along the Arrow lakes (about 400 km downstream) had fresh salmon. Near the turn of the century, chinook salmon were still common in the upper Columbia River (Lyons 1969); shortly thereafter salmon were rarely seen above Kettle Falls, and by 1921 no adult salmon were returning to Lake Windermere (Lorraine 1924). The Similkameen River in BC, tributary to the Okanagan River in Washington, apparently supported a salmon run as far as 180 km upstream to the Similkameen Falls (150 km north of the US-Canada border) where salmon were frequently harvested according to oral histories of Native Americans (Fanning 1985). The current site of Enloe Dam, 29 km south of the US-Canada border, was also an important location for the harvest of adult salmon (probably chinook) by Native Americans in late prehistoric times. Because Enloe Dam was constructed without adult fish passage provisions, no salmon have migrated above it since its completion in 1923. The Okanagan River in BC supports one of the main upper Columbia River stocks of sockeye salmon, along with a few chinook salmon. No historical records of stock size are available, but there is little doubt that abundance of both has decreased greatly over the last century. Thus, in the three upper Columbia River tributaries in BC that historically supported Pacific salmon, stocks have become extinct in two and have seriously declined in the other.

For many rivers and streams of Vancouver Island and the south-central coast drainages of BC, it is only possible to make very rough estimates of historical abundance for coho and chinook salmon (Table 3). These stocks have shown up to twofold or more decreases between historical and recent periods. Very rough estimates of historical levels of salmon abundance can be made for Queen Charlotte Islands rivers and streams (Table 3) from early escapements given by Shirvell and Charbonneau (1984) and the review by Northcote et al. (1989). Escapements of pink, chum, and chinook salmon appear to have decreased substantially, and coho and sockeye to a lesser extent in these waters. In addition to long-term declining escapements of Pacific salmon to Queen Charlotte Islands spawning streams, virtual extinction of some stocks has

Table 3. Average annual abundance (escapement) of Pacific salmon estimated for historical (1800s) and recorded for recent (1953–92) periods for major river systems (excluding the Fraser River) of British Columbia.

River systems	Species	Historical	Recent	Change
Columbia				
1. Upper mainstem	Chinook	Probably >1,000	0	Extinction
2. Okanagan River	Chinook	Probably <1,000	<100	Decrease
	Sockeye	Probably >100,000	19,000	Decrease
3. Similkameen R.	Chinook	Probably <1,000	0	Extinction
Vancouver Island	Pink	Unknown	329,000	Unknown
	Chum	Unknown	891,000	Unknown
	Coho	Probably > 500,000	237,000	Decrease
	Chinook	Probably > 200,000	66,000	Decrease
	Sockeye	Unknown	322,000	Unknown
South-Central Coast	Pink	Unknown	3,489,000	Unknown
	Chum	Unknown	1,107,000	Unknown
	Coho	Probably > 500,000	290,000	Decrease
	Chinook	Probably > 100,000	71,000	Decrease
	Sockeye	Unknown	701,000	Unknown
Queen Charlotte Islands	Pink	Probably > 1,000,000	651,000	Decrease
	Chum	Probably > 600,000	326,000	Decrease
	Coho	Probably > 100,000	80,000	Decrease
	Chinook	Probably > 3,000	1,900	Decrease
	Sockeye	Probably > 50,000	39,000	Decrease
Skeena-Nass	Pink	Unknown	1,334,000	Unknown
	Chum	Unknown	18,000	Unknown
	Coho	Probably > 100,000	69,000	Decrease
	Chinook	Probably > 100,000	44,000	Decrease
	Sockeye	Probably < 1,000,000	1,041,000	Increase
Transboundary	Pink	Unknown	536,000	Unknown
	Chum	Unknown	87,000	Unknown
	Coho	Unknown	80,000	Unknown
	Chinook	Unknown	39,000	Unknown
	Sockeye	Unknown	236,000	Unknown
Yukon	Chum	Unknown	Present but no estimates	Unknown
	Chinook	Unknown	Present but no estimates	Unknown
Mackenzie	Chum	Unknown	Probably < 1,000	Unknown
	Chinook	Unknown	A few strays	Unknown

occurred (no recorded escapement or other evidence of the species for 10 consecutive years). Between 1947 and 1983 some 32 salmon stocks were said to have become "extinct" (13 for pink, 14 for coho, 4 for chum and 1 for sockeye salmon; Shirvell and Charbonneau 1984). Regional staff now consider that extinctions are questionable in 17 of these streams (R. Sjolund, Canada Dep. Fisheries and Oceans, Queen Charlotte City, BC, pers. comm.). In the Skeena and Nass watersheds, coho and chinook salmon probably have declined substantially from their historical abundance whereas sockeye salmon may well have increased (Table 3). Historical estimates of stock abundance are not available for the transboundary rivers of northwestern BC, and there are virtually no data other than the indication of the presence of fish for Yukon or Mackenzie system stocks (in the latter, several hundred chum were captured in the late 1970s and early 1980s [McLeod and O'Neil 1983]).

Short-Term Trends

Total escapement levels of Pacific salmon in Vancouver Island rivers have not shown any significant trends since the early 1950s even though there have been declines in escapement of even-year pink salmon from the mid-1960s to mid-1980s and significant declining trends of coho salmon (Fig. 6). Sockeye salmon escapements over the 4-decade period have shown a significant increasing trend but with sharp fluctuations since the 1980s. Similarly for south-central coast drainages, overall escapement records show a significant increasing trend despite the opposite and significant trends for coho and chinook salmon (Fig. 7), which are presumably "balanced" by the increasing trend for even-year pink and for sockeye salmon escapements.

No significant trend exists in overall salmon escapement to Queen Charlotte Islands rivers (Fig. 8), where "saw-toothed" fluctuations result largely from the strong influence of even-year pink escapements. Escapement records for odd-year pink, coho, chinook, and sockeye salmon all show significant declining trends in recent decades (Fig. 8).

For the Skeena-Nass systems, total escapements have clearly, though irregularly, increased over the 4 decades owing to the significantly increasing trends shown for even- and odd-year pink populations as well as for those of sockeye salmon (Fig. 9). The trend for coho is one of significant decline but that for chinook salmon is not significant.

Escapement data for the transboundary rivers are currently lacking, but a reasonable catch per unit effort (CPUE) time series exists for a number of terminal, commercial gillnet fisheries on the Stikine, Taku, and Alsek rivers (Fig. 10). During this period, CPUE for all salmon species increased irregularly and significantly even though no single species showed a similar trend (Fig. 10). The overall increase is likely due to a trend ($p = 0.053$) towards increases in sockeye and coho salmon catches. Chinook salmon catches have declined significantly; however, this is attributed to the implementation of restrictive harvest policies during this period (J. Muir, Alaska Dep. Fish and Game, Douglas, pers. comm.). Available escapement index data suggests that chinook, coho, and possibly sockeye salmon stocks may be rebuilding while chum salmon escapements are declining. Escapement estimates for pink salmon are scarce but trends in CPUE apparently reflect current escapement levels (J. Muir, Alaska Dep. Fish and Game, Douglas, pers. comm.).

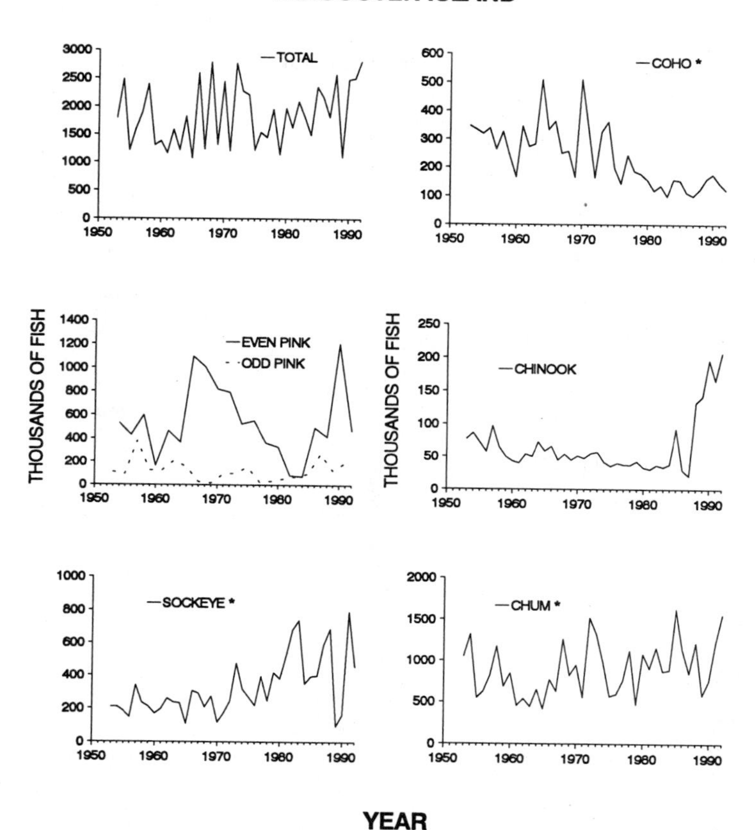

Figure 6. Annual Pacific salmon escapement to recorded rivers and streams of Vancouver Island, 1953–92. * = trend statistically significant at $p \leq 0.05$.

Overall Abundance in British Columbia

Estimated escapements of all species of Pacific salmon to the nine major drainage areas (excluding the transboundary rivers) of BC have increased from ~10 million fish in the 1950s to well over 20 million during most years since the mid-1980s (Fig. 11). This significant trend is driven mainly by increases in pink and sockeye salmon escapements, as there is no significant trend for chum and chinook salmon escapements and a highly significant decreasing trend for coho salmon. (Estimated total escapement for the latter was <300,000 in 1992, the lowest on record over the 40-year span!)

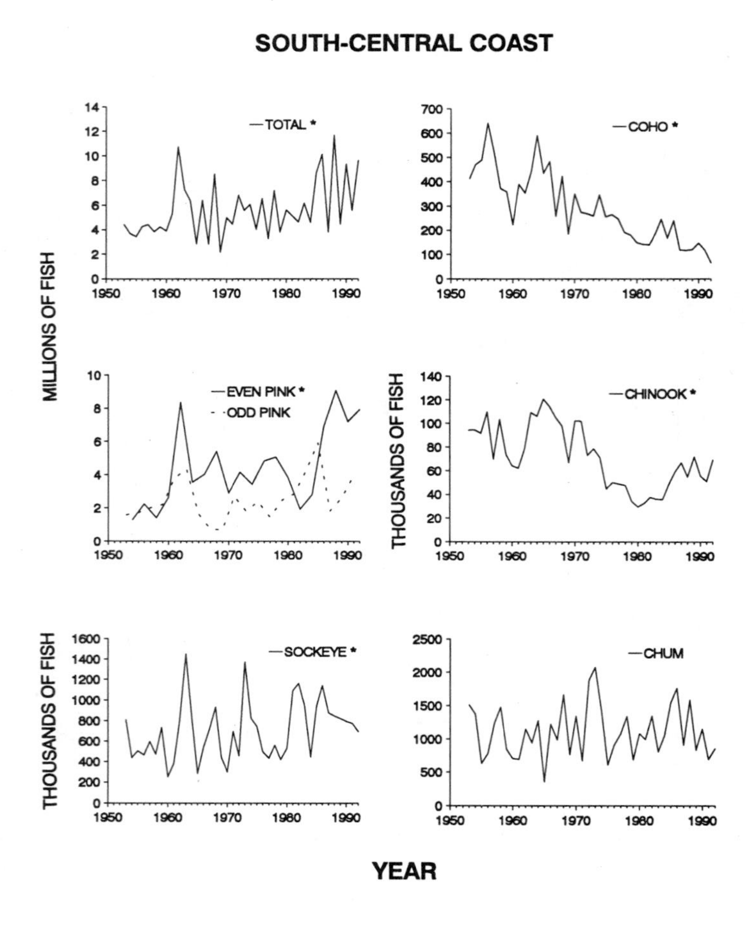

Figure 7. Annual Pacific salmon escapement to recorded rivers and streams of the south-central coast, 1953–92. * = trend statistically significant at $p \leq 0.05$.

Discussion

This overview of long- and short-term trends in abundance of Pacific salmon in BC is heavily dependent on records of escapement to its major rivers and streams made in large part by Fishery Officers of the Canada Department of Fisheries and Oceans. Not surprisingly, the accuracy of these data varies considerably not only between species but also between regions, rivers, and streams. Species differences in accuracy are considered briefly by Northcote and Burwash (1991),

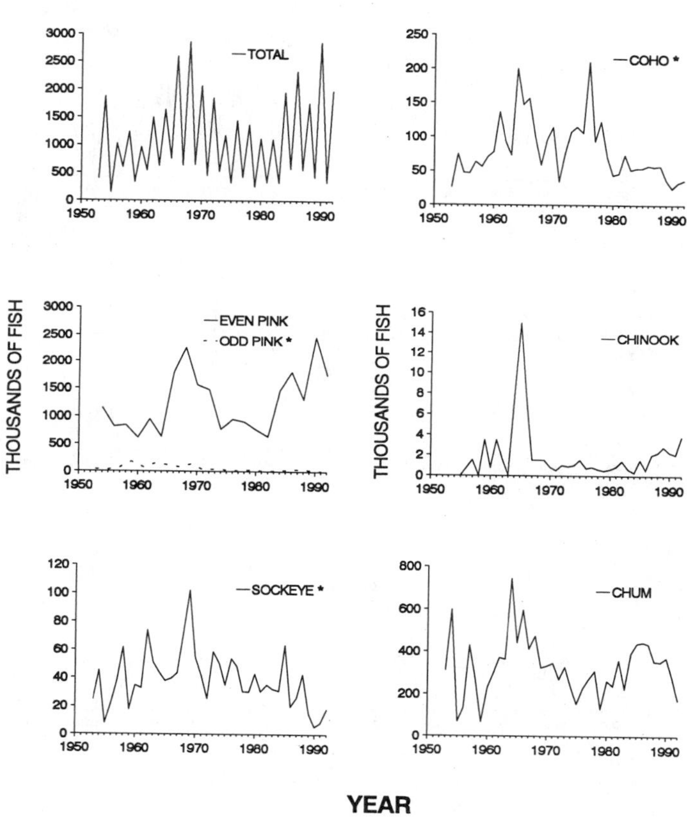

Figure 8. Annual Pacific salmon escapement to recorded rivers and streams of the Queen Charlotte Islands, 1953–92. * = trend statistically significant at p ≤0.05.

based on discussions with senior fishery managers. Escapement estimates for sockeye and pink salmon are considered to be the most reliable, followed by those for chum, chinook, and coho salmon in that order. Furthermore, accuracy in escapement estimates has improved over the recent 4 decades, and we point out where such change may seriously bias the indicated trends.

The long-term trends for major decreases in abundances of all species of Pacific salmon in the Fraser River watershed (Table 2) appear to have been followed by most species in most of the other drainage areas of the province where historical data are sufficient to make comparisons

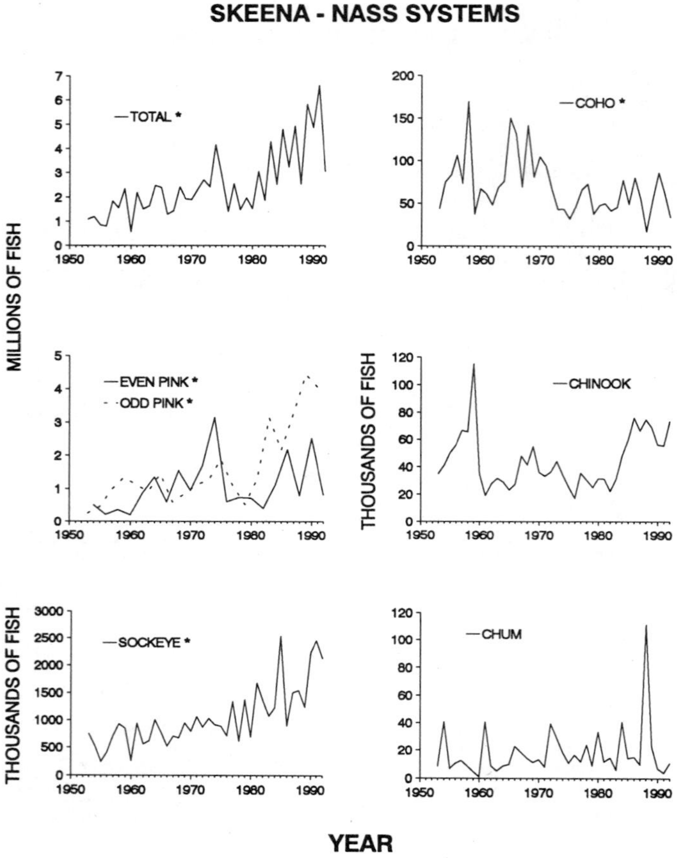

Figure 9. Annual Pacific salmon escapement to recorded rivers and streams of the Skeena-Nass systems, 1953–92. * = trend statistically significant at $p \leq 0.05$.

(Table 3). The only clear exception is in the Skeena-Nass area where sockeye salmon escapements, averaging about 4X those to the Nass River, have shown a considerable increase, probably resulting from major improvement in spawning habitat by enhancement activities. Despite this, the estimated average level of historical escapement at <1 million sockeye salmon annually is far from firm for the Skeena River.

Causes of the general long-term decline (and in some cases extinction) in most salmon stocks in BC are multiple, complex, and confounded. Certainly, permanent impoundments on some

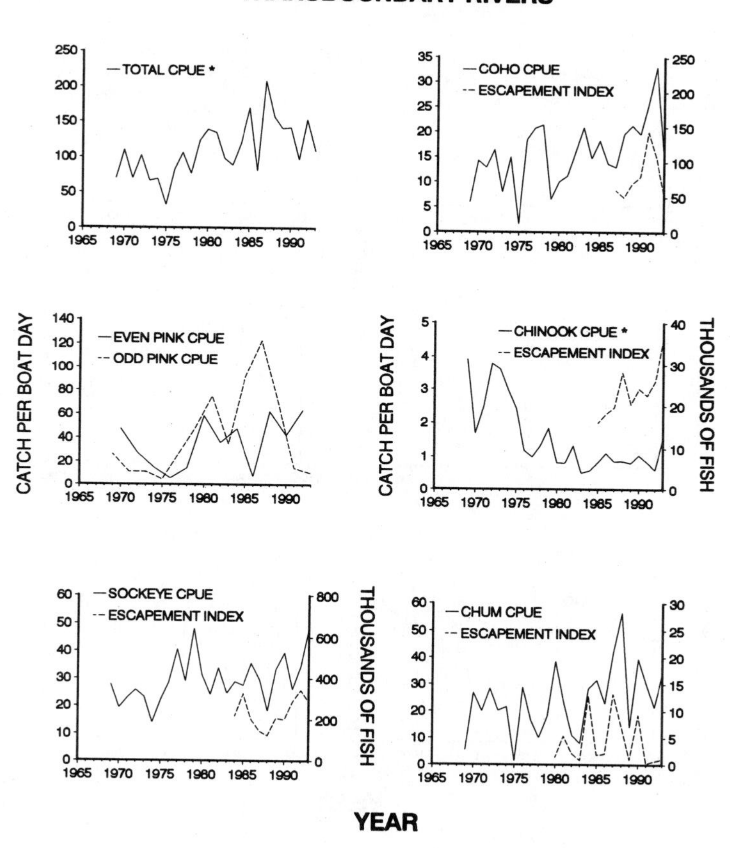

Figure 10. Annual Pacific salmon catch per unit effort (1969–93) and escapement indices for transboundary rivers (Stikine, Taku, and Alsek rivers). * = trend statistically significant at $p \leq 0.05$; no trend analysis for escapement data.

of the smaller river systems and on tributaries to some of the major ones blocked salmon from reaching their historical spawning areas. The temporary but devastating blockage by railway construction in 1913 on the middle mainstem of the Fraser River exterminated the large upriver stocks of pink salmon and sharply reduced those of sockeye salmon for decades. The multitude of mainstem and tributary dams on the Columbia River system must have had similar effects on stock components originating within BC. Degradation of spawning and early rearing habitat for

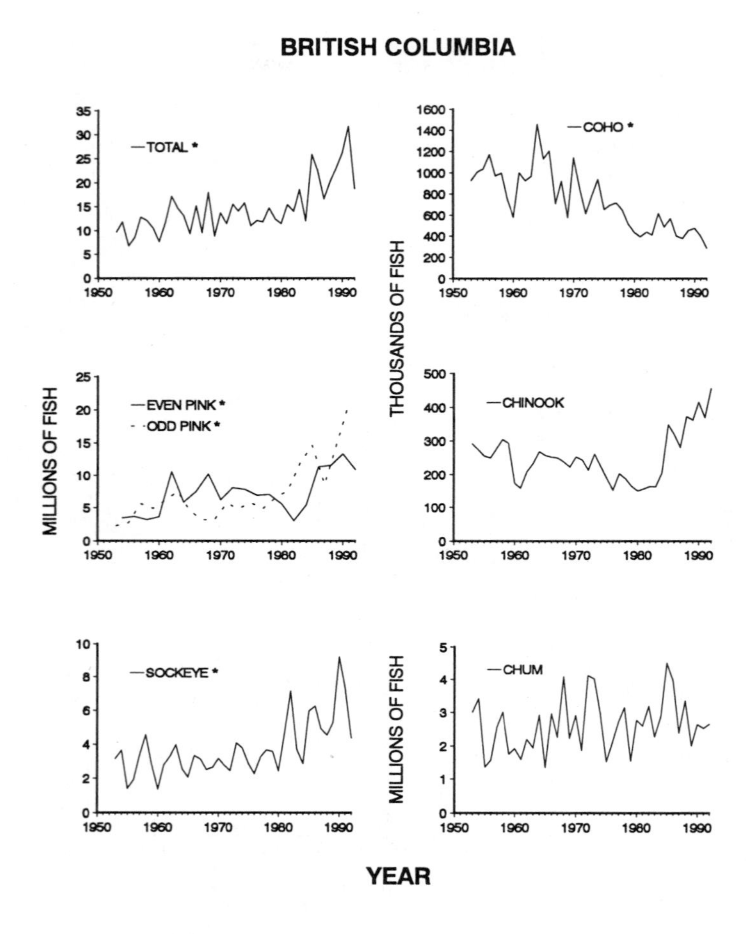

Figure 11. Annual Pacific salmon escapement to all recorded rivers and streams of British Columbia, 1953–92. * = trend statistically significant at $p \leq 0.05$. Does not include transboundary rivers.

salmon as a result of early mining, agricultural, and logging activities also contributed to the decline of stock abundance from historical levels, as did overfishing (mainly commercial up to the mid-1900s).

The Fraser River is an important producer of Pacific salmon in BC (and indeed in the world) not only because of its large drainage area (Northcote and Larkin 1989) but also because of the high availability, diversity, and productivity of spawning and juvenile rearing habitat within many of the watershed sub-basins that conditions such output. There are no dams on the mainstem Fraser River nor on many of its major tributaries, so salmon access upstream and downstream is good. Though urbanization and industrial activity have been intense in its lowermost reaches, these apparently have had little effect on successful access of salmon to and from the less developed major upriver spawning and rearing grounds. Furthermore, effective, stock-specific fisheries management has been successfully practiced on pink, sockeye, and chum salmon for several decades. The Skeena system contains several of the previously mentioned attributes of the Fraser, and it should be expected to rank second only to the Fraser system in BC Pacific salmon production.

Trends in BC salmon abundance over the recent decades show some reason for cautious optimism but also for conditional pessimism. We discuss this in reference to the six major drainage areas in BC that contain all five species of Pacific salmon and accommodate the bulk of total salmon escapement—namely the Fraser, Vancouver Island, south-central coast, Queen Charlotte Islands, and Skeena-Nass and transboundary systems (Fig. 1). In four of these areas, total salmon escapements show a significantly increasing trend over the recent decades, driven in large part by increasing trends in pink and sockeye salmon escapements. Fishery managers seem reasonably confident in the accuracy of escapement data for these two species and do not suggest that the trends are simply a result of better and more complete enumeration in the most recent years. In the Queen Charlotte Islands, however, there is no trend for increased total escapement of salmon (Fig. 8). Though pink salmon occur in over 60% of some 224 streams surveye in the area (Northcote et al. 1989), only a few of these support sizable escapements. Sockeye salmon, which occur in <5% of these streams and have no large escapements, have declined sharply since 1970 (Fig. 8).

Chum salmon escapements in four of the six areas under consideration (south-central coast, Queen Charlotte Islands, Skeena-Nass, and transboundary) show no significant trend. Fraser chum escapements have increased significantly (Fig. 3), but have declined in the most recent years. The large annual fluctuations in escapements of chum salmon, at least to the Fraser, Vancouver Island and south-central coastal drainages, are believed to be largely a result of commercial fisheries management actions (M. Joyce, Canada Dep. Fisheries and Oceans, New Westminster, BC, pers. comm.).

Coho salmon escapements have declined sharply in all areas except possibly the transboundary rivers. Though coho spawners are notoriously difficult to enumerate in streams and the escapement counts therefore are less reliable than those for the other species, there should be little reason for enumeration to become less accurate in recent years. Coho salmon often use small tributaries for spawning, and most young remain at least a year rearing in such waters before migrating to sea. Thus, of the five species of Pacific salmon, coho would seem to be most vulnerable to the various activities of man, which usually influence small streams most severely. Several hypotheses have been proposed for the observed coast-wide declines in coho abundance, including overfishing, freshwater habitat loss, marine carrying capacity limitations, and changes in ocean conditions (Walters 1993).

Chinook salmon escapements in the six areas show a variety of trend changes. In the Fraser River, they have a significant and gradually increasing trend over the 4 decades under discussion; however, recent increases (1984–present) are likely a result of conservation measures implemented in the early 1980s (N. Schubert, Canada Dep. Fisheries and Oceans, New Westminster, BC, pers. comm.). In the south-central coastal drainages, chinook salmon escapements have been declining since the mid-1960s. Vancouver Island and transboundary chinook stocks have slowly declined from the 1950s and 1960s, respectively, until the late 1980s when escapements increased probably in response to conservation and enhancement measures. Escapements in the Queen Charlotte Islands showed little overall change (perhaps some increase since the mid-1980s) whereas those in the Skeena-Nass rivers were highly erratic. The increased escapements of recent years in some of these areas may be, in part, an artifact of improved enumeration techniques being applied to this species.

Clearly, any optimism about the overall trends for four of the five species in these areas or for BC generally needs to be well cautioned. Those trends for coho salmon surely should signal pessimism, but pessimism well laced with a determination for correction. More restrictive harvest limitations (to reduce exploitation rates), better habitat protection and restoration, responsible enhancement activities, and the implementation of large adaptive management experiments (see Walters 1993 for details) are necessary to reverse these trends.

The escapement trends generally are not sensitive to extinction of small stocks, which use small streams or tributaries. Although such losses may not show up in the escapement aggregates from a watershed or region, they can be very important in terms of genetic diversity. So, while the short-term trends for Pacific salmon abundance are certainly not as alarming as in some other regions of the eastern Pacific, there is little room for complacency. One only has to recall the long-term trend of decline in BC salmon stock abundance, a decline from which we are not yet recovering, and also remember that many small stocks have already been lost irrevocably.

Acknowledgments

We are especially grateful for the extensive help provided on short notice by a number of Canada Department of Fisheries and Oceans staff, including I. Boyce, A. von Finster, M. Joyce, P. Milligan, G. Serbic, N. Schubert, R. Sjolund, and T. Whitehouse; additionally, thanks are extended to J. Muir, Alaska Department of Fish and Game. Helpful information was given by K.D. Adam, Biological Sciences, University of Waikato, Hamilton, New Zealand; P. Dill, Okanagan University College, Kelowna, BC, Canada; and by G. Haas and C. Walters, Department of Zoology, University of British Columbia, Vancouver, BC, Canada. Support for manuscript preparation and travel for D.Y. Atagi came from NSERC Grant 583454 awarded to the senior author and the Canada Department of Fisheries and Oceans. Symposium sponsors kindly provided travel support to the senior author. D. McPhail prepared Figure 1. J. Howard prepared Figs. 4 and 5.

Literature Cited

Beacham, T.D. 1984a. Catch, escapement, and exploitation of pink salmon in British Columbia, 1951-1981. Canadian Technical Report of Fisheries and Aquatic Sciences 1276.

Beacham, T.D. 1984b. Catch, escapement, and exploitation of chum salmon in British Columbia, 1951-1981. Canadian Technical Report of Fisheries and Aquatic Sciences 1270.

Cox, D.R., and A. Stuart. 1955. Some quick sign tests for trend in location and dispersion. Biometrika 42: 80-95.

Fanning, M.L. 1985. Enloe Dam passage project, Annual Report 1984, Volume 1. IEC Beak Consultants Ltd., Richmond, British Columbia, Canada.

Fraser, F.J., P.J. Starr, and A.Y. Fedorenko. 1982. A review of the chinook and coho salmon of the Fraser River. Canadian Technical Report of Fisheries and Aquatic Sciences 1126.

Hopwood, V.G. (ed.). 1971. David Thompson travels in western North America, 1784-1812. Macmillan of Canada. Toronto, Ontario, Canada.

Lorraine, M.J. 1924. The Columbia unveiled. The Times-Mirror Press. Los Angeles, California, USA.

Lyons, C. 1969. Salmon: our heritage. Mitchell Press Limited. Vancouver, British Columbia, Canada.

McLeod, C.L., and J.P. O'Neil. 1983. Major range extensions of anadromous salmonids and first record of chinook salmon in the Mackenzie River drainage. Canadian Journal of Zoology 61: 2183-2184.

McPhail, J.D., and R. Carveth. 1992. A foundation for conservation: the nature and origin of the freshwater fish fauna of British Columbia. Fish Museum, Department of Zoology, University of British Columbia. Vancouver, British Columbia, Canada.

Nehlsen, W., J.E. Williams, and J.A. Lichatowich. 1991. Pacific salmon at the crossroads: stocks at risk from California, Oregon, Idaho, and Washington. Fisheries 16: 4-21.

Northcote, T.G., and M.D. Burwash. 1991. Fish and fish habitats of the Fraser River basin, p. 117-141. *In* A.H.J. Dorcey and J.R. Griggs (eds.), Water in sustainable development: exploring our common future in the Fraser River basin. Westwater Research Centre, University of British Columbia. Vancouver, British Columbia, Canada.

Northcote, T.G., and P.A. Larkin. 1989. The Fraser River: a major salmonine production system, p. 172-204. *In* D.P. Dodge (ed.), Proceedings of the International Large River Symposium. Canadian Special Publication of Fisheries and Aquatic Sciences 106.

Northcote, T.G., A.E. Peden, and T.E. Reimchen. 1989. Fishes of the coastal marine, riverine and lacustrine waters of the Queen Charlotte Islands, p. 147-174. *In* G.G.E. Scudder and N. Gesler (eds.), The Outer Shores. Based on the proceedings of the Queen Charlotte Islands First International symposium, University of British Columbia, August, 1984. Vancouver, British Columbia, Canada.

Palmer, R.N. 1972. Fraser River chum salmon. Canadian Department of Environment. Fisheries Service Technical Report 1972-1.

Ricker, W.E. 1950. Cyclic dominance among Fraser River sockeye. Ecology 31: 6-26.

Ricker, W.E. 1987. Effects of the fishery and of obstacles to migration on the abundance of Fraser River sockeye salmon (*Oncorhynchus nerka*). Canadian Technical Report of Fisheries and Aquatic Sciences 1522.

Ricker, W.E. 1989. History and present state of the odd-year pink salmon runs of the Fraser River region. Canadian Technical Report of Fisheries and Aquatic Sciences 1702.

Sachs, L. 1978. Angewandte Statistik. Springer-Verlag. Berlin, West Germany.

Shirvell, C.S., and C. Charbonneau. 1984. Can salmon stream indexing improve salmon escapement estimates?, p. 51-65. *In* P.E.K. Symons and M. Waldichuck (eds.), Proceedings of the workshop on stream indexing for salmon escapement estimation. Canadian Technical Report of Fisheries and Aquatic Sciences 1326.

Starr, P.J., A.T. Charles, and M.A. Henderson. 1984. Reconstruction of British Columbia sockeye salmon (*Oncorhynchus nerka*) stocks: 1970-1982. Canadian Manuscript Report of Fisheries and Aquatic Sciences 1780.

Walters, C.J. 1993. Where have all the coho gone?, p. 1-8. *In* L. Berg and P.W. Delaney (eds.), Proceedings of the Coho Workshop. The Association of Professional Biologists of British Columbia and the North Pacific International Chapter of the American Fisheries Society, Nanaimo, British Columbia, Canada.

Factors Contributing to Stock Declines

Genetic Factors Contributing to Declines of Anadromous Salmonids in the Pacific Northwest

Reginald R. Reisenbichler

Abstract

Deleterious genetic change in wild anadromous salmonids (*Oncorhynchus* spp.) is expected from fisheries differentially harvesting fish that spawn at particular times within a season, mature at particular sizes or ages, or grow at particular rates. Other sources include overfishing, habitat degradation or destruction, and interactions with hatchery fish, particularly when these phenomena severely reduce population size. Gene flow from hatchery to wild fish populations also is deleterious because hatchery populations genetically adapt to the unnatural conditions of the hatchery environment at the expense of adaptation for living in natural streams. This domestication is significant even in the first generation of hatchery rearing. Spawner-recruit theory serves as a framework for discussing the consequences of deleterious genetic change. This theory can illustrate how the fitness or productivity of a population is reduced and whether genetic change is largely offset by natural selection within one generation, or accumulates over many generations. Although our knowledge is far from complete, sufficient information exists to demand actions to reduce or avoid deleterious genetic change. As additional information becomes available, these actions may be changed, perhaps relaxed. One suggested action is to establish or maintain "refuge" populations of wild fish that are to be protected from habitat degradation, selective or intense fishing, and interactions with hatchery fish.

Introduction

Any serious attempt to maintain or increase sustainable production of an anadromous salmonid species requires maintenance of the species' genetic diversity (Berst and Simon 1981; Ryman 1981, 1991). This diversity exists both among populations (representing, in part, adaptation of populations to local environments) and within populations, and is fundamental to a species' ability to adapt and persist in the face of changing environmental conditions (e.g., Schonewald-Cox et al. 1983). The among-population component of diversity is preserved by conserving the various populations composing each species; however, we are failing at this task. Harvest, habitat degradation, or hatchery programs may be decreasing the genetic diversity among extant populations of anadromous salmonids (Reisenbichler and Phelps 1989, Waples 1991, Reisenbichler et al. 1992). Recent listings of sockeye salmon (*Oncorhynchus nerka*) and two groups of chinook salmon (*O. tshawytscha*) from Idaho under the US Endangered Species Act, and other data (e.g., Konkel and McIntyre 1987, Nehlsen et al. 1991, Riddell 1993), indicate widespread decline or extirpation of populations.

Genetic adaptation and diversity within populations compose the primary theme of this paper. I focus on genetic changes in populations of Pacific salmon (*Oncorhynchus* spp.) that are expected to reduce productivity (number of recruits per spawner; here used interchangeably with "fitness"). Current knowledge is inadequate to provide accurate estimates of the actual loss in productivity from such genetic problems (e.g., Law 1991). Nevertheless, existing data and population genetics theory strongly suggest that genetic change has reduced the productivity of salmon and steelhead (*O. mykiss*) populations and, in conjunction with degraded environments, may be threatening the persistence of some populations, and ultimately some species.

Fishing and hatcheries are sources of deleterious genetic change in fish populations. I present data that suggest how such genetic change occurs and how it contributes to stock decline. I develop a simple framework for interpreting the effects of genetic change and discuss means to avoid such change. Degraded habitat, which also causes genetic change in fish populations, is not discussed because it often persists for long periods, and genetic adaptation to degraded conditions benefits the population. Also, fisheries managers generally have less control of habitat than of fisheries and hatcheries. Problems from small population size (e.g., inbreeding) may result from fishing or habitat degradation; they are discussed under fishing.

This discussion is focused on the conservation of wild populations of fish. I recognize that populations of hatchery fish (those fish reared in a hatchery during any portion of their life, generally during the first few months to 1 year for Pacific salmon and steelhead in the Pacific Northwest) represent a portion of the genetic resources of each species, and therefore often merit protection and concern similar to that increasingly displayed for wild populations. However, I consider hatchery populations of secondary importance, partially because their sustainability is unproven. Collectively, wild populations have persisted since long before the Pleistocene epoch (Neave 1958), and many individual populations have persisted at least since European man's early explorations of western North America. In contrast, hatchery populations of Pacific salmon have persisted only since the 1950s (Lichatowich and McIntyre 1987). Indeed, genetic manipulation of hatchery stocks may be necessary to avoid serious consequences to wild stocks when the two types interbreed (Reisenbichler and McIntyre 1986, Waples 1991).

A basic premise of my discussion is the principle that natural selection has worked to produce near-optimal distributions of genetically determined traits (Darwin 1859, Hartl 1980). For example, the distribution of spawning time for an undisturbed population provides, on average, nearly the highest possible fitness (survival and reproductive success) for the population, given historical, ecological, and other constraints. Any significant change in that distribution (unmatched by serendipitous environmental change) caused by fishing or interbreeding with hatchery fish should reduce the population's productivity (Nelson and Soulé 1987).

Genetic Changes from Fisheries

Fisheries often harvest fish nonrandomly and may thereby genetically alter a fish population and reduce its productivity (Law and Grey 1989). Often the earliest, biggest, or fastest-growing fish are most intensively harvested. Generally when a fishery is selective (the fishery preferentially harvests fish with a given [select] trait) and the trait is genetically determined (exhibiting additive genetic variance), the population will change genetically. Such change will likely alter the population's adaptiveness for its environment and reduce its productivity. (Some changes in genetically determined traits may not affect productivity [e.g., one that affects only mate selection]; selectively neutral changes should be rare.) The degree to which genetic change accumulates over generations will depend on how intensely natural selection opposes those genetic changes (e.g., Rose 1983, Riddell 1986). When the change accumulates over many generations, genetic recovery after cessation of selective fishing may require many more generations, or may never be complete if alleles (alternative forms of a gene) or particular gene combinations (Emlen 1991) have been lost.

In general, unequivocal examples of deleterious genetic change caused by fishing or measures of a resultant decrease in productivity for wild populations are lacking. Such changes have been difficult to document for anadromous salmonids, in part, because they are confounded with environmental variability and the direct (nongenetic) effects of fishing (e.g., Nikolskii 1969, Riddell 1986, McAllister et al. 1992). Intense natural selection on the trait of interest, or on genetically correlated traits, may largely offset genetic change from a fishery (e.g., Ricker 1981, Riddell 1986) or other sources of mortality (see Lande 1982 and Rose 1983 for examples from nonfish species). Even where the genetic composition of a population recovers within one generation (i.e., no cumulative change), I argue that the additional mortality associated with such counter-balancing selection reduces the productivity of the population. Appropriate controlled experiments to document genetic change from fishing have not been conducted with wild fish (McAllister et al. 1992). Genetic changes in life histories, however, have been shown to arise from age- or size-specific cropping in laboratory populations of invertebrates (references in Law and Grey 1989) and *Tilapia mossambica* (Silliman 1973).

Lack of absolute demonstrations notwithstanding, I proceed by discussing several life-history traits of anadromous salmonids that may be altered by selective fishing. In addition, the indirect effects of fishing associated with reduced population size are discussed briefly.

Changes in Run Timing or Time of Spawning

Terminal fisheries harvest salmon as they approach or reside in the stream of origin, and often differentially or selectively harvest some temporal portion of the run. Such selective harvest occurs when a certain number of the fish returning each year are harvested, after which fishing stops. Other situations leading to selection for specific, temporal portions of a run often occur when several populations overlap in their return time. For example, a fishery may be closed during part of the run to avoid harvesting certain protected populations, or the fishery may target one population but incidentally harvest a commingling portion of another population (e.g., Ricker 1981, Gharrett et al. 1993).

Partial overlap can occur among populations of different species, naturally spawning (wild) populations from a single species, or hatchery and wild populations from a single species. One example of mixed species overlap occurs in the Columbia River, where winter steelhead primarily enter the river from November to April, and lower-river spring chinook salmon primarily enter from February to April (Oregon Department of Fish and Wildlife and Washington Department of Fish and Wildlife 1993). A net fishery directed at the spring chinook salmon incidentally harvests some fish from the late portion of the steelhead run. Currently, however, only a small proportion of the steelhead are caught, and thus selection should be light (Oregon Department of Fish and Wildlife and Washington Department of Fish and Wildlife 1993).

In many streams of Oregon and Washington, hatchery steelhead return to their home stream, on average, 6 wk before wild steelhead, yet substantial overlap occurs (Fig. 1). Fisheries targeted on the hatchery fish may select against (i.e., harvest more of) the early portion of the wild run.

Alexandersdottir (1987) described a substantial shift in return time to the spawning grounds and spawning time for the Sashin Creek (Baranof Island in northern Southeast Alaska) population of pink salmon (*O. gorbuscha*). This shift from mid-August in 1934 to mid-September during the 1940s in large part appeared to be a genetic effect, which apparently resulted from a corresponding shift in fishing effort during 1930-42. In response to later season openings and reduced fishing after 1950, the population returned to its original timing within five generations. Data from Sashin Creek relating freshwater survival of juvenile pink salmon to the arrival date of their parents (references in Alexandersdottir 1987) suggested that the shift in spawning time had resulted in a >50% reduction in freshwater survival. This also may have reduced the productivity of the population by ~50%. Of course some of the shift in timing was a direct (non-genetic) effect of fishing; however, a substantial genetic component was indicated by the multi-generation lag in return to the original spawning time.

A shift in return time for wild coho salmon (*O. kisutch*) in the Clackamas River, Oregon, has been attributed to selective fishing in the commercial gillnet fishery that occurs downstream in the Columbia River during September-November (Cramer and Cramer 1994, Alan McGie,

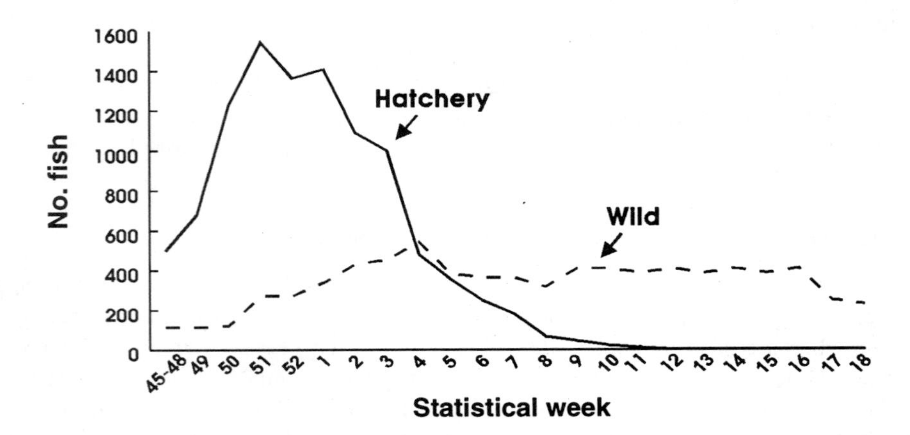

Figure 1. Typical run timing for hatchery and wild steelhead returning to the Quinault River, Washington; numbers were those expected in winter 1992–1993 (Quinault Indian Nation and Washington Dep. Wildlife, unpublished data). Fisheries that target the hatchery fish can selectively harvest the early portion of the wild run. Statistical week 1 is the first week in January.

Oregon Dep. Fish and Wildlife [ODFW], Corvallis, pers. comm.). Fishing on the early and middle portions of the run more than doubled from the 1960s through the 1970s and seemed responsible for a substantial shift in the peak time of return to the North Fork Dam (Cramer and Cramer 1994). A genetic basis for much of this shift is indicated by a coincident shift for jacks (small mature males, having spent only 0.5 year at sea rather than the more typical 1.5 years), which are not harvested in the gillnet fishery (Fig. 2). The shift in run timing corresponded with later spawning time, shortened rearing period for the resulting juvenile fish, and a decline in stock productivity of ~50% (Cramer and Cramer 1994).

Although situations leading to selective harvest commonly occur, a resultant genetic response of the affected populations seldom has been documented. Nevertheless, time of return and time of spawning have been shown to be heritable (e.g., Nosho and Hershberger 1976, Garrison and Rosentreter 1980, Taylor 1980, Gharrett and Smoker 1993). Thus, differential harvest with respect to either trait can be expected to genetically change the selected populations and reduce productivity (e.g., Miller and Brannon 1982, Brannon 1987). Frissell and Hirai (1988) analyzed spawning and incubation conditions for chinook salmon in relation to patterns of flooding along the south coast of Oregon, and concluded that fish spawning in certain months typically have substantially greater egg-to-fry survival for their progeny than do fish spawning in other months. Their results suggest that changes in spawning time induced by a fishery can substantially reduce the productivity of a population, and that productivity will be lost if populations are not allowed to adjust their timing to respond to changes in environmental conditions.

Genetic Changes in Size or Age at Maturity

Many salmon fisheries differentially harvest the largest or oldest fish (Ricker 1981). This occurs in gillnet fisheries where the large (usually old) fish are more likely to be captured than

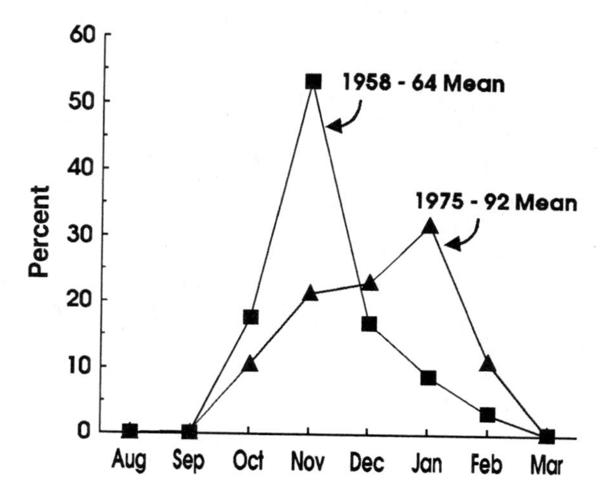

Figure 2. Estimated proportion of jack (age-2) coho salmon returning each month to the North Fork Clackamas River, Oregon, for 1958–1964 and 1975–1992 (Oregon Dep. Fish and Wildlife, unpublished data). The shift presumably reflects genetic effects from the Columbia River gillnet fishery for age-3 fish.

are the small (usually young) fish, or vice versa. Two other situations, not mutually exclusive, also result in smaller size or earlier age at maturity. First, in some fisheries the fast-growing fish become vulnerable to harvest earlier in the season than do slow-growing fish. Fast-growing fish may be subject to harvest for weeks or months longer than the slow-growing fish. This situation occurs when fish become vulnerable to harvest at a particular size and remain vulnerable until they mature at a common age with smaller fish, or until the fishing season ends. Second, in some fisheries the fish destined to mature at older ages can be vulnerable to harvest for 1 or more years longer than are fish that mature at early ages; hence, a much smaller proportion of the former survive to mature. This situation occurs where all fish from a cohort become vulnerable to the fishery at the same age and remain vulnerable until they mature. Few data are available to test for genetic changes in these traits; however, inasmuch as size and age at maturity are heritable (Donaldson and Menasveta 1961, Donaldson 1970, Garrison and Rosentreter 1980, Gjedrem 1984, Hankin et al. 1993), differential selection by fisheries can be expected to effect genetic change in these traits.

Ricker (1981) summarized data showing that the size of all five species of Pacific salmon in British Columbia had declined during 1951-1975. Pink salmon declined, on average, 19% and 34% in weight for odd and even years, respectively; chum (*O. keta*), sockeye, coho, and chinook salmon declined 12%, 8%, 24%, and 33%, respectively. The decrease for chinook salmon was at least partially caused by a decline in mean age, but the decline for chum salmon occurred despite an increase in mean age. Except for sockeye salmon, decreases in size were not correlated with ocean temperature or salinity data, and Ricker (1981) proposed genetic change from selective fishing as the most likely explanation for the observed changes in growth rate or size at age in pink, chum, and coho salmon. Ricker (1981) noted an increase in the size difference between sockeye salmon having three and four growing seasons at sea. He judged this increase to result from a genetic effect attributed to greater exploitation of fish near the median size (large ocean age-4 fish and small ocean age-3 fish were more likely to escape the fishery; e.g., Todd and Larkin 1971) and accomplished through at least partial "segregation of the two ages in time or place of spawning."

Ricker's (1981) work seemed a vivid testimony of the undesirable genetic effects of selective fishing. Healey (1986) subsequently evaluated the data set with additional years of information, and noted that changes in sea-surface temperature in the central Gulf of Alaska matched "the temporal pattern of changing size of pink, coho, and chinook salmon." He stated that changes in size could have resulted from genetic selection for slower growth, but "recent increases in size in some populations and failure of other populations to demonstrate expected changes in size are contrary to this hypothesis." McAllister et al. (1992) emphasized that Ricker's (1981) estimates of heritability from historical catch data on pink salmon were "potentially confounded with selective depletion of larger stocks, or oceanographic and density-dependent effects." McAllister et al. (1992) also noted that such estimates for hatchery fish probably were overestimates for wild populations. Of Ricker's (1981) conclusions on the likelihood of genetic change, only that for the divergence in mean size of sockeye having three and four growing seasons remains undisputed.

As Ricker (1981), Hankin and Healey (1986), and others have pointed out, the commercial troll and ocean sport fisheries select for earlier age at maturity in chinook salmon. For example, the mean ages for spawners have declined at least 0.7 year since the first part of the century in the Klamath (Hankin and McKelvey 1985) and Sacramento (Reisenbichler 1986) rivers in

California. Since age at maturity is heritable in chinook salmon (Donaldson and Menasveta 1961, Hankin et al. 1993), the genetic composition of the spawners must be different from the genetic composition of the same cohort before becoming vulnerable to the fishery. I suggest that a resultant genetic effect is not obvious (Healey 1986), in part because natural selection is sufficiently intense to offset most of the genetic change within one generation.

Older (and larger) females carry more eggs (Healey and Heard 1984) and may dig deeper redds (van den Berghe and Gross 1984) in larger substrate (Ricker 1981), which may be less susceptible to damage during high flows (Frissell and Hirai 1988). Larger females also may be less prone to superimposition by smaller females (Hankin and McKelvey 1985) and more suited for long-distance migrations within rivers (Schaffer and Elson 1975). This is not to say that all spawners should be old; there are advantages to small size also (e.g., Healey and Heard 1984) and presumably niches for both. A balance of ages similar to the pre-1950 age compositions probably achieves, on average, the highest possible productivity. A large range in age at maturity also should lead to increased stability and greater total production over several decades, particularly for populations subject to very high mortality in some years such as in southwestern Oregon (Frissell and Hirai 1988). This reasoning suggests that selection for early age at maturity reduces the productivity of chinook salmon (Hankin and Healey 1986) whether or not substantial genetic change accumulates over generations. Data are not available to test for a loss in productivity with high statistical power (e.g., Reisenbichler 1986, Peterman 1990), and a loss from such selection has not been demonstrated.

Thorpe (1993) summarized data and conclusions from Asian populations of sockeye salmon where "directional selection [by the fishery] . . . appear[ed] to drive these populations towards a preponderance of smaller, earlier maturing, but highly heterozygous types" (also see Smith et al. 1991a). Data from Asian pink salmon (Thorpe 1993, citing Altukhov 1989) showed that high heterozygosity was associated with increased individual growth rates, but that "the progeny of parents of average heterozygosity survived better than those of high or low [heterozygosity]." Collectively these results suggested that such selective fishing reduced the productivity of Asian sockeye salmon.

Schaffer and Elson (1975) concluded that fishing had caused genetic change for early age at maturity in (sea-run) Atlantic salmon (*Salmo salar*) from eastern North America (but see Riddell 1986). However, a substantial increase in the incidence of mature parr was more interesting. For example, the incidence of mature male parr increased from 31% of the population in the Matamek River (Quebec) in 1967 to 75% in 1982 (Caswell et al. 1984). Myers (1984) reported that ~80% of the male salmon in the Codroy River, Newfoundland, matured as parr, and estimated a resultant loss of 60% of the adult male salmon production to that stream.

As Ricker (1981) and others had done for Pacific salmon, Caswell et al. (1984) prepared a compelling story for genetic change in response to a fishery; however, as with Pacific salmon, further analysis (Myers et al. 1986) suggested that the observed shift in age structure could be explained by environmental factors alone. Reduced density of juvenile salmon was the predominant environmental factor resulting from intense fishing that substantially reduced the number of spawners and offspring. Juvenile salmon grow faster at reduced densities, which causes an increase in the proportion of males that mature as parr (i.e., without going to sea). Such environmental change also should cause a genetic response, but intuition as well as models developed from first principles of life-history theory (e.g., Stearns and Koella 1986) suggest that the phenotypic response of individual organisms to environmental change will overshadow the genetic response of the population, at least in the first few generations. Myers et al. (1986) concluded that no evidence existed for genetic changes in the Matamek River population, and the issue remains unresolved.

Reduced Population Size

Fishing (and habitat degradation or loss) can reduce populations to low levels, which can effect genetic change by leading to the demise of a population and loss of the associated genome, by inbreeding (mating between relatives), or by random genetic drift (chance fluctuations in allele frequencies). Small populations are prone to extinction because of random fluctuations in survival or reproduction; unpredictable, detrimental changes in the environment such as reduced food supply or bad weather; or depensatory effects (decreased survival or reproduction as the population declines; Soulé 1987). The actual population sizes that place the population at substantial risk from these factors (estimated by an exercise called population viability analysis) depend on the biology of the population and the peculiarities of its environment (Soulé 1986, 1987). Depending on the circumstances, even populations substantially larger than 500 individuals may be at risk of extinction from these factors. Rigorous viability analyses and specific guidelines generally are not available for salmon and steelhead.

Small populations (less than 25 breeding males and 25 breeding females for populations with discrete generations, and in some circumstances less than 250 of each sex [Soulé 1980; also see Waples and Teel 1990]) are prone to loss of genetic diversity and productivity from genetic drift and inbreeding (Hartl 1980). Inbreeding depression and the loss of alleles from genetic drift are commonly recognized as problems to be avoided (Nelson and Soulé 1987).

Declines in the productivity of wild populations from inbreeding or random drift have not been demonstrated for salmon or steelhead, which is perhaps partly because of lack of adequate data or from being confounded with other factors. Another possible reason is that sufficient animals migrate between populations to mitigate the deleterious genetic effects of small population size. Successful reproduction (gene flow) by just one stray (a fish originating from another population) per generation may offset the effects of inbreeding (Wright 1969, Allendorf 1983). However, sufficient gene flow in the past does not ensure sufficiency now or in the future (Rieman and McIntyre 1993). Widespread loss or serious depletion of populations (e.g., Konkel and McIntyre 1987, Nehlsen et al. 1991, Riddell 1993) reduces the number of fish available to stray from one population to another and may exacerbate problems from small population size.

Genetic Change from Interbreeding with Hatchery Fish

Hatchery programs can induce genetic change in wild populations even with no interbreeding of hatchery and wild fish. This can occur when selection regimes are altered (e.g., selective effects from a fishery targeted on hatchery fish) or through reduction in the size of wild populations via competition, predation, or other factors (Reisenbichler 1984, Nickelson et al. 1986, Reisenbichler and McIntyre 1986, Hindar et al. 1991, Waples 1991). Here I simply note that such change generally is undesirable for wild populations (see end of Introduction).

Mature hatchery fish commonly occur in streams with mature wild fish (e.g., Labelle 1992). This often is the intent of managers who release hatchery fish away from the hatchery as juveniles so that they home to their release site as adults (e.g., Wagner 1969). Interbreeding with wild fish has been observed even where the mean time of spawning for hatchery and wild fish differed by >1 mo (Leider et al. 1984). Indeed, interbreeding of hatchery and wild fish is the intended result of "outplanting" or "supplementation" hatchery programs, which are designed to increase the number of salmon or steelhead produced in streams by increasing the number of naturally spawning fish (Reisenbichler and McIntyre 1986, Waples 1991). If the hatchery fish are genetically different from wild fish, particularly if they are less well-adapted for rearing in natural streams than are wild fish, interbreeding will reduce the fitness of the wild population (Reisenbichler 1984, Emlen 1991, Byrne et al. 1992). Even if hatchery fish are not genetically different, interbreeding from poorly planned supplementation of small populations can increase deleterious effects from small population size by vastly magnifying the reproductive success of only a small portion of the population (i.e., by substantially increasing the variance in family size; Ryman and Laikre 1991).

Hynes et al. (1981), Verspoor (1988), Swain and Riddell (1990), Johnsson and Abrahams (1991), and others reviewed or reported evidence for domestication in fishes (see also Allard 1988 for plants). Experience with other aquatic organisms also has shown that domestication can lead to rapid genetic changes in a population (Doyle 1983). Ruzzante and Doyle (1993) found distinct genetic divergence in agonistic and schooling behavior of medaka (*Oryzias latipes*) after only two generations under artificial conditions that differed only in the level of social interaction (concentrated vs. dispersed food). Other studies with resident salmonids (e.g., Kohane and Parsons 1988, and eight studies summarized by Wohlfarth [1993]) suggest that genetic changes associated with captivity generally are disadvantageous for natural rearing. Frankham and Loebel (1992) found substantial domestication in *Drosophila melanogaster* after eight generations in captivity (reproductive fitness doubled) and inferred reduced fitness for natural conditions.

Four studies of anadromous salmonids (three with steelhead [Reisenbichler and McIntyre 1977, Leider et al. 1990, and National Biological Service, unpubl. data; Fig. 3] and one with coho salmon [Nickelson et al. 1986]) have compared the reproductive success of hatchery and wild fish, or the growth and survival of offspring of hatchery and wild fish in natural streams. Each study provided strong evidence that hatchery fish are substantially inferior genetically to wild fish in natural streams. Genetic differences between hatchery and wild fish could arise from the following:

1. initiation of the hatchery population with wild fish from a genetically distinct population,
2. random effects due to small population size in the hatchery (the same effects discussed previously for wild fish),
3. artificial selection (selective breeding) by hatchery personnel,
4. relaxed selection (the sheltered environment in the hatchery allows fish that are poorly adapted for rearing in natural streams to survive), or
5. natural selection for fish that are well adapted to the hatchery environment and subsequent release (domestication selection).

If only the first three factors apply, managers should be able to avoid differences between hatchery and wild fish (Gall 1993). Genetic differences from relaxed selection, in large part, may be unavoidable consequences of improved survival, which is the foundation of successful artificial propagation. If domestication selection is a substantial cause of the genetic differences, changes

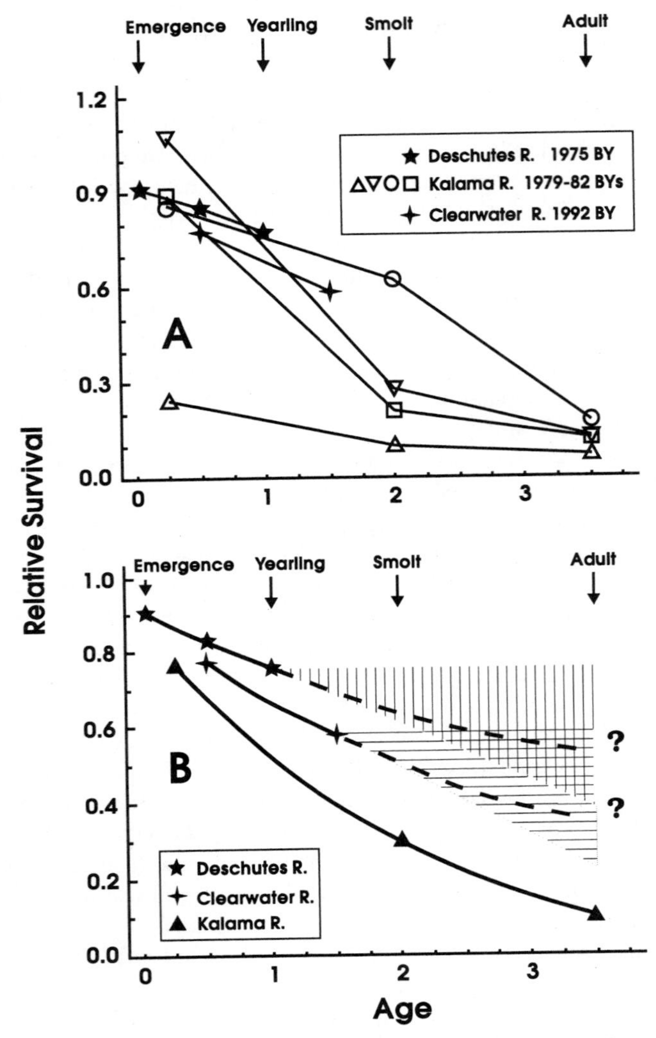

Figure 3. Survival for offspring of hatchery steelhead rearing in natural streams (or at sea) relative to that of offspring for wild steelhead at various ages or life-history stages. Relative survivals were evaluated from the eyed-embryo stage in Oregon's Deschutes River (★; Reisenbichler and McIntyre 1977), unfertilized eggs in Washington's Kalama River (Δ ∇ O □; Leider et al. 1990), and swim-up fry in Idaho's Clearwater River (+; National Biological Service, unpubl. data). (A) Data are given for each year-class separately. (B) Data for the Kalama River (▲) are arithmetic means from the four year-classes. Curves were fitted by eye. Dashed lines represent extrapolation of the data from the Deschutes and Clearwater rivers following a trajectory similar to that for the Kalama River. Shaded areas represent reasonable limits for these extrapolations. BY = brood year.

in hatchery environments will be necessary to avoid or reduce it (Doyle 1983). These changes may substantially increase the cost of operating hatcheries.

Whether domestication selection is important can be tested rather easily. If any or all of the first four factors above are the sole cause(s) of the genetic difference between hatchery and wild fish, the offspring of wild fish should grow and survive better, or at least no worse, than do the offspring of hatchery fish in any environment, including the hatchery. Reisenbichler and McIntyre's (1977) study in Oregon's Deschutes River system, and an ongoing study (National Biological Service, unpubl. data) in Idaho's Clearwater River system compared offspring of hatchery and wild steelhead in the respective hatcheries. Both studies found significantly faster growth and higher survival over the first year of life (i.e., to release from the hatchery as smolts) for the offspring of hatchery fish. The hatchery population evaluated in the former study had been initiated from the local wild population (the one used in the comparison) only two generations before the study. Apparently, domestication selection is an important cause of the genetic differences between hatchery and wild fish, at least for steelhead.

Important differences, in addition to number of generations in the hatchery, existed among the three studies of steelhead. For example, hatchery environments and stock of origin differed. Nevertheless, comparison of the results for the three studies offers a provisional basis for designing and managing hatchery programs until additional information becomes available. The decreasing relationship between the survival of hatchery fish in natural streams (relative to that for wild fish) and the number of generations that the hatchery population had been in the hatchery (Fig. 4) depicts a substantial change with the first generation in the hatchery, and a declining

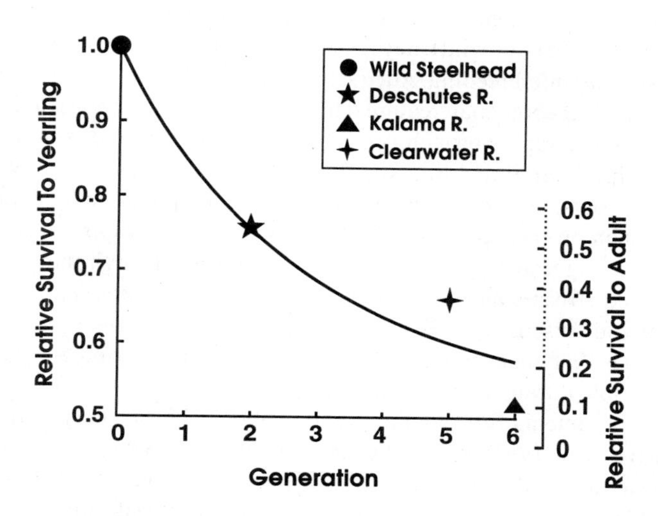

Figure 4. Survival for offspring of hatchery steelhead relative to that of offspring for wild steelhead versus number of generations the hatchery populations were in the hatchery. Hatchery and wild fish were reared together in natural streams until termination of the experiment or until migrating to sea. Survival was from the eyed-embryo or swim-up fry stage for Deschutes River fish (★; Reisenbichler and McIntyre 1977), the swim-up fry stage for Clearwater River fish (+; National Biological Service, unpubl. data), and unfertilized eggs (prior to being spawned naturally) for Kalama River fish (▲; Leider et al. 1990). The value for zero generations is one by definition.

rate of change with additional generations in the hatchery. Extrapolation of the data from the Deschutes River, following a trajectory similar to that for the Kalama River (Fig. 3B), suggests that the egg-to-adult survival for naturally rearing fish may be halved after only two generations in the hatchery. If the decline in fitness for natural rearing is of this magnitude, the likelihood of success for supplementation programs may be very low. Fitness for natural rearing may decline even faster for farmed salmonids because they rear in artificial environments for their entire lives (although they sometimes escape to interbreed with wild fish [e.g., Hansen 1991]).

The apparent degree and rate (reflected by a substantial effect after only two generations) of domestication selection is surprising to some people because they believe that the typically high survival of fish in the hatchery precludes rapid change from natural selection. Survival from fertilization to release from salmon hatcheries typically exceeds 60% (compared with less than 15% during the same time interval for wild populations; Howell et al. 1985). One should note, however, that much of the selective mortality probably occurs after the fish are released from the hatchery, when mortality of hatchery fish often exceeds 99% (Howell et al. 1985). Among fish that are genetically equivalent for survival from smolt to adult life stages, those fish that are genetically well adapted for rearing from egg to smolt stages under hatchery conditions (crowding, abundant food, little or no shelter, etc.) on average should be physiologically and morphologically the best prepared for the subsequent rigors after release. Fish that are genetically well adapted for rearing in hatcheries should grow faster and be larger at the time of release than should poorly adapted fish. Larger fish at the time of release usually survive better after release (Johnson 1970, Senn and Noble 1968, Buchanan and Wade 1982, Mathews and Ishida 1989, but see Bilton et al. 1984).

Emlen (1991) presented a genetic model that indicated interbreeding between genetically distinct populations (hatchery or wild) may result in substantially reduced productivity for more than 10 generations after interbreeding. Adequate genetic similarity to avoid protracted losses in productivity from interbreeding may require a history of gene flow between the populations (in part, stemming from geographic proximity); ecological equivalence or similarity apparently is not sufficient in fish (Gharrett and Smoker 1991) or other species (King 1955). It is unclear whether differences between hatchery and wild fish exclusively caused by domestication selection can result in such protracted losses of productivity. The possibility that protracted losses may occur is at least suggested by the substantial differences in environmental conditions between hatcheries and streams and the associated selective pressures (e.g., Hjort and Schreck 1982, and the previous discussion). Hatchery rearing over many generations may disrupt coadapted sets of genes that enhance fitness in natural streams (Gharrett and Smoker 1991) and may not be recoverable during a humaan lifetime (Emlen 1991).

In contrast to arguments for avoiding hybridization with hatchery fish (e.g., this paper, Waples 1991, Gharrett et al. 1993), Wohlfarth (1993) argues for "stocking first generation crossbreds between domestic and wild strains" to "restore" wild populations. A comparable argument by Moav et al. (1978) for using hatchery fish to "genetically improve" wild populations was refuted by Nelson and Soulé (1987). Wohlfarth's (1993) argument was based on a summary of recovery rates (number of fish recovered/number of fish released) or weights from eight published studies of resident trout and char (*Salvelinus* spp.), and two of anadromous salmonids. Those data, however, were insufficient for one to confidently predict even the direction of change in fitness or productivity because data on reproductive success for both the F_1 and the more informative F_2 generation (Emlen 1991, Gharrett and Smoker 1991) were missing. Indeed, both studies cited by Wohlfarth (1993) for anadromous fish indicated harm from releasing hybrids.

Wohlfarth (1993) cites Bams (1976) as showing a higher recovery rate for hybrid (local x introduced) pink salmon than for purebred salmon; however, he fails to point out that those purebred fish were introduced, not local fish. Bams (1976) reported equal survival for purebred and hybrid fish and went on to present evidence that returns to the spawning grounds were high for purebred local fish, low for hybrid fish, and lowest for purebred introduced fish. Apparently, hybrid fish homed to the natal stream only about one-half as well as did local fish. Such exaggerated straying should substantially reduce the productivity of a population. The other study of anadromous fish (steelhead, Reisenbichler and McIntyre 1977) showed significantly lower survival for hatchery x wild fish than for wild x wild fish, which also suggested that adding hybrid fish would lower the productivity of a wild population. Except in some situations where a population is likely to go extinct within a few generations, Wohlfarth's (1993) recommendation for releasing hybrids should be held in abeyance.

A Simple Model

Reproductive relations or spawner-recruit models (Ricker 1954) serve to illustrate changes in the fitness or productivity of a population. These simple models describe the number of adult offspring (recruits) expected from any given number of spawners (Fig. 5). The models typically reflect high survival for the offspring at low densities of spawners (hence, low densities of juvenile offspring), and declining survival as the number of spawners increases. At high densities, additional spawners produce few or no additional recruits, and at even higher densities, the number of recruits may decline. The "equilibrium population size" occurs where the curve crosses the replacement line (the line where the number of recruits always equals the number of spawners). "Harvestable surplus" or "surplus production" are fish produced in excess of replacement and are indicated by the vertical distance between the curve and the replacement line. In reality (and to the chagrin of many fishers and managers), the actual relation varies from year to year in response to environmental conditions.

Loss in productivity for a fish population is represented by a decrease in the height, and sometimes a change in the shape, of the reproductive relationship (Junge 1970, Reisenbichler 1986). The typical result is that the stable population size and surplus production for the population decline; fewer fish are available to spawn or for harvest, and the population recovers more slowly from short-term environmental deterioration (the population loses resiliency). Consider, for example, a hypothetical fish population in a seriously degraded habitat (Fig. 5A). The population substantially declines in abundance and has little surplus production. Any substantial and sustained harvest from this population will drive it to extinction. Even without fishing, it is unclear to observers whether the population will persist.

Declines in population productivity from fishing or hatcheries shift the reproductive relationship downward, similar to the effect from habitat degradation. Habitat degradation and the non-genetic effects of overfishing, in general, have caused greater losses in productivity or resilience than has genetic degradation. In the long term (e.g., over scores of generations), however, the harmful effects of accumulated genetic degradation within populations, loss of populations and the associated genetic diversity, and the accompanying hindrance of genetic adaptation to changing environmental conditions may equal or exceed the effects of habitat degradation and overfishing (e.g., Murphy 1968; Schaffer and Elson 1975; Warren and Liss 1980). Even a modest loss of

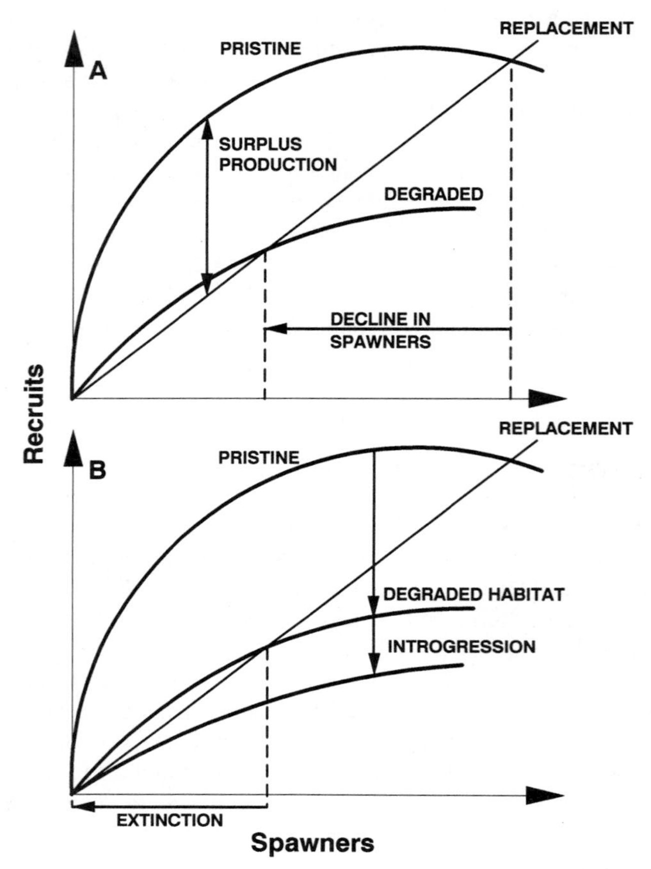

Figure 5. Hypothetical spawner-recruit relations. (A) A drop from the pristine relation owing to habitat degradation is shown. The equilibrium number of spawners (indicated by the vertical, dashed lines) declines to less than one-half the former level, and very little surplus production remains (i.e., the "degraded" curve barely exceeds the replacement line). The population may go extinct from fluctuations in environmental conditions or demographic parameters for lack of resiliency. Many contemporary populations of salmon and steelhead appear to be well represented by this situation (Konkel and McIntyre 1987, Nehlsen et al. 1991, Riddell 1993). (B) An additional drop from interbreeding with hatchery fish or selective fishing is shown. Even a modest drop is sufficient to lower the curve below the replacement line with the result that the population goes extinct.

adaptiveness for already degraded populations may cause extinction in the absence of rapid genetic recovery or favorable human intervention (Fig. 5B).

Genetic degradation may either be offset almost completely within one generation by natural selection (case 1) or accumulate over several or many generations (case 2). As discussed previously, a decrease in productivity may be the same for either type of genetic change; however, for case 1, productivity should recover almost immediately upon elimination of problems from fishing or hatcheries, while for case 2 (cumulative genetic change), recovery will require several to many generations. So for equal losses in productivity, cumulative genetic change poses the greatest threat to a population. Upon elimination of the problems with fishing or hatcheries (and under constant environment), the spawner-recruit relation for case 1 moves from a depressed level to nearly the original level in one generation. Under the same conditions, the relation for case 2 increases more slowly, passing along a family of progressively higher but closer curves, one per generation, until reaching the original level (or if important genetic variation has been lost, until reaching a stable relation below the original level). Even if managers adjust fishing and hatchery practices to avoid further deleterious genetic change, the population suffering from cumulative genetic degradation may not recover productivity in time to avoid extinction.

Approaches for Reducing Deleterious Genetic Change

Means are available to reduce the deleterious genetic effects of fishing, and such means have been applied, or at least proposed, when managers have perceived such problems. Clearly, inbreeding and random drift are reduced by lowering harvest rates to allow more fish to spawn. These effects also are avoided or reduced by preserving neighboring, conspecific populations, which are the populations most likely to provide strays. But there can be too much straying. Kapuscinski and Lannon (1984) suggested that increasing the gene flow beyond natural levels, even if migrants originate from hatchery populations, would genetically enhance wild populations. However, Hindar et al. (1991) and Ryman (1991) have pointed out the fallacy in that recommendation, which also should be apparent from the discussion here. Some steelhead populations in Oregon and Washington may suffer from too much gene flow from hatchery populations (Reisenbichler and Phelps 1989, Reisenbichler et al. 1992). Maintaining natural levels of straying, usually from genetically similar populations, should be a goal of managers, not exaggerating the numbers of strays.

Reducing harvest rates can reduce any type of selection by a fishery. Size selectivity also has been reduced by modifying gear or using alternative gear. For example, gillnet mesh sizes have been regulated. In addition, gillnets, which are well known for size selectivity, largely have been replaced by purse seines in southeast Alaska for harvesting pink salmon (Alexandersdottir 1987). Incidental harvest in mixed-stock fisheries is reduced by increasing the proportion of the catch taken in terminal fisheries. Marking all fish before release from the hatchery and requiring fishers to release unmarked (wild) fish can reduce the harvest of wild fish in some fisheries while allowing high exploitation of hatchery fish. This approach is being used for steelhead in many river systems, and has been proposed for coho and chinook salmon for harvest in ocean troll fisheries.

Inadvertent selection for size or age at maturity or for return time often is reduced when populations or species are harvested in terminal fisheries. For example, the move from long-duration ocean troll fisheries could substantially reduce selection for age at maturity in chinook salmon (Hankin and Healey 1986). Theoretically, maximum size limits also could be used to reduce the selection by troll and ocean sport fisheries for early age at maturity in chinook salmon; however, many people would judge this method as unacceptable for fear that a substantial proportion of the released fish would die (Ricker 1976). Selection for return time or spawning time can be reduced by holding the exploitation rate constant while a population returns to its home stream. Of course, this procedure is most acceptable when forecasts of run size are sufficiently accurate to avoid substantial overharvest or underharvest.

Two strategies are available for reducing the deleterious genetic effects from hatchery fish interbreeding with wild fish: (1) reduce the amount of interbreeding, or (2) reduce the genetic difference between hatchery fish and wild fish (before interbreeding). Reisenbichler and McIntyre (1986), Nelson and Soulé (1987), Goodman (1990), Hindar et al. (1991), Waples (1991), and others have provided more detailed discussions of various approaches (Table 1) for implementing these strategies.

Managers avoid or reduce competition (e.g., Nickelson et al. 1986, Riddell and Swain 1991), predation (e.g., Sholes and Hallock 1979), or other interactions between hatchery and wild fish by judiciously selecting the times, sizes, and locations of release for hatchery fish, and by restricting the hatchery program to a numerical scale in balance with associated wild populations and the carrying capacities of the freshwater, estuarine, and marine habitats (e.g., Lichatowich and McIntyre 1987). For example, hatchery fish generally are released at times when they will

Table 1. Two strategies for reducing the deleterious genetic effects from hatchery fish interbreeding with wild fish, and means used to implement the strategies. The two strategies are not mutually exclusive.

Reduce the amount of interbreeding between hatchery fish and wild fish	Reduce the genetic difference between hatchery fish and wild fish (before interbreeding)
Release fewer or no hatchery fish.	Use only wild fish, randomly selected from throughout the population to be affected, to initiate the hatchery population.
Release hatchery fish only at the hatchery or at locations where they are unlikely to interbreed with wild fish when they return as adults.	Initiate the hatchery program with wild fish genetically similar to the targeted wild population when fish are not available from the latter.
Rear sterile hatchery fish (appropriate for aquaculture operations [Hindar et al. 1991, Donaldson et al. 1993]).	Include additional wild fish in the hatchery population each generation (if accomplished without significantly depleting the wild population).
Hold hatchery fish in acclimation ponds for a period before release to increase the precision with which they home to the release area (which should be located away from areas where wild fish spawn).	Employ appropriate breeding protocols to avoid problems from inbreeding, genetic drift, and selective breeding in the hatchery (e.g., Simon et al. 1986, Allendorf and Ryman 1987, Gall 1993).
Advance or retard the time of spawning for hatchery fish so that there is little or no overlap between wild fish and hatchery fish that spawn naturally.	Use only wild fish for broodstock in the hatchery each year to reduce the level of domestication.
	Alter the hatchery environment to reduce domestication selection and loss of fitness for natural rearing.
	Selectively breed hatchery fish to mitigate domestication selection (e.g., Gall 1993).
	Release hatchery fish at early life-history stages to avoid extended exposure to domestication selection. (Early releases are appropriate only where other deleterious impacts [competition, predation, etc.] on wild fish are avoided.)

go directly to sea, rather than compete with wild fish in streams for extended periods of time; they also are released when wild fish are not vulnerable to predation from the hatchery fish.

We are at a crossroads for conserving our salmon and steelhead (Nehlsen et al. 1991). Although our knowledge is far from complete, sufficient information exists to demand actions to reduce or avoid deleterious genetic change. As additional information becomes available, these actions may be changed, perhaps relaxed. Because the suggested actions will be implemented slowly or not at all for many populations, and because some of the suggested actions are untested, a significant number of wild populations throughout the Pacific Northwest should be protected from habitat degradation, selective or intense fishing, and interactions with hatchery fish. The various management alternatives employed with the remaining populations (including no change in current practices) should be evaluated and should include adequate safeguards. Much of this evaluation will require large-scale, adaptive management approaches (Walters 1986). Of course, public acceptance and implementation of proposed actions to conserve wild populations will be enhanced by disseminating existing information.

Acknowledgments

P. Bisson, G. Brown, S. Chitwood, R. Cooper, D. Cramer, S. King, L. Luckinbill, M. Lynch, A. McGie, J. McIntyre, and others shared their data or thoughts with me. J. McIntyre and B. Rieman reviewed a preliminary draft of this manuscript with a very short deadline. G. Brown assembled data and prepared the figures.

Literature Cited

Allard, R.W. 1988. Genetic changes associated with the evolution of adaptedness in cultivated plants and their wild progenitors. Journal of Heredity 79: 225-238.

Allendorf, F.W. 1983. Isolation, gene flow, and genetic differentiation among populations, p. 51-65. *In* C. Schonewald-Cox, S. Chambers, B. MacBryde, and W. Thomas (eds.), Genetics and Conservation: A Reference for Managing Wild Animal and Plant Populations. Benjamin-Cummings, Menlo Park, California.

Allendorf, F.W. and N. Ryman. 1987. Genetic management of hatchery stocks, p. 141-159. *In* N. Ryman and F. Utter (eds.), Population Genetics and Fishery Management. University of Washington Press, Seattle, Washington.

Alexandersdottir, M. 1987. Life history of pink salmon *(Oncorhynchus gorbuscha)* in southeastern Alaska and implications for management. Ph.D. thesis. University of Washington, Seattle.

Altukhov, Y.P. 1989. Genetic processes in populations. Nauka, Moscow.

Bams, R.A. 1976. Survival and propensity for homing as affected by presence or absence of locally adapted paternal genes in two transplanted populations of pink salmon (*Oncorhynchus gorbuscha).* Journal of the Fisheries Research Board of Canada 33: 2716-2725.

Berst, A.H. and R.C. Simon (eds). 1981. Proceedings of the Stock Concept Symposium. Canadian Journal of Fisheries and Aquatic Sciences 38(12).

Bilton, H.T., R.B. Morely, A.S. Coburn, and J. Van Tine. 1984. The influence of time and size at release of juvenile coho salmon (*Oncorhynchus kisutch*) on returns at maturity: results of releases from Quinsam River Hatchery, B.C. in 1980. Canadian Technical Report in Fisheries and Aquatic Sciences 1306.

Brannon, E.L. 1987. Mechanisms stabilizing salmonid fry emergence timing. Canadian Special Publication in Fisheries and Aquatic Sciences 96: 120-124

Buchanan. D.V. and M.G. Wade. 1982. Development and assessment of steelhead in the Willamette River Basin. Annual Progress Report, Contract F-117-R-2. Oregon Department of Fish and Wildlife, Fish Division, Portland, Oregon.

Byrne, A., T.C. Bjornn, and J.D. McIntyre. 1992. Modeling the response of native steelhead to hatchery supplementation programs in an Idaho River. North American Journal of Fisheries Management 12: 62-78.

Caswell, H., R.J. Naiman, and R. Morin. 1984. Evaluating the consequences of reproduction in complex salmonid life cycles. Aquaculture 43: 123-134.

Cramer, D.P. and S.P. Cramer. 1994. Status and population dynamics of coho salmon in the Clackamas River. Technical Report. Portland General Electric Company, 121 S.W. Salmon St., Portland, Oregon 97204.

Darwin, C. 1859. Origin of Species. John Murray, London.

Donaldson, L.R. 1970. Selective breeding of salmonoid fishes, p. 65-74. *In* Marine Aquiculture. Oregon State University Press, Corvallis.

Donaldson, L.R. and D. Menasveta. 1961. Selective breeding of chinook salmon. Transactions of the American Fisheries Society 90: 160-164.

Doyle, R.W. 1983. An approach to the quantitative analysis of domestication selection in aquaculture. Aquaculture 33: 167-185.

Emlen, J.M. 1991. Heterosis and outbreeding depression: a multi-locus model and an application to salmon production. Fisheries Research 12: 187-212.

Frankham, R. and D.A. Loebel. 1992. Modeling problems in conservation genetics using captive *Drosophila* populations: rapid genetic adaptation to captivity. Zoo Biology 11: 333-342.

Frissell, C.A., and T. Hirai. 1988. Life history patterns, habitat change, and productivity of fall chinook stocks of southwest Oregon. *In* Proceedings of the Northeast Pacific Chinook and Coho Workshop, North Pacific International Chapter of the American Fisheries Society. Canada Department of Fish and Oceans, Vancouver, British Columbia.

Gall, G.A.E. 1993. Genetic change in hatchery populations, p. 81-92. *In* J.G. Cloud and G.H. Thorgaard (eds.), Genetic Conservation of Salmonid Fishes. Plenum Press, New York.

Garrison, R.L. and N.M. Rosentreter. 1980. Stock assessment and genetic studies of anadromous salmonids. Annual Progress Report, Project number AFS-73-3. Oregon Department of Fish and Wildlife, Portland.

Gharrett, A.J. and W.W. Smoker. 1991. Two generations of hybrids between even-and odd-year pink salmon (*Oncorhynchus gorbuscha)*: a test for outbreeding depression? Canadian Journal of Fisheries and Aquatic Sciences 48: 1744-1749.

Gharrett, A.J., and W.W. Smoker. 1993. Genetic components in life history traits contribute to population structure, p. 197-202. *In* J.G. Cloud and G.H. Thorgaard (eds.), Genetic Conservation of Salmonid Fishes. Plenum Press, New York.

Gharrett, A.J., B.E. Riddell, J.E. Seeb, and J.H. Helle. 1993. Status of the genetic resources of Pacific Rim salmon, p. 295-302. *In* J.G. Cloud and G.H. Thorgaard (eds.), Genetic Conservation of Salmonid Fishes. Plenum Press, New York.

Gjedrem, T. 1984. Genetic variation in age at maturity and its relation to growth rate, p. 52-61. *In* R.N. Iwamoto and S. Sower (eds), Salmonid Reproduction. University of Washington, Seattle.

Goodman, M.L. 1990. Preserving the genetic diversity of salmonid stocks: a call for federal regulation of hatchery programs. Environmental Law 20: 111-166.

Hankin, D.G. and M.C. Healey. 1986. Dependence of exploitation rates for maximum yield and stock collapse on age and sex structure of chinook salmon (*Oncorhynchus tshawytscha*) stocks. Canadian Journal of Fisheries and Aquatic Sciences 43: 1746-1759.

Hankin, D.G. and R. McKelvey. 1985. Comment on fecundity of chinook salmon and its relevance to life history theory. Canadian Journal of Fisheries and Aquatic Sciences 42: 393-395.

Hankin, D.G., J.W. Nicholas, and T.W. Downey. 1993. Evidence for inheritance of age of maturity in chinook salmon (*Oncorhynchus tshawytscha).* Canadian Journal of Fisheries and Aquatic Sciences 50: 347-358.

Hansen, L.P. (ed.) 1991. Interactions between cultured and wild Atlantic salmon. Aquaculture 98: 1-324.

Hartl, D.L. 1980. Principles of Population Genetics. Sinauer Associates, Sunderland, Massachusetts.

Healey, M.C. 1986. Optimum size and age at maturity in Pacific salmon and effects of size-selective fisheries, p. 39-52. *In* D.J. Meerburg (ed.), Salmonid Age at Maturity. Canadian Special Publication in Fisheries and Aquatic Sciences 89.

Healey, M.C. and W.R. Heard. 1984. Inter- and intrapopulation variations in the fecundity of chinook salmon (*Oncorhynchus tshawytscha*) and its relevance to life history theory. Canadian Journal of Fisheries and Aquatic Sciences 41: 476-483.

Hindar, K., N. Ryman, and F. Utter. 1991. Genetic effects of cultured fish on natural fish populations. Canadian Journal of Fisheries and Aquatic Sciences 48: 945-957.

Hjort, R.C. and C.B. Schreck. 1982. Phenotypic differences among stocks of hatchery and wild coho salmon (*Oncorhynchus kisutch*) in Oregon, Washington, and California, USA. Fishery Bulletin 80: 105-120.

Howell, P., K. Jones, D. Scarnecchia, L. Lavoy, W. Kendra, and D. Ortmann. 1985. Stock assessment of Columbia River anadromous salmonids. Vol. I; Chinook, coho, chum, and sockeye salmon stock summaries. Final Report, Project 83-335. Bonneville Power Administration, Portland, Oregon.

Hynes, J.D., E.H. Brown, Jr., J. Helle, N. Ryman, and D. Webster. 1981. Guidelines for the culture of fish stocks for resource management. Canadian Journal of Fisheries and Aquatic Sciences 38: 1867-1876.

Johnson, A.K. 1970. The effect of size at release on the contribution of 1964-brood Big Creek hatchery coho salmon to the Pacific Coast sport and commercial fisheries. Research Reports of the Fish Commission of Oregon 2: 64-76.

Johnsson, J.L. and M.V. Abrahams. 1991. Interbreeding with domestic strain increases foraging under threat of predation in juvenile steelhead trout (*Oncorhynchus mykiss*): an experimental study. Canadian Journal of Fisheries and Aquatic Sciences 48: 243-247.

Junge, C.O. 1970. The effects of superimposed mortalities on reproduction curves. Research Reports of the Fish Commission of Oregon 2: 56-63.

Kapuscinski, A.R.D. and J.E. Lannan. 1984. Application of a conceptual fitness model for managing Pacific salmon fisheries. Aquaculture 43: 135-146.

King, J.C. 1955. Evidence for the integration of the genepool from studies of DDT resistance in *Drosophila*. Cold Spring Harbor Symposium on Quantitative Biology 20: 311-317.

Kohane, M.J. and P.A. Parsons. 1988. Domestication: evolutionary change under stress. Evolutionary Biology 23: 31-48.

Konkel, G.W. and J.D. McIntyre. 1987. Trends in spawning populations of Pacific anadromous salmonids. US Fish and Wildlife Service, Fish and Wildlife Technical Report 9. Washington, DC.

Labelle, M. 1992. Straying patterns of coho salmon (*Oncorhynchus kisutch*) stocks from southeast Vancouver Island, British Columbia. Canadian Journal of Fisheries and Aquatic Sciences 49: 1843-1855.

Lande, R. 1982. A quantitative genetic theory of life history evolution. Ecology 63: 607-615.

Law, R. 1991. Fishing in evolutionary waters. New Scientist 2: 35-37.

Law, R., and D.R. Grey. 1989. Evolution of yields from populations with age-specific cropping. Evolutionary Ecology 3: 343-359.

Leider, S.A., M.W. Chilcote, and J.J. Loch. 1984. Comparative life history characteristics of hatchery and wild steelhead trout (*Salmo gairdneri*): evidence for partial reproductive isolation. Canadian Journal of Fisheries and Aquatic Sciences 41: 1454-1462.

Leider, S.A., P.L. Hulett, J.J. Loch, and M.W. Chilcote. 1990. Electrophoretic comparison of the reproductive success of naturally spawning transplanted and wild steelhead trout through the returning adult stage. Aquaculture 88(3-4): 239-252

Lichatowich, J.A. and J.D. McIntyre. 1987. Use of hatcheries in the management of Pacific anadromous salmonids. American Fisheries Society Symposium 1: 131-136.

Mathews, S.B. and Y. Ishida. 1989. Survival, ocean growth, and ocean distribution of differentially timed releases of hatchery coho salmon (*Oncorhynchus kisutch*). Canadian Journal of Fisheries and Aquatic Sciences 46: 1216-1226.

McAllister, M.K., R.M. Peterman, and D.M. Gillis. 1992. Statistical evaluation of a large-scale fishing experiment designed to test for a genetic effect of size-selective fishing on British-Columbia pink salmon (*Oncorhynchus gorbuscha*). Canadian Journal of Fisheries and Aquatic Sciences 49: 1294-1304.

Miller, R.J. and E.L. Brannon. 1982. The origin and development of life history patterns in Pacific salmonids, p. 296-309. *In* E.L. Brannon and E.O. Salo (eds.), Proceedings of the Salmon and Trout Migratory Behavior Symposium. University of Washington, Seattle.

Moav, R., T. Brody, and G. Hulata. 1978. Genetic improvement of wild fish populations. Science 201: 1090-1094.

Murphy, G.I. 1968. Pattern in life history and the environment. American Naturalist 102(927): 391-403.

Myers, R.A. 1984. Demographic consequences of precocious maturation of Atlantic salmon (*Salmo salar*). Canadian Journal of Fisheries and Aquatic Sciences 41: 1349-1353.

Myers, R.A., J.A. Hutchings, and R.J. Gibson. 1986. Variation in male parr maturation within and among populations of Atlantic salmon, *Salmo salar*. Canadian Journal of Fisheries and Aquatic Sciences 43: 1242-1248.

Neave, F. 1958. The origin and speciation of *Oncorhynchus*. Transactions of the Royal Society of Canada 52(3,5): 25-39.

Nehlsen, W., J.E. Williams, and J.A. Lichatowich. 1991. Pacific salmon at the crossroads: stocks at risk from California, Oregon, Idaho, and Washington. Fisheries 16(2): 4-21.

Nelson, K., and M. Soulé. 1987. Genetical conservation of exploited fishes, p. 345-368. *In* N. Ryman and F. Utter (eds.), Population Genetics and Fishery Management. University of Washington Press, Seattle.

Nickelson, T.E., M.F. Solazzi, and S.L. Johnson. 1986. Use of hatchery coho salmon (*Oncorhynchus kisutch*) presmolts to rebuild wild populations in Oregon coastal streams. Canadian Journal of Fisheries and Aquatic Sciences 43: 2443-2449.

Nikolskii, G.V. 1969. Theory of Fish Population Dynamics as the Biological Background for Rational Exploitation and Management of Fishery Resources. Translated by J.E.S. Bradley. Oliver and Boyd, Edinburgh.

Nosho, T.Y. and W.K. Hershberger (eds.). 1976. Salmonid Genetics. Division of Marine Resources, University of Washington, Seattle.

Oregon Department of Fish and Wildlife and Washington Department of Fish and Wildlife. 1993. Status report: Columbia River fish runs and fisheries, 1938-1992. Joint Columbia River Management Staff. Clackamas, Oregon/Battleground, Washington.

Peterman, R.M. 1990. Statistical power analysis can improve fisheries research and management. Canadian Journal of Fisheries and Aquatic Sciences 47: 2-15.

Reisenbichler, R.R. 1984. Outplanting: potential for harmful genetic change in naturally spawning salmonids, p. 33-39. *In* J.M. Walton and D.B. Houston (eds), Proceedings of the Olympic Wild Fish Conference. Peninsula College, Port Angeles, Washington.

Reisenbichler, R.R. 1986. Use of spawner-recruit relations to evaluate the effect of degraded environment and increased fishing on the abundance of fall-run chinook salmon, *Oncorhynchus tshawytscha*, in several California streams. Ph.D. dissertation. University of Washington, Seattle.

Reisenbichler, R.R. and J.D. McIntyre. 1977. Genetic differences in growth and survival of juvenile hatchery and wild steelhead trout, *Salmo gairdneri*. Journal of the Fisheries Research Board of Canada 34: 123-128.

Reisenbichler, R.R. and J.D. McIntyre. 1986. Requirements for integrating natural and artificial production of anadromous salmonids in the Pacific Northwest, p. 365-374. *In* R.H. Stroud (ed.), Fish Culture in Fisheries Management. American Fisheries Society, Bethesda, Maryland.

Reisenbichler, R.R., J.D. McIntyre, M.F. Solazzi, and S.W. Landino. 1992. Genetic variation in steelhead of Oregon and northern California. Transactions of the American Fisheries Society 121: 158-169.

Reisenbichler, R.R. and S.R. Phelps. 1989. Genetic variation in steelhead (*Salmo gairdneri*) from the north coast of Washington. Canadian Journal of Fisheries and Aquatic Sciences 46: 66-73.

Ricker, W.E. 1954. Stock and recruitment. Journal of the Fisheries Research Board of Canada 11: 559-623.

Ricker, W.E. 1976. Review of the rate of growth and mortality of Pacific salmon in salt water, and noncatch mortality caused by fishing. Journal of the Fisheries Research Board of Canada 33: 1483-1524.

Ricker, W.E. 1981. Changes in the average size and average age of Pacific salmon. Canadian Journal of Fisheries and Aquatic Sciences 38: 1636-1656.

Riddell, B.E. 1986. Assessment of selective fishing on the age at maturity in Atlantic salmon (*Salmo salar*): a genetic perspective, p. 102-109. *In* D.J. Meerburg (ed.), Salmonid Age at Maturity. Canadian Special Publication in Fisheries and Aquatic Sciences 89.

Riddell, B.E. 1993. Spatial organization of Pacific salmon: what to conserve? p. 23-42. *In* J.G. Cloud and G.H. Thorgaard (eds.), Genetic Conservation of Salmonid Fishes. Plenum Press, New York.

Riddell, B.E., and D.P. Swain. 1991. Competition between hatchery and wild coho salmon (*Oncorhynchus kisutch)*: genetic variation for agonistic behavior in newly-emerged wild fry. Aquaculture 98: 161-172.

Rieman, B.E., and J.D. McIntyre. 1993. Demographic and habitat requirements for conservation of bull trout. General Technical Report INT-302. USDA-Forest Service, Ogden, Utah.

Rose, M.R. 1983. Theories of life-history evolution. American Zoologist 23: 15-23.

Ruzzante, D.E. and R.W. Doyle. 1993. Evolution of social behavior in a resource-rich structured environment: selection experiments with medaka (*Oryzias latipes)*. Evolution 47: 456-470.

Ryman, N. (ed.) 1981. Fish gene pools. Ecological Bulletins 34. The Editorial Service/FRN, Stöckholm.

Ryman, N. 1991. Conservation genetics considerations in fishery management. Journal of Fish Biology 39(A): 211-224.

Ryman, N. and L. Laikre. 1991. Effects of supportive breeding on the genetically effective population size. Conservation Biology 5: 325-329.

Schaffer, W.M. and P.F. Elson. 1975. The adaptive significance of variation in life history among local populations of Atlantic salmon in North America. Ecology 56: 577-590

Schonewald-Cox, C.M., S.M. Chambers, B. MacBryde, and W.L. Thomas (eds.). 1983. Genetics and Conservation: A Reference for Managing Wild Animal and Plant Populations. Benjamin-Cummings, Menlo Park, California.

Senn, H.G. and R.E. Noble. 1968. Contribution of coho salmon (*Oncorhynchus kisutch*) from a Columbia River hatchery. Washington Department of Fisheries Research Paper 3(1): 51-62.

Sholes, W.H. and R.J. Hallock. 1979. An evaluation of rearing fall-run chinook salmon, *Oncorhynchus tshawytscha*, to yearlings at Feather River Hatchery, with a comparison of returns from hatchery and downstream releases. California Fish and Game 64: 239-255.

Silliman, R.P. 1973. Selective and unselective exploitation of experimental populations of *Tilapia mossambica*. Fishery Bulletin 73: 495-507.

Simon, R.C., J.D. McIntyre, and A.R. Hemmingsen. 1986. Family size and effective population size in a hatchery stock of coho salmon (*Oncorhynchus kisutch)*. Canadian Journal of Fisheries and Aquatic Sciences 43: 2434-2442.

Smith, P.J., R.I.C.C. Francis, and M. McVeagh. 1991. Loss of genetic diversity due to fishing pressure. Fisheries Research 10(3-4): 309-316.

Soulé, M.E. 1980. Thresholds for survival: maintaining fitness and evolutionary potential, p. 151-170. *In* M.E. Soulé and B.A. Wilcox (eds.), Conservation Biology: An Evolutionary-Ecological Perspective. Sinauer Associates, Sunderland, Massachussets.

Soulé, M.E. (ed.). 1986. Conservation Biology: The Science of Scarcity and Diversity. Sinauer, Sunderland, Massachusetts.

Soulé, M.E. (ed.). 1987. Viable Populations for Conservation. Cambridge University Press, Cambridge, Great Britain.

Stearns, S.C. and J.C. Koella. 1986. The evolution of phenotypic plasticity in life-history traits: predictions of reaction norms for age and size at maturity. Evolution 40: 893-913.

Swain, D.P. and B.E. Riddell. 1990. Variation in agonistic behavior between newly emerged juveniles from hatchery and wild populations of coho salmon (*Oncorhynchus kisutch*). Canadian Journal of Fisheries and Aquatic Sciences 47: 566-571.

Taylor, S.G. 1980. Marine survival of pink salmon fry from early and late spawners. Transactions of the American Fisheries Society 109: 79-82.

Thorpe, J.E. 1993. Impacts of fishing on genetic structure of salmonid populations, p. 67-80. *In* J.G. Cloud and G.H. Thorgaard (eds.), Genetic Conservation of Salmonid Fishes. Plenum Press, New York.

Todd, I.S. and P.A. Larkin. 1971. Gillnet selectivity in sockeye (*Oncorhynchus nerka*) and pink salmon (*O. gorbuscha*) of the Skeena River system, British Columbia. Journal of the Fisheries Research Board of Canada 28: 821-842.

van den Berghe, E.P. and M.R. Gross. 1984. Female size and nest depth in coho salmon (*Oncorhynchus kisutch*). Canadian Journal of Fisheries and Aquatic Sciences 41: 204-206.

Verspoor, E. 1988. Reduced genetic variability in first-generation hatchery populations of Atlantic salmon (*Salmo salar*). Canadian Journal of Fisheries and Aquatic Sciences 45: 1686-1690.

Wagner, H.H. 1969. Effect of stocking location of juvenile steelhead trout, *Salmo gairdneri*, on adult catch. Transactions of the American Fisheries Society 98: 27-34.

Walters, C.J. 1986. Adaptive Management of Renewable Resources. Macmillan, New York.

Waples, R.S. 1991. Genetic interactions between hatchery and wild salmonids: lessons from the Pacific Northwest. Canadian Journal of Fisheries and Aquatic Sciences 48: 124-133

Waples, R.S. and D.J. Teel. 1990. Conservation genetics of Pacific salmon. I. Temporal changes in allele frequency. Conservation Biology 4: 144-156.

Warren, C.E. and W.J. Liss. 1980. Adaptation to aquatic environments, p. 15-40. *In* R.T. Lackey and L.A. Nielsen. (eds.), Fisheries Management. Halsted Press, New York.

Wohlfarth, G.W. 1993. Genetic management of natural fish populations, p. 227-230. *In* J.G. Cloud and G.H. Thorgaard (eds.), Genetic Conservation of Salmonid Fishes. Plenum Press, New York.

Wright, S. 1969. The Theory of Gene Frequencies. Evolution and the Genetics of Populations, Volume 2. Chicago University Press, Chicago.

The Role of Competition and Predation in the Decline of Pacific Salmon and Steelhead

Kurt L. Fresh

Abstract

In this paper, I examine the role of competition and predation in the decline of Pacific salmon and steelhead (*Oncorhynchus* spp.) populations along the Pacific coast of North America. Few studies have clearly established the role of competition and predation in anadromous population declines, especially in marine habitats. A major reason for the uncertainty in the available data is the complexity and dynamic nature of competition and predation; a small change in one variable (e.g., prey size) significantly changes outcomes of competition and predation. In addition, large data gaps exist in our understanding of these interactions. For instance, evaluating the impact of introduced fishes is impossible because we do not know which nonnative fishes occur in many salmon-producing watersheds. Most available information is circumstantial. While such information can identify where inter- or intraspecific relationships may occur, it does not test mechanisms explaining why observed relationships exist. Thus, competition and predation are usually one of several plausible hypotheses explaining observed results.

Competition and predation should not be considered primary causes of population declines. For competition and predation to contribute to anadromous population declines, something must occur to alter the outcomes of these interactions (e.g., predation mortality increases). Competition and predation are altered as a result of the following: introductions of nonnative, non-salmonid fishes, introductions of artificially produced salmonids, environmental changes, and non-environmental changes in predator or competitor populations (e.g., from fishing). Efforts to restore salmon populations must direct action at identifying and eliminating primary causes of population declines and not simply treating secondary effects (i.e., competition and predation) of these causes.

Introduction

Many populations of naturally spawning Pacific salmon (*Oncorhynchus* spp.) and steelhead trout (*O. mykiss*) along the west coast of North America have declined to critically low levels (Konkel and McIntyre 1987, Nehlsen et al. 1991, Brown et al. 1994). One recently published assessment of the status of salmon and steelhead in California, Oregon, Washington, and Idaho

(The Wilderness Society 1993) concluded that salmon were extinct in 40% of their combined ranges and threatened or endangered in 27%.

To restore depressed populations and maintain viable populations in the future, we must understand why declines have occurred. Overfishing, freshwater habitat loss, water quality alterations, loss of genetic integrity of wild fish, and biological interactions (e.g., competition and predation) have been identified as factors causing anadromous population declines (Nehlsen et al. 1991, Hilborn 1992, The Wilderness Society 1993, Brown et al. 1994, Botkin et al. 1995). The purpose of this paper is to examine the role that competition and predation in freshwater and marine habitats have had in the decline of Pacific salmon and steelhead. Competition among adults for space and mates during reproduction was not included as part of this review.

I first summarize what is known about competition and predation for anadromous salmonids. Second, I discuss how these biological interactions have contributed to decreases in abundance of anadromous populations. Finally, I consider the effects of competition and predation within the context of restoration of salmonid populations.

Competition

Competition is the demand by two or more individuals of the same or different species for a resource that is actually or potentially limiting (Larkin 1956). As a result of competition, some competing individuals obtain less of the scarce resource than is optimum. These individuals may experience declines in reproductive rates, they may die, or they may be forced to emigrate from where they are living.

STREAMS

Stream-dwelling juvenile salmonids appear to compete primarily for space rather than for food or other resources (Chapman 1966, Hearn 1987). Individuals compete for positions based upon their importance for food acquisition and as cover (Fausch and White 1981). Competition for space is most critical during seasonally occurring periods of low flow (late spring to early fall) while in winter space is less critical owing to the fishes' lower metabolic requirements and levels of aggression (Hartman 1965, Glova 1986). Intraspecific competition is often of greater intensity than interspecific competition (Fraser 1969, Allee 1974, Lonzarich 1994). Aggressive interactions between individuals of the same species (i.e., interference competition) result in the formation of social hierarchies. The dominant individuals occupy preferred positions and are less likely to be displaced from territories, thereby having the highest growth rates (Chapman 1962, Mason and Chapman 1965, Allee 1974, Nielsen 1992).

Interspecific competition occurs between non-salmonids and salmonids as well as between different species of salmonids. Competition between salmonids and non-salmonids occurs infrequently (Moyle 1977, Brown and Moyle 1981, Baltz and Moyle 1984, Reeves et al. 1987, Lonzarich 1994), and is probably most significant in larger streams where non-salmonids are more abundant (Li et al. 1987). Interspecific competition is one mechanism used to partition scarce resources in streams (e.g., Hartman 1965, Glova 1986, Fausch and White 1986, Hearn 1987). Effects of aggressive interactions between competing individuals are highly localized and may result in shifts in microhabitat use by one or both of the interacting species (Li et al. 1987). Outcomes of interspecific competitive interactions vary with the species involved (Fausch and White 1986, Li et al. 1987), size differences among competing individuals (Fausch and White 1986), and numerous environmental factors such as temperature and streamflow (Hearn 1987, Reeves et al. 1987, Fausch 1988). Competition among sympatric salmonid species is minimized by species-specific differences in habitat preference, emergence timing, body morphology, environmental tolerances, or a combination of these factors (Hearn 1987, Bisson et al. 1988, Dolloff and Reeves 1990).

Lakes

The most extensive use of lakes by anadromous salmonids along the west coast of North America is by juvenile sockeye salmon (*O. nerka*), which rear for ≤ 3 years in lakes before migrating to sea. Intraspecific competition is considered the most important interaction involving sockeye juveniles. Burgner (1991) concluded that intraspecific competition for food among juvenile sockeye occurs when there are large numbers of sockeye in one year class, two or more year classes use the same resources, or species other than sockeye utilize the same resources. For example, in the Wood River Lakes, Alaska, the size (mean weight in g on September 1) of sockeye salmon fry rearing in the nursery lakes is inversely related to the number of parent spawners per rearing lake area (Fig. 1). One explanation for this relationship is that food supplies are limiting and, as a result, growth declines as density of rearing fry increases (indexed by changes in numbers of parent spawners per nursery lake area).

Marine Habitats

Food is the most limiting resource in marine habitats. Because measuring the amount of food available in coastal or open ocean habitats is difficult, evaluations of when, where, and what stocks might encounter food limitations during marine life have relied upon studies of food habits and dietary overlap, measures of stomach fullness and daily ration, simulation models, and evaluations of abundance, survival, and age composition from salmon management databases (Walters et al. 1978; Healey 1980; McCabe et al. 1983; Nielsen et al. 1985; Nickelson 1986; Peterman 1984, 1987; Fisher and Pearcy 1988; Emlen et al. 1990; Pearcy 1992; Beamish and Bouillon 1993; Cooney 1993; Rogers and Ruggerone 1993). One frequently used approach to identify where competition occurs is to test for density-dependent growth and survival (Peterman 1984, Emlen et al. 1990, Cooney 1993, Rogers and Ruggerone 1993). Evidence of a carrying capacity effect (e.g., decreased growth or survival as salmon densities increase) suggests food is limiting and competition is occurring. Regardless of the difficulty in proving that food is limiting, the carrying capacity of the ocean for salmonids is not unlimited; at some point, food supplies will be limiting.

Inter- and intraspecific competition involving juvenile salmonids is more likely to occur in estuarine and nearshore coastal areas. In these habitats, juvenile salmon can encounter food limitations that reduce growth and survival because they are spatially and temporally concentrated and have similar diets (e.g., Healey 1980; Peterman 1982, 1987; Nielsen et al. 1985; Pearcy and Fisher 1988; Brodeur and Pearcy 1990; Brodeur et al. 1992; Thomas and Mathisen 1993). Adult salmon also may encounter carrying capacity limitations when passing through nearshore coastal areas during their return migrations to natal streams (Rogers and Ruggerone 1993).

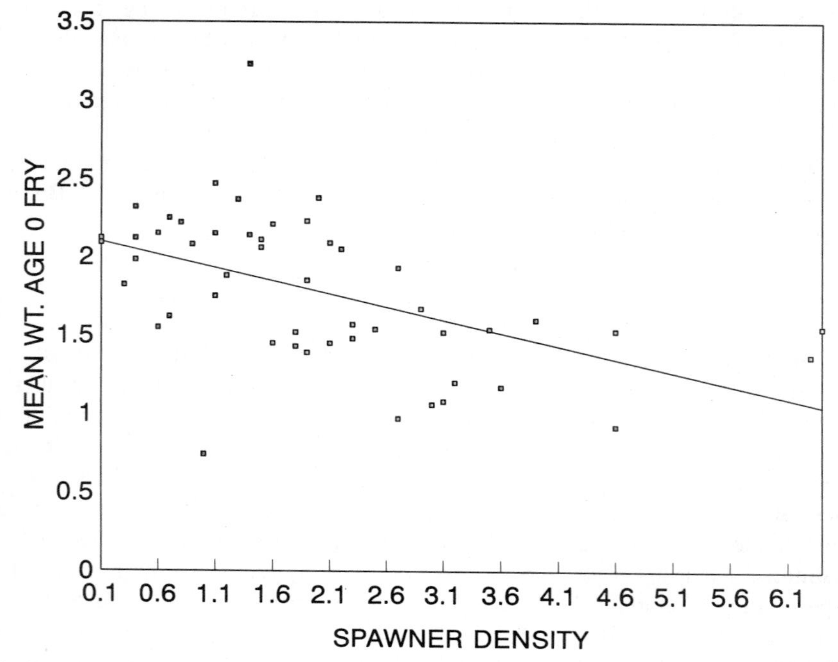

Figure 1. Relationship between mean size of sockeye salmon fry (weight in grams on September 1) and the relative density of parent spawners (numbers of spawners per km^2 of nursery lake area) from the previous year in the Wood River Lakes, Alaska. Data represent different lakes and years combined into one plot (after Burgner 1991). Source: Rogers (1977).

Predation

A large number of species eat salmon throughout their life cycle. The available literature indicates that 33 fish species, 13 bird species, and 16 marine mammal species are predators of juvenile and adult salmon (Table 1).

FRESHWATER

Numerous estimates of predation mortality in freshwater have been made (Table 2). Freshwater predators often consume large numbers of pink (*O. gorbuscha*) and chum (*O. keta*) salmon fry during their downstream migration. Up to 85% of pink and chum fry are eaten in some systems (e.g., Hooknose Creek, British Columbia; see Hunter 1959) even though fry need only migrate a short distance to reach the estuary (several km) and can accomplish this within one night. The small size of newly emerged pink, and chum fry (typically <40 mm) and their minimal avoidance capabilities make them especially vulnerable to predators.

Sockeye salmon juveniles are eaten by predators throughout their freshwater life (Foerster 1968, Burgner 1991). Similar to pink and chum fry, sockeye fry are particularly vulnerable to

Table 1. Predators of juvenile and adult Pacific salmon in freshwater and estuarine/marine habitats.

Species		Freshwater	Marine	Selected
Common name	Scientific name	predator	predator	references
Fish predators				
River lamprey	*Lampetra ayresi*	x	x	Beamish and Neville (1995)
Spiny dogfish	*Squalus acanthias*		x	Beamish et al. (1992)
American shad	*Alosa sapidissima*	x		Wendler (1967)
Pacific herring	*Clupea harengus pallasi*		x	Ito and Parker (1971)
Coho salmon	*Oncorhynchus kisutch*	x	x	Fresh and Schroder (1987), Hargreaves (1988)
Chinook salmon	*O. tshawytscha*	x	x	Dunford (1975), Sholes and Hallock (1979)
Cutthroat trout	*O. clarki*	x	x	McCart (1967), Fresh et al. (1981)
Rainbow trout	*O. mykiss*	x	x	Fresh et al. (1981), Fresh and Schroder (1987)
Arctic char	*Salvelinus alpinus*	x		Meacham and Clark (1979)
Dolly Varden	*S. malma*	x	x	Lagler and Wright (1962), Foerster (1968)
Lake trout	*S. namaycush*	x		Gilhousen and Williams (1989)
Lake whitefish	*Coregonus clupeaformis*	x		Gilhousen and Williams (1989)
Mountain whitefish	*Prosopium williamsoni*	x		Gilhousen and Williams (1989)
Northern squawfish	*Ptychocheilus oregonensis*	x		Rieman et al. (1991), Tabor et al. (1993)
Sacramento squawfish	*P. grandis*	x		Brown and Moyle (1981)
Channel catfish	*Ictalurus punctatus*	x		Poe et al. (1994)
Pacific cod	*Gadus macrocephalus*		x	Simenstad et al. (1979)
Tomcod	*Microgadus proximus*		x	Cooney et al. (1978)
Walleye pollock	*Theragra chalcogramma*		x	Armstrong and Winslow (1986)
Pacific hake	*Merluccius productus*		x	Hargreaves et al. (1990, cited by Wood et al. 1993)
Burbot	*Lota lota*	x		Gilhousen and Williams (1989)
Striped bass	*Morone saxatilis*	x	x	Stevens (1966), Johnson et al. (1992)
Smallmouth bass	*Micropterus dolomieui*	x		Tabor et al. (1993)
Largemouth bass	*Micropterus salmoides*	x		Poe et al. (1994)
Yellow perch	*Perca flavescens*	x		Dahle (1979)
Walleye	*Stizostedion vitreum vitreum*		x	Rieman et al. (1991), Poe et al. (1994)
Chub mackerel	*Scomber japonicus*		x	Washington Dep. Fish and Wildlife (unpubl. data)

Table 1—cont.

Species		Freshwater predator	Marine predator	Selected references
Common name	Scientific name			
Fish predators—cont.				
Coastrange sculpin	*Cottus aleuticus*	x		Hunter (1959)
Prickly sculpin	*C. asper*	x		Hunter (1959)
Shorthead sculpin	*C. confusus*	x		K. Fresh (Washington Dep. Fish and Wildlife, unpubl. data)
Reticulate sculpin	*C. perplexus*	x		Patten (1972)
Torrent sculpin	*C. rhotheus*	x		Patten (1972)
Staghorn sculpin	*Leptocottus armatus*		x	Dunford (1975)
Bird predators				
Double crested cormorant	*Phalacrocorax auritus*		x	Robertson (1974)
Harlequin duck	*Histrionicus histrionicus*	x		McCart (1967)
Common merganser	*Mergus merganser*	x	x	Simenstad et al. (1979), Wood (1987a, b)
Bald eagle	*Haliaeetus leucocephalus*	x	x	Simenstad et al. (1979), Angell and Balcomb (1982)
Short-billed gull (mew)	*Larus canus*	x		McCart (1967)
Ring-billed gull	*L. delawarensis*	x		Ruggerone (1986)
Glaucous-winged gull	*L. glaucescens*	x		Meacham and Clark (1979)
Bonaparte's gull	*L. philadelphia*	x		Meacham and Clark (1979)
Black tern	*Chlidonias niger*	x		McCart (1967)
Arctic tern	*Sterna paradisaea*	x		Meacham and Clark (1979)
Caspian tern	*Sterna caspia*		x	Simenstad et al. (1979)
Common murre	*Uria aalge*		x	Varoujean and Mathews (1983)
Rhinoceros auklet	*Cerorhinca monocerata*		x	D. Manuwal, University of Washington, Seattle, unpubl. data)

Table 1—cont.

Species		Freshwater predator	Marine predator	Selected references
Common name	Scientific name			
Mammal predators				
River otter	*Lutra canadensis*	x		Dolloff (1993)
California sea lion	*Zalophus californianus*		x	Simenstad et al. (1979), Fiscus (1980)
Northern sea lion	*Eumetopias jubatus*		x	Simenstad et al. (1979), Fiscus (1980)
Northern fur seal	*Callorhinus ursinus*		x	Simenstad et al. (1979), Fiscus (1980)
Harbor seal	*Phoca vitulina*	x	x	Beach et al. (1985), Olesiuk (1993)
Larga seal	*P. largha*		x	Fiscus (1980)
Fin whale	*Balaenoptera physalus*[a]		x	Fiscus (1980)
Humpback whale	*Megaptera novaeangliae*[a]		x	Fiscus (1980)
Pacific whiteside dolphin	*Lagenorhynchus obliquidens*		x	Fiscus (1980)
False killer whale	*Pseudorca crassidens*[a]		x	Fiscus (1980)
Killer whale	*Orcinus orca*		x	Fiscus (1980)
Harbor porpoise	*Phocoena phocoena*		x	Fiscus (1980)
Dall's porpoise	*Phocoenoides dalli*		x	Fiscus (1980)
Beluga whale	*Delphinapterus leucas*		x	Meacham and Clark (1979), Fiscus (1980)
Sperm whale	*Physeter catodon*[a]		x	Fiscus (1980)
Bear	*Ursus* spp.	x	x	Gard (1971)

[a]Considered by Fiscus (1980) to be an accidental occurrence.

Table 2. Some estimates of freshwater predation mortality. Examples were selected to show the range of life history stages, locations, and time intervals where predation mortality has been quantified. Outmig. = period of outmigration.

Prey, stage	Habitat	Time period	Mean loss (%)	Range (%)	Source
Chum, fry	Stream	Outmig.	47	35–62	Neave (1953)
Chum, fry	Stream	1 week	58	33–55	Hunter (1959)
Chum, fry	Stream	Outmig.	37	2–68	Semko (1954)
Pink, fry	Stream	Outmig.	34	5–86	Neave (1953)
Pink-chum, fry	Stream	Outmig.	45	23–85	Hunter (1959)
Sockeye, fry	Stream	Migrate to lake	84	67–98	Foerster (1968)
Sockeye, smolt	Lake	Outmig.	63	—	Rogers et al. (1972)
Sockeye, fingerlings	Lake	May-Sept.	59	—	Ruggerone and Rogers (1992)
Chinook and steelhead, smolts	Lake	Outmig.	14	9–19	Rieman et al. (1991)
Coho, fingerlings	Stream	Outmig.	—	24–65	Wood (1987b)
Coho, smolts	River	Outmig.	<1	—	Fresh (Washington Dep. Fish and Wildlife, unpubl. data)

predators as they emerge from the gravel and migrate downstream, but in this case, fry are migrating to rearing lakes (Table 2); some estimates of predation mortality during this period exceed 95%. Predators also consume large numbers of juvenile sockeye while they rear in lakes and as the smolts leave rearing lakes (Foerster and Ricker 1941, Eggers 1978, Burgner 1991). In the Chignik Lakes, Alaska, Ruggerone and Rogers (1992) found that 59% of the average population of sockeye fry was consumed by juvenile coho (*O. kisutch*) between May and September. In the Agulowak River, Alaska, Arctic char (*Salvelinus alpinus*) ate 33–66% of the outmigrating sockeye smolts in 1 year (Rogers et al. 1972).

Compared with pink, chum, and sockeye, there are few estimates of predation mortality of chinook (*O. tshawytscha*) and coho salmon in freshwater. Available data suggest that predation rates on wild populations are low under natural conditions (Patten 1971; Buchanan et al. 1981; Brown and Moyle 1981; Wood 1987a, b) but are much higher in non-natural situations, such as around dams and diversions during downstream migration (Brown and Moyle 1981). In the John Day Reservoir on the Columbia River, predators ate an annual average of 14% of migrating juvenile salmonids (mostly chinook) (Rieman et al. 1991); 21% of this loss occurred in a small area immediately below McNary Dam. Because both hatchery and wild fish were mixed, it was not possible to determine predation on wild populations.

Marine

The impact of predation on salmonids during marine life is poorly understood. A significant portion of the total natural marine mortality of salmon occurs during the estuarine and early marine period of juvenile life (Parker 1968, Ricker 1976, Bax 1983, Fisher and Pearcy 1988);

the extent to which this loss is due to predators is unclear. Few studies have estimated predation mortality of juvenile salmon during early marine life (Wood 1987a, Beamish et al. 1992, Wood et al. 1993). Instead, most assessments of predation during marine life have been limited to analyses of predator stomach contents (Fresh et al. 1981, Hargreaves 1988). Interpreting stomach contents data alone without accompanying data on the size of predator and prey populations can lead to erroneous conclusions. A low incidence of salmon smolts in predator stomachs can reflect a high predation mortality if the predator population is large and the smolt population size is low.

The magnitude of predation mortality of salmon in freshwater and marine habitats is a function of characteristics of predators, prey (i.e., the salmon), and their environment. Examples of these characteristics include predator and prey abundance, the size and number of individuals, condition or health of predators and prey, light intensity, and water temperature (Mead and Woodall 1968, Coutant et al. 1979, Hargreaves and LeBrasseur 1986, Fresh and Schroder 1987, Gregory 1993, Mesa 1994, Mesa et al. 1994). Depending on the direction of change of these factors, the magnitude of predation mortality increases or decreases. For instance, as prey size decreases, predation increases (Taylor and McPhail 1985, Hargreaves and LeBrasseur 1986). Similarly, predation increases as salmon become more visible to predators because of increased light intensity, decreased streamflows, or decreased turbidity (Ginetz and Larkin 1976).

Declines in Salmonid Stocks: Role of Competition and Predation

Competition and predation are natural processes that have influenced the abundance of salmon and steelhead throughout their evolutionary history. Anadromous salmonids have evolved characteristics that minimize both predation mortality and the loss of fitness due to effects of competition; otherwise, they would have gone extinct long ago. For competition and predation to contribute to the decline of an anadromous population, something must alter the outcome of biological interactions (e.g., predation mortality must increase). Moreover, this change must result in decreased numbers of reproducing adults. Ways in which this can happen include the following: introductions of nonnative, non-salmonid fish; introductions of artificially produced salmonids; environmental changes; and changes in population sizes of predators and competitors caused by non-environmental factors such as fishing. Although each of these factors is discussed separately in the following sections, several may operate simultaneously on any one salmonid stock. I have also focused on human-induced changes because understanding their effects will be most useful in helping to restore declining populations.

Introductions of Non-Salmonid Fishes

Overview

Since the arrival of Europeans, numerous species of plants, invertebrates, and fish have been introduced throughout North America (Taylor et al. 1984, Moyle et al. 1986). While ecosystem effects of plant and invertebrate introductions can be severe (Li and Moyle 1981, Nichols et al. 1990, Northcote 1991), I focus here on introductions of non-salmonid fishes.

The successful establishment of a nonnative fish alters the physical or biological nature of the receiving environment (Taylor et al. 1984, Brown and Moyle 1991). As a result, growth, survival, or abundance of native species can decline owing to parasites and diseases, inhibition of reproduction, changes in the nature of existing biological interactions, or environmental alterations (Moyle 1976, Taylor et al. 1984, Moyle et al. 1986, Ross 1991). Effects on native species are difficult to predict because they depend on two factors: (1) physiological, behavioral, and ecological potentialities of the introduced species; and (2) physical and biological properties of the ecosystem (Taylor et al. 1984, Herbold and Moyle 1986, Ross 1991). In general, the successful establishment of nonnative species and subsequent displacement of native fish is greater in habitats that have been extensively modified, especially by anthropogenic factors (Ross 1991, Baltz and Moyle 1993).

To my knowledge, there is no comprehensive assessment for fish species introductions into anadromous salmonid-producing watersheds of the North American west coast. Introductions into some areas, such as California (Moyle 1976) and the Columbia and Sacramento rivers (Moyle 1976, Li et al. 1987, Bisson et al. 1992, Poe et al. 1994), have been well documented. However, in most areas, little or no information exists on the extent to which nonnative fish introductions have negatively impacted native salmon and steelhead because of the effects of competition and predation (Table 3).

Competition

Considerable concern has been expressed about potential impacts on salmonids of American shad (*Alosa sapidissima*) because they are present in various river systems along the North American west coast (Table 3) that also have depressed populations of salmon and steelhead. However, in most West Coast rivers American shad are not abundant, suggesting that they have not caused population declines in those river systems.

American shad are abundant in two rivers, the Sacramento and Columbia, although in the Sacramento River, the numbers of shad have declined considerably from peak levels in the early 1900s (Stevens et al. 1987). Conversely, in the Columbia River, numbers of American shad recently increased to their highest historical levels (Fig. 2) at the same time as some salmon and steelhead runs declined to critically low levels. As a result, competition between American shad and juvenile salmon and steelhead has been hypothesized as one cause of anadromous population declines in the Columbia River system (Kaczynski and Palmisano 1992, Bevan et al. 1994). Other than the inverse relation between shad abundance and salmon runs, evidence supporting this hypothesis results primarily from high dietary overlaps between shad and juvenile salmonids in some habitats (e.g., estuaries; see McCabe et al. 1983). Even though dietary overlaps are often cited as evidence of interspecific competition, they are inconclusive without accompanying information on food supplies and spatial and temporal overlap of the potentially interacting species. To my knowledge, estimates of the amount of food available do not exist for the estuary or reservoirs associated with the Columbia River. Some overlap of shad and several anadromous species in reservoirs occurs, but overlap in the estuary appears to be minimal (Dawley et al. 1986).

Table 3. Examples of non-salmonid fish species introduced into anadromous salmonid-producing watersheds along the Pacific coast of North America. Species selected were those where available information suggested they could be predators or competitors of naturally produced salmon and steelhead. (Note: Lake Washington sockeye salmon were included as a naturally reproducing population even though they have both native and introduced portions. Since their introduction, the non-native portion has been sustained largely by natural production.)

Species/watershed	Species potentially impacted			
	Chinook	Coho	Steelhead	Sockeye
Potential Competitors				
American shad[a]				
Sacramento River, California	x	x	x	
Russian River, California	x	x		
Klamath River, Oregon	x		x	
Coos River, Oregon	x	x		
Columbia River	x	x	x	x
Chehalis River, Washington	x	x	x	
Threadfin shad				
Sacramento River, California	x	x	x	
Longfin smelt[b]				
Lake Washington, Washington				x
Bluegill				
Russian River, California	x	x		
Columbia River	x	x	x	
Lake Washington, Washington				x
Yellow perch				
Columbia River	x	x	x	
Lake Washington, Washington				x
Potential Predators				
American shad				
Sacramento River, California	x			
Klamath River, Oregon	x			
Columbia River	x			
Yellow perch				
Lake Washington, Washington				x
Channel catfish				
Columbia River	x	x	x	
Striped bass				
Sacramento River, California	x	x	x	
Coos River, Oregon	x			
Smallmouth bass				
Russian River, California	x	x		
Umpqua River, Oregon	x		x	
Mid-Columbia River	x	x	x	x
John Day River, Oregon	x			
Lake Osoyoos, Washington				x
Lake Washington, Washington				x

Table 3—cont.

Species/watershed	Species potentially impacted: Chinook	Coho	Steelhead	Sockeye
Potential Predators—cont.				
Largemouth bass				
Lake Washington, Washington				x
Russian River, California	x	x		
Walleye				
Columbia River	x	x	x	x
White bass/striped bass hybrid				
Tenmile Lakes, Oregon		x		

[a]Shad are present in many North American west coast rivers; these were selected to show the range of systems in which they are found.

[b]The exact origin of longfin smelt in Lake Washington is unknown, but they were probably strays from an adjacent watershed.

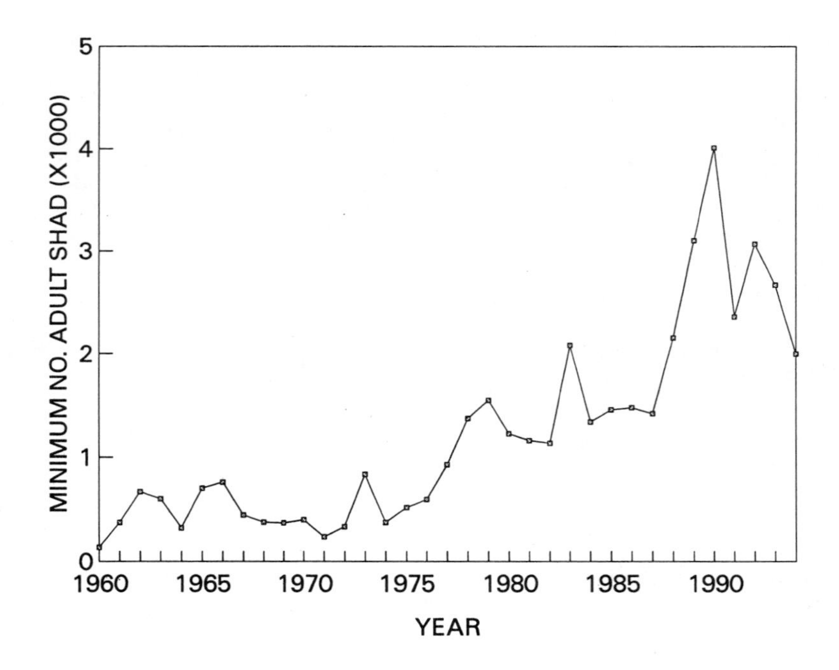

Figure 2. Minimum numbers of adult American shad entering the Columbia River. Source: Washington Department of Fish and Wildlife and Oregon Department of Fish and Wildlife, unpubl. data.

Predation

Introduction of a piscivorous species can impact native fish directly as a result of consumption or indirectly by affecting habitat use and competitive interactions (Brown and Moyle 1991, Rieman et al. 1991). Various salmon- and steelhead-producing river systems have introduced piscivores (Table 3). For instance, striped bass (*Morone saxatilis*) were introduced into the Sacramento River in 1879. Even though striped bass prey on salmon (Stevens 1966) and have historically been abundant in the system, I am unaware of any study that evaluates their impact on native anadromous populations in the basin.

In 1914, striped bass were introduced into the Coos River, Oregon. Since their introduction, the abundance of striped bass and fall chinook in the system has been inversely related, suggesting striped bass were negatively impacting native chinook (Johnson et al. 1992). However, changes in habitat occurred simultaneously, making it difficult to isolate effects of the striped bass. A predation model developed by Johnson et al. (1992) estimated that striped bass in this system could consume between 42,000 and 383,000 juvenile salmonids or an equivalent of 1,000 to 46,000 adult salmon. Despite this information, a decision was made to enhance the striped bass fishery to a level of 20,000 adults, potentially resulting in the loss of 15,000 adult salmonids (Johnson et al. 1992).

The most rigorous evaluation of predation by nonnative species on anadromous populations has been in the Columbia River. Smallmouth bass (*Micropterus dolomieui*), walleye (*Stizostedion vitreum vitreum*), and adult American shad are introduced species known to prey on juvenile salmon (Poe et al. 1994). Rieman et al. (1991) calculated that smallmouth bass and walleye in the John Day Reservoir ate ~3% of the outmigrating salmonids annually (a mixture of hatchery and wild). Although overall predation appears low, some specific salmon populations may be more heavily impacted by nonnative species. Tabor et al. (1993) found a high incidence of subyearling chinook in smallmouth bass stomachs just below the last natural spawning area of wild salmon in the mainstem Columbia River. They speculated that many of the chinook being eaten were juveniles from wild spawners.

Introduction of Artificially Produced Salmonids

Overview

The use of artificially produced fish is widely believed to be a major factor contributing to the decline in abundance of salmon and steelhead along the North American west coast (Marnell 1986, Nehlsen et al. 1991, Hilborn 1992, Brown et al. 1994). Competition and predation between native and cultured fish are two types of impacts that may result from artificially cultured fish introductions (Marnell 1986, Steward and Bjornn 1990, Nehlsen et al. 1991, Hilborn 1992, Brown et al. 1994).

Artificially produced salmonids are equivalent to nonnative species introductions even when conspecifics are already present in the receiving environment. As mentioned previously, impacts of nonnative species will depend upon physiological, ecological, and behavioral characteristics of the introduced fish. For artificially cultured fish, these characteristics are a function of their genetic origin and how they are reared and released (Mead and Woodall 1968, Fenderson and Carpenter 1971, Hume and Parkinson 1987, Swain et al. 1991). Cultural practices influence many attributes of hatchery fish that affect outcomes of biological interactions, including the following: size and morphology (Swain et al. 1991), behavior (Fenderson and Carpenter 1971, Mesa 1991, Nielsen 1994), habitat utilization patterns (Dickson and MacCrimmon 1982, Levings et al. 1986, Petrosky and Bjornn 1988), and movements (Levings et al. 1986, Hume and Parkinson 1987).

Competition

Competition between hatchery-produced and wild salmonids is often cited as a mechanism to explain how hatchery fish introductions have impacted native salmonids in streams (Marnell 1986, Fausch 1988, Brown et al. 1994). Although studies show that hatchery fish can disrupt the growth, survival, and abundance of native salmonid communities in streams (Bjornn 1978, Nickelson et al. 1986), competition's role in causing these changes is unclear (Fausch 1988, Steward and Bjornn 1990). For instance, Nickelson et al. (1986) reported a 44% decline in abundance of wild juvenile coho salmon in Oregon coastal streams following the release of hatchery-produced coho juveniles. The authors speculated that hatchery coho were able to outcompete wild coho because they were larger and thus could evict wild fish from their territories.

Flagg et al. (1995) concluded that the over tenfold reduction in densities of wild coho spawners in the lower Columbia River resulted, in part, from the large size of juvenile coho used in hatchery programs as well as stocking hatchery coho at densities ≤7X the carrying capacity of the receiving environment. As a result, wild juvenile coho were competitively displaced by the larger-sized (hence, competitively superior) and abundant hatchery coho. In the Eagle River, British Columbia, Perry (1995) found that the return per spawner of wild coho and the survival of hatchery coho fry declined as hatchery coho fry abundance was increased; he attributed these trends to the effects of competition between the hatchery and wild coho in freshwater. Similarly, Bjornn (1978) and Tripp and McCart (1983) concluded that competitive interactions between native and hatchery fish caused the decline in native salmonid populations they studied. Even though competition was a plausible explanation for the results of the previous studies, none of them specifically measured mechanisms responsible for changes in the native salmonid populations.

Both intra- and interspecific competition can occur as a result of hatchery fish introductions (Allee 1974, Fausch and White 1986, Kennedy and Strange 1986, Spaulding et al. 1989, Nielsen 1994). Nielsen (1994) provides an excellent example of the effects of intraspecific competition between hatchery and wild salmonids. Following the introduction of hatchery coho salmon in the Noyo River, California, Nielsen (1994) found that agonistic encounters between hatchery and wild coho resulted in the displacement of 83% of the wild coho from their usual microhabitats. Foraging behavior of the wild fish was also altered as a result of aggressive encounters. Production of wild coho salmon in the Noyo River declined following the introduction of the hatchery coho although this decline did not appear to be significant when compared with concurrent changes in other wild populations where hatchery fish were not used.

Interspecific competition between artificially produced and native salmonids has been the focus of a great deal of research (Allee 1974, Kennedy and Strange 1986, Marnell 1986, Moyle et al. 1986, Fausch 1988, Nielsen 1994). Unfortunately, we lack the ability to reliably predict which combinations of sympatry will produce intense interspecific competition and which will not (Fausch 1988). Because species that have not co-evolved do not possess behavioral and morphological mechanisms to reduce competition, intense interspecific competition can occur when an introduction brings together species that are not naturally sympatric (Hearn 1987, Fausch 1988). Introductions that produce sympatry in species that have co-evolved elsewhere but not in the system where the introduction occurs may also result in interspecific competition (Hearn 1987). This occurs because in allopatry, a species can exhibit ecological release, a process whereby the niche of the species expands in the absence of competition. When a new species is introduced, intense interspecific competition occurs because the two species in that particular system did not co-evolve. Thus, they lack mechanisms to reduce effects of competition, even though they co-occur elsewhere.

In marine habitats, evidence that competition occurs between hatchery and wild fish consists primarily of density-dependent declines in survival or growth of wild salmonids simultaneously with increasing numbers of hatchery fish (Thomas and Mathisen 1993, Perry 1995). For example, Hilborn (1992) suggested that sockeye salmon juveniles produced by the Babine Lake, British Columbia, spawning channels depressed smolt-to-adult survival of wild sockeye owing to competition (Fig. 3). While this suggests a density-dependent interaction (e.g., competition for food), no specific evaluation of causative mechanisms was conducted.

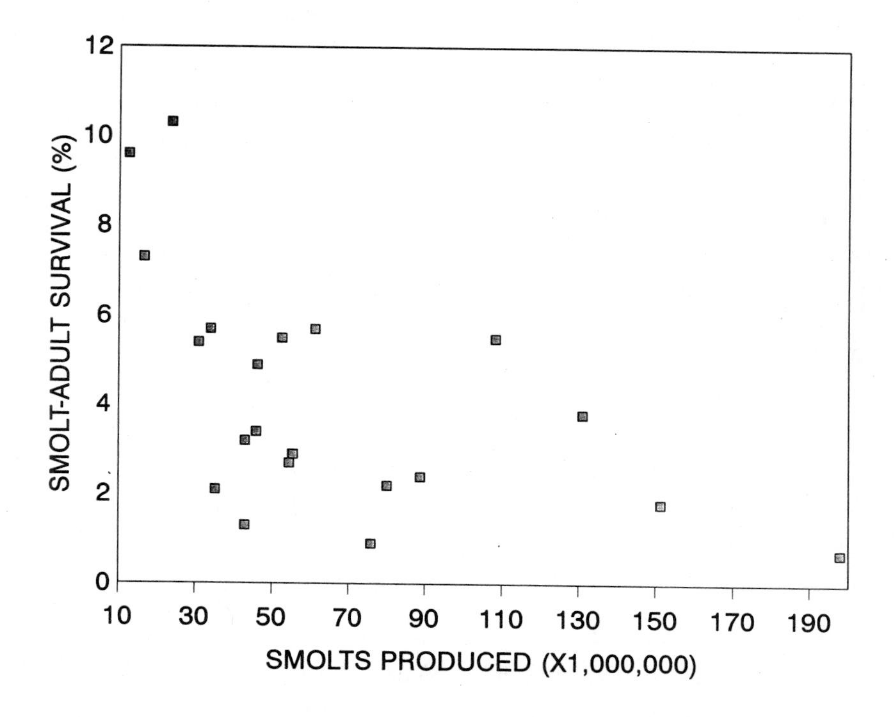

Figure 3. The relationship between total sockeye salmon smolt production and smolt-to-adult survival from Babine Lake, British Columbia (after Hilborn 1992). Smolt production represents the total number of juvenile sockeye produced in a year and includes both wild and artificially produced fish. Survival declines as more artificially produced fish are released. Source: McDonald and Hume (1984) and Macdonald et al. (1987).

One situation where competition between hatchery and wild salmonids during the marine life phase likely occurs is in populations of pink and chum salmon in Washington and southern British Columbia (Gallagher 1979, Peterman 1987, Beachum 1993). Many chum populations in this region exhibit strong odd-even year cyclicity in abundance and age composition (Gallagher 1979). This between-year variability has been hypothesized to result from chum competing with pink salmon during early marine life (Belford 1978, Gallagher 1979, Peterman 1987, Beachum 1993). Such an interaction is plausible because the number of adult pink salmon that return to spawn in the region is much greater in odd-numbered years; thus, pink fry are abundant in even-numbered years, but scarce in odd-numbered years. Moreover, the two species have similar distributions and food requirements during early marine life (Simenstad et al. 1980, Peterman 1987).

Evidence of the pink salmon effect on chum is exhibited in Puget Sound where chum salmon survival declines as pink salmon escapement increases (Fig. 4). For Fraser River pink and chum salmon, Peterman (1987) concluded, based on survival rate correlations, that most of the between-year variability in marine survival of chum salmon occurs during early ocean life when pink and chum juveniles are sympatric for a prolonged period. If competition between these two species occurs during early marine life, then increasing numbers of pink salmon juveniles with

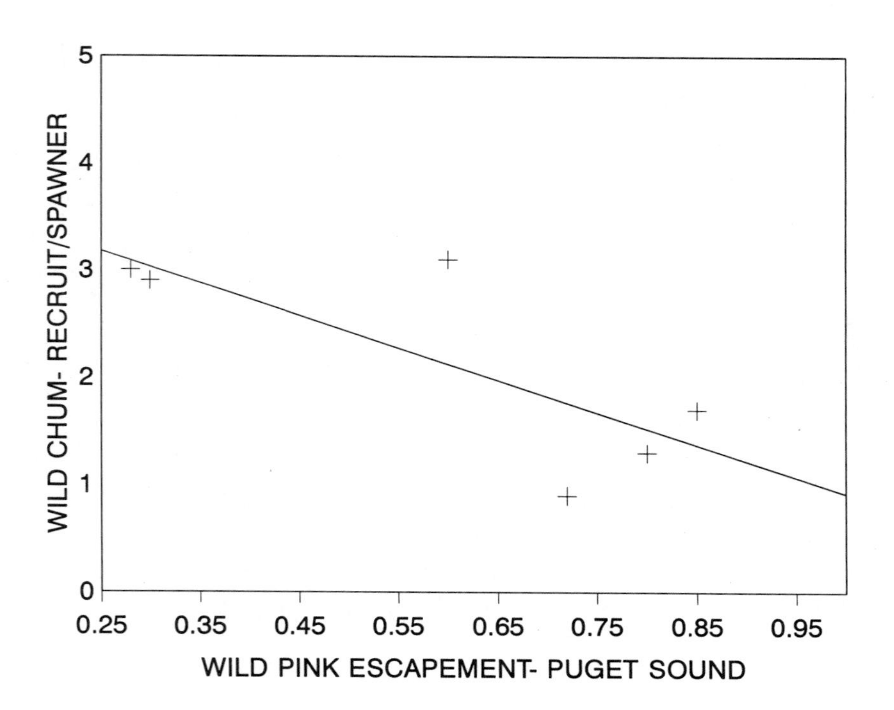

Figure 4. Relationship between pink salmon escapement and chum salmon survival (as measured by recruit per spawner) in Puget Sound, Washington ($r^2 = 0.62$). Each data point is a separate year from 1959 to 1979. Source: Washington Department of Fish and Wildlife (WDFW), unpubl. data; J. Ames, WDFW, pers. comm.

enhancement programs will depress chum survival. This can occur if high densities of hatchery fish depress growth rates of wild fish and increase the period when wild fry remain vulnerable to predators (Belford 1978).

Predation

Compared with competition, less is known about predation by hatchery fish on wild salmonids (Fausch 1988). Several studies have used data on stomach contents, predator abundance, etc., to estimate consumption of juvenile wild salmon by hatchery-produced fish (Table 4). For example, Evenson et al. (1981) calculated that the average annual loss of wild chinook and steelhead over a 3-year period in the Rogue River, Oregon, due to predation by hatchery fish was 9.7%. Martin et al. (1993) evaluated predation on juvenile chinook in the Tucannon River, Washington, from a release of 119,082 juvenile steelhead; they estimated ~10,000 of these fish remained in the study area and ate 456 wild fish (95% CI = 4–3,117) in the first 4.5 months following their release in April.

Much of the information on predation by hatchery fish on wild salmonids is circumstantial in nature, such as that obtained from analyses of fishery management databases (Table 4). One example of this type of data is from Washington, where Johnson (1973) proposed that predation by hatchery coho salmon had caused the decline of several wild pink and chum populations. His conclusion was based on circumstantial evidence such as a comparison of adult returns of chum and coho at various hatcheries. Johnson (1973) observed that at a number of hatcheries, adult chum returns declined dramatically shortly after the initiation of hatchery coho programs. Further, he found that wild chum populations did not decline in areas where hatchery coho programs had not been established (e.g., South Puget Sound or on larger streams). This and other circumstantial evidence led Johnson (1973) to conclude that hatchery coho were responsible for the declines of a number of chum populations.

Environmental Changes

Considerable attention has been focused on the role that human-induced environmental changes have had in the decline of Pacific salmon and steelhead populations (e.g., Nehlsen et al. 1991). For >150 years, humans have altered or eliminated access to freshwater habitats (Li et al. 1987, Raymond 1988); modified the quantity, type, and quality of freshwater and marine habitats (Scott et al. 1986, Hicks et al. 1991, Bisson et al. 1992, Simenstad et al. 1992, Reeves et al. 1993); and changed the physiochemical nature of waters the fish use (e.g., Seiler 1989, Hicks et al. 1991). Many authors regard environmental changes, especially impacts to freshwater habitats, as a major cause of the decline of anadromous populations in the Pacific Northwest (Nehlsen et al. 1991, The Wilderness Society 1993, Brown et al. 1994).

One way environmental changes cause salmon abundance to decline is by altering outcomes of biological interactions, such as by changing availability of resources, abundance of predator and competitor populations, condition (e.g., stress levels) of interacting individuals, and sizes of interacting individuals (Ginetz and Larkin 1976, Coutant et al. 1979, Fisher and Pearcy 1988, Brodeur and Pearcy 1990, Gregory 1993, Mesa 1994). Changes in water temperatures (e.g., caused by removing riparian vegetation along streams) altered competitive interactions between redside shiner (*Richardsonius balteatus*) and juvenile steelhead trout (Reeves et al. 1987). At

Table 4. Examples of situations where hatchery salmonids are potentially preying upon wild salmonids. Two general types of data analysis were used: (1) fishery management databases, and (2) stomach contents.

Wild population	Predator	Evidence	Source
Fishery database analyses			
PUGET SOUND CHUM AND PINK SALMON			
Samish River	Coho salmon	Decline in rack returns	Johnson (1973)
Skykomish River	Coho salmon	Decline in rack returns	Johnson (1973)
Green River	Coho salmon	Spawner abundance index	Johnson (1973)
Minter Creek	Coho salmon	Decline in rack returns	Johnson (1973)
Nemah River	Coho salmon	Decline in rack returns	Johnson (1973)
Willapa River	Coho salmon	Decline in rack returns	Johnson (1973)
Stillaguamish River	Coho salmon	Decline in terminal run size	J. Ames (Washington Dep. Fish and Wildlife, unpubl. data)
Stomach contents analyses			
CHINOOK SALMON			
Rogue River, Oregon	Chinook, steelhead	$\overline{X}$ = 9.7% salmon juveniles eaten per year	Evenson et al (1981)
Salmon River, Idaho	Steelhead	~13% of juveniles eaten in 1 year	Bevan et al. (1994)
Feather River, California	Chinook, steelhead	7.5 million wild fish eaten in 1 year	Sholes and Hallock (1979)[a]
Tucannon River, Washington	Steelhead	Mean predation loss = 456 fish	Martin et al. (1993)
Nicola River, British Columbia	Chinook	No predation observed	Levings and Lauzier (1989)
SOCKEYE			
Lake Washington, Washington	Rainbow trout	2% of sockeye eaten per year	Beauchamp (1987)

[a]No method was provided for how the predation loss was derived.

higher temperatures, the redside shiner were able to competitively displace steelhead from their territories. A higher incidence of disease also led to higher mortality among the less competitive species while the dominant competitor was unaffected.

Another example of how environmental changes alter biological interactions is in the Columbia River where damming and impounding the river has created favorable habitat conditions for a number of nonnative piscivorous species, most of which originated from the midwestern United States (US) (Li et al. 1987). These nonnative piscivores prey on native salmonids and have altered the native fish community structure and food web in the Columbia River system (Li et al. 1987, Tabor et al. 1993, Poe et al. 1994).

While it is clear that human-induced environmental changes can alter competition and predation, such changes will not always cause the abundance of salmonid populations to decline. Salmon have lived in naturally changing freshwater and marine environments throughout their evolutionary history and, thus, have adapted to some level of environmental changes (e.g., Taylor 1991). For instance, juvenile salmon in some rivers move in the fall to habitats that provide greater protection from winter flood conditions (Bustard and Narver 1975, Peterson 1982, Nickelson et al. 1992). It is when salmonids cannot adapt to environmental changes that abundance levels can decline.

One example of how environmental changes, biological interactions, and declines in salmonid populations can be linked is predation around dams and diversions, such as in the Columbia and Sacramento rivers (Hall 1979, Rieman et al. 1991). In the Columbia River, predation by fish and birds is a particularly significant cause of dam-related mortality (Ruggerone 1986, Rieman et al. 1991). Detailed studies of predation in one reservoir (John Day Reservoir) demonstrated that the average annual loss of salmonid smolts due to fish predators, especially northern squawfish (*Ptychocheilus oregonensis*), was 14% (Rieman et al. 1991). High predation losses of smolts are also associated with other mainstem dams and reservoirs in the Columbia River system (Uremovich et al. 1980).

I speculate that losses of salmonids to northern squawfish and other fish predators in the Columbia River are higher now than they were historically. What was once a free-flowing system has been converted into a series of dams and impoundments. Squawfish predation is especially high around dams, diversions, and in lakes; in free-flowing rivers, predation rates are typically lower (Brown and Moyle 1981, Rieman et al. 1991). It is not possible to quantify how predation losses in the Columbia River may have changed as a result of damming the river because historical data on pre-dam predation rates do not exist. Also, without knowledge of stock-specific mortality rates, it is impossible to know how mortality varies among stocks (e.g., is it more severe on certain stocks?).

OTHER FACTORS CAUSING CHANGES IN PREDATOR AND COMPETITOR POPULATIONS

Factors other than environmental variability (e.g., diseases, parasites, and fishing) can also alter biological interactions. For example, since the early 1970s, pinniped abundance, particularly harbor seals (*Phoca vitulina*) and California sea lions (*Zalophus californianus californianus*), has increased to near historical levels because the harvest of these animals was prohibited by the Canadian and US governments (Olesiuk et al. 1990a, Calambokidis and Baird 1994). Because marine mammals have recently increased in abundance, are known to prey on adult and juvenile salmon and steelhead, and tend to aggregate at river mouths (Beach et al. 1985, Olesiuk et al. 1990b), it is believed that these animals are contributing to declines in some anadromous populations (Kaczyinksi and Palmisano 1992). The most comprehensive evaluation of marine mammal predation on salmon is a study by Olesiuk (1993) of harbor seals in the Strait of Georgia, British Columbia. Olesiuk (1993) calculated that nearly 386 mt of salmon or 3% of the mean annual escapement in the area in recent years was consumed by harbor seals in 1988.

Because information used by Olesiuk (1993) is unavailable for other areas of the North American west coast, comparable estimates cannot be computed for these regions. In a thorough review of the available information on salmon and steelhead in northern California and western Oregon, Botkin et al. (1995) concluded that, despite serious weaknesses in existing data and analyses, "marine mammals are a minor factor in the harvest of salmon." However, in some specific cases where stocks are already at critically low levels of abundance or where conditions exist that enhance predation, marine mammal predation can have a significant impact (Calambokidis and Baird 1994). For example, marine mammal predation adjacent to Lake Washington, Washington, is affecting winter-run steelhead. In recent years, California sea lions have annually consumed ≤65% of the returning adults at the Hiram Chittendon Locks in Seattle (R. Leland, WDFW, Mill Creek, pers. comm.) and are considered a major factor responsible for the decline of this steelhead run.

General Discussion

Evaluation of Available Information

Evaluating the role of competition and predation in the decline of salmon and steelhead populations depends upon the quantity and quality of available information. I found few instances where the contribution of these interactions to population declines could be clearly established. One reason for this was the existence of significant data gaps. For example, the lack of comprehensive surveys of nonnative fish introductions in western North American watersheds makes it impossible to fully assess effects of nonnative fishes on native salmonid communities.

Many of the data on competition and predation were circumstantial in nature, particularly in evaluations of interactions in marine habitats. Examples of such data include abundance or age composition data obtained from fishery management agencies, food habits and diet overlap (Peterman 1984, 1987). Analyses of this type of data identifies the potential for inter- or intraspecific relationships (Perry 1995) but does not directly measure or test potential mechanisms. Thus, competition and predation can be one of several plausible hypotheses explaining observed results. For example, the decline in the chum salmon populations observed by Johnson (1973) could have been due to predation by juvenile coho salmon from hatcheries on chum fry or habitat loss occurring at the same time as coho programs expanded.

Much of the uncertainty in the data stems from the inherent complexity and dynamic nature of biological interactions. Effects of competition and predation on a particular salmonid population depend upon a whole suite of variables (e.g., size of predators and competitors, environmental conditions, and abundance levels); a small change in one may dramatically change mortality. This is particularly apparent when evaluating effects of hatchery fish introductions. As noted previously, numerous variables associated with rearing and releasing hatchery fish (e.g., numbers of hatchery fish that remain in the area, size of hatchery fish that are released) affect competition and predation with native salmonids. Thus, predicting what types of biological interactions may result from introducing hatchery fish and the associated impacts on native salmonids remains complex.

Data were especially ambiguous when assessing effects of interspecific competition. Competition is a difficult interaction to study and measure even in the best of conditions; for instance, effects of competition can be easily masked by a number of environmental factors (Taylor et al. 1984, Fausch 1988). Clear demonstration that interspecific competition is occurring requires evidence that a niche shift in one species occurs in the presence of another species (Hearn 1987). This necessitates manipulative experiments in either natural or controlled laboratory settings (Fausch 1988). These types of experiments have been successfully accomplished in stream environments while in marine habitats they have only been accomplished at very small scales (e.g., laboratory tanks). Stream habitats represent a better opportunity for competition experiments because we know more about the biology and ecology of the system, thus making it easier to design and conduct experiments; they can be more easily modeled in laboratory settings (e.g., laboratory stream channels); and sampling methods are relatively inexpensive. It is no accident that we know more about competition in streams than about competition in marine habitats. The reliance on circumstantial data to assess effects of biological interactions in marine habitats is undoubtedly a reflection of the difficulty associated with measuring biological interactions in such large systems.

Biological Interactions and the Restoration of Salmonid Populations

Reviews by Nehlsen et al. (1991), Konkel and McIntyre (1987), The Wilderness Society (1993), Brown et al. (1994), and others have drawn attention to the depressed status of many Pacific salmon and steelhead populations. Clearly, once we have identified where depressed populations occur, we must next take action to restore populations and better maintain their future health. A major step in accomplishing this goal must be to identify reasons for population declines so that the appropriate corrective actions can be taken. In deciding what actions should be taken, it is important that we recognize that competition and predation are a secondary effect or symptom of other changes that have occurred. For biological interactions to have a role in the decline of an anadromous population, something must occur that alters the outcome of competition or predation (e.g., predation mortality must increase). For example, altering abundance levels of populations of salmon predators by changing harvest levels can affect overall predation mortality. In the Columbia and Sacramento rivers, dams and diversions, which kill salmon directly (e.g., as a result of passage through turbines) or indirectly (e.g., by increasing predation risk), are the underlying cause of salmon mortality while predation represents a secondary effect of the dams.

To effectively restore depressed populations, we must identify and eliminate the underlying causes of population declines and not simply treat secondary effects (Meffe 1992, Black 1994). This is analogous to a doctor treating a patient (Meffe 1992). To cure the patient, the doctor must identify the disease that is responsible for the symptoms that he observes. If the physician treats only symptoms without understanding what is causing the symptoms, the patient may fail to recover because the doctor treats the wrong disease. Similarly, to restore a salmon population, the factors responsible for the deleterious biological interactions must be identified and corrective action directed at these factors. For example, while reducing the abundance of northern squawfish around dams in the Columbia River may increase abundance of salmon (Rieman and Beamesderfer 1990), it avoids the issue of why predation became a problem in the first place (the damming and impounding of the river). In this case, to cure the depressed salmon runs, a solution should involve the dams.

Establishing whether and how biological interactions have contributed to declines of specific salmonid populations will not be easy. The available data will rarely, if ever, be unequivocal. Uncertainty in decision making is not a unique feature of salmon populations or of competition and predation (Ludwig et al. 1993). It requires a willingness to use innovative, adaptive, and experimental management approaches in situations where many of our answers will come from experience (Walters 1986). A key part of these types of approaches must be monitoring both biological interactions and changes in salmon population abundance. For example, abundance levels of wild populations associated with hatchery programs should be carefully monitored; depending on the outcome of the monitoring efforts, managers can decide whether to proceed with additional enhancement (Olson et al. 1995, Perry 1995).

In some situations, decision makers will require additional data or a reanalysis of existing data using new methods before determining how to proceed. Such additional studies can be time-consuming and costly, may do little to reduce uncertainty, and might delay initiation of recovery actions, perhaps by many years. Such delays must be balanced against the likelihood that further work will reduce uncertainty and the status of the salmonid population. If no action is taken, already depressed salmonid populations may decline further or become extinct as a result of even a small change in predation or competition. Even populations that do not become extinct can be trapped at low levels of abundance by depensatory mortality (e.g., predation, Peterman 1987).

Acknowledgments

The manuscript was greatly improved by the comments of P. Moyle, S. Schroder, E. Volk, and two anonymous reviewers, and by discussions with J. Ames.

Literature Cited

Allee, B. J. 1974. Spatial requirements and behavioral interactions of coho salmon (*Oncorhynchus kisutch*) and steelhead trout (*Salmo gairdneri*). PhD dissertation, University of Washington. Seattle.

Angell, T. and K.C. Balcomb III. 1982. Marine birds and mammals of Puget Sound. Washington Sea Grant, Seattle.

Armstrong, R.H. and P.C. Winslow. 1968. An incidence of walleye pollock predation on salmon young. Transactions of the American Fisheries Society 97: 202-203.

Baltz, D.M. and P.B. Moyle. 1984. Segregation by species and size classes of rainbow trout, *Salmo gairdneri*, and Sacramento sucker, *Catostomus occidentalis*, in three California streams. Environmental Biology of Fishes 10: 101-110.

Baltz, D.M. and P.B. Moyle. 1993. Invasion resistance to introduced species by a native assemblage of California stream fishes. Ecological Applications 3: 246-255.

Bax, N. 1983. Early marine mortality of marked juvenile chum salmon (*Oncorhynchus keta*) released into Hood Canal, Puget Sound, Washington in 1980. Canadian Journal of Fisheries and Aquatic Sciences 40: 426-435.

Beach, R.J., A.C. Geiger, S.F. Jefferies, S.D. Treacy, and B. Troutman. 1985. Marine mammals and their interactions with fisheries of the Columbia River and adjacent waters, 1980-1982. Northwest and Alaska Fisheries Center, NMFS, NWAFC Processed Report 85-04. Seattle, Washington.

Beachum, T.D. 1993. Competition between juvenile pink (*Oncorhynchus gorbuscha*) and chum salmon (*Oncorhynchus keta*) and its effect on growth and survival. Canadian Journal of Zoology 71: 1270-1274.

Beamish, R. and D.R. Bouillon. 1993. Pacific salmon production trends in relation to climate. Canadian Journal of Fisheries and Aquatic Sciences 50: 1002-1016.

Beamish, R.J. and C. M. Neville. 1995. Pacific salmon and Pacific herring mortalities in the Fraser River plume caused by river lamprey (*Lamptera ayresi*). Canadian Journal of Fisheries and Aquatic Sciences 52: 644-650.

Beamish, R., B.L. Thomson, and G.A. McFarlane. 1992. Spiny dogfish predation on chinook and coho salmon and the potential effects on hatchery-produced salmon. Transactions of the American Society 121: 444-455.

Beauchamp, D.A. 1987. Ecological relationships of hatchery rainbow trout in Lake Washington. PhD dissertation, University of Washington. Seattle.

Belford, D.L. 1978. Simulation of coho smolt predation on pink and chum fry. The importance of relative size and growth rate. MS thesis, University of British Columbia. Vancouver.

Bevan, D.E., J. Harville, P. Bergman, T. Bjornn, J. Crutchfield, P. Klingeman, and J. Litchfield. 1994. Snake River Salmon Recovery Team: Final Recommendations to the National Marine Fisheries Service. Portland, Oregon.

Bisson, P.A., T.P. Quinn, G.H. Reeves, and S.V. Gregory. 1992. Best management practices, cumulative effects, and long-term trends in fish abundance in Pacific Northwest River systems, p. 189-232. *In* R.J. Naiman (ed.), Watershed Management: Balancing Sustainability and Environmental Change. Springer-Verlag, New York.

Bisson, P. A., K. Sullivan, and J.L. Nielsen. 1988. Channel hydraulics, habitat use, and body form of juvenile coho salmon, steelhead, and cutthroat trout in streams. Transactions of the American Fisheries Society 117: 262-272.

Bjornn, T. C. 1978. Survival, production, yield of trout and chinook salmon in the Lemhi River, Idaho. University of Idaho, College of Forestry, Wildlife, and Range Sciences Bulletin Number 27. Moscow.

Black, M. 1994. Recounting a century of failed fishery policy toward California's Sacramento River salmon and steelhead. Conservation Biology 8: 892-894.

Botkin, D., K. Cummins, T. Dunne, H. Reiger, M. Sobel, L. Talbot, and L. Simpson. 1995. Status and future of salmon of Western Oregon and Northern California: Findings and options. Center for the Study of the Enviornment, Report No. 8. Santa Barbara, California.

Brodeur, R.D., R.C. Francis, and W.G. Pearcy. 1992. Food consumption of juvenile coho (*Oncorhynchus kisutch*) and chinook salmon (*O. tshawytscha*) on the continental shelf off Washington and Oregon. Canadian Journal of Fisheries and Aquatic Sciences 49: 1670-1685.

Brodeur, R.D. and W.G. Pearcy. 1990. Trophic relations of juvenile Pacific salmon off the Oregon and Washington coast. Fisheries Bulletin (US) 88: 617-636.

Brown, L.R. and P.B. Moyle. 1981. The impact of squawfish on salmonid populations: a review. North American Journal of Fisheries Management 1: 104-111.

Brown, L.R. and P.B. Moyle. 1991. Changes in habitat and microhabitat partitioning within an assemblage of streamfishes in response to predation by Sacramento squawfish (*Ptychocheilus grandis*). Canadian Journal of Fisheries and Aquatic Sciences 48: 849- 856.

Brown, L.R., P.B. Moyle, and R.M. Yoshiyama. 1994. Historical decline and current status of coho salmon in California. North American Journal of Fisheries Management 14: 237-261.

Buchanan, D.V., R.M. Hooton, and J.R. Moring. 1981. Northern squawfish (*Ptychocheilus oregonensis*) predation on juvenile salmonids in sections of the Willamette River basin, Oregon. Canadian Journal of Fisheries and Aquatic Sciences 38: 360-364.

Burgner, R. 1991. Life history of sockeye salmon, p. 1-117. *In* C. Groot and L. Margolis (eds.), Pacific Salmon Life Histories. University of British Columbia Press, Vancouver.

Bustard, D.R. and D.W. Narver. 1975. Aspects of the winter ecology of juvenile coho salmon (*Oncorhynchus kisutch*) and steelhead trout (*Salmo gairdneri*). Journal of the Fisheries Research Board Canada 32: 667-680.

Calambokidis, J. and R.W. Baird. 1994. Status of marine mammals in the Strait of Georgia, Puget Sound, and the Strait of Georgia, and potential impacts, p. 282-300. *In* R.C.H. Wilson, R.J. Beamish, F. Aitkens, and J. Bell (eds.), Review of the Marine Environment and Biota of the Strait of Georgia, Puget Sound, and Juan de Fuca Strait. Canadian Department of Fisheries and Oceans, Technical Report No. 1948.

Chapman, D.W. 1962. Aggressive behavior of juvenile coho salmon as a cause of emigration. Journal of the Fisheries Research Board of Canada 19: 1047-1080.

Chapman, D.W. 1966. Food and space as regulators of salmonid populations in streams. American Naturalist 100: 345-357.

Cooney, R.T. 1993. A theoretical evaluation of the carrying capacity of Prince William Sound, Alaska for juvenile Pacific salmon. Fisheries Research 18: 77-88.

Cooney, R.T., D. Urquhart, R. Neve, J. Hilsinger, R. Clasby, and D. Barnard. 1978. Some aspects of the carrying capacity of Prince Williams Sound, Alaska for hatchery-released pink and chum salmon fry. University of Alaska, Fairbanks. Sea Grant Report 78-4, IMS Report R78-3.

Coutant, C.C., R.B. McLean, and D.L. DeAngelis. 1979. Influences of physical and chemical alterations on predator-prey interactions, p. 57-68. *In* H. Clepper (ed.), Predator-Prey Systems in Fisheries Management. Sport Fish. Institute, Washington, DC.

Dahle, T.F. 1979. Observations of fingerling chinook salmon in the stomachs of yellow perch from the Klamath River, California. California Fish and Game 65: 168.

Dawley, E.L., C.W. Sims, R.D. Ledgerwood, T. Blahm, C.W. Sims, J. Durkin, and others. 1986. Migrational characteristics, biological observations, and relative survival of juvenile salmonids entering the Columbia River Estuary, 1966-1983. Final Report, Bonneville Power Administration, US Department of Energy. Portland, Oregon.

Dickson, T.A. and H.R. MacCrimmon. 1982. Influence of hatchery experience on growth and behavior of juvenile Atlantic salmon (*Salmo salar*) within allopatric and sympatric stream populations. Canadian Journal of Fisheries and Aquatic Sciences 39: 1453- 1458.

Dolloff, C.A. 1993. Predation by river otters (*Lutra canadensis*) on juvenile coho salmon (*Oncorhynchus kisutch*) and Dolly Varden (*Salvelinus malma*) in Southeast Alaska. Canadian Journal of Fisheries and Aquatic Sciences 50: 312-315.

Dolloff, D.A. and G.H. Reeves. 1990. Microhabitat partitioning among stream-dwelling juvenile coho salmon, *Oncorhynchus kisutch*, and Dolly Varden, *Salvelinus malma*. Canadian Journal of Fisheries and Aquatic Sciences 47: 2297-2306.

Dunford, W.E. 1975. Space and food utilization by salmonids in marsh habitats of the Fraser River Estuary. MS thesis, University of British Columbia. Vancouver, British Columbia.

Eggers, D.M. 1978. Limnetic feeding behavior of juvenile sockeye salmon in Lake Washington and predator avoidance. Limnology and Oceanography 23: 1114-1125.

Emlen, J.M., R.R. Reisenbichler, A.M. McGie, and T.E. Nickelson. 1990. Density dependence at sea for coho salmon (*Oncorhynchus kisutch*). Canadian Journal of Fisheries and Aquatic Sciences 47: 1765-1772.

Evenson, M.D., R.D. Ewing, E. Birks, A.R. Hemmingsen, and J. Dentler. 1981. Coles River Hatchery evaluation. Oregon Department of Fish and Wildlife, Fish Division, Progress Report 1981.

Fausch, K.D. 1988. Tests of competition between native and introduced salmonids in streams. Canadian Journal of Fisheries and Aquatic Sciences 45: 2238-2246.

Fausch, K.D. and R.J. White. 1981. Competition between brook trout and brown trout for positions in a Michigan stream. Canadian Journal of Fisheries and Aquatic Sciences 38: 1220-1227.

Fausch, K.D. and R.J. White. 1986. Competition among juveniles of coho salmon, brook trout, and brown trout in a laboratory stream, and implications for Great Lakes tributaries. Transactions of the American Fisheries Society 115: 363-381.

Fenderson, O.C. and M.R. Carpenter. 1971. Effects of crowding on the behavior of juvenile hatchery and wild landlocked Atlantic salmon (*Salmo salar* L.). Animal Behavior 19: 439-447.

Fiscus, C.H. 1980. Marine mammal-salmonid interactions: a review, p. 121-131. *In* W.J. McNeil and D.C. Himsworth (eds.), Salmonid Ecosystems of the North Pacific. Oregon State University Press and Oregon State University Sea Grant Program, Corvallis.

Fisher, J.P. and W.G. Pearcy. 1988. Growth of juvenile coho salmon (*Oncorhynchus kisutch*) off Oregon and Washington, USA, in years of differing coastal upwelling. Canadian Journal of Fisheries and Aquatic Sciences 45: 1036-1044.

Flagg, T.A., F.W. Waknitz, D.J. Maynard, G.B. Milner, and C.V.W. Mahnken. 1995. The effect of hatcheries on native coho salmon populations in the lower Columbia River. American Fisheries Society Symposium 15: 366-375.

Foerster, R.E. 1968. The sockeye salmon (*Oncorhynchus nerka*). Fisheries Research Board of Canada Bulletin 162.

Foerster, R.E. and W.E. Ricker 1941. The effect of reduction of predaceous fish on survival of young sockeye salmon at Cultus Lake. Journal of the Fisheries Research Board of Canada 5: 315-336.

Fraser, F.J. 1969. Population density effects on survival and growth of juvenile coho salmon and steelhead trout in experimental stream channels, p. 253-268. *In* T.G. Northcote (ed.). Symposium on Salmon and Trout in Streams. H.R. MacMillan Lectures in Fisheries, University of British Columbia, Vancouver.

Fresh, K.L., R.D. Cardwell, and R.R. Koons. 1981. Food habits of Pacific salmon, baitfish, and their potential competitors and predators in the marine waters of Washington, August 1978 to September 1979. Washington Department of Fisheries, Progress Report No. 145. Olympia.

Fresh, K.L. and S.L. Schroder. 1987. Influence of the abundance, size, and yolk reserves of juvenile chum salmon (*Oncorhynchus keta*) on predation by freshwater fishes in a small coastal stream. Canadian Journal of Fisheries and Aquatic Sciences 44: 236-243.

Gallagher, A.F., Jr. 1979. An analysis of factors affecting brood year returns in the wild stocks of Puget Sound chum (*Oncorhynchus keta*) and pink salmon (*Oncorhynchus gorbuscha*). MS thesis, University of Washington. Seattle.

Gard, R. 1971. Brown bear predation on sockeye salmon at Karluk Lake, Alaska. Journal of Wildlife Management 35: 193-204.

Gilhousen, P. and I.V. Williams. 1989. Fish predation on juvenile Adams River sockeye salmon in the Shuswap Lakes in 1975 and 1976, p. 83-100. *In* Studies of the Lacustrine Biology of the Sockeye Salmon (*Oncorhynchus nerka*) in the Shuswap System. Bulletin XXIV, International Pacific Salmon Fisheries Commission.

Ginetz, R. and P.A. Larkin. 1976. Factors affecting rainbow trout (*Salmo gairdneri*) predation on migrant fry of sockeye salmon (*Oncorhynchus nerka*). Journal of the Fisheries Research Board Canada 33: 19-24.

Glova, G.J. 1986. Interaction of food and space between experimental populations of juvenile coho salmon (*Oncorhynchus kisutch*) and coastal cutthroat trout (*Salmo clarki*) in a laboratory stream. Hydrobiologia 132: 155-168.

Gregory, R.S. 1993. Effect of turbidity on the predator avoidance behavior of juvenile chinook salmon (*Oncorhynchus tshawytscha*). Canadian Journal of Fisheries and Aquatic Sciences 50: 241-246.

Hall, F.A. 1979. An evaluation of downstream migrant chinook salmon (*Oncorhynchus tshawytscha*) losses at Hallwood-Cordura fish screen. California Department of Fish and Game, Anadromous Fisheries Branch, Administrative Report No. 79-5. Stockton.

Hargreaves, N.B. 1988. A field method of determining prey preferences of predators. Fishery Bulletin 86: 763-772.

Hargreaves, N.B., and R.J. Lebrasseur. 1986. Size selectivity of coho (*Oncorhynchus kisutch*) preying on juvenile chum salmon (*O. keta*). Canadian Journal of Fisheries and Aquatic Sciences 43: 581-586.

Hartman, G.F. 1965. The role of behavior in the ecology and interaction of underyearling coho salmon (*Oncorhynchus kisutch*) and steelhead trout (*Salmo gairdneri*). Journal of the Fisheries Research Board of Canada 22: 1035-1081.

Healey, M.C. 1980. The ecology of juvenile salmon in Georgia Strait, British Columbia, p. 203-229. *In* W.J. McNeil and D.C. Himsworth (eds.), Salmonid Ecosystems of the North Pacific. Oregon State University Press, Corvallis.

Hearn, W.E. 1987. Interspecific competition and habitat segregation among stream-dwelling trout and salmon: A review. Fisheries 12: 24-31.

Herbold, B. and P.B. Moyle. 1986. Introduced species and vacant niches. American Naturalist 128: 751-760.

Hicks, B.J., J.D. Hall, P.A. Bisson, and J.R. Sedell. 1991. Responses of salmonids to habitat changes, p. 483-518. *In* W.R. Meehan (ed.) Influences of Forest and Rangeland Management on Salmonid Fishes and Their Habitats. American Fisheries Society Special Publication 19. Bethesda, Maryland.

Hilborn, R. 1992. Hatcheries and the future of salmon in the Northwest. Fisheries 17: 5-8.

Hume, J.M.B. and E.A. Parkinson. 1987. Effect of stocking density on the survival, growth, and dispersal of steelhead trout fry (*Salmo gairdneri*). Canadian Journal of Fisheries and Aquatic Sciences 44: 271-281.

Hunter, J.G. 1959. Survival and production of pink and chum salmon in a coastal stream. Journal of the Fisheries Research Board Canada 16: 835-886.

Ito, J. and R.R. Parker. 1971. A record of Pacific herring (*Clupea harengus pallasi*) feeding on juvenile chinook salmon (*Oncorhynchus tshawytscha*) in a British Columbia estuary. Journal of the Fisheries Research Board of Canada 28: 1921.

Johnson, J.H., A.A. Nigro, and R. Temple. 1992. Evaluating enhancement of striped bass in the context of potential predation on anadromous salmonids in Coos Bay, Oregon. North American Journal of Fisheries Management 12: 103-108.

Johnson, R.C. 1973. Potential interspecific problems between hatchery coho smolts and juvenile pink and chum salmon. Puget Sound stream studies, pink and chum investigations. Washington Department of Fisheries. Olympia.

Kaczynski, V.W. and J.F. Palmisano. 1992. A review of management and environmental factors responsible for the decline and lack of recovery of Oregon's wild anadromous salmonids. Technical Report, Oregon Forest Industries Council. Salem, Oregon.

Kennedy, G.J.A., and C.D. Strange. 1986. The effects of intra- and interspecific competition on the survival and growth of stocked juvenile Atlantic salmon, *Salmo salar* L., and resident trout, *Salmo trutta* L., in an upland stream. Journal of Fish Biology 28: 479- 489.

Konkel, G.W. and J.D. McIntyre. 1987. Trends in spawning populations of Pacific anadromous salmonids. US Fish and Wildlife Service Fish and Wildlife Technical Report 9. Washington, DC.

Lagler, K.F. and A.T. Wright. 1962. Predation of the Dolly Varden, *Salvelinus malma*, on young salmons, *Oncorhynchus* spp., in an estuary of Southeastern Alaska. Transactions of the American Fisheries Society 91: 90-93.

Larkin, P.A. 1956. Interspecific competition and population control in freshwater fish. Journal of the Fisheries Research Board of Canada 13: 327-342.

Levings, C.D. and R. Lauzier. 1989. Migration patterns of wild and hatchery-reared juvenile chinook salmon (*Oncorhynchus tshawytscha*) in the Nicola River, British Columbia, p. 267-275. *In* B. Shepherd (rapporteur), Proceedings of the 1988 Northwest Pacific Chinook and Coho Salmon Workshop. North Pacific International Chapter, American Fisheries Society. Bethesda, Maryland

Levings, C.D., C.C. McAllister, and E.D. Chang. 1986. Differential use of the Campbell River Estuary, British Columbia, by wild and hatchery-reared juvenile chinook salmon (*Oncorhynchus tshawytscha*). Canadian Journal of Fisheries and Aquatic Sciences 43: 1386-1397.

Li, H.W. and P.B. Moyle. 1981. Ecological analysis of species introductions into aquatic systems. Transactions of the American Fisheries Society 110: 772-782.

Li, H.W., C.S. Schreck, C.E. Bond, and E. Rexstad. 1987. Factors influencing changes in fish assemblages of Pacific Northwest streams, p. 193-202. *In* W.J. Matthews and D.C. Heins (eds.), Community and Evolutionary Ecology of North American Stream Fishes. University of Oklahoma Press, Norman and London.

Lonzarich, D.C. 1994. Stream fish communities in Washington: patterns and processes. PhD dissertation, University of Washington. Seattle.

Ludwig, D., R. Hilborn, and C. J. Walters. 1993. Uncertainty, resource exploitation, and conservation: lessons from history. Science 260: 17, 36.

Macdonald, P.D.M., H.D. Smith, and L. Jantz. 1987. The utility of Babine smolt enumerations in management of Babine and other Skeena River sockeye salmon (*Oncorhynchus nerka*) stocks, p. 280-295. *In* H.D. Smith, L. Margolis, and C.C. Wood (eds.). Sockeye Salmon (*Oncorhynchus nerka*) Population Biology and Future Management. Canadian Special Publication in Fisheries and Aquatic Sciences 96.

Marnell, L. 1986. Impacts of hatchery stocks on wild fish populations, p. 339-347. *In* R.H. Stroud (ed.), Fish Culture in Fisheries Management. American Fisheries Society, Bethesda, Maryland.

Martin, S.W., A.E. Viola, and M.L. Schuck. 1993. Investigations of the interactions among hatchery-reared summer steelhead, rainbow trout, and wild spring chinook salmon in southeast Washington. Washington Department of Wildlife, Fisheries Management Division Report #93-4. Olympia.

Mason, J.C. and D.W. Chapman. 1965. Significance of early emergence, environmental rearing capacity, and behavioral ecology of juvenile coho salmon in stream channels. Journal of the Fisheries Research Board Canada 22: 173-190.

McCabe, G.T., Jr., W.D. Muir, R.L. Emmett, and J.T. Durkin. 1983. Interrelationships between juvenile salmonids and non-salmonid fish in the Columbia River estuary. Fishery Bulletin 81: 815-826.

McCart, P. 1967. Behavior and ecology of sockeye salmon fry in the Babine River. Journal of the Fisheries Research Board Canada 24: 375-428.

McDonald, J. and J.M. Hume. 1984. Babine Lake sockeye salmon (*Oncorhynchus nerka*) enhancement program: testing some major assumptions. Canadian Journal of Fisheries and Aquatic Sciences 41: 70-92.

Meacham, C.P. and J.H. Clark. 1979. Management to increase anadromous salmon production, pp. 377-386. *In* H. Clepper (ed.), Predator-Prey Systems in Fisheries Management. Sport Fishing Institute, Washington DC.

Mead, R.W. and W. Woodall. 1968. Comparison of sockeye salmon fry produced by hatcheries, artificial channels, and natural spawning areas. International Pacific Salmon Fish Commission, Progress Report No. 20. New Westminster, British Columbia.

Meffe, G.K. 1992. Techno-arrogance and halfway technologies: salmon hatcheries on the Pacific Coast of North America. Conservation Biology 6: 350-354.

Mesa, M.G. 1991. Variation in feeding, aggression, and position choice between hatchery and wild cutthroat trout in an artificial stream. Transactions of the American Fisheries Society 120: 723-727.

Mesa, M.G. 1994. Effects of multiple acute stressors on the predator avoidance ability and physiology of juvenile chinook salmon. Transactions of the American Fisheries Society 123: 786-793.

Mesa, M.G., T.P. Poe, D.M. Gadomski, and J.H. Petersen. 1994. Are all prey created equal? A review and synthesis of differential predation on prey in substandard condition. Journal of Fish Biology 45(Supplement A): 81-96.

Moyle, P.B. 1976. Fish introductions in California: history and impact on native fishes. Biological Conservation 9: 101-118.

Moyle, P.E. 1977. In defense of sculpins. Fisheries 2: 20-23.

Moyle, P.B., H.L. Li, and B.A. Barton. 1986. The Frankenstein effect: impacts of introduced fishes on native fishes in North America, p. 415-426. *In* R.H. Stroud (ed.), Fish Culture in Fisheries Management. American Fisheries Society, Bethesda, Maryland.

Neave, F. 1953. Principles affecting the size of pink and chum populations in British Columbia. Journal of the Fisheries Research Board of Canada 9: 450-491.

Nehlsen, W., J.E. Williams, and J.A. Lichatowich. 1991. Pacific salmon at the crossroads: stocks at risk from California, Oregon, Idaho, and Washington. Fisheries 16: 4-21.

Nichols, F.H., J.K. Thompson., and L.E. Schemel. 1990. Remarkable invasion of San Francisco Bay (California, USA) by the Asian clam *Potamocorbula amurensis*. II. Displacement of a former community. Marine Ecology Progress Series 66: 95-101.

Nickelson, T.E. 1986. Influences of upwelling, ocean temperature, and smolt production on marine survival of coho salmon (*Oncorhynchus kisutch*) in the Oregon production area. Canadian Journal of Fisheries and Aquatic Sciences 43: 527-535.

Nickelson, T.E., J.D. Rogers, S.L. Johnson, and M.F. Solazzi. 1992. Seasonal changes in habitat use by juvenile coho salmon (*Oncorhynchus kisutch*) in Oregon coastal streams. Canadian Journal of Fisheries and Aquatic Sciences 49: 783-789.

Nickelson, T. E., M. F. Solazzi, and S.L. Johnson. 1986. Use of hatchery coho salmon (*Oncorhynchus kisutch*) to rebuild wild populations in Oregon coastal streams. Canadian Journal of Fisheries and Aquatic Sciences 48: 2443-2449.

Nielsen, J.D., G.H. Geen., and D. Bottom. 1985. Estuarine growth of juvenile chinook salmon (*Oncorhynchus tshawytscha*) as inferred from otolith microstructure. Canadian Journal of Fisheries and Aquatic Sciences 42: 899-908.

Nielsen, J.L. 1992. Microhabitat-specific foraging behavior, diet, and growth of juvenile coho salmon. Transactions of the American Fisheries Society 121: 616-634.

Nielsen, J.L. 1994. Invasive cohorts—impacts of hatchery-reared coho salmon on the trophic, developmental, and genetic ecology of wild stocks, p. 361-385. *In* D.J. Stouder, K.L. Fresh, and R. Feller (eds.). Theory and Application in Fish Feeding Ecology. The Belle Baruch Library in Marine Science, University of South Carolina, Belle Baruch Press. Columbia.

Northcote, T.G. 1991. Success, problems, and control of introduced mysid populations in lakes and reservoirs. American Fisheries Society Symposium 9: 5-16.

Olesiuk, P.F. 1993. Annual prey consumption by harbor seals (*Phoca vitulina*) in the Strait of Georgia, British Columbia. Fishery Bulletin 91: 491-515.

Olesiuk, P.F., M.A. Bigg, and G.M. Ellis. 1990a. Recent trends in the abundance of harbour seals, *Phoca vitulina*, in British Columbia. Canadian Journal of Fisheries and Aquatic Sciences 47: 992-1003.

Olesiuk, P.F., M.A. Bigg, G.M. Ellis, S.J. Crockford, and R.J. Wigen. 1990b. An assessment of the feeding habits of harbour seals (*Phoca vitulina*) in the Strait of Georgia, British Columbia, based on scat analysis. Canadian Technical Report of Fisheries and Aquatic Sciences No. 1730.

Olson, D.E., B.C. Cates, and D.H. Diggs. 1995. Use of a national fish hatchery to complement wild salmon and steelhead production in an Oregon stream. American Fisheries Society Symposium 15: 317-328.

Parker, R.R. 1968. Marine mortality schedules of pink salmon of the Bella Coola River, central British Columbia. Journal of the Fisheries Research Board Canada 25: 757- 794.

Patten, B.G. 1971. Predation by sculpins on fall chinook salmon, *Oncorhynchus tshawytscha*, fry of hatchery origin. NOAA National Marine Fisheries Service, Special Scientific Report Fisheries No. 621. Washington, DC.

Patten, B.G. 1972. Predation, particularly by sculpins, on salmon fry in freshwaters of Washington. National Technical Information Service No. COM-72-10527.

Pearcy, W.G. 1992. Ocean Ecology of North Pacific Salmonids. Washington Sea Grant Program, Seattle and London.

Pearcy, W.G. and J.P. Fisher. 1988. Migrations of coho salmon, *Oncorhynchus kisutch*, during their first summer in the ocean. Fishery Bulletin 86: 173-195.

Perry, E.A. 1995. Salmon stock restoration and enhancement: strategies and experiences in British Columbia. American Fisheries Society Symposium 15: 152-160.

Peterman, R.M. 1982. Non-linear relation between smolts and adults in Babine Lake sockeye salmon (*Oncorhynchus nerka*) and implications for other salmon populations. Canadian Journal of Fisheries and Aquatic Sciences 39: 904-913.

Peterman, R.M. 1984. Density-dependent growth in early ocean life of sockeye salmon (*Oncorhynchus nerka*). Canadian Journal of Fisheries and Aquatic Sciences 41: 1825- 1829.

Peterman, R.M. 1987. Review of the components of recruitment of Pacific salmon. American Fisheries Society Symposium 1: 417-429.

Peterson, N.P. 1982. Immigration of juvenile coho salmon (*Oncorhynchus kisutch*) into riverine ponds. Canadian Journal of Fisheries and Aquatic Sciences 39: 1308-1310.

Petrosky, C.E., and T.C. Bjornn. 1988. Response of wild rainbow (*Salmo gairdneri*) and cutthroat trout (*S. clarki*) to stocked rainbow trout in fertile and infertile streams. Canadian Journal of Fisheries and Aquatic Sciences 45: 2087-2105.

Poe, T.P., R.S. Shively, and R.A. Tabor. 1994. Ecological consequences of introduced piscivorous fishes in the lower Columbia and Snake rivers, p. 347-360. *In* D.J. Stouder, K.L. Fresh, and R. Feller (eds.), Theory and Application in Fish Feeding Ecology. The Belle Baruch Library in Marine Science, University of South Carolina, Belle Baruch Press. Columbia.

Raymond, H.L. 1988. Effects of hydroelectric development and fisheries enhancement on spring and summer chinook salmon and steelhead in the Columbia River Basin. North American Journal of Fisheries Management 8: 1-24.

Reeves, G.H., F.H. Everest, and J.D. Hall. 1987. Interactions between the redside shiner (*Richardsonius balteatus*) and the steelhead trout (*Salmo gairdneri*) in western Oregon: the influence of water temperature. Canadian Journal of Fisheries and Aquatic Sciences 44: 1602-1613.

Reeves, G.H., F.H. Everest, and J.R. Sedell. 1993. Diversity of juvenile salmonid assemblages in coastal Oregon basins with different levels of timber harvest. Transactions of the American Fisheries Society 122: 309-317.

Ricker, W.E. 1976. Review of the rate of growth and mortality of Pacific salmon in saltwater and non-catch mortality caused by fishing. Journal of the Fisheries Research Board Canada 33: 1483-1524.

Rieman, B.E. and R.C. Beamesderfer. 1990. Dynamics of a northern squawfish population and the potential to reduce predation on juvenile salmonids in a Columbia River reservoir. North American Journal of Fisheries Management 10: 228-241.

Rieman, B.E., R.C. Beamesderfer, S. Vigg, and T.P. Poe. 1991. Estimated loss of juvenile salmonids to predation by northern squawfish, walleyes, and smallmouth bass in John Day Reservoir, Columbia River. Transactions of the American Fisheries Society 120: 448-458.

Robertson, I. 1974. The food of nesting double-crested and pelagic cormorants at Mandrate Island, British Columbia with notes on feeding ecology. Condor 76: 346-348.

Rogers, D.E. 1977. Collection and analysis of biological data from the Wood River Lake System, Nushagak, Bristol Bay, Alaska. Will fertilization increase growth and survival of juvenile sockeye salmon in the Wood River Lakes? University of Washington, College of Fisheries, Fisheries Research Institute, FRI-UW-7617-C. Seattle.

Rogers, D.E., L. Gilbertson, and D. Eggers. 1972. Predator-prey relationship between Arctic char and sockeye smolts at the Agulowak River, Lake Aleknagik, in 1971. University of Washington, College of Fisheries, Fisheries Research Institute, Circular No. 72-7. Seattle.

Rogers, D.E. and G.T. Ruggerone. 1993. Factors affecting marine growth of Bristol Bay sockeye salmon. Fisheries Research 18: 89-104.

Ross, S. T. 1991. Mechanisms structuring stream fish assemblages: are there lessons from introduced species? Environmental Biology of Fishes 30: 359-368.

Ruggerone, G.T. 1986. Consumption of migrating juvenile salmonids by gulls foraging below a Columbia River Dam. Transactions of the American Fisheries Society 115: 736-742.

Ruggerone, G.T. and D.E. Rogers. 1992. Predation on sockeye salmon fry by juvenile coho salmon in the Chignik Lakes, Alaska: implications for salmon management. North American Journal of Fisheries Management 12: 87-102.

Scott, J.B., C.R. Steward, and Q.J. Stober. 1986. Effects of urban development on fish population dynamics in Kelsey Creek, Washington. Transactions of the American Fisheries Society 115: 555-567.

Seiler, D. 1989. Differential survival of Grays Harbor basin anadromous salmonids: water quality implications, p. 123-135. *In* C.D. Levings, L.B. Holtby, and M.A. Henderson (eds.), Proceedings of the National Workshop on effects of Habitat Alteration on Salmonid Stocks. Canadian Special Publication of Fisheries and Aquatic and Sciences 105.

Semko, R.S. 1954. A method for determining the consumption of predators of the young Pacific salmon during early stages of development. Trudy Soveshchaniya 6: 124-134. (Fisheries Research Board Canada, Translation No. 215.)

Sholes, W.H. and R.J. Hallock. 1979. An evaluation of rearing fall-run chinook salmon, *Oncorhynchus tshawytscha*, to yearlings at Feather River Hatchery with a comparison of returns from hatchery and downstream releases. California Fish and Game 65: 239-255.

Simenstad, C.A., D. Jay, and C.R. Sherwood. 1992. Impacts of watershed management on land-margin ecosystems: The Columbia River estuary, p. 266-306. *In* R.J. Naiman (ed.), Watershed Management: Balancing Sustainability and Environmental Change. Springer-Verlag, New York.

Simenstad, C.A., W.J. Kinney, S.S. Parker, E.O. Salo, J.R. Cordell, and H. Buechner. 1980. Prey community structure and trophic ecology of outmigrating juvenile chum and pink salmon in Hood Canal, Washington. A synthesis of three years' studies, 1977-1979. Fisheries Research Institute Report FRI-UW-8026, University of Washington, College of Fisheries. Seattle.

Simenstad, C.A., B.S. Miller, C.F. Nybalde, K. Thornburgh, and L.J. Bledsoe. 1979. Food web relationships of Northern Puget Sound and the Strait of Juan de Fuca. A synthesis of available knowledge. US Environmental Protection Agency, EPA-600/7-79-259.

Spaulding, J.S., T.W. Hillman, and J.S. Griffith. 1989. Habitat use, growth, and movement of chinook salmon and steelhead in response to introduced coho salmon, p. 157-208. *In* Summer and Winter Ecology of Juvenile Chinook Salmon and Steelhead Trout in the Wenatchee River, Washington. Don Chapman Consultants, Inc., Boise, Idaho.

Stevens, D.E. 1966. Food habits of striped bass, *Roccus saxatilis*, in the Sacramento-San Joaquin Delta, p. 68-96. *In* J.L. Turner and D. W. Kelly (eds.), Ecological Studies of the Sacramento-San Joaquin Delta. California Fish and Game Bulletin 136. Sacramento.

Stevens, D.E., H.K. Chadwick, and R.E. Painter. 1987. American shad and striped bass in California's Sacramento-San Joaquin River system, p. 66-78. *In* M.J. Dadswell, R.J. Klanda, C.M. Moffitt, R.L. Saunders, R.A. Rulifson, and J.E. Cooper (eds.) Common Strategies of Anadromous and Catadromous Fishes. American Fisheries Society Symposium 1. Bethesda, Maryland.

Steward, C.R. and T.C. Bjornn. 1990. Supplementation of salmon and steelhead stocks with hatchery fish: a synthesis of published literature. Part 2. Analysis of salmon and steelhead supplementation. US Department of Energy, Bonneville Power Administration, Division of Fish and Wildlife, Tech Rep. 90-1. Portland, Oregon.

Swain, D.P., B.E. Riddell, and C.B. Murray. 1991. Morphological differences between hatchery and wild populations of coho salmon (*Oncorhynchus kisutch*): environmental versus genetic origin. Canadian Journal of Fisheries and Aquatic Sciences 48: 1783-1791.

Tabor, R.A., R.S. Shively, and T.P. Poe. 1993. Predation on juvenile salmonids by smallmouth bass and northern squawfish in the Columbia River near Richland, Washington. North American Journal of Fisheries Management 13: 831-838.

Taylor, E.B. 1991. A review of local adaptation in Salmonidae, with particular reference to Pacific salmon and Atlantic salmon. Aquaculture 98: 185-207.

Taylor, E.B., and J.D. McPhail. 1985. Burst swimming and size-related predation of newly emerged coho salmon (*Oncorhynchus kisutch*). Transactions of the American Fisheries Society 114: 546-551.

Taylor, J.N., W.R. Courtney, Jr., and J.A. McCann. 1984. Known impacts of exotic fishes in the continental United States, p. 322-373. *In* W.R. Courtenay, Jr., and J.R. Stauffer, Jr. (eds.), Distribution, Biology, and Management of Exotic Fishes. The Johns Hopkins University Press, Baltimore.

Thomas, G.L. and O.A. Mathisen. 1993. Biological interactions of natural and enhanced stocks of salmon in Alaska. Fisheries Research 18: 1-18.

Tripp, D. and P. McCart. 1983. Effects of different coho stocking strategies on coho and cutthroat trout production in isolated headwater streams. Canadian Department of Fisheries and Aquatic Sciences Technical Report 1212.

Uremovich, B.L., S.P. Cramer, C.F. Willis, and C.O. Junge. 1980. Passage of juvenile salmonids through the ice-trash sluiceway and squawfish predation at Bonneville Dam. Oregon Department of Fish and Wildlife, Fisheries Division, Progress Report. Clackamas.

Varoujean, D.H. and D.R. Mathews. 1983. Distribution, abundance, and feeding habits of seabirds off the Columbia River, May-June, 1982. University of Oregon, Institute of Marine Biology, Report No. OIMB 83-1.

Walters, C.J. 1986. Adaptive Management of Renewable Resources. Macmillan, New York.

Walters, C.J., R. Hilborn, R.M. Peterman, and M.J. Staley. 1978. Model for examining early ocean limitation of Pacific salmon production. Journal of the Fisheries Research Board Canada 35: 1301-1315.

Wendler, H.O. 1967. The American shad of the Columbia River with a recommendation for management of the fishery. Washington Department of Fisheries, Olympia.

Wood, C.C. 1987a. Predation of juvenile Pacific salmon by the common merganser (*Mergus merganser*) on eastern Vancouver Island. I: Predation during the seaward migration. Canadian Journal of Fisheries and Aquatic Sciences 44: 941-949.

Wood, C.C. 1987b. Predation of juvenile Pacific salmon by the common merganser (*Mergus merganser*) on eastern Vancouver Island. II: Predation of stream-resident juvenile salmon by merganser broods. Canadian Journal of Fisheries and Aquatic Sciences 44: 950-959.

Wood, C.C., N.B. Hargreaves, D.T. Rutherford, and B.T. Emmett. 1993. Downstream and early marine migratory behaviour of sockeye salmon (*Oncorhynchus nerka*) smolts entering Barkley Sound, Vancouver Island. Canadian Journal of Fisheries and Aquatic Sciences 50: 1329-1337.

The Wilderness Society. 1993. Pacific salmon and federal lands. The Wilderness Society. Washington, DC.

Degradation and Loss of Anadromous Salmonid Habitat in the Pacific Northwest

Stanley V. Gregory and Peter A. Bisson

Abstract

Many stocks of Pacific salmon (*Oncorhynchus* spp.) have become extinct over the last century, and other stocks currently are declining and risk extinction. Habitat degradation has been associated with >90% of the documented extinction or declines of salmon stocks. Surveys of public and private lands in Washington, Oregon, Idaho, and northern California indicate that freshwater habitats for salmonids are in poor to fair condition. Fundamental aspects of habitat in the Pacific Northwest include known trends in habitat change, institutional histories and land uses, and principles for restoration of streams and rivers.

Declines of Pacific salmon stocks closely followed Euro-American settlement of the Pacific Northwest in the 19th century. Causes for declining stocks are complex, but habitat degradation has been explicitly identified as a factor in the declines of most stocks. The historical land use on both public and private lands in the Pacific Northwest has left us with a legacy of altered habitats that will require considerable time for recovery, and return to historical conditions will never occur on a large proportion of the landscape. Loss of floodplain habitats in both montane and lowland riparian forests has been one of the most pervasive and unregulated forms of habitat use.

Inconsistent development of environmental management issues or guidelines for land-use practices presents a major obstacle to managing freshwater habitats in the Pacific Northwest. Currently, Pacific Northwest states have a fragmented and uncoordinated collection of statutes relating to different land-use types, zoning, and different resource users. Effective habitat management at a landscape scale requires incorporation of the entire landscape relevant to salmon life histories and integration of management policies for both public and private lands.

The goal of restoration is to reestablish an ecosystem's ability to maintain its function and organization without continued human intervention. It does not mandate returning to some arbitrary prior state. Any restoration program should be nested within a larger program of landscape management that protects, maintains, and restores ecosystem structure and function. The most critical questions related to aquatic habitat restoration include the following: the degree to which the habitat can be repaired or restored, priorities for locations where restoration efforts will be beneficial, and ecologically sound approaches for habitat restoration. The major agent of aquatic ecosystem restoration in the Pacific Northwest is periodic flooding, and the challenge for human efforts is to supplement natural processes of restoration. In the future, our success in ecosystem management will be measured by the degree to which we are able to decrease the need for restoration programs.

Introduction

Though the causes have been strenuously debated, there is no question that anadromous salmonids in the Pacific Northwest have declined or that considerable aquatic and riparian habitat has been altered or lost since the mid-1850s (National Research Council [NRC] 1996). Recent evaluations of the status of Pacific salmon (*Oncorhynchus* spp.) have concluded that many stocks have become extinct over the last century and that many other stocks currently are declining and risk extinction (Walters and Cahoon 1985, Chapman 1986, Konkel and McIntyre 1987, Nehlsen et al. 1991, Frissell 1993, Washington Department of Fisheries et al. 1993). Habitat degradation has been associated with >90% of the documented extinctions or declines of these stocks (Nehlsen et al. 1991). No survey of public or private lands in Washington, Oregon, Idaho, or northern California has concluded that freshwater habitats for salmonids are in good to excellent condition (General Accounting Office 1988, USDI Bureau of Land Management [BLM] 1991, Nickelson et al. 1992, Thomas et al. 1993, Forest Ecosystem Management Assessment Team [FEMAT] 1993). Forest practices, agriculture, livestock grazing, road building, urbanization, and dams have diminished the ability of freshwater habitats to support anadromous stocks of salmon and trout. Factors not related to habitat (e.g., excessive commercial and sport salmon harvest, hatchery practices, disease or predation, and ocean conditions) also have contributed to the decline of Pacific salmon and are addressed in other papers (Fresh 1996, Mundy 1996, Pearcy 1996, Reisenbichler 1996).

Causes and rates of habitat change have varied across the Pacific Northwest over the last 150 years, involving many types of human activity on forested land, croplands, rangelands, residential areas, and industrial developments. Generalizations are difficult and may be misleading because the degree of habitat change ranges from short-term, localized modification to large-scale, long-term loss of habitat. In some cases, habitats have been destroyed through diking and filling, land draining, channelization, or stream rerouting. Other forms of habitat alteration significantly reduce major aspects of salmonid habitat, such as pools, wood accumulations, side channels and other lateral habitats, or floodplains. Studies often focus on immediate changes in habitat structure and composition, but alteration of ecosystems processes (e.g., hydrologic regimes, delivery of sediment, thermal loading) and ecosystem structure (e.g., riparian forests, beaver populations, wetlands) may influence habitat conditions over much larger areas and time periods (Naiman 1992).

Attempts to improve survival in other aspects of salmon life history such as upstream adult migrations or harvest reductions may have little effect if freshwater habitat for salmon is inadequate. Recovery of salmon in response to natural improvements in climate or environmental conditions in the north Pacific Ocean will be limited if freshwater habitats pose limits to the early phase of their life histories (Francis and Sibley 1991, Pearcy 1992). Freshwater streams and riparian areas are critical to the life history of the fish (e.g., clean gravels for spawning; an open gravel environment from which newly hatched fish can emerge; low velocity, shallow habitat along stream margins for rearing of young fish; overwintering habitat; refuges to survive natural flooding; deep pools; and cold water).

Environmental factors and human activities contributing to the decline of salmon differ from basin to basin. Attempts to single out practices responsible for the decline of anadromous salmonids or to rank their impacts are likely to be misleading and ultimately to weaken collective or integrated approaches for managing the common resources and landscapes of the Pacific

Northwest (Botkin et al. 1995). It is impossible to ensure the survival of the stocks of anadromous salmonids of the Pacific Coast without providing high-quality freshwater habitat for spawning, rearing, and passage of juvenile fish, and migration of returning adults. In this paper, we identify known trends in habitat change in the Pacific Northwest, describe institutional histories and land uses that influence habitat conditions and future alternatives, and explore principles and applications for restoring streams and rivers.

Historical Patterns of Habitat Alteration and Loss

Humans have occupied the Pacific Northwest for ~18,000 years, strongly influencing the distribution and abundance of native plants and animals. For example, burning of prairies and forests by Native Americans modified vegetation in local areas, such as the prairies of the Willamette Valley, Oregon, and the Puget Trough, Washington, maintaining more of an oak savannah or grassland than a forested environment (Boyd 1990). Many Native Americans subsisted on anadromous salmon as part of their food base for much of that time. Fish traps, spearing, and net fisheries at falls and other constrictions to migratory pathways had the potential to strongly alter salmon and trout populations (Stewart 1977), but Native Americans changed relatively little of the overall habitat of salmon.

Declines of Pacific salmon stocks closely followed Euro-American settlement of the Pacific Northwest in the 19th century. In 1832, Captain Nathaniel Wyeth established a commercial fishing and salt packing operation on Sauvies Island at the confluence of the Willamette River and Columbia River, but competition with the Hudson Bay Company proved too great, and he sailed back to the east coast of the United States (US) in 1835 with only half a cargo load of salmon (Cobb 1922). The first successful commercial harvest in the Pacific Northwest did not occur until 1861, when H.N. Nice and Jotham Reed began a commercial salting operation in the Columbia River 96 km below Portland, Oregon. William Hume built the first cannery at Eagle Cliff, Washington, in 1866, 2 years after he helped establish the first salmon cannery on the Sacramento River, California (Adams 1885). In that year, all returning adult salmon to the Sacramento River were eliminated by massive sedimentation caused by gold mining (Stone 1897). Forty canneries were operating in the Columbia River by 1885. Abundances of all species of salmon in the Columbia River declined by the mid-1880s, prompting Marshall McDonald (1895), the US Commissioner of Fish and Fisheries, to note:

> It is not a matter of wonder that, under the existing conditions, there has been serious deterioration in the value of the fisheries. It is, indeed, a matter of surprise that any salmon have been able to elude the labyrinth of nets which bar their course to the Upper Columbia.

Declines of salmon were attributed first to overharvest of fish in the terminal fisheries and later to habitat degradation. In 1876, hatchery development in Oregon began as an attempt to overcome the perceived loss of the natural system's ability to support natural production. McDonald (1875) further noted:

> In 1888 the U.S. Fish Commission, by direction of Congress, established a salmon hatching station on the Clackamas River, Oregon. . . . This work was undertaken on the urgent solicitation of those concerned in the salmon fisheries of the Columbia River, who realized that their fisheries were being exhausted, and it was hoped that some compensation for the deficiency in natural reproduction could be made by artificial stocking and breeding.

It quickly became evident that artificial propagation was not a simple panacea for overly aggressive commercial fishing and that land-use practices were degrading aquatic habitat conditions. Livingstone Stone (1897), founder of the first hatcheries on the Pacific coast, observed:

> When it [the Clackamas hatchery] first passed into the hands of the US Fish Commission it yielded 5,000,000 salmon eggs a year, but it was too near civilization to prosper long as a salmon-breeding station, and gradually mills and dams, timber cutting on the upper waters of the Clackamas, and logging in the river, together with other adverse influences, so crippled its efficiency that it was given up this year as a collecting point for salmon eggs. . . .

Even the value and success of the McCloud River hatchery in California, the first hatchery on the US west coast, was related to the extensive destruction of habitat in the basin by mining. Stone (1897) also noted:

> McCloud River . . . is the only cold tributary of the Sacramento that has not been roiled by gold mining, in consequence of which the salmon come into the McCloud to breed in the summer, not only from choice, but also from necessity.

Mining after the Gold Rush of 1862 destroyed critical habitat for salmonids throughout the Northwest, and water quality was impaired through degradation of ambient chemistry (e.g., oxygen, suspended sediment) and introduction or re-exposure of toxic substances (e.g., mercury, cyanide). Many laws governing mining operations have changed little over the last 100 years, as evidenced by the federal Mining Act of 1872 (Nelson et al. 1991).

Routes of early settlers into the Pacific Northwest followed the major river drainages (Fig. 1), moving westward along the Snake River system, down the Columbia River or across the Cascades into the western valleys of Oregon and Washington, and south into northern California (Nolan 1993). These same routes became the major arteries for today's transportation infrastructure and for the major centers of present human populations (Fig. 2, 3; Northam 1993). These corridors undoubtedly will dictate the pattern of future human activity and concomitant ecological changes in the region. Aquatic ecosystems along these transportation routes and urban centers will be altered more intensively than more remote habitats in rural lands or mountainous areas.

Development of towns and cities along the major rivers brought almost unbridled degradation of water quality, and by the early 1920s, John Cobb (1922) noted:

> Next to the fishing operations of man, the gravest danger to the salmon fisheries of the Pacific coast lies in the pollution of the rivers which the salmon ascend for spawning purposes. . . . The large increase in the population of the coast States within recent years, with the resulting increase of mills and factories, has greatly increased the amount of sewage from cities and towns and the waste of the manufacturing plants.

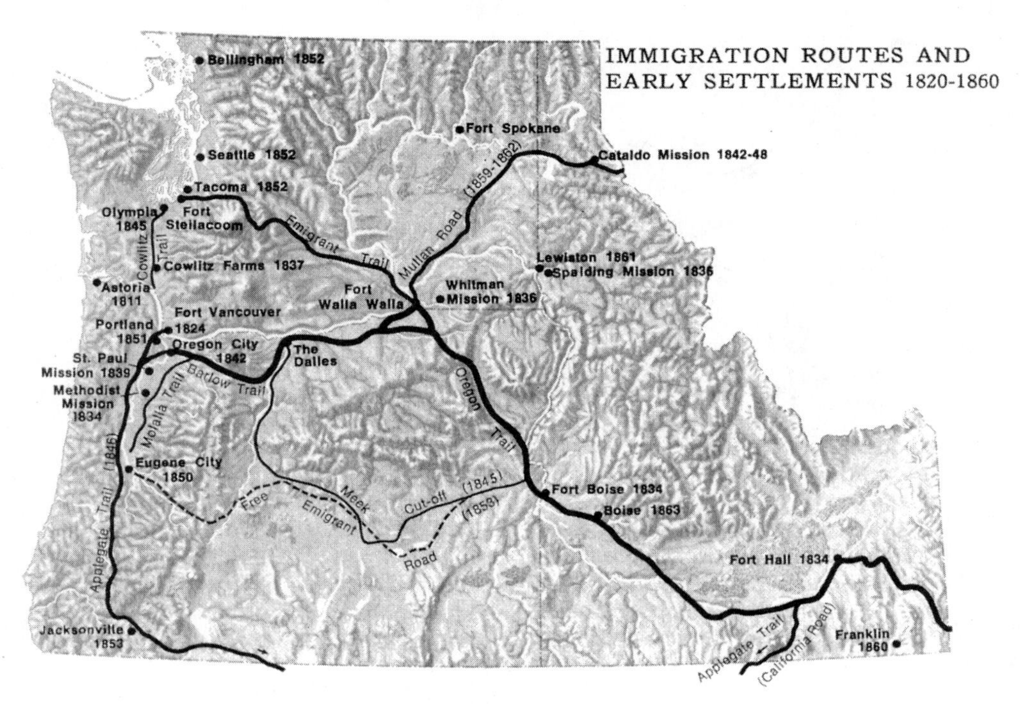

Figure 1. Routes of Euro-American settlers moving into the Pacific Northwest. Source: Nolan (1993); reprinted by permission of Oregon State University Press.

The first large dams of the Northwest were built in the 1930s, but settlers had been damming streams for power generation and mill operations since the mid-1800s. Dams were constructed with little or no regulation on their construction or locations. Their impacts on the fisheries were recognized before the turn of the century (Smith 1895). Fish passage was rarely accommodated by early dams, and even Bonneville Dam, constructed in 1933, originally was designed with no provision for passage of salmon into the Columbia River basin above Portland. Growth of factories, agriculture, and mining required enormous consumption of water, and laws on water withdrawal reflected the pioneering nature of the expansion of western society, first come, first served. Modern water allocations in the western US are still based on this principle, and water rights from the late 1800s remain in effect. Most water withdrawals were not considered in their consequences either on habitat in the streams or on fish that were diverted with the water, as noted by Cobb (1922):

> The irrigation ditch, a comparatively new product on this coast, while of great benefit in developing the arid lands in certain sections, as at present operated is a considerable menace to the salmon fisheries. But few ditches have screens at their head, and as a result many thousands of young salmon slowly making their way to the ocean home pass into and down these to an early doom.

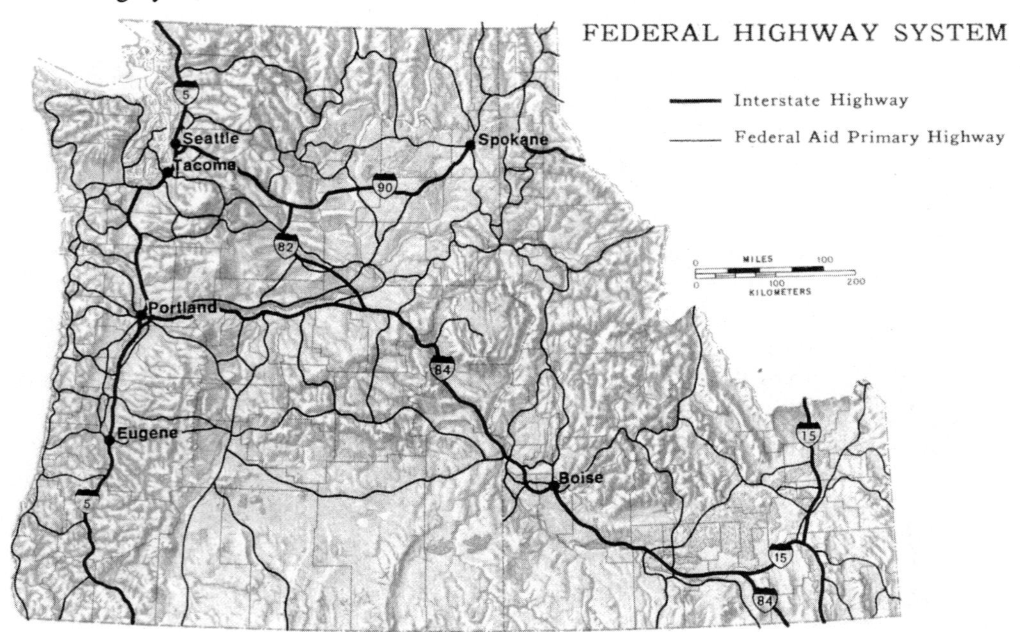

Figure 2. Modern highways and transportation routes in the Pacific Northwest. Source: Nolan (1993); reprinted by permission of Oregon State University Press.

Pollution of major rivers during the early 20th century made extensive reaches devoid of fish and most invertebrate life because of elevated temperatures, lack of oxygen, and toxic substances (Gleeson 1972). A survey of fish and water quality in the Willamette River in 1944 indicated few live fish in the 69-km reach of river from Newberg to Portland (Dimick and Merryfield 1945). The survey also noted that cutthroat trout fry died within 2 minutes after being placed in the South Fork of the Santiam River. Stream temperatures in major tributaries reached 31.7°C (89°F), and oxygen concentrations were observed as low as 0.0 mg L^{-1} and ranged from 0.6 to 1.4 mg L^{-1} in the lower river below Newberg, Oregon. Water quality in mainstem lowland rivers diminished or eliminated the access to headwater streams for adult salmon and created an almost impassable gauntlet for smolts migrating downstream.

After little more than a century of Euro-American settlement, the states and regulatory agencies of the Pacific Northwest were forced to acknowledge the destruction of freshwater habitat and water quality throughout the region. The last half of the 20th century marked a period of land-use laws, pollution regulation, mandated sewage treatment, and establishment of water rights for fish and aquatic ecosystems. Each step toward restoring the aquatic ecosystems of the Pacific Northwest has been seen as a potential infringement on the rights of individuals and continues to be fiercely debated. The momentum that drives the development of environmental regulation, land zoning, and alliances of private citizens is the recognition that our waters and anadromous salmonids are a common resource and regional heritage.

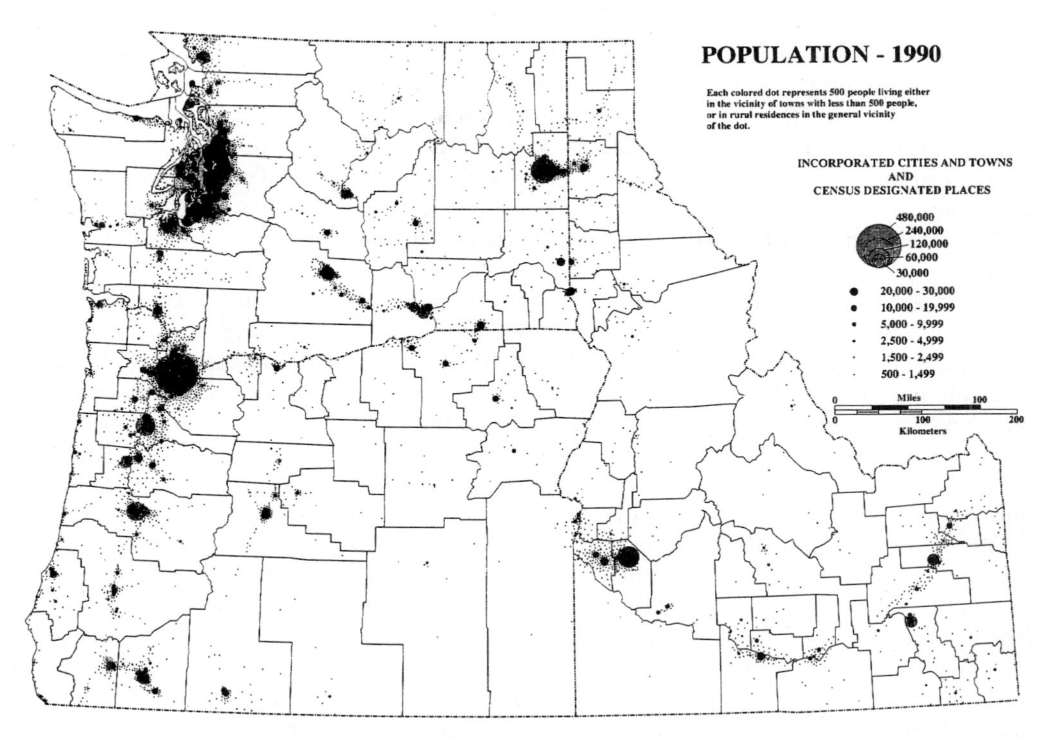

Figure 3. Major urban areas and population distribution in the Pacific Northwest. Source: Matzke (1993); reprinted by permission of Oregon State University Press.

Status of Salmon and Patterns of Their Decline

In the late 1980s and early 1990s, concerns over dwindling numbers of returning adult salmon stimulated regional evaluations of local extinctions and the status of existing stocks (Konkel and McIntyre 1987, Nehlsen et al. 1991, The Wilderness Society 1993; P. Higgins, S. Dobush, and D. Fuller, Humboldt Chapter, American Fisheries Society unpubl. data). These studies echoed 100 years of reports on habitat loss and salmon declines by regional aquatic biologists, and also attempted to quantify historical extinctions and populations at risk of extinction over broad areas, something many of the earlier reports had not done. According to Nehlsen et al. (1991), >106 stocks of anadromous salmonids have become extinct in Washington, Oregon, Idaho, and California, and 214 stocks considered to be at risk of extinction were further identified. Generally, anadromous salmonids were found to be at greater risk near the southern portions of their ranges than more northerly populations, and interior populations (e.g., upper Sacramento River, middle and upper Columbia River) were at greater risk than populations in systems draining the Coast Range (NRC 1996).

Causes for extinctions or declining stocks are complex and differ from basin to basin, but habitat degradation (including loss caused by dams) was explicitly identified as a factor in the declines of 194 of the 214 stocks and was believed to be the principal factor in the declines of 51 at-risk stocks (Nehlsen et al. 1991). No evidence was found for fishing salmon stocks to extinction even when weirs were used as a terminal fishery. Analyses of the status of individual species, such as steelhead (*O. mykiss*) (Cooper and Johnson 1992), coho (*O. kisutch*) and chinook (*O. tshawytscha*) salmon (Lichatowich 1989), and cutthroat trout (*O. clarki*) (Trotter et al. 1993) have all identified habitat loss as a widespread, significant contributor to stock declines. Protection and restoration of existing salmon stocks will require integrated efforts to address the many sources of mortality, but the quantity and quality of habitats clearly will be central issues in the future of Pacific salmon (Pacific Rivers Council 1993, The Wilderness Society 1993, NRC 1996).

The history of extinctions and threats to the survival of existing salmon stocks is not uniform across the Pacific Northwest. Regional patterns reveal the nature of past habitat change and indicate areas where greater numbers of stocks depend on future protection and restoration. The Columbia River basin, with its numerous dams and water withdrawals, accounts for 63% of the stocks that are known to have become extinct in Washington, Oregon, Idaho, and California, but 20% of the extinct stocks originally occurred in California (Nehlsen et al. 1991). This reflects both the intensity of habitat alteration in these areas and the harsher environmental conditions at the edge of these species' distributions. Stocks on the edges of the geographic range of the Pacific salmon must tolerate habitat conditions that can be marginal for the species, and even moderate levels of habitat alteration may be adequate to eliminate entire stocks.

According to Nehlsen et al. (1991), relatively few recent extinctions have been documented for stocks inhabiting the Oregon coast (10 stocks) and the Washington coast and Puget Sound (8 stocks), but each of these geographic areas contains as many stocks at risk of extinction as currently occur in the Columbia River basin (58 in Oregon coast, 60 in Washington coast and Puget Sound, 57 in Columbia River basin). The lower extinction rate of salmon in coastal areas illustrates the potential to save a greater proportion of the original species populations in coastal Oregon and Washington, but the large number of at-risk stocks reflects our recent history of land management and fisheries regulation in those parts of the Pacific Northwest with the greatest potential to support anadromous salmonids. The extent of stock declines in coastal Oregon and Washington emphasizes the need to reverse the current trend in habitat alteration in the Pacific Northwest (Moyle and Williams 1990, Frissell 1993).

Alteration of Salmonid Habitats by Land-Use Practices

Modification of aquatic habitats generally affects one or more of six fundamental components of stream ecosystems: channel structure, hydrology, sediment input, environmental factors, riparian forests, and exogenous material (Table 1). Actions that change channel structure, hydrology, or sediment delivery essentially alter the physical habitat that potentially can be occupied by anadromous salmonids. Environmental factors change either the physical environment or water chemistry, which either directly affect the physiology of salmonids or indirectly influence their food resources. Riparian forests influence numerous processes such as flood routing, sediment trapping, nutrient uptake, allochthonous inputs, wood, shade, stream temperature, and root strength (Naiman et al. 1988, Gregory et al. 1991), but a critical aspect of altering riparian forests

Table 1. Types of habitat alteration and effects on salmonid fishes in the Pacific Northwest. Based in part on Hicks et al. (1991b), Swanston (1991), and National Research Council (1996).

Ecosystem feature	Altered component	Effects on salmonid fishes and their ecosystems	Selected references
Channel structure	Floodplains	Loss of overwintering habitat, loss of refuge from high flows, loss of inputs of organic matter and large wood	Peterson and Reid (1984), Sedell and Froggatt (1984), Brown and Hartman (1988), Booth (1991)
	Pools and riffles	Shift in the balance of species, loss of deep water cover and adult holding areas, reduced rearing sites for yearling and older juveniles	Hawkins et al. (1983), Bisson and Sedell (1984), Sullivan et al. (1987), Bisson et al. (1988), Moore and Gregory (1988), Sedell and Everest (1991), Ralph et al. (1994)
	Large wood	Loss of cover from predators and high flows, reduced sediment and organic matter storage, reduced pool-forming structures, reduced organic substrate for macroinvertebrates, formation of new migration barriers, reduced capacity to trap salmon carcasses	Narver (1971), Swanson and Lienkaemper (1978), Bryant (1983), Cederholm and Peterson (1985), Harmon et al. (1986), Bisson et al. (1987), Andrus et al. (1988), Murphy and Koski (1989), Aumen et al. (1990), Gregory et al. (1991), Naiman et al. (1992)
	Substrate	Reduced survival of eggs and alevins, loss of interstitial spaces used for refuge by fry, reduced macroinvertebrate production, reduced biodiversity	Burns (1972), Murphy et al. (1981), Hawkins et al. (1982), Everest et al. (1987), Swanson et al. (1990), Montgomery and Buffington (1993)
	Hyporheic zone	Reduced exchange of nutrients between surface and subsurface waters and between aquatic and terrestrial ecosystems, reduced potential for recolonizing disturbed substrates	Stanford and Ward (1988, 1992), Triska et al. (1989, 1990)
Hydrology	Discharge	Altered timing of discharge-related life cycle cues (e.g., migrations), changes in availability of food organisms related to timing of emergence and recovery after disturbance, altered transport of sediment and fine particulate organic matter, reduced biodiversity	Swanson et al. (1982), Bilby and Bisson (1987), Chamberlin et al. (1991), Naiman et al. (1992)
	Peak flows	Scour-related mortality of eggs and alevins, reduced primary and secondary productivity, long-term depletion of large wood and organic matter, involuntary downstream movement of juveniles during freshets, accelerated erosion of streambanks	Cristner and Harr (1982), Berris and Harr (1987), Culp (1988), Anderson (1992), Borchardt (1993), Luchetti and Fuerstenburg (1993)
	Low flows	Crowding and increased competition for foraging sites, reduced primary and secondary productivity, increased vulnerability to predation, increased fine sediment deposition	Smoker (1955), Mason and Chapman (1965), Chapman (1966), Wissmar and Swanson (1990), Hicks et al. (1991a)

Table 1—cont.

Ecosystem feature	Altered component	Effects on salmonid fishes and their ecosystems	Selected references
Hydrology—cont.	Rapid fluctuations	Altered timing of discharge-related life cycle cues (e.g., migrations), stranding, intermittent connections between mainstem and floodplain rearing habitats, reduced primary and secondary productivity	Poff and Ward (1989), Reice et al. (1990)
Sediment	Surface erosion	Reduced survival of eggs and alevins, reduced primary and secondary productivity, interference with feeding, behavioral avoidance and breakdown of social organization, pool filling	Cordone and Kelly (1961), Burns (1972), Iwamoto et al. (1978), Noggle (1978), Bisson and Bilby (1982), Berg and Northcote (1985), Everest et al. (1987), Chapman (1988), Platts et al. (1989)
	Mass failures and landslides	Reduced survival of eggs and alevins, reduced primary and secondary productivity, behavioral avoidance, formation of upstream migration barriers, pool filling, addition of new large structure to channels	Beschta (1978), Cederholm et al. (1981), Everest et al. (1987), Swanson et al. (1987), Chapman (1988), Benda (1990), Chamberlin et al. (1991), Megahan et al. (1992)
Water quality	Temperature	Altered adult migration patterns, accelerated development of eggs and alevins, earlier fry emergence, increased metabolism, behavioral avoidance at high temperatures, increased primary and secondary production, increased susceptibility of both juveniles and adults to certain parasites and diseases, altered competitive interactions between species, mortality at sustained temperatures >23-29°C, reduced biodiversity	Brett (1952), Averett (1969), Hall and Lantz (1969), Brown and Krygier (1970), Bisson and Davis (1976), Wurtsbaugh and Davis (1977), Newbold et al. (1980), Tschaplinski and Hartman (1983), Hughes and Davis (1986), Beschta et al. (1987), Reeves et al. (1987), Holtby (1988), Platts and McHenry (1988), Bjornn and Reiser (1991), Karr (1991)
	Dissolved oxygen	Reduced survival of eggs and alevins, smaller size at emergence, increased physiological stress, reduced growth	Davis (1975), Ringler and Hall (1975), Brett and Blackburn (1981), Scrivener (1988), Bjornn and Reiser (1991)
	Nutrients	Increased primary and secondary production, possible anoxia during extreme algal blooms, increased eutrophication rate of standing waters, certain nutrients (e.g., non-ionized ammonia, some metals) possibly toxic to eggs and juveniles at high concentrations	Dimick and Merryfield (1945), Warren et al. (1964), Triska et al. (1984), Gregory et al. (1987), Bothwell (1989), Nelson et al. (1991), Bisson et al. (1992)

Table 1—cont.

Ecosystem feature	Altered component	Effects on salmonid fishes and their ecosystems	Selected references
Riparian forest	Production of large wood	Loss of cover from predators and high flows, reduced sediment and organic matter storage, reduced pool-forming structures, reduced organic substrate for macroinvertebrates	Narver (1971), Swanson and Lienkaemper (1978), Bryant (1983), Grette (1985), Harmon et al. (1986), Bisson et al. (1987), Andrus et al. (1988), Murphy and Koski (1989), Aumen et al. (1990), Shirvell (1990), Van Sickle and Gregory (1990), Bilby and Ward (1991), Gregory et al. (1991)
	Production of food organisms and organic matter	Reduced heterotrophic production and abundance of certain macroinvertebrates, reduced surface-drifting food items, reduced growth in some seasons	Mispagel and Rose (1978), Naiman and Sedell (1979), Vannote et al. (1980), Minshall et al. (1985), Gregory et al. (1987), Wissmar and Swanson (1990), Gregory et al. (1991), Bilby and Bisson (1992), Naiman et al. (1992)
	Shading	Increased water temperature, increased primary and secondary production, reduced overhead cover, altered foraging efficiency	Murphy et al. (1981), Shortreed and Stockner (1983), Wilzbach (1985), Gregory et al. (1987), Culp (1988), Bilby and Bisson (1992)
	Vegetative rooting systems and streambank integrity	Loss of cover along channel margins, decreased channel stability, increased streambank erosion, increased landslides	Burroughs and Thomas (1977), Beschta (1991), Platts (1991), National Research Council (1992), Forest Ecosystem Management Assessment Team (1993)
	Nutrient modification	Altered nutrient inputs from terrestrial ecosystems, altered primary and secondary production	Berg and Doerksen (1975), Minshall et al. (1985), Gregory et al. (1991), Stanford and Ward (1992)
Exogenous material	Chemicals	Reduced survival of eggs and alevins, toxicity to juveniles and adults, increased physiological stress, altered primary and secondary production, reduced biodiversity	Dimick and Merryfield (1945), Seiler (1989), Karr (1991), Nelson et al. (1991), Norris et al. (1991)
	Exotic organisms	Increased mortality through predation, increased interspecific competition, introduction of diseases	Li et al. (1987), Karr (1994), National Research Council (1996)

is the time required for recovery of mature forest conditions (Agee 1988). This successional process creates a context for other types of habitat modification and limits rates of recovery. Exogenous materials, including dissolved chemicals, particulate material, and exotic organisms, represent factors that commonly are not part of the evolutionary history of the aquatic ecosystems. Responses can be severe and may persist as long as the material remains in the ecosystem.

Conversion of lowland forests, coastal tide lands, floodplains, and headwater forests, as well as alteration of water quality, have affected anadromous salmonids and aquatic ecosystems throughout the Pacific Northwest (Frenkel and Morlan 1991, Lucchetti and Fuerstenberg 1993). Attention to forestry-related issues in recent years has focused the public's attention on land-use policies in the upland forests, which are predominantly public lands. Historical loss of estuaries and lowland freshwater habitats (Boulé and Bierly 1987) has been considerable. For example, Simenstad et al. (1982) identified five major estuaries in Puget Sound in which >70% of the available habitat has been lost. Much of the habitat of lower main rivers is no longer in forest lands but instead in areas zoned for agriculture, urban, and industrial development. Many of these lands have been converted from coniferous forests to grasslands, meadows, deciduous forests, or paved surfaces.

As a consequence of settlement, many historical lowland or floodplain forests have been eliminated, and recent society has little memory of the conditions of those riparian forests and the roles that they played (Sedell et al. 1990). Riparian forests in lower valley floodplains, particularly secondary channels and off-channel ponds, were particularly critical for survival of rearing salmon during winter floods and provided cold-water refuges during warmer periods of the year (Ward et al. 1982, Peterson and Reid 1984, Brown and Hartman 1988).

Floodplains also provide coarse beds of alluvial sediments through which subsurface river flow passes much like a trickle filter in wastewater treatment plants (Stanford and Ward 1992). This hyporheic zone, the subsurface flow between surface water and the water table, serves as a filter for nutrients and maintains high water quality (Triska et al. 1989). Human activities have altered lowland rivers incrementally in small patches by numerous practices such that existing channels and floodplains are minor relicts of original conditions. As a result of these "diffuse" alterations over space and time, the degree and consequences of habitat alteration are rarely recognized.

Few quantitative studies of salmonid habitats were conducted prior to World War II, and historical reconstructions of riverine conditions are scarce; thus, accurate assessment of habitat loss is difficult. Comparison of current conditions of the upper Willamette River with maps constructed by the cadastral land survey of the 1850s reveals extensive simplification (Sedell and Froggatt 1984). Sections of the river that originally were braided and contained side channels and floodplain lakes are now single channels with little or no lateral connections. Lowland streams and rivers have been simplified and channelized so extensively that it is rare to find reaches that resemble natural channels and floodplain forests. A survey of 43,000 km of streams in Oregon indicated that 55% were either moderately or severely impacted by nonpoint source pollution (Edwards et al. 1992).

Despite the scarcity of quantitative historical studies, it is clear that habitat availability and quality have significantly declined and that current environmental protection and resource management policies have not been able to reverse that trend (Hicks et al. 1991b, Bisson et al. 1992). Land-use practices differ in their impacts and the portions of the landscape and river drainages that are altered. Forested lands make up 46% of the land cover of Washington, Oregon, and

Idaho, and the federal government manages or supervises ~60% of the forest lands (Table 2; Jackson and Kimerling 1993). Rangelands account for 32% of the land base, and croplands and pasture make up another 20%. Only 2% of the Pacific Northwest is represented by urban or developed lands. These general trends in land use are consistent throughout the states of the region, but proportions of federal and non-federal lands within a land-use type differ substantially between states (Table 2).

Habitat Loss Associated with Forest Management

Forest practices (e.g., timber harvest, yarding, road building) alter many components and processes of aquatic ecosystems and the land-water interface. These interactions have been evaluated and synthesized in several major symposia, reports, and books (Krygier and Hall 1971, Karr and Schlosser 1977, Iwamoto et al. 1978, Salo and Cundy 1987, Raedeke 1988, Hartman and Scrivener 1990, Meehan 1991, Naiman 1992, Peterson et al. 1992). These works provide detailed reviews of the effects of forest practices on aquatic ecosystems, and the following section simply highlights some of the major changes related to habitat alteration on forest lands (Table 1).

Habitat Change

When commercial logging began in the mid-19th century, there were no roads for moving logs to the sawmills, and rivers served as the early routes for transportation (Sedell and Luchessa 1982). Splash dams were constructed to generate sufficient flows for moving the logs down

Table 2. Areas of different land-use types in three Pacific Northwest states. Modified from Jackson and Kimerling (1993).

		Oregon		Washington		Idaho	
Land use	Ownership	km^2	%	km^2	%	km^2	%
Forests	Federal	75,669	30	38,340	22	67,745	30
	Nonfederal	47,984	19	51,128	29	16,475	7
Rangeland	Federal	53,160	21	6,750	4	63,463	28
	Nonfederal	37,037	15	22,557	13	26,693	12
Cropland	Nonfederal	24,682	10	37,968	22	40,590	18
Pasture	Nonfederal	7,754	3	5,747	3	5,480	2
Urban land	Nonfederal	3,808	2	6,329	4	1,930	1
National parks	Federal	683	<1	7,329	4	393	<1
Total landbase	Federal	129,512	51	52,419	30	131,600	59
	Nonfederal	121,265	49	123,729	70	91,168	41
	Combined	250,777	100	176,148	100	222,768	100

stream channels. During relatively low flow conditions, a slurry of water and logs was suddenly released, destroying riparian zones and aquatic communities as it moved downstream. Structurally complex habitats within these streams were channelized and cleared to facilitate transportation. The techniques of splash damming and log driving down rivers had been used in timber harvest across the North American continent as settlers moved west and had also been used in Europe for centuries. At the same time that log drives were first appearing in the Pacific Northwest, detrimental effects of log drives were being documented in Sweden (Malmgren 1885). Splash damming and log drives from the 1870s through the 1920s altered streams and rivers to such an extent that they have not yet fully healed (Sedell et al. 1991).

The history of logging on both public and private lands in the Pacific Northwest left a legacy of altered habitats that will require considerable time for recovery (Cordone and Kelly 1961), and the return to historical conditions will probably never occur on a large proportion of the forested landscape. Stream surveys by federal agencies have shown that habitat is in fair to poor condition (BLM 1991, FEMAT 1993, Hessburg 1993, Thomas et al. 1993). The BLM estimated that 64% of the riparian areas on their lands in Oregon and Washington and 45% of their riparian areas in Idaho did not meet the objectives of their management policies (BLM 1991). FEMAT (1993) concluded that "aquatic ecosystems in the range of the northern spotted owl exhibit signs of degradation and ecological stress. . . . Although several factors are responsible for declines of anadromous fish populations, habitat loss and modification are major determinants of their current status." In addition, evaluations sponsored by the forest industry acknowledged the overall decrease in stream habitat quality on forest lands (Kaczynski and Palmisano 1992, Palmisano et al. 1993).

One of the few quantitative studies of habitat change was based on a survey of pools in Pacific Northwest streams, conducted by the US Fish and Wildlife Service from 1934 to 1946 (Rich 1948). The Pacific Northwest Research Station of the USDA Forest Service (Forest Service) and its cooperators resurveyed the same streams ~50 years later to determine changes in channel conditions (Sedell and Everest 1991). All streams are dynamic and channel change is a natural process, but overall trends may reflect large-scale responses to human activities. Frequencies of very large pools (based on criteria of >1.8 m deep and >42 m^2 surface area) in 658 km of stream in 13 basins in Washington and Oregon decreased by an average of 58%, ranging from a loss of 94% in the Coweeman River basin to a gain of 10% in the Wind River basin (McIntosh et al. 1993). The gain in pool habitat in the Wind River was a result of channel restoration efforts that followed the Yacolt burn and subsequent log drives in the 1910s, which had reduced the amount of large wood in the Wind River prior to the original survey. Decreases in large pool habitat on private forest lands in coastal Oregon averaged >80%. Pool habitat in largely unmanaged sub-basins of the Wenatchee River, Washington, and the Willamette River, Oregon, over the same period increased 212% and 400%, respectively (J. Sedell and B. McIntosh, USDA Forest Service, Corvallis, Oregon, pers. comm.). On the basis of habitat surveys from 1934 to 1946, McIntosh et al. (1993) concluded that the frequency of large pools in watersheds with forest management in eastern Oregon and Washington declined by an average of 31%, while pools in unmanaged basins increased by 200%. These changes have occurred since 1934, which followed more than 80 years of extensive habitat alteration in all of the surveyed basins.

Loss of large-pool habitat has been caused by various forest management-related factors, including the removal of large wood and large boulders, an increase in the amount of fine sediment (sand and gravel) deposited in pool bottoms, and in some instances, by channelization

(FEMAT 1993). Large pools do not provide the full range of habitat conditions needed by all adult and juvenile salmonids, but they are important holding areas for adults migrating upstream and serve as rearing sites for the juveniles of certain species (Hicks et al. 1991b). Loss of large pools reduces the availability of holding areas for adult salmon. This may be particularly important in streams where large pools with inflowing groundwater provide cold-water refuges during adult migrations in summer months (Berman and Quinn 1991). Additionally, large pools often provide cover for both juvenile and adult salmonids from terrestrial predators. The observation that >50% of the large pools have been lost from many streams over the last half century reveals the extent of physical habitat alteration, especially given that many of these streams had already been changed by human activities when the original pool surveys were initiated.

A study of streams in old growth forests, forests with moderate harvest (<50% harvested within the last 40 years), and forests with intensive harvest (>50% harvested within the last 40 years) in western Washington documented significant changes in pool habitat and amounts of large wood (Ralph et al. 1994). Pool areas and depths were significantly lower in streams in old-growth forests than in harvested basins, and pools >1 m in depth were almost eliminated in harvested basins. A reduction in the abundance of large pieces of wood was also related to logging.

Studies of the effects of past timber harvest over the last several decades frequently are criticized because forest practices have changed during the interim. Unfortunately, the landscape is filled with lands changed by historical practices that no longer occur, and anadromous salmon return to habitats that reflect very little of the improvement in recent forest practices. In Oregon, the majority of private forest lands were harvested at least once prior to the development of the Forest Practices Act (Oregon Department of Forestry 1988). Current habitat conditions on forest land in the Pacific Northwest have been shaped by roughly 150 years of timber harvest, and recent land-use regulations are designed to allow some degree of recovery in the future.

Channel Structure

One of the most profound changes in habitat related to forest practices is alteration of channel structure. Channel structure may be affected directly by sedimentation, mass failure, changes in rooting and vegetative cover, and direct channel modification by heavy equipment (Cederholm et al. 1981, Chamberlin et al. 1991). Changes in hydrologic regimes and loss of in-channel wood may cause indirect, long-term modification of channel structure. Channel changes frequently are evaluated at the scale of a stream reach, but the most important scale for analysis of land-use practices on channel structure is the drainage basin (Sullivan et al. 1987, Ryan and Grant 1991). Channels may respond differently to physical change depending on geology, climate, sediment loading, vegetation, slope, and basin position (Montgomery and Buffington 1993). Decreased heterogeneity of channel units and loss of pool habitat are common responses to forest practices in the Pacific Northwest (McIntosh et al. 1993), but fisheries managers must be cautious about basing management efforts on simplistic assumptions of channel dynamics (Sullivan et al. 1987, Montgomery and Buffington 1993).

The 1970s marked the first well-documented recognition of the role of wood in stream ecosystems (Swanson and Lienkaemper 1978, Bilby and Likens 1980, Harmon et al. 1986, Bisson et al. 1987). Numerous studies have demonstrated that clearcutting, often in combination with stream clean-up, have dramatically reduced the volumes and types of wood in streams throughout the region (Harmon et al. 1986, Andrus et al. 1988, Bilby and Ward 1991). Removal of mature trees along streams reduces natural loading rates for centuries (McDade et al. 1990, Van Sickle and Gregory 1990). Loss of wood from channels directly influences the distribution and abundance of fish populations and is one of the longest lasting effects of forest harvest on anadromous salmonids (Murphy and Koski 1989, Hicks et al. 1991b).

The wide array of changes in channel structure (e.g., loss of pools, reduction in wood, sedimentation, decreased heterogeneity) influences all freshwater stages of anadromous salmonid life histories (Hicks et al. 1991b). Responses of fish populations and other members of the aquatic community are complex; many different responses have been noted in streams throughout the region (Murphy et al. 1981, Hawkins et al. 1983, Murphy et al. 1986, Hartman and Scrivener 1990, Nickelson et al. 1991, Bisson et al. 1992). Forestry operations may also lead to reduced macroinvertebrate populations, which serve as the food base for anadromous salmonids (Newbold et al. 1980, Hawkins et al. 1982, Culp 1988), although under certain circumstances the organisms that feed on algae may benefit from increased autotrophic production associated with removal of forest canopy (Erman et al. 1977, Bilby and Bisson 1992).

Floodplains are fundamental and often overlooked components of stream channels and alluvial valleys (Gregory et al. 1991). Secondary channels provide important refugia in moderate- to high-gradient streams during floods (Seegrist and Gard 1972, Tschaplinski and Hartman 1983). Seasonally flooded channels and riverine ponds support a major component of the populations of coho salmon and other fish species during winter months (Peterson 1982, Peterson and Reid 1984, Brown and Hartman 1988). Loss of floodplain habitats in both montane and lowland riparian forests has been one of the most pervasive and unregulated forms of habitat loss in the Pacific Northwest (NRC 1996).

Stream Cleaning

Recent policies for maintaining and enhancing large wood have caused considerable frustration for those who recall the period in the 1950–70s when fishery agencies required removal of logging-related woody debris from streams. Contradictions between earlier recommendations and more recent policies actually are not as contradictory as they first appear. Practices that led to degraded habitat conditions in the late 1950s often caused the introduction of large volumes of sediment directly into stream channels. Roads were located immediately adjacent to streams, road fill was side-cast directly into channels, logs were yarded along stream corridors, and trees were felled directly into the channel. Tremendous volumes of sediment and slash were left in streams at the end of logging operations. These practices resulted in unstable conditions during subsequent winter floods, high demand for oxygen by the decomposing debris, stagnant pools, and increased direct solar radiation on the channel, which led to high temperatures, low oxygen, unstable channels, poor quality spawning gravel, and apparent blocks to migration (Bisson et al. 1987).

Fishery biologists noted the mortality of fish and called for preventing erosion of sediment and slash into streams, as well as for removing slash deposits; however, the recommendations acknowledged the need to retain wood in the streambed that existed at the site prior to logging (Hall and Lantz 1969). Even earlier efforts to remove debris accumulations also recognized potential adverse effects of wood removal. In 1949, the Oregon Fish Commission removed 170 log jams and 32 beaver dams in 27 miles of the Clatskanie River, but Merrell (1951) noted:

> Some criticism has been made of the clearing in this particular area, it having been suggested that overclearing ruined the stream by eliminating pools and exposing the gravel to shifting and scouring.

In many efforts to remove excess slash and debris, almost all wood was removed in an attempt to clear the stream for fish passage and reduce the demands on oxygen supply in the water. Stream cleaning practices often became overly zealous in that it was assumed that if some removal was good, total removal was better.

Research in the 1970s in the Fraser Experimental Forest of Colorado (Heede 1972), H.J. Andrews Experimental Forest in Oregon (Swanson and Lienkaemper 1978), and Hubbard Brook Experimental Forest in New Hampshire (Bilby and Likens 1980) identified the physical and ecological functions of large wood in streams. Aquatic scientists were soon calling for maintaining large wood and even restoring large wood to historical levels. This was not so much a contradiction of earlier policies as a recognition that proper management required maintaining natural channel functions and avoiding the types of practices that caused excessive loading of logging debris and sediment. By the 1980s, understanding of the role of large wood in ecosystems rapidly expanded to both terrestrial and aquatic ecosystems and their interfaces (Swanson et al. 1982, Harmon et al. 1986).

Habitat Loss Associated with Agriculture and Livestock Grazing

Agricultural lands (including croplands and pastures) make up ~20% of the land base of the region, and rangelands account for >30% of the land. In combination, these lands used for production of crops or livestock account for ~50% of the northwestern states (Pease 1993). These lands are located in the lower portions of the river basins where stream gradients are low and valleys are formed primarily by alluvial deposition. Agricultural and range lands usually contain more species of fish than steeper headwater streams in forests (Hughes and Gammon 1987) and often some of the more productive aquatic habitat within the basin (Li et al. 1987). These lands also contain the mainstem reaches that are essential for migration of anadromous salmonids.

Land-use practices on agricultural and range lands have greatly reduced the availability and quality of salmonid habitat (Platts 1991), and analysis of habitat conditions and development of legislation or Best Management Practices (BMPs) on private agricultural lands have been notably lacking. Agricultural lands generally occur in lowland valleys that historically contained the majority of floodplains and wetlands within the region (Sedell and Froggatt 1984). Most of these aquatic habitats were eliminated by channelization, draining, road building, and filling operations prior to World War II, and many of these changes occurred before 1900 (Bowen 1978, Boag 1992). Fishery biologists have no quantitative measures of the degree to which the elimination of lowland aquatic systems affected salmon, but recent evidence indicates that these were some of the most productive habitats within the landscape. Studies of effects of livestock grazing on aquatic ecosystems and salmonids generally have observed responses consistent with studies of habitat relationships on forest lands (Chapman and Knudsen 1980, Kauffman and Krueger 1984, Platts 1991). Where riparian vegetation is heavily grazed and channel structure is changed, populations of some fish species decline, the balance of species is altered, and stream flows are negatively affected (Elmore and Beschta 1987, Beschta 1991, Elmore 1992).

In contrast to policies for forests, land-use regulations pertaining to streams for agricultural and range lands are less protective. The few that do exist apply to a small fraction of the lands and do not explicitly identify BMPs (Kauffman 1988). Many state and federal programs have relied on voluntary compliance (Oregon Agricultural Practices - ORS 568.900-933), and there is little evaluation of attempts to protect or restore aquatic habitat. The lack of consistency in development of environmental management issues or guidelines for land-use practices is a major obstacle to managing freshwater habitats in the Pacific Northwest.

Habitat Loss Associated with Urbanization

Urban lands make up only 2% of the land base of the Pacific Northwest (Pease 1993), but they exert disproportionate influences on salmonid production because urban areas are frequently located in important salmonid migration corridors and wintering sites. In spite of their relatively small area, >70% of the population of the region lives in cities and towns (76%, 70%, 57%, 93% for Washington, Oregon, Idaho, and California, respectively [American Almanac: Statistical Abstracts of the United States 1994]). Regional resource management is dictated primarily by the urban sector, but constraints on land use are borne almost entirely by the rural sector. Increases in the proportion of the urban population will only create greater conflicts between interests of the general public, private landowners, and natural resource agencies that manage the majority of the land base.

Though total urban area may be small, cities and towns are located at critical positions on major rivers, tributary junctions, and estuaries. The confluences of major rivers in the Pacific Northwest (the Willamette and Columbia rivers, Puget Sound and its tributaries) are centers of major regional metropolitan areas (Lucchetti and Fuerstenberg 1993, Nolan 1993). Aquatic habitats in urban areas are more highly altered than in any other land-use type in the Pacific Northwest, and the proportion of the streams within the urban areas that are degraded is greater than the proportion of highly altered streams on agricultural, range, or forested lands (Booth 1991).

Most urban areas are located on historical wetlands, but drainage requirements for residences and urban centers have eliminated ≥90% of these productive aquatic habitats in some drainage systems (Boulé and Bierly 1987). Water quality and habitat conditions in these critical migration pathways within river networks potentially restrict movement of salmonid smolts from their natal streams, survival in winter rearing areas, or return of adult salmon to the headwaters. In addition, habitat degradation and direct effects on invertebrate communities reduce food supplies for fish assemblages (Hachmoller et al. 1991, Borchardt 1993). Lossess of wetlands, tidal sloughs, and estuaries in heavily urbanized or industrialized river basins have been extensive; in some areas of Puget Sound, >95% of estuarine and coastal wetland habitats have been eliminated since the 19th century (Sherwood et al. 1990, Simenstad et al. 1992). Though forest practices and, to a much lesser degree, agricultural practices have drawn intense scrutiny resulting in more protective land-use regulations, urbanization and industrial development tend to cause the most extensive alteration of aquatic ecosystems. Future population increases in the Pacific Northwest will expand the spatial extent of this source of habitat loss.

Legal History and Role of Habitat Regulation

Public attention and legislative regulation have been focused primarily on management of aquatic habitat in forests on both public and private lands. Timber harvest practices were not regulated in riparian zones until the 1970s; thus, there were >120 years of human activity and ~50–70 years of intensive harvest on public lands prior to mandated consideration of streamside protection. Some logging of federal forests occurred during the first half of the 20th century, but timber harvest accelerated dramatically after World War II—an era with little or no riparian protection. From 1940 to 1960, federal lands accounted for 19% of the total forest harvest in Oregon, but from 1960 to 1990, federal forest provided 52% of the total harvest (Dave Steere, Oregon Dep. Forestry, Salem, Oregon, pers. comm.). Even during the recent development of the Northwest Forest Plan, the federal forests accounted for 35% of the harvest in Oregon in 1992. The current debate over regional forest management has raised concerns about the abrupt shift in the proportions of federal and non-federal harvest, but previous shifts in harvest levels within the forest industry in the two decades after World War II rival those of today. Much of the logging in the Pacific Northwest currently occurs on private forest lands, which have less stringent regulations for streamside protection.

Environmental guidelines for forest practices first called for riparian protection on federal lands in the late 1960s and early 1970s. Riparian management was addressed directly in the forest planning process of the National Forest Management Act of 1976, but National Forests were encouraged to develop individual standards and guidelines, which were not coordinated and differed substantially from forest to forest (Gregory and Ashkenas 1990). In 1971, Oregon was the first state to enact a Forest Practices Act (FPA) for private forest lands. Initial legislation was aimed primarily at maintaining shade over the streams, decreasing erosion and sediment inputs, and providing for replanting after logging. The original FPA provided relatively little riparian protection around small streams and allowed for essentially complete removal of merchantable timber. Recent revisions develop more protective measures for maintaining ecological functions of aquatic ecosystems. Similar changes have occurred in Washington, Idaho, and California, but substantial differences exist in the protection requirements of neighboring states. The state of Washington has developed a watershed analysis approach that offers the potential to develop watershed-specific guidelines based on local resources and watershed conditions (Washington Forest Practices Board 1993).

Floodplains are not directly addressed in any of the statutes of the Pacific Northwest states except for protection of streamside wetlands. One-hundred-year floodplains are not recognized in the protection requirements for private commercial or state forestlands. Floodplain conditions are examined under Washington's watershed assessment approach, but no specific regulations or guidelines require management actions. Lack of consideration of floodplains in regional forest management of state and private lands reveals a fundamental weakness in the regulations, given the certainty of floods and the land-water interactions that occur during such events.

Recently, some National Forests in the Pacific Northwest have begun to implement a new aquatic conservation strategy based on a set of recommendations called PACFISH (Sedell et al. 1994). This strategy involves interim guidelines requiring functional riparian protection with no timber harvest within the streamside management zone, full floodplain protection, and even protection of small ephemeral channels that do not contain fish. The strategy also formally establishes key watersheds (drainages with at-risk species and other important aquatic resources) where protection of fish habitat is given top priority, and it encourages the development of regional conservation strategies based on thorough watershed assessment. Modification of forest management policies by the federal government in 1993 under the development of FEMAT (1993) incorporated substantially greater riparian protection than had been previously required and addressed floodplains on federal forestlands.

In general, society has called for high standards of environmental protection on public and private forest lands, but management activities in public forests are restricted to a greater extent than in private forests (Robinson 1987). Approximately 80% of the anadromous salmonid stocks identified by Nehlsen et al. (1991) and state agency reports as at risk of extinction spend a substantial portion of their life history on federal lands (FEMAT 1993, Thomas et al. 1993), but some stocks, particularly those inhabiting coastal lowlands, spend most of their time in freshwater on state and privately owned lands (NRC 1996). Currently, Pacific Northwest states have a fragmented and uncoordinated collection of statutes relating to different land-use types, zoning, and different resource users. Regulations for riparian protection on private non-forest lands are often minimal. Mature trees are required to be maintained along streams on lands zoned for forestry, but riparian forests along streams that pass through land zoned for agricultural, residential, or industrial use are allowed to be almost completely removed. These lowland streams historically were some of the most productive habitats in river drainages for anadromous salmonids, and only fragments of these habitats and their stocks are still in existence (Sedell and Froggatt 1984, Naiman et al. 1991). There is a need for greater consistency in the levels of environmental protection applied to different land-uses. We suggest that an integrated land-use practices act for each state would promote management practices that would not have to be identical for each type of land use or zoning but would ensure the ecological considerations that form the basis for management would be consistent.

Future Directions in Riparian Management and Habitat Protection

The future of anadromous and resident salmonids of the Pacific Northwest requires protecting existing intact, healthy aquatic ecosystems, restoring degraded systems, and developing sustainable resource management policies (Frissell 1993, Moyle and Yoshiyama 1994, Sedell et al. 1994, NRC 1996). Any discussion of habitat loss in the Pacific Northwest that did not call for these actions would be deficient, but these goals are not simple and each contains ecological and social traps that have the potential to impede rather than accelerate habitat recovery.

Protecting Existing Aquatic Ecosystems

One of the major tools in landscape management at a regional scale is development of systems of watershed reserves for aquatic ecosystems (FEMAT 1993, Frissell 1993, Pacific Rivers Council 1993, Moyle and Yoshiyama 1994). This reserve system in the Pacific Northwest is based primarily on public lands managed by the Forest Service and BLM. Recognition of watersheds as major functional elements of regional landscapes and of their role in protecting entire stocks or local populations is one of the major advances in ecosystem management in the Pacific Northwest. The approach is essential for sound ecosystem management, but it is constrained by land ownership patterns in the region. Lowland rivers and estuaries predominantly occur on private lands, and steeper mountainous terrain occupies the majority of federally owned lands. Approximately 20% of the stocks of anadromous salmon identified as at risk by the American Fisheries Society and state agencies in the Pacific Northwest do not occur on federal lands (FEMAT 1993). In addition, many of the stocks that occur on public lands must pass through lowland rivers during their migrations. Effective habitat management at a landscape scale requires incorporation of the entire landscape relevant to salmon life histories and integration of management policies for both public and private lands (NRC 1996). Exclusive reliance on public lands for aquatic habitat protection will jeopardize the future of many anadromous stocks and accelerate the loss of major large-river and floodplain habitats in the lower portions of river basins.

Restoring Degraded Systems

The most critical questions related to aquatic habitat restoration include the following: the degree to which it can be repaired or restored, priorities for locations where restoration efforts will be beneficial, and ecologically sound approaches for habitat restoration. As with sustainability, restoration is a term that finds almost unanimous acceptance, but misguided or ineffective restoration programs can undermine public confidence and even cause additional ecological damage (NRC 1992, Hilborn and Winton 1993). The goal of restoration is to reestablish an ecosystem's ability to maintain its function and organization without continued human intervention. It does not mandate returning to some arbitrary prior state. Any restoration program should be nested within a larger program of landscape management that protects, maintains, and restores ecosystem structure and function (Wissmar and Swanson 1990, Sedell et al. 1991, Moyle and Yoshiyama 1994). Resource analysis should precede any restoration effort, starting at the scale of entire river basins, focusing down to specific watersheds, and finally addressing local reach characteristics.

Ecosystems are dynamic and changing; thus, restoration to a previous condition often is impossible or even ecologically undesirable. Ecosystem restoration is based on restoring systems to the point that they can provide the natural materials and ecological functions that create habitat. Artificially constructing habitats does not constitute ecological restoration. Many practices commonly are mentioned within the context of restoration and often are used interchangeably with the term restoration, but their differences are important. Rehabilitation involves the reestablishment of specific components or processes to some degree of their previous state, but not complete recovery of ecological function. The term "habitat improvement" is widely used, but it has the misleading connotation that habitats are increasing in quality or function. In most cases, the habitat has been severely degraded and only a small fraction of its potential function has been restored. Ecological repair is more limited and may focus on a few limited characteristics of the ecosystem and reestablish relatively few of historical conditions. Mitigation is a substitution of systems, habitats, artificial processes, or simple economic value for the loss of natural habitat or ecological functions.

Stream restoration practices often attempt to reestablish specific geomorphic features or channel structure (Reeves et al. 1991a, Gregory and Wildman 1994) or increase densities of selected species (House and Boehne 1986, Nickelson et al. 1992). Delivery of geomorphic elements that are locally missing or deficient, facilitating natural hydrologic processes, or reestablishment of riparian plant communities can accelerate the rate of recovery of ecological processes or communities (Reeves et al. 1991b). Inaccurate assessment of physical processes, riparian plant ecology, causes of habitat degradation, or factors that limit populations or communities can result in either ineffective restoration efforts or habitat degradation or ecological damage (Frissell and Nawa 1992). Design of restoration requires thorough ecological and landscape analysis. Appropriate criteria for evaluation of the performance of restoration efforts are based on achieving the overall goals in a dynamic environment rather than simple persistence in the original location.

Several principles guide future restoration efforts from a habitat viewpoint. Choices exist between restoring areas that have been severely altered versus areas that have been only slightly changed and would recover quickly with less input of material and effort (Pacific Rivers Council 1993). Potential for recovery is greatly diminished where proportionately more natural processes, structures, and aquatic communities have been lost. However, even these severely altered habitats become appropriate candidates for restoration where there are important resources, critical habitats, or unique opportunities. Sound ecological restoration considers landscape pattern and connectivity. Setting priorities involves not just a matter of locating good or poor habitat, but considering how these areas are spatially arrayed. Restoration of river basins should be built upon nodes of high quality habitat that serve as refuges and provide sources of biotic colonists to rebuild connectivity throughout the basin. Efforts must extend into estuaries to include habitats that are critical for several life-history stages of salmonids (Shreffler et al. 1990).

One of the most important challenges of restoration is to change practices that altered habitat in the first place (Beschta et al. 1991). If environmental degradation continues, restoration efforts will be impeded or ineffective. Aquatic ecosystems should be allowed to recover naturally before habitat improvements are undertaken unless heroic efforts are required to save resources from extinction or prevent catastrophic habitat change. It is difficult to wait when the need for ecological recovery is great, but the potential for restoration efforts to be ineffective or inappropriately located is far greater in rapidly changing systems (Sedell and Beschta 1991).

Habitat restoration or rehabilitation commonly utilizes engineering approaches to erect permanent structures in streams (NRC 1992). In addition, administration of projects by agencies frequently identifies habitat targets so that funds can be allocated efficiently and project performance can be evaluated (Frissell and Nawa 1992). Unfortunately, natural processes and long-term dynamics of stream channels and communities are largely ignored, and rehabilitation projects may provide little or no benefit and may cause ecological damage (Sedell and Beschta 1991). Sound restoration of aquatic ecosystems is based on a solid foundation of ecological principles and a clear recognition of the dynamic nature of streams, rivers, wetlands, lakes, and riparian forests (Naiman et al. 1993).

Ecological restoration facilitates the reestablishment of natural physical and ecological processes that occur in the local area. Use of native species and locally adapted stocks maintains the integrity of the genetic characteristics of local populations. Some of the most important components of habitat restoration include protection or restoration of floodplains and riparian plant communities. The agents of habitat change (identified previously in Table 1) are also the basis for ecological recovery (Table 3). Environmental and biological uncertainty should be recognized and management options should emphasize alternatives that offer flexibility and opportunities to learn. Knowledge of the behavior of the environment will always be limited, and the range of possible outcomes of management actions in aquatic ecosystems introduces a large degree of uncertainty (Ludwig et al. 1993). Given the uncertainty inherent in ecosystem management and restoration, actions should be reversible either by natural processes or by human correction, if possible.

Restoration of riparian forests may be accelerated by silvicultural practices (Bilby and Bisson 1991). The goal of silvicultural management in riparian management zones should be to provide the natural ecological functions of riparian vegetation where previous practices have diminished the diversity of riparian plant communities. Riparian silviculture should encourage natural patterns of succession and create diverse and structurally complex riparian plant communities (Agee 1988). Reestablishment of shade over stream channels can be accelerated by protecting remaining streamside vegetation, especially young trees. However, in areas dominated by shrub cover, underburning may encourage regeneration of desired tree species (Agee 1993). In riparian areas where short-term canopy recovery is required for temperature protection, hardwood species may be used to rapidly reestablish vegetative cover. In many riparian zones of the Pacific Northwest, conifers are needed for long-term shading and inputs of large wood. More research is needed to determine the most effective techniques of restoring conifers to hardwood or shrub-dominated riparian zones. Pre-commercial thinning of small trees from upslope forests can provide material for placing directly into streams for short-term improvements, particularly in small streams lacking large wood.

The time frame for ecosystem restoration is constrained by the processes that shape stream channels, riparian plant communities, and aquatic communities (Table 3). In almost all cases, ecological recovery will require decades before natural systems can maintain themselves without human intervention, and centuries will be required for complete restoration of certain ecosystem components or processes. Resource management agencies should explicitly describe the time frame for restoration and clearly identify anticipated patterns of recovery.

The major agent of aquatic ecosystem restoration in the Pacific Northwest is periodic natural disturbance. Natural disturbances create and maintain the structural and ecological characteristics of riparian areas (Resh et al. 1988, Aumen et al. 1990, Bayley 1995). Disturbances in riparian areas include floods, windthrow, fire, insect outbreaks, and disease, which in combination create complex habitats and diverse plant and animal communities (Reice et al. 1990, Gregory et al. 1991, Reice 1994). Floods are essential for the sustained productivity of rivers. Streams are shaped by floods, and many rivers throughout the world are more productive after flooding (Junk et al. 1989). Flooding is a renewal process that creates pools, cleans gravel, and delivers dissolved and particulate nutrients (Elwood and Waters 1969, Bayley 1995). Fish and invertebrate communities in streams are resilient and often respond rapidly to disturbance, but the availability of refuges accelerates recovery of invertebrate community structure (Lamberti et al. 1991, Anderson 1992) and fish populations (Bisson et al. 1988, Sedell et al. 1990, Lamberti et al. 1991). Small-scale refuges during floods include deep pools, debris dams, boulders and logs, off-channel habitats on floodplains, and stems and roots of streamside forests. Floodplain habitats, large wood, and pool habitats have declined substantially in recent years and are among the major habitat losses related to the decline of Pacific salmon (Sedell et al. 1991).

Table 3. Sources of habitat modification, active restoration approaches, and estimated time scales for recovery. All of these responses vary according to the degree of landscape and stream alteration, natural disturbance events, and the magnitude of restoration efforts.

Ecosystem feature	Alteration characteristics	Ecosystem processes that restore structure or function	Recovery period (yr)
Channel structure	Floodplains	Reconnect floodplain to main channel. Silvicultural planting revegetates floodplain surface.	10–100
	Pools and riffles	Pools can be dug and riffles can be deposited, but new bedforms will not be stable if inconsistent with natural channel structure and hydrologic regimes.	1–10
	Large wood	Large wood can be placed in streams. Attention to natural dynamics and distributions is necessary to prevent further habitat degradation through the restoration effort. Natural succession may require centuries to restore inputs of wood.	5–25
	Substrate	Sediments can be placed in stream reaches if deficient. Artificial flushing of excessive sediment loads is largely ineffective.	5–20
	Hyporheic zones	Processes that reestablish bedforms and bed composition create new distribution of subsurface flow. Reestablishment of hydrologic sources may restore subsurface flow.	5–20
Hydrology	Discharge	If silvicultural acceleration of upslope and riparian vegetation is possible, recovery of evapotranspiration rates will accompany forest recovery. Dams prevent recovery of natural discharge patterns.	10–50
	Low flows	If silvicultural acceleration of conifer regeneration vegetation is possible, replacement of second-growth deciduous vegetation reduces evapotranspiration rates.	25–50
	Rapid fluctuations	After cessation, colonization by aquatic organisms and revegetation would require decades.	1–10
Sediment	Surface erosion	Revegetation of the watershed and riparian areas will diminish inputs of soil from surface erosion.	10–50
	Mass failures and landslides	Mass failures will diminish if road systems are reduced or upgraded, but failure rates will remain elevated as long as roads and culverts alter local water movements on steep slopes.	20–200
Water quality	Temperature	Reestablishment of canopy cover over streams reduces solar inputs and stream temperature.	10–40
	Dissolved oxygen	Temperature effects on oxygen will be related to shade recovery, and organic demands will be reduced as material decomposes and redistributes.	1–10

Table 3—cont.

Ecosystem feature	Alteration characteristics	Ecosystem processes that restore structure or function	Recovery period (yr)
Water quality—cont.	Nutrients	Revegetation of the watershed and riparian areas will retain nutrients and diminish inputs to surface waters.	10–40
Riparian forest	Production of large wood	Development of mature to old-growth riparian forest contributes new wood. Transport of wood and boulders from upstream or landslides delivers large structural elements to local reaches.	100–300
	Production of food organisms and organic matter	Development of mature to old-growth riparian forests restores natural inputs from terrestrial ecosystems.	40–80
	Shading	Reestablishment of canopy cover over streams reduces solar inputs and primary production declines to predisturbance levels. Riparian vegetation influences stream temperature.	5–20
	Rooting systems	Development of woody vegetation along streams and streambank integrity strengthens banks as root systems develop. Grasses and forbs provide similar functions along natural meadows.	20–80
	Nutrient modification	Successional development of riparian plant communities restores nutrient filtering capacity.	20–80
Exogenous material	Chemicals	Anthropogenic chemicals vary greatly in persistence in the environment.	Unknown
	Exotic organisms	Once established, exotic species likely will not be eliminated from the regional assemblage of species. Exotic plant communities may become less abundant as forests return to mature forest conditions.	Unknown

Land-use practices should be designed to maintain natural disturbance processes and to retain the beneficial effects of disturbance events to the degree possible. Analysis of the consequences of disturbances includes explicit assessment of short-term effects, local site-specific effects, long-term effects, and basin and landscape effects. At present, no resource management agency has a formal "after the disturbance" policy designed to protect beneficial changes caused by natural disturbances; too often the management response has been to fix the changes caused by the disturbance. Recognition of the role of floods and related natural disturbances in streams and riparian areas will reduce the tendency for disaster relief efforts that simply repeat previous resource management mistakes.

Habitat restoration is no substitute for appropriate environmental protection, and approaches built solely upon rehabilitation cannot maintain ecosystem health. The growth of habitat restoration programs in state and federal agencies is the most undeniable and well-documented evidence for our failure to effectively manage aquatic ecosystems of the Pacific Northwest. Our success in the future will be measured by the degree to which we are able to decrease the need for restoration programs.

Human Population

Discussion of habitat alteration and future alternatives is fundamentally a question of human population and rates of resource consumption. Salmon stocks and the ecosystems that support them have been altered so extensively that there is no question about the outcome if current land-use practices and population growth rates continue without change. Unfortunately, the time frame for response is shorter than the historical trajectory that brought us to this point. The combined population of Oregon, Washington, and Idaho was just >1 million people in 1900 and is currently >9 million (Fig. 4) (American Almanac: Statistical Abstract of the United States 1994). The population of the Pacific Northwest is projected to double to ~17 million by the year 2025 (assuming an average annual population increase of 1.9%), requiring only 30 years to attain the same absolute increase in numbers that required 90 years previously.

Projected increases in human population in the Pacific Northwest will be accompanied by increased demand not only for forest products but also for land, water, and energy. Only 2% of the Pacific Northwest is urban or residential lands, but the additional millions of people anticipated for this region in the next few decades will either live in those urban areas or will consume current forest and agricultural land to build homes and communities. Even considering a low domestic water consumption of 750 L d^{-1} per person (excluding agricultural and industrial water requirements), population growth in the Pacific Northwest will impact an additional 6 billion L d^{-1} of water in only 30 years. This consumption of water also will be reflected in the delivery of sewage and waste water to the region's streams and rivers.

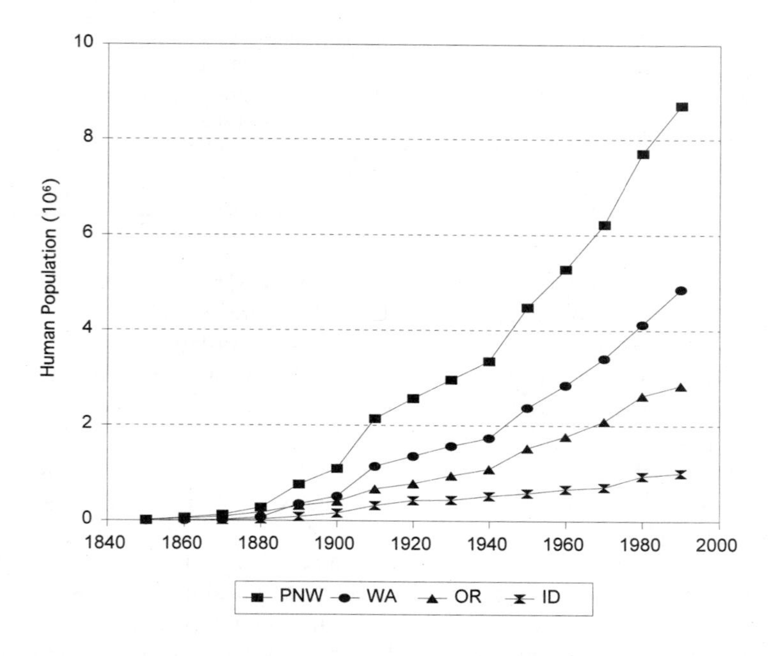

Figure 4. Trends in population of Oregon, Washington, Idaho, and the Pacific Northwest from 1850 to 1990. Source: American Almanac: Statistical Abstracts of the United States (1994).

The major source of water for future communities of the Pacific Northwest comes from forest lands, primarily designated federal forest lands. The Forest Service has allocated ~440,000 ha of federal land in Washington and 1,112,000 ha in Oregon as designated water supplies for local municipalities (B. McCammon, Region 6, USDA Forest Service, Portland, Oregon, pers. comm.). Federal lands provide the domestic water supply for 43% of the population in Oregon and for 34% of the population of Washington. In many ways, water will be the most valuable product coming from federal lands in the near future, and public forest lands will be a critical component in the supply of water for the region in the coming century.

The people of the Pacific Northwest must evaluate the success of our efforts to manage ecosystems based on our ability to deal with ecological and institutional change rather than our static performance at any point in time. In 1902, Overton Price, the Assistant Forester for the newly formed Forest Service, noted, "It is the history of all great industries directed by private interests that the necessity for modification is not seen until the harm has been done and its results are felt." The Pacific Northwest finds itself repeating the lessons of other regions. Future management of the ecosystems of the Pacific salmon will require ecologically sound approaches for protection and restoration of aquatic habitats, effective regulations and human incentive systems, and long-term resource monitoring programs.

Acknowledgments

The authors acknowledge the support of the H.J. Andrews Long-Term Ecological Research Program of the National Science Foundation, the Coastal Oregon Productivity Enhancement Program, and the Department of Fisheries and Wildlife and Agricultural Experiment Station of Oregon State University in the research and preparation of this paper.

Literature Cited

Adams, E. H. 1885. Salmon canning in Oregon. Bulletin of the United States Fish Commission: 362-365.

Agee, J. K. 1988. Successional dynamics of forest riparian zones, p. 31-43. *In* K.J. Raedeke (ed.), Streamside Management: Riparian Wildlife and Forestry Interactions. Contribution no. 59, Institute of Forest Resources, University of Washington. Seattle.

Agee, J. K. 1993. Fire Ecology of Pacific Northwest Forests. Island Press, Washington, DC.

American Almanac 1993-1994: Statistical Abstract of the United States. 113th Edition. The Reference Press Inc., Austin, Texas.

Anderson, N.H. 1992. Influence of disturbance on insect communities in Pacific Northwest streams. Hydrobiologia 248: 79-92.

Andrus, C.W., B.A. Long, and H.A. Froehlich. 1988. Woody debris and its contribution to pool formation in a coastal stream 50 years after logging. Canadian Journal of Fisheries and Aquatic Sciences 45: 2080-2086.

Aumen, N.G., C.P. Hawkins, and S.V. Gregory. 1990. Influence of woody debris on nutrient retention in catastrophically disturbed streams. Hydrobiologia 190: 183-192.

Averett, R.C. 1969. Influence of temperature on energy and material utilization by juvenile coho salmon. Ph.D. thesis, Oregon State University. Corvallis.

Bayley, P. B. 1995. Understanding large river-floodplain ecosystems. BioScience 45(3): 153-158.

Benda, L. 1990. The influence of debris flows on channels and valley floors in the Oregon Coast Range, USA. Earth Surface Processes and Landforms 15: 457-466.

Berg, A. and A. Doerksen. 1975. Natural fertilization of a heavily thinned Douglas-fir stand by understory red alder. Research Note 56, School of Forestry, Oregon State University. Corvallis, Oregon.

Berg, L. and T.G. Northcote. 1985. Changes in territorial, gill-flaring, and feeding behavior of juvenile coho salmon (*Oncorhynchus kisutch*) following short-term pulses of suspended sediment. Canadian Journal of Fisheries and Aquatic Sciences 42: 1410-1417.

Berman, C.H. and T.P. Quinn. 1991. Behavioral thermoregulation and homing by spring chinook salmon, *Oncorhynchus tshawytscha* (Walbaum), in the Yakima River. Journal of Fish Biology 39: 301-312.

Berris, S.N. and R.D. Harr. 1987. Comparative snow accumulation and melt during rainfall in forested and clearcut plots in the western Cascades of Oregon. Water Resources Research 23: 135-142.

Beschta, R. L. 1978. Long-term patterns of sediment production following road construction and logging in the Oregon Coast Range. Water Resources Research 14: 1011-1016.

Beschta, R.L. 1991. Stream habitat management for fish in the northwestern United States: the role of riparian vegetation. American Fisheries Society Symposium 10: 53-58.

Beschta, R.L., R. Bilby, G. Brown, L.B. Holtby, and T.D. Hofstra. 1987. Stream temperature and aquatic habitat: fisheries and forestry interactions, p. 191-232. *In* E.O. Salo and T.W. Cundy (eds.). Streamside Management: Forestry and Fishery Interactions. Contribution no. 57, Institute of Forest Resources, University of Washington. Seattle.

Beschta, R.L., W.S. Platts, and B. Kauffman. 1991. Field review of fish habitat improvement projects in the Grande Ronde and John Day River basins of eastern Oregon. Project No. 91-069, Report to Bonneville Power Administration, Division of Fish and Wildlife, US Department of Energy. Portland, Oregon.

Bilby, R.E. and P.A. Bisson. 1987. Emigration and production of hatchery coho salmon (*Oncorhynchus kisutch*) stocked in streams draining an old-growth and a clear-cut watershed. Canadian Journal of Fisheries and Aquatic Sciences 44: 1397-1407.

Bilby, R.E. and P.A. Bisson. 1991. Enhancing fisheries resources through active management of riparian areas, p. 201-209. *In* B. White and I. Guthrie (eds.), Proceedings of the 15th Northeast Pacific Pink and Chum Salmon Workshop. Pacific Salmon Commission, Vancouver, British Columbia, Canada.

Bilby, R.E. and P.A. Bisson. 1992 Allochthonous versus autochthonous organic matter contributions to the trophic support of fish populations in clear-cut and old-growth forested streams. Canadian Journal of Fisheries and Aquatic Sciences 49: 540-551.

Bilby, R.E. and G.E. Likens. 1980. Importance of organic dams in the structure and function of stream ecosystems. Ecology 61: 1107-1113.

Bilby, R.E. and J.W. Ward. 1991. Characteristics and function of large woody debris in streams draining old-growth, clear-cut, and 2nd-growth forests in southwestern Washington. Canadian Journal of Fisheries and Aquatic Sciences 48: 2499-2508.

Bisson, P.A. and R.E. Bilby. 1982. Avoidance of suspended sediment by juvenile coho salmon. North American Journal of Fisheries Management 2: 371-374.

Bisson, P.A., R.E. Bilby, M.D. Bryant, C.A. Dolloff, G.B. Grette, R.A. House, M.L. Murphy, KV. Koski, and J.R. Sedell. 1987. Large woody debris in forested streams in the pacific northwest: past, present, and future, p. 143-190. *In* E.O. Salo and T.W. Cundy (eds.). Streamside Management: Forestry and Fishery Interactions. Contribution no. 57, Institute of Forest Resources, University of Washington. Seattle.

Bisson, P.A. and G.E. Davis. 1976. Production of juvenile chinook salmon, *Oncorhynchus tshawytscha*, in a heated model stream. Fishery Bulletin 74: 763-774.

Bisson, P.A., G.G. Ice, C.J. Perrin, and R.E. Bilby. 1992. Effects of forest fertilization on water quality and aquatic resources in the Douglas-fir region, p. 179-193. *In* H.N. Chappell, G.F. Weetman, and R.E. Miller (eds.), Forest Fertilization: Sustaining and Improving Nutrition and Growth of Western Forests. Contribution 73, Institute of Forest Resources, University of Washington. Seattle.

Bisson, P.A., J.L. Nielsen, and J.W. Ward. 1988. Summer production of coho salmon stocked in Mount St. Helens streams 3-6 years after the 1980 eruption. Transactions of the American Fisheries Society 117: 322-335.

Bisson, P.A., T.P. Quinn, G.H. Reeves, and S.V. Gregory. 1992. Best management practices, cumulative effects, long-term trends in fish abundance in Pacific Northwest river systems, p. 189-232. *In* R.J. Naiman (ed.), Watershed Management: Balancing Sustainability and Environmental Change. Springer-Verlag, New York.

Bisson, P.A. and J.R. Sedell. 1984. Salmonid populations in streams in clearcut vs. old-growth forests of western Washington, p. 121-129. *In* W.R. Meehan, T.R. Merrell, Jr., and T.A. Hanley (eds.), Fish and Wildlife Relationships in Old-Growth Forests. American Institute of Fisheries Research Biologists, Juneau, Alaska.

Bisson, P.A., K. Sullivan, and J.L. Nielsen. 1988. Channel hydraulics, habitat use, and body form of juvenile coho salmon, steelhead, and cutthroat trout in streams. Transactions of the American Fisheries Society 117: 262-273.

Bjornn, T.C. and D.W. Reiser. 1991. Habitat requirements of salmonids in streams. American Fisheries Society Special Publication 19: 83-138.

Boag, P.G. 1992. Environment and Experience: Settlement Culture in Nineteenth Century Oregon. University of California Press, Berkeley.

Booth, D.B. 1991. Urbanization and the natural drainage system—impacts, solutions and prognoses. Northwest Environmental Journal 7: 93-118.

Borchardt, D. 1993. Effects of flow and refugia on drift loss of benthic macroinvertebrates: implications for habitat restoration in lowland streams. Freshwater Biology 29: 221-227.

Bothwell, M.L. 1989. Phosphorus-limited growth dynamics of lotic periphytic diatom communities: areal biomass and cellular growth rate responses. Canadian Journal of Fisheries and Aquatic Sciences 46: 1293-1301.

Botkin, D., K. Cummins, T. Dunne, H. Regier, M. Sobel, L. Talbot, and L. Simpson. 1995. Status and future of salmon in western Oregon and northern California: status and future options. Report Number 8, The Center for the Study of the Environment. Santa Barbara, California.

Boulé, M.E. and K.F. Bierly. 1987. History of estuarine wetland development and alteration: what have we wrought? Northwest Environmental Journal 3: 43-61.

Bowen, W.A. 1978. The Willamette Valley: Migration and Settlement on the Oregon Frontier. University of Washington Press, Seattle.

Boyd, R. T. 1990. Demographic history, 1774-1874. *In* W. Suttles (ed.), Northwest Coast. Smithsonian Institution, Washington, DC.

Brett, J.R. 1952. Temperature tolerances in young Pacific salmon, *Oncorhynchus*. Canadian Journal of Fisheries and Aquatic Sciences 9: 268-323.

Brett, J.R. and J.M. Blackburn. 1981. Oxygen requirements for growth of young coho (*Oncorhynchus kisutch*) and sockeye (*O. nerka*) salmon at 15°C. Canadian Journal of Fisheries and Aquatic Sciences 38: 399-404.

Brown, G.W. and J.T. Krygier. 1970. Effects of clear-cutting on stream temperature. Water Resources Research 6: 1133-1139.

Brown, T.G. and G.F. Hartman. 1988. Contribution of seasonally flooded lands and minor tributaries to the production of coho salmon in Carnation Creek, British Columbia. Transactions of the American Fisheries Society 117: 546-551.

Bryant, M.D. 1983. The role and management of woody debris in west coast salmonid nursery streams. North American Journal of Fisheries Management 3: 322-330.

Burns, J.W. 1972. Some effects of logging and associated road construction on northern California streams. Transactions of the American Fisheries Society 101: 1-17.

Burroughs, E.R., Jr., and B.R. Thomas. 1977. Declining root strength in Douglas-fir after felling as a factor in slope stability. USDA Forest Service Research Paper INT-190. Ogden, Utah.

Cederholm, C.J. and N.P. Peterson. 1985. The retention of coho salmon (*Oncorhynchus kisutch*) carcasses by organic debris in small streams. Canadian Journal of Fisheries and Aquatic Sciences 42: 1222-1225.

Cederholm, C.J., L. M. Reid, and E.O. Salo. 1981. Cumulative effects of logging road sediment on salmonid populations in the Clearwater River, Jefferson County, Washington, p. 38-74 *In* Proceedings of a Conference on Salmon Spawning Gravel: A Renewable Resource in the Pacific Northwest. Report 39, State of Washington Water Resource Center. Washington State University, Pullman, Washington.

Chamberlin, T.W., R.D. Harr, and F.H. Everest. 1991. Timber harvesting, silviculture, and watershed processes. American Fisheries Society Special Publication 19: 181-205.

Chapman, D. W. 1966. Food and space as regulators of salmonid populations in streams. American Naturalist 100: 345-357.

Chapman, D. W. 1986. Salmon and steelhead abundance in the Columbia River in the nineteenth century. Transactions of the American Fisheries Society 115: 662-670.

Chapman, D. W. 1988. Critical review of variables used to define effects of fines in redds of large salmonids. Transactions of the American Fisheries Society 117:1-21.
Chapman, D.W. and E. Knudsen. 1980. Channelization and livestock impacts on salmonid habitat and biomass in western Washington. Transactions of the American Fisheries Society 109: 357-363.
Cobb, J.N. 1922. Pacific salmon fisheries. Report of the United States Commissioner of Fisheries for 1921, United States Commission of Fish and Fisheries. Washington, DC.
Cooper, R. and T.H. Johnson. 1992. Trends in steelhead (*Oncorhynchus mykiss*) abundance in Washington and along the coast of North America. Washington Department of Wildlife Report 92-20. Olympia.
Cordone, A.J. and D.W. Kelley. 1961. The influence of inorganic sediment on the aquatic life of streams. California Fish and Game 47: 189-228.
Cristner, J. and R. D. Harr. 1982. Peak streamflows from the transient snow zone, western Cascades, Oregon. Proceedings of the 50th Western Snow Conference 50: 27-38.
Culp, J. M. 1988. The effect of streambank clearcutting on the benthic invertebrates of Carnation Creek, British Columbia, p. 75-80 *In* T.W. Chamberlin (eds.), Proceedings of the Workshop: Applying 15 Years of Carnation Creek Results. Carnation Creek Steering Committee, Pacific Biology Station, Nanaimo, British Columbia.
Davis, J.C. 1975. Minimal dissolved oxygen requirements of aquatic life with emphasis on Canadian species: a review. Journal of the Fisheries Research Board of Canada 32: 2295-2332.
Dimick, R.E. and F. Merryfield. 1945. The fishes of the Willamette River system in relation to pollution. Bulletin Series No. 20, Engineering Experiment Station, Oregon State College. Corvallis.
Edwards, R., G. Portes, and F. Fleming. 1992. Regional nonpoint source program summary. Environmental Protection Agency, Region 10, Seattle, Washington.
Elmore, W. 1992. Riparian responses to grazing practices, p. 442-457 *In* R. J. Naiman (ed.). Watershed Management: Balancing Sustainability and Environmental Change. Springer-Verlag, New York.
Elmore, W. and R.L. Beschta. 1987. Riparian areas: perceptions in management. Rangelands 9: 260-265.
Elwood, J.W. and T.F. Waters. 1969. Effects of floods on food consumption and production rates of a stream brook trout population. Transactions of the American Fisheries Society 98: 253-262.
Erman, D.C., J.D. Newbold, and K.B. Roby. 1977. Evaluation of streamside bufferstrips for protecting aquatic organisms. Contribution Number 165, California Water Resources Center, University of California. Davis.
Everest F.H., R.L. Beschta, J.C. Scrivener, KV. Koski, J.R. Sedell, and C.J. Cederholm. 1987. Fine sediment and salmonid production: a paradox, p. 98-142. *In* E.O. Salo and T.W. Cundy (eds.), Streamside Management: Forestry and Fishery Interactions. Contribution no. 57, Institute of Forest Resources, University of Washington. Seattle.
Forest Ecosystem Management Assessment Team. 1993. Forest ecosystem management: An ecological, economic, and social assessment. US Government Printing Office, Washington, DC.
Francis, R.C. and T.H. Sibley. 1991. Climate change and fisheries: what are the real issues. The Northwest Environmental Journal 7: 295-307.
Frenkel, R.E. and J.C. Morlan. 1991. Can we restore our salt marshes?, lessons from the Salmon River, Oregon. Northwest Environmental Journal 7: 119-135.
Fresh, K.L. 1996. The role of competition and predation in the decline of Pacific salmon and steelhead, p. 245-275. *In* D.J. Stouder, P.A. Bisson, and R.J. Naiman (eds.), Pacific Salmon and Their Ecosystems: Status and Future Options. Chapman and Hall, New York.
Frissell, C.A. 1993. Topology of extinction and endangerment of native fishes in the Pacific Northwest and California, USA. Conservation Biology 7: 342-354.
Frissell, C.A. and R.K. Nawa. 1992. Incidence and causes of physical failure of artificial habitat structures in streams of western Oregon and Washington. North American Journal of Fisheries Management 12: 182-197.
General Accounting Office. 1988. Public rangelands; some riparian areas restored but widespread improvement will be slow. US General Accounting Office GAO/RCED-88-105. Washington, DC.
Gleeson, G.W. 1972. Return of a river: the Willamette River, Oregon. Water Resources Research Institute WRRI-13, Oregon State University. Corvallis.

Gregory, S.V. and L.R. Ashkenas. 1990. Riparian management guidelines for the Willamette National Forest. US Forest Service Technical Report, Willamette National Forest, US Forest Service. Eugene, Oregon.

Gregory, S.V., G.A. Lamberti, D.C. Erman, KV. Koski, M.L. Murphy, and J.R. Sedell. 1987. Influence of forest practices on aquatic production, p. 233-255. *In* E.O. Salo and T.W. Cundy (eds.), Streamside Management: Forestry and Fishery Interactions. Contribution no. 57, Institute of Forest Resources, University of Washington. Seattle.

Gregory, S.V., F.J. Swanson, W.A. McKee, and K.W. Cummins. 1991. An ecosystem perspective of riparian zones. Bioscience 41: 540-551.

Gregory, S.V. and R.C. Wildman. 1994. Aquatic ecosystem restoration project. Quartz Creek, Willamette National Forest fifth year progress report, Willamette National Forest. Eugene, Oregon.

Grette, G.B. 1985. The abundance and role of large organic debris in juvenile salmonid habitat in streams in second growth and unlogged forests. M.S. thesis, University of Washington. Seattle.

Hachmoller, B., R.A. Matthews, and D.F. Brakke. 1991. Effects of riparian community structure, sediment size, and water quality on the macroinvertebrate communities in a small, suburban stream. Northwest Science 65: 125-132.

Hall, J.D. and R.L. Lantz. 1969. Effects of logging on the habitat of coho salmon and cutthroat trout in coastal streams, p. 355-375. *In* T.G. Northcote (ed.), Salmon and Trout in Streams. H. R. MacMillan Lectures in Fisheries. University of British Columbia, Vancouver.

Harmon, M.E., J.F. Franklin, F.J. Swanson, P. Sollins, S.V. Gregory, J.D. Lattin, N.H. Anderson, S.P. Cline, N.G. Aumen, J.R. Sedell, G.W. Lienkaemper, K. Cromack, Jr., and K.W. Cummins. 1986. Ecology of coarse woody debris in temperate ecosystems. Advances in Ecological Research 15: 133-302.

Hartman, G.F. and J.C. Scrivener. 1990. Impacts of forestry practices on a coastal stream ecosystem, Carnation Creek, British Columbia. Canadian Bulletin of Fisheries and Aquatic Sciences 223.

Hawkins, C.P., M.L. Murphy, and N.H. Anderson. 1982. Effects of canopy, substrate composition, and gradient on the structure of macroinvertebrate communities in Cascade range streams of Oregon. Ecology 63: 1840-1856.

Hawkins, C.P., M.L. Murphy, N.H. Anderson, and M.A. Wilzbach. 1983. Density of fish and salamanders in relation to riparian canopy and physical habitat in streams of the northwestern United States. Canadian Journal of Fisheries and Aquatic Sciences 40: 1173-1185.

Heede, B.H. 1972. Influences of a forest on the hydraulic geometry of two mountain streams. Water Resources Bulletin 8: 523-530.

Hessburg, P.F. 1993. Eastside forest ecosystem health assessment. Volume III, Assessment. US Forest Service, US Department of Agriculture. Wenatchee, Washington.

Hewlett, J. D., and A. R. Hibbert. 1967. Factors affecting the response of small watersheds to precipitation in humid areas, p. 275-290. *In* W. E. Sopper and H. W. Lull (eds.). Forest hydrology. Pergamon, Oxford.

Hicks, B.J., R.L. Beschta, and R.D. Harr. 1991a. Long-term changes in streamflow following logging in western Oregon and associated fisheries implications. Water Resources Bulletin 27: 217-225.

Hicks, B.J., J.D. Hall, P.A. Bisson, and J.R. Sedell. 1991b. Responses of salmonids to habitat change, p. 483-518. *In* W.R. Meehan (ed.), Influences of Forest and Rangeland Management on Salmonid Fishes and Their Habitats. American Fisheries Society Special Publication 19, Bethesda, Maryland.

Hilborn, R. and J. Winton 1993. Learning to enhance salmon production: lessons from the salmonid enhancement program. Canadian Journal of Fisheries and Aquatic Sciences 50: 2043-56.

Holtby, L.B. 1988. Effects of logging on stream temperatures in Carnation Creek, British Columbia, and associated impacts on the coho salmon (*Oncorhynchus kisutch*). Canadian Journal of Fisheries and Aquatic Sciences 45: 502-515.

House, R.D. and P.L. Boehne. 1986. Effects of instream structures on salmonid habitat and populations in Tobe Creek, Oregon. North American Journal of Fisheries Management 6: 38-46.

Hughes, R.M. and G.E. Davis. 1986. Production of coexisting juvenile coho salmon and steelhead trout in heated model stream communities. American Society for Testing and Materials Special Technical Publication 920: 322-337.

Hughes, R.M. and J.R. Gammon. 1987. Longitudinal changes in fish assemblages and water quality in the Willamette river, Oregon. Transactions of the American Fisheries Society 116: 196-209.

Iwamoto, R.N., E.O. Salo, M.A. Madej, and R.L. McComas. 1978. Sediment and water quality: a review of the literature including a suggested approach for water quality criteria. EPA 910/9-78-048, Environmental Protection Agency Region X. Seattle, Washington.

Jackson, P.L. and A.J. Kimerling. 1993. Atlas of the Pacific Northwest. Oregon State University Press, Corvallis.

Junk, W.J., P.B. Bayley, and R.E. Sparks. 1989. The flood pulse concept in river-floodplain systems, p. 110-127. *In* D.P. Dodge (ed.), Proceedings of the International Large River Symposium. Canadian Journal of Fisheries and Aquatic Sciences, Special Publication 106.

Kaczynski, V.W. and J.F. Palmisano. 1993. Oregon's wild salmon and steelhead trout: a review of the impact of management and environmental factors. Oregon Forest Industries Council, Salem.

Karr, J.R. 1991. Biological integrity: a long-neglected aspect of water resource management. Ecological Applications 1: 66-84.

Karr, J. R. 1994. Restoring wild salmon: we must do better. Illahee 10: 316-319.

Karr, J.R. and I. Schlosser, 1977. Impact of nearstream vegetation and stream morphology on water quality and stream biota. Environmental Protection Agency - 600 3-77 097. Washington, DC.

Kauffman, J.B. 1988. The status of riparian habitats in Pacific Northwest forests, p. 45-55. *In* K. J. Raedeke (ed.), Streamside Management: Riparian Wildlife and Forestry Interactions. Contribution no. 59, Institute of Forest Resources, University of Washington. Seattle.

Kauffman, J.B. and W.C. Krueger. 1984. Livestock impacts on riparian ecosystems and streamside management implications. Journal of Range Management 37: 430-437.

Konkel, G.W. and J.D. McIntyre. 1987. Trends in spawning populations of Pacific anadromous salmonids. Fish and Wildlife Technical Report 9, United States Department of Interior, Fish and Wildlife Service. Washington, DC.

Krygier, J.T. and J.D. Hall. 1971. Forest land uses and stream environment. Proceedings of a Symposium. College of Forestry and Department of Fisheries and Wildlife, Oregon State University, Corvallis.

Lamberti, G.A., S.V. Gregory, L.R. Ashkenas, R.C. Wildman, and K.M.S. Moore. 1991. Stream ecosystem recovery following a catastrophic debris flow. Canadian Journal of Fisheries and Aquatic Sciences 48: 196-208.

Li, H., C.B. Schreck, C.E. Bond, E. Rexstad. 1987. Factors influencing changes in fish assemblages of Pacific northwest streams, p. 193-202 *In* W.J. Matthews and D.C. Heins (eds.), Community and Evolutionary Ecology of North American Stream Sishes. University of Oklahoma Press, Norman, Oklahoma.

Lichatowich, J.A. 1989. Habitat alteration and changes in abundance of coho (*Oncorhynchus kisutch*) and chinook salmon (*O. tshawytscha*) in Oregon's coastal streams, p. 92-99. *In* C.D. Levings, L.B. Holtby, and M.A. Anderson (eds.). Proceedings of the National Workshop on Effects of habitat Alteration on Salmonid Stocks. Canadian Special Publication in Fisheries and Aquatic Sciences 105. 199 p.

Lucchetti, G. and R. Fuerstenberg. 1993. Management of coho salmon habitat in urbanizing landscapes of King County, Washington, USA, p. 308-317. *In* L. Berg and P. Delaney (eds.), Proceedings of the 1992 Coho Workshop, Nanaimo, British Columbia. North Pacific International Chapter, American Fisheries Society, and Association of Professional Biologists of British Columbia. Vancouver, British Columbia.

Ludwig, D., R. Hilborn, and C. Walters. 1993. Uncertainty, resource exploitation, and conservation: Lessons from history. Science 260: 35-36.

Malmgren, A. J. 1885. The injurious effects of rafting on the lake and river fisheries of Sweden. Bulletin of the United States Fish Commission for 1885. p. 264-266.

Mason, J.C. and D.W. Chapman. 1965. Significance of early emergence, environmental rearing capacity, and behavioral ecology of juvenile coho salmon in stream channels. Journal of the Fisheries Research Board of Canada 22: 173-190.

Matzke, G. 1993. Population, p. 18-24. *In* P.L. Jackson and A.J. Kimerling (eds.), Atlas of the Pacific Northwest. Oregon State University Press, Corvallis.

McDade, M.H., F.J. Swanson, W.A. McKee, and J. Van Sickle. 1990. Source distances for coarse woody debris entering small streams in western Oregon and Washington. Canadian Journal of Forest Research 20: 326-330.

McDonald, M. 1895. The salmon fisheries of the Columbia River. Bulletin of the United States Fish Commission for 1894: 153-207.

McIntosh, B.A., J.R. Sedell, J.E. Smith, R.C. Wissmar, S.E. Clarke, G.H. Reeves, and L.A. Brown. 1993. Management history of eastside ecosystems: changes in fish habitat over 50 years, 1935 to 1992, p. 291-483. *In* P.F. Hessburg (ed.), Eastside Forest Ecosystem Health Assessment. Volume III, Assessment. U S Forest Service, US Department of Agriculture. Wenatchee, Washington.

Meehan, W.R. (ed.). 1991. Influences of Forest and Rangeland Management on Salmonid Fishes and Their Habitats. American Fisheries Society Special Publication 19. Bethesda, Maryland.

Megahan, W.F., J.P. Potyondy, and K.A. Seyedbagheri. 1992. Best management practices and cumulative effects from sedimentation in the South fork Salmon River: an Idaho case study, p. 401-414. *In* R.J. Naiman (ed.), Watershed Management: Balancing Sustainability and Environmental Change. Springer-Verlag, New York.

Merrell, T.R. 1951. Stream improvement as conducted in Oregon on the Clatskanie River and tributaries. Fish Commission Research Briefs, Fish Commission of Oregon 3: 41-47.

Minshall, G.W., K.W. Cummins, R.C. Petersen, C.E. Cushing, D.A. Burns, J.R. Sedell, and R.L. Vannote. 1985. Developments in stream ecosystem theory. Canadian Journal of Fisheries and Aquatic Sciences 42: 1045-1055.

Mispagel, M.E. and S.D. Rose. 1978. Arthropods associated with various age stands of Douglas-fir from foliar, ground, and aerial strata. Coniferous Forest Biome Bulletin 13, University of Washington. Seattle.

Montgomery, D.R. and J.M. Buffington. 1993. Channel classification, prediction of channel response, and assessment of channel condition. Report TFW- SH10-93-002, Washington State Department of Natural Resources. Olympia.

Moyle, P.B. and J.E. Williams. 1990. Biodiversity loss in the temperate zone: decline of the native fish fauna of California. Conservation Biology 4: 275-284.

Moore, K.M.S. and S. V. Gregory. 1988. Response of young-of-the-year cutthroat trout to manipulation of habitat structure in a small stream. Transactions of the American Fisheries Society 117: 162-170.

Moyle, P.B. and R.M. Yoshiyama. 1994. Protection of aquatic biodiversity in California: a five-tiered approach. Fisheries 19: 6-18.

Mundy, P.R. 1996. The role of harvest management in the future of Pacific salmon populations: shaping human behavior to enable the persistence of salmon, p. 315-329. *In* D.J. Stouder, P.A. Bisson, and R.J. Naiman (eds.), Pacific Salmon and Their Ecosystems: Status and Future Options. Chapman and Hall, New York.

Murphy, M.L., C.P. Hawkins, and N.H. Anderson. 1981. Effects of canopy modification and accumulated sediment on stream communities. Transactions of the American Fisheries Society 110: 469-478.

Murphy, M.L., J. Heifetz, S.W. Johnson, KV. Koski, and J.F. Thedinga. 1986. Effects of clear-cut logging with and without buffer strips on juvenile salmonids in Alaskan streams. Canadian Journal of Fisheries and Aquatic Sciences 43: 1521-1533.

Murphy, M.L. and KV. Koski. 1989. Input and depletion of woody debris in Alaska streams and implications for streamside management. North American Journal of Fisheries Management 9: 427-436.

Naiman, R.J. (ed.). 1992. Watershed Management: Balancing Sustainability and Environmental Change. Springer Verlag, New York.

Naiman, R.J., T.J. Beechie, L.E. Benda, D.R. Berg, P.A. Bisson, L.H. MacDonald, M.D. O'Connor, P.L. Olson, and E.A. Steel. 1992. Fundamental elements of ecologically healthy watersheds in the Pacific Northwest Coastal Ecoregion, p. 127-189. *In* R.J. Naiman (ed.), Watershed Management—Balancing Sustainability and Environmental Change. Springer-Verlag, New York.

Naiman, R.J., H. Decamps, J. Pastor, and C.A. Johnston. 1988. The potential importance of boundaries to fluvial ecosystems. Journal of the North American Benthological Society 7: 289-306.

Naiman, R.J., H. Decamps, and M. Pollock. 1993. The role of riparian corridors in maintaining regional biodiversity. Ecological Applications 3: 209-212.

Naiman, R.J., D.G. Lonzarich, T.J. Beechie, and S.C. Ralph. 1991. General principles of classification and the assessment of conservation potential in rivers, p. 93-123. *In* P.J. Boon and G.E. Petts (eds.), River Conservation and Management. John Wiley and Sons, Inc., Chichester, United Kingdom.

Naiman, R.J. and J.R. Sedell. 1979. Relationships between metabolic parameters and stream order in Oregon. Canadian Journal of Fisheries and Aquatic Science 37: 834-847.

Narver, D.W. 1971. Effects of logging debris on fish production, p. 100-111. *In* J.T. Krygier and J.D. Hall (eds.), Proceedings of a Symposium on Forest Land Uses and Stream Environment, October 19-20, 1970. Oregon State University, Corvallis.

National Research Council. 1992. Restoration of aquatic ecosystems. National Academy Press, Washington, DC.

National Research Council. 1996. Upstream: Salmon and Society in the Pacific Northwest. National Academy Press, Washington, DC.

Nehlsen, W., J.E. Williams, and J.A. Lichatowich. 1991. Pacific salmon at the crossroads: stocks at risk from California, Oregon, Idaho and Washington. Fisheries 16: 4-21.

Nelson, R.L., M.L. McHenry, and W.S. Platts. 1991. Mining, p. 425-457. *In* W.R. Meehan (ed.), Influences of Forest and Rangeland Management on Salmonid Fishes and Their Habitats. American Fisheries Society Special Publication 19, Bethesda, Maryland.

Newbold, J.D., D.C. Erman, and K.B. Roby. 1980. Effects of logging on macroinvertebrates in streams with and without buffer strips. Canadian Journal of Fisheries and Aquatic Sciences 37: 1076-1085.

Nickelson, T.E., J.D. Rodgers, S.L. Johnson, and M.F. Solazzi. 1991. Seasonal changes in habitat use by juvenile coho salmon (*Oncorhynchus kisutch*) in Oregon coastal streams. Canadian Journal of Fisheries and Aquatic Sciences 49: 783-789.

Nickelson, T. E., M. F. Solazzi, S. L. Johnson, and J. D. Rodgers. 1992. Effectiveness of selected stream improvement techniques to create suitable summer and winter rearing habitat for juvenile coho salmon in Oregon coastal streams. Canadian Journal of Fishery and Aquatic Sciences 49:790-794.

Nickelson, T.E., J.W. Nicholas, A.M. McGie, R.B. Lindsay, D.L. Bottom, R.J. Kaiser, and S.E. Jacobs. 1992. Status of anadromous salmonids in Oregon coastal basins. Report to the Governor, Oregon Department of Fish and Wildlife. Corvallis.

Noggle, C.C. 1978. Behavioral, physiological and lethal effects of suspended sediment on juvenile salmonids. M.S. thesis, University of Washington. Seattle.

Nolan, M.L. 1993. Historical geography, p. 6-17. *In* P.L. Jackson and A.J. Kimerling (eds.), Atlas of the Pacific Northwest. Oregon State University Press, Corvallis.

Norris, L.A., H.W. Lorz, and S.V. Gregory. 1991. Forest chemicals, p. 207-296. *In* W.R. Meehan (ed.), Influences of Forest and Rangeland Management on Salmonid Fishes and Their Habitats. American Fisheries Society Special Publication 19, Bethesda, Maryland.

Northam, R.M. 1993. Transportation, p. 25-30. *In* P.L. Jackson and A. J. Kimerling (eds.). Atlas of the Pacific Northwest. Oregon State University Press, Corvallis.

Oregon Department of Forestry. 1988. Assessment of Oregon's forests. A collection of papers published by the Oregon State Department of Forestry. Salem.

Pacific Rivers Council. 1993. Entering the watershed: an action plan to protect and restore America's river ecosystems and biodiversity. Pacific Rivers Council. 319 p.

Palmisano, J.F., R.H. Ellis, and V.W. Kaczynski. 1993. The impact of environmental and management factors on Washington's wild anadromous salmon and trout. Washington Forest Protection Association and Washington Department of Natural Resources, Olympia.

Pearcy, W.G. 1992. Ocean Ecology of North Pacific Salmonids. Washington Sea Grant Program, University of Washington Press, Seattle.

Pearcy, W.G. 1996. Salmon production in changing ocean domains, p. 331-352. *In* D.J. Stouder, P.A. Bisson, and R.J. Naiman (eds.), Pacific Salmon and Their Ecosystems: Status and Future Options. Chapman and Hall, New York.

Pease, J.R. 1993. Land use and ownership, p. 31-39. *In* P.L. Jackson and A.J. Kimerling (eds.), Atlas of the Pacific Northwest. Oregon State University Press, Corvallis.

Peterson, N.P. 1982. Immigration of juvenile coho salmon (*Oncorhynchus kisutch*) into riverine ponds. Canadian Journal of Fisheries and Aquatic Sciences 39: 1308-1310.

Peterson, N.P. and L.M. Reid. 1984. Wall-base channels: their evolution, distribution, and use by juvenile coho salmon in the Clearwater River, Washington, p. 215-226. *In* J.M. Walton and D.B. Houston (eds), Proceedings of the Olympic Wild Fish Conference, Port Angeles, Washington, USA, March 23-25, 1983.

Peterson, N., A. Hendry, and T.P. Quinn. 1992. Assessment of cumulative effects on salmonid habitat: some suggested parameters and target conditions. Timber, Fish, and Wildlife TFW-F3-92-001. Olympia, Washington.

Platts, W.S. 1991. Livestock grazing, p. 389-423. *In* W.R. Meehan (ed.), Influences of Forest and Rangeland Management on Salmonid Fishes and Their Habitats. American Fisheries Society Special Publication 19, Bethesda, Maryland.

Platts, W.S. and M.L. McHenry. 1988. Density and biomass of trout and char in western streams. General Technical Report INT-241, USDA Forest Service, Intermountain Research Station. Ogden, Utah.

Platts, W.S., R.J. Torquemada, M.L. McHenry, and C.K. Graham. 1989. Changes in salmon spawning and rearing habitat from increased delivery of fine sediment to the South Fork Salmon River, Idaho. Transactions of the American Fisheries Society 118: 274-283.

Poff, N.L. and J.V. Ward. 1989. Implications of streamflow variability and predictability for lotic community structure: a regional analysis of streamflow patterns. Canadian Journal of Fisheries and Aquatic Sciences 46: 1805-1818.

Price, O. 1902. The influence of forestry upon the lumber industry, p. 309-312. *In* Yearbook of Agriculture, US Department of Agriculture. Government Printing Office, Washington, DC.

Raedeke, K.J. (ed.). 1988. Streamside Management: Riparian Wildlife and Forestry Interactions. Contribution no. 59, Institute of Forest Resources, University of Washington. Seattle.

Ralph, S.C., G.C. Poole, L.L. Conquest, and R.J. Naiman. 1994. Stream channel condition and in-stream habitat in logged and unlogged basins of western Washington. Canadian Journal of Fisheries and Aquatic Sciences 51: 37-51.

Reeves, G.H., K.M. Burnett, F.H. Everest, J.R. Sedell, D.B. Hohler, and T. Hickman. 1991a. Responses of anadromous salmonid populations and physical habitat to stream restoration in Fish Creek, Oregon, 1983-1990. US Department of Energy, Bonneville Power Administration, Division of Fish and Wildlife. Portland, Oregon.

Reeves, G. E., J. D. Hall, T. D. Roelofs, T. L. Hickman, and C. O. Baker. 1991b. Rehabilitating and modifying stream habitats, p. 519-558. *In* W.R. Meehan (ed.) Influences of forest and rangeland management on salmonid fishes and their habitats. American Fisheries Society Special Publication 19, Bethesda, Maryland. 751 p.

Reice, S. R. 1994. Nonequilibrium determinants of biological community structure. American Scientist 82: 424-435.

Reice, S.R., R.C. Wissmar, and R.J. Naiman. 1990. Disturbance regimes, resilience, and recovery of animal communities and habitats in lotic ecosystems. Environmental Management 14: 647-659.

Reisenbichler, R.R. 1996. Genetic factors contributing to declines of anadromous salmonids in the Pacifc Northwest, p. 223-244. *In* D.J. Stouder, P.A. Bisson, and R.J. Naiman (eds.), Pacific Salmon and Their Ecosystems: Status and Future Options. Chapman and Hall, New York.

Resh, V.H., A.V. Brown, A.P. Covich, M.E. Gurtz, H.W. Li, G.W. Minshall, S.R. Reice, A.L. Sheldon, J.B. Wallace, and R.C. Wissmar. 1988. The role of disturbance in stream ecology. Journal of the North American Benthological Society 7: 433-455.

Ringler, N.H. and J.D. Hall. 1975. Effects of logging on water temperature and dissolved oxygen in spawning beds. Transactions of the American Fisheries Society 104: 111-121.

Rich, W.H. 1948. A survey of the Columbia River and its tributaries with special reference to the management of its fishery resources. Special Scientific Report 51, US Fish and Wildlife Service. Washington, DC.

Robinson, W. L. 1987. An industrial forester's perspective on streamside management, p. 289-294. *In* E.O. Salo and T.W. Cundy (ed.). Streamside Management: Forestry and Fishery Interactions. Contribution no. 57, Institute of Forest Resources, University of Washington. Seattle.

Ryan, S.E. and G.E. Grant. 1991. Downstream effects of timber harvesting on channel morphology in the Elk River basin, Oregon. Journal of Environmental Quality 20: 60-72.

Salo, E.O. and T.W. Cundy (eds.), 1987. Streamside Management: Forestry and Fishery Interactions. Contribution no. 57, Institute of Forestry Resources, University of Washington. Seattle.

Scrivener, J.C. 1988. Changes in composition of the streambed between 1973 and 1985 and the impacts on salmonids in Carnation Creek, p. 59-65. *In* T.W. Chamberlin (ed.), Proceedings of the Workshop: Applying 15 Years of Carnation Creek Results. Pacific Biological Station, Nanaimo, British Columbia.

Sedell, J.R. and R.L. Beschta. 1991. Bringing back the "bio" in bioengineering. American Fisheries Society Symposium 10: 160-175.

Sedell, J.R. and K.J. Luchessa. 1982. Using the historical record as an aid to salmonid habitat enhancement, p. 210-223. *In* N.B. Armantrout (ed.), Symposium on Acquisition and Utilization of Aquatic Habitat Inventory Information. American Fishery Society, Western Division.

Sedell, J.R. and J.L. Froggatt. 1984. Importance of streamside forests to large rivers: the isolation of the Willamette river, Oregon, USA, from its floodplain by snagging and streamside forest removal. Verhandlungen Internationale Vereinigen Limnologie 22: 1828-1834.

Sedell, J.R., G.H. Reeves, and K.M. Burnett. 1994. Development and evaluation of aquatic conservation strategies. Journal of Forestry 92: 28-31.

Sedell, J.R., G.H. Reeves, F.R. Hauer, J.A. Stanford, and C.P. Hawkins. 1990. Role of refugia in recovery from disturbance: modern fragmented and disconnected river systems. Environmental Management 14: 711-724.

Sedell, J.R., R.J. Steedman, H.A. Regier, and S.V. Gregory. 1991. Restoration of human impacted land-water ecotones, p. 110-129. *In* M.M. Holland, P.G. Risser, and R.J. Naiman (eds.), Ecotones: the Role of Landscape Boundaries in the Management and Restoration of Changing Environments. Chapman Hall, New York.

Sedell, J.R. and F.H. Everest. 1991. Historic changes in pool habitat for Columbia River Basin salmon under study for TES listing. United States Forest Service, Pacific Northwest Research Station, Draft General Technical Report. Portland, Oregon.

Seegrist, D.W. and R. Gard. 1972. Effects of floods on trout in Sagehen Creek, California. Transactions of the American Fisheries Society 101: 478-482

Seiler, D. 1989. Differential survival of Grays Harbor basin anadromous salmonids: water quality implications, p. 123-135 *In* C.D. Levings, L.B. Holtby, and M.A. Henderson (eds.), Proceedings of the National Workshop on Effects of Habitat Alteration on Salmonid Stocks. Canadian Special Publication of Fisheries and Aquatic Sciences 105.

Sherwood, C.R., D.A. Jay, R.B. Harvey, P. Hamilton, and C.A. Simenstad. 1990. Historical changes in the Columbia River estuary. Progressive Oceanographer 25: 299-357.

Shirvell, C.S. 1990. Role of instream rootwads as juvenile coho salmon (*Oncorhynchus kisutch*) and steelhead trout (*O. mykiss*) cover habitat under varying streamflows. Canadian Journal of Fisheries and Aquatic Sciences 47: 852-861.

Shortreed, K.R.S. and J.G. Stockner. 1983. Periphyton biomass and species composition in a coastal rainforest stream in British Columbia: effects of environmental changes caused by logging. Canadian Journal of Fisheries and Aquatic Sciences 40: 1887-1895.

Shreffler, D.K., C.A. Simenstad, and R.M. Thom. 1990. Temporary residence by juvenile salmon of a restored estuarine wetland. Canadian Journal of Fisheries and Aquatic Sciences 47: 2079-2084.

Simenstad, C.A., K.L. Fresh, and E.O. Salo. 1982. The role of Puget Sound and Washington coastal estuaries in the life history of Pacific salmon: an unappreciated function, p. 343-364. *In* V.S. Kennedy (ed.), Estuarine Comparisons. Academic Press, Toronto.

Simenstad, D.A., D.A. Jay, and C.R. Sherwood. 1992. Impacts of watershed management on land-margin ecosystems: the Columbia River Estuary as a case study, p. 266-306. *In* R.J. Naiman (ed.), Watershed Management: Balancing Sustainability and Environmental Change. Springer-Verlag, New York.

Smith, H. M. 1895. Notes on a Reconnaissance of the Fisheries of the Pacific Coast of the United States in 1984. Bulletin of the United States Fish Commission for 1894. p. 223-288.

Smoker, W.A. 1955. Effects of streamflow on silver salmon production in western Washington. Ph.D. dissertation, University of Washington. Seattle.

Stanford, J.A. and J.V. Ward. 1988. The hyporheic habitat of riverine ecosystems. Nature 335: 64-66.

Stanford, J.A. and J.V. Ward. 1992. Management of aquatic resources in large catchments: recognizing interactions between ecosystem connectivity and environmental disturbance, p. 91-124. *In* R.J. Naiman (ed.), Watershed Management: Balancing Sustainability and Environmental Change. Springer-Verlag, New York.

Stewart, H. 1977. Indian fishing: early methods on the Northwest coast. Douglas and McIntyre, Vancouver, British Columbia.

Stone, L. 1897. The artificial propagation of salmon on the Pacific Coast of the United States, with notes on the natural history of the Quinnault salmon. Bulletin of the United States Fish Commission for 1896.

Sullivan, K., T.E. Lisle, C.A. Dolloff, G.E. Grant, and L.M. Reid. 1987. Stream channels: the link between the forests and fishes, p. 39-97. *In* E.O. Salo and T.W. Cundy (eds.). Streamside Management: Forestry and Fishery Interactions. Contribution no. 57, Institute of Forest Resources, University of Washington. Seattle.

Swanson, F.J. and G.W. Lienkaemper. 1978. Physical consequences of large organic debris in Pacific Northwest streams. General Technical Report PNW-69, USDA Forest Service, Pacific Northwest Forest and Range Experiment Station. Portland, Oregon.

Swanson, F.J., R.L. Fredrickson, and F.M. McCorison. 1982. Material transfer in a western Oregon forested watershed, p. 231-266. *In* R.L. Edmonds (ed.), Analysis of Coniferous Forest Ecosystems in the Western United States. Hutchinson Ross Publishing Co., Stroudsburg, Pennsylvania.

Swanson, F.J., S.V. Gregory, J.R. Sedell, and A.G. Campbell. 1982. Land-water interactions: the riparian zone, p. 267-291. *In* R.L. Edmonds (ed.), Analysis of Coniferous Forest Ecosystems in the Western United States. Hutchinson Ross Publishing Co., Stroudsburg, Pennsylvania.

Swanson, F.J., L.E. Benda, S.H. Duncan, G.E. Grant, W.F. Megahan, L.M. Reid, and R.R. Ziemer. 1987. Mass failures and other sediment production in Pacific Northwest landscapes, p. 9-38. *In* E.O. Salo and T.W. Cundy (eds.), Streamside Management: Forestry and Fishery Interactions. Contribution no. 57, Institute of Forest Resources, University of Washington. Seattle.

Swanson, F.J., J.F. Franklin, and J.R. Sedell. 1990. Landscape patterns, disturbance, and management in the Pacific Northwest, USA, p. 191-213. *In* I.S. Zonneveld and R.T. Forman (eds.), Trend in Landscape Ecology. Springer-Verlag, New York.

Swanston, D.N. 1991. Natural processes. American Fisheries Society Special Publication 19: 139-179.

Thomas, J.W., M.G. Raphael, R.G. Anthony, E.D. Forsman, A.G. Gunderson, R.S. Holthausen, B.G. Marcot, G.H. Reeves, J.R. Sedell, and D.M. Solis. 1993. Viability assessments and management considerations for species associated with late-successional and old-growth forests of the Pacific Northwest. Report of the Scientific Analysis Team, US Forest Service, US Department of Agriculture. Portland, Oregon.

Triska, F.J., V.C. Kennedy, R.J. Avazino, G.W. Zellweger, and K.E. Bencala. 1989. Retention and transport of nutrients in a third order stream: hyporheic processes. Ecology 70: 1893-1905.

Triska, F.J., V.C. Kennedy, R.J. Avanzino, G.W. Zellweger, and K.E. Bencala. 1990. In situ retention-transport response to nitrate loading and storm discharge in a third-order stream. Journal of the North American Benthological Society 9: 229-239.

Triska, F.J., J.R. Sedell, K. Cromack, Jr., S.V. Gregory, and F.M. McCorison. 1984. Nitrogen budget for a small coniferous forest stream. Ecological Monographs 54: 119-140.

Trotter, P.C., P.A. Bisson, and B.R. Fransen. 1993. Status and plight of the searun cutthroat trout, p. 203-212. *In* J.G. Cloud and G.H. Thorgaard (eds.), Genetic Conservation of Salmonid Fishes. Plenum Press, New York.

Tschaplinski, P.J. and G.F. Hartman. 1983. Winter distribution of juvenile coho salmon (*Oncorhynchus kisutch*) before and after logging in Carnation Creek, British Columbia, and some implications for overwinter survival. Canadian Journal of Fisheries and Aquatic Sciences 40: 452-461.

US Department of the Interior Bureau of Land Management. 1991. Riparian-wetland initiative for the 1990's. BLM/WO/GI-91/001+4340. Washington, DC.

Vannote, R.L., G.W. Minshall, K.W. Cummins, J.R. Sedell, and C.E. Cushing. 1980. The river continuum concept. Canadian Journal of Fisheries and Aquatic Sciences 37: 130-137.

Van Sickle, J. and S.V. Gregory. 1990. Modeling inputs of large woody debris to streams from falling trees. Canadian Journal of Forest Research 20: 1593-1601.

Walters, C.J. and P. Cahoon. 1985. Evidence of decreasing spatial diversity in British Columbia salmon stocks. Canadian Journal of Fisheries and Aquatic Sciences 42: 1033-1037.

Ward, G.W., K.W. Cummins, R.W. Speaker, A.K. Ward, S.V. Gregory, and T.L. Dudley. 1982. Habitat and food resources for invertebrate communities in South Fork Hoh River, Olympic National Park, Washington, p. 9-14. *In* E.E. Starkey, J.F. Franklin, and J.W. Matthews (eds.), Ecological Research in National Parks of the Pacific Northwest: Proceedings. Oregon State University Forest Research Laboratory, Corvallis.

Warren, C.E., J.H. Wales, G.E. Davis, and P. Doudoroff. 1964. Trout production in an experimental stream enriched with sucrose. Journal of Wildlife Management 28: 617-660.

Washington Department of Fisheries, Washington Department of Wildlife, and Western Washington Treaty Indian Tribes. 1993. 1992 Washington State salmon and steelhead stock inventory. Available from the Washington Department of Fisheries and Wildlife, Olympia, WA. 212 p.

Washington Forest Practices Board. 1993. Standard methodology for conducting watershed analysis. Version 2.0. Washington Department of Natural Resources. Olympia, Washington.

Wilderness Society. 1993. The living landscape, Volume 2. Pacific Salmon on Federal Lands. Bolle Center for Forest Ecosystem Management, The Wilderness Society, Seattle, Washington.

Wilzbach, M.A. 1985. Relative roles of food abundance and cover in determining the habitat distribution of stream-dwelling cutthroat trout (*Salmo clarki*). Canadian Journal of Fisheries and Aquatic Sciences 42: 1668-1672.

Wissmar, R.C. and F.J. Swanson. 1990. Landscape disturbances and lotic ecotones, p. 65-89. *In* R.J. Naiman and H. Decamps (eds.), Ecology and Management of Aquatic-Terrestrial Ecotones. Parthenon Press, London.

Wurtsbaugh, W.A. and G.E. Davis. 1977. Effects of temperature and ration level on the growth and food conversion efficiency of *Salmo gairdneri* Richardson. Journal of Fish Biology 11: 87-98.

Ziemer, R.R., J. Lewis, T.E. Lisle, and R.M. Rice. 1991. Long-term sedimentation effects of different patterns of timber harvesting. International Association of Hydrologic Science 203:143-150.

The Role of Harvest Management in the Future of Pacific Salmon Populations: Shaping Human Behavior to Enable the Persistence of Salmon

Phillip R. Mundy

Abstract

A harvest management paradigm suitable for sustainable use of Pacific salmon (*Oncorhynchus* spp.) has been developed by observing and contrasting interactions between harvest, habitat, and salmon productivity in three large salmon-producing areas that have long-term records of commercial salmon landings. The drainages of Bristol Bay (Alaska), the Fraser River (British Columbia, Canada), and the Columbia River of the northwestern United States have three things in common: (1) long historical records of salmon exploitation, (2) a history of unlimited exploitation of salmon followed by decreases in abundance, and (3) a contemporary harvest management regime limiting exploitation. These localities differ distinctly in the nature of the freshwater spawning and rearing habitat available to salmon, which ranges from pristine to restored to badly damaged. The present status of the principal commercially exploited salmon species in each locality also differs, with populations in the pristine and restored habitat returning to states of high productivity while some of those in the badly damaged habitat in the Columbia River are currently listed as federally endangered species. Human activities in the Columbia River Basin, such as agriculture and hydroelectric development, both precede and parallel the development of chinook harvest. Harvest management actions in Bristol Bay and the Fraser River have proven effective in restoring salmon in the presence of suitable habitat. The qualitative attributes of the effective salmon harvest management paradigm within an ecosystem context include biological diversity, transboundary scope, and interdisciplinary staffing. The quantitative attributes of the effective harvest management paradigm are escapement goals, geographic gradients in fishing mortalities, and zero-sum mortality allocation.

Introduction

Natural resource management is an often humbling exercise in shaping human behavior to enable the persistence of those parts of our ecosystem that we all exploit. In salmon fisheries management, I am constantly reminded that we do not manage fish, we manage people. In looking for the origins of, and solutions to, the problems afflicting salmon and their ecosystems, we must remember that humans are but small parts of a larger natural-cultural system that owes its existence to the health of its ecosystems. In particular, salmon fisheries are human activities that are directly dependent on healthy functioning ecosystems.

Some of our ecosystems have deteriorated to the point where a number of the fisheries for Pacific salmon (*Oncorhynchus* spp.) are sitting squarely on the train tracks of history, and the sound of the train whistle is growing louder. The warning whistle is being blown by the incontrovertible evidence of the decline and extirpation of Pacific salmon throughout their range in the contiguous United States (US) (Nehlsen et al. 1991, Norman and King 1992). Harvesting and human activities contributing to freshwater habitat degradation have dealt many Pacific salmon populations a series of damaging, and often fatal, blows in North America (Netboy 1974, Northwest Power Planning Council [NPPC] 1986). At present, four groups of salmon are listed as endangered or threatened by the US government: (1) Sacramento River winter chinook (*O. tshawytscha*), (2) Snake River basin sockeye (*O. nerka*), and Snake River basin fall chinook, (3) Snake River basin spring chinook, and (4) Snake River summer chinook. The actual number of listed salmon populations represents but a small fraction of the number of populations in California (Mills et al. 1996), Oregon (Kostow 1996), and Washington (Johnson et al. 1996) that may qualify for consideration for federal listing as endangered or threatened species (Nehlsen et al. 1991, Lichatowich 1993). Given that Pacific salmon continue to decline despite the experience and efforts of more than a century, it is reasonable to look for ways to improve our abilities to protect and restore Pacific salmon.

Understanding the interactions and dependencies between harvest, habitat health, and salmon population productivity (Ricker 1954) is essential to improving our abilities to identify and implement salmon restoration efforts. I address three key questions: (1) What is the role of harvest management in the decline and extirpation of Pacific salmon populations, (2) what is the role of harvest management in recovery of the salmon resource, and (3) is it possible for harvest management actions to compensate for losses to salmon populations due to other human activities (e.g., habitat degradation)?

In search of answers to these questions, I compare the population status of two salmon species in three geographic localities in North America. The localities have three things in common: (1) long historical records of salmon exploitation, (2) a history of unlimited exploitation followed by declining abundance, and (3) a contemporary harvest management regime limiting exploitation. However, these localities are distinctly different in the nature of the freshwater spawning and rearing habitat available to salmon, which ranges from pristine to restored to badly damaged, for the drainages of Bristol Bay (Alaska), Fraser River (British Columbia, Canada), and the Columbia River of the northwestern US, respectively.

In Bristol Bay and the Fraser River, recent annual harvests of the principal commercial species, sockeye salmon (Figs. 1 and 2) have reached or exceeded the levels seen in the peak of the earliest commercial fisheries (Northcote and Atagi 1996, Wertheimer 1996). At the opposite end of the harvest spectrum, landings of the principal commercial species in the Columbia River, chinook salmon (Fig. 3), have reached record low levels despite production from hatcheries.

Examination of the three localities shows that habitat loss and degradation, and unlimited fishing emerge as parallel companions of the initial decline in population numbers of the principal commercial salmon species in the Fraser and Columbia rivers, whereas only unlimited harvest was associated with the initial declines in Bristol Bay. Effective control of harvest figured

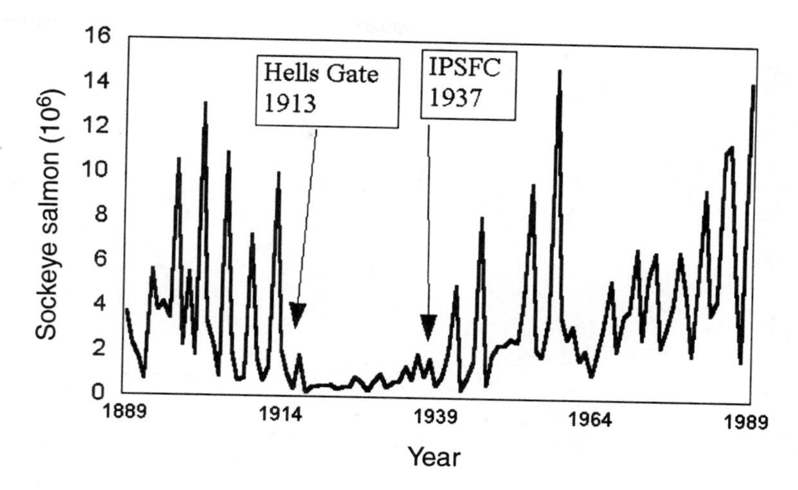

Figure 1. Numbers landed (10^6) of Fraser River (Canada) sockeye salmon, 1889–1989. Years of initial habitat obstruction at Hells Gate and the creation of the International Pacific Salmon Fisheries Commission (IPSFC) are indicated. Sources: Gilhousen (1992, Table 1, column III, p. 11) and Roos (1991, Appendix H2, p. 413).

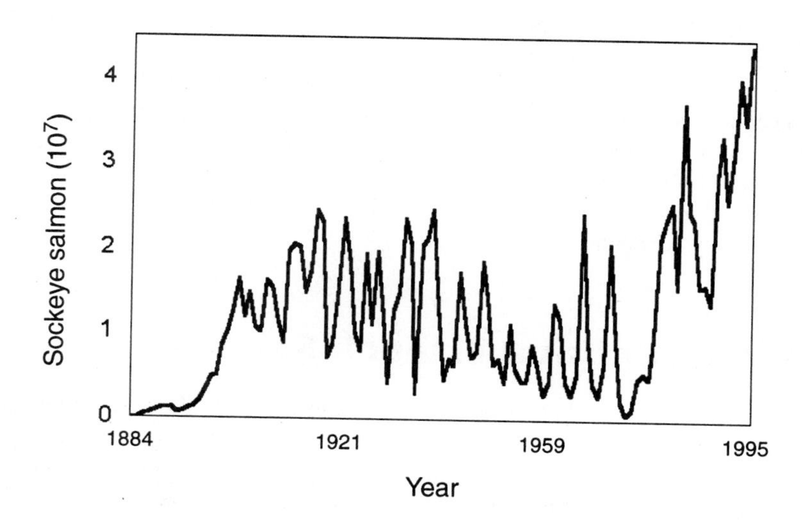

Figure 2. Total numbers landed (10^7) of Bristol Bay (Alaska) sockeye salmon, 1884–1995. Source: Alaska Department of Fish and Game, Juneau.

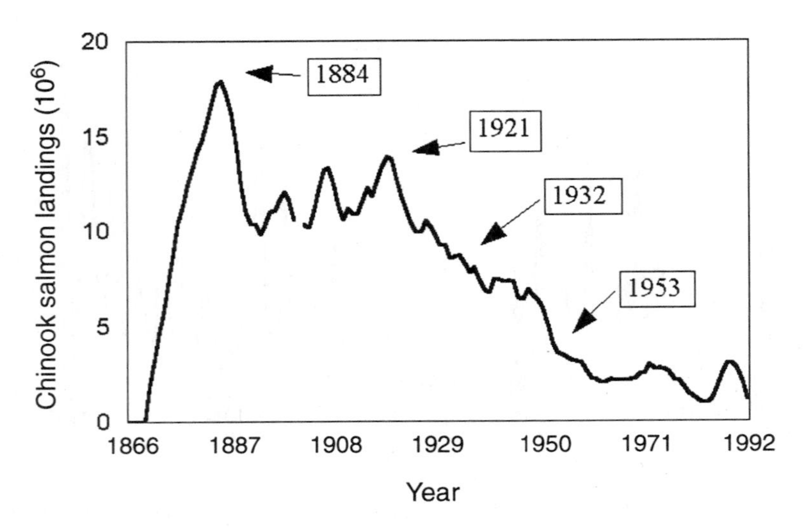

Figure 3. Five-year moving averages of total annual landings (10^6 kg) of Columbia River (Washington, Oregon, Idaho, Canada) chinook salmon, 1866–1992. Landing sources: 1866–1936, Craig and Hacker (1940); 1937, Cleaver (1951); 1938–91, Norman and King (1992).

prominently in the recoveries of sockeye salmon in both Bristol Bay and the Fraser River, and recovery of lost habitat was a prominent feature of Fraser River sockeye recovery. In the Columbia River, continuing habitat losses combine with ineffective harvest regulation as probable causes for the continuing failure of Columbia River chinook salmon. As an object lesson in the impacts of human development on all salmon-bearing ecosystems, the Columbia River Basin is discussed in detail. Following the historical analysis, I explore the beginnings of an effective harvest management paradigm suitable to sustainable use of Pacific salmon through management within an ecosystem context.

Columbia River Basin Chinook Salmon

By 1975, the end of the major dam construction era, the total length of spawning and rearing habitat accessible to chinook salmon in the Columbia River Basin had declined from original values by 48% (from 29,051 km to 15,117 km). This loss of habitat was more severe for spring and summer chinook than for fall chinook salmon (Table 1). Please note that length of freshwater habitat accessible is only a crude measure of productive capacity for salmon since not every portion of every accessible stream and river is equally suitable for rearing and spawning. Similarly, the percent reduction in length of habitat accessible is a minimum measure of loss in productive capacity. Much of the habitat that remains accessible has been degraded by human activities.

Agricultural irrigation, logging, and commercial fishing developed simultaneously in the Columbia River Basin, followed closely by mining and hydroelectric development. The critical human mass necessary to produce statehood for Oregon came in 1859, for Washington in 1889,

Table 1. Changes in amount of chinook salmon habitat in the Columbia River as length of spawning and rearing habitat accessible in kilometers for three life-history types. Source: Northwest Power Planning Council (1986).

Type	Original	1975	% original	% lost
Spring	17,088	8,718	51	49
Summer	8,002	3,650	46	54
Fall	3,961	2,749	69	31
Totals	29,051	15,117	52	48
Average			55	45

and for Idaho in 1890. Each admission to the Union was preceded by substantial influxes of settlers, agricultural development, and other extractive uses of natural resources. The use of irrigation dates to 1840, with organized water projects for Washington (Walla Walla) and Oregon (Hood River, John Day, and Umatilla) being in place by the 1860s (NPPC 1986). The role of irrigation in the reduction and extirpation of salmon populations is well known (Corely 1963, Stober et al. 1979, Whitney and White 1984, NPPC 1986). When irrigated agricultural enterprise in the Columbia River Basin blossomed at the beginning of the 20th century (Table 2), it was the culmination of more than half a century of steady growth.

Logging reduces salmon productivity by rendering the streams and rivers unsuitable for salmon (Reeves et al. 1993, Ralph et al. 1994). The chronology of logging in the Columbia River Basin follows much the same pattern seen in the development of irrigated agriculture. Logging dates to 1827, with 37 saw mills active by 1850. Intensive logging and attendant construction of splash dams and other instream modifications for log transportation date to about 1880 (NPPC 1986).

The growth in logging from ~1880 was stimulated in part by the growth of the mining industry. Damage to salmon habitat by displacement of spawning gravel from stream beds by placer mining was extensive in the lower Snake River basin and in the John Day and Powder rivers of Oregon (NPPC 1986).

Table 2. Hectares under irrigation, and cubic meters (x 10^3) of water delivered to agricultural enterprises by the Bureau of Reclamation, Columbia River Basin. Modified from Northwest Power Planning Council (1986), Appendix D, pages 178 and 181.

Year	ha	m^3 x 10^3
1889	161,874	
1900	202,343	
1910	930,777	
1925	1,173,588	
1947		3,255
1966	2,670,925	
1967		10,343
1979		14,374
1980	3,075,611	
1981		13,227

Chronologically trailing other sources of salmon habitat loss only slightly, the first hydroelectric project was constructed in the Columbia River Basin on the Willamette River, Oregon, in 1888, and two more were built before the beginning of the 20th century (NPPC 1986). Hydroelectric dams decrease salmon productivity by denying access to spawning and rearing habitat, and by rendering the river unsuitable for migration of adults and juveniles (Raymond 1988, Bell 1991). Owing to the locations of these first dams, loss of salmon habitat to hydroelectric development was relatively minor prior to 1900. Substantial loss of salmon habitat from Columbia River tributaries occurred because of hydroelectric development from 1900 until 1933, when the first of the large hydroelectric dams was put in service on the main Columbia River at Rock Island, Washington (Public Power Council [PPC] 1995). Prior to 1933, a total of 20 hydroelectric projects >10 MW were constructed on Columbia River tributaries (one federal, three publicly owned, and 16 investor-owned) (PPC 1995). In addition, two federal hydroelectric projects <10 MW (PPC 1995), and numerous privately funded small hydroelectric projects (NPPC 1986) were built during the same time period. As of 1986, 136 hydroelectric projects of all types were operating in the Columbia River Basin (NPPC 1986).

Harvests impact salmon productivity directly by reducing numbers in the spawning populations, and indirectly by reducing phenotypic diversity, which affects factors important to basic productivity (e.g., average number of eggs per female) (Russell 1931 in Cushing 1983, Miller 1957, Ricker 1981). When fishing removes enough spawners from a population to cause it to decline, overfishing has occurred.

Observations consistent with overfishing of the salmon runs by the commercial fishery of the lower Columbia River occurred during the 1870s when Native American harvesters, who fished upriver from the commercial fisheries, found they could no longer meet their basic subsistence needs for salmon (letter to the Commissioner of Indian Affairs from J. Simms, US Indian Agent, Colville Agency, August 23, 1877). Seventeen years later, biologists examined the lower river fisheries in order to find explanations for the sharp declines in salmon (possibly spring chinook and sockeye) returns to the Yakima River in Washington State (McDonald 1894). Information collected from the commercial fisheries of the lower Columbia River, which would have permitted a quantitative assessment of its impacts on the salmon populations of individual tributaries, was unavailable during the time of McDonald (1894). Thus, putting numerical values on the relative roles of fishing and habitat in calculating salmon productivity remains problematic.

As late as 1936, salmon fisheries were an important part of the regional economy, employing 3,820 harvesters and generating $10 million annually (Craig and Hacker 1940). Although Craig and Hacker (1940) recognized that preventing overfishing was important, the authors emphasized that maintaining suitable spawning and nursery grounds was of paramount importance to the success of salmon fishing in achieving conservation. Craig and Hacker (1940) list human population growth, logging, mining, hydroelectric power, and flood control and navigation as causes for the decline in salmon resources during the 19th and early 20th centuries.

Factors contributing to the first major Columbia River chinook salmon harvest declines from 1884 to 1889 include the reduction of late spring and early summer chinook by fishing, and reductions in fishing effort resulting from falling demand for the relatively highly priced Columbia River salmon (Craig and Hacker 1940). Species identification of the early landings was not particularly accurate, which suggests the largest reported landing of Columbia chinook in 1883 could have included species other than chinook (Craig and Hacker 1940).

A contemporary of Craig and Hacker, Rich (1941), linked habitat declines to fishing pressure as a source of decline: "The way in which the Chinook salmon runs have held up under the excessive exploitation and a constant reduction in the available spawning area is remarkable." In the same paper, Rich (1941) issued a prophetic warning to fishers, laymen, and administrators about the futility of trying to replace lost salmon spawning and rearing habitats with hatcheries.

Other contemporaries of Craig and Hacker also recognized the interaction between habitat loss and the effects of fishing in determining salmon population size. With regard to Columbia River blueback (sockeye) salmon, Johnson et al. (1948) stated, "The blueback is . . . in an advanced stage of depletion . . . a very intense fishery, coupled with elimination of the majority of the important spawning grounds, has reduced the populations to a fraction of their former abundance." However, Johnson et al. (1948) were not concerned about trends in escapement of chinook as of 1935.

Since 1936, several authors have concluded that overfishing contributed to the decline of Columbia River chinook salmon. Thompson (1951) documented declines in nominal landings per unit effort of spring and summer chinook between 1876 and 1919 that were clearly associated with declines in actual chinook population size (Fig. 3, Chapman 1986). In a comprehensive review of the historical evidence for overfishing of Columbia River salmon, Chapman (1986) joined Thompson (1951) in concluding that overfishing was a factor in the decline of chinook.

Historian Netboy (1974) reported that chinook salmon runs of the Columbia River were overfished and in radical decline after 1885. In addition, Netboy (1974) recognized the role of habitat losses in salmon declines by citing the US Army Corps of Engineers "308 Report" of 1948, which documented the existence of over 300 dams of all types in the Columbia River Basin at that time.

After 1941, observations supported the negative impact of fishing on Columbia River chinook salmon. For example, Van Hyning (1973) documented the increase of ocean fishing as the main contributor to the decline of Columbia River fall chinook from 1938 to 1959. By this time, fall chinook had become the dominant race of chinook in the Columbia River drainages. The ocean fishery clearly had a negative effect on run sizes during this period. It is noteworthy that Van Hyning's (1973) analysis incorporated indirect measures of the effects of habitat degradation.

Examination of trends in the rise and fall of Columbia River Basin chinook salmon is facilitated by a 5-year moving average of the annual landings, which is used to remove the short-term noise (Fig. 3). Trends are grouped into five eras: 1866, 1884, 1921, 1932, and 1953. From its inception to about 1883, the fishery was probably reaping the benefits of harvesting relatively lightly exploited populations of chinook salmon. Although Craig and Hacker (1940) estimated annual aboriginal harvest at 8.263 million kg of chinook (~910,000 fish), many of the aboriginal peoples had perished in epidemics prior to the growth of the commercial fisheries.

As an apparent response to exploitation, populations declined to lower levels during the second era (~1884). This decline coincided with declining salmon markets, reduced fishing effort, and substantial loss and degradation of spawning and rearing habitat. Annual landings during the last 5 years of this era were on the order of 1.5–2 million chinook salmon, based on a nominal average weight of 9.08 kg per chinook. Chapman (1986) used 10.45 kg for spring/summer chinook salmon. From 1884 until the end of the second era in 1920, the fishery was working at an apparent annual equilibrium landings level on the order of 1.25 million chinook salmon.

The economics of World War I caused an increase in fishing effort in the river, in the mouth of the river, and on the ocean (Craig and Hacker 1940). Increased demand for salmon products resulted in the final peak of the fishery. The year 1921, as fixed by the point where the 5-year moving average of chinook landings dropped below an annual harvest of 13.62 million kg (~1.5 million chinook), clearly marked the point where the Columbia River Basin chinook populations started the slide toward extirpation (Fig. 3). At this time, the sum of the effects of accelerating habitat loss and degradation and ineffective harvest management regimes converged to drive salmon population numbers below the critical point where they would have been able to replace their numbers from one generation to the next.

For the next three eras (1921-present), it is likely that overfishing joined forces with rapidly accelerating habitat degradation to cause lasting reductions in chinook salmon population levels (Craig and Hacker 1940, Rich 1941, Van Hyning 1973). During the third era (1921–31), Columbia River chinook landings experienced a decline as sharp as that marking the beginning of the second era in 1884. Despite an increase in fishing effort during this time period, the decline in landings in the third era apparently resulted from decreased productive capacity of the populations (Craig and Hacker 1940).

The 5-year moving average of landings dropped below 9.1 million kg (~1 million chinook salmon) at the beginning of the fourth era in 1932, a year which also witnessed the beginning of large hydroelectric dam development in the mainstem of the Columbia River. Operations began at Rock Island in 1933, Bonneville in 1938, and Grand Coulee in 1941. The combination of large hydroelectric development projects and fishing pressure led to the third collapse of chinook landings starting about 1941 (Fig. 3; see Van Hyning 1973).

In 1953, when McNary Dam went into operation on the Columbia River, the 5-year moving average for chinook salmon landings dropped to <4.6 million kg (~500,000 chinook). Although the Columbia River chinook harvest in 1988 was 4.785 million kg (489,000 chinook), the 5-year moving average remained low owing to lower landing figures before and after 1988. The current era has seen most of the big river dam construction, with 15 dams being built on the Columbia and Snake rivers from 1953 to 1975, a 500% numerical increase over the preceding era (PPC 1995).

Harvest management of Columbia River chinook populations remains ineffective because the two principal harvest control entities do not provide harvest regulations that explicitly facilitate chinook salmon spawning escapements to individual tributaries. Salmon harvest regulations under the Columbia River Fisheries Management Plan (US Federal District Court, Portland, Oregon) provide for aggregated spawning escapements to hydroelectric dams, not to tributary spawning grounds. The harvest management of Columbia River chinook under the Pacific Salmon Commission (PSC) (Jensen 1986) has two fundamental flaws. First, with the exception of one natural spawning population, impacts of PSC harvests on naturally spawning stocks are not directly measured. Second, the PSC bases harvest regulations on the number of salmon landed rather than the numbers caught. Catch measures the number of salmon actually killed, whereas landings measure the number of salmon actually kept on the vessel. Landings may be only a fraction of the number killed. For example, it is estimated that the PSC sports fishery in the Strait of Georgia caught 18 salmon for every 10 salmon reported as landed in 1993. Furthermore, the combined Canadian and US PSC fisheries annually caught, but did not land, the equivalent of 411,130 to 479,104 adult chinook between 1990 and 1993 (PSC 1994).

The Fraser River Sockeye Salmon

The history of sockeye salmon recovery in the Fraser River, British Columbia, is very complex, being composed of many biologically different spawning populations, habitat loss, fishing across international boundaries, and the pride and ingenuity of nations (Roos 1991). Reduced to its essence, fish lost access to a majority of the spawning grounds as a result of activities related to construction of a railroad bed in the Hells Gate Canyon in 1913, combined with a series of uncoordinated fisheries that allowed unlimited harvest prior to 1937 (Fig. 1). Thus, a major loss of available habitat interacted with fishing to cause a decline in the Fraser River sockeye populations. This decrease was accompanied by declines in the landings (Roos 1991). A treaty between Canada and the US established the International Pacific Salmon Fisheries Commission in 1937. As a consequence of the treaty, a fishway was built to bridge the obstruction of the Fraser River at Hells Gate, a major program was established to provide vital information on the biological attributes of sockeye populations, and management measures necessary to recover the salmon populations were implemented (Roos 1991). The successful management measures coordinated the regulations of fisheries harvesting the Fraser River sockeye, identified the origin of the landings by major spawning ground, and set the fishing regulations for the fisheries in order to achieve spawning requirements (e.g., adult spawner escapement).

Bristol Bay Sockeye Salmon

Many times in the past hundred years, Bristol Bay, Alaska, has been the site of the world's largest salmon fishery. Since at least as early as 1961, the management program has attempted to provide a minimum level of spawning escapement to each of the watersheds. Spawner counts are made near the outlets of the nursery lakes and verified by qualitative aerial surveys of distributions of the spawners within each watershed (Fried and Yuen 1987). Stock identification is practiced by restricting the fishery to relatively small marine harvest areas nearshore by timing harvest periods to correspond with the timing of adult returns to the various watersheds (Mundy and Mathisen 1981, Eggers 1992) and by scale pattern analysis (Cross and Stratton 1991). Although some annual migrations (as represented by annual catch) reached extremely low levels during the early 1970s (Fig. 2), the control and elimination of harvest in years of low returns (Eggers 1992, 1993) and the presence of pristine spawning and rearing habitats have permitted the populations to sustain unprecedented levels of harvest during the 1980s (Fig. 2).

Conclusions

On the basis of the preponderance of evidence and a century of experience, the three salmon populations presented above serve to illustrate four key points relevant to understanding the three questions I posed on the relationship among harvest, habitat, and salmon productivity. First, what is the role of harvest management in the decline and extirpation of Pacific salmon populations? The histories of the three populations demonstrate that unlimited exploitation of salmon contributed to reductions in the production of salmon. Second, what is the role of harvest management in recovery of the salmon resource? After the imposition of limits on exploitation, two of the three salmon populations returned to levels of production similar to or greater than the period of unlimited exploitation. Third, is it possible for harvest management actions to compensate for losses to salmon populations due to other human activities (e.g., habitat degradation)? Perhaps not. The two populations that recovered did so in the presence of pristine or recovered freshwater habitat, while the population that continues to decline does so in the presence of severely degraded freshwater habitat.

In summary, while I do not claim that one watershed or salmon population can be used as a "control" for another, the comparisons permit some conclusions that are applicable to all species and life-history types of salmon (Karr 1992). The three populations described herein indicate that salmon production is closely related to human population density and land-use practices. Salmon recovery appears to be impossible in the absence of sufficient freshwater habitat, even in the presence of the admittedly suboptimal harvest limitations practiced on Columbia River chinook salmon. A history of unlimited exploitation does not necessarily preclude the future productivity of a salmon population. Fishing activity, limited or unlimited, has the potential to impose genetically heritable changes in characteristics of the populations (Miller 1957, Ricker 1981). However, this cannot be currently evaluated.

An Effective Pacific Salmon Harvest Management Paradigm

Clearly, the question of why certain salmon-bearing ecosystems are apparently functioning normally while others appear to be dying, is of paramount importance to the proper design of harvest management programs. The comparison of the three populations presented here demonstrates that the limits on salmon exploitation rates appropriate to conservation are ultimately dependent on the status of the habitat from which the populations originate. Hence, salmon harvest managers need to look beyond the fishing fleet and its docks to the impacts of civilization on habitat. Ecosystem management is not merely a stylish term. The long-term persistence of all salmon species throughout their ranges depends upon implementing a salmon harvest management paradigm that applies exploitation rates consistent with the status of the salmon-bearing ecosystems.

Combining an explicit recognition of the role of habitat in determining salmon productivity (Mundy et al. 1995) with the basic harvest regimes of the sockeye salmon fisheries of the Fraser River, Canada (Roos 1991), and Bristol Bay, Alaska (Mundy and Mathisen 1981, Eggers 1992), provides the entry point to the paradigm. An effective harvest management paradigm for Pacific salmon may be defined in terms of its objectives and the information necessary to attain those objectives. Effective harvest management in an ecosystem context has the same objectives as traditional single-species harvest management: conservation, public safety, and product quality. However, the minimum information necessary to achieve these objectives goes well beyond that required to achieve the same objectives under the old single-species management (Mundy 1985, Fried and Yuen 1987, Hilborn 1987, Walters and Collie 1988, Eggers 1992, McAllister and Peterman 1992). Information requirements are greater because the assumptions permitted in a productive, stable habitat are no longer valid given that the sources of harvest are numerous and because some harvests remain unidentified. In this paradigm, the inadvertent taking of salmon by humans is recognized as incidental harvest. This taking results from other human activities during the salmon's life cycle (e.g., logging, road building, agricultural cultivation and irrigation, pollution, hydroelectric power generation, fishing for other species, and by directed fishing for the same and for other life-cycle stages of salmon).

The hard-won and still debated concept that future spawning stock size is ultimately dependent on present spawning stock size (Ricker 1954, Cushing 1983, Walters 1986, Hilborn and Walters 1992, Hilborn et al. 1995) needs to be enlarged to include other indicators of ecosystem health. The parameters of a stock-recruitment function appropriate to effective harvest management in an ecosystem context can be drawn from relationships incorporating the density and species composition of riparian vegetation, the percent of fine sediment in the spawning substrate, the abundance of critical life-history stages of at least one prey and one predator species, and the abundance of one species used as an alternative by the salmon's predators. If it were possible to explicitly include at least one of the preceding abiotic or biotic variables in the salmon stock-recruitment function, it would remind harvest managers of the ephemeral nature of the environment's productive capacity, especially in areas of high human population density.

The qualitative attributes of effective harvest management include biological diversity, transboundary scope, and interdisciplinary staffing. As an example of the need for more substantial auxiliary information on biological diversity, consider the following speculation. The decline in salmon in the Snake River basin appears to have been preceded by a sharp decline in Pacific lamprey (*Lampetra tridentata*) (K.L. Witty, Oregon Dep. Fish and Wildlife, pers. comm.). Juvenile Pacific lamprey are eaten by the juvenile salmon's native predator, northern squawfish (*Ptychocheilus oregonensis*). The decline in Pacific lamprey, coupled with the introduction of exotic predators such as smallmouth bass (*Micropterus dolomieu*), may have substantially increased natural mortalities of salmon in freshwater. And then again, it may not have. Given that salmon resource management agencies only recently began to collect and publish abundance data on the lamprey and northern squawfish, the evidence available to answer questions of the role of prey shifting in the decline of Pacific salmon in the Snake River basin remains anecdotal and circumstantial. Similarly lacking are quantitative data on riparian vegetation and stream bed condition in relation to surveys of spawning adults and rearing juveniles.

A transboundary scope for effective harvest management is essential to indefinitely sustain Pacific salmon and their ecosystems. In the course of their life cycles, Columbia River chinook salmon, along with most other Pacific salmon populations, migrate through a range of harvest management regimes of differing capabilities. Obviously, limiting effective harvest management regimes to areas close to the spawning grounds is only likely to prevent extirpation in relatively pristine areas. In those cases where stocks from damaged freshwater habitats interact extensively with ineffective harvest management regimes, extirpation seems likely. The Pacific Salmon Treaty (Jensen 1986) and its predecessor, the International Pacific Salmon Fisheries Commission (Roos 1991), embody the principles, if not always the practice, of international cooperation in management for salmon conservation.

Effective harvest management requires interdisciplinary staffing beyond the disciplines of fish biology and mathematics ordinarily found in the old single-species harvest management. It is essential to develop "a framework for integrating predictable and observable features of flowing water systems with the physical-geomorphologic environment" (Vannote et al. 1980). The hydrology and geomorphology of the watersheds, as well as the consequences of riparian vegetation for salmon production, need to be a part of Pacific salmon harvest management. These factors are especially important for conservation of stocks originating from damaged habitat. As Rich (1941) and Craig and Hacker (1940) wrote more than half a century ago, understanding habitat is essential to sustainable salmon production.

The quantitative attributes of effective harvest management include escapement goals, geographic gradients in fishing mortalities, and zero-sum mortality allocation. Effective management strategies are designed to provide adequate spawning escapements to all spawning grounds and to accurately measure the attainment of these goals on an annual basis. Without monitoring there is no harvest management because salmon harvest management runs on information (Walters 1986). The escapement goals also apply to the riparian vegetation even though an escapement of willows may seem an odd concept to a fisheries biologist. But appropriate habitat is as important as providing spawning escapements when trying to meet a conservation objective. Escapement goals under effective harvest management are quantifiable objectives. Information collected by locality and life-history stage can include spawning numbers, habitat, and associated species. Escapement goals must be accompanied by monitoring programs to be meaningful.

The concept that fishing mortalities need to decrease as distance from the spawning grounds increases is essential to reduce the risk of extirpation for salmon populations originating in damaged habitat. The farther that harvest occurs from the spawning grounds, the less likely accurate stock identification becomes, and the lower the likelihood that effective harvest management can be achieved. The concept that the magnitude of salmon fishing mortality should be inversely proportional to the distance of harvest from the spawning grounds is especially critical when distant mixed-stock fisheries harvest populations from damaged habitat.

The concept of zero-sum mortality holds that when one source of mortality increases, another source of mortality must decrease in order to keep the population size from decreasing. A basic law of biology is that each Pacific salmon population can bear only a certain average total mortality before it starts to decline (Ricker 1954). As a conservative approximation, if the 5-year average total annual mortality from egg to spawner for a chinook salmon population reaches the level where one female chinook cannot be expected to produce two spawners in the next generation, the population will necessarily decline in numbers. When a population is declining, the probability of extinction is 100% if the trend is not halted or reversed (Rieman and McIntyre 1993). For populations at critically low levels, such as those threatened or endangered, anthropogenic influences on mortality need to decrease; otherwise, the population will be extirpated. Implementation of the zero-sum principle requires the measurement and maintenance of survival at each life-history stage and that controls be implemented when possible on sources of mortality.

Finally, neither effective harvest management nor any other harvest management regime can work unless there are consequences for the humans involved in salmon-consuming activities when survival standards and escapement goals for salmon are not met. Both Bristol Bay, Alaska, and the Fraser River, Canada, support thriving sockeye salmon populations today, resulting from implementation of effective harvest management regimes. Whenever spawning populations have reached critically low levels, fishing has been reduced or stopped. For example, harvesters and processors in Bristol Bay lost an entire year's income in 1973 when biologists allocated nearly all of the adult return of 2.3 million sockeye to the spawning escapements. Forgoing the harvesting led to large returns of the sockeye salmon in 1978.

By contrast, in the Columbia River Basin, when the El Niño Southern Oscillation reduces ocean productivity and drought reduces freshwater survival, human behavior remains unaltered. The hydroelectric system, commercial barge transportation system, irrigation systems, and timber industry continue to consume or perturb salmon as they operate. Without direct consequences to all those sectors of the economy that take salmon as these animals dwindle, it remains unlikely that the wide-scale geographic effort necessary to prevent salmon from being extirpated can be successfully mounted.

Acknowledgments

I thank R. Whitney, D.W. Chapman, D. Eggers, and S. Fried for critical reviews. I acknowledge the hard work of two anonymous peer reviewers who were most helpful. Discussions with J. Lichatowich, W. Liss, and C. Warren made material contributions to the paper. Thanks to D. Stouder for her editorial work on this paper. Thanks also to N. Mundy for help with the editing.

Literature Cited

Bell, M.C. 1991. Fisheries handbook of engineering requirements and biological criteria. Fish Passage Development and Evaluation Program, North Pacific Division, US Army Corps of Engineers. Portland, Oregon.

Chapman, D.W. 1986. Salmon and steelhead abundance in the Columbia River in the nineteenth century. Transactions of the American Fisheries Society 115: 662-670.

Cleaver, F.C. (ed.). 1951. Fisheries Statistics of Oregon. Oregon Fish Commission Contribution No. 16. Portland, Oregon.

Corely, D. 1963. Salmon, steelhead . . . and fish screens. Idaho Wildlife Review, May-June: 3-7.

Craig, J.A. and R.L. Hacker. 1940. The history and development of the fisheries of the Columbia River. Bulletin of the Bureau of Fisheries No. 32, Volume 49. Washington, DC.

Cross, B.A. and B.L. Stratton. 1991. Origins of sockeye salmon in East Side Bristol Bay fisheries in 1988 based on linear discrimination function analysis of scale patterns. Alaska Department of Fish and Game Technical Fishery Report No. 91-09. Juneau.

Cushing, D.H. 1983. Key papers on fish populations. IRL Press Limited, Washington, DC.

Eggers, D.M. 1992. The costs and benefits of the management program for natural sockeye salmon stocks in Bristol Bay, Alaska. Fisheries Research 14: 159-177.

Eggers, D.M. 1993. Robust harvest policies for Pacific salmon fisheries, p. 85-106. In G. Kruse, D.M. Eggers, R.J. Marasco, C. Pautzke, and T.J. Quinn II (eds.), Proceedings of the International Symposium on Management Strategies for Exploited Fish Populations. Alaska Sea Grant College Program Report No. 93-02. University of Alaska, Fairbanks.

Fried, S.M. and H.J. Yuen. 1987. Forecasting sockeye salmon (*Oncorhynchus nerka*) returns to Bristol Bay, Alaska: a review and critique of methods, p. 273-279. In H.D. Smith, L. Margolis, and C. Wood (eds.), Sockeye Salmon (*Oncorhynchus nerka*) Population Biology and Future Management. Canadian Special Publication in Fisheries and Aquatic Sciences 96.

Gilhousen, P. 1992. Estimation of Fraser River sockeye salmon escapements from commercial harvest data, 1892-1944. International Pacific Salmon Fisheries Commission Bulletin XXVII. Vancouver, Canada.

Hilborn, R. 1987. Living with uncertainty in resource management. North American Journal of Fisheries Management 7: 1-5.

Hilborn, R. and C.J. Walters. 1992. Quantitative Fisheries Stock Assessment. Chapman and Hall, New York.

Hilborn, R., C.J. Walters, and D. Ludwig. 1995. Sustainable exploitation of renewable resources. Annual Review of Ecology and Systematics 26: 45-67.
Jensen, T.C. 1986. The United States-Canada Pacific salmon interception treaty: an historical and legal overview. Environmental Law, Northwestern School of Law of Lewis and Clark College 16(3): 363-422.
Johnson, D.R., W.M. Chapman, and R.W. Schoning. 1948. The effects on salmon populations of the partial elimination of fixed fishing gear on the Columbia River in 1935. Oregon Fish Commission Contribution No. 11. Portland.
Johnson, T.H., R. Lincoln, G.R. Graves, and R.G. Gibbons. 1996. Current status and future of wild salmon and steelhead in Washington state, p. 127-144. *In* D.J. Stouder, P.A. Bisson, and R.J. Naiman (eds.), Pacific Salmon and Their Ecosystems: Status and Future Options. Chapman and Hall, New York.
Karr, J.R. 1992. Ecological integrity: protecting earth's life support systems, p. 223-238. *In* R. Costanza, B.G Norton, and B.D. Haskell (eds.), Ecosystem Health. New Goals for Environmental Management. Island Press, Washington, DC.
Kostow, K. 1996. The status of salmon and steelhead in Oregon, p. 145-178. *In* D.J. Stouder, P.A. Bisson, and R.J. Naiman (eds.), Pacific Salmon and Their Ecosystems: Status and Future Options. Chapman and Hall, New York.
Lichatowich, J. 1993. Historical perspective of Pacific Northwest salmon. Report to the Wilderness Society. Portland, Oregon.
McAllister, M.K. and R.M. Peterman. 1992. Experimental design in the management of fisheries: a review. North American Journal of Fisheries Management 12: 1-18.
McDonald, M. 1894. The salmon fisheries of the Columbia River basin, p. 3-18. *In* Report of the Commissioner of Fish and Fisheries Investigations in the Columbia River Basin in Regard to the Salmon fisheries. Senate Miscellaneous Document No. 200, 53rd Congress 2d Session, and House Miscellaneous Document No. 86, 53d Congress, 3d Session. Washington, DC.
Miller, R.B. 1957. Have the genetic patterns of fishes been altered by introductions or selective fishing? Journal of the Fisheries Research Board of Canada 14: 797-806.
Mills, T.J., D. McEwan, and M.R. Jennings. 1996. California salmon and steelhead: beyond the crossroads, p. 91-111. *In* D.J. Stouder, P.A. Bisson, and R.J. Naiman (eds.), Pacific Salmon and Their Ecosystems: Status and Future Options. Chapman and Hall, New York.
Mundy, P.R. 1985. Harvest control systems for commercial marine fisheries management: theory and practice, p. 1-34. *In* P.R. Mundy, T.J. Quinn, and R.B. Deriso (eds.), Fisheries Dynamics, Harvest Management and Sampling. Washington Sea Grant Technical Report 85-1. University of Washington, Seattle.
Mundy, P.R. and O.A. Mathisen. 1981. Abundance estimation in a feedback control system applied to the management of a commercial salmon fishery, p. 81-98. *In* K. Brian Haley (ed.), Applied Operations Research in Fishing. Plenum Publishing Corp., New York.
Mundy, P.R., T.W.H. Backman, and J.M. Berkson. 1995. Selection of conservation units for Pacific salmon: lessons from the Columbia River, p. 28-38. *In* J.M. Nielson (ed.), Evolution and The Aquatic Ecosystem: Defining Unique Units in Population Conservation. American Fisheries Society, Bethesda, Maryland.
Nehlsen, W., J.E. Williams, and J.A. Lichatowich. 1991. Pacific salmon at the crossroads: stocks at risk from California, Oregon, Idaho, and Washington. Fisheries 16(2): 4-21.
Netboy, A. 1974. The Salmon Their Fight for Survival. Houghton Mifflin Co., Boston, Massachusetts.
Norman, G. and S. King. 1992. Status report, Columbia River fish runs & fisheries, 1938-91. Washington Department of Fisheries, Battleground, and the Oregon Department of Fish and Wildlife, Portland.
Northcote, T.G. and D.Y. Atagi. 1996. Pacific salmon abundance trends in the Fraser River watershed compared with other British Columbia systems, p. 000-000. *In* D.J. Stouder, P.A. Bisson, and R.J. Naiman (eds.), Pacific Salmon and Their Ecosystems: Status and Future Options. Chapman and Hall, New York.
Northwest Power Planning Council. 1986. Council staff compilation of information on salmon and steelhead losses in the Columbia River Basin, and appendices. Portland, Oregon.
Public Power Council. 1995. Public Power Fundamentals. Portland, Oregon.
Pacific Salmon Commission. 1994. Joint Chinook Technical Committee. 1993 Annual Report. Pacific Salmon Commission Report TCCHINOOK (94)-1. Vancouver, British Columbia.

Ralph, S.C., G.C. Poole, L.L. Conquest, and R.J. Naiman. 1994. Stream channel morphology and woody debris in logged and unlogged basins of western Washington. Canadian Journal of Fisheries and Aquatic Sciences 51: 37-51.

Raymond, H.L. 1988. Effects of hydroelectric development and fisheries enhancement on spring and summer chinook salmon and steelhead in the Columbia River basin. North American Journal of Fisheries Management 8: 1-24.

Reeves, G.H., F.H. Everest, and J.R. Sedell. 1993. Diversity of juvenile anadromous salmonids assemblages in coastal Oregon basins with different levels of timber harvest. Transactions of the American Fisheries Society 122: 309-317.

Rich, W.H. 1941. The present state of the Columbia River salmon resources. Proceedings of the Sixth Pacific Science Congress, University of California Press, Berkeley. Oregon Fish Commission Contribution No. 3. Portland.

Ricker, W.E. 1954. Stock and recruitment. Journal of the Fisheries Research Board of Canada 11: 559-623.

Ricker, W.E. 1981. Changes in the average size and age of Pacific salmon. Canadian Journal of Fisheries and Aquatic Sciences 38: 1636-1656.

Rieman, B.E. and J.D. McIntyre. 1993. Demographic and habitat requirements for conservation of bull trout. US Forest Service Intermountain Research Station, General Technical Report INT-302. Ogden, Utah.

Roos, J. F. 1991. Restoring Fraser River salmon. Pacific Salmon Commission. Vancouver, British Columbia.

Stober, Q.J., M.R. Criben, R.V. Walker, A.L. Setter, I. Nelson, J.C. Gislason, R.W. Tyler, and E.O. Salo. 1979. Columbia River irrigation withdrawal environmental review: Columbia River fishery study. Final Report to the US Army Corps of Engineers. University of Washington, College of Fisheries, Fisheries Research Institute FRI-UW-7919. Seattle.

Thompson, W.F. 1951. An outline for salmon research in Alaska. University of Washington, Fisheries Research Institute Circular No. 18. Seattle.

Van Hyning, J.M. 1973. Factors affecting the abundance of fall chinook salmon in the Columbia River. Research Reports of the Fish Commission of Oregon 4(1).

Vannote, R.L., G.W. Minshall, K.W. Cummins, J.R. Sedell, and C.E. Cushing. 1980. The river continuum concept. Canadian Journal of Fisheries and Aquatic Sciences 37: 130-137.

Walters, C.J. 1986. Chapter 4. Adaptive Management of Renewable Resources. MacMillan Publishing Company, New York.

Walters, C.J. and J.S. Collie. 1988. Is research on environmental factors useful to fisheries management? Canadian Journal of Fisheries and Aquatic Sciences 45: 1848-1954.

Wertheimer, A.C. 1996. Status of Alaska salmon, p. 179-197. *In* D.J. Stouder, P.A. Bisson, and R.J. Naiman (eds.), Pacific Salmon and Their Ecosystems: Status and Future Options. Chapman and Hall, New York.

Whitney, R.R. and S.T. White. 1984. Estimating losses caused by hydroelectric development and operation, and setting goals for the Fish and Wildlife Program of the Northwest Power Planning Council. Completion report. Appendix A. University of Washington, School of Fisheries. Seattle.

Salmon Production in Changing Ocean Domains

William G. Pearcy

Abstract

The ocean's carrying capacity for anadromous salmonids is dynamic in time and space. It is constantly changing on interannual, decadal, centennial and millennial time scales. Since 1976 a major change has occurred in the Northeast Pacific Ocean, with unfavorable ocean conditions for salmonids in the Coastal Upwelling Domain, and highly favorable conditions farther north in the Coastal Downwelling and Central Subarctic domains and the Bering Sea. High sea levels and warm temperatures along the coast, an intense Aleutian Low, and weak upwelling are associated with these recent changes. During the 1960s and early 1970s, when hatchery releases of smolts were increased to compensate for loss of freshwater habitat, the opposite trend prevailed, with good ocean survival in the Coastal Upwelling Domain and lower survival in the Gulf of Alaska. Although the exact mechanisms that affect high or low salmon production are still speculative, ocean climate is clearly implicated and should be considered in management decisions. Favorable ocean conditions will be required for full recovery of many depressed stocks.

Introduction

To understand trends in the production or survival of anadromous salmonids we must learn about all aspects of their life histories, including both freshwater and marine. Salmon survival and production are dependent on the ocean environment, where nearly all of their growth and much of their mortality occurs (see Groot and Margolis 1991, Pearcy 1992 for reviews). Although we cannot control ocean conditions, such conditions modify survival rates and should be major considerations in managing commercially harvested stocks.

Both growth and survival are dependent upon ocean productivity. Production of salmon, in turn, depends on the following factors:

- primary productivity, modified by solar radiation, availability of nutrients, stability of the water column, temperature, etc.;
- number and transfer efficiency of trophic links in the food web from primary producers to the trophic level that includes salmonids;
- fraction of the suitable prey produced that are actually consumed by salmonids versus their competitors;

- conversion and growth efficiency of salmonids;
- predation rates on salmonids; and
- summation of geographic areas of differing salmonid productivities.

These factors, which determine the carrying capacity of the ocean for salmonids (i.e., the biomass of salmonids that can be supported by their ecosystems), vary in both space and time. Moreover, they interact strongly. A change in any factor may directly affect one or more trophic processes and may indirectly modify interspecific interactions by favoring a competitor or predator within a salmonid ecosystem.

In this paper I address two main questions: First, where are the different salmonid production ecosystems in the northeastern Pacific Ocean and how do they differ in physical and biological characteristics? Second, what temporal changes in these regions have affected salmonid production?

The Subarctic Pacific Ocean

Salmonid production ecosystems in the North Pacific include the waters of subarctic origin north of the Subarctic Boundary (Fig. 1). This huge water mass is the ocean feeding grounds for anadromous salmonids. It is characterized by low salinity and temperature, high primary and secondary productivity, large standing stocks of zooplankton and nekton, pronounced seasonal variations in production, and low species richness compared with waters to the south (Pearcy 1991).

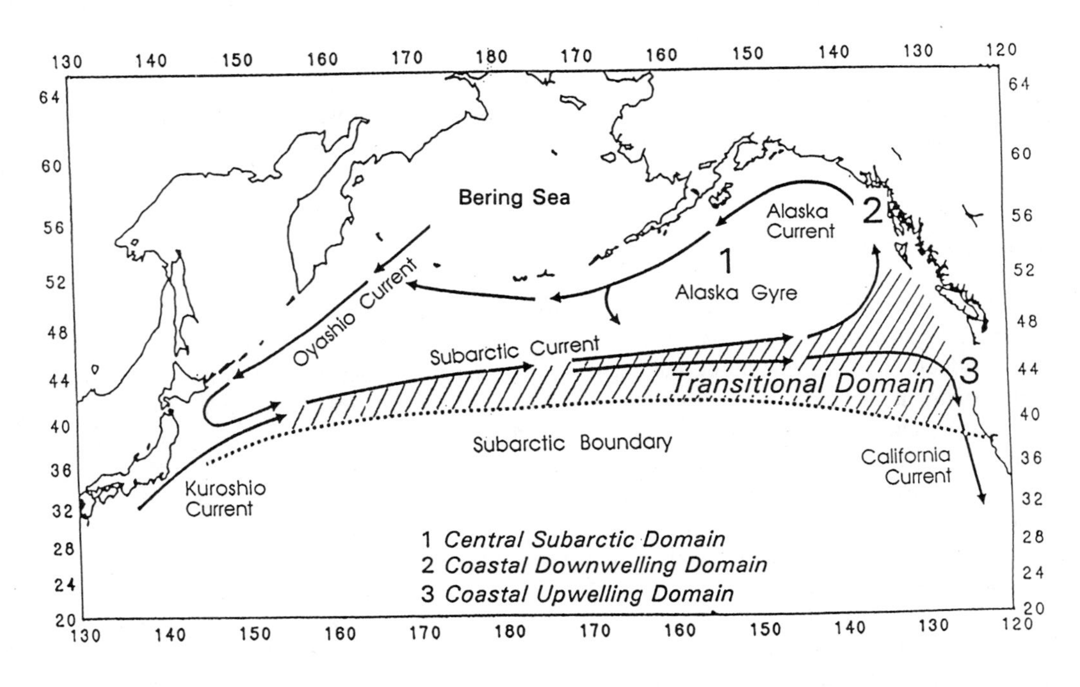

Figure 1. General circulation and salmonid domains of the Northeast Pacific.

The eastern North Pacific area of the subarctic Pacific has been divided into three domains that have different physical and biological characteristics. These include the Central Subarctic Domain, the Coastal Downwelling Domain, and the Coastal Upwelling Domain (Ware and McFarlane 1989). In addition, the Transitional Domain of Dodimead et al. (1963) and the Bering Sea are important regions of salmonid production (Fig. 1).

THE OCEANIC DOMAINS

The Central Subarctic and the Transitional domains provide for offshore foraging for salmonids. The Central Subarctic Domain includes waters of the Subarctic Gyre, usually characterized by a subsurface temperature minimum, whereas the Transitional Domain, between the Central Subarctic Domain and the Subarctic Boundary, includes the easterly flowing Subarctic Current (Fig. 1; Dodimead et al. 1963). Both domains have a permanent halocline (a rapid increase in salinity) between 100 and 200 m.

These oceanic waters of the Central Subarctic and Transitional domains are highly productive. Nutrients are replenished in surface waters by wind mixing from intense winter storms to the depth of the permanent halocline. Cyclonic circulation around the Gulf of Alaska and Ekman pumping from wind stress curl cause upwelling of nutrient-rich water in the Alaskan Gyre. Primary production during the spring and summer is high when a seasonal thermocline limits the depth of mixing, light intensity is high, and major nutrients are replete in the euphotic zone (where primary production occurs). Early estimates of production rates were ~50–100 g $Cm^{-2} yr^{-1}$ (McAllister 1969, Sambrotto and Lorenzen 1987). Recent measurements, however, are about three times higher, 140–200 g $Cm^{-2} yr^{-1}$, among the highest reported for any open ocean region; these higher values may be linked to real changes in primary productivity or to different procedures for measuring productivity (Welschmeyer et al. 1993). Most of the production is apparently by small (<3 μm) nanoplankton (Booth 1988).

Even though major nutrients are always plentiful (Anderson et al. 1969), a spring phytoplankton bloom does not develop in the subarctic Pacific, either because of intense grazing by microzooplankton (Parsons and Lalli 1988, Miller et al. 1991, Welschmeyer et al. 1993) or because of lack of essential micronutrients such as iron (Martin and Fitzwater 1988, Martin et al. 1991, Zhuang et al. 1992). Annual zooplankton production at Ocean Station 'P' (OSP; 50°N, 145°W) was ~13 g C m^{-2} during 1961–66 (McAllister 1969, Parsons and Lalli 1988).

Because of the relatively high primary productivity, efficient grazing by zooplankton, and the large size and high lipid content of many of the carnivorous macrozooplankton and micronekton, the Central Subarctic and Transitional domains provide a rich feeding ground for salmonids and for other nekton that migrate across the Subarctic Boundary during spring and summer (Neave and Hanavan 1960, Taniguchi 1981, Brodeur 1988, Pearcy 1991). The Central Subarctic Domain, however, is the domain of the salmon. Here, Pacific salmon are the dominant top trophic level carnivore in the relatively simple epipelagic food web (Brodeur 1988, Pearcy 1991). Maturing salmonids feed opportunistically on euphausiids, amphipods, squids, fishes, copepods, pteropods and various zooplankton (Ito 1964, LeBrasseur 1966, Pearcy et al. 1988). Salmon comprised 90% or more of the fishes and squids caught in drift gillnets during the summer in waters where the sea-surface temperature is ~7–12°C and where the depth of the 6° C isotherm is <100 m (W.G. Pearcy and J.P. Fisher, unpubl. data). The Transitional Domain, alternatively, is inhabited by many other species of epipelagic fishes as well as salmonids

(Pearcy 1991, McKinnell and Waddell 1993). Some of these fishes, such as Pacific pomfret (*Brama japonica*) and Pacific saury (*Cololabis saira*), dominate the numbers of nekton during the summer and may compete with salmon for food (Pearcy et al. 1988, Pearcy 1991).

Coastal Downwelling Domain

The eastward flowing Subarctic Current bifurcates hundreds of kilometers offshore, forming the northward flowing Alaska Current and the southward flowing California Current. These two eastern boundary currents define the Coastal Downwelling and Coastal Upwelling domains. Downwelling prevails along the northern coast of British Columbia and along the coast of Alaska as a result of freshwater inputs and onshore wind forcing (Royer 1981, Schumacher and Reed 1986, Ware and McFarlane 1989). Primary production in this coastal region is high (300 g Cm^{-2} yr^{-1}) (Sambrotto and Lorenzen 1987). Productivity here is probably several times that in the Central Subarctic Domain, especially along the shelf break, as a result of entrainment of nutrient-rich water into the surface by freshwater movement offshore, wind-induced upwelling during the summer, tidal mixing, and mixing caused by topographic features (Parsons 1987). Zooplankton biomass is high during the summer. Estimates of zooplankton production range from 27 to 50 g $Cm^{-2}yr^{-1}$ over the shelf and inside waters (Cooney 1987). Neritic zooplankton may have several generations a year and are augmented by meroplankton produced on the continental shelf as well as large macrozooplankton, such as euphausiids and *Neocalanus* copepods, which are advected onshore from oceanic waters during periods of downwelling. These oceanic zooplankters often dominate the shelf community and are important food for juvenile salmonids (Manzer 1960, Cooney 1984, Hartt and Dell 1986).

The Coastal Downwelling Domain is the primary migratory corridor for many stocks of juvenile salmon (especially pink [*Oncorhynchus gorbuscha*], sockeye [*O. nerka*], and chum [*O. keta*] salmon), as they migrate along the coast to the north and northwest during the summer and fall en route to the oceanic Subarctic Pacific (Hartt and Dell 1986). (Alternatively, juvenile steelhead [*O. mykiss*] migrate directly offshore into the Gulf of Alaska.) Coastal regions of this domain are also important as a feeding area for stocks of coho (*O. kisutch*) and chinook (*O. tshawytscha*) salmon, including those that are not highly migratory. This region also supports a large biomass of benthic and pelagic fishes, such as walleye pollock (*Theragra chalcogramma*), Pacific cod (*Gadus macrocephalus*), Pacific herring (*Clupea pallasi*), rockfishes (*Sebastes* spp.) and sablefish (*Anoplopoma fimbria*). Many of these pelagic species consume the same prey as salmon (e.g. euphausiids and small fishes), and fluctuations in their abundances may impact stocks of salmonids.

The Coastal Upwelling Domain

The Coastal Upwelling Domain is located along the equatorward flowing California Current, from southern British Columbia to Baja California. The productivity of this domain is influenced by advection of water from the north and by coastal upwelling. Water entering the California Current from the north is largely subarctic in origin, cool, fresh, and rich in nutrients and plankton. Wickett (1967), Chelton et al. (1982), and Roesler and Chelton (1987) concluded that interannual variations in the biomass of zooplankton off California are primarily driven by transport into the California Current from the Subarctic Current to the north, with increased equatorward transport resulting in high standing stocks.

Coastal upwelling, induced by the prevailing northerly and northwesterly winds from April through September, results in offshore transport of surface waters, vertical advection of deeper waters (often rich in nutrients), and high primary and secondary productivity. Satellite imagery reveals bands of cool, chlorophyll-rich water along the coast following episodic upwelling events (Abbott and Zion 1985, Thomas and Strub 1989). Blooms of phytoplankton usually occur during the spring and summer off Washington and Oregon (Landry et al. 1989, Strub et al. 1990). The Columbia River plume is often a prominent feature extending offshore and to the south along the Oregon coast during the summer (Barnes et al. 1972). During the winter, winds are generally from the south and southwest in the northern part of this domain, resulting in onshore and northerly surface currents and downwelling along the coast.

Salmonids are not a major component of the pelagic fish communities in this domain (Ware and McFarlane 1989). Dominant species of commercially important fishes are Pacific hake (*Meluccius productus*), northern anchovy (*Engraulis mordax*), Pacific sardine (*Sardinops sagax*), Pacific herring, Pacific mackerel (*Scomber japonicus*), and jack mackerel (*Trachurus symmetricus*). Again, these species may play a important role in the production of salmonids, either as predators (Pacific hake and Pacific mackerel; N.B. Hargreaves, Pacific Biological Station, Nanaimo, British Columbia, pers. comm.) or as alternatives to juvenile salmonids as prey (e.g., northern anchovy, smelt (*Hypomesus* spp.), Pacific herring [Ware and McFarlane 1988, Holtby et al. 1990]).

Comparisons Among Domains

Ware and McFarlane's (1989) estimates of the mean biomass and catches of commercially important fishes in the Central Subarctic, Coastal Downwelling and Coastal Upwelling domains (1974–1983) indicate that the total production in the Central Subarctic Domain is low compared with the coastal domains (Fig. 2A). However, the biomass of salmon in the Central Subarctic Domain is about five times that in the other two domains of the Northeast Pacific.

I tabulated the average commercial catches of salmon from the coastal states and British Columbia for the period 1982–1991. During this period total catches in Alaska, including the Bering Sea (307,000 tons), were 76% of the total for the west coast of North America (Fig. 2B). Assuming all pink, chum, and sockeye were produced in oceanic waters (as did Ware and McFarlane 1989 in Fig. 2A) and all coho and chinook salmon production was from coastal domains, the catches produced in oceanic waters are about eight times larger than the combined coastal regions. These estimates indicate the great importance of oceanic waters to salmon production in the Northeast Pacific.

Temporal Variations

We are all aware of the large fluctuations in salmonid production that have occurred in the Pacific Ocean during recent years. These changes are correlated with remarkable fluctuations in ocean climate. Thus, in the following I first discuss changes in production of coho salmon in the Coastal Upwelling Domain in relation to upwelling and El Niño Southern Oscillation (ENSO) events, and then the large-scale shift in ocean climate in all domains that began in 1976 and persisted through 1993.

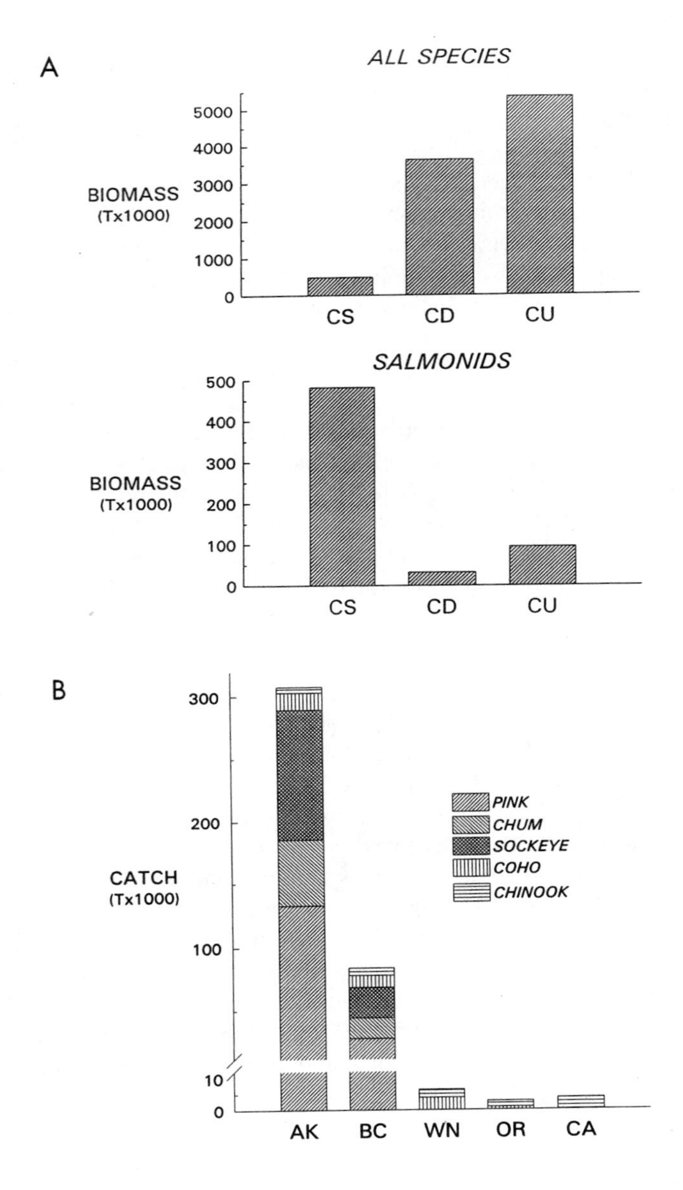

Figure 2. (A) Estimates of average biomass (metric tons x 10^3) of fishes and salmon in the Central Subarctic (CS), Coastal Downwelling (CD) and Coastal Upwelling (CU) domains, 1974–83 (from Ware and McFarlane 1989). (B) Estimates of average catch (metric tons x 10^3) of salmon in Alaska (AK), British Columbia (BC), Washington (WA), Oregon (OR), and California (CA), 1982–1991. Catch numbers were converted to weight using average weights by species from British Columbia catches (M. Kostner, Dep. Fisheries and Oceans, Vancouver, British Columbia, pers. comm.). Source: Alaska Department of Fish and Game (1991); M. Kostner, Dep. Fisheries and Oceans (pers. comm.); Pacific Fishery Management Council (1988, 1992).

Oregon Coho Salmon

Year-to-year variations in the production (catch and escapement) of adult hatchery coho salmon in the Oregon Production Index (OPI) area (south of Willapa Bay, Washington) were high between 1960 and 1992 (Fig. 3). Production of hatchery adults, as well as indices of wild coho catch and escapement, decreased dramatically between the 1975 and 1976 smolt release years. Since 1976 the total production of adults in low years, including 1991 and 1992, is similar to that in the early 1960s, when hatchery release of smolts was just beginning, with <1 million smolts being released. The long-term trend in OPI public hatchery smolt-to-adult survival is negative during this period. Poor production of OPI coho after 1976 coincided with major changes in ocean conditions that have continued into the 1990s. These include weak coastal upwelling (Fig. 4A), elevated sea surface temperatures (Fig. 4B), and high sea levels.

The correlation between coastal upwelling during the summer and smolt-to-adult survival of hatchery OPI coho salmon was positive and significant ($p < 0.05$) in earlier years (1960–1981; see Scarnecchia 1981, Nickelson 1986). The correlation between upwelling at 42°N and survival, including recent years (1960–1992), is still statistically significant ($p < 0.05$) even after removal of autocorrelation in the time series. Trends at 45°N, where upwelling intensities are about half those at 42°N (Fig. 4), are similar. However, even though 45°N is closer to the

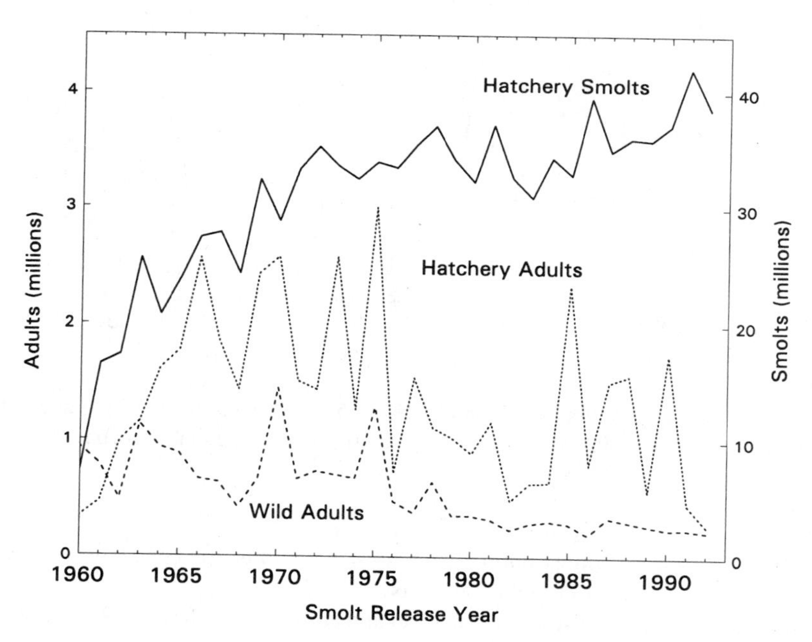

Figure 3. Production (catch + escapement) of Oregon Production Index (OPI) public hatchery and wild coho salmon, and releases of OPI public hatchery smolts by smolt release year, 1960–92. Wild production is overestimated. Source: Oregon Dep. Fish and Wildlife.

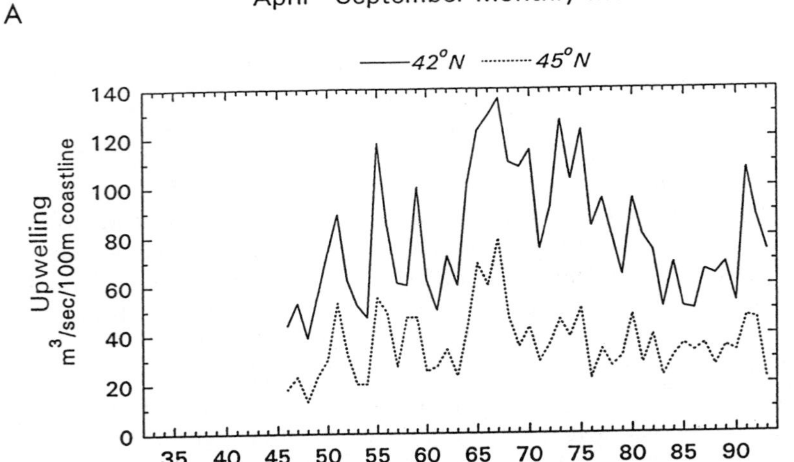

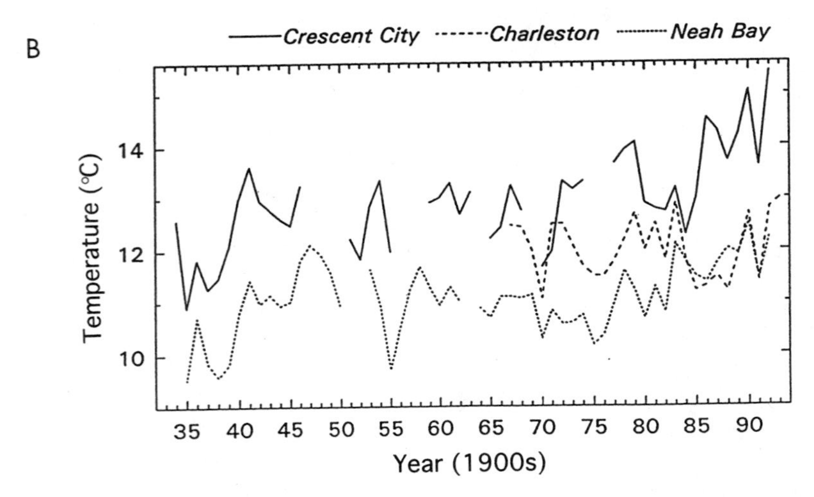

Figure 4. (A) Mean monthly upwelling at 42° and 45°N, 1945–1993, in m³ min⁻¹ 100 m⁻¹ of coastline. (B) Sea surface temperature, 1934–93, at Crescent City, California; Charleston, Oregon; and Neah Bay, Washington.

Columbia River, the source of most public hatchery coho smolts, the correlations between upwelling and survival are weaker here than at 42°N.

Relationships between smolt-to-adult survival and coastal upwelling at 42°N between April and September vary by smolt migration years for two periods, 1960–75 and 1976–92 (Fig. 5). A major change in the ocean climate occurred in the North Pacific Ocean after 1975 (Pearcy 1992, Beamish and Bouillon 1993, Francis and Hare 1994), hence the division into these two time series. Survival was high in strong upwelling years of 1964–70, 1973, and 1975, and low

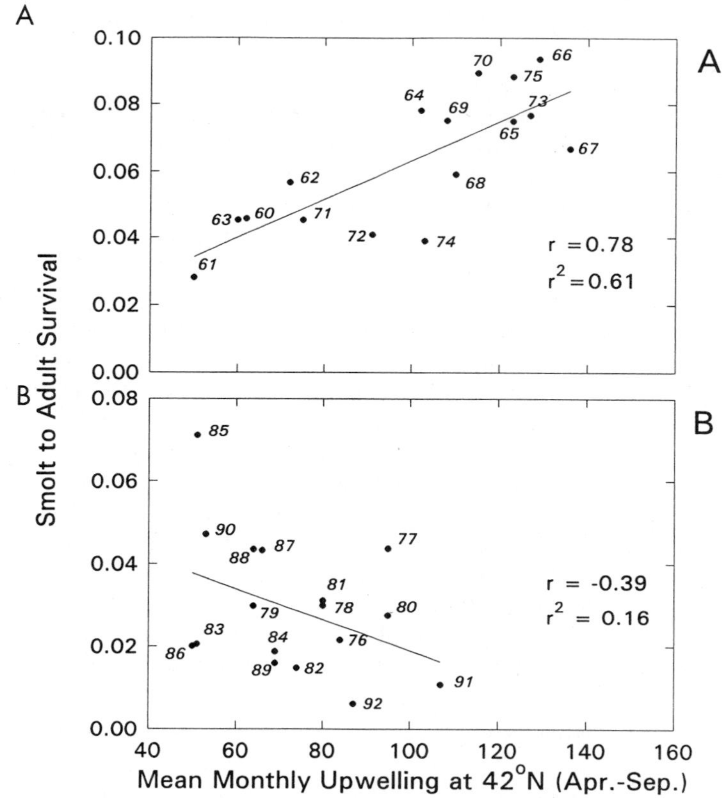

Figure 5. Smolt-to-adult survival of OPI public hatchery coho salmon versus mean monthly upwelling (m^3 min^{-1} 100 m^{-1} of coastline) during April-September at 42°N by smolt migration year: (A) 1960–75; (B) 1982–92.

during the other moderate or weak upwelling years from 1960 to 1975. Of the last 23 years, only 7 years had strong summer upwelling values over 110 units, all of which occurred during the 1960–75 period; the correlation between ocean survival and upwelling during this period was positive and significant ($p < 0.05$, Fig. 5A), even after correction for autocorrelation from the upwelling time series.

Surprisingly, after the regime change of 1976, the relationship between upwelling and survival is negative, though not statistically significant ($p > 0.05$, Fig 5B). In 1991 and 1992, upwelling was strong and survival was poor, whereas in 1985 upwelling was weak and survival was strong. The deterioration of the positive correlation between survival and upwelling suggests that factors other than the quantity of water upwelled along the coast are related to coho survival. Besides intensity of upwelling, the quality of upwelled water may be critical. Upwelling may be ineffective in injecting cool, nutrient-laden water into the euphotic zone, especially during El Niño years, such as 1982–83 and 1991–93, when upwelled water originated above a deep thermocline and nutricline (Brodeur et al. 1985, Miller et al. 1985, Hayward 1993), or during years when upwelling was below the nutricline but the percentage of nutrient-rich subarctic water was low (Pearcy 1992). Unfortunately, we do not have sufficient data to evaluate this nutrient-limitation idea.

In the past, the returns of coho hatchery jacks have been a good predictor of ocean catch and return runs of adult OPI coho during the following year (Pacific Fishery Management Council 1993). This suggests that the adult population was usually determined by the time of the jack return (~6 mo in the ocean) and survival rates during the final ocean year were fairly constant. However, from the large return of jack coho salmon in 1982, adult production in 1983 was severely overestimated because adult mortality was exceptionally high during the severe El Niño of 1983. During 1983, sea temperatures were high, phytoplankton and zooplankton concentrations were low (Fiedler 1984, Miller et al. 1985, Thomas and Strub 1989), and the few coho adults that survived were small and emaciated (Pearcy and Schoener 1987, Johnson 1988). Since this big El Niño event, deviations from the relationship between jack returns in year n and adults in year $n+1$ have increased markedly (Fig. 6), and return of jacks has become less reliable as a predictor of adults. This could be caused by variable rates of survival during the final ocean year, or interannual variation in the proportion of jacks that mature and return to hatcheries.

What is the linkage between upwelling and ocean conditions and coho survival? Mortality of salmon is usually greatest and most variable early in ocean life, when fish are small and subject to high rates of predation (see Pearcy 1992 for review). High rates of survival in years of favorable ocean productivity may be due to fast growth rates and reduced size selective mortality (Scarnecchia 1981, Holtby et al. 1990). Alternatively, predation may be reduced in years of strong upwelling when juveniles are dispersed offshore and when alternative prey are more available (Fisher and Pearcy 1988). Fisher and Pearcy (1988) found little evidence for reduced growth or poor condition of coho smolts during the first 2 mo of ocean life—the time that they suggested was a "critical period"—during the poor upwelling and survival years of 1983–84

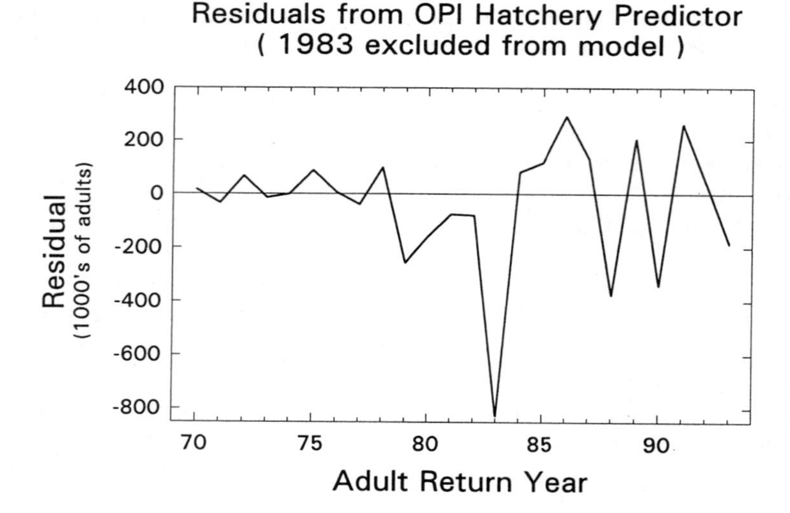

Figure 6. Residuals (in 1000s of adults) from OPI jack predictor versus adult return year 1970–93 (1983 was excluded from the model). Source: P. Lawson, Oregon Dep. Fish and Wildlife.

versus the better survival years of 1982 and 1985. However, Holtby et al. (1990) concluded that marine survival of much smaller coho smolts was positively correlated with early ocean growth. Apparently, fast growth has a survival advantage up to some threshold size, presumably because larger size and faster swimming speeds enhance predator avoidance.

Predation may be the most important factor determining smolt survival. During warm years in the Coastal Upwelling Domain (e.g., 1983–1984), large numbers of predatory fishes, such as Pacific mackerel and jack mackerel, invaded coastal waters (Pearcy and Schoener 1987). These fishes often consume euphausiids, fishes, and decapod larvae and likely compete with juvenile salmon (Brodeur et al. 1987; Brodeur and Pearcy 1990, 1992). But these migratory species may be more important as predators on juvenile salmonids. Pacific hake and Pacific mackerel prey intensively on salmon smolts and young Pacific herring (Ware and McFarlane 1988; N.B. Hargreaves, Pacific Biological Station, Nanaimo, British Columbia, pers. comm.). Abundance of age-1 and age-2 herring is positively correlated with the early marine survival of Carnation Creek coho, suggesting that herring may be alternative prey to coho salmon smolts (Holtby 1988, Holtby et al. 1990). Therefore, predation on juvenile salmon may be more severe during years of poor ocean productivity, when mainstay forage animals, such as euphausiids, northern anchovy, Pacific herring, and various species of smelts, are less available and predatory fishes, birds, and marine mammals feed more on salmonids (Pearcy 1992).

The Changing Ocean Climate

Since 1976, a major change in ocean climate and a geographic shift in salmon production has occurred in the Northeast Pacific Ocean. While production of OPI coho salmon in the Coastal Upwelling Domain has declined drastically, catches of salmon in the Coastal Downwelling Domain and the Bering Sea have increased to record levels. The spectacular increases in the catches of salmon in Alaska are shown by Wertheimer (1996; see also Fig. 7). Rogers (1987) calculated that the biomass of Alaskan salmon increased 1.7 times between 1956–62 and 1980–84. Brodeur and Ware (1995) estimated that the biomass of salmonid fishes increased by 2.4 times between the late 1950s and the 1980s in the Gulf of Alaska. According to Beamish and Bouillon (1993), the long-term trends in the catches of pink, chum, and sockeye salmon from United States, Canada, Japan, and Russia were often similar, with major increases beginning in the late 1970s. This suggests that the production of salmon has increased over an enormous area of the subarctic North Pacific Ocean.

These changes in salmonid production are correlated with large-scale variations in ocean climate. Warm temperatures and high sea levels have prevailed along the North American coast during winters since 1976 (Cole and McLain 1989; Royer 1989, 1992; Folland and Parker 1990; Freeland 1990; Trenberth and Hurrell 1995). This large increase is reflected in the winter air temperatures from Bristol Bay and the winter sea-surface temperatures from Kodiak (Rogers and Ruggerone 1993; Fig. 8). The Alaska Current has apparently become stronger, while the California Current has become weaker (Chelton 1984, Hollowed and Wooster 1992).

The physical forcing model, first proposed by Chelton and Davis (1982), postulates that the Alaska and California currents vary inversely in response to interannual variations in North Pacific winds—when one current is strong the other is weak and vice versa. Increased cyclonic

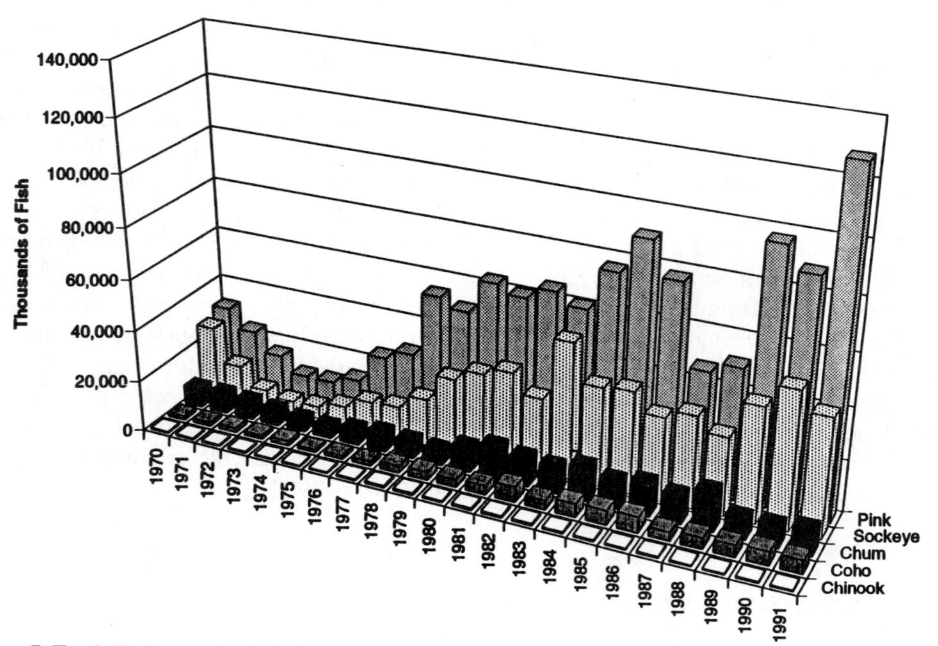

Figure 7. Total Alaskan commercial salmon harvest by species (in 1,000s of fish), 1970–91. Source: Alaska Dep. of Fish and Game 1991.

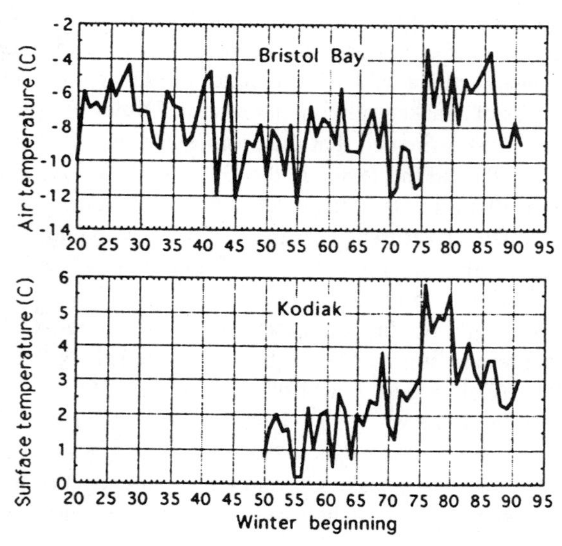

Figure 8. Annual winter air temperatures in Bristol Bay, 1919–20 to 1990–91, and sea surface temperature at Kodiak, Alaska, 1950–51 to 1990–91. Source: Rogers and Ruggerone (1993).

wind circulation in the Gulf of Alaska alters the bifurcation of the Subarctic or North Pacific Current to increase transport into the Alaska Current at the expense of the California Current (Fig. 9). Intensification and easterly shift in location of the Aleutian Low during the winter results in a strong flow of warm, moist air into Alaska from the south (and anomalously cool air into the central and western Pacific), increased wind mixing, an intensified Alaska Current transporting warm waters from the south, and in increased Ekman pumping in the cyclonic circulation (Emery and Hamilton 1985, Hollowed and Wooster 1992, Trenberth and Hurrell 1995).

Periods of very low pressures during the winter in the North Pacific have been unusually frequent during the late 1970s and the 1980s (Trenberth and Hurrell 1995). Although the low pressure weakened during 1988–1991 and near-surface temperatures decreased in the North Pacific (Tabata 1989, Royer 1992), sea level pressure subsequently dropped again in 1992 and temperatures have remained abnormally high in the California Current region or increased in the Gulf of Alaska (T.C. Royer, University of Alaska, Fairbanks, pers. comm.).

Climate changes associated with fluctuations in the intensity of the Aleutian Low in the North Pacific Ocean may be linked with ENSO events in the tropical Pacific (Neibauer 1988, Wallace et al. 1990). The correlation between winter sea level pressures and the Southern Oscillation Index is significant (Trenberth and Hurrell 1995). Both indices have been low for an unprecedented duration since 1976. During this interval, three El Niño events occurred (1982–83, 1986–87, and 1992–93), with few intervening anti-El Niño or La Niña periods (Trenberth and Hurrell 1995, Wooster and Hollowed 1995). Warm temperatures in the eastern North Pacific suggest that this warm climate event persisted through 1993.

Zooplankton standing stocks have apparently increased in the Gulf of Alaska in response to these climatic changes. Brodeur and Ware (1992) hypothesized that intensification of circulation in the Alaskan Gyre and onshore advection resulted in increased zooplankton production and subsequent advection of nutrients and plankton into coastal regions, where they would enhance the growth and survival of juvenile salmonids. The large increase in the abundance of copepods at Ocean Station 'P' beginning in 1976 was also associated with intensification of the Aleutian Low (Beamish and Bouillon 1993).

Although macrozooplankton apparently increased in the Gulf of Alaska since 1976, the reasons for the increase are not clear (Brodeur and Ware 1992). Increased Ekman pumping may inject essential micronutrients into the euphotic zone (major nutrients are not limiting in the subarctic Pacific [Anderson et al. 1969]), or intensified westerly winds could increase aeolian transport of iron, an essential micro-element for phytoplankton growth (Martin and Fitzwater 1988; Zhuang et al. 1990, 1992; Duce and Tindale 1991). Additionally, the tropho-dynamics and efficiency of the food web could change without increased primary productivity (e.g., changes in the sizes and composition of prey organisms or the number of links in the food chain; Parsons 1987). Also, changes in zooplankton distribution in the Gulf of Alaska since 1976 may affect salmonid production as much as increases in zooplankton production.

Elevated temperatures and increased production of salmon are closely correlated, making it difficult to separate effects of ocean temperatures. During recent years, elevated temperatures in the Gulf of Alaska have been correlated with increased growth or survival of some stocks. For example, Cooney (University of Alaska, Fairbanks, pers. comm.) found a positive relationship between survival of hatchery pink salmon in Prince William Sound and spring ocean temperatures. Rogers (1984) and Rogers and Ruggerone (1993) showed that major trends in the catches of Alaskan sockeye, pink, and chum salmon were generally correlated with changes in winter

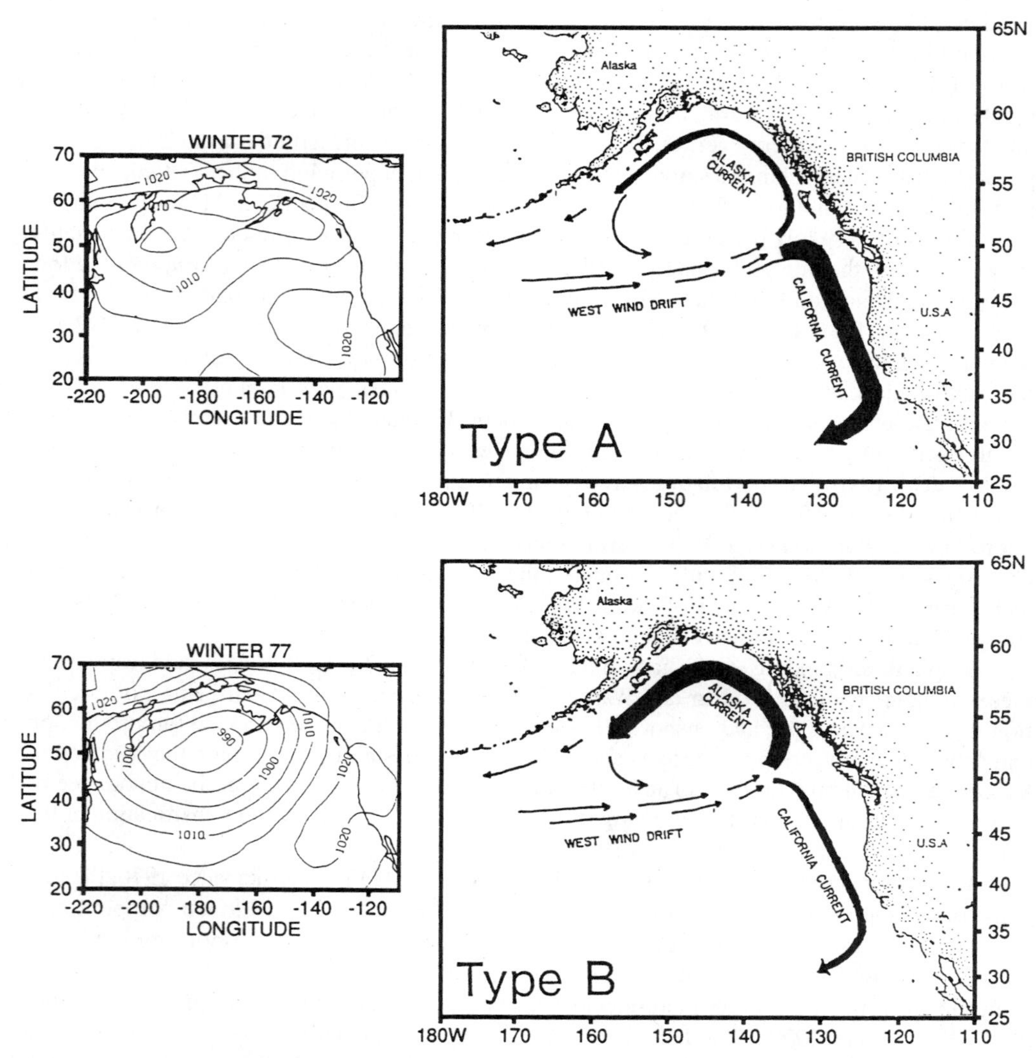

Figure 9. Two hypothesized states of winter atmospheric and oceanic circulation in the North Pacific Ocean. Source: Emery and Hamilton (1985) and Hollowed and Wooster (1992), Figure 4.

temperatures. Temperatures increased rapidly between the winters of 1975–76 and 1976–77 (Fig. 8), coincident with a dramatic increase in salmon catches. They concluded from scale measurements of Nushagak sockeye salmon that growth was positively correlated with temperature during the first 2 years at sea, but during the last year at sea, growth was positively correlated with adult length rather than temperature. Rogers (1984) believed that winter temperatures could modify the range of salmon at sea and hence their susceptibility to predation. If food is abundant, elevated water temperatures may increase food conversion efficiency and growth rate (Brett et al. 1969); increased growth, in turn, may reduce size-selective mortality.

Major shifts in ocean climate, as in 1976, are recurring phenomena. Oscillations in the marine climate of the North Pacific are now widely recognized. Episodes of good and poor production of salmon apparently have opposite trends in the California Current and the Alaska Current systems, with warm periods favoring high production of pink and sockeye salmon in the Alaska Current system and cool periods favoring high production of coho salmon in the California Current (Francis and Sibley 1991; Fig. 10). However, neither north-migrating nor south-migrating stocks of chinook salmon from the California Current region follow these trends (J.P. Fisher and W.G. Pearcy, unpubl. data).

Hare and Francis (1995) and Francis and Hare (1994) identified coherent shifts in Kodiak winter air temperature, a North Pacific sea level pressure index and catches of Alaska pink and sockeye salmon. The timing of these linkages indicated that salmon production was affected during the first year of ocean life. These sudden changes separated three climate regimes: 1920–47 (warm), 1948–76 (cool), and 1977–90 (warm). High catches of Alaskan pink and sockeye salmon were associated with the warm episodes.

Hollowed and Wooster (1992) and Wooster and Hollowed (1995) reported shorter period fluctuations within these regimes. They found that sea surface temperature anomalies during the winter alternated between warm and cool periods, each lasting 6 to 12 years, and that the average time from the start of one warm era to the next was about 17 years. Synchronous recruitment of strong year-classes of groundfish (e.g., Pacific hake, Pacific cod, and some rockfishes) occurred during warm years, when circulation in the Gulf of Alaska was strong. Three of the warm eras since 1932 were ushered in by ENSO events. More recently, Royer (1992) reports an 18.6-year periodicity in Alaska temperatures that he believes is caused by lunar tides.

Long-term fluctuation in the populations of pelagic fishes in the Northeast Pacific have been linked to 40–60 years oscillations in both coastal upwelling and primary and secondary productivity (Ware and Thomson 1991). Fish scales deposited in laminated sediments of anoxic basins provide evidence for interdecadal fluctuations in the abundance of pelagic fishes even further into the past. Over the past 1,700 years, scale deposition rates in the Santa Barbara Basin varied with periods of ~60 years and 60–100 years for Pacific sardine and northern anchovy, respectively (Baumgartner et al. 1992), indicating large variations in the productivity of the California Current ecosystem (Fig. 11). Apparently, the productivity of pelagic fishes was many times higher at the start of this century (see also Smith and Moser 1988). On even longer geological time scales, ocean productivity during the late Pleistocene was much less than that of the Holocene, when summer upwelling in the northern California Current and off Peru was stronger (DeVries and Pearcy 1982, Lyle et al. 1992, Sancetta et al. 1992).

This historical perspective suggests that salmonid stocks waxed and waned in different parts of their range with interdecadal, centenary and millenary fluctuations in past ocean climates. Since anadromous salmonids of the genus *Oncorhynchus* have persisted during the past 50–100 million years (McPhail 1996), they must have evolved mechanisms to enable them to adapt to large-scale climatic changes.

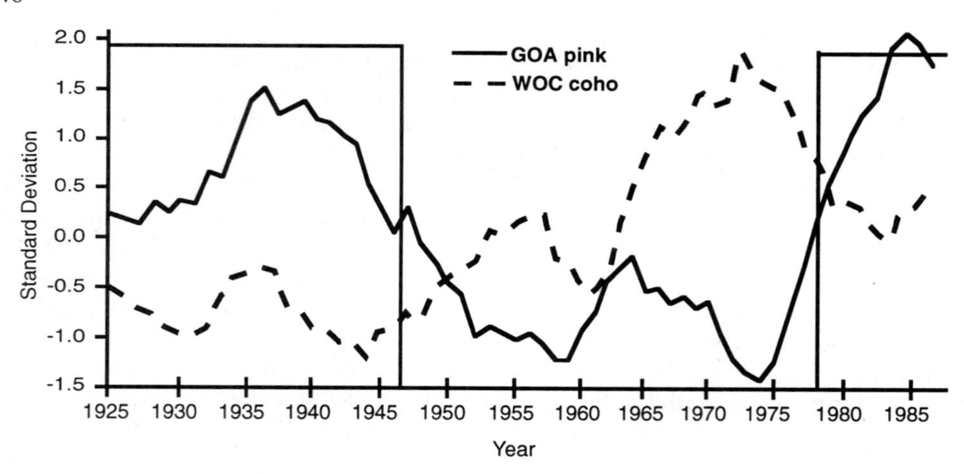

Figure 10. Normalized and smoothed (5-year running mean) plot of pink salmon catch in the Gulf of Alaska (GOA) and coho salmon catch in Washington, Oregon, and California (WOC). Source: Modified from Francis and Sibley (1991), Figure 2.

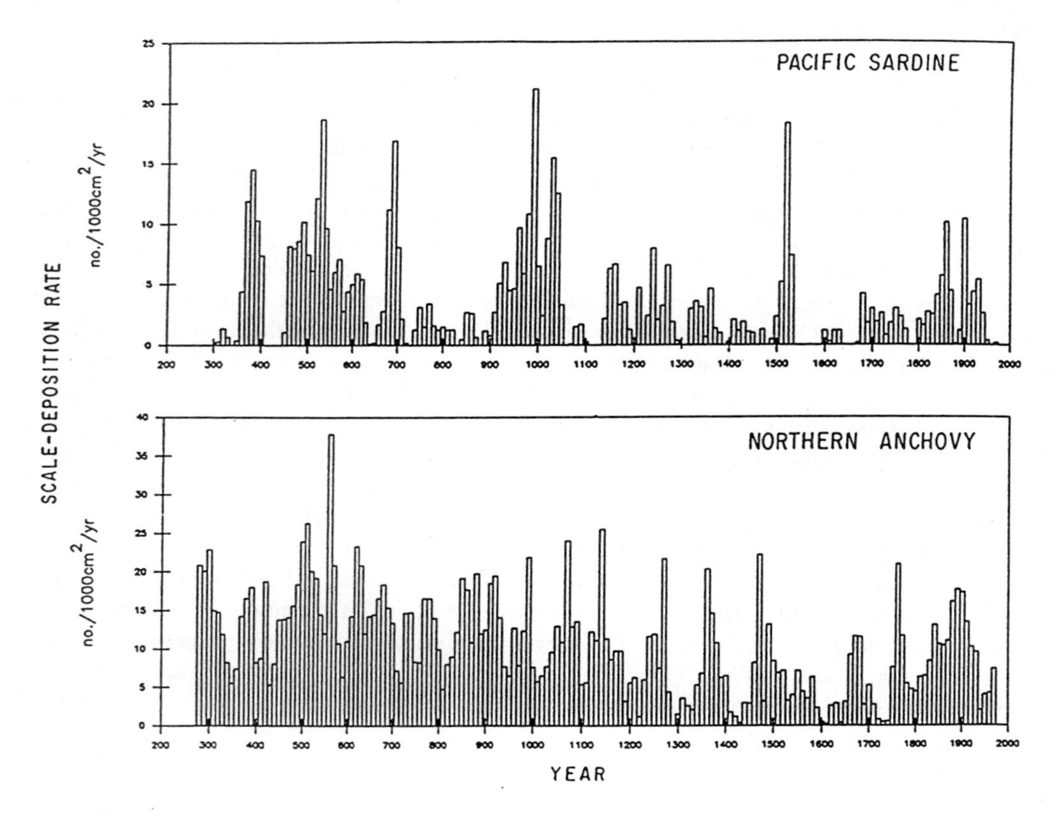

Figure 11. Scale deposition rates of Pacific sardine and northern anchovy in the sediments of the Santa Barbara basin over the last two millennia. Source: Baumgartner et al. (1992).

Epilogue

Salmon catches have exceeded historical levels in the northern North Pacific since the regime shift of 1976. Because climate changes are inevitable, the future shift to a lower productivity regime in this region will be a grand experiment to test the carrying capacity of the oceanic "commons" of the North Pacific Ocean. As ocean productivity decreases in the northern North Pacific Ocean, density-dependent responses may become generally evident for many enhanced and naturally produced stocks of salmonids. Global warming may either extend the present period of high production in the subarctic Pacific, or conversely, exacerbate a future decline.

The best evidence for a limited ocean carrying capacity for salmon comes from negative relationships between numbers of fish and their rates of growth (see Pearcy 1992 for review). Density-dependent growth of stocks of some species has already been convincingly documented (Rogers 1980; Peterman 1984, 1987, 1991; Kaeriyama 1989; Ishida et al. 1993; Rogers and Ruggerone 1993). A change to less favorable ocean conditions will undoubtedly provide more evidence for both intraspecific and interspecific interactions affecting growth. Density-dependent survival may be less obvious since early marine stages of many stocks inhabit coastal oceans, and mortality rates are thought to be high and variable during early life–history stages.

Hatchery enhancement has contributed to increased salmon production, especially in Japan and Alaska. If we are approaching the ocean's carrying capacity, increased hatchery releases may not increase the biomass of salmon produced. Furthermore, releases by one country along the Pacific Rim may affect the size, numbers, and economic value of adult salmon, both hatchery and wild, returning to other countries (Peterman 1991). This is a scientific and management problem of international concern that needs immediate study.

Acknowledgments

This research was supported by NOAA, National Marine Fisheries Service (NA37FE0186-01) and by the Oregon State University Sea Grant College Program (NA89AA-D-SG108, Project S60). Thanks to R. Francis, P. Lawson, and J. Fisher, and three anonymous reviewers for helpful comments on the manuscript.

Literature Cited

Abbott, M.R. and P.M. Zion. 1985. Satellite observations of phytoplankton variability during an upwelling event. Continental Shelf Research. 4: 661-680.

Alaska Department of Fish and Game. 1991. Alaska commercial salmon catches, 1978-1991. Regional Information Report No 5J91-16.

Anderson, G.C., T.R. Parsons, and K. Stephens. 1969. Nitrate distribution in the subarctic northeast Pacific Ocean. Deep-Sea Research 16: 329-334.

Barnes, C.A., A.C. Duxbury, and B.A. Morse. 1972. Circulation and selected properties of the Columbia River effluent at sea, p. 41-80. *In* A.T. Pruter and D.L. Alverson (eds.), Columbia River Estuary and Adjacent Ocean Waters. University of Washington Press, Seattle, Washington.

Baumgartner, T.R., A. Soutar, and V. Ferreira-Bartrina. 1992. Reconstruction of the history of Pacific sardine and northern anchovy populations over the past two millennia from sediments of the Santa Barbara basin, California. California Cooperative Oceanic Fisheries Investigations Reports 33: 24-40.

Beamish, R.J. and D.R. Bouillon. 1993. Pacific salmon production trends in relation to climate. Canadian Journal of Fisheries and Aquatic Sciences 50: 1002-1016.

Booth, B.C. 1988. Size classes and major taxonomic groups of phytoplankton a two locations in the subarctic Pacific Ocean in May and August, 1984. Marine Biology 97: 275-286.

Brett, R.J., J.E. Shelbourn, and C.T. Shoop. 1969. Growth rate and body composition of fingerling sockeye salmon, *Oncorhynchus nerka*, in relation to temperature and ration size. Journal of the Fisheries Research Board of Canada 26: 2363-2394.

Brodeur, R.D. 1988. Zoogeography and trophic ecology of the dominant epipelagic fishes in the northern North Pacific. Bulletin of the Ocean Research Institute, University of Tokyo, Japan, no. 26(Pt.II): 1-27.

Brodeur, R.D., D.M. Gadomski, W.G. Pearcy, H.P. Batchelder, and C.B. Miller. 1985. Abundance and distribution of ichthyoplankton in the upwelling zone off Oregon during anomalous El Niño conditions. Estuarine, Coastal and Shelf Science 21: 365-378.

Brodeur, R.D., H.V. Lorz, and W.G. Pearcy. 1987. Food habits and dietary variability of pelagic nekton off Oregon and Washington, 1979-1984. NOAA Technical Report NMFS 57.

Brodeur, R.D. and W.G. Pearcy. 1990. Trophic relations of juvenile Pacific salmon off the Oregon and Washington coast. Fishery Bulletin 88: 617-636.

Brodeur, R.D. and W.G. Pearcy. 1992. Effects of environmental variability on trophic interactions and food web structure in a pelagic ecosystem. Marine Ecology Progress Series 84: 101-119.

Brodeur, R.D. and D.M. Ware. 1992. Long-term variability in zooplankton biomass in the subarctic Pacific Ocean. Fisheries Oceanography 1: 32-38.

Brodeur, R.D. and D.M. Ware. 1995. Interdecadal variability in distribution and catches of epipelagic nekton in the Northeast Pacific Ocean, p. 329-356. *In* R.J. Beamish (ed.), Symposium on Climate Change and Northern Fish Populations. Canadian Special Publication of Fisheries and Aquatic Sciences 121.

Chelton, D.B. 1984. Short-term climate variability in the Northeast Pacific Ocean, p. 87-99. *In* W.G. Pearcy (ed.), The Influence of Ocean Conditions on the Production of Salmonids in the North Pacific. Oregon State Univ. Sea Grant College Program, Corvallis, Oregon.

Chelton, D.B., P.A. Bernal, and J.A. McGowan. 1982. Large-scale interannual physical and biological interaction in the California Current. Journal of Marine Research 40: 1095-1125.

Chelton, D.B. and R.E. Davis. 1982. Monthly mean sea-level variability along the West Coast of North America. Journal of Physical Oceanography 12: 757-784.

Cole, D.A. and D.R. McLain. 1989. Interannual variability of temperature in the upper layer of the North Pacific eastern boundary region, 1971-87. NOAA Technical Memo NMFS-SWFC-125. La Jolla, California. 19 p.

Cooney, R.T. 1984. Some thoughts on the Alaska coastal current as a feeding habitat for juvenile salmon, p. 256-268. *In* W.G. Pearcy (ed.), The Influence of Ocean Conditions on the Production of Salmonids in the North Pacific. Oregon State Univ. Sea Grant Program, Corvallis, Oregon.

Cooney, R.T. 1987. Zooplankton, p. 285-303. *In* D.W. Hood and S.T. Zimmerman (eds.), The Gulf of Alaska Physical Environment and Biological Resources. NOAA, Ocean Assessment Division, Alaska Office.

DeVries, T.J. and W.G. Pearcy. 1982. Fish debris in sediments of the upwelling zone off central Peru: a late Quaternary record. Deep-Sea Research 28: 87-109.

Dodimead, A.J., F. Favorite, and T. Hirano. 1963. Salmon of the North Pacific Ocean. Part II. Review of oceanography of the subarctic Pacific region. International North Pacific Fisheries Commission, Bulletin 13.

Duce, R.A. and N.W. Tindale. 1991. Atmospheric transport of iron and its deposition in the ocean. Limnology and Oceanography 36: 1715-1726.

Emery, W.J. and K. Hamilton. 1985. Atmospheric forcing of interannual variability in the Northeast Pacific Ocean; connections with El Niño. Journal of Geophysical Research 90(Cl): 857-868.

Fiedler, P.C. 1984. Satellite observations of the 1982-83 El Niño along the US Pacific coast. Science 224: 1251-1254.

Fisher, J.P. and W.G. Pearcy. 1988. Growth of juvenile coho salmon (*Oncorhynchus kisutch*) in the ocean off Oregon and Washington, USA, in years of differing coastal upwelling. Canadian Journal of Fisheries and Aquatic Sciences 45: 1036-1044.

Folland, C.K., and D. Parker. 1990. Observed variations of sea surface temperature, p. 21-52. *In* M.E. Schlesinger (ed.), Climate-Ocean Interaction. Kluwer, Dordrecht, Netherlands.

Francis, R.C. and T.H. Sibley. 1991. Climate change and fisheries: What are the real issues? Northwest Environmental Journal 7: 295-307.

Francis, R.C. and S.R. Hare. 1994. Decadal-scale regime shifts in large marine ecosystems of the North-east Pacific: a case for historical science. Fisheries Oceanography 3: 279-291.

Freeland, H.J. 1990. Sea surface temperatures along the coast of British Columbia: Regional evidence of a warming trend. Canadian Journal of Fisheries and Aquatic Sciences 47: 346-350.

Groot, C. and L. Margolis (eds.). 1991. Pacific Salmon Life Histories. University of British Columbia Press. Vancouver, Canada.

Hare, S.R. and R.C. Francis. 1995. Climate change and salmon production in the Northeast Pacific, p. 357-372. *In* R.J. Beamish (ed.), Symposium on Climate Change and Northern Fish Populations. Canadian Special Publication of Fisheries and Aquatic Sciences 121.

Hartt, A.C. and M.B. Dell. 1986. Early oceanic migrations and growth of juvenile Pacific salmon and steelhead trout. International North Pacific Fisheries Commission Bulletin, 46.

Hayward, T.L. 1993. Preliminary observations of the 1991-1992 El Niño in the California Current. California Cooperative Oceanic Fisheries Investigations Reports 34: 21-29.

Hollowed, A.B. and W.S. Wooster. 1992. Variability of winter ocean conditions and strong year classes of marine fishes and winter northeast Pacific groundfish. ICES Marine Science Symposium 195: 433-444.

Holtby, L.B. 1988. The importance of smolt size to the marine survival of coho salmon. Pages 211-219 *in* 1988 Northeast Pacific Chinook and Coho Salmon Workshop, North Pacific Inter-national Chapter American Fisheries Society.

Holtby, L.B., B.C. Andersen, and R.K. Kadawaki. 1990. Importance of smolt size and early ocean growth to interannual variability in marine survival of coho salmon (*Oncorhynchus kisutch*). Canadian Journal of Fisheries and Aquatic Sciences 47: 2,181-2,194.

Ishida, Y., S. Ito, M. Kaeriyama, S. McKinnell, and K. Nagasawa. 1993. Recent changes in age and size of chum salmon (*Oncorhynchus keta*) in the North Pacific Ocean and possible causes. Canadian Journal of Fisheries and Aquatic Sciences 50: 290-295.

Ito, J. 1964. Food and feeding of Pacific salmon (Genus *Oncorhynchus*) in their oceanic life. Hokkaido Regional Fisheries Research Bulletin 29: 85-97.

Johnson, S.L. 1988. The effects of the 1983 El Niño on Oregon's coho (*Oncorhynchus kisutch*) and chinook (*O. tshawytscha*) salmon. Fisheries Research 6: 105-123.

Kaeriyama, M. 1989. Aspects of salmon ranching in Japan, p. 625-638. *In* H. Kawabe, F. Yamazaki, and D.L.G. Noakes (eds.), Proceedings of the International Symposium on Charrs and Masu Salmon, Physiology and Ecology Japan, Vol. 1, Hokkaido University, Sapporo, Japan, Oct. 1988.

Landry, M.R., J.R. Postel, W.K. Peterson, and J. Newman. 1989. Broad-scale distributional patterns of hydrographic variables on the Washington/Oregon shelf, p. 1-40. *In* M.R. Landry and B.M. Hickey (eds.), Coastal Oceanography of Washington and Oregon. Elsevier Oceanography Series 47, Amsterdam.

LeBrasseur, R.J. 1966. Stomach contents of salmon and steelhead trout in the northeastern Pacific Ocean. Journal of the Fisheries Research Board of Canada 23: 85-100.

Lyle, M., R. Zahn, F. Prahl, J. Dymond, R. Collier, N. Pisias, and E. Suess. 1992. Paleoproductivity and carbon burial across the California Current: The multitracers transect, 42°N. Paleoceanography 7: 251-272.

Manzer, J.I. 1960. Stomach contents of juvenile Pacific salmon in Chatham Sound and adjacent waters. Journal of the Fisheries Research Board of Canada 26: 2,219-2,223.

Martin, J.H. and S.W. Fitzwater. 1988. Iron deficiency limits phytoplankton growth in the northeast Pacific subarctic. Nature 331: 341-343.

Martin, J.H., R.M. Gordon, and S.E. Fitzwater. 1991. The case for iron. Limnology and Oceanography 36: 1793-1802.

McAllister, C.D. 1969. Aspects of estimating zooplankton production from phytoplankton production. Journal of the Fisheries Research Board of Canada 26: 199-220.

McKinnell, S. and B. Waddell. 1993. Associations of species caught in Japanese large scale pelagic squid driftnet fishery in the central North Pacific Ocean: 1988-1990. International North Pacific Fisheries Commission, Bulletin 53 (II): 91-109.

McPhail, J.D. 1995. The origin and speciation of *Oncorhynchus* revisited, p. 29-38. *In* D.J. Stouder, P.A. Bisson, and R.J. Naiman (eds.), Pacific Salmon and Their Ecosystems: Status and Future Options. Chapman and Hall, New York.

Miller, C.B., H.P. Batchelder, R.D. Brodeur, and W.G. Pearcy. 1985. Response of zooplankton and ichthyoplankton off Oregon to the El Niño event of 1983, p. 185-187. *In* W.S. Wooster and D.L. Fluharty (eds.), El Niño North. Washington Sea Grant Program, WSG-WO-85-3, Seattle, Washington.

Miller, C.B., B.W. Frost, B. Booth, P.A. Wheeler, M.R. Landry, and N. Welschmeyer. 1991. Ecological processes in the subarctic Pacific: Iron limitation cannot be the whole story. Oceanography 4: 71-78.

Neave, F. and M.G. Hanavan. 1960. Seasonal distribution of some epipelagic fishes in the Gulf of Alaska region. Journal of the Fisheries Research Board of Canada 17: 221-233.

Nickelson, T.E. 1986. Influences of upwelling, ocean temperature, and smolt abundance on marine survival of coho salmon (*Oncorhynchus kisutch*) in the Oregon Production Area. Canadian Journal of Fisheries and Aquatic Sciences 43: 527-535.

Niebauer, H.J. 1988. Effects of El Niño—Southern Oscillation and North Pacific weather patterns on internal variability in the subarctic Bering Sea. Journal of Geophysical Research 93(C5): 5,051-5,068.

Pacific Fishery Management Council. 1988. A review of the 1987 ocean salmon fisheries. Pacific Fishery Management Council. Portland, Oregon.

Pacific Fishery Management Council. 1992. A review of the 1991 ocean salmon fisheries. Pacific Fishery Management Council. Portland, Oregon.

Pacific Fishery Management Council. 1993. Preseason report I. Stock abundance analysis for 1993 ocean salmon fisheries. Pacific Fishery Management Council. Portland, Oregon.

Parsons, T.R. 1987. Ecological relations, p. 561-570. *In* D.W. Hood and S.T. Zimmerman (eds.), The Gulf of Alaska. Physical Environment and Biological Resources. NOAA, Ocean Assessment Division, Alaska Office.

Parsons, T.R. and C.M. Lalli. 1988. Comparative oceanic ecology of the plankton communities of the subarctic Atlantic and Pacific Oceans. Oceanograpy and Marine Biology, Annual Review 26: 317-359.

Pearcy, W.G. 1991. Biology of the Transition Region, p. 39-55. *In* J.A. Wetherall (ed.), Biology, Oceanography and Fisheries of the North Pacific Transition Zone and Subarctic Frontal Zone. NOAA Technical Report NMFS 105.

Pearcy, W.G. 1992. Ocean Ecology of North Pacific Salmonids. University of Washington Press. Seattle, Washington.

Pearcy, W.G., R.D. Brodeur, J.M. Shenker, W.W. Smoker, and Y. Endo. 1988. Food habits of Pacific salmon and steelhead trout, midwater trawl catches and oceanographic conditions in the Gulf of Alaska, 1980-1985. Bulletin of the Ocean Research Institute, University of Tokyo no. 26(Pt. 2): 29-78.

Pearcy, W.G., and A. Schoener. 1987. Changes in the marine biota coincident with the 1982-1983 El Niño in the northeastern subarctic Pacific Ocean. Journal of Geophysical Research 92(C13): 14417-14428.

Peterman, R.M. 1984. Density-dependent growth in early ocean life of sockeye salmon (*Oncorhynchus nerka*). Canadian Journal of Fisheries and Aquatic Sciences 41: 1825-1829.

Peterman, R.M. 1987. Review of the components of recruitment of Pacific salmon, p. 417-429. *In* M.J. Dodswell, R.J. Klauda, C.M. Moffitt, R.L. Saunders, R.A. Rulifson and J.E. Cooper (eds.), Common Strategies of Anadromous and Catadromous Fishes. American Fisheries Society, Symposium 1, Bethesda, Maryland.

Peterman, R.M. 1991. Density-dependent marine processes in North Pacific salmonids: lessons for experimental design of large-scale manipulations of fish stocks. ICES Marine Science Symposium 192: 69-77.

Roesler, C.S. and D.B. Chelton. 1987. Zooplankton variability in the California Current, 1951-1982. California Cooperative Oceanic Fisheries Investigations Reports 28: 59-96.

Rogers, D.E. 1980. Density-dependent growth of Bristol Bay sockeye salmon, p. 267-283. *In* W.J. McNeil and D.C. Himsworth (eds.), Salmonid Ecosystems of the North Pacific. Oregon State University Press, Corvallis, Oregon.

Rogers, D.E. 1984. Trends in abundance of northeastern Pacific stocks of salmon, p. 100-127. *In* W.G. Pearcy (ed.), The Influence of Ocean Conditions on Production of Salmonids in the North Pacific. Oregon State University Sea Grant Program, Corvallis, Oregon.

Rogers, D.E. 1987. Pacific salmon, p. 461-476. *In* D.W. Hood and S.T. Zimmerman (eds.), The Gulf of Alaska. Physical Environment and Biological Resources. NOAA, Ocean Assessment Division, Alaska Office.

Rogers, D.E. and G.T. Ruggerone. 1993. Factors affecting marine growth of Bristol Bay sockeye. Fisheries Research 18: 89-103.

Royer, T.C. 1981. Baroclinic transport in the Gulf of Alaska. Part I: Seasonal variations of the Alaska Current. Part II: A freshwater driven coastal current. Journal of Marine Research 39: 239-266.

Royer, T.C. 1989. Upper ocean temperature variability in the Northeast Pacific Ocean: Is it an indicator of global warming? Journal of Geophysical Research 94(C12): 18175-18183.

Royer, T.C. 1992. High latitude oceanic variability associated with the 18.6 year nodal tide. Journal of Geophysical Research 98(C3): 4639-4644.

Sambrotto, R.N. and C.J. Lorenzen. 1987. Phytoplankton and primary productivity, p. 249-282. *In* D.W. Hood and S.T. Zimmerman (eds.), The Gulf of Alaska. Physical Environment and Biological Resources. NOAA, Ocean Assessment Division, Alaska Office.

Sancetta, C., M. Lyle, and L. Heusser. 1992. Late-glacial to Holocene changes in winds, up-welling, and seasonal production in the northern California Current System. Quaternary Research 38: 359-370.

Scarnecchia, D.L. 1981. Effects of streamflow and upwelling on yield of wild coho salmon (*Oncorhynchus kisutch*) in Oregon. Canadian Journal of Fisheries and Aquatic Sciences 38: 471-475.

Schumacher, J.D. and R.K. Reed. 1986. On the Alaska Coastal Current in the western Gulf of Alaska. Journal of Geophysical Research 91(C8): 9655-9661.

Smith, P.E. and H.G. Moser. 1988. CalCOFI time-series: An overview of fishes. California Cooperative Oceanic Fisheries Investigations Reports 29: 66-77.

Strub, P.T., C. James, A.C. Thomas, and M.R. Abbott. 1990. Seasonal and nonseasonal variability of satellite-derived surface pigment concentration in the California Current. Journal of Geophysical Research 95(C7): 11503-11530.

Tabata, S. 1989. Trends and long-term variability of ocean properties at Ocean Station P in the northeast Pacific and Western Americas, p. 113-132. *In* D.H. Peterson (ed.), Aspects of Climate Variability in the Pacific and Western Americas. (American Geophysical Union, Washington, DC), Geophysical Monographs 55.

Taniguchi, A. 1981. Plankton productivities in the Pacific subarctic boundary zone: food conditions of the migrating pelagic fishes. Research Institute North Pacific Fisheries, Hokkaido University Special Volume 23-35.

Thomas, A.C. and P.T. Strub. 1989. Large-scale patterns of phytoplankton pigment distribution during the spring transition along the west coast of North America. Journal of Geophysical Research 94(C12): 18,095-18,117.

Trenberth, K.E. and J.W. Hurrell. 1995. Decadal coupled atmosphere-ocean variations in the North Pacific Ocean, p. 15-24. *In* R.J. Beamish (ed.), Symposium on Climate Change and Northern Fish Populations. Canadian Special Publication of Fisheries and Aquatic Sciences 121.

Wallace, J.M., C. Smith, and Q. Jiang. 1990. Spatial patterns of atmosphere-ocean interaction in the northern winter. Journal of Climate 3: 990-998.

Ware, D.M. and G.A. McFarlane. 1988. Relative impact of Pacific hake, sablefish and Pacific cod on west coast of Vancouver Island herring stocks. International North Pacific Fisheries Commission Bulletin 47: 67-77.

Ware, D.M. and G.A. McFarlane. 1989. Fisheries production domains in the Northeast Pacific Ocean. Canadian Special Publication Fisheries and Aquatic Sciences 108: 359-379.

Ware, D.M. and R.E. Thomson. 1991. Link between long-term variability in upwelling and fish production in the Northeast Pacific Ocean. Canadian Journal of Fisheries and Aquatic Sciences 48: 2296-2306.

Welschmeyer, N.A., S. Strom, R. Goericke, G. DiTullio, M. Belvin, and W. Peterson. 1993. Primary production in the Subarctic Pacific Ocean: Project SUPER. Progress in Oceanography 32: 101-135.

Wertheimer, A. 1996. Status of Alaska salmon, p. 179-197. *In* D.J. Stouder, P.A. Bisson, and R.J. Naiman (eds.), Pacific Salmon and Their Ecosystems: Status and Future Options. Chapman and Hall, New York.

Wickett, W.P. 1967. Ekman transport and zooplankton concentration in the North Pacific. Journal of the Fisheries Research Board of Canada 24: 581-594.

Wooster, W.S. and A.B. Hollowed. 1995. Decadal-scale variations in the eastern subarctic Pacific: A. Winter ocean conditions, p. 81-85. *In* R.J. Beamish (ed.), Symposium on Climate Change and Northern Fish Populations. Canadian Special Publication of Fisheries and Aquatic Sciences 121.

Zhuang, G., R.A. Duce, and D.R. Kester. 1990. The dissolution of atmospheric iron in surface seawater of the open ocean. Journal of Geophysical Research 95(C9): 16,207-16,216.

Zhuang, G., Z. Yi, R.A. Duce, and P.R. Brown. 1992. Link between iron and sulfur cycles suggested by detection of Fe (II) in remote marine aerosols. Nature 355: 537-539.

Salmon Policies and Politics

Salmon Fisheries in the Pacific Northwest: How Are Harvest Management Decisions Made?

Larry G. Rutter

Abstract

The growing scarcity of salmon (*Oncorhynchus* spp.) in the Pacific Northwest has spurred investigations into the reasons for their decline. Because people in the region perceive salmon to be an important part of their identity as a community, they want the problems corrected. Critical eyes are being directed towards harvest management by some who are genuinely worried about the salmon resource and by others who simply are looking for ways to limit their own responsibility for the problems.

Within these politically murky waters wade the salmon managers, working in processes established in federal, state, and tribal laws, in judicial rulings arising from Native American fishing rights, and by international treaty. Though the resource has problems, many stocks still are healthy and are producing fish that people want to catch for food, recreation, their livelihood, or to sustain their culture. Accordingly, decisions must be made every year to set allowable harvest levels.

The rules and the processes are extremely complex and poorly understood by most people. They must and can be improved, but tough, pragmatic choices will have to be made about which stocks to manage. The appropriate role that harvest management should play in the recovery of depressed stocks must be determined. We need to obtain and apply new information while we learn, in this age of declining public funding, how to reduce the cost of management. A better balance between participatory, democratic decision making and the effective exercise of government authority must be found. How and where salmon are harvested will have to change. We will have to find more creative ways to accommodate the needs and legal rights of treaty Native American tribes. In short, we must adapt to a future in which salmon management and salmon fisheries will be much different from what they are today.

Introduction

Many salmon (*Oncorhynchus* spp.) stocks are in serious trouble in the Pacific Northwest. Some already have been lost, and others soon may be. Although legitimate disagreements exist as to whether entire species are at risk, many naturally spawning populations in Washington, Oregon,

Idaho, California, and southern British Columbia clearly have been greatly diminished (Canada Department of Fisheries and Oceans 1989, Nehlsen et al. 1991, Washington Department of Fisheries et al. 1993). A lot of controversy surrounds the question of what should be done about the declines. What specific human activities and natural phenomena have caused the problem? Is habitat destruction the major culprit? Or is overfishing to blame? How about recurrent droughts? Ocean survivals? The questions are more than academic; the extent of culpability implies a level of responsibility for remedial actions, and some of these actions will be very costly.

For analytical and jurisdictional convenience, the factors contributing to the decline of salmon have been categorized into three "H's" representing harvest, hatcheries, and habitat. In the context of the Columbia River, one speaks of four "H's" because hydropower development has so greatly changed the ecology of the river that it is evaluated separately from other habitat perturbations.

In this paper, I address harvest. Because fishing in its many forms represents the only human activity which intentionally kills fish, it has come under great scrutiny in this era of declining natural production. To some degree, the fishing industry has become a scapegoat for problems it did not cause, yet has been denied credit for sacrifices already made in the form of greatly curtailed fisheries. In some cases, harvest has contributed significantly to the problem (Canada Department of Fisheries and Oceans 1990, Pacific Fishery Management Council 1992) or has not responded to changing circumstances (Chinook Technical Committee 1993). Therefore, it is appropriate that we as a society reexamine how fisheries management, acting through various resource management institutions, currently balances conservation and exploitation goals so that we might identify needed reforms.

The first step in this reexamination must be to understand how the system currently works. Accordingly, I describe how a typical fishing season is planned and implemented in the Pacific Northwest. Key institutions and processes are then identified and critiqued in greater detail. I focus particularly on the Pacific Salmon Commission, arguably the least understood of the major institutions affecting salmon harvests. Last, I offer a few modest prescriptions for changes.

An Overview of the Annual Salmon Management Process

Most harvest management decisions in the Pacific Northwest are affected by the Pacific Salmon Commission (PSC). The PSC develops harvest regimes for major ocean fisheries adjacent to Southeast Alaska, British Columbia, and Washington State. These fisheries greatly affect the number of chinook (*O. tshawytscha*) and coho (*O. kisutch*) salmon that return to Northwest fisheries and streams. Different kinds of limits apply for different fisheries; for example, the major Canadian fisheries that affect Northwest stocks of coho and chinook salmon are constrained by maximum annual catch limits known as "ceilings." These ceilings, which are often set in advance and apply for a period of 2 to 4 years, are not responsive to annual fluctuations in the size of the runs, nor have they been modified yet to reflect recent downward trends in production and survival. When the regimes expire (as nearly all have as of this writing), one country or both usually will seek changes in harvest limits. The ensuing bilateral negotiations generally occur in January and February. In the past, if the PSC failed to agree to new regimes, the expired agreement was simply extended without modification.

In early March of each year, the Pacific Fishery Management Council (PFMC) begins its preseason management planning for ocean fisheries off Washington, Oregon, and California. By this time, preliminary forecasts of chinook and coho stocks for the upcoming fishing season usually are available. Assuming that the PSC has reached agreement, catch limits for the major intercepting fisheries in Southeast Alaska and Canada will be known. Options for the various sport, commercial, and treaty Native American fisheries are developed by the Council at its March meeting, analyzed with complicated computer models, and then distributed for public review and comment. The options define a range of ocean harvests within which a final plan will be adopted. The final plan must comply with national standards prescribed in the Magnuson Fishery Management and Conservation Act and the regional framework management plan developed earlier by the PFMC for the United States (US) west coast salmon fishery. In developing and considering options, the PFMC receives scientific advice from its technical teams and hears testimony from the public and state and tribal management agencies. At its April meeting, the PFMC votes to adopt a single preseason plan that sets the seasons for each ocean fishery. This plan is forwarded to the US Secretary of Commerce, who may modify it before adopting it and promulgating federal regulations. The regional office of the National Marine Fisheries Service (NMFS) monitors the ocean fisheries during the season and, if circumstances warrant, may adjust the regulations.

While participating in the PFMC process to set ocean regulations, state and tribal managers also are developing preseason fishing plans for the "terminal" fisheries, those located closer to the streams where the salmon originate (e.g., the Columbia River system, coastal Washington rivers, and Puget Sound). Some of this planning occurs concurrently with the PFMC meetings, in what is popularly known as the "North of Falcon" process, because ocean harvest levels affect the number of fish available for the terminal fisheries and escapements. The plans and regulations developed by the states and tribes must comply with conservation and allocation standards set in federal court orders implementing treaty Native American fishing rights, unless the affected states and tribes agree otherwise on a case-by-case basis. Strict compliance with these orders sometimes could spell disaster for important commercial and recreational fisheries because there is almost always one or more stocks present in the fishing areas for which both escapement and fishery objectives cannot be fully met. This provides a powerful incentive to the states and tribes to negotiate compromises that allow some level of fishing to occur. Intense, sometimes heated negotiations occur between managers, and between managers and affected user groups, to find an acceptable balance among competing fisheries while meeting at least minimal escapement requirements. These negotiations rely heavily, sometimes too heavily, on computer simulation models that estimate the impacts of various fishery options on catches, allocations, and escapements. Once an agreed plan is developed, it is implemented by the states and tribes with regulations promulgated according to their respective laws. State and tribal managers then monitor harvests and update their estimates of run sizes as the season progresses, making changes in the regulations as necessary in light of the changing information. After the season, final estimates of catches and escapements result, and the process begins again for the next season.

A Closer Look At Key Institutions

As noted above, all significant harvest management decisions in the Pacific Northwest originate directly or indirectly from the PSC, the PFMC, or state and tribal fishery management agencies. To understand how harvest management decisions are made in the Northwest, one has to understand how and why these institutions function.

THE PACIFIC SALMON COMMISSION

In 1985, the Pacific Salmon Commission was created by the Treaty Between the Government of Canada and the Government of the United States of America Concerning Pacific Salmon (better known and cited hereafter as the Pacific Salmon Treaty). After more than a decade of bilateral negotiations, the two countries had agreed to the following key principles, stated succinctly in Paragraph 1 of Article III:

1. With respect to stocks subject to this Treaty, each Party shall conduct its fisheries and its salmon enhancement programs so as to:
   (a) prevent overfishing and provide for optimum production; and
   (b) provide for each Party to receive benefits equivalent to the production of salmon originating in its waters.

Subparagraph (a) is the basic "conservation principle" of the Treaty. The Treaty defines overfishing as "fishing patterns which result in escapements significantly less than those required to produce maximum sustainable yields." Note that, in addition to avoiding overfishing, the two countries assumed an affirmative duty to "provide for optimum production," a level of production far greater than the amount necessary merely to sustain the existence of the stocks. Subparagraph (b), the so-called "equity principle," defines the fundamental (and, so far, elusive) standard for allocating between the two countries (see Annex IV of the Pacific Salmon Treaty) for actual harvest levels and management agreements applicable to specific fisheries.

Purpose, Structure, and Membership of the Pacific Salmon Commission

The PSC exists to implement the Pacific Salmon Treaty, a task which it has undertaken rather narrowly to date by focusing on the negotiation of fishery harvest regimes that expire at different times. Indeed, Todd and Jensen (1988) characterized the PSC process as an institutionalization of the negotiations that occurred between the two countries since 1971. Whether these interminable negotiations have produced results in proportion to the effort invested over the last several years legitimately can be questioned.

Although "Pacific Salmon Commission" is a term sometimes used to describe the entire PSC bureaucracy, the PSC itself comprises eight individuals, four each from Canada and the US. Each commissioner has an alternate to serve in his or her absence. In actual practice, alternate commissioners function in a manner almost indistinguishable from regular commissioners, so bilateral deliberations of the PSC involve up to 16 individuals. The Chair of the bilateral PSC rotates annually between the US and Canada.

Agreements made by the PSC are not self-executing; they must be approved by the federal governments of the two countries and implemented by their respective domestic management agencies. Until recently, that task in Canada fell exclusively to the Department of Fisheries and

Oceans, the federal agency responsible for managing salmon fisheries in Canada. Now, as a result of recent court rulings regarding aboriginal fishing rights, some management responsibilities are being shared with Canada's aboriginal people.

The situation is more complicated in the US, where authority for managing fisheries is much more dispersed. Four states (Alaska, Washington, Oregon, and Idaho) and at least two federal agencies (NMFS and the US Fish and Wildlife Service) share responsibilities with 24 treaty Native American tribes for managing fisheries or stocks affected by the Pacific Salmon Treaty.

The four commissioners that make up the US Section of the PSC are appointed by and serve at the pleasure of the US President for terms up to 4 years. One represents Alaska, one represents Washington and Oregon, collectively, and one represents the treaty Native American tribes of Washington, Oregon, and Idaho. The fourth commissioner must be an official of the federal government, and does not have a vote. Decision making within the US Section must comply with the Pacific Salmon Treaty Act, the US federal legislation that implements the Pacific Salmon Treaty, which provides that any decision taken by the US Section must be by consensus among the three voting US commissioners. The Pacific Salmon Treaty Act requires that the chairmanship of the US Section generally rotate annually among the four commissioners, unless they agree otherwise.

The Canadian commissioners are appointed by and serve at the pleasure of the Minister of Fisheries, a member of Canada's Parliament and the top fisheries official in Canada. The Canadian Section usually includes two federal officials; one will chair the Canadian Section and the other will be his or her alternate. Because many Pacific Salmon Treaty provisions have expired and require renegotiation, Ottawa recently appointed a new commissioner (an ambassador with extensive experience in international negotiations) to its federal position on the PSC. This commissioner's alternate is a high-ranking official in Canada's Department of Fisheries and Oceans. Canada's non-federal commissioners and alternate commissioners represent the commercial, recreational, and native sectors, and one represents the Province of British Columbia. Although consensus is sought within the Canadian Section, it is not a requirement; lacking consensus, the Canadian federal commissioners have the authority to decide on behalf of the Canadian Section (B. Graham, Canada Dep. Fisheries and Oceans, Vancouver, British Columbia, pers. comm.).

The Pacific Salmon Treaty created three "panels" to provide advice to the PSC. Each country may appoint up to six people, plus alternates, to each panel. The Northern Panel has responsibility for salmon originating in rivers in Southeast Alaska and northern British Columbia. The Southern Panel has responsibility for salmon originating in rivers south of Cape Caution, except Fraser River sockeye (*O. nerka*) and pink (*O. gorbuscha*) salmon, which are the purview of the Fraser River Panel. The Fraser River Panel assumed most of the harvest management duties of the International Pacific Salmon Fisheries Commission, which was terminated as part of the Pacific Salmon Treaty agreement. Its inseason management role makes it unique among the panels; it meets at least weekly throughout the summer to manage the lucrative Fraser sockeye and pink salmon fisheries in southern British Columbia and Washington State.

Because of the extensive migratory range of chinook salmon, fisheries in the Northern Panel area harvest chinook originating in the Southern Panel area, so the Pacific Salmon Treaty provides for joint meetings of those two panels to deal with chinook issues. These joint panel meetings proved to be unwieldy, so the PSC created a much smaller Chinook Working Group to deal with chinook issues. Comprising selected members of the Southern and Northern panels and a few advisors, the Chinook Working Group functions essentially as another panel.

Several joint technical committees exist to assist the PSC and the panels. They include the Transboundary River, Northern Boundary, Chinook, Coho, Chum, Fraser River, and Data Sharing technical committees. Members of the technical committees are not employees of the PSC. They are fishery biologists and technical experts employed by the two countries' domestic management agencies. The number of members for the Fraser River Technical Committee currently is limited to three from each country; none of the other committees has official limits. The largest of all, the Chinook Technical Committee, by recent count had 32 members (Chinook Technical Committee 1993). A perusal of the most recent membership list disclosed that >100 individuals are official members of one or more of the PSC's technical committees (PSC 1993).

The administrative office of the PSC (the Secretariat) is located in Vancouver, British Columbia, and has a permanent staff of ~20 people. The Secretariat is funded equally by the US and Canada to provide administrative and certain technical support to the PSC. It handles the logistics of scheduling and coordinating meetings, publishing and distributing technical reports, and maintaining the records and library of the PSC. Approximately half of the staff consists of fishery biologists who assist the Fraser River Panel in managing the Fraser sockeye and pink salmon fisheries (PSC 1993).

Including the PSC, the panels, technical committees, working groups, support staff, and the Secretariat, the PSC is a large organization indeed. Well over 100, and sometimes as many as 200, people may be involved in an official capacity at a meeting of the PSC.

Normally, the PSC and panels meet three or four times annually to review fisheries, negotiate replacements for expired chapters of Annex IV, and attempt to resolve problems that inevitably arise. The technical committees often meet concurrently with these meetings, but also meet several more times each year to carry out their responsibilities. A week-long meeting of the PSC and panels occurs in late November to discuss the previous season's fisheries and to review catch, escapement, and other fishery data compiled by the two countries' management agencies. If a fishery agreement has expired, the panels may begin in November to negotiate a new agreement, usually after the PSC has provided some basic guidance on the scope of the discussions. Negotiations typically continue through another week of panel meetings in January and are supposed to conclude at the PSC's week-long annual meeting in February.

In recent years, the PSC has been unsuccessful in reaching agreement within the scheduled time frame and has had to schedule extra sessions well into the springtime. In 1993, for example, despite 10 days of panel meetings in January and 10 more days of PSC meetings in February, no agreements were reached. A 3-day special PSC meeting in April also ended in a stalemate. The Commission finally agreed to a stop-gap, 1-year arrangement in late June, but only after the two federal governments had met for 2 days in Montreal and narrowed the scope of the issues considerably (US Department of State 1993). The situation deteriorated even further in 1994; although all of the fishing regimes had expired, the PSC failed to reach even a 1-year agreement. PSC negotiations again were supplanted, at least temporarily, by direct negotiations between the two countries. Dissatisfied with the pace of these negotiations, Canada stopped all talks, imposed a transit fee on American fishing vessels traveling between Washington and Alaska through Canadian waters, and announced plans for an aggressive fishery designed to minimize US harvests of Fraser River sockeye salmon. Canada agreed to resume the talks and lift the transit fee only after receiving assurances from US Vice President Al Gore that the US was committed to addressing all of the issues. However, the ensuing negotiations also failed to produce an agreement, and the 1994 fishing seasons were managed independently by the two countries with a decided lack of cooperation. As of this writing, it is uncertain when the PSC process might return to something resembling a more normal pattern.

Pacific Salmon Commission Decision Making

Unfortunately, an analysis of PSC decision making largely is a study in how and why decisions are not made. Because each country's section has only one vote in the PSC, no decision can be taken unless both sides agree (Pacific Salmon Treaty Act 1985). The same decision rule applies to each of the bilateral panels. As originally envisioned, the national sections of each panel would develop positions on the various fisheries under their jurisdiction, making internal decisions along the way in accordance with applicable domestic laws. Then the panels would meet bilaterally to negotiate agreements that would be approved by the PSC. If an issue proved to be unresolvable at the panel level or involved broad policy questions beyond the purview of any one panel, it would be presented to the PSC for resolution.

However, in actual practice the process has not worked as expected. Panels often fail to resolve their differences, either because of an unwillingness to compromise or because issues are linked to matters beyond the panel's jurisdiction. By the time the issue arrives before the PSC, which conducts its deliberations almost exclusively in executive session, the sides are so polarized and wedded to their respective positions that solutions, for a number of reasons, are nearly impossible.

First, individual commissioners usually see things the same way as their respective regional panel sections. Indeed, they probably participated in national caucuses to help shape the positions before they were advocated at the panel level. Second, many of the issues are mired in technical uncertainties. Although both countries have a strong science ethic, science cannot always provide the answers. However, scientists often can, and do, find interpretations of the data to help arm highly interested commissioners with enough information to keep the debate alive indefinitely. Third, commissioners are under intense political pressure not to make any decisions contrary to their panels, which are "closer" to the constituents. Fourth, the status quo has a built-in advantage. As Hardin (1968) observed, proposed reforms are unfairly held to a higher standard than the status quo. In the PSC context, this almost always favors one vote-wielding faction over another. To be approved by the PSC, a proposal must be acceptable to everyone because everyone effectively has the ability to veto an agreement. As a result, few changes are made and existing arrangements continue unchanged.

These problems are particularly acute for the US Section because of its disparate interests and legal requirement for total consensus among the voting commissioners. By the time the US Section reaches a position, it may have very little flexibility left to bring to the bilateral negotiations, having used it all up to achieve agreement among its own competing constituencies. On several occasions, dissension within the US Section made it impossible even to table a position to negotiate with Canada. More often than not, the disagreement involves chinook salmon. Geography and chinook biology lead to difficulty for the US: Alaska wants to take more fish, the southern states and tribes want Alaska and Canada to reduce their catches, and chinook salmon insist on crossing the borders.

The Effect of Equity on Commission Decisions

More than any other issue, equity has threatened the PSC's ability to function effectively to decide fishery harvest matters. It now threatens even the continued survival of the PSC. Canada believes strongly that it is not receiving benefits equivalent to its own production as promised by the Pacific Salmon Treaty's equity principle. Canada also believes that its situation is worsening, pointing to declining salmon production from Washington and Oregon, which reduces Canada's opportunity to intercept American fish, and to increasing US interceptions of Canadian salmon in Alaska. Given this view of its equity status, Canada sometimes conditions its agreement to fishing regimes based on progress on the equity issue (Chamut 1992). In practical terms, this means that Canada will not agree to fishery regimes unless the US reduces its interceptions of Canadian salmon to address the alleged equity imbalance. In particular, Canada will continue to seek some combination of reductions in sockeye and coho salmon interceptions in Alaska and Fraser River sockeye in Washington.

The US Section has not agreed with Canada's contention that an imbalance exists. It also has argued that Canada cannot legitimately demand compensation for lost interception opportunities that stem from management actions required for conservation. However, the US commissioners often disagree over how best to address the equity issue, making it impossible to effectively deal with it at the PSC table. At some risk of oversimplification, the disparate views can be summarized in the following manner. The Alaskan delegation worries about the potential consequences of an equity calculation. They see the outcome of such a process as a potential threat to their fisheries. The tribes and southern states worry about escalating Canadian interceptions of southern stocks and about the lack of progress on conservation issues as a result of Canada's insistence that such progress be linked to equity.

Because the US Section could not reach agreement internally, the US government decided to take the issue outside the PSC process, and agreed with Canada to address equity at a "government-to-government" level (US Department of State 1993). Since that agreement, struck in June 1993, several high-level meetings occurred between the two governments, but these also failed to produce an agreement. More than a year has passed since the equity issue was taken out of the PSC's hands, and yet it remains unclear as to when the PSC will be able to resume its assigned task of managing an ever-changing resource.

Why Chinook Salmon Problems Often Lead to Gridlock

A coastwide decline in natural chinook salmon stocks in the late 1970s and early 1980s provided perhaps the most important motive to the two countries to conclude negotiations and sign the Pacific Salmon Treaty in 1985. Because of their extensive migratory range, chinook are vulnerable to a veritable gauntlet of US and Canadian fisheries. With each side unable to control the total harvest, many stocks were overfished. The Pacific Salmon Treaty provided the first opportunity to manage chinook throughout their migratory range, thus ensuring that conservation actions taken in one fishery would not be negated by increased catches in other fisheries.

The intent of the Pacific Salmon Treaty's chinook rebuilding program is to rebuild the depressed stocks by 1998. A relatively simple, four-stock simulation model that relied on a number of key assumptions about chinook salmon productivity, survival rates, maturation schedules, and migration patterns was used to estimate appropriate harvest limits necessary to accomplish the rebuilding goal. These limits took the form of maximum catch levels, ceilings, imposed on the landed catches in major ocean fisheries in Southeast Alaska, northern British Columbia, the west coast of Vancouver Island, and the Strait of Georgia. The ceilings were intended to immediately reduce harvest rates and thereby increase escapements. Over time, increased escapements would lead to greater production, so adherence to the ceilings would effect a declining trend in harvest rates and further accelerate rebuilding of the depressed stocks. Other fisheries affecting many of these same stocks, including those in the ocean off Washington and Oregon, in Puget Sound, and in chinook-producing rivers like the Columbia, would not be managed by fixed ceilings. Instead, these would be managed to ensure that the bulk of the depressed stocks saved by the ceilings would pass through and accrue to spawning escapement.

For several stock groupings, survival rates were higher than anticipated for the first few years of the program. Additionally, the fish were concentrated in northern fishing areas, and thus were caught at a faster rate. Seasonal limits were reached earlier in the year, particularly in Southeast Alaska (Chinook Technical Committee 1992). Although these were largely anticipated outcomes of the fixed catch policy, they caused a lot of dissatisfaction among fishers, particularly in Alaska. These fishers, facing increasingly shortened chinook seasons, felt they were being unfairly denied the opportunity to harvest fish that were so abundant in their fishing areas. Reports of record-level fishing in certain terminal area fisheries, notably in the Columbia River, fueled this discontent with the ceilings. In more recent years, survival for many stocks, including those from the Columbia River, has declined significantly, so much so that the Chinook Technical Committee recently concluded that some stocks, for a number of reasons not all linked to harvest, will not rebuild by 1998 (Chinook Technical Committee 1993). Southern states and tribes began to call for further reductions in catches in the northern fisheries.

These problems place conflicting demands on the PSC to change the regimes. Northern fishing interests argue that the ceilings should be raised to allow them to harvest stocks that are abundant in their areas and ease problems caused by shortened fishing seasons. Conversely, southern interests seek to lower the ceilings to further reduce harvest of the stocks that are failing to rebuild. The PSC has spent much of its time trying to resolve this conflict through refinement of the chinook rebuilding program. An enormous amount of effort has been invested in developing an abundance-based approach to adjusting ceilings, one that would, more or less, automate adjustments to the ceilings based on predefined abundance criteria and the status of stocks affected by a fishery. Despite these efforts, the original program remains largely unchanged. In fact, through 1993, the ceilings for the ocean fisheries still were fixed at the same levels that applied in 1985. For 1994, all of the fishery regimes had expired and the two countries failed to agree to new regimes, though they have tended to manage the chinook fisheries as though the old regimes were still in place. Meanwhile, production has declined so far from the levels that prevailed when the original agreements were made that the ceilings no longer effectively reduce harvest rates on some stocks to the extent necessary to rebuild them by 1998 (Chinook Technical Committee 1993).

Perhaps the most frustrating cost of the continuing impasse has been lost opportunity. Rather than developing and using its influence as a high-profile international advocate of the fishery resource, the PSC has squandered several years mired in internal, divisive, and unproductive debate over chinook harvest management. This has happened despite a growing recognition that harvest management alone, particularly as manifested in the PSC's chinook management regimes, will not be adequate to rebuild some of the most depressed stocks (Chinook Technical Committee 1993).

The effort to refine the chinook program continues today, largely at the technical level. Relatively little progress has occurred on the political level in 1994, despite high-level pressures to do so. For example, in June 1993, representatives of the federal governments of the two countries agreed to meet in March 1994 to evaluate the PSC's performance with respect to the goals of the chinook rebuilding program (US Department of State 1993). At about the same time, on behalf of the US Department of State's Office of the Assistant Secretary of State for Oceans and International Environmental and Scientific Affairs, E.G. Constable gave notice to the US Section that its failure to develop an agreed plan to rebuild depressed chinook stocks by 1998 placed the US in jeopardy of not fulfilling its international obligations under the Pacific Salmon Treaty. This was the first step in an ill-defined process which could lead to the US government "taking over" for the US Section of the PSC (as mandated by the Pacific Salmon Treaty Act of 1985). Adding to the pressure is the fact that several chinook stocks have been afforded special protection pursuant to the Endangered Species Act (ESA) of 1973. Before any chinook agreement reached by the PSC can be accepted by the US government, NMFS must first determine that it complies with the ESA.

Despite the June 1993 agreement, the bilateral review of the PSC's progress on rebuilding chinook salmon never occurred. Furthermore, no bilateral agreement was reached on any of the chinook management regimes for 1994, largely because of the impasse in negotiations that stemmed from the equity issue.

New Pressures for Action: The Emerging Coho Crisis

The latest major conservation problem to confront the PSC involves natural coho stocks in Washington, Oregon, and southern British Columbia. In the late 1980s, Canada became increasingly concerned about the status of natural production from Strait of Georgia stocks; escapements and catches were "declining at a disturbing rate" (Canada Department of Fisheries and Oceans 1990). Canada began an ambitious study to identify causes for the decline and possible measures for reversing it. Although the decline can be related to a number of factors, such as loss of freshwater habitat, poor ocean conditions, and overzealous hatchery projects, the major factor was overfishing. Accordingly, the key prescription proposed was a reduction in fishery exploitation rates. Because few fish from these stocks are intercepted in US fisheries, Canada can solve the problem utilizing domestic management actions; it really does not need, and did not seek, bilateral cooperation to address the Strait of Georgia coho problem.

However, the same is not true for depressed natural coho stocks originating in Washington and Oregon, which also have shown an alarming decline in production, but which migrate extensively into Canadian waters. Some of these stocks have declined so dramatically that petitions have been filed to extend protection to them under the ESA. For some of these stocks, particularly those that originate in Washington, the Canadian troll fishery off the west coast of Vancouver Island is the single largest harvester, taking well over half the total harvest of some stocks (PFMC 1992). With such high interception rates, too few fish return to Washington to meet escapement objectives and provide minimal fishery needs.

The west coast of Vancouver Island troll fishery has been managed under a fixed ceiling of 1.8 million coho since 1987, except in 1993, when it was reduced for 1 year to 1.7 million. As it turned out, abundance was down so low in 1993 that the fishery was only able to harvest less than a million fish, far below the agreed ceiling level (Canada Department of Fisheries and Oceans 1993). For several years, the US has tried to convince Canada that it should reduce the ceiling significantly to help address the conservation problem. However, as noted above, Canada repeatedly has characterized such a reduction as a further erosion in its equity position and wants compensation for it (Chamut 1992). Canada has expressed a willingness to consider fish trades for compensation. For example, Canada might agree to reduce the west coast troll fishery in exchange for reduced interceptions of Canadian salmon by US fisheries in Alaska or northern Puget Sound (Chamut 1992). This approach has supporters in both countries, but they have not succeeded in effecting a deal.

To date, the US Section has advanced no proposals to trade species for two reasons. First, many people in the US Section believe that having to pay for conservation actions is inconsistent with the Pacific Salmon Treaty. They point to the willingness of the US, without compensation, to reduce harvests in its intercepting fisheries when salmon abundance is down. Second, given the decision-making rules that apply to the US Section, it is almost impossible to decide to make such a trade. Consider, for example, a treaty Native American tribe whose economy is almost entirely dependent on the Fraser River sockeye salmon fishery, or Alaskan fishers dependent on fishing in Southeast Alaska for their income. Under what circumstance is it likely that either of these groups would agree to reduce its harvest for the possible benefit of someone else, or to provide a solution to a problem not of its making? These are just a few of the very complex resource problems and political pressures that face the US Section and the bilateral PSC in 1994 and beyond.

The Pacific Fishery Management Council

The Pacific Fishery Management Council is one of eight regional management councils created by the Magnuson Fishery Conservation and Management Act of 1976 (the Magnuson Act). Congress passed this law for the following reasons: (1) to assert jurisdiction over fisheries within the exclusive economic zone of the US, (2) to control foreign fishing in the zone, (3) to establish conservation-based management of these resources, and (4) to promote development of the US fishing industry. The regional councils were created to develop fishery plans and regulations consistent with national standards established in the Magnuson Act. Among other things, the plans must prevent overfishing and achieve optimum yield, utilize the best scientific information available, provide for management throughout the range of the stocks, and promote efficient utilization of fishery resources.

Purpose, Structure, and Membership

The PFMC develops plans for fisheries in the ocean seaward of the territorial limits of California, Oregon, and Washington. Although it has responsibility for many non-salmonid species, management of salmon is a major part of the PFMC's focus, reflecting the great importance of salmon to the region. Fishery plans developed by the PFMC are forwarded to the US Secretary of Commerce, who reviews them for compliance with national standards of the Magnuson Act and other applicable federal laws. The Secretary may modify the plans pursuant to certain procedures, and then publishes the regulations in the Federal Register. Once the fisheries begin, the regional NMFS office cooperates with the affected states and tribes to monitor harvest levels and resource conditions. If necessary, in accordance with the Magnuson Act, emergency inseason regulations may be promulgated under authority of the Secretary to close or otherwise modify the ocean fisheries.

The PFMC comprises 13 voting and 5 non-voting members. Each of the four states' principal fishery management officials and the NMFS Regional Director, or their designees, have a permanent position. The eight other voting members are appointed by the Secretary of Commerce for 3-year terms from lists of individuals nominated by the governors of the states of Washington (one obligatory and one at-large position), Oregon (one obligatory and one at-large position), California (one obligatory and two at-large positions), and Idaho (one obligatory position). The Magnuson Act requires that the nominees be "knowledgeable and experienced" about fishing or fisheries management in the region. The Magnuson Act also requires that the five non-voting PFMC members consist of the Area Director of the US Fish and Wildlife Service, regional commander of the US Coast Guard, Executive Director of the Pacific States Marine Fisheries Commission, a representative of the Department of State, and one non-voting member appointed by and serving at the pleasure of the Governor of Alaska.

Several permanent advisory subpanels provide advice to the PFMC on specific fisheries. One of these, the Salmon Advisory Subpanel, has ~20 members representing commercial and recreational organizations, consumers, and Native American fishers (PFMC 1993a). The subpanel serves as an organizational link between the PFMC and key interest groups affected by its decisions.

Two committees provide technical advice to the PFMC on issues related to salmon management. The Scientific and Statistical Committee provides general technical advice and analysis relative to methodologies and scientific questions involved in managing salmon and other species. It comprises mostly federal, state, tribal, and university scientists specializing in biostatistical analysis and economics. The Salmon Technical Team, a much smaller group of six fishery biologists, is responsible for providing annual forecasts of salmon abundance and analyzing the various fishery options under consideration by the PFMC. Members of the Salmon Technical Team are employees of state, federal, or tribal management agencies, and are assigned to assist the PFMC each year with the preseason planning process.

Council Decision Making

The PFMC's process for developing and adopting annual management plans for salmon begins at a week-long meeting in early March. Like all PFMC meetings, and unlike the PSC, this meeting is open to the public and can have as many as a hundred or more observers who come to give or hear testimony. At their March meeting, the PFMC reviews a report prepared by the Salmon Technical Team (the Team) on the previous season's catches, run sizes, and escapements (PFMC 1993b). The Team also presents a report containing forecasts of ocean abundance for the upcoming season's chinook and coho runs. This report includes an analysis of the consequences of maintaining the harvest limits of the previous season, thereby providing a general sense of the condition of the various runs relative to the previous year. It also may include particular concerns the Team has regarding management of the fisheries or the status of the resource (PFMC 1993c).

The PFMC then begins developing options for the ocean fishing season. Beginning early in the week, it receives a report from its Salmon Advisory Subpanel and takes public testimony from representatives of the states, tribes, commercial and recreational fishing groups, environmental groups, and any other interested members of the public. By mid-week, the PFMC adopts three preliminary options that define a range of harvests within which it expects to set actual seasons and harvest limits for the ocean commercial and recreational salmon fisheries. A separate but interrelated set of options is adopted for two major regions of the coast: (1) north of Cape Falcon, Oregon (located ~50 km south of the mouth of the Columbia River), and (2) south of Cape Falcon. The stock compositions of fisheries differ enough between these two regions to define reasonably discrete sets of fisheries to manage.

The Salmon Technical Team then begins organizing the options for analysis, which can be quite complicated because of the interrelationship among fisheries, beginning and ending dates, daily bag limits, and other details; this process allows the Team to identify any additional refinements that may be necessary to facilitate impact analysis. Following yet another round of public testimony late in the week, the PFMC votes to adopt specific options for public review and comment, which are later analyzed by the Salmon Technical Team and a staff economist for biological and socioeconomic impacts. Within a week or so following the meeting, the Team's analyses are published in a report and made available to the public (PFMC 1993d). Aided by the PFMC's administrative staff, subsets of the PFMC conduct several smaller regional hearings over the next few weeks to collect public input on the options. Testimony taken at these smaller meetings, which usually occur in coastal communities, is summarized, along with any written testimony received by mail, and made available to the full PFMC at its meeting in early April.

Like the March meeting, the April meeting may be attended by hundreds of individuals involved in the management process or affected by the decisions. Early in the week, the PFMC will again hear public testimony debating the options. Some of the most influential testimony comes from state and tribal representatives reporting on negotiations among the various managers that have occurred and continue to occur throughout the week. Finally, on Friday, the PFMC formally votes to adopt seasons for each of the various salmon fisheries. The Magnuson Act specifies that the PFMC's decisions require concurrence among the majority of the voting members present. The adopted seasons are forwarded to the Secretary of Commerce to begin the next step in the review and adoption process.

In contrast to the PSC process, the PFMC process always results in a timely decision on a fishing plan. Occasionally, these decisions may be successfully challenged in court, or modified by the Secretary of Commerce as a result of the post-PFMC review process, but eventually a decision is made that establishes the fishing seasons.

Despite its success at making decisions, the PFMC process has its detractors. Some argue that it favors the needs of the fishing industry over those of the resources it manages (American Fisheries Society 1993). Others take a more parochial view and are simply critical of the PFMC for not showing enough favor to their particular interest. In recent years, some environmental and consumer organizations have become increasingly critical of the regional councils in general and decisions affecting salmon in particular. Major criticisms include the PFMC's membership. Because some members are personally affected by decisions made by the PFMC, this creates at least the appearance of conflict of interest. Others are critical of the PFMC's failure to use its influence more effectively to protect salmon habitat; however, the PFMC's influence over land management issues currently is limited by the Magnuson Act to commenting on other agencies' decisions.

Some people still harbor resentment over the effect that Native American treaty rights have had on their fisheries and the influence of tribes on the process. In the late 1970s and early 1980s, the PFMC treated certain groups of stocks within a species as a single unit for the purpose of managing the ocean salmon fisheries. The intent was to provide the 50% allocation to treaty tribes required by the 1974 ruling in *US v. Washington* (the Boldt Decision) in the aggregate, rather than on a run-by-run basis. This practice allowed for larger ocean fisheries, but because of differences in the relative strengths of the runs, often returned too few fish to coastal rivers to allow fishing in the rivers where the tribes are located. In the 1981 *Hoh v. Baldrige* ruling, three coastal tribes, the Hoh, Quileute, and Quinault, won a lawsuit against the Secretary of Commerce to stop this practice. The tribes also challenged the specific escapement goals set by the state of Washington for coastal streams, which served to constrain the tribes' fisheries in the river. As a result of their victory in that lawsuit, the tribes now work as equals with the state to establish escapement objectives. Fishery plans adopted by the PFMC must conform to these objectives and, unless the state and tribes otherwise agree, provide the 50% Native American treaty allocation for each species on a river-by-river basis.

Arguably, the *Hoh v. Baldrige* decision changed management of salmon fisheries north of Cape Falcon more than any other single event. It forced weak stock management on mixed-stock fisheries and greatly increased the tribes' influence in ocean fishery management. In recognition of the treaty tribes' role as co-managers, for more than a decade the state of Washington has nominated a treaty Native American representative to one of its at-large positions on the PFMC. As the Magnuson Act comes up for reauthorization by Congress, the tribes will seek to amend the act to better reflect and acknowledge their standing as co-managers of the resource, further institutionalizing their role in the process. Specifically, testimony was presented in 1993 by a tribal member, J. Harp, to the US Senate wherein the tribes indicated they will advocate the addition of up to three voting positions to the PFMC for Native American representation.

Tribal and State Fishery Management Entities

Each of the several coastal states has an agency empowered by state law to regulate the sport and commercial fisheries within its borders (except tribal-regulated treaty Native American fisheries). These agencies employ fishery biologists, statisticians, resource planners, field technicians, hatchery personnel, and others to carry out their missions of managing state fishery resources. Because the treaty tribes manage their own fisheries, they also have fishery departments employing similar albeit fewer experts. They promulgate and enforce fishing regulations according to tribal laws and procedures established by the federal courts. Many tribes own and operate hatcheries; others operate facilities in cooperation with the state. For efficiency and other reasons, some tribes have created multi-tribal organizations, to whom they have delegated a broad range of management, regulatory, enhancement, enforcement, and habitat management functions (e.g., Skagit System Cooperative, Point No Point Treaty Council).

Most of the treaty tribes in Washington and Oregon are also members of either the Northwest Indian Fisheries Commission (Puget Sound and Washington coastal treaty tribes) or the Columbia River Intertribal Fish Commission (Columbia River Basin treaty tribes). Although they differ in specifics, these two organizations employ fishery biologists, biometricians, computer specialists, statisticians, policy analysts, public relations experts, and others to provide a broad range of specialized services to their member tribes. They coordinate among tribes, and between tribes and other agencies, and serve as forums for tribes to develop coordinated policies to protect, enhance, and implement their fishing rights.

Effect of Treaty Native American Fishing Rights on Harvest Management Decisions

Although the historical and well-known 1974 decision in *US v. Washington* dramatically changed fisheries management in the Pacific Northwest, most people really do not understand much about it. Basically, the courts ruled that tribes have rights reserved by them when they signed treaties with the US government in the mid-1850s. These rights entitle them collectively to harvest one-half of the harvestable surplus of each run of fish returning to or passing through their historical fishing areas. The 1974 decision recognizes the tribes' right to manage their own fisheries; the state may regulate tribal fishing only under very limited, specific circumstances.

Fishing opportunities vary widely among tribes because each tribe may fish only in its historical fishing areas, which must be established through the courts. For example, some tribes (e.g., the Upper Skagit Tribe) have fishing rights limited to only a single river system, so their fishery is limited to, and completely dependent upon, fish returning to that river. Other tribes have access to a large number of fish runs passing through their area. The Makah Tribe, for example, has fishing rights in the western Strait of Juan de Fuca and the Pacific Ocean off northwest Washington. As specified by the 1974 ruling in *US v. Washington*, this gives the Makah Tribe rights to a share of all of the many runs that migrate through these waters.

The geographical limits of tribal fishing rights, and in particular the differences among tribes, have enormous implications for how fisheries are managed and harvests allocated in the Pacific Northwest. It means that large "mixed stock" fisheries (both Native American and non-Native American) must be managed to ensure that some of the harvestable fish return to the terminal areas where some tribes are located. It also means that tribes must negotiate with each other to determine intertribal shares because one tribe's catch can affect the number available to other tribes, and collectively they are limited to 50% of each run. A great deal of fishery data and technical analyses are needed to manage the fisheries to achieve these allocation goals.

The State/Tribal Harvest Management Process

Much of the state and tribal annual management process already has been described; tribal and state representatives are key participants in both the PSC and PFMC processes. However, those two processes primarily address management of ocean fisheries, whereas the states and tribes also must develop and implement plans for fisheries in inland marine areas (e.g., Puget Sound) and rivers. The states and tribes also have responsibility for setting escapement objectives for the various stocks within their jurisdictions; these objectives must be accommodated by the ocean management processes. The states and tribes are the primary sources of fishery and resource data upon which almost all fishery decisions are made. Lastly, these groups are responsible for setting harvest and production policies for the various stocks and fisheries.

For the past 10 years, the states of Washington, Oregon, and Idaho and the treaty tribes have practiced cooperative management. In the late 1970s and early 1980s, nearly every fishery management issue resulted in a dispute requiring resolution by the federal courts. This changed dramatically in the mid-1980s; since that time, the states and tribes have managed their respective fisheries with much less conflict. Now, with relatively few exceptions, they successfully and routinely cooperate to develop annual fishing plans, coordinate fisheries and enhancement programs, share data collection and analysis responsibilities, and attempt to cooperatively resolve problems that arise.

Fishery management procedures, escapement policies, artificial production plans, and policy and technical dispute resolution mechanisms are provided in regional management plans developed by the states, tribes, and federal resource agencies, and filed with the federal courts. These plans include the *Hoh v. Baldrige* Framework Management Plan of 1987, the Puget Sound Salmon Management Plan of 1985, and the Columbia River Fisheries Management Plan of 1988 (associated with the 1985 ruling in *US v. Oregon*), and they apply to fisheries and stocks originating in coastal Washington streams, Puget Sound, and the Columbia River, respectively. In addition to these regional plans, the states and tribes have developed more detailed plans addressing specific river systems or stocks.

As noted earlier, the states and tribes participate fully in the PFMC process not only because of their respective constituents' fisheries in the ocean, but also because the annual decisions made in that process affect the number of fish escaping the ocean for the other fisheries. To accommodate planning for ocean and terminal fisheries simultaneously, state and tribal managers initiated the North of Falcon process in the mid-1980s. State, tribal, and federal fishery managers interact with affected constituencies in an open public forum to attempt to reach consensus on an acceptable package of ocean and terminal fisheries. They strive to find a combination of fishery management actions that fairly allocates fish among competing fisheries, satisfies treaty Native American allocation requirements, and best meets escapement objectives under that year's forecast of resource conditions.

In mixed-stock areas, where hatchery and natural stocks are intermingled, the harvest must be constrained to provide adequate escapement for key stocks managed for natural production, taking into account also the needs of subsequent terminal fisheries. For example, nearly all mixed-stock coho fisheries in Washington are managed to meet the needs of eight key natural coho stocks, namely those originating in the Queets, Hoh, Quileute, Skagit, Stillaguamish, and Snohomish rivers, and tributaries of Grays Harbor and Hood Canal. Because there always is one weak stock that is most limiting on mixed-stock fisheries, the state (on behalf of its non-Native American constituents) and the tribe in whose terminal area the weakest stock originates have powerful incentives to reach agreement. Failure to reach agreement, for example, to lower the escapement level below established goals, could trigger strict application of court-adopted rules that would severely limit fishing for everyone.

The process relies heavily on complex computer models developed and used by state, tribal, and federal agency biologists. Largely on the basis of data collected over many years from coded-wire tagging studies, these models employ algorithms to estimate interactions between stock abundances, harvest, and escapement. Frequently, the models are expected to provide scientifically insupportable levels of precision and accuracy. Enormous legal, political, and economic pressures exist to find that perfect solution, one that maximizes allowable harvests, reaches an acceptable (albeit sometimes arbitrary) escapement level for that year, and complies with allocation requirements.

Generally, the North of Falcon caucus conducts its business in two 2-day public meetings between the PFMC's March and April meetings. Key participants, especially the state and tribal managers, will continue to meet on an ad-hoc basis throughout the week of the April PFMC meeting to try to resolve any outstanding differences.

When the North of Falcon process is successful, agreements reached among the state and tribal managers are presented to the PFMC for its consideration. Although the PFMC must still create a public record to support its decisions, the PFMC gives substantial weight to agreements proffered by consensus of the affected tribes and states (J. Harp, PFMC, Portland, Oregon, pers. comm.). When no consensus is reached, representatives of the states and tribes must independently present testimony to the PFMC, trying to persuade it to decide in their favor. Finally, on Friday of the April meeting, the PFMC will cast its votes and make the decisions in the presence or absence of consensus.

Because the PFMC and the North of Falcon processes focus almost exclusively on chinook and coho salmon, the state and tribal managers need to complete their preseason planning for the other species, such as sockeye, pink, and chum salmon fisheries in Puget Sound, and sockeye and steelhead *(O. mykiss)* fisheries in the Columbia River. These plans generally are developed by the states and tribes in accordance with procedures established in the Puget Sound Salmon Management Plan or the Columbia River Management plan, respectively. The timing of some of these runs overlaps with chinook and coho runs; thus, much of the planning for these fisheries already may have occurred in concert with planning for those species. When completed, the preseason plans are usually recorded in various memoranda of understanding between the affected states and tribes and, depending on the applicable rules, may be filed with the federal court pursuant to 1974 ruling in *US v. Washington* or the 1985 ruling in *US v. Oregon.*

Implementing the Fishing Plans

The last step in the annual harvest management process is to implement the fishing seasons. The tribes and states must carefully monitor the fisheries and returning fish runs because the entire preseason process is based on forecasts, which are far from perfect. For most commercial fisheries north of Cape Falcon, Oregon, catch monitoring programs operated by the states and tribes provide inseason data that can be used to update the size of the runs. On the basis of these updates, inseason regulations are promulgated to implement the terms of the agreed preseason plans. Compared with commercial fisheries, fewer inseason adjustments are made to recreational fisheries. These fisheries are planned such that they avoid the need for inseason regulatory changes, which can be difficult to convey to the public.

Some Prescriptions for Changes

Any complicated management system has opportunities for improvement, and salmon harvest management is no exception. The preceding description of how things currently work illustrates that reforms are needed; some are procedural, others are substantive. In no order of priority, I offer the following suggestions, which reflect only my opinions.

Clearly, we need an overall management framework that explicitly identifies which natural stocks we are going to manage and, by implication, which we are not going to manage. The framework should include fishing levels that will apply under predetermined conditions of abundance. Decisions as to which stocks should drive which fisheries should not be made on an ad-hoc basis each year by fishery managers under intense pressure to respond to the loudest or most powerful interest groups advocating or opposing fishing. Nor should the range of possible decisions be entirely preempted by the ESA. Excessively restrictive scientific and legal interpretations of how this law should be applied to salmon have removed nearly all the discretion of fishery managers. Admittedly, some of the these managers' past decisions may have been wrong and may have contributed to the poor condition of natural stocks that we now want to sustain. If so, we should develop recovery plans for those stocks and change our harvest management regimes to sustain them at productive levels.

The necessary choices should be guided by science, drawing from positive examples of sustained resource exploitation (Rosenberg et al. 1993), using all past lessons about salmon management, and incorporating promising new technologies (Waples et al. 1990). However, we need to recognize that some of these choices involve balancing among competing values, which is the stuff of politics (Barke 1986).

Fishery managers need to establish criteria for judging how well our harvest management plans and regulations are working (Miller 1990). Where such criteria already exist, they should be reviewed and updated as necessary to reflect our current knowledge. Have we selected and are we achieving exploitation rates that will sustain the stocks over a broad range of conditions? Are we meeting our responsibilities to rebuild formerly overexploited natural salmon populations? Though we know that many factors have contributed to the decline of salmon populations, there will always be dissension and public confusion over how to parcel out the blame. As fishery managers, it is incumbent upon us to have objective, scientifically defensible criteria for judging our harvest plans and regulations. Highly interested fisheries critics will be watching, and some will be looking for ways to dodge their own responsibilities.

We need to develop fishing plans that are more robust to unforeseen circumstances or errors in our information systems and knowledge base. We need to confront uncertainties and build them into our plans (Ludwig et al. 1993). Almost certainly, this will require a general reduction of harvests in mixed-stock fisheries, at least until technology either improves our ability to predict abundance and monitor catches or makes it feasible to more selectively harvest the stronger stocks.

We need to simplify fisheries management and lower its cost. One of the ironies of salmon management is that complexity and costs climb as harvests decline. As any fisheries manager knows, pressures exist to squeeze every possible fish out of the system; witness the complicated set of regulations contained in the PFMC's annual plans for the last several years (PFMC 1993d). Unfortunately, while pressures increase to extract maximum benefits from the fisheries, budgets for management activities are declining in response to the growing reluctance of taxpayers to pay for government. Reductions in real-time fishery monitoring and resource assessment programs must be reflected in simpler, more conservative harvest regimes.

We need to work with treaty Native American tribes to find more creative ways to meet our obligations with respect to their treaty fishing rights. Their economies and cultures are inextricably tied to the harvest of salmon, yet in some cases these have been sacrificed for the excesses of non-Native American development. Currently, their treaty right is quantified in terms of a percentage of the allowable harvest. That may be appropriate in the context of relative plenty, but may be less so when allowable harvests fall below some minimal levels. Perhaps it is time to consider that the larger society has allocated some of its share in the harvest to alternative uses, such as dams, irrigation, pollution, urbanization, and forest practices. Perhaps instead of thinking in terms of maximum treaty harvest rights, we should consider defining minimal harvest rights for treaty fisheries, at least until allowable fishery harvests are restored to some semblance of their former levels. This concept might also provide greater flexibility for the state, allowing it to use its remaining shares of various runs as it sees fit to maximize benefits to the state.

We need to improve our ability to make decisions. An obvious candidate for improvement is the US Section of the PSC, where the participatory decision-making process has run amok. The true cost of institutionalized stalemate in the PSC process may be in lost opportunities to direct that organization's considerable resources and potential stature towards more positive actions to benefit the salmon resource. If status quo harvest regimes are the best we can do in the international context (and I do not believe they are), then we need to become more efficient. It is unlikely that our government, or our public, will forever tolerate the cost of sending a hundred people or more to 3 weeks of meetings each year simply to perpetuate the status quo.

When decision rules for the US Section were being developed and debated in 1985, people were fearful that one group or region might use the Pacific Salmon Treaty process to trample on the interests of another. The resulting compromise, in essence a decision to give everyone a veto, was lauded as a delicate compromise, and credited with making the Pacific Salmon Treaty possible. The architects of this delicate balance knew that it would force difficult negotiations within the Section (Jensen 1986). Surely, however, they did not foresee, nor intend, the extent of the resulting gridlock. They also did not foresee that the US commissioners, for the most part, would refuse to settle differences among the panels. With the benefit of hindsight, we can find a better balance between participatory decision making and the potential abuse of the authority to govern.

Conclusion

Because salmon and salmon fishing are so important to the people of the Pacific Northwest, a great deal of effort goes into managing salmon harvest. It is essential that the stocks be managed throughout their entire range because of the highly migratory nature of the salmon. By necessity, many political jurisdictions, many organizations, and large numbers of people must be involved in the processes for determining harvest levels. Though sometimes unwieldy, these processes help ensure that those who are most affected by the status of the salmon resource can participate in the decisions. Society has been willing to invest significant resources into research and management systems because the decisions rely upon a great deal of scientific information. As a result, we can better our understanding and apply our knowledge about the effects of exploitation on salmon stocks. It would be wrong, therefore, to leave the impression that salmon fishery management has been entirely unsuccessful. Indeed, many stocks of salmon in the Pacific Northwest remain quite productive, despite the myriad pressures placed upon them by their human neighbors.

However, improvement is possible and necessary. We need to review, update, and clarify the objectives of our fishing plans to ensure consistency with our current knowledge of the status of natural stocks. Objective, explicit criteria for assessing plan performance are necessary. In addition, the plans have to be more resilient to uncertainties and changing circumstances. We need to be more creative in how we accommodate the treaty fishing rights of Native American tribes. Fisheries management must be simplified to lower its costs, and we need to improve the efficacy of some of our decision-making processes.

Salmon are too important to the Pacific Northwest not to do these things. They are important largely because people want to catch them. When too few salmon remain to allow fishing, our quality of life will be diminished. We need to rebuild the salmon runs back to fishable, sustainable levels. Harvest management can contribute to this effort, but with the human population continually increasing, this will be a never-ending struggle. Perhaps the most important thing fishery managers must do is find better ways to affect decisions about land and water uses, because those decisions have much greater influence on the future of the salmon resource than does harvest management.

Literature Cited

American Fisheries Society. 1993. AFS legislative briefing statement: reauthorization of the Magnuson Act. Fisheries 18(10): 20-26.

Barke, R. 1986. Science, Technology, and Public Policy. C.Q. Press, Washington, DC.

Canada Department of Fisheries and Oceans. 1989. Biological advice on Pacific Salmon. Pacific Stock Assessment Review Committee Advisory Document 89-1. Vancouver, BC, Canada.

Canada Department of Fisheries and Oceans. 1990. Strait of Georgia coho salmon resource status and management planning process. Vancouver, BC, Canada.

Canada Department of Fisheries and Oceans. 1993. 1993 post season report for Canadian treaty limit fisheries. Report prepared for Nov. 29-Dec. 3, 1993 meeting of the Pacific Salmon Commission. Vancouver, BC, Canada.

Chamut, P. 1992. Plenary statement regarding the Canadian position. Given in plenary session of Pacific Salmon Commission, December 3, 1992. Reproduced in the Eighth Annual Report of the Pacific Salmon Commission, p. 5-10. Vancouver, BC, Canada.

Chinook Technical Committee. 1992. 1991 Annual report (TCCHINOOK(92)-4). Pacific Salmon Commission, Vancouver, BC, Canada.

Chinook Technical Committee. 1993. 1992 Annual report (TCCHINOOK(93)-2). Pacific Salmon Commission, Vancouver, BC, Canada.

Hardin, G. 1968. The tragedy of the commons. Science 162:1243-48.

Jensen, T.C. 1986. The United States-Canada Pacific Salmon Interception Treaty: an historical and legal overview. 16 Environmental Law 363: 369-370.

Ludwig, D., R. Hilborn, and C. Walters. 1993. Uncertainty, resource exploitation, and conservation: lessons from history. Science 260: 17, 36.

Miller, R.J. 1990. Properties of a well-managed nearshore fishery. Fisheries 15(5): 7-12.

Nehlsen, W., J.E. Williams, and J.A. Lichatowich. 1991. Pacific salmon at the crossroads: stocks at risk from California, Oregon, Idaho and Washington. Fisheries 16(2):4-21.

Pacific Fishery Management Council. 1992. Assessment of the status of five stocks of Puget Sound chinook and coho. Technical Report prepared by the Puget Sound Stock Review Group. Portland, Oregon.

Pacific Fishery Management Council. 1993a. Roster. Portland, Oregon.

Pacific Fishery Management Council. 1993b. Review of 1992 ocean salmon fisheries. Portland, Oregon.

Pacific Fishery Management Council. 1993c. Preseason report I: stock abundance analysis for 1993 ocean salmon fisheries. Prepared by the Pacific Fishery Management Council Salmon Technical Team. Portland, Oregon.

Pacific Fishery Management Council. 1993d. Preseason report II: analysis of proposed regulatory options for 1993 ocean salmon fisheries. Prepared by the Pacific Fishery Management Council Salmon Technical Team and Staff Economist. Portland, Oregon.

Pacific Salmon Commission. 1993. Pacific Salmon Commission 1992/93 eighth annual report. Vancouver, BC, Canada.

Rosenberg, A.A., M.J. Fogarty, M.P. Sissenwine, J.R. Beddington, and J.G. Shepherd. 1993. Achieving sustainable use of renewable resources. Science 262: 828-829.

Todd, I. and T. Jensen. 1988. An assessment of the Pacific Salmon Commission's first 3 years; frustration and disappointment—but optimism for the future. Lewis and Clark School of Law, Anadromous Fish Law Memo 47: 13-15. Portland, Oregon.

US Department of State. 1993. Record of Agreement (of meeting between representatives of the Government of the United States and the Government of Canada concerning the Pacific Salmon Treaty in Montreal, Canada, June 2-4, 1993). Bureau of Oceans and International Environmental and Scientific Affairs, US Department of State. Washington, DC.

Waples, R.S., G.A. Winans, F.M. Utter, and C. Mahnken. 1990. Genetic approaches to the management of Pacific salmon. Fisheries 15(5): 19-25

Washington Department of Fisheries, Washington Department of Wildlife, and Western Washington Treaty Indian Tribes. 1993. 1992 Washington State salmon and steelhead stock inventory. Olympia.

Habitat Policy for Salmon in the Pacific Northwest

James R. Sedell, Gordon H. Reeves, and Peter A. Bisson

Abstract

Earlier in the 20th century, habitat decisions were based on a belief that aquatic habitats could be manipulated with technology to benefit salmon, especially in terms of fish passage. The importance of riparian zones and biophysical watershed processes to salmon productivity was poorly appreciated. Recent events, coupled with an awareness of widespread habitat simplification, have changed this perspective. Spurred by passage of state forest practices acts, federal clean air and clean water acts, and the Boldt tribal fishing rights decisions in the 1970s, federal and state agencies recognized the importance of riparian zones as critical links between aquatic and terrestrial ecosystems. The listing of the northern spotted owl and several stocks of salmon (*Oncorhynchus* spp.) under the federal Endangered Species Act in the early 1990s prompted a team of scientists under a mandate from US President Clinton to suggest an ecosystem-based approach to habitat management that relied less on engineered habitat substitution and more on streamside buffers that preserved land-water interactions. This approach constituted a landscape-scale application of the principles of adaptive management in which conservative interim buffer guidelines (i.e., large buffers) could be adjusted if watershed analysis showed that smaller buffers would not be likely to harm aquatic resources.

While federal forest lands are expected to anchor the recovery of some salmon stocks in the future, the location of these lands regionally and within river basins prevents them from serving as refugia for many stocks that inhabit coastal lowlands containing urban, agricultural, and non-federal forests. Comprehensive, region-wide improvement in aquatic ecosystems can only occur when habitat policy decisions are shared among affected natural resource users and when watershed-scale strategies are implemented that identify and protect remaining nodes of productive habitat and seek to restore riparian corridors and greenway systems which reduce habitat fragmentation are implemented.

Looking Backward

The decision-making process for managing habitat has evolved from being a marginal issue in salmon restoration to being a major factor in recovery plans for salmon (*Oncorhynchus* spp.) in the Pacific Northwest. It has been a mixture of good and bad logic and serendipitous politics; it involves relationships between land-use activities and issues of watershed health and resource sustainability (Naiman et al. 1992). In order to understand how this mixture came about, we must retrace the history of habitat management in the Pacific Northwest. In this way, we can see how decisions were made in the past and better understand the basis for how current decisions are being made. Most of the examples we cite are related to forestry activities, but other land and water uses have had histories of similar decisions. We conclude our review with a discussion of how recent developments in forest planning are evolving into a general set of regional protocols for addressing the very complex habitat problems related to salmon and their ecosystems in Pacific Northwest watersheds.

Habitat loss has always been a concern of salmon conservation. Even in the 19th century there were papers in the fisheries journals, often anecdotal, suggesting that habitat was being lost at an alarming rate (Stone 1892). The earliest known stock extinction (a sockeye salmon population in the Puget Sound area) related to post-settlement land development was caused by a farmer building a dam across a spawning stream (Nehlsen et al. 1991). Before the turn of the 20th century, the notion of a salmon reserve, a large area where habitat would forever be protected from human development, had been put forward (Stone 1892). Although hatcheries figured prominently in the fisheries management programs of federal and state agencies even in the early part of the 20th century, the extent of habitat loss and knowledge that large hydroelectric dams were about to be built led to some of the first stream surveys in the 1920s and 1930s (Craig and Hacker 1940, Rich 1948). These surveys constitute important historical benchmarks against which current conditions can be measured.

After World War II and the Korean War, there were infusions of surplus heavy equipment that fisheries agencies used to modify stream channels in an attempt to improve habitat. The perception that drove these activities was that migrating salmon were often blocked by log jams and waterfalls; these and other obstructions were the principal culprits in habitat loss. Agency policies viewed impediments to the upstream migration of adult salmon as sources of mortality or reproductive failure (Evermann and Meek 1898, Gharrett and Hodges 1950, McKernan et al. 1950). In addition, fish biologists worked with the United States (US) Army Corps of Engineers to straighten and smooth stream channels so that fish would not be stranded with the rise and fall of the hydrograph. Stream cleaning was believed to benefit navigation and efficient log transport, as well as reduce the risk of fish becoming stranded (Sedell and Luchessa 1982).

The very large flood of late 1964 caused widespread habitat changes in many coastal watersheds of Oregon and Washington, including the creation of many log jams. Repairing the effects of this 100-year event dominated state and federal fish habitat restoration programs for 15 years. Many of these programs involved using heavy equipment or hand cleaning methods to remove log jams. Thus, there was much enthusiasm for engineered habitat solutions that began in the post-war years. The period from 1950 to the mid-1970s also witnessed the construction of many of the region's salmon hatcheries and other large-scale enhancement facilities such as rearing ponds and spawning channels (Hilborn and Winton 1993). Most projects were directed toward increasing the production of one or two species and did not closely resemble the complexity of natural stream environments (Bisson et al. 1992). This was consistent with management's philosophy for a commodity-production orientation towards other natural resources in the region. Protection of ecological processes and conservation of entire aquatic communities were not high priorities.

The impact of the federal court decisions in the mid-1970s on fisheries management in the Pacific Northwest cannot be overestimated (Lee 1993). By affirming the Native American treaty tribes' rights to half of the available salmon harvest in Boldt Decision Phase I and to protection of habitat in which salmon were produced in Boldt Phase II, the courts put fisheries agencies, fishers, and land owners on notice that the tribes would be co-managers of the resource and that harvest and habitat management would have to be linked together more effectively than in the past. The 1970s were also a time when the federal government enacted the Clean Water Act, Clean Air Act, and Endangered Species Act, and many western states were passing the first forest practices acts for state and private forests. Although somewhat cursory, the new laws included requirements to protect streams and riparian zones.

In combination, the Native American fishing rights decisions and new environmental laws moved salmon management organizations to take habitat issues much more seriously in the 1970s and 1980s. Natural resource agencies and timber companies formed fish-forestry groups, the US Congress created the Northwest Power Planning Council whose charter included a mandate to restore salmon runs in the Columbia River basin (Lee 1993), universities held symposia devoted to habitat problems (Krygier and Hall 1971, Salo and Cundy 1987), and stream habitat protection assumed greater importance in land-use plans.

The late 1980s and early 1990s produced a final development that set the stage for the current habitat crisis. The decision of the US Fish and Wildlife Service to list the northern spotted owl and of the National Marine Fisheries Service to list Sacramento River winter chinook (*Oncorhynchus tshawytscha*) and Snake River sockeye (*O. nerka*) and chinook salmon under the Endangered Species Act (ESA) had profound implications for the region's economic underpinnings. Following the spotted owl listing, the Seattle Chapter of the Audubon Society successfully petitioned the courts to halt logging on federal lands until the government could come up with a plan having a reasonable certainty of protecting fish and wildlife. In directing the USDA Forest Service (Forest Service) to solve the problem of managing multiple threatened and endangered species on federal lands, the US Congress demanded that a plan be prepared to provide sustainable fish, wildlife, and timber resources. A small team of Forest Service and university scientists was charged by a congressional subcommittee chairman with solving the timber, old-growth forest, and owl problems, and to include fish because of their critical importance to the region.

FEMAT and Habitat Decisions

The administrative gridlock accompanying the 1991 and 1992 injunctions against logging federal timber in the range of the spotted owl prompted the first visit by a US President and most of his domestic cabinet members to the Pacific Northwest expressly to help solve a natural resource problem. When President Clinton came to Portland, Oregon, in 1993, it was telling that he rarely mentioned owls but he mentioned fish a number of times. One of the main outcomes of the President's Forest Conference was the formation of an interdisciplinary scientific group, the Forest Ecosystem Management Assessment Team (FEMAT), whose charge was to consider both the long-term health of Pacific Northwest ecosystems and human socio-economic systems while developing plans that were scientifically sound, ecologically credible, and legally responsible (FEMAT 1993). The charge of legal responsibility was particularly important because court decisions indicated that federal agencies had not adhered to the principles of the National Forest Management Act (NFMA) in the preceding decade, and it was clear that future logging on federal lands could not take place without legal endorsement.

The FEMAT was asked to produce a plan that yielded a predictable and sustainable timber harvest and that incorporated fish habitat protection as a major component. All federal agencies were directed to cooperate. To protect salmon, the plan had to include the variability and availability of both freshwater and marine environments and develop alternatives employing an ecosystem-based approach. However, the plan was to minimize, to the extent possible, the economic impact of listed organisms on non-federal lands. Large patches of habitat for threatened and endangered organisms (Fig. 1) were set aside in federal forest lands so that the burden of conserving these species would be reduced for states or private land owners. Balancing this objective with the legal requirements of the ESA and the NFMA challenged team members because they were scientists, not lawyers, and little case law pertaining to similar situations existed that could help guide them. The NFMA required that forests be managed for "viable populations, well distributed across the landscape," but FEMAT understood that human presence in watersheds would obviate this goal in some instances by creating significant gaps in species' distributions.

The FEMAT also understood that whatever forest management option was finally adopted could not guarantee protection sufficient to prevent future salmon listings under ESA because federal forest lands supported only a portion of the life cycles of anadromous species (FEMAT 1993). Factors operating outside federal forests influenced the abundance of different stocks. On a region-wide basis, federal forest lands made up only about one-third the total land area within the range of the northern spotted owl. However, the position was taken that the Forest Service and USDI Bureau of Land Management (BLM) had an obligation to manage watersheds in such a way that productive fish habitat would be maintained regardless of the abundance of particular fish populations. Additionally, management plans had to account for uncertainty in such a way that forestry and related activities would not harm streams and riparian zones even when unforeseen natural disturbances occurred.

What finally emerged was an aquatic conservation strategy (Sedell et al. 1994) that included riparian reserves, watershed analysis, habitat restoration, and monitoring built around a system of drainages called "key watersheds" (Fig. 1). These were 4th- to 6th-order watersheds that possessed aquatic habitat considered to be in good condition and to have known populations of potentially at-risk anadromous or resident salmonids. It was assumed that fish from these productive watersheds would eventually be available to colonize surrounding areas as habitat in other watersheds of a basin improved over a period of 1 to 2 centuries. Key watersheds were relatively large (15–1,000 km^2), and the riparian buffer requirements within them (100–140 m wide, each side of the stream) were quite stringent. However, the aquatic conservation strategy specified that riparian reserves could be changed if a detailed watershed analysis had been conducted and a valid ecological rationale for redefining the boundaries existed. Thus, in a very real sense, the strategy utilized principles of adaptive learning (Walters and Holling 1990, Lee 1993) but began with a conservative protection approach that would be likely to cause minimal harm to important aquatic resources. The burden of proof for being allowed to alter riparian buffers so that more timber could be harvested from watersheds was on land managers, who were required to conduct a detailed analysis to justify proposed changes. This strategy of adaptive management became a central theme in President Clinton's Forest Plan (FEMAT 1993). The FEMAT report also recognized that most watersheds would benefit from some sort of restoration, and that watershed analysis would be quite helpful in identifying restoration priorities.

Figure 1. Distribution of key watersheds within the range of the northern spotted owl. Tier 1 watersheds contain at-risk anadromous or resident salmonids (bull trout, *Salvelinus confluentus*). Tier 2 watersheds are important water sources for drainages containing at-risk stocks. Source: Forest Ecosystem Management Assessment Team (1993).

The distribution of key watersheds and wilderness reserves throughout the range of the northern spotted owl was not uniform (Fig. 1). Some regions were fairly well covered while others had few key watersheds in which fish habitat protection and restoration would become a high priority. Large reserves were located along the central axis of the Cascade Mountains, in Washington's Olympic Peninsula, and in the Siskiyou Mountains of southern Oregon and northern California. Other regions such as southwest Washington, western and central Oregon, and many coastal lowlands, had fewer key watersheds. Absence of an extensive network of key watersheds in these regions resulted from the majority of these lands belonging to state and private ownership.

Key watersheds were not meant to be areas where all land uses, including timber harvest, were prohibited. Rather, they were meant to be drainages in which protection of aquatic habitat was given the primary emphasis as other management activities took place. In an idealized watershed, FEMAT requirements provided 100-m buffers for permanently flowing streams, wetlands, and hillslopes with highly unstable soil (Fig. 2). While buffers are present on all permanently flowing streams, many of the ephemeral streams do not have wide buffer zones. Although there may be certain ecological gains from maintaining late-successional buffers along ephemeral streams, there may be even greater benefits (e.g., reduced forest fragmentation) to integrating riparian management objectives of these numerous small drainages with those of adjacent hillslopes. Protection measures around these small streams will still be necessary during logging, but watershed analysis may indicate that some ephemeral streams may not require wide buffers in order for adequate ecological interactions to occur.

Designation of riparian reserves for permanently flowing and ephemeral small streams, as well as unstable hillslopes, distinguished the interim guidelines of the FEMAT from existing forest management plans (Table 1). At the time of the legal injunction against logging late-successional and old-growth federal forests within the range of the northern spotted owl, standard BLM riparian prescriptions in a drainage the size of Augusta Creek, an 11,200-ha watershed in the McKenzie River system of western Oregon, would have accounted for about 8% of the total watershed area. The riparian management guidelines of the Willamette National Forest of Oregon were somewhat more conservative and would have encumbered ~14% of the drainage area (Table 1).

The FEMAT concluded that neither of these riparian reserve systems would have been sufficient to prevent continued degradation of aquatic habitat over a century or more (FEMAT 1993). Application of interim buffer guidelines (100–140 m) for fish-bearing streams and smaller buffers (50 m) for non-fish bearing streams to watersheds such as Augusta Creek would have maintained ~30–40% of the landscape in riparian reserves—several times the amount of land existing forest plans had required. Augusta Creek drained the west slope of the Cascade Mountains and had a moderate drainage density—about 4 km of stream length per km^2 watershed area. However, some streams in the Coast Range of Oregon and Washington had drainage densities of about 6–8 km of stream length per km^2 watershed area. Application of interim FEMAT buffers to these watersheds would have placed ~65–80% of the total area in riparian reserves, a very large and politically unpopular increase over existing protection standards. But the intent of President Clinton's Forest Plan was that, given the precarious status of many stocks of Pacific salmon, a stringent set of protection guidelines should be followed until a thorough analysis indicated that buffer requirements could be relaxed in some parts of the drainage (Sedell et al. 1994). The really significant change was that the FEMAT proposal placed the burden of proof on watershed analysis to show where timber harvest in riparian areas could occur without the likelihood of significant damage to aquatic communities over time.

The other factor that prompted FEMAT to suggest wide vegetative buffers was the knowledge that a number of other, less well-known plant and animal species spent most or all of their lives in riparian zones (FEMAT 1993). Most of the ecological functions affecting aquatic species (Fig. 3, top) were believed to be achieved within a distance from the edge of the channel equal

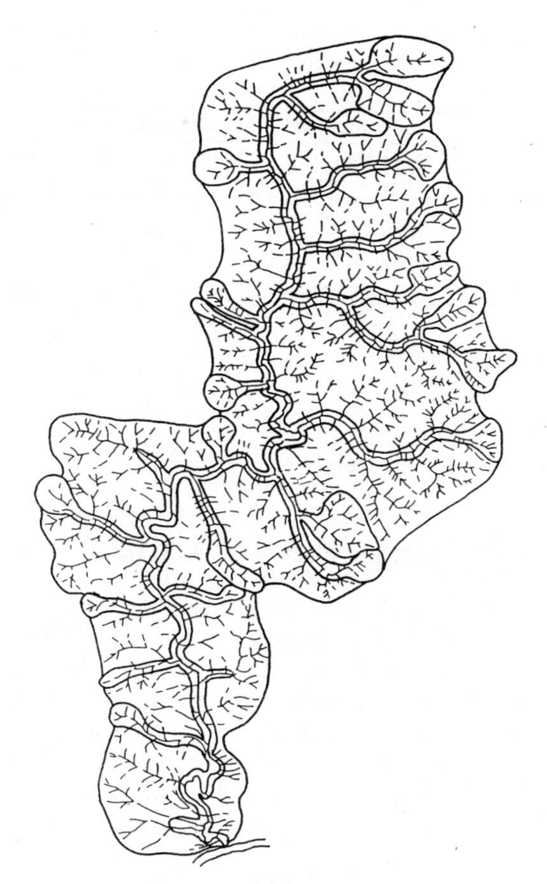

Figure 2. Schematic diagram of how riparian reserves might be arrayed after modification by watershed analysis. The channel network is depicted as follows: —— = permanently flowing streams; ---- = ephemeral streams.

to the height of a mature tree at that site (one "site-index" tree height), a width of ~45–75 m. However, a buffer this wide would be insufficient to protect all aspects of the interior microclimate associated with riparian zones in old-growth forests (Chen 1991). Interim buffer guidelines for permanently flowing streams were therefore extended to the width of two site-index tree heights (90–150 m), which FEMAT (1993) concluded would have a greater likelihood of producing microclimate conditions similar to what would be found in unmanaged forests (Fig. 3, bottom). Additionally, the wider buffers would provide a margin for error owing to scientific uncertainty or to the occurrence of infrequent large, natural disturbances. The overall impact of applying the riparian reserve guidelines to federal forest lands would lead to a significant reduction in watershed area available for timber harvest. The FEMAT believed this reduction alone would promote the recovery of ecological processes in watersheds where habitat had been damaged, as well as protect those functions in watersheds where habitat existed in a natural condition.

Table 1. Riparian reserve widths (one side of stream). Percent of total drainage area in riparian reserves are from Augusta Creek, Oregon, an 11,200-ha watershed. Modified from Thomas et al. (1993).

	Approximate widths (m) of riparian reserves and riparian management areas			
Stream category	Bureau of Land Management	Willamette National Forest	Reserves in non-key watersheds	Reserves in key watersheds
High value, permanently flowing, fish bearing	75	65	110	110
Lower value, permanently flowing, fish bearing	50	30	110	110
Permanently flowing, non-fish bearing	30	30	55	55
Intermittent	0	8	28	55
Area in riparian reserves	8.5%	14%	36%	53%

Habitat Policy in Non-Federal Lands

State and privately owned forest lands in the Pacific Northwest have been regulated by state forest practices rules since the 1970s, whereas until recently federal forests have been regulated through regional forest plans. The approximate percent of key ecological functions included in streamside management zones of different widths can be assessed by comparing the width of the riparian corridor needed for complete functioning, based on available scientific information, with the width of the buffer prescribed by rules and regulations (Fig. 3). Most state forest practices rules prescribe buffers that provide less than complete protection for some riparian functions. For fish-bearing streams, state rules prescribe buffers up to ~30 m, or ~20–30% of a dominant site-index tree height.

Compared with state rules, the interim guidelines recommended by FEMAT provide a much higher level of protection. Therefore, habitat protection policies for federal forests specify a very low risk approach while state policies apply an approach that accepts a greater risk of long-term habitat alteration. It is possible to get a general idea of how effectively state forest practices rules protect aquatic resources using approximate indicators of the completeness of ecological processes as a function of riparian width. For example, if microclimate functions important to riparian-associated wildlife species such as certain amphibians, birds, or bats are considered, we can examine cumulative effectiveness curves for soil moisture, temperature, etc., to suggest riparian management widths needed to protect these species (Fig. 3). More importantly, the curves provide a basis for predicting how reductions in buffer widths are likely to affect different ecological characteristics.

Habitat management policies in the past decade have benefited from information on the location of high-quality habitats and from scientific data suggesting incremental gains or losses associated with varying riparian management zones. Along with this new information has come the political will to protect and restore riparian zones as an important component of the restoration of at-risk salmonids. It is generally acknowledged that state and privately owned lands will not be held to the same low risk standard as that which will be applied to federal forests, but how much risk is acceptable on these lands has not yet been resolved.

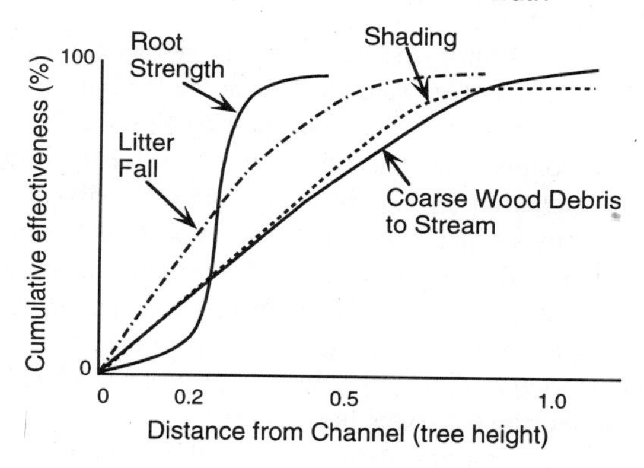

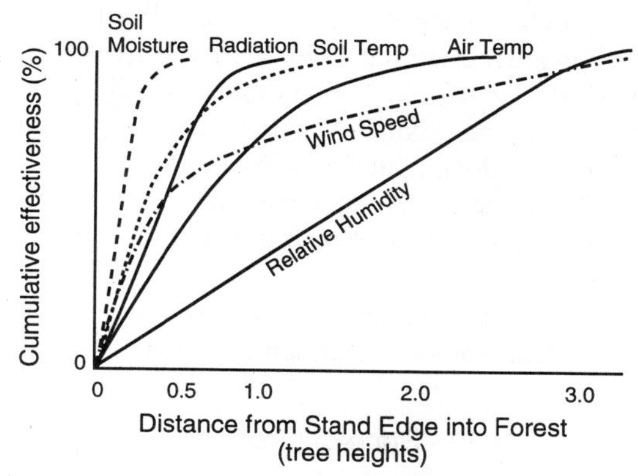

Figure 3. Top: generalized curves indicating percent of certain functions or processes affecting interactions between streams and adjacent riparian zones achieved within varying distances (as indexed to the height of a dominant tree) from the edge of the stream channel. Bottom: generalized curves indicating percent of microclimatic attributes achieved within varying distances from the edge of a riparian forest stand. Source: FEMAT (1993), based in part on Chen (1991).

Looking Forward

The overall effect of the President's Forest Plan and subsequent Record of Decision, if ultimately upheld in court, will be to remove a substantial portion of federal forest lands from the annual timber harvest base. Volumes of timber available for harvest that are believed to be sustainable over time while protecting fish and wildlife habitat are likely to be as little as 10–30% of the timber volumes logged annually during the 1980s. Federal forests and wilderness areas will, in effect, become the mainstays of a refuge system for salmon and other forest-dependent species at risk of extinction in the Pacific Northwest. The policy decision to set aside significant lands for habitat conservation, whether driven by the ESA or other laws and treaties, will be very costly for the region.

Will this be sufficient to prevent at-risk salmon populations from becoming extinct? We believe in some instances it will not. Key watersheds in some federal forests are located around high-elevation montane environments with much of their land above those parts of river basins that support anadromous salmonids. State and privately owned forests in the Pacific Northwest tend to be located at lower elevations, which are more heavily utilized by salmon. Drainages with the greatest potential use by all species of salmon often occur in coastal lowlands and broad river valleys between the Coast Range and Cascade Mountains, areas usually dominated by agricultural and urban land uses. While having improved substantially over the last decade, habitat protection guidelines for state and privately owned forests in California, Oregon, and Washington are still less restrictive than those proposed by FEMAT (FEMAT 1993), and at present none of the Pacific Northwest states have an agricultural practices act that explicitly recognizes and protects riparian functions. In the absence of sweeping reforms of land-use laws to provide more even protection along all watercourses, continued degradation of aquatic habitat on non-federal lands remains likely.

Growth of major urban centers in the Pacific Northwest illustrates the dynamic population increases that have occurred in the region after 1950. Similar increases are seen in the number of automobile registrations, kilometers of roads, fishing license sales, and national forest recreational use (Fig. 4). Pressures on the productive capabilities of the region's natural resources have reached all-time high levels, along with the virtual certainty that they will further increase with continued immigration to the area. Given the inevitability of additional population growth and economic expansion, what can be done to help protect habitat currently in good condition and to rehabilitate habitat that is degraded?

A key to designing and implementing effective habitat conservation programs at the local level (i.e., the scale of individual watersheds or sub-basins) is to involve potentially affected

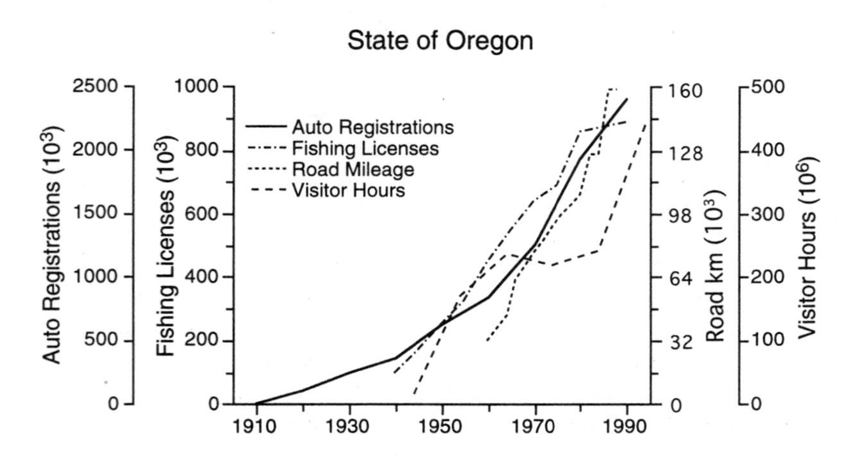

Figure 4. Trends in the number of automobile registrations, kilometers of roads, fishing license sales, and national forest recreational use during the 20th century.

natural resource agencies, land owners, tribes, local governments, and environmental interest groups in policy decisions. For the most part, previous attempts to bring diverse interests to the table have failed (e.g., the "salmon summit" of the early 1990s that attempted, unsuccessfully, to forge a consensus agreement to resolve the crisis of declining Columbia River salmon). Whether more widespread salmon listings, or the threat of listings, under the ESA will be sufficient to catalyze successful co-management agreements remains to be seen, but unless the political will is present to develop consensus programs that have sound ecological underpinnings (Volkman and Lee 1994), prospects for non-litigated solutions seem remote.

Urban and industrial sites, highways, and other permanent structures will prevent restoration of riparian zones in heavily developed areas and some agricultural lands. In those areas, buffers along major river systems will not be continuous and riparian corridors will remain fragmented. Habitat improvement plans will need to identify locations of healthy riparian zones and opportunities for reestablishing corridors of riparian vegetation between them where possible. However, productivity will be lost where development prevents interactions between streams and riparian zones from occurring. Such losses may be unavoidable, but it will be important to locate remaining nodes of good-quality habitat so that they can be managed in a way consistent with the habitat needs of salmon and other aquatic resources.

Watershed analysis procedures for federal, state, and private forest lands are currently being refined; however, the analytical tools should be applicable to agricultural and urban lands as well as commercial forests and should assist in locating nodes of remaining good-quality habitat as well as identifying land-use activities that could cause habitat degradation. Although the stringent standards being applied to watersheds in federal forests will probably not be applied to non-federal land owners, it is essential that habitat management planning take place jointly. This will allow for identifying clear habitat goals, locating remaining productive habitats, and implementing reasonable conservation measures throughout the watershed. A basic premise is that habitat requirements should be met as salmon move through different parts of the river basin across different ownerships, and small, fragmented populations should receive the protection necessary to conserve genetic diversity wherever possible (Reeves et al. 1996).

We believe a sound technical underpinning based on ecological processes has now been established for managing the freshwater habitat of Pacific salmon. We now need better integration of habitat policy with regional policies affecting the harvest and propagation of salmon and other aquatic resources. Management policies involving resident and anadromous fishes are not always complementary. For example, a decision to enhance exotic game fishes or non-native salmonid stocks may affect both the habitat and genetic integrity of native salmonid populations. Harvest rates of commercially and recreationally important salmon may be so great as to deprive streams of the fish needed to occupy available habitats, and the loss of salmon carcasses may itself degrade the productivity of aquatic ecosystems (Bilby et al. 1996).

The future of habitat policy will also involve better coordination between publicly and privately owned land, a coordination enhanced by watershed-based planning groups in which all interested organizations are represented. We cannot return to completely pristine river systems; the Pacific Northwest landscape will consist of permanently degraded as well as productive habitats. Our ability to maintain and strengthen healthy salmon populations in high quality environments in spite of permanent habitat losses in some areas will depend on our ability to work together and to develop systems of practical incentives for land and water managers to increase environmental stewardship across all land uses. The extent to which such cooperative efforts succeed in improving habitat for salmon will depend largely on a willingness to change, a commitment to research and monitoring, and an ability to learn from past mistakes.

Acknowledgments

We thank D. Stouder, R. Naiman, and two anonymous referees for constructive suggestions on the manuscript, and N. Thomas for graphical assistance.

Literature Cited

Bilby, R.E., B.R. Fransen, and P.A. Bisson. 1996. Incorporation of nitrogen and carbon from spawning coho salmon into the trophic system of small streams: evidence from stable isotopes. Canadian Journal of Fisheries and Aquatic Sciences 53: 164-173.

Bisson, P.A., T.P. Quinn, G.H. Reeves, and S.V. Gregory. 1992. Best management practices, cumulative effects, and long-term trends in fish abundance in Pacific Northwest river systems, p. 189-232. *In* R.J. Naiman (ed.), Watershed Management: Balancing Sustainability and Environmental Change. Springer-Verlag, New York.

Chen, J. 1991. Edge effects: microclimatic pattern and biological responses in old-growth Douglas-fir forests. Ph.D. dissertation, University of Washington. Seattle.

Craig, J.A. and R.L. Hacker. 1940. The history and development of the fisheries of the Columbia River. US Bureau of Fisheries Bulletin No. 32: 133-216. Washington, DC.

Evermann, B.W. and S.E. Meek. 1898. A report upon salmon investigations in the Columbia River Basin and elsewhere on the Pacific coast in 1896. Bulletin of the United States Fish Commission for 1897, Volume XVIII. US Government Printing Office, Washington, DC.

Forest Ecosystem Management Assessment Team. 1993. Forest ecosystem assessment: an ecological, economic, and social assessment. Report of the Forest Ecosystem Management Assessment Team to President Clinton. USDA Forest Service. Portland, Oregon.

Gharrett, J.T. and J.I. Hodges. 1950. Salmon fisheries of the coastal rivers of Oregon south of the Columbia. Oregon Fish Commission Contribution 13. Portland.

Hilborn, R. and J. Winton. 1993. Learning to enhance salmon production: lessons from the Salmonid Enhancement Program. Canadian Journal of Fisheries and Aquatic Sciences 50: 2043-2056.

Krygier, J.T. and J.D. Hall (eds). 1971. Forest Land Uses and Stream Environment. Oregon State University Press, Corvallis.

Lee, K.N. 1993. Compass and Gyroscope. Integrating Science and Politics for the Environment. Island Press, Washington, DC.

McKernan, D.L., D.R. Johnson, and J.I. Hodges. 1950. Some factors influencing the trends of salmon populations in Oregon. Transactions of the 15th North American Wildlife Conference 1950: 427-449.

Naiman, R.J., T.J. Beechie, L.E. Benda, P.A. Bisson, L.H. MacDonald, M.D. O'Connor, C. Oliver, P. Olson, and E.A. Steel. 1992. Fundamental elements of ecologically healthy watersheds in the Pacific Northwest Coastal Ecoregion, p. 127-188. *In* R.J. Naiman (ed.), Watershed Management: Balancing Sustainability and Environmental Change. Springer-Verlag, New York.

Nehlsen, W., J.E. Williams, and J.A. Lichatowich. 1991. Pacific salmon at the crossroads: stocks at risk from California, Oregon, Idaho and Washington. Fisheries 16: 4-21.

Reeves, G.H., L.E. Benda, K.M. Burnett, P.A. Bisson, and J.R. Sedell. 1995. A disturbance-based ecosystem approach to maintaining and restoring freshwater habitats of evolutionarily significant units of anadromous salmonids in the Pacific Northwest. American Fisheries Society Symposium 17: 334-349.

Rich, W.H. 1948. A survey of the Columbia River and its tributaries with special reference to the management of its fishery resources. US Fish and Wildlife Service Special Scientific Report 51. Portland, Oregon.

Salo, E.O. and T.W. Cundy (eds.). 1987. Streamside Management: Forestry and Fishery Interactions. University of Washington Institute of Forest Resources Contribution Number 57. Seattle.

Sedell, J.R. and K.J. Luchessa. 1982. Using the historical record as an aid to salmonid habitat enhancement, p. 210-223. *In* N.B. Armantrout (ed.), Acquisition and Utilization of Aquatic Habitat Inventory Information. Proceedings of a symposium, Portland, Oregon, October 28-30, 1981. The Hague Publishing, Billings, Montana.

Sedell, J.R., G.H. Reeves, and K.M. Burnett. 1994. Development and evaluation of aquatic conservation strategies. Journal of Forestry 92(4): 28-31.

Stone, L. 1892. A salmon national park, p. 149-162. *In* Transactions of the American Fisheries Society, Proceedings of the 21st Annual Meeting, Holland House, New York, New York, May 25, 1892.

Volkman, J. M. and K. N. Lee. 1994. The owl and Minerva: ecosystem lessons from the Columbia. Journal of Forestry 92(4): 48-52.

Walters, C.J. and C.S. Holling. 1990. Large-scale management experiments and learning by doing. Ecology 71(6): 2060-2068.

Water-Management and Water-Quality Decision Making in the Range of Pacific Salmon Habitat

Charles F. Gauvin

Abstract

In the United States, water-resources and water-quality decision making have evolved as essentially separate processes. Compartmentalization of the two processes has fostered incremental decision making and frustrated efforts to manage resources and human activities on an ecosystem scale. Natural resources crises like that involving Pacific salmon (*Oncorhynchus* spp.) are the virtually certain outcomes of our nation's failure to overcome this compartmentalization and manage the aquatic environment on a holistic basis.

Water-resources management encompasses two subsystems (resource allocation and development), both of which value water as a commodity and share a common resource exploitation objective. Throughout the western United States, state laws having a lineage in the legal doctrine of prior appropriation govern the private use of water resources. Federal laws and policies govern the development and use of some water resources located on federal land and the development and use of water resources generally for purposes such as flood control, navigation, hydropower, and irrigation. With a few notable exceptions, both state appropriation laws and federal management laws emphasize human subsistence and commercial water uses and de-emphasize or ignore outright water-resource values like fish habitat.

Our primary water-quality protection authority, the federal Clean Water Act, recognizes the importance of the biological integrity of water resources but has not led to effective regulatory mechanisms for protecting and restoring biological integrity. Despite nationwide progress to control certain pollutant sources, other pollutant sources and aquatic ecosystem disturbances remain largely beyond the Clean Water Act's reach. In addition to the Clean Water Act, federal laws of more limited jurisdiction like the National Forest Management Act and the Federal Land Policy and Management Act have not been effective in preventing the pervasive forms of habitat degradation that have depleted many salmon stocks.

After reviewing the major legal and policy influences on water-resources management and water-quality decision making, I examine the need for a new, integrative paradigm to conserve the watersheds that serve as Pacific salmon habitat. Consideration of such a paradigm is timely because of ongoing federal and state efforts to prevent the extinction of salmon stocks and because of pending congressional reauthorization of the Clean Water Act.

Introduction

In this paper, I address the following question: How are water-management and water-quality decisions made by governmental authorities whose jurisdiction encompasses the habitat of Pacific salmon? I have chosen to focus only on what I view as the key mechanisms in water-resources and water-quality decision making as they affect Pacific salmon (*Oncorhynchus* spp.). Analyzing those decision-making mechanisms repeatedly forced me to recall a statement that United States (US) Supreme Court Justice Oliver Holmes once made about the law. According to Justice Holmes, when it comes to understanding the law, "an ounce of experience is worth a pound of logic." Justice Holmes's statement did not specifically contemplate water law, but it easily could have applied to the way our nation's legal and political system allocates and manages water resources in the watersheds that serve as Pacific salmon habitat.

In the US, water policy and law have evolved into two essentially different processes: water-resources management and water–quality management. In the jurisdictions that contain Pacific salmon habitat, water-resources management involves (1) the allocation of private rights to use water, and (2) the development and operation of water-resources projects (e.g., hydroelectric power generation, flood control, irrigation, navigation, and water supply). Until recently, water-resources management has been virtually blind to considerations of environmental quality. Water-quality management is focused on controlling and limiting the discharge of pollutants and mitigating their effects on human health and the environment. This process has always had as its endpoint some broad notion of environmental quality, whether human health or ecosystem health.

To return to Justice Holmes for a moment, consider how logic might have determined the rules of the game in water law and policy. Even the narrowest logic would have viewed water as a renewable resource and created a system that secured its availability for whatever use (industry, hydropower, irrigation) that was perceived as the most important. That, in fact, was the logic that drove the reclamation movement in the early years of this century: Water resources, properly husbanded, would be forever available for human use. Somewhere along the way, however, logic ran hard aground on the shoals of experience. In part, experience has revealed the flaws in the logic of reclamation and other resource development programs: Water resources, if exploited to the point of plunder, are not infinite. Experience also has demonstrated the strong role that changes in societal values can play in water law and policy.

Thus, it is experience, not logic, that has delivered the crisis in which we today find Pacific salmon and their watersheds. Nehlsen et al. (1991) identify habitat destruction, hydropower, and stream dewatering as the principal causes of decline for ~197 of 214 at-risk salmon stocks. This is a ringing indictment, a prima facie case, that our nation's water laws and policies have failed to protect key elements of our ecosystems. Although the Pacific salmon crisis is one of the greatest natural resource disasters in American history, it is also the modern-day repetition of a similar 19th century experience, namely, the destruction of New England's Atlantic salmon stocks.

We must bear in mind the larger role of water policy and law in both creating, or failing to halt, the destruction of the Pacific Northwest's salmon stocks and reversing that destruction to rebuild the remaining stocks. In the complex of causes that has brought some Pacific salmon stocks to the brink of extinction and pushed many others beyond it, it is often difficult to inculpate site-specific actions. The road to extinction is quite unlike the scene of a crime drama; there usually is no corpus delicti, and we seldom learn the villain's identity. The decisive role of large-scale historical factors, such as the damming of the Columbia and Snake rivers and the destruction of riparian systems that began when European settlers logged most of the region's forests, rather than individual catastrophic events, suggests that changes in water law and policy could make a decisive difference in the outcome of the salmon crisis.

From the standpoint of Pacific salmon conservation, the most glaring need for change in water law and policy is to overcome the compartmentalization of water-resources management and water-quality management. Separation of the two processes has meant that water-resources management could proceed largely without consideration of environmental quality, and that water-quality management would lack all of the tools necessary to achieve environmental quality. It also has fostered incremental decision making and frustrated efforts to manage human activities on an ecosystem scale. Natural resources crises like the current one involving Pacific salmon are the virtually certain outcomes of our failure to overcome this compartmentalization and manage the aquatic environment holistically. With Justice Holmes's admonition in mind, I examine the compartments of water law and policy and see what they hold for the future of Pacific salmon and their ecosystems.

Overview of Policy and Legal Framework

Water-Resources Management

Private Water Use

Appropriation rights to water are the subject of many treatises and the primary source of income for a significant sector of the legal profession west of the Mississippi River. Appropriation law governing private water use derives from the customs and usage of mining camps and other early settlement in the western US (the West). Commentators on the appropriation system have identified three basic policies underlying appropriation law: (1) "free" water, (2) capture, and (3) water as property (Butro et al. 1993). The three policies interact in the following manner. From the earliest settlement onward, water resources were free for human exploitation. For water to be free, it had to be readily accessible. Claims to water became synonymous with claims to real property with one important distinction. Persons claiming the right to use water were not required to own or occupy streamside property but only to assert and perfect a right to appropriate the water itself. To facilitate access, western legislatures enacted laws early on providing rights-of-way to non-riparian property owners.

Along with ready access, the policy of free water developed pricing mechanisms that valued water at only its commercial sale price. External costs like damage to fish and wildlife habitat did not and, to a large extent, still have not entered the pricing equation. The obvious bias in the system is to underprice water so that there is no economic penalty for waste. This means that there is little or no incentive for users to conserve water.

The policy of capture meant that one claiming a water right had to capture (i.e., divert and remove) the entire amount underlying the claim. This continues to be the way in which water-rights claimants perfect their rights. Failure to capture the entire amount carries the risk of forfeiture to other claimants, a risk that causes water-rights holders to waste water by annually appropriating all the water to which they are entitled regardless of actual need. Once captured, water was, and largely remains, available for any use without the stipulation (common to the riparian rights systems of the eastern US) that the user return it to its source. This has allowed some notorious interbasin water transfers such as that associated with the water in California's Owens River Valley, which the City of Los Angeles captured for its own purposes beginning in the early decades of this century.

Once perfected, water rights are enforceable in order of priority. The seniormost right holders must have satisfied their water rights before the "juniors" can take their water. To administer water rights, states have established water-resources boards or similar authorities as well as specialized water courts.

The policy of water as property anointed the appropriation system with a special, nearly inviolate status under the law. Once water became property, rights to water received federal and state constitutional protection against government infringement. Water-rights holders could and still do claim that, just as government cannot take a tract of land, a house, or an automobile without due process and just compensation, government cannot take a water right without due process and just compensation. More than any other policy underlying the appropriation system, this one continues to confound state law reform efforts while spilling over into and clouding the debate surrounding private rights to water on federal lands and those lands that the federal government reserved for Native Americans.

Many of the Pacific Northwest's streams are over-appropriated, meaning that water rights exceed mean annual flow. Along with rights to water on non-federal lands, irrigators and municipalities have laid claim to waters on federal lands administered by the USDA Forest Service (Forest Service) and the USDI Bureau of Land Management (BLM). In the past several years, disputes have arisen regarding the obligation of water diversion and storage operations to release bypass flows, which are minimally necessary to maintain aquatic life in the diverted river and tributary reaches on federal lands. Property-rights claims of irrigators and municipalities clash with the federal land management agencies' resource stewardship responsibilities. In 1991, Agriculture Secretary Madigan decreed that the Forest Service would not require bypass flows in special-use permits for water storage and diversion projects on Forest Service lands. Madigan's decision, although in response to a dispute involving water storage and diversion projects in Colorado's Front Range, is part of a large controversy involving federal versus state authority over instream flows on federal lands throughout the West.

The net effects of the appropriation system have been disastrous for Pacific salmon. Waste and overuse have resulted in widespread dewatering of spawning and rearing habitat. Diversion works used to capture water have blocked migratory fish passage. The status of water as property has made governmental efforts to obtain water for instream uses like fish habitat especially burdensome.

The past several decades have seen some diminution of the appropriation system's potency. State constitutional and statutory proscriptions against waste have brought some increases in use efficiency. Programs to purchase or lease-back water rights or reserve unappropriated waters for instream uses have met with limited success. Some states have used public trust theory (the idea that such public values in water resources as fish and wildlife habitat and recreational uses limit private water rights) to limit existing appropriations.

Despite limitations on the appropriations system, keeping water in streams for fishery habitat conservation purposes has proven difficult. Even if a state has legislation providing for purchase or lease-back of water rights, state water resources or fishery authorities must negotiate with individual water-rights holders regarding important details like price and timing of water releases and withdrawals. Water-rights holders have tended to value their rights at astronomical levels, making it nearly impossible to consummate water transfers for instream purposes. Even where state law clearly provides for public reservation of unappropriated water, the amount of water reserved has been the subject of time-consuming and costly litigation.

Water-Resources Development

There is a distinction between the system of private rights to appropriate water use laws and the policies that have led to widespread and often large-scale development of rivers and watersheds in the range of Pacific salmon. Although the appropriations system is manifestly private in character and reflective of the creed of rugged individualism, its beneficiaries accepted the beneficent hand of government intervention when it came to water-resources development. Physical limitations on individual farmers' ability to capture and store water gave rise to the reclamation movement, which enlisted government in the creation of storage and distribution projects. Energy needs of emerging industries and expanding municipalities (or at least the congressional sponsors' and development agencies' perception of such needs) drove the creation of hydroelectric storage and generation facilities. The need for waterborne transportation of commodities and finished goods fueled the construction of navigation projects.

Whereas the appropriation system establishes under state law individual rights to water in a private (albeit not really free) market setting, federal water resource development policies and programs have a manifestly public works character. Aside from this significant distinction, the two systems have shared common resource exploitation objectives. Through the US Army Corps of Engineers (the Corps), the Bureau of Reclamation (BuRec), and other agencies, federal water resource development projects have replicated on an macrocosmic ecosystem scale what the appropriation system wrought microcosmically for headwater streams and mainstem river reaches.

Reclamation, the earliest federal water resource development program, aimed to create public works projects that transformed the arid West into a veritable garden. The program's ostensible purpose was to facilitate settlement and economic growth. After a slow beginning in the early part of this century, federal reclamation projects by the Corps and BuRec during the 1930s set out to harness the mighty Columbia River, the continent's greatest Pacific salmon nursery, for hydropower, flood control, and irrigation. Skillful and sometimes deceitful shepherding of congressional appropriation bills for individual projects enabled sponsors to effect water resource development with great dispatch. Reisner (1986) describes the Depression-era process to authorize construction of the Grand Coulee Dam: "What had begun as an emergency program to put the country back to work, to restore its sense of self-worth, to settle the refugees of the Dust Bowl, grew into a nature-wrecking, money-eating monster that our leaders lacked the courage or ability to stop."

Loosed of their moorings in the policy of reclamation, major federal multipurpose water-resources projects became programs in search of policies and largely have remained so. This has been incremental decision making at its apogee, aptly illustrating the truth of Justice Holmes's prescription about the prevalence of experience over logic as well as the law of unintended consequences. Ecosystem destruction like that which has occurred in the Columbia-Snake River

Basin and California's Central Valley has prompted an array of mitigation and enhancement programs as add-ons. As a result, basic operating decisions affecting water allocation within project systems have become staggeringly cumbersome, complex, and costly. Consider, for example, the host of governmental authorities, private user groups, and other interests that now seek to influence operation of the Columbia-Snake River hydropower system through the Northwest Power Planning Council (the Power Council): Bonneville Power Administration, public utility districts, state water-resources boards, irrigation districts, municipal and other wastewater treatment systems, Native American governance and business entities, federal and state land management agencies, port authorities and private shipping organizations, federal and state environmental and fishery and wildlife agencies, private manufacturing industry organizations, commercial and recreational fishery organizations, and conservation organizations.

A similar host of governmental agencies and interest groups now plays a role in the operating decisions of BuRec regarding California's Central Valley project. Moreover, by virtue of its listing decisions for Snake River salmon stocks (two chinook salmon runs [*O. tshawytscha*] and sockeye salmon [*O. nerka*]) and one Sacramento River salmon stock (winter-run chinook), the National Marine Fisheries Service (NMFS) and other federal agencies have assumed a direct role in project operating and water allocation decisions for both the Columbia-Snake River Basin projects and the Central Valley Project.

Thus, what began as a public works program that required nothing more than passage of appropriations legislation for individual water projects has grown incrementally as project impacts on fish and wildlife have stimulated demand for interest group participation. Although this is undoubtedly an understandable result, its efficacy remains open to question. Implementation of project operation and water allocation measures to conserve and restore depleted salmon stocks encounters not only the physical obstacle of the projects themselves but also a formidable set of entrenched economic and political interests.

Water-Quality Management

Whereas the appropriation system and policy grew out of settlement of the rural West, our nation's system of water-quality management had its origins in the urban environments of the eastern US and Western Europe. From early on, the objective of water-quality management has been to limit or control the chemical, physical, and biological changes in the aquatic environment that occur as a result of wastewater discharges from factories and sewage treatment facilities. Water allocation issues rarely have entered into water-quality management decision making. The exception occurs when it has been necessary to ensure sufficient dilution capability in a water body receiving one or more wastewater discharges. Only recently, in the technical and policy literature and in a few regulatory decisions, has water-quality management begun to consider instream flow issues and their effects on the aquatic environment.

The federal Clean Water Act (the Act) governs both federal water-quality management efforts and, through minimum standards and delegation agreements, state water-quality management efforts. Its enactment in 1972 took the federal government on a new course. Although previous water-quality legislation had provided funding for pollution control projects and had given federal authorities some enforcement authority, it had not comprehensively defined water pollution, or created a nationwide system for reducing and eliminating water pollution, or provided any reasonably certain or effective enforcement mechanisms (Gauvin 1993a). Section 101 of the Act proclaimed a national objective "to restore and maintain the chemical, physical, and biological integrity of the Nation's waters" by eliminating the discharge of pollutants into navigable waters by 1985 and by achieving an interim water-quality level that would protect fish, shellfish, and wildlife while providing for recreation in and on the water. The Act's most significant mechanisms to date for accomplishing its objective are its discharge permits and dredge and fill permits programs.

The discharge permit program regulates point sources, which Section 502 of the Act defines as "any discernible, confined, discrete conveyance, including, but not limited to, any pipe, ditch, channel, [or] tunnel." Section 502 expressly excludes from the definition of point source agricultural stormwater discharges and irrigation return flows, thus exempting significant sources of water pollution within the range of Pacific salmon. Section 402 authorizes the US Environmental Protection Agency (EPA) and state water-quality agencies to issue discharge permits for point sources, which include limits on the chemical or physical characteristics of a point source's effluent. Discharge limits reflect both technological (i.e., pollutant removal capability) and water-quality considerations.

Much of the regulatory effort on the part of EPA and the states during the past decade has centered on the development of so-called effluent limitations based on water-quality standards, which the Clean Water Act requires states to promulgate and which are the Act's benchmark for determining water quality. A water-quality standard has two parts: (1) a designated use or uses (e.g., coldwater fish habitat, warmwater fish habitat, drinking water, or non-contact industrial cooling water) for the water body to which it applies, and (2) a water-quality criterion that is either a narrative expression (e.g., no toxics in toxic amounts) or a numeric unit of measurement (e.g., ≤ 3 μg L^{-1} of free cyanide or ≥ 5 μg DO). A water-quality criterion represents a more or less precise scientific estimate regarding the level of water quality that is necessary to protect the designated use(s) of a particular water body.

In addition to discharge permitting for point sources, the Clean Water Act's Section 404 dredge and fill permitting program controls some physical alterations of wetlands and, incidentally, riparian corridors. The EPA and Corps jointly administer the program at the federal level and oversee state permitting programs.

Section 208 of the Clean Water Act attempted to broaden the focus of state regulators beyond the sources of aquatic impairment that came within Sections 402 and 404. Section 208 directed states to develop management plans and Best Management Practices (BMPs) to prevent or reduce loadings of such pollutants as sediment, pathogens, nutrients, and toxics from such non-point sources as agriculture, timber harvest, and land development activities.

The move by some states to develop BMPs grew out of the recognition that the Clean Water Act's federally enforceable jurisdiction under Sections 402 and 404 represented only a small slice of the total aquatic environment. In much of the nation's rural headwater drainages, including most of its Pacific salmon habitat, there are few if any point sources. Land disturbances, like those associated with timber harvest and grazing, seldom if ever come within Section 404's permitting process. Data collected by the EPA during the 1980s showed tens of thousands of river km not attaining designated uses for fishery habitat. Impairments such as siltation, nutrients, and pathogens contributed to >40% of the failure to attain water-quality standards (EPA 1990). More than 50% of the impairment was due to non-point source pollution from agriculture, resource extraction, and silviculture (EPA 1990).

With the Water Quality Act of 1987, Congress attempted to address non-point source pollution in the form of runoff that washes from farm fields and timber harvesting, mining, grazing, and other land-disturbing activities. The 1987 law added Section 319 to the Clean Water Act, which required states to prepare and submit to EPA a management plan and schedule for implementing BMPs to control polluted runoff. Section 319 merely required preparation and submittal of a management plan and implementation schedule. Unlike the Clean Water Act's provisions for point source limitations, Sections 208 and 319 did not enable EPA to impose on sources of polluted runoff its own BMPs where states failed to adopt or enforce their own.

While disappointing from a national perspective, Sections 208 and 319 have had some positive results in states with Pacific salmon habitat: Oregon, Idaho, and Washington now have forest practices laws and regulations that impose BMPs on timber harvest and associated activities (McDonald et al. 1991). The BMPs involve such requirements as buffer strips in riparian areas, restrictions on road building and equipment handling, removal of slash debris from riparian areas, and limitations on stream crossings (Doppelt et al. 1993).

Although BMPs are a step in the right direction, questions remain concerning their effectiveness. Oregon's forest practices law and regulations, often cited as an example of a strong state program to control polluted runoff, are "almost completely inadequate to protect riverine-riparian ecosystems and biodiversity" (Doppelt et al. 1993). Idaho's forest practices law, which imposes a minimal buffer strip along rivers and streams, offers at best token protection for aquatic habitats. States are making some effort to monitor the implementation and, to a lesser extent, the effectiveness of their BMPs (Adler et al. 1993). The cookbook nature of most BMPs is well suited to implementation monitoring; audit teams visiting private timberlands can make reasonably accurate and cost-effective determinations of compliance. Effectiveness monitoring, however, raises more difficult questions about the extent to which the uniform prescriptions in the forest practices regulations conform to site-specific conditions. Doppelt et al. (1993) suggest that "the ideal way to identify 'riparian management areas' is by using ecological definitions on a site-specific basis, but most state agencies—if they address this at all—specify minimum widths of these zones, leave-tree numbers, and percent of required shading on an across-the-board basis which ignores, and is most often at odds with, ecological realities."

Despite nationwide progress on controlling point sources and some movement by the states on controlling polluted runoff, the Clean Water Act has other serious shortcomings when it comes to protecting and restoring salmonid habitat. There is no statutory directive for anti-degradation, which could protect the nation's truly pristine and high-quality waters. The EPA views its current anti-degradation policy as unenforceable owing to the lack of an express statutory foundation and has not used it to force states to comply. In a comprehensive study of state anti-degradation programs, the National Wildlife Federation (1992) found great disparity among the requirements of state programs and concluded that all but 15 fell short of EPA's standards. Thus, like the polluted runoff control program, EPA's anti-degradation program relies heavily on the willingness of states to develop and enforce programs that are critical to attaining the Clean Water Act's objectives.

Clean Water Act Section 313 requires federal agencies "having jurisdiction over any property...or...engaged in any activity resulting, or which may result, in the discharge or runoff of pollutants..." to comply with all federal and state requirements under the Act. Despite the clear mandate of that section, the EPA has not enforced its own anti-degradation policy on federal lands (Adler et al. 1993). Instead, the agency has relied on states to enter cooperative agreements with federal land managers to apply state anti-degradation standards. Under one such memorandum of agreement, Idaho now applies its anti-degradation program to waters on federal lands. However, the EPA's timidity in imposing its anti-degradation policy on states and in applying it directly to federal lands means that some of the last undisturbed salmon habitat in the Pacific Northwest is receiving less protection than it should.

Where it concerns many riparian zone alterations and direct fishery habitat impacts, Section 404 of the Clean Water Act is virtually powerless. There are no express requirements that state water-quality criteria address fluvial geomorphology or protection of riparian zone inputs, like large woody debris, to stream channels. EPA guidance under Section 404 defines "wetlands" narrowly to exclude many riparian areas and exempts numerous activities, including normal agricultural, silvicultural, and ranching activities that individually or cumulatively can degrade riparian zones and streams.

Even when viewed from a strictly water-quality perspective, state water-quality standards and criteria often address only water column concentrations of pollutants. Thus, they can fail to account for the long-term mass loadings of persistent pollutants in the aquatic environment. Water-quality criteria also address primarily the human health effects of pollutants. Where aquatic species criteria exist, they often rely on only single-species bioassays and are imprecise predictors of real-world impacts. The EPA has sought to remedy the deficiencies in state water-quality criteria and make them better suited to application in controlling polluted runoff. In 1989, the agency issued protocols for biological assessments of rivers and streams (EPA 1989). Since then, the agency has encouraged states to incorporate the protocols in their water-quality standards and to use them in assessing water-quality impairments and evaluating the effectiveness of such control measures as BMPs (McDonald et al. 1991). If rigorously applied, the EPA bioassessment protocols are valuable tools in protecting sensitive headwater streams such as those that serve as spawning and rearing habitat for Pacific salmon.

What is perhaps the greatest flaw in the Clean Water Act from a holistic management standpoint is a seldom-noticed proviso in Section 101, which Western-state members of Congress succeeded in adding to the Act as part of the 1977 reauthorization package. Section 101(g) plainly states that "the authority of each state to allocate quantities of water within its jurisdiction shall not be superseded, abrogated, or otherwise impaired." This provision underscores the compartmentalization of water-resources management and water-quality management and has limited efforts to realize the Act's objective of restoring and maintaining the chemical, physical, and biological integrity of our nation's waters.

Some states now include instream flow requirements in their water-quality standards which, as creatures of state law, are immune from Section 101(g). The United States Supreme Court's recent decision in *PUD of Jefferson County and City of Tacoma v. Washington Department of Ecology,* 511 US_ _, 128 L.Ed.2d 716, 114 S.Ct. 1900 (1994), upholds the authority of states to impose numeric and narrative water-quality standards on federally licensed hydropower projects. The decision is a major development in fishery conservation. Whether the EPA can force the state of California to include in its water-quality standards for salinity sufficient instream flows for Sacramento River winter-run chinook and other aquatic species has become a contentious state-federal issue.

Given the Clean Water Act's shortcomings, one is entitled to wonder about its utility in protecting Pacific salmon and their habitat. Yet the Act remains our only water-quality law of general jurisdiction and, as such, offers the potential to address water quality on an ecosystem-wide, comprehensive basis across federal and non-federal lands. The Act's legislative history suggests that an ecosystem-wide approach is what chief sponsor Senator Edmund Muskie and others had in mind when, in 1972, they crafted the words "to restore and maintain the chemical, physical, and biological integrity" of our nation's waters as the primary objective of the new law (Gauvin 1993a). Whether the Clean Water Act can accomplish that objective is a timely question given the ongoing congressional effort to reauthorize it. In the following section, I argue that the reauthorization process must make some inroads in the compartmentalization of water-resources management and water-quality decision making if it is to achieve its objectives, and that doing so would help to avert, for other species, the crisis that now confronts Pacific salmon.

The Role of Fisheries Conservation in Managing Water Resources and Water Quality

Fisheries conservation still does not figure prominently in the general framework of water-resources management or water-quality management. Several federal laws incorporate fisheries conservation objectives in their respective jurisdictions, and thereby influence management of water resources and water quality in the range of Pacific salmon. With the exception of the Federal Power Act, which Congress originally intended as a means of developing water-resources and later amended to control the effects of such development, the laws and policies discussed below represent efforts to compensate for the failure of existing water-resources and water-quality management to conserve aquatic resources like Pacific salmon.

The Endangered Species Act

The Endangered Species Act of 1973 (ESA) implements the policy of Congress "that all Federal departments and agencies shall seek to conserve endangered species and threatened species and shall utilize their authorities in furtherance of [such] purpose..." (ESA, Section 2(c)(1)). The ESA's requirements apply to federal agency actions. Where state water laws and policies are concerned, Congress refrained from applying the Act's enforceable requirements to state and local government actions not directly involving the conservation of fish and wildlife and specifically directed that federal agencies "cooperate with state and local agencies to resolve water resource issues in concert with conservation of endangered species" (ESA, Section 2(c)(1)).

In the context of Pacific salmon conservation, the ESA's definition of species applies to individual stocks that are reproductively isolated and evolutionarily significant (56 Fed. Reg. 58612 (November 20, 1991)). To date, NMFS has listed the Sacramento River winter-run chinook and Snake River sockeye as endangered and the Snake River spring-summer chinook and the Snake River fall chinook as threatened. In 1994, NMFS changed the listing for the two Snake River chinook stocks from threatened to endangered.

The listing of any species can trigger the ESA's Section 7 consultation requirements. Whenever any action that may affect a listed fish species requires a federal license, permit, or other authorization, the federal agency with decision-making authority must consult with NMFS, the lead agency under the ESA for managing Pacific salmon, to ensure that any agency action "is not likely to jeopardize the continued existence of any endangered species or threatened species or result in the destruction or adverse modification of [critical] habitat" (ESA, Section 7 (a)(2)). Consultation results in a biological opinion "detailing how the agency action affects the species or its critical habitat" (ESA, Section 7 (6)(3)(A)). If the opinion finds species jeopardy or adverse habitat modification, NMFS must suggest "reasonable and prudent alternatives" to the agency action (ESA, Section 7 (6)(3)(A)).

A federal court decision rejected the rationale for the 1993 NMFS determinations that Columbia-Snake River hydrosystem dams posed no jeopardy for the listed Snake River salmon stocks (*Idaho Department of Fish and Game v. National Marine Fisheries Service*, 56 F.3d 1071 (D. Or. 1994)). That decision could set new standards for application of the ESA to the listed Snake River stocks and the scores of other Pacific salmon stocks that have been petitioned for listing under the ESA. Ultimately, it and other judicial decisions under the ESA may induce federal hydropower system operators, land managers, and fishery management agencies to avoid replication of the conditions that have given rise to the raft of listing petitions for Pacific salmon.

Despite several years of interagency consultation, the issuance of several biological opinions, and the development of a draft by the NMFS-appointed recovery team for Pacific salmon in the Columbia-Snake River system, NMFS has not yet issued a final recovery plan for listed salmon stocks. Continuing controversy involving the extent to which the Bonneville Power Administration's budget should support salmon recovery and the appropriate role of using barges to transport salmon smolts around hydropower dams thus remains officially unresolved, despite the passing of more than four years since the listing of Snake River salmon stocks under the ESA.

In the case of actions that do not trigger Section 7's consultation requirements, the ESA's ban on the "take" of listed species may prohibit the adverse modification of habitat, such as might occur in connection with timber harvest or other land-disturbing activities on non-federal lands. However, in *Sweet Home Chapter of Communities for a Better Oregon v. Babbitt*, 17 F.3d 1463 (DC Cir. 1994), the court held that the ESA's ban on taking of listed species does not apply to activities that adversely modify habitat but do not involve a physical taking of the species. The US Supreme Court's 1995 decision in *Sweet Home* resolved the split among the federal courts in favor of including adverse habitat modifications within the definition of take, and it affirmed the validity of prohibiting adverse habitat modification on non-federal lands. This established one of the ESA's strongest species protection tools (*Sweet Home Chapter of Communities for a Better Oregon v. Babbitt*, __US_ _ 130 L.Ed. 2d 621, 115 S.Ct. 714 (1995)).

Whatever its virtues as a safety net for critically depleted fish and wildlife species, Congress never intended the ESA to be a first line of defense against species extinction. Yet the ESA has assumed just that role where resource management laws have failed to achieve their intended purpose. Congress is considering a number of proposals to reauthorize and amend the ESA. Several would substantially weaken the law and undermine its ability to recover Pacific salmon stocks. Pending reauthorization, new listings of salmon stocks have been halted by a congressional moratorium on listings enacted in the DOI appropriations legislation.

The Pacific Northwest Power Planning and Conservation Act of 1980

The Northwest Power Planning Act arose largely from the failure of the federal Fish and Wildlife Coordination Act of 1958 to compel the Corps to develop effective salmon passage facilities at the mainstem Columbia and Snake River hydropower projects (Bean 1983). In the Northwest Power Planning Act, Congress authorized creation of the Northwest Power Planning Council (the Power Council) and directed it to develop a regional electric power and conservation plan "to protect, mitigate, and enhance fish and wildlife, including related spawning grounds and habitat, on the Columbia River and its tributaries" (Pacific Northwest Power Planning and Conservation Act, Section 4(a)). The Northwest Power Planning Act requires the Bonneville Power Administration (BPA), the Corps, BuRec, and the Federal Energy Regulatory Commission (FERC) to take into account "to the fullest extent practicable" the Power Council's fish and wildlife program in the operation of existing hydroelectric facilities and the development of new ones (Pacific Northwest Power Planning and Conservation Act, Sections 4(d)(3) and 6). Its most immediate and perhaps lasting effect has been to increase public participation in power planning and fish and wildlife conservation through the proceedings of the Power Council.

Several Power Council programs under the Northwest Power Planning Act directly affect water-resources management in the Columbia-Snake River Basin. The first is the Power Council's Water Budget, which seeks to enhance hydropower system flows during peak juvenile salmon outmigration periods. However, the failure of the hydropower system's chief operating agency, the Corps, to view the Water Budget as anything more than a non-binding "cooperative agreement" has frustrated implementation of the Water Budget (Wood 1993) and was a significant contributing factor in the filing of ESA listing petitions for the Snake River salmon stocks. In response, the Power Council adopted Fish and Wildlife Program Amendments in December 1991 that attempted to clarify the Water Budget's flow requirements, but these have been viewed as ineffectual (Wood 1993).

The Power Council's 1991 Fish and Wildlife Program Amendments prompted a lawsuit by salmon conservationists, who challenged the Amendments as insufficient to meet the Northwest Power Planning Act's salmon restoration objectives. Specifically, the lawsuit addressed the Power Council's unwillingness to make immediate changes in Columbia-Snake River hydrosystem operations to facilitate downstream passage of juvenile salmon and establish firm biological objectives for rebuilding depleted stocks of Snake River salmon. In a September 1994 opinion, the US Court of Appeals for the Ninth Circuit made the following statement:

> The Council's approach seems largely to have been from the premise that only small steps are possible, in light of entrenched river user claims of economic hardship. Rather than asserting its role as regional leader, the Council has assumed the role of a consensus builder, sometimes sacrificing the Act's fish and wildlife goals for what is, in essence, the lowest common denominator acceptable to power interests. . . . (*Northwest Resource Information Center, et al. v. Northwest Power Planning Council*, 35 F.3d 1371 (9th Cir. 1994)).

The court remanded the Fish and Wildlife Program Amendments to the Power Council and instructed it to revise them to incorporate more rigorous fish passage measures based on firm biological objectives. In 1994, the Power Council issued new amendments that responded to the most significant concerns about the hydrosystem's impacts on salmon. Nevertheless, because of changes in the region's political leadership that occurred in the 1994 elections, those amendments have not been effectively implemented.

The Power Council's Protected Areas Program has met with limited success. The program's purpose is to set aside from future hydropower development certain rivers and river reaches having high-quality fish and wildlife habitat values. Its recommendation to protect 15% of the 350,000 river miles (560,000 km) in the Basin resulted in BPA's decision to block access to power lines for future hydropower projects and has played a role in FERC licensing decisions for non-federal hydropower projects (Palmer 1991).

The Power Council's view that its role in water resource management is largely advisory and its unwillingness to enforce its directives have severely limited the Northwest Power Planning Act's effectiveness (Wood 1993). The Power Council's 1994 Fish and Wildlife Program Amendments sought to reassert its prerogative over water resource management matters in the mainstem of the Columbia and Snake rivers, but the Council has done little since adopting the amendments to implement them.

The Central Valley Project Improvement Act of 1992

As part of the Reclamation Project Authorization and Adjustment Act of 1992, Congress directed BuRec to allocate 987,000 ha-m (800,000 acre-ft) of water (740,000 ha-m [600,000 acre-ft] in dry years) for fish and wildlife habitat improvements in California's Central Valley. The Project Improvement Act also established a $50 million annual fish and wildlife habitat restoration fund. Providing additional water for fisheries should help to recover the depleted winter-run chinook salmon in the Sacramento River, which is now listed as an endangered species. Already controversial before its enactment, the Central Valley Project legislation faces strong opposition in its implementation from agricultural interests.

The National Forest Management Act of 1976

National Forest lands in the Pacific Northwest contain many streams that serve as spawning and rearing habitat for Pacific salmon stocks. Management of activities like timber harvest, grazing, and mining on National Forest lands has been a subject of longstanding concern to Pacific salmon conservationists. Sedimentation and changes in stream morphology from extractive land uses are primary factors in salmon stock extinctions and depletions (Nehlsen et al. 1991, Forest Ecosystem Management Assessment Team [FEMAT] 1993; The Wilderness Society, unpubl. data).

In response to a growing conflict between conservationists and the timber industry regarding the environmental consequences of clearcutting on National Forests, the US Congress enacted the National Forest Management Act of 1976 (NFMA) (Wilkinson 1992). The NFMA establishes land and resource planning for individual forest units. The object of such planning is to establish allowable timber harvest levels and harvest modes. The NFMA also authorizes the Forest Service to develop a riparian management policy and expressly prohibits timber harvesting where it is likely "to seriously and adversely affect water conditions for fish habitat" (NFMA Section 1604 (g)).

Under NFMA, National Forests develop 10-yr forest plans that seek to "provide for multiple use and sustained yield of goods and services from the National Forest System in a way that maximizes long term net public benefits in an environmentally sound manner" (36 C.F.R. Section 219). Although multiple use has been and continues to be a problematical goal in the context of fishery conservation (Trout Unlimited 1994), NFMA created new opportunities for public involvement in the management of National Forest resources. Fishery conservation interests now have a direct role in specific land-use decisions for all units within the National Forest system, both at the forest planning level and in commenting on and appealing specific management decisions such as timber sales.

The NFMA has been less effective in promoting sustainable riparian area management. Forest Service regulations require only that "special attention shall be given to land and vegetation for approximately 100 feet [31 m] from the edges of all perennial streams, lakes and other bodies of water" (36 C.F.R. Section 219.27(e)). They also require soil and water conservation measures to mitigate damage and enhance productivity on a site-specific basis. The 31-m "special attention" rule is subject to wide variations in application and, even if vigorously applied, will provide only marginal protection and will be ineffective in steep headwater areas or in preventing siltation and sediment transport from ephemeral streams. Forest Service personnel monitor soil and water conservation measures for implementation purposes, but follow-up monitoring for effectiveness purposes varies both among regions of the National Forest System and among individual National Forest Units (Gauvin 1993a).

Forest Service regulations under NFMA also attempt more rigorously to protect minimum viable populations of fish from adverse effects of timber harvest (36 C.F.R. Section 219). Implementation, however, is biologically flawed because the Forest Service "typically fails to distinguish between species and individual fish stocks (or populations)" (Doppelt et al. 1993, The Wilderness Society, unpubl. data). Thus, Forest Plans do not preserve habitat on an ecosystem basis and can allow considerable habitat degradation for reproductively isolated stocks (Doppelt et al. 1993).

While allowing for greater public participation in National Forest planning and management, the Forest Service's implementation of NMFA has not been effective in preventing the region-wide depletion of Pacific salmon stocks (The Wilderness Society, unpubl. data, FEMAT 1993). This failure in NFMA's implementation led to the emergency intervention effort on the part of US President Clinton's administration and the old growth management plan that emerged from the administration's April 1993 Forest Conference.

The Federal Land Policy and Management Act of 1976

Like the Forest Service, the BLM administers vast tracts of federal land containing spawning and rearing habitat for Pacific salmon. The BLM manages most of its lands on a multiple use basis and, until the US Congress passed the Federal Land Policy and Management Act of 1976 (FLPMA), it lacked any comprehensive definition of its responsibilities and authorities.

The FLPMA attempts to guide BLM actions in much the same way that NFMA applies to the Forest Service. Thus, BLM must develop "land use plans which provide by tracts or areas for the use of public lands" and "use a systematic, interdisciplinary approach to achieve integrated consideration of physical, biological, economic and other sciences" and "give priority to the designation and protection of areas of critical environmental concern" (FLPMA, Section

202(a) and(c)). Unlike NFMA, FLPMA lacks detailed standards and guidelines, which means that there is little guidance for the federal courts in interpreting FLPMA's requirements, and control of BLM's management discretion depends primarily on congressional oversight (Doppelt et al. 1993).

The FLPMA's definition of areas of critical environmental concern includes "areas where special management attention is required to protect and prevent irreparable damage to important . . . fish and wildlife resources. . ." (FLPMA, Section 103(a)). Broad discretion has been granted to BLM to designate areas of critical environmental concern but the bureau has not developed regulations specifically protecting riparian area. Indeed, BLM regulations under FLPMA do not provide even the minimal riparian area protections found in Forest Service regulations. As a result of its lack of specific standards and management goals, FLPMA has not been effective in preventing or mitigating damage to Pacific salmon habitat. As with NFMA, FLPMA's failure to protect fish and wildlife habitat on federal lands was a significant contributing factor in US President Clinton's administration's 1993 special intervention efforts in the management of old growth forest ecosystems.

FEMAT, PACFISH, AND THE 1995 TIMBER SALVAGE RIDER

The growing crisis over the decline of the Pacific Northwest's salmon and its relationship to the larger decline of the region's forest ecosystems prompted two significant federal agency responses, in the form of FEMAT's report and management plan for the Northwest's late-successional forests and the Forest Service-BLM Pacific salmon and steelhead (PACFISH) management strategy. Both FEMAT and PACFISH are significant advances in forest and fishery resource management throughout the range of the Pacific Northwest's salmon (see Sedell et al. 1996 and Williams and Williams 1996 for analysis of FEMAT and PACFISH).

Both FEMAT and PACFISH demonstrated the potential for scientifically based ecosystem management to achieve a workable balance among timber and other natural resource interests. In 1995, however, Congress upset that balance when it enacted the so-called timber salvage rider in Section 2001 of the Rescissions Law. Although purportedly for the purpose of allowing limited logging to salvage dead, diseased, and insect-infested stands of timber and alleviate fire hazards on federal lands, in a number of areas the timber salvage rider has effectively undermined the fishery conservation benefits of FEMAT, PACFISH, and applicable land management and water-quality laws.

Proponents of the timber salvage rider succeeded in tacking it onto an omnibus law the main purpose of which was to make budget cuts that met fiscal targets. As part of an appropriations bill, the timber salvage rider never received full congressional scrutiny or public discussion. To support its passage by Congress, the rider's proponents pointed to the need to remove dead and dying timber and invoked the example of the widespread forest fires that had occurred on public lands in summer 1994. Yet the rider went even further than its proffered rationale by allowing the harvest of green timber associated with salvage and by targeting for harvest timber sales that land managers had previously withdrawn or that were canceled by injunction, including sales that were not consummated for environmental reasons. In addition, although its Senate sponsor claimed that Congress did not intend to authorize timber harvest in violation of the FEMAT plan, the timber industry has contended in subsequent litigation that the rider allows harvest of timber stands permitted to be harvested by the FEMAT plan, but without the necessity of complying with the plan's provisions or applicable environmental laws (Kriz 1996). The extent to which the rider undermines or nullifies applicable requirements of FEMAT is the subject of ongoing litigation. Regardless of the outcome of that litigation, the timber salvage rider already has earned the pejorative name that its opponents have given it: logging without laws.

Although the timber salvage rider will have impacts on lands within federal ownership in some 10 states, much of the logging that the rider authorizes has occurred or will occur within the range of Pacific salmon. One advocacy group involved in contesting timber sales in the Pacific Northwest estimates that sales in Washington and Oregon will encompass 10,608 ha of public lands (all data from court documents submitted by the federal government in *Northwest Forest Resources Council vs. V. Glickman*, no. 95–6244 [D. Or.]). By the time it expires on December 31, 1996, in Oregon alone the salvage rider will have allowed harvest of otherwise off-limits timber stands in salmon-bearing drainages within the BLM's districts of Coos Bay, Medford, and Roseburg, and within the Siskiyou, Rogue River, and Umpqua National Forests. In Washington, the rider has permitted clearcutting of old-growth timber stands in the Dosewallips River watershed under the original terms of a 1990 timber sale contract, and not under the more rigorous requirements of the FEMAT plan. In Idaho, the timber salvage rider has made possible a salvage sale in the drainage of the Salmon River's South Fork, critical habitat for the endangered Snake River chinook. In the words of the federal judge reviewing the sale, "but for the Rescissions Act, the salvage sale could not be implemented" (Kriz 1996).

Where FEMAT and PACFISH demonstrated that elected officials were capable of receiving the message that scientists had sent concerning the degraded condition of the Pacific Northwest's watersheds and the negative consequences of excessive logging and roading for the region's salmon, in adopting the timber salvage rider the US Congress said, in effect, "This time, let's kill the messenger." The result, as one commentator has described it, is that "the Forest Service is defying the expert wildlife agencies and permitting logging that would not withstand scrutiny under our environmental laws. Before the rider, the Forest Service took some care in developing salvage sales because it knew the public was watching and that it might have to answer for its actions in a court of law. ...Under the logging without laws rider, the Forest Service is abandoning that sense of caution" (Goldman 1996).

The timber salvage rider is a dramatic example of how, in the last analysis, political decision making can undermine attempts to apply science to conserving and restoring the Pacific Northwest's salmon. The cruel irony is that US President Clinton, who had leveraged his political capital in facilitating the FEMAT plan and PACFISH as solutions to the Pacific Northwest's ecosystem crisis, proceeded to sign the Rescissions Bill into law after initially vetoing it, and subsequently claimed that he and his administration were unaware of the timber salvage rider's impacts on the FEMAT plan and PACFISH.

The Federal Power Act

Whereas the development of the nation's federal water-resources projects has occurred largely outside of a single policy framework, the development of non-federal hydropower projects since 1935 has proceeded under the policy framework of the Federal Power Act (FPA). Congress passed the FPA to centralize the planning and regulation of non-federal hydropower projects in a single agency, FERC, and to encourage the development of river resources for more power (Echeverria et al. 1989). In response to heightened environmental concerns about hydropower development, Congress amended the FPA in 1986 by enacting the Electric Consumers Protection Act (ECPA). The ECPA strengthened the authority of federal and state fish and wildlife agencies and required FERC to give equal consideration to power and non-power values.

Licensing of a non-federal hydropower project begins with interagency consultation, whereby a license applicant, before filing with FERC, must discuss its development plan with federal and state resource agencies. The intent is to identify resource issues early in the process, to ensure that the applicant performs the necessary project studies (including instream flow and fishery habitat analyses), and to allow both the applicant and agencies to resolve conflicts before filing of the application. This process allows public involvement, but public participation rights are unclear and public involvement in the consultation stage is minimal.

The filing of an application triggers FERC's environmental analysis. At this stage, FERC decides whether to prepare an Environmental Impact Statement (EIS) or an Environmental Assessment (EA) for the project. Until recently, FERC seldom prepared EISs and preferred to subject project applications to the far less rigorous EA process. Recent pressure from conservation interests has caused FERC to re-examine this practice and to begin using a more comprehensive analysis of the cumulative effects of hydropower development within watersheds.

During FERC's environmental analysis, intervening parties have the right to submit recommendations regarding license terms and conditions. In some cases, however, FERC has required submittal of recommendations before environmental analysis is complete and before FERC has completed an EA or an EIS. Intervening parties in those cases have had less than full disclosure of environmental information that may be relevant to their recommendations.

Recommendations of fish and wildlife agencies and license conditions of state environmental agencies related to water quality have special status under the FPA. Section 30(c) of the FPA requires FERC to include as license conditions the recommendations of fish and wildlife agencies with respect to non-power conduit projects (which are exempt from licensing), small projects of <5 MW, and certain projects where developers are seeking benefits under the Public Utility Regulatory Policies Act of 1978. Section 10(j) applies to all hydropower projects that require licenses and allows fish and wildlife agencies to propose license conditions that "protect, mitigate damage to, and enhance fish and wildlife..." (Echeverria et al. 1989). If it rejects a 10(j) recommendation, FERC must state its reasons in writing (Federal Power Act, as amended by the Electric Power Consumers Act of 1986, Section 10(j)).

State environmental agencies have the authority, under Clean Water Act Section 401, to issue water-quality certifications for FERC-licensed projects. Water-quality certifications usually involve conventional parameters like dissolved oxygen, but a number of states now are including requirements for fishery habitat and aesthetics in their water-quality standards. The US Supreme Court's recent decision in *PUD of Jefferson County and City of Tacoma v. Washington Department of Ecology* (511 US_ _, 128 L.Ed.2d 716, 114 S.Ct. 1900 (1995)) affirmed the authority of states to impose narrative water-quality standards under Clean Water Act Section 401.

Even with the strictures that ECPA placed on FERC's discretion, in the past the agency has been notoriously insensitive to fishery and environmental concerns (Gauvin 1993b). In 1992, FERC issued new licensing regulations that defined the term fishway to include only facilities for upstream fish passage and limited the authority of fish and wildlife agencies to prescribe fishways at FERC projects. The FERC continues to view state river protection plans skeptically and holds an unduly narrow view of the restrictions that the plans impose on its licensing mandate. US President Clinton's appointees to the commission have begun to change the agency's policies and practices for the better, but it remains to be seen whether FERC will make lasting changes in its strongly pro-development approach to licensing non-federal hydropower projects.

Constructing a New Paradigm for Water-Resources and Water-Quality Management

The foregoing discussion and survey show two historical trends in water law and policy. The first is compartmentalization, whereby water-resources management and water-quality management have evolved as essentially separate decision-making processes. The second is incrementalism, whereby state and federal laws have attempted to compensate for ecosystem dysfunction with special-purpose remedies like instream flow reservations, the public trust doctrine, endangered species protections, and regional and watershed-specific planning and management programs. The sum of those increments is positive, as state and federal authorities are beginning to take a more holistic approach to management decisions and overcome the historical compartmentalization of water-resources management and water-quality management.

In the long term, effective management of aquatic ecosystems and undoing the destruction wrought in the Pacific Northwest's watersheds will require a more integrative approach. From the perspective of a juvenile salmon, survival depends as much on effective water-resources management as it does water-quality management. Nehlsen et al. (1991) correctly emphasized that survival and recovery of the region's salmonid biodiversity require dealing comprehensively and coherently with the complex of causes (e.g., instream flows, water quality, gene transfers and disease, and over-harvest) that has depleted and extirpated some fish stocks. While integrating water-resources management and water-quality management will not produce a complete solution, it would give managers a common ground from which to address the other factors necessary to rebuild weakened stocks. It also would reduce the opportunities for finger-pointing among the resource user groups, each of which attempts to minimize its own responsibility by exposing the role of others in depleting salmon stocks. The forensic approach is bankrupt; it is time to abandon it and focus on the factors that produce healthy aquatic ecosystems.

There are several conceptual frameworks that could help resource managers to focus on the factors that produce healthy aquatic ecosystems and could play a large role in the recovery of Pacific salmon. On the forest management front, FEMAT and PACFISH (Sedell et al. 1996, Williams and Williams 1996) hold considerable promise for reforming decision making and offer an ecosystem-based approach that may be adaptable to water-management decision making. Bioassessment protocols developed by EPA address the ability of rivers and streams to support all forms of aquatic life and are useful in identifying sources of impairment, evaluating the effectiveness of control measures, and in other contexts. Although they address impairments from a number of potential sources, application to date has focused on water-quality management. Further evaluation will be necessary to determine their usefulness in water-management decision making.

Similarly, others have developed broad biological goals for the health of aquatic ecosystems. Karr (1991) defined biological integrity to encompass the ability of an aquatic ecosystem to support and maintain a balanced, integrated, adaptive community of organisms having a species composition, diversity, and functional organization comparable to that of natural habitat of the region. To guide management decisions toward protection and restoration conditions inherent in the definition of biological integrity, Karr (1991) created the Index of Biotic Integrity (IBI). This set of metrics includes species richness and composition, trophic composition, and fish abundance and condition. The IBI facilitates assessment of current instream conditions as well as identification and control of anthropogenic factors leading to ecological stress. It does not replace the need for physical and chemical water-quality monitoring, but it offers a site-specific assessment and management tool that incorporates professional judgment and quantitative analysis.

Reauthorization of the Clean Water Act offers the opportunity to add substance to the Act's biological integrity objectives, to create effective mechanisms to realize biological integrity in water-quality management, and to begin integrating water-quality and water-resources management within a common biological integrity framework. Despite the Act's mandate against abrogating state water allocation laws, Congress could take a more integrative approach by incorporating the following items in the reauthorization legislation:

1. an effective, working definition of the Act's biological integrity objective;
2. a directive that EPA develop and the states adopt standards and criteria that encompass all components of the aquatic ecosystem and address the significant forms of ecological stress, including those that the EPA bioassessment protocols and the IBI encompass and address;
3. federally enforceable mechanisms to control polluted runoff in watersheds on federal and non-federal lands, to be implemented within a reasonable, but certain, time frame;
4. a federally enforceable statutory anti-degradation policy that protects both pristine and non-pristine, but biologically important, aquatic habitat;
5. a stronger, more expansive Section 404 permit program that covers both wetlands and critical riparian areas, and that addresses, in addition to currently regulated activities, biologically significant activities that are now exempted; and
6. incentive funding for states that implement the standards and criteria described in item 2.

Unfortunately, despite some positive movement in the 103rd Congress, the 104th Congress has approached Clean Water Act reauthorization in a manner that would set back, by decades, progress toward the above goals. In 1995, the US House enacted a Clean Water Act reauthorization bill that would weaken existing water-quality law and make it less responsive to ecological parameters; would postpone implementation of controls on polluted runoff; would relax anti-degradation standards; and would emasculate current protections for wetlands. Although unlikely to receive approval in the US Senate, the House legislation sets an unacceptably low standard for compromise and makes it unlikely that the US Congress will be responsive to the concerns that this paper has identified as critical to the recovery of Pacific salmon and their ecosystems.

Whatever the future holds for federal water-quality management law, it is important to recognize the unique role of state law. Political experience has shown that Western water interests are loath to accept a forceful approach to changing the appropriation system. As I noted previously, while reform has been modest to date, the appropriation system is becoming more responsive to changes in societal values. The challenge is to encourage the appropriation system to continue changing and to direct change toward the protection of healthy aquatic ecosystems and the restoration of unhealthy ones. Compromise legislation adopted by the US Congress as part of the 1996 Farm Bill establishes a special commission to develop recommendations to reconcile conflicts between state water laws and federal land management law under FLPMA. If developed and implemented in a manner consistent with FLPMA's broad objectives, such recommendations could help to restore Pacific salmon habitat on federal lands and may serve as an example of positive change in the direction of restoring salmon habitat on non-federal lands.

Neither an amended Clean Water Act nor an environmentally conscious evolution of the appropriation system will effect the immediate, system-wide transformation that is necessary to rebuild salmon stocks that federal water projects have depleted. For the near term, the fate of the Snake River stocks is in the hands of the NMFS recovery team and, in all likelihood, the federal courts. On a longer-term basis, if NMFS can use the ESA effectively to begin rebuilding depleted Pacific salmon stocks, it would be best for management authority to revert to the Power Council, provided that the Power Council can implement system operating changes at least as strong as those in the 1994 Fish and Wildlife Program Amendments.

Beyond the Columbia-Snake River Basin, BuRec projects control flows on many rivers that contain Pacific salmon habitat. To help rebuild weakened stocks, BuRec must restore instream flows and require water conservation measures on the part of water users. In addition, BuRec should enforce contracts with users that prohibit spreading water outside water conservation district boundaries and comply with the National Environmental Policy Act (NEPA) by subjecting changes in contractual water use to rigorous environmental impact analysis (Curtis 1993). By requiring conservation measures and enforcing its contracts, BuRec could make a significant amount of water available for instream uses like fish habitat (Curtis 1993). Overall, BuRec should develop biologically based principles for operation of all its projects and, through NEPA, enter into contracts that are consistent with those principles.

As long as government agencies fail to recognize water-resources and water-quality management as vitally related processes, our watersheds will remain ecologically dysfunctional. Amending the Clean Water Act to give impetus to biological integrity as an objective for managing the nation's aquatic environment would signal a major long-term change in federal policy, and other federal programs like the Power Council's 1994 Fish and Wildlife Program Amendments would bring immediate relief to depleted Pacific salmon stocks. State appropriation laws, already in flux, will move in the right direction if given proper federal example and incentive. To borrow the words that helped put US President Clinton in the White House in 1992, "the future can be ours if only we have the courage to change."

Literature Cited

Adler, R., D. Cameron, and J. Landman. 1993. The Clean Water Act 20 Years Later. Island Press, Washington, DC.

Bean, M.J. 1983. The Evolution of National Wildlife Law. Praeger Publishing, New York, New York.

Butro, S., D. Getches, L. MacDonnell, and C. Wilkerson. 1993. Searching Out the Headwaters: Change and Rediscovery in Western Water Policy. Island Press, Washington, DC.

Curtis, J. 1993. Water, water, everywhere. Trout 34(3): 41-68.

Doppelt, B., C. Frissell, J. Karr, and M. Scurlock. 1993. Entering the Watershed: A New Approach to Save America's River Ecosystems. Island Press, Washington, DC.

Echeverria, J., P. Barrow, and R. Roos-Collins. 1989. Rivers at Risk. Island Press, Washington, DC.

Forest Ecosystem Management Assessment Team. 1993. Forest ecosystem management: an ecological, economic, and social assessment. USDA Forest Service. Portland, Oregon.

Gauvin, C. 1993a. How clean is clean enough: making the Clean Water Act work for trout. Trout 34(2): 24-71.

Gauvin, C. 1993b. FERC: the rogue agency. Fly Fisherman 24(5): 21-31.

Goldman, P. 1996. The logging without laws rider: Lawlessness fuels the industry's greed at the expense of our nation's forests. Unpubl. report prepared for Sierra Club Legal Defense Fund.

Karr, J. 1991. Biological integrity: a long-neglected aspect of water resource management. Ecological Applications 1(1): 66-84.

Kriz, M. 1996. Timber! National Journal 25(5): 252-257.

MacDonald, L.H., A. Smart, and R.C. Wissmar. 1991. Monitoring guidelines to evaluate effects of forestry activities on streams in the pacific northwest and Alaska. US Environmental Protection Agency Rep. EPA/910/9-91-001. Region X, Seattle.

National Wildlife Federation. 1992. Waters at risk: keeping clean waters clean. Washington, DC.

Nehlsen, W., J.W. Williams, and J.H. Lichatowich. 1991. Pacific salmon at the crossroads: stocks at risk from California, Oregon, Idaho, and Washington. Fisheries 16(2): 4-21.

Palmer, T. 1991. The Snake River: Window to the West. Island Press, Washington, DC.

Reisner, M. 1986. Cadillac Desert. Penguin Books, New York, New York.

Sedell, J.R., G.H. Reeves, and P.A. Bisson. 1996. Habitat policy for salmon in the Pacific Northwest, p. 375-387. *In* D.J. Stouder, P.A. Bisson, and R.J. Naiman (eds.), Pacific Salmon and Their Ecosystems: Status and Future Options. Chapman and Hall, New York.

Trout Unlimited. 1994. Technical support document for conserving salmonid biodiversity on federal lands: Trout Unlimited's policy on mining, grazing, and timber harvest. Arlington, Virginia.

United States Environmental Protection Agency. 1989. Rapid bioassessment protocols for use in stream and rivers: benthic macroinvertebrates and fish. US Environmental Protection Agency Technical Report EPA-440/5-88-089.

United States Environmental Protection Agency. 1990. The quality of the nation's water: results from the 1988 national water quality inventory. Washington, DC.

Wilkinson, C.F. 1992. Crossing the Next Meridian: Land, Water, and the Future of the West. Island Press, Washington, DC

Williams, J.E. and C.D. Williams. 1996. An ecosystem-based approach to management of salmon and steelhead habitat, p. 541-556. *In* D.J. Stouder, P.A. Bisson, and R.J. Naiman (eds.), Pacific Salmon and Their Ecosystems: Status and Future Options. Chapman and Hall, New York.

Wood, C. 1993. Implementation and evaluation of the water budget. Fisheries 18(11): 6-17.

A Resource in Crisis: Changing the Measure of Salmon Management

Christopher A. Frissell, William J. Liss, Robert E. Gresswell, Richard K Nawa, and Joseph L. Ebersole

Abstract

An overriding focus on extraction of biomass and numerical goals in fishery management has promoted the depletion and biotic impoverishment of the Pacific salmon (*Oncorhynchus* spp.) resource. The prevalence of mechanistic thinking has marginalized or excluded critical ecological and cultural functions that sustain the resource and embody much of what humans value about it. This approach to salmon management has led to its own demise. We now face the task of forging a new approach to resource management, necessarily founded on restoration and long-term conservation of salmon, their ecosystems, and their non-commodity cultural contexts. We advocate a more contextual perspective of management and science, emphasizing that numerical performances are outcomes of underlying causal processes and relationships that determine the realized capacities of resources and ecosystems. Rather than striving to directly control resources to produce desired states or yields, management should foster and maintain the processes and relationships that determine the consumptive and nonconsumptive values of salmon and their ecosystems. We define restoration as the identification and alteration of activities and processes that presently constrain a biological resource and its environment, allowing the resource to re-express its intrinsic capacity. Management success is gauged by different empirical measures, including trends in spatial and developmental diversity, temporal continuity and spatial connectivity, life-history patterns and strategies, integrity of colonization pools, the reversibility of biological and physical processes, and the quality and sustainability of human experiences and communities.

Introduction—The Conceptual Core of the Salmon Crisis

The natural-cultural systems that encompass natural resources evolve over time as a function of human aspirations, values, and investments within some limits set by natural ecological capacities and vagaries of climate and geological events. When human expectations are perceived to be in conflict with environmental capacities, a resource crisis happens (Warren 1989). Crisis may be triggered when new information calls into question the sustainability of an exploitative practice, or when the undesirable ancillary consequences of cumulative management actions become widely known.

Empirical measures are one critical means by which we monitor and gauge the success of a resource management program. Performance measures and monitoring data are sometimes used to fine tune a management system (e.g., through harvest regulation to maintain escapement goals in a fish population), but more frequently they serve to legitimate a prevailing management paradigm. Therefore, in this paper we argue that we must evaluate measures of the performance of salmon (*Oncorhynchus* spp.) management from the viewpoint of the broader conceptual perspective in which they reside.

Facts and data gain their meaning only in the context of a set of conceptual and theoretical assumptions about how the world works. As Kuhn (1970) has forcefully argued, "the answers you get depend on the questions you ask." Questions will depend on the cultural traditions and often implicit assumptions that underlie a specific discipline and the institutional organization of the investigator (McEvoy 1986; Bella 1987, 1992). Many debates about the success of a scientific theory or management program are focused on arguments about data (or often, the apparent lack of it) when in fact the roots of the conflict lie in a deep-seated clash of world views and theoretical assumptions about how the world is organized; the roles of the scientists, managers, citizens, and salmon that inhabit it; and what data are relevant or what they mean. The challenge of interdisciplinary research and environmental management can be viewed as less of a data gap than a conceptual gap that prevents data from having transdisciplinary meaning.

Understanding that data get their meaning from a conceptual context implies that the solutions to our management problems will not necessarily emerge from gathering more or better data. McEvoy (1986) documents the tragic and inveterate failure of what was once the world's most elaborate collection of research and data-gathering institutions to solve California's fisheries problems during this past century. In McEvoy's (1986) example, fisheries research institutions originated and evolved to meet the political needs of extractive industrial interests. Capture of management institutions by vested interests, whether overt or insidious, is perhaps inevitable, but its consequences remain unrecognized or unacknowledged except in the light of critical examination of the conceptual assumptions of management conventions and the science that supports them (Warren 1989, Livingston 1991, Bella 1992). Perhaps more often than we suspect, we will resolve longstanding resource crises only through critical reexamination of our conceptual assumptions about how the world works and through creative analysis and synthesis of existing or easily available new data from an alternative perspective.

We argue that only by explicitly rendering the conceptual underpinnings of resource science and management, subjecting these concepts to critical examination and discourse, and innovatively revising them, will human cultures progress toward long-term conservation, restoration, and sustainable management of salmon ecosystems. The decline of salmon and the ecosystems upon which they depend is clearly an ecological and cultural issue that requires a conceptual framework integrating many disciplines of science and management. In this paper, we discuss two contrasting approaches to resource science and management. The prevailing approach is a predominantly mechanistic and specialized world view focused on extractive use of resources. We describe a conceptual framework that redefines management as a restorative rather than extractive venture. A brief case study illustrates how such an approach can improve understanding and help frame the conditions necessary for reversal of declines in salmon runs.

Mechanism in Resource Science and Management

Pepper (1970) describes four general world views or coherent systems of assumptions that underlie human thought. Each world view emphasizes different aspects of metaphysics, with different assumptions about the nature of objects or systems and their relationship to each other. No human endeavor lies wholly within a single one of these world views. However, one view can come to dominate the others in any cultural sphere, and we suggest that the prevailing perspective of science, resource management, and the relationship between science and management has become predominantly mechanistic in its origin and aspirations (Warren 1989). Discrete mechanism views the machine as a metaphor for how the world works; it is exemplified in Newtonian physics (Pepper 1942) but is evident in more subtle forms throughout the biological, physical, and resource management sciences. This view rests on a systems theory with three major metaphysical assumptions (Warren et al. 1983, Warren 1989): (1) behavior of a system is determined by interaction of its elements largely independent of its environment or field of location, (2) a system is analyzable into discrete atomic elements, and (3) interactions of elements are only superficial and do not lead to changes in their nature.

This view is evident in the pervasive assumption that logging and management of forests proceed independently of the harvest and management of salmon, at least until scientific consensus is reached on the specific mechanistic linkages between these resources and the principles that govern their operation. Moreover, any responses of fish to forest management are assumed to affect only the number of fish, not their qualitative ability to persist. Because in this view changes in fish populations are seen simply as numerical, not evolutionary, recovery of fish populations is assumed to be simple and speedy once the proximal cause of their decline is reversed. Mechanistic emphasis on primary causality explains the dominant tactics prescribed for stock recovery (i.e., curtailment of fishing, elimination of wild predators on fish, addition of juvenile hatchery fish of the same species, or installation of structural devices assumed to promote fish production).

The mechanistic world view dominates science and management (and politics) because it fits well the industrial-capitalistic principles commonly presumed to drive our culture and economy (Warren 1989, Hall 1990). The emphasis of this kind of resource science and management is on extracted products (i.e., the yield of marketed goods from ecosystems rather than their intrinsic functions and non-market services). Ecosystems are defined and managed as if they were factories for the efficient production and harvest of valued products. The success of management is gauged primarily by measures of catch, yield, biomass (as an indicator primarily of potential harvest), and net economic efficiency (or in the case of privatized resources, short-term profit).

Fisheries, forestry, agriculture, and other resource management disciplines have been guided by roughly the same set of root assumptions and core values, and all are presently engaged in parallel ethical, epistemological, and social crises. Most recent attempts to define ecosystem management (e.g., Salwasser 1991) and to apply adaptive management to resource systems (e.g., Walters 1986) remain largely rooted in the mechanistic, exploitative view of resource management. Most of the "new" programs promoted by agencies and industry so far emphasize aggressive control of the ecosystem to produce yields of several selected commodity resources simultaneously rather than continuing to exploit one or two at the expense of all others (Frissell et al. 1992, Grumbine 1994). Intrinsic values of ecosystems (e.g., biodiversity and genetic heritages; scenic, cultural, or spiritual values; even water and air quality) continue to be viewed chiefly as amenities or non-market values that are significant primarily because they constrain the business of extracting commodities that have more immediate and preconceived market value (Hall 1990, Dixon and Sherman 1990). Freed from the grip of entrenched political and industrial interests, ecosystem management could be the precursor to a pragmatic perspective that fundamentally redefines management goals and strategies so that many apparent conflicts of values, side effects of extractive activities, the legacies or threats of resource depletion, changing economics, and the role of restoration are rationally addressed and better resolved. However, it seems unlikely that any such resolution could come without some diminishment or major alteration of many vested exploitative activities (Frissell et al. 1992, Ludwig et al. 1993, Grumbine 1994, Noss and Cooperrider 1994).

Mechanistic Management and the Salmon Crisis

The present crisis in salmon management has arisen not because we have failed to achieve our goals nor because society has suddenly changed its goals. Crisis has arrived precisely because we have succeeded in achieving many of the goals embodied in our past view of management, and we have achieved them on an unprecedented global scale. Fragmentation and simplification of ecosystems, deterioration of the long-term productive capacity of resources, depletion of natural biodiversity, impairment of air and water quality, and loss of employment and cultural integrity in resource-based communities are predictable although not all explicitly intended outcomes of natural research management under a mechanistic paradigm (Warren 1989). Norms of efficiency, production, control, domestication, and privatization of resources assume and require the fundamental simplification of natural and cultural diversity. Hall (1990) says that "neoclassical economics argues implicitly for the destruction of the natural (as opposed to the developed) world, and as such assists in the destruction of many existing non-market economies. . . ." Highly co-evolved, self-sustaining natural-cultural systems (invaluable as living models of what we today call sustainable development) are destroyed each time an indigenous people and its resources are captured or obliterated by the tide of centralized, western-industrial culture and economy (Suzuki and Knudtson 1992). The loss of a long-standing resource-dependent subculture within our own society can similarly impoverish resource systems (McEvoy 1986, Warren 1989).

We contend that today's salmon crisis arises from two separate but related sources. The first is an internal failing, a consequence of the fundamental mismatch between an overly mechanistic paradigm and the inherent capacities of natural resource systems. This shortcoming is manifest in the present depletion of salmon fisheries, an outcome presumably unanticipated in official quarters (e.g., some managers and their scientists believed that hatchery fish could entirely replace wild fish for sustainable fisheries). Because the organization of natural-cultural systems in fact violates many assumptions of the machine metaphor, it has not proven possible to manage ecosystems or specific resources for the simultaneous purposes of production, optimization, and control (Ludwig et al. 1993). In general, management for production (maximum or optimum, however defined) and efficiency leads inevitably to structural and functional simplification of naturally complex systems, pushing them into new, often unanticipated (and unproductive) domains of behavior. Attempts to actively manage for steady state as the primary goal in natural resource systems have generally proven unsuccessful, sometimes with catastrophic consequences (Holling 1973). For example, simplified production systems are often more sensitive than natural systems to vagaries of climate and weather, or disease. The goal of ecosystem control, implying stability or predictability of outputs, retreats ever farther from our grasp even as aggregate measures of production and efficiency may indicate near-term success.

The problem of control is exemplified in interannually unstable and generally declining survival rates among hatchery salmon populations (e.g., Hilborn 1992). The optimal and controlled spawning and rearing environment of hatchery fish almost certainly results in diminished diversity of genotypes and life histories within populations (e.g., Verspoor 1988, Leary et al. 1989), whereas the diversity of life histories and habitats within wild populations spreads risks of mortality, increasing the likelihood of year-to-year and long-term persistence (den Boer 1968, MacLean and Evans 1981, Walters and Cahoon 1985). Although hatchery salmon may increase aggregate productivity measured as an average over a period of years, the volatile nature of this production and unanticipated side effects impacting wild populations (Hindar et al. 1991) can discourage the evolution and persistence of stable fishing economies and communities.

The second underlying cause of the salmon crisis is external, or residual to past goals of resource management. It results from the inability of the mechanistic world view to successfully capture or represent much of human experience, and from subsequent devaluation of many values and ideals that human cultures still hold dear (Warren 1989, Hall 1990). These include the so-called environmental amenities and human virtues such as diversity, integrity, and other intrinsic values of natural ecosystems; quality of life; free access to self-sustaining natural systems as a source of inspiration, spiritual and emotional renewal, or material for innovation; confidence in the endurance of existing ways of life and continuity of community traditions; intergenerational equity; diversity of cultures and cultural knowledge; freedom of religion; and rights of social justice. These values, we argue, are not new. They are deeply rooted in our cultural and natural histories. That such values may strike some resource managers and scientists as new, threatening, or illegitimate is a testament to the inordinate degree to which mechanistic and industrial conceptual frameworks and ideologies have grown to dominate our profession (Warren 1989, Leopold 1990).

A Contextual View of Natural Resources

If the salmon crisis is, as we have argued, the culmination of both internal and external failings of mechanistically oriented natural resource management, alleviation of the crisis will entail a major change in perspective. The popular term for such a change is paradigm shift, or what Leopold (1990) calls "a basic metamorphosis." We believe such a shift is already well underway although by no means manifest. Changing priorities and values are implied by new professional and political dialogue on biodiversity, conservation, biotic integrity, ecosystem approaches to management, environmental ethics, and restoration. However, articulation of the theoretical underpinnings and empirical benchmarks for a shift in perspective is lacking, and this is reflected in disparate perceptions and intense but often fruitless arguments about defining these new priorities. Recent, vehement political backlash against the implementation of longstanding environmental laws indicates the need for broader cognizance of and public discourse on the assumptions and perspectives that underlie resource management.

We suggest that recent changes in resource management can be viewed as a shift to a less mechanistic and more contextualistic view of resource management. On the basis of Pepper (1942), Warren et al. (1983) identified three major metaphysical assumptions underlying contextual systems theory: (1) behavior of a system is determined jointly by its elements and its context; (2) a system, its context, and its elements interpenetrate in such ways as to make abstraction and analysis of the system fundamentally distorting; and (3) interactions of a system, its context, and its elements lead to qualitative changes at each of these levels.

Note that these core assumptions of the contextual view stand in direct opposition to the precepts of mechanism described in this paper. Proposed solutions to the present resource crisis fail to embrace more contextualistic modes of thought and may hinder, not further, a shift in perspective. Meffe (1992) characterized such efforts as "techno-arrogance and halfway technologies" because they fail to address the root causes of ecosystemic and cultural problems. The role of science and technology in resolving resource crises is substantially changed and more limited within a contextual perspective, and the organization and function of the human elements of resource systems assume much greater importance (Warren 1989, Ludwig et al. 1993, Gresswell and Liss 1995).

In the following sections, we describe a theoretical framework for a more contextualistic view of resource management and discuss the implications of such a shift for the management of salmon ecosystems.

Ecosystem Restoration and Salmon Resources: A Conceptual Framework

Owing to widespread and persistent declines in runs of wild salmon in the US Pacific Northwest (Nehlsen et al. 1991, Frissell 1993a), managers must focus increasingly on salmon restoration. Restoration is not a new concept; moreover, the failure of past salmon management could largely be explained as the failure of past approaches to restoration. We suggest that successful, ecologically meaningful restoration calls for a fundamentally different approach to ecosystems and their management. There now exists widespread agreement that restoration is a necessary premise for managing salmon resources under virtually any model. While a few managers continue to argue that the present crisis results solely from changing public expectations, we believe this view denies the obvious and well-documented evidence of large-scale collapse of salmon fisheries (and other natural resources and ecosystem functions) in the Pacific Northwest. In large part, expectations are changing because the public is directly affected and moved by the accelerating and widespread deterioration of natural resources (consumptive and non-consumptive) that they deeply value.

Our view of resource restoration is derived from the conceptual framework or general systems theory developed by Warren et al. (1979). In this conceptual framework, a system and its environment are co-extensive (in both space and time) and co-determining. Performances are the observable characteristics of a system, but in this view, systems are defined by their potential and realized capacities to exhibit suites of performances (Fig. 1). System performances at each stage of development are jointly determined by system realized capacity and environmental performances. A system with a given realized capacity will manifest different performances in different environments (Fig. 1). Thus, realized capacity can be conceptualized as all possible performances of a system given its current state in all of its possible environments.

As a system develops, both its realized capacities and performances change over time and space. The pattern or course of development is determined by system potential capacity and the environment in which the system develops (Fig. 1). Potential capacity can be conceptualized as all possible patterns of development, or changes in realized capacities and performances, in all possible developmental environments. Potential capacity thus incorporates the possible future (and theoretical past) developmental pathways of a system through time. Theoretically, then, patterns of development and associated performances of a system are expressions of that system's potential capacity in interacting with the environment.

We conceptualize a resource, R, as being embedded in a natural-cultural system (Fig. 2; Gregor 1982, Eldridge 1988). This system forms the biophysical (climate, water, soils, and biota) and cultural environment to which the resource must adapt in order to persist (Fig. 2A). In a

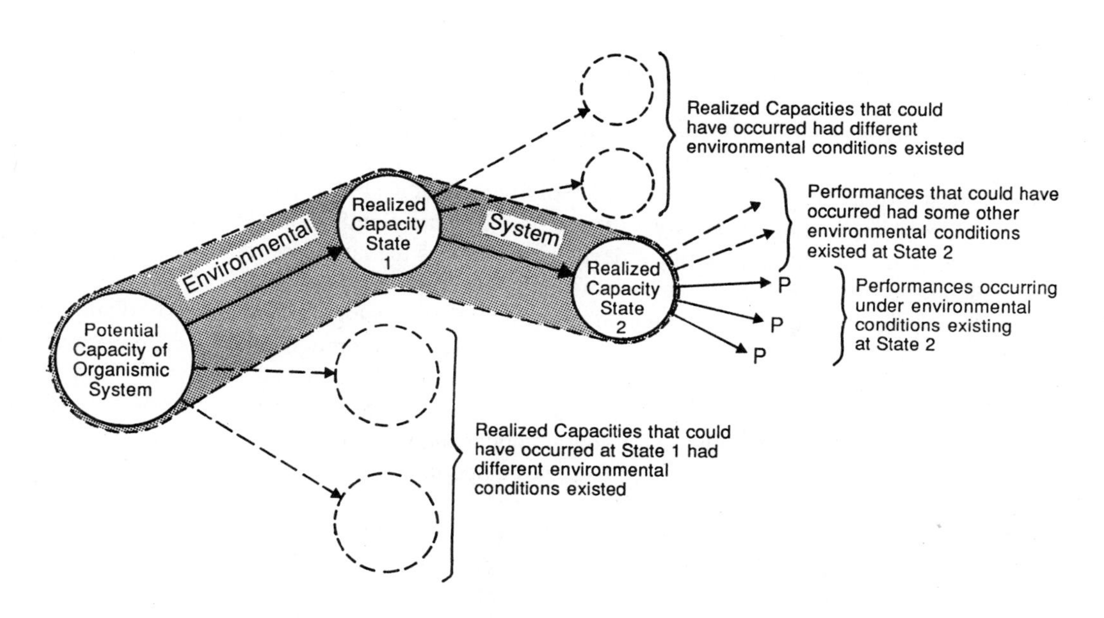

Figure 1. Conceptual model of the organization and development of a biological system. From some initial potential capacity, the system passes through a particular sequence of realized capacities, jointly determined by its potential capacity and the prevailing environmental conditions. Source: Warren et al. (1979).

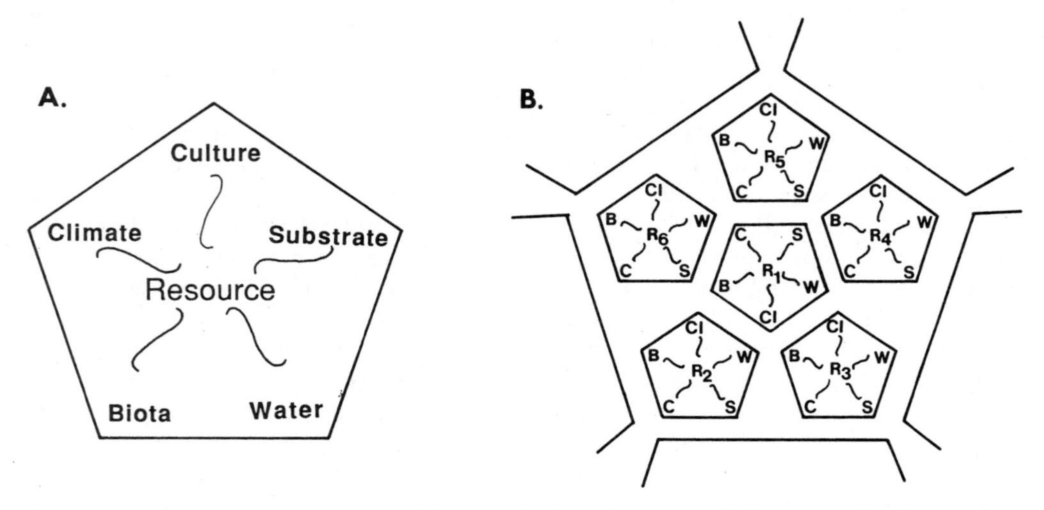

Figure 2. A. Conceptual representation of the organization of a resource (*R*) as a natural cultural system, with interacting culture (*C*), substrate (geology and soils, *S*), water (*W*), biota (*B*), and climate (*Cl*) subsystems. B. The landscape is represented as a mosaic of spatially organized and interrelated resource systems (R_1-R_6) embedded in a hierarchy of similarly organized systems. Source: Modified from Warren et al. (1983).

broad sense, the biophysical aspect of the natural-cultural system can be viewed as the habitat of the resource. Resource habitat varies from geographic location to location (Fig. 2B) and, at each location, habitat conditions change through time. This spatial-temporal array of habitat developmental patterns across the landscape is an expression of the potential capacity of the environmental habitat mosaic (Fig. 3; Frissell et al. 1986, Ebersole et al. 1996) and, together with the cultural environment (e.g., harvest), forms the template (sensu Southwood 1977) for resource development. The diversity of patterns of development across the landscape can be termed developmental diversity (represented as the vertical array in Fig. 3).

Resource potential capacity is expressed through time as development in response to temporal changes in habitat conditions at any geographic location and through space as differences in resource developmental patterns among geographic locations (Fig. 3). Thus, across the landscape, resource developmental diversity is concordant with developmental diversity of the habitat or the ecosystem as a whole. Our definition of habitat encompasses the effects of dams and diversions, as well as changes in watersheds as expressed in acute, chronic, or cumulative responses in aquatic environments.

Human activities can constrain expression of resource capacity (Fig. 4A). One example is constraint of expression of habitat capacity (Fig. 4B; Ebersole et al. 1996). Habitat developmental pathways are either eliminated (the habitat at a particular location is no longer suitable for resource persistence) or they converge on a few patterns characteristic of disrupted habitats (Bisson et al. 1991). Developmental diversity of both the resource and its habitat is reduced and resource performances (e.g., abundance, distribution, yield) are adversely affected (Schlosser 1991). For example, Reeves et al. (1993) found that the diversity of anadromous salmonid assemblages declined with logging in basins of coastal Oregon, and this simplification of the fish community corresponded

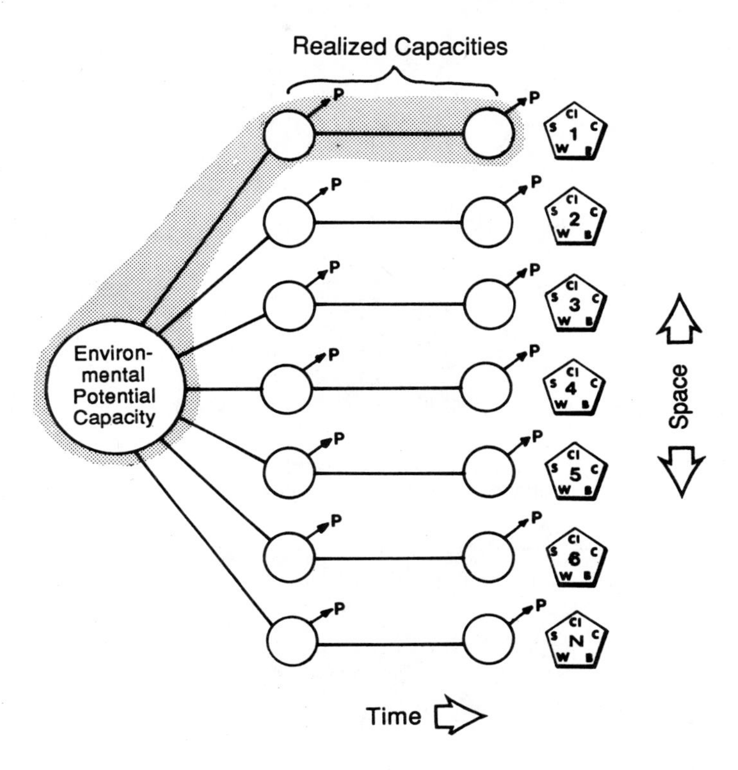

Figure 3. The influence of a regional environment on the developmental trajectories, realized capacities, and local performances of a spatial mosaic of *N* resource systems. The outcome is developmental diversity of resource systems over time and space, observed at a given time as spatial diversity of resources across a landscape. *P* = observable performance of resources at particular developmental states. Other symbols as in Figure 2. Shaded area is equivalent to Figure 1; vertical space dimension at right is equivalent to Figure 2.

with declining habitat complexity. Highly altered habitats, or the introduction of novel habitats such as large reservoirs, are associated with a major shift in ecosystem domain, such as transformation from a salmonid-dominated community in the mainstem of the Columbia River to a community dominated by non-native and warm-water fishes (Li et al. 1987). Cultural activities such as fishing and stocking of hatchery fish can also constrain expression of resource capacity through direct effects on the ability of fish populations to respond and adapt to their changing environments.

In this perspective, the goal of resource restoration is to enable the resource to express a currently suppressed capacity by relaxing or removing environmental constraints (Fig. 4A). This involves allowing habitat to re-express its capacity (Figure 4B; Ebersole et al. 1996) and, thus, restoring the template for resource recovery. The outcome is developmental diversity of

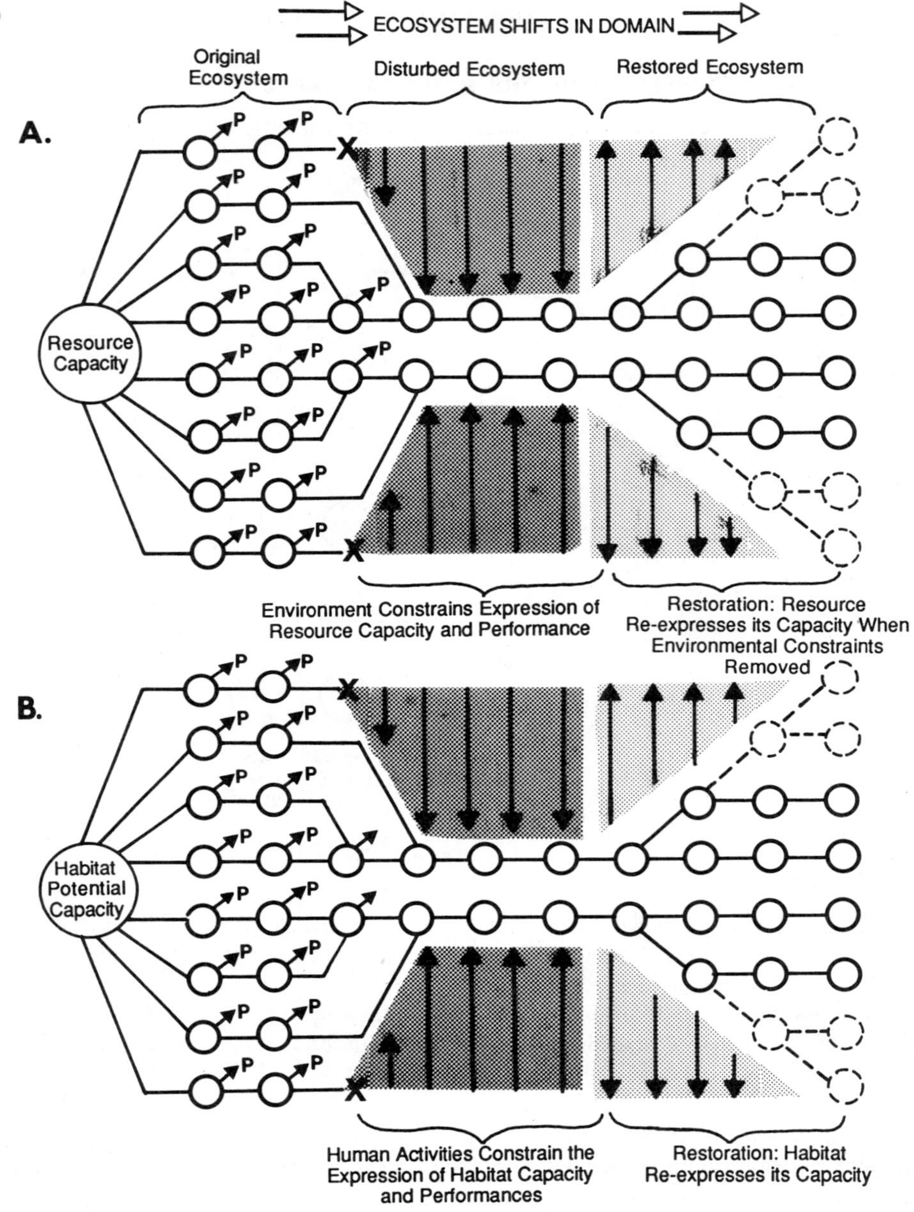

Figure 4. Changes in the realized capacity of a resource (A) and its environment or habitat (B) over time in response to human disturbances that constrain developmental pathways. Restoration is depicted as the relaxation or removal of constraints, allowing the self-regeneration of developmental diversity of the resource and its environment. --- = developmental pathways not yet expressed, but which could potentially reappear under suitable conditions; X = subsystems that were extinguished or otherwise eliminated from the system during its development; P = observable performances of subsystems at various developmental stages.

habitats and biotic resources. Numerical performances such as abundance, distribution, and yield are outcomes of the interaction of resource capacity and environment. These performance measures will be intrinsically protected and enhanced if the environment allows the capacities of resources to more fully express themselves. In some cases, restoration or re-expression of resource and habitat capacity may not occur (dashed recovery patterns, Fig. 4) either because constraints have not been entirely eliminated or because resource or habitat capacity (or both) has been so fundamentally altered that full expression is no longer possible.

Salmon populations and the ecosystems they inhabit are viewed not as static structures but as vital integrated systems with self-regenerative inclinations. Regier et al. (1989) similarly describe riverine rehabilitation as recovering and fostering the self-organizing capability of ecosystems. Restoration or rehabilitation is accomplished not by supplementing ecosystems with the addition of more fish or new habitat elements but rather by reducing or removing environmental constraints imposed by degraded habitat, fishing, and hatchery practices. Restoration can be approached on spatial scales ranging from individual river basins to physiographic regions and on biological scales ranging from communities to individual species or stocks.

At the salmonid community level, the potential capacity for the development of the biological community in any river basin resides in the species pool. This pool of potential colonists is maintained in surrounding river basins. Given time (decades to centuries?), species from the regional pool can recolonize a river basin and restore species diversity if habitat conditions in the basin are suitable and barriers to movement through the basin can be overcome. The spatial extent of, and time required for, recovery of species diversity in a river basin will depend on the degree to which habitat in the basin has been restored and the integrity, or biological and spatial diversity, of the species pool.

Species-pool integrity may be disrupted by large-scale or regionally pervasive habitat changes that cause persistent decline or elimination of species from surrounding basins (Frissell 1993a). This habitat fragmentation, in effect, may increase the distance (MacArthur and Wilson 1967, Wilcox 1980) of the source pool of colonists in a particular basin and so may delay recolonization and prevent full restoration of species diversity. Large-scale changes in species pools, which have occurred along the Pacific coast of North America over the past century (Frissell 1993a), can eliminate opportunities for successful restoration of species diversity for time spans that may exceed several human lifetimes. For example, there are few sources of potential founders or colonists to reestablish pink (*O. gorbuscha*) or chum salmon (*O. keta*) populations in California, even assuming available suitable habitat. Forms of coho salmon (*O. kisutch*) adapted to life in interior rivers (Taylor and McPhail 1985), once widely distributed in the Columbia Basin and possibly the Sacramento River in California, are now rare or absent south of the Fraser River in British Columbia (Frissell 1993a). The potential for re-expression of this lost diversity within existing remnant populations of such forms might exist should appropriate habitat redevelop.

Establishment of non-native fishes, mysid shrimp, and other introduced organisms in the regional species pool has also contributed to potentially irreversible ecosystem changes, such as the development of large warm-water fish communities in the Columbia River and its tributaries (Li et al. 1987) and the reorganization of food webs in oligotrophic lakes (Spencer et al. 1991). Because many introduced species are able to establish self-sustaining populations, changes in the regional species pool pose ecological and evolutionary consequences that can be far less reversible than are changes in the quality and distribution of habitats within river systems, which if allowed may recover over periods of decades or centuries. For example, although it appears technically feasible, if not politically expedient, to remove some large dams from the Columbia River system, it is unlikely that populations of centrarchids, ictalurids, percids, and other introduced fishes will ever be eliminated from this ecosystem. A single small lake or backwater harboring these species can serve as a colonization source for many thousands of kilometers of riverine habitat.

Potential capacity can be viewed as all possible performance trajectories in all possible environments. Accordingly, the life-history capacity of a species of salmon (all possible life-history types) is partially represented by all life-history types present within the range of the species (all possible environments) (Fig. 5). Some subset of this life-history pool originally occurred in individual river basins, consistent with available habitat mosaics, and is an expression of life-history capacity within each basin. Human activities can constrain expression of life-history capacity and reduce life-history diversity either through direct biological effects

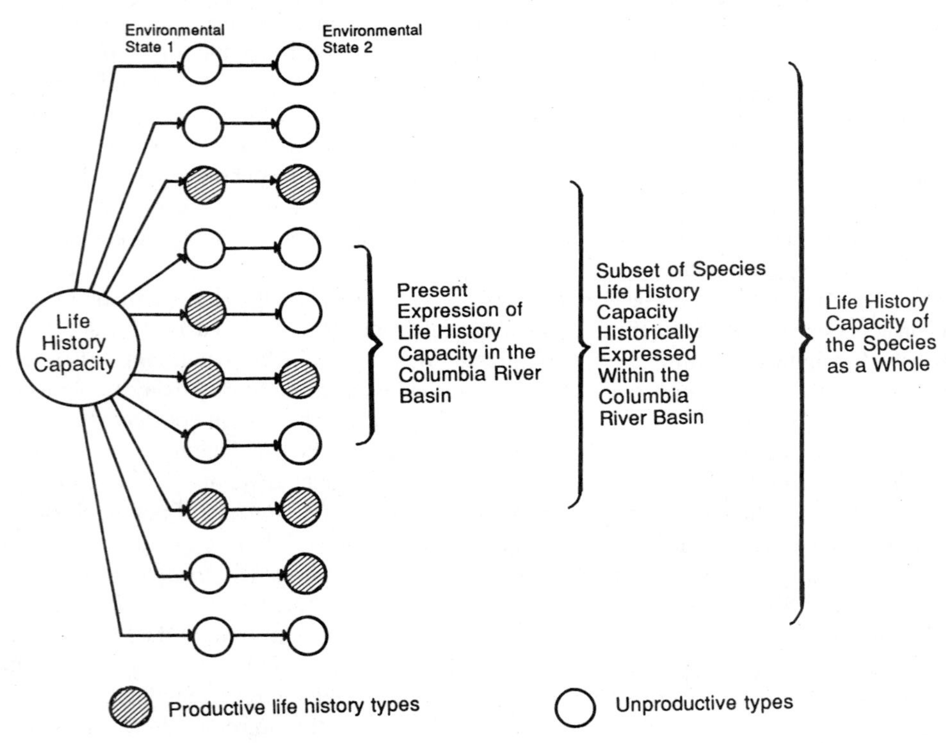

Figure 5. A species defined in terms of its potential capacity to express an array of life-history types in an array of environments. Local regions within the range of the species encompass a subset of the possible life-history diversity, and only certain of these types are productive under prevailing environmental conditions, which change over time. The effects of management have often been to constrain the life-history diversity of populations, with attendant loss or developmental submergence of many of the historically and potentially productive types. Recovery of a population or species entails the re-expression of its historical life-history diversity and adaptive capacity.

such as size- or growth-selective fishing (Ricker 1981) or through disruption of habitat mosaics (Lichatowich et al. 1995) by introduction of barriers to migration, by deterioration of key habitat patches (Regier et al. 1989), or by introduction of non-native species. Life-history types that once may have been productive can be selectively reduced or eliminated. At the species level, the goal of restoration should be to provide for re-expression of life-history and stock diversity within a drainage basin or physiographic region (Ebersole et al. 1996).

Resource diversity can buffer performances against environmental change and thereby enhance resource persistence and sustainability (den Boer 1968, Andrewartha and Birch 1984, Opdam 1991). Loss of diversity can increase variability in resource performances and enhance the probability of resource extinction (den Boer 1968, Nehlsen et al. 1991); moreover, increased variation in the availability of resources generates uncertainty and conflict among resource users (Walters 1986). We believe that biological diversity confers productivity upon resources, and thus it is diversity (the expression of population-habitat relationships) more than simply mass quantities of fish that must be conserved to sustain some level of resource exploitation. Diversity of biological resources, and the ecological resilience such diversity promotes, provides the necessary bioeconomic and geographic basis for long-term development and maintenance of social and cultural diversity in human society as well, fostering resilience and sustainability of resource systems in the face of changing values (Suzuki and Knudtson 1992, Smith 1994, Gresswell and Liss 1995).

This view of restoration is essentially one of resource stewardship (sensu Leopold 1949). It does not necessarily imply that natural or historical conditions were static or in perfect balance, nor that they were unaffected by humans. It is neither intrinsically biocentric nor wholly anthropocentric. It does acknowledge and embrace variation, complexity, pattern, connectivity, and sustainability in natural-cultural ecosystems. It focuses on the self-maintaining capacity, or resilience, that has co-evolved in virtually all organisms, populations, and ecosystems to some degree. When recent human alterations rapidly or extensively constrain the capacity of ecosystems and biological resources to recover to historical conditions and dynamics, effective restoration must involve removing the causes of these constraints. Restoration, then, primarily focuses on the management of humans and their activities in ways that allow valued biological and ecological systems and processes to continue and propagate—thus, ultimately sustaining the many services, conditions, and materials that humans directly and indirectly value.

Adaptive Capacity, Diversity, and Restoration Goals

Since habitat restoration involves allowing habitat capacity to physically and biologically re-express itself, it must proceed from an integrated watershed or ecosystem perspective that addresses all important components of habitat complexity and connectivity (Regier et al. 1989, Stanford and Ward 1991, Frissell and Nawa 1992, Reeves and Sedell 1992, Frissell 1993b, Doppelt et al. 1993). Projects that focus on mechanistic tuning or supplementation of a few selected habitat components may not provide for and can even preclude the full re-expression of resource capacity; unanticipated, adverse side effects are common (Frissell and Nawa 1992, Doppelt et al. 1993). A population and its immediate habitat, and the relationship with their larger environment (e.g., a drainage basin, and the northeast Pacific Ocean) should be monitored to guide and assess the progress of restoration (Stanford and Ward 1991, Frissell and Nawa 1992). Frissell (1993b) and Doppelt et al. (1993) provide some specific concepts and terms for developing ecologically sound strategies for restoring riverine systems at the scale of whole drainage basins.

Often salmon restoration goals are explicitly directed at improving specific numerical performances. An example is the Northwest Power Planning Council's goal of doubling the run of salmon in the Columbia River (Northwest Power Planning Council 1993). Our view is that numerical performances such as abundance and yield are the product of the interaction of resource capacity and environment. Consequently, from a theoretical standpoint, management efforts should be focused primarily on resource and environmental capacities rather than particular performances (Warren 1979, Warren et al. 1983). This view does not denigrate the importance of numerical performances but merely suggests that numerical goals may best be achieved by protecting the intrinsic capacities of resources rather than by directly manipulating resource components in an attempt to produce a specific state. Further, it suggests that short-term numerical performances may not be the most reliable indicator of the well-being of a resource. Thus, goals should address restoration not simply of numerical performances but of developmental diversity (which reflects the spatial and temporal re-expression of capacity) at the levels of ecosystems, species, populations, and life-history types within populations.

For example, in traditional approaches to coho salmon, managers have collapsed the spawner counts from many streams into regional average estimates of escapement representing administrative harvest regulation zones. In contrast, to assess the success of a management program, we would be more concerned with among-stream variability in spawning escapement and whether geographical distribution and temporal diversity of run timing have changed. An examination of stream-by-stream time series of spawner counts can provide early indications of unreversed local population collapses that in the long term undermine the productive capacity and sustainability of a regional salmon fishery (e.g., Walters and Cahoon 1985). Such declines also indicate ecosystematic dysfunction where the once abundant food resources and nutrient pathways provided for coho runs are diminishing. A spatially explicit analysis can reveal areas of the landscape where salmon decline is associated with particular kinds of land and water resource alterations (Frissell 1993a). Although both mechanistic and contextualistic approaches in this case might rely on much the same raw data, they employ different hypotheses, analyses, statistics, and inferences. They ask different questions of the data.

Fishery managers in recent years have placed increased emphasis on protecting genetic diversity, an integral part of adaptive capacity; however, maintenance of genetic diversity alone (as it is typically defined and measured, e.g., using allozyme electrophoresis) may be insufficient to prevent resource depletion. Even with adequate genetic diversity, life-history diversity (and thus, the many resource performances dependent on such expressed diversity) can be constrained by environmental forces. An environmental mosaic or habitat template must be available to make possible the expression of life-history diversity (Balon 1993, Healey 1994). From the standpoint of ecosystem functions, cultural significance, or economic values, we value fish not because of the genes they carry, but because of the full expression of their biological diversity and adaptive capacity that develops within a complex array of natural environments (Gresswell and Liss 1995). Beyond maintaining existing resources and the biological diversity from which they are derived, the future evolution of new subspecies or species can occur only if a diversity of environments is maintained to exert divergent selective pressure (Balon 1993) and to afford opportunities for reproductive isolation.

Management approaches intended to circumvent specific environmental constraints may actually impose other constraints, sometimes unanticipated or unacknowledged, on the expression of adaptive capacity. Hatcheries provide a good example of a very mechanistic approach to management in which the solution to the lack of salmon is assumed to be the artificial addition of more young salmon. However, even with adequate genetic diversity incorporated into broodstocks, hatchery populations face an artificially constrained spawning and early rearing environment and may never be able to express the life-history diversity found within and among wild salmon stocks. Homogeneous fertilization, incubation, and early rearing environments may preclude opportunities for developmental and life-history diversity at later life stages (Balon 1993). This is one reason that programs intended to supplement wild stocks with hatchery-reared fish have rarely (if ever) benefited a natural population (Steward and Bjornn 1990, Waples 1991). Management that is directed at maintaining resources and environments in particular optimal or controlled states (e.g., a protected incubation and rearing environment with artificial matings and a relaxed mortality regime) is equivalent to intentionally constraining the expression of capacity, and it will inevitably lead to loss of developmental diversity and corresponding side effects.

Resource recovery, and management success in general, we believe should be monitored and evaluated on the basis of increases in developmental diversity and distribution of habitats and resources, with less emphasis on shorter-term changes in abundance or harvest (the post-hoc fruits of resource extraction). Expanding distribution can indicate the reappearance of life-history types that are colonizing new habitats or reemerging in newly restored habitats. Such changes in behavioral, developmental, and spatial diversity indicate probable increased resilience of populations, confirm that the resource system maintains some capacity for self-regeneration and adaptive evolution, and may presage future trends in productivity or abundance. We will illustrate these points with a brief case study and then return to them in our recommendations and conclusions.

Case Study: The Chinook Salmon of the Klamath Mountains, Oregon

We briefly examine the history and status of fall chinook salmon (*O. tshawytscha*) populations that inhabit the coastal rivers of southwest Oregon in the northern extreme of the Klamath Mountains physiographic province. This region encompasses the coastal river basins from just south of the Coquille River in Oregon south to the Klamath River area in northern California. We focus on Euchre Creek, a drainage of 62 km^2 located immediately north of the Rogue River in Curry County, Oregon (Fig. 6), as a typical historical example. The story of these fish and their habitats, which we have studied since 1985, serves to illustrate both the problems created by past approaches to resource management and the potential utility of more contextualistic approaches to salmon conservation. We focus on two arenas of resource exploitation—timber and fish—that have interacted over time to cause the long-term alterations of aquatic ecosystems and loss of salmon resources.

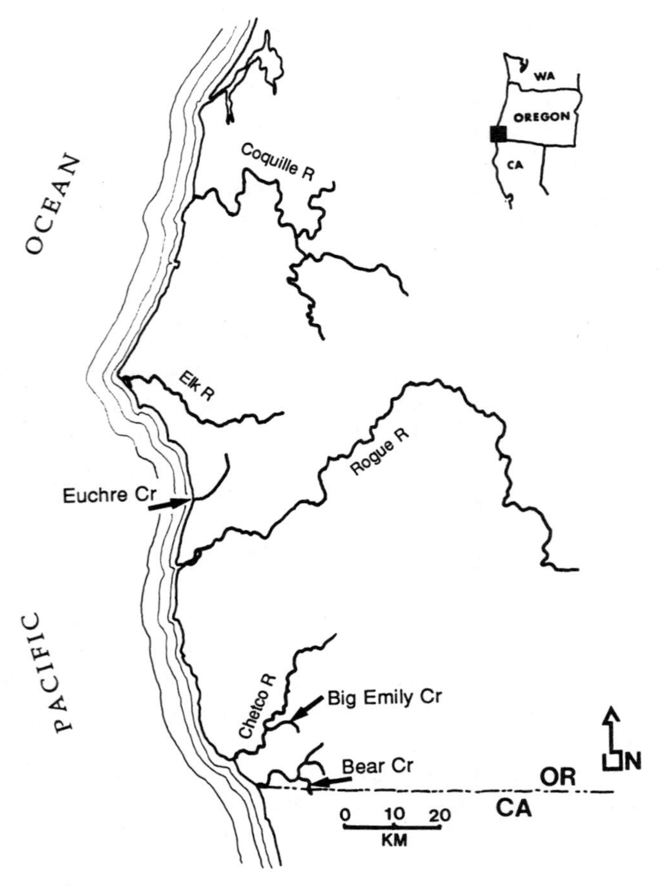

Figure 6. Map of southwest Oregon, USA, showing locations of the streams mentioned in text.

Changes in Salmon Populations and Fisheries

Before the arrival of European colonists in the late 19th century, each coastal stream in southern Oregon and northern California harbored a permanent band of native people. Subsisting on salmon for much of their food and trade economy, these societies existed as relatively stable, continuous, and interconnected entities for thousands of years. McEvoy (1986) details the complex social and ecological innovations evolved by Klamath River tribes for salmon harvest and conscientious conservation. These cultures sustained salmon fisheries that apparently equaled or exceeded the largest catches known in Euro-American commercial fisheries. Although native fishers were buffeted by many of the same environmental fluctuations that afflict today's fisheries, the history of coastal tribes stands as testimony that sustainable, natural-resource-based economies and societies involving the long-term coexistence of abundant salmon and humans are possible (Suzuki and Knudtson 1992).

Between 1850 and 1920, immigrants of European and Asian descent colonized the coastal region and extirpated most indigenous settlements, whose people had already been decimated by disease. Commercial fishing was established on all the larger rivers and streams in the region during this period. Peak commercial catches, including large numbers of coho and chinook salmon and some steelhead (*O. mykiss*), occurred in most southern Oregon and northern California rivers prior to 1920 (Cobb 1930). In the early decades of this century, chum salmon were observed to be widely distributed in the area (Snyder 1908), and occasional char (*Salvelinus* spp.) and small numbers of large-bodied, spring- or summer-run fish (*Oncorhynchus* spp.) were also reported to be present in small coastal streams (Ziesenhenne 1935), but these fishes were not targeted in commercial fisheries.

Declines in salmon catch are often attributed to overfishing; however, prior to 1940 substantial habitat change had already occurred in southern Oregon rivers. Human alterations of the landscape included extensive trapping of beaver, clearance of lowland and floodplain forests, log drives, and channelization of many of the most productive streams and wetlands in low elevation areas. Habitat changes probably contributed to much of the early decline in fish runs, particularly coho salmon, and to the apparent extinction of chum salmon, summer-run salmon or steelhead, and char populations from most southern Oregon coastal streams by mid-century.

Since 1950, salmon populations and harvest have continued to ebb, but the timing of decline varies among streams. Streams draining basins comprising a large proportion of lower-elevation lands in private ownership experienced dramatic collapses in salmon runs during the late 1950s and 1960s. For example, in Euchre Creek, coho were apparently extinct by ~1970, and the fall chinook salmon run had dwindled from an estimated annual run size of >2,000 fish in the 1950s (R. Riikula, Oregon Dep. Fish and Wildlife, Gold Beach, unpubl. data) to a low of ≤100–200 fish from the 1960s through the 1980s (an average spawning density of just 2.7 fish km^{-1}) (Nawa et al. 1988). By contrast, chinook salmon populations in drainages dominated by federal forest reserves (e.g., Bear Creek, Big Emily Creek; Fig. 6) did not dramatically decline until the late 1970s and 1980s (data in Cooney and Jacobs 1990), ~5–15 years after extensive logging activities began on the federal lands. Commercial and sport harvests of salmon in southern Oregon have been greatly curtailed or eliminated in recent years.

The historical pattern of change in the fishery resources of these basins is one of incremental depletion of diversity, which is expressed in the progressive loss of some species and decline in abundance and distribution of others. These trends are reflected in the cumulative simplification and shrinkage of the fisheries themselves and in the attendant loss of connectivity between humans and the salmon resource. For perhaps the first time in the past 9,000 years, humans can no longer support themselves by fishing in the rivers of southern Oregon. Consequently, conservation and management is focused not on sustaining an exploitable resource but on salvaging the last relict populations to maintain the possibility of future restoration of harvestable salmon runs.

Changes in Freshwater Habitat

Although large-scale clearcutting of private, large-corporate forest lands in southwest Oregon began in the 1950s, extensive logging and road construction did not begin on federal lands until the late 1960s and early 1970s. Large storms in 1955, 1964, and 1972 triggered mass erosion and channel changes, the effects of which were concentrated in basins where slopes had been extensively disturbed during the previous few decades (Lisle 1982, McHugh 1986, Ryan and Grant 1991). These observations suggest that, irrespective of fluctuations in abundance related to changes in the marine environment or the effects of ocean fishing, factors affecting freshwater habitat play a significant role in determining or mediating the causes of salmon decline. Deterioration of multiple facets of the freshwater environment reduces the ability of salmon populations to adjust to the stresses and mortality sources they encounter elsewhere in their life cycle.

The land-use history of Euchre Creek, a basin of mixed private and public ownership, illustrates patterns common to many basins in southwest Oregon. The gentle slopes of small, private holdings were cleared as homesteads near the turn of the century (Fig. 7A). Floodplain forests,

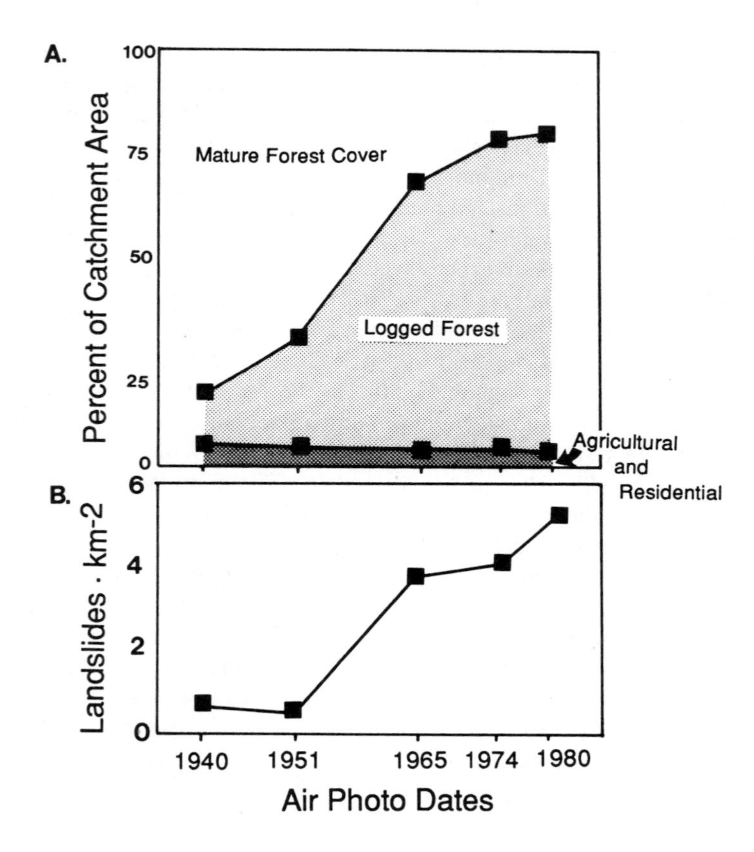

Figure 7. A. Change in land use patterns in Euchre Creek basin since 1940 expressed as percent of the basin area in each category. Mature forest category includes a mosaic of unlogged forests of various ages. Logged forest includes areas cut at least once by clearcut or partial-cut methods. Based on interpretation of aerial photographs for the indicated years. B. Increase in number of large, active landslides in the Euchre Creek basin over the same time period, based on air photograph interpretation. Includes easily observed and classified (>100 m^2 surface area) features in both natural and disturbed areas of the basin, unvegetated and actively eroding, or recently failed at the photograph date.

though selectively logged, were largely left in place. In the early decades of this century, more accessible lands now under industrial forest ownership began to be logged. Prior to the 1950s, loggers used selective cutting and light-impact transportation methods, such as horses or small tractors.

With the availability of heavy equipment and increased demand for Douglas fir and other tree species after World War II, clearcutting and the construction of extensive road networks and skid trails became the favored methods of industrial forest managers. Following the mid-1960s, the rate of conversion of primary forest to cut-over lands slowed, largely owing to the exhaustion of timber supplies on corporate ownerships in the basin. Although relatively smaller increments of the basin were logged each year after 1965, this activity did represent a major change in management of the steep and remote federal land base. Prior to 1950, timber volumes sold from the Siskiyou National Forest in southwest Oregon as a whole were $<$24,000-94,000 $m^3 y^{-1}$ (Price 1978). By the 1970s and 1980s, $\geq$470,000 $m^3 y^{-1}$ were sold annually from this national forest. By contrast, agricultural and residential development showed a small decline caused by afforestation of some lands that had proved marginal for crops and grazing (Fig. 7A). For example, sheep numbers declined by ~20% in Curry County between 1940 and 1970 (Price 1978).

Prior to the mid-1970s, the incidence of mass erosion in the basin, measured by the occurrence of large landslides on aerial photographs, increased with the increase in cumulative area of the basin subject to logging activities (Fig. 7B). However, despite deceleration in rate of total (public and private) forest removal after 1965, the incidence of landslides continued to climb. Close inspection of changes in landslide types and their spatial associations with management activities reveals several factors that contribute to this cumulative effect. First, the incidence of road failure and debris slides remained high on industrial forest lands for $\geq$20 years after they were logged. Road networks continued to fail and chronically erode, and some large, logged areas with slump-earthflow topography showed signs of increased rates of debris slides long after second-growth vegetation became established. High slope-failure rates in some areas may reflect the persistent or lagged effects of accelerated creep or surged flow deformation that typically follow vegetation removal on extensive, metastable hillslopes in the Klamath Mountains (e.g., Swanston et al. 1988). Any possible trend toward stabilization of slopes on industrial forest lands will have been obscured or reversed by a new round of logging that began in the late 1980s. Much of the latest logging involves complete removal of 20- to 30-year-old second-growth stands for wood pulp. Such a management regime appears to afford little respite between cycles of deforestation for recovery of slope stability.

A second reason for continued cumulative increases in erosion over time was the disproportionate sensitivity of the basin to disturbance of headwater lands managed by federal agencies (Frissell 1992). Despite evidence that federal managers employed more cautious logging practices than had been used on most corporate lands in Euchre Creek, landslide rates increased more dramatically on federal lands following logging than they did on most private lands. However, federal ownerships were concentrated in the steepest and most geologically unstable portions of the basin where each acre of forest disturbance causes far more erosion than a similar activity on more stable lands (McHugh 1986; C. Frissell, unpubl. data). Moreover, more slope failures propagated into large, in-channel debris flows. By 1974, the occurrence of debris-flow initiation sites in headwater channels on federal lands increased more than fiftyfold in streams disturbed by roads and logging compared with adjacent, undisturbed federal forest; 21 large debris flows originating from roads and clearcut units on federal lands were observed in 1974 photographs of Euchre Creek. The incidence of debris flows increased further by the 1980s.

Many of these debris flows triggered channel changes and secondary streamside debris slides that contributed large masses of sediment directly to the stream network (C. Frissell, unpubl. data).

Changes in erosion and sediment loads triggered massive and persistent changes in the mainstem of Euchre Creek (Fig. 8). This has been the primary spawning and rearing habitat for fall chinook salmon in the basin. In 1940, the stream was dominated by reaches with partially or completely closed forest canopies. Riparian forests were well-developed even though much of the stream had been subject to selective logging and livestock grazing since early in the century, and the basin had burned extensively in the late 1800s. By 1965, large storms and infusions of sediment from disturbed slopes and tributary streams triggered dramatic channel widening and aggradation. The spatial extent of channels with open riparian canopies increased, and many reaches shifted from sinuous or anastomosing channels with forested banks and islands to wide unvegetated gravel bars with braided low-flow channels. Some reaches of Euchre Creek have shown signs of incipient recovery in subsequent decades, but these areas remain vulnerable to renewed widening and lateral shifting even during small storm events. A storm in February 1986, which produced a peak flow with a 2- to 5-year recurrence interval in area rivers (Frissell and Nawa 1992) and produced or reactivated many landslides, aborted recovery in many reaches of Euchre Creek. Alluvial valley segments that dominate the lower half of the mainstem, once

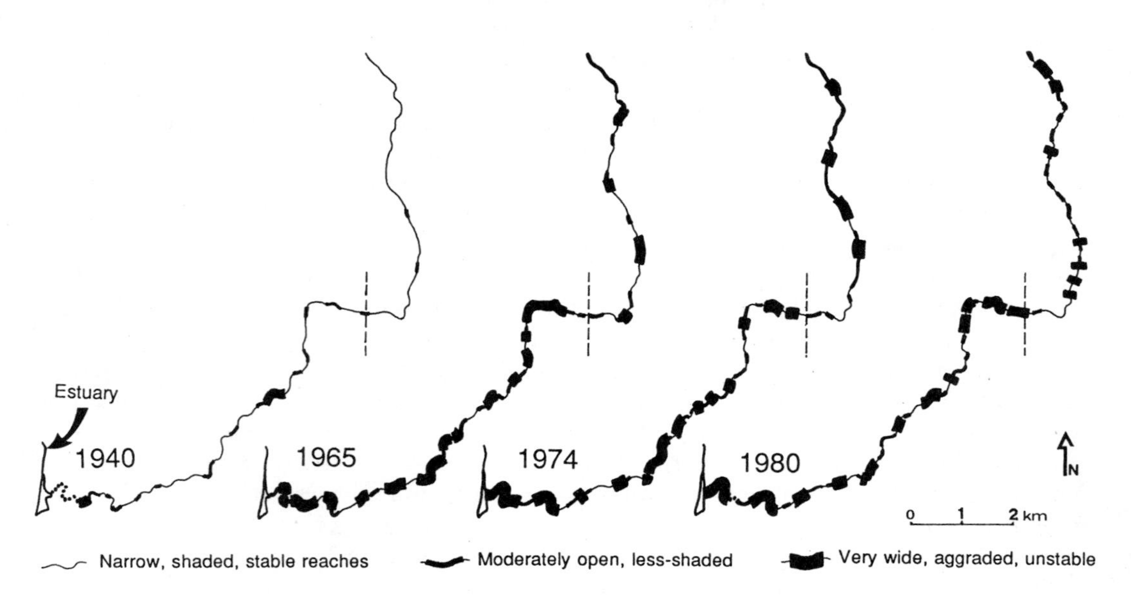

Figure 8. Change in channel conditions in the mainstem of Euchre Creek at intervals since 1940, mapped from air photographs. — = channels almost entirely obscured by forest canopy; ▬ = reaches lacking canopy cover over the low-flow water surface; ■ = highly aggraded and unstable reaches with wide gravel bars and exposed multi-thread or braided channels. --- = transition from narrow, higher-gradient canyon segments upstream to wide, gently sloping alluvial valley segments downstream.

probably the most productive spawning and rearing areas for chinook salmon, show little sign of persistent recovery to pre-1965 channel conditions (Fig. 8). Recovery in some reaches is further hampered by livestock grazing. Primary riparian forests apparently resisted the impacts of grazing for many decades, but once these forests were destroyed or fragmented by large, sediment-laden floods and related channel changes, grazing effectively discouraged reestablishment of woody vegetation at some sites.

The transformation of the habitat mosaic in Euchre Creek and similar streams had many negative effects on summer and winter habitat conditions; these effects are well known to reduce salmonid survival. For example, highly aggraded channels in disturbed drainage basins in southwest Oregon have less summer habitat (shallower riffles and reduced pool volumes) available than comparable, less disturbed streams (Frissell 1992). The exposed, shallower surface flows in such streams are more vulnerable to heating than are deeper, more shaded channels (McSwain 1987). Summer rearing densities of all species of anadromous salmonids decline with increasing maximum temperature in these streams, and temperature increases of just a few degrees can substantially depress fish production and survival (Frissell 1992). Overwinter survival of salmon eggs and fry is almost certainly lower and more interannually variable in drainage basins extensively disturbed by human activities than in relatively undisturbed basins. The effectiveness of large woody debris (LWD) in stabilizing channel morphology and providing cover for fish has been reduced. Streambed surfaces are more heavily dominated by fine sediments in relatively disturbed drainages (Frissell 1992). Scour and fill of lethal magnitude is far more frequent and widespread in salmon spawning reaches of highly disturbed streams compared

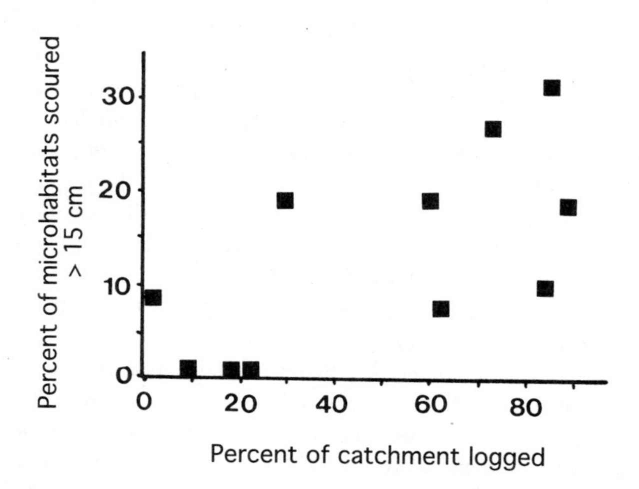

Figure 9. Percentage of salmon spawning microhabitats subject to lethal-magnitude scour during winter storms in south coast Oregon streams as a function of the cumulative percentage of the contributing drainage area logged. For this analysis, we assumed scour of ≥15 cm was sufficient to destroy at least some eggs of smaller female salmon. Data aggregated over 2-4 years with multiple reaches sampled in most of the 11 study streams (Nawa and Frissell 1993; C. Frissell and R. Nawa, unpubl data). During 3 of these years, annual peak flow was less than the 2-year recurrence interval, and highest peak did not exceed the 5-year recurrence interval in magnitude.

with similar habitats in streams draining more intact watersheds (Fig. 9; Nawa and Frissell 1993; C. Frissell and R. Nawa, unpubl. data). These differences appear to be reflected in larger and more diverse populations of adult and juvenile chinook and coho salmon remaining in drainage basins where hillslopes and channels are more stable (Frissell 1992).

Fishing and the Adaptive Capacity of Southern Oregon Salmon

The persistence and productivity of salmon stocks are affected by changes in habitat but also by the direct and indirect effects of other human interventions, especially fishing. Older female salmon tend to be larger in body size and fecundity (Healey 1991); however, each year a salmon remains at sea it remains vulnerable to catch in the ocean troll fishery. During the 1970s and 1980s, cumulative mortalities from such fisheries were high enough that fishing constituted a strong selective force against maturation at older ages and larger body sizes. Although such selective harvest can directly lead to collapse of some populations (Hankin and Healey 1986), the more immediate effect is to constrain the adaptive capacity of populations by selectively removing certain life-history types that might prove critical for survival of the population (Warren and Liss 1980). For example, changes in the habitat of southern Oregon rivers, including reduced stability of spawning habitats, most likely select for large-bodied fish, which bury eggs deeply enough to survive the effects of many storm flows (Holtby and Healey 1986). Smaller, younger female salmon dig redds that are shallower and, therefore, their eggs and fry are far more vulnerable to scour in unstable streams.

Apparent shifts in the spawning time of chinook salmon populations corroborate the significance of scour-related mortality in the collapse of populations. In relatively stable drainage basins of southern Oregon where large populations are still present (e.g., Dry Creek, Elk River, Big Emily Creek; see Fig. 6), adult salmon tend to return and spawn in late fall and early winter (peak of spawning in December) (Nicholas and Hankin 1988; Oregon Dep. Fish and Wildlife, Gold Beach, Oregon, unpubl. data). In Euchre Creek and other heavily disturbed streams, spawning populations are much smaller, and the peak of spawning usually occurs about mid- or even late-January (C. Frissell and R. Nawa, unpubl. data; Oregon Dep. Fish and Wildlife, Gold Beach, Oregon, unpubl. data). Few salmon return earlier than late November or December to these streams. This is significant because on average the largest winter storms strike this portion of the coast during the first two weeks of January (Pyke 1972). In unstable systems, fall and early-winter spawning life histories may experience heavy mortality because their progeny must survive in the streambed during mid-winter storms. Winter steelhead, which spawn after the January storm period, appear to have comparatively high and consistent reproductive success in Euchre Creek. Older local residents have reported to us that prior to collapse of the Euchre Creek salmon runs in the 1960s, large numbers of fish entered the stream in fall and spawning activity peaked in late November or early December.

Why did the apparent loss of early spawning life-history types correspond with collapse of salmon populations rather than shifting to a population of similar abundance with a later spawning peak? Some other factor obviously limits productivity of late-spawning salmon in these streams. In stream systems that are stable enough to support early spawning, progeny of early spawners may benefit from growth and survival advantages during the spring months (Reimers

1971), exploiting opportunities for early smolting that are not available to progeny of later spawners. Fish that reach sufficient size and maturity to smolt by June can avoid the crowded, warm, poor-growth conditions in these streams during midsummer (Reimers 1971). Progeny of late-spawning parents emerge later and might miss much of the spring growth and June smoltification windows. To attain size for successful smolting, progeny of later spawners are probably required to remain in the river until cooler, wetter conditions in fall are again conducive to good growth (Reimers 1971). In this way, offspring of later-spawning fish could be forced to contend with the adverse summer and winter habitat conditions induced by human disturbance of the drainage basin.

Thus, the effects of anthropogenic habitat change and fishing pressure have been synergistic in several ways. Fishery-driven selection against older, larger females probably exacerbates the effects of spawning habitat deterioration by constraining the population from shifts in age and size structure that might compensate for reduced bed stability as the watershed changes. Conversely, habitat changes reduce the ability of salmon populations to sustain fishing pressure because habitat alterations depress survival and may select against younger age-at-maturity and early-spawning life-history types, which we suspect in most years are the most productive components of the population from the harvest standpoint.

This example illustrates how changes in habitat and adult life histories, such as loss of early spawning, early ocean entrance, and late-maturing phenotypes within populations, can cause complex developmental and ecological consequences that constrain the ability of fish populations to adapt to other sources of environmental variation. Frequently, the consequences of this loss of life-history diversity and adaptive capacity are seen as increased fluctuations of recruitment or population size followed by collapse (den Boer 1968, Walters and Cahoon 1985). Many kinds of direct, indirect, and interactive environmental effects can cause such changes, which diminish intrapopulation diversity and constrain adaptive capacity (Warren and Liss 1980).

The Ghost of Restoration Past

The primary response of management agencies to the depletion of salmon populations in Euchre Creek and other southern Oregon rivers has been to attempt to supplement fall chinook salmon runs with fish of hatchery origin. For example, fry and fingerling salmon have been repeatedly released into the middle reaches of Euchre Creek in an effort to establish a reproducing population in habitat that has been perceived by the state management agency as chronically underseeded. This program has not produced the anticipated large returns of adult salmon, but in some years a small number of fish do return to release sites and spawn, primarily in early winter. In one year, we monitored streambed scour and fill at redd sites of several of these fish, and observed streambed changes of lethal magnitude in nearly every instance (Nawa et al. 1988). The fish, putatively of hatchery origin, homed faithfully and spawned at or near their release site, but the release site lies within the most severely aggraded and unstable segment of the entire stream system. Homing behavior of the stocked fish did not promote spawning success because their reproduction was curtailed by the same processes that have depressed or eliminated wild fish from this segment of Euchre Creek. Most salmon believed to be of wild origin spawn consistently at sites far upstream from this segment in much more stable reaches of the stream. The success of hatchery fish in a challenging reproductive environment is further compromised by the common pattern of inadvertent but strong artificial selection in hatchery environments for fish that return to spawn early in the run and at younger ages, a pattern that is apparently maladaptive in Euchre Creek and similarly altered streams. This example illustrates serious flaws in the notion that hatchery stocking can supplement a depressed wild population whose habitat is impaired. Since ~1991, biologists have taken steps to curtail such stocking programs in southern Oregon rivers, but such programs are prone to cyclical reinvention as management personnel and political pressures change.

The second major response of resource managers has been to spend large amounts of time and money building instream structures presumed to improve habitat conditions. Such projects have experienced high failure rates and have not overcome the physical processes driving habitat decline (e.g., sedimentation, loss of channel stability, and increased water temperature) (Frissell and Nawa 1992). Despite the lack of verified success, state and federal agencies continue to direct massive fiscal and personnel resources to promote and construct instream projects.

The Future of Salmon Restoration in Southwest Oregon

To effectively reverse declines in salmon populations in southwest Oregon, it is necessary to identify the processes thought to be driving the declines, their spatial and temporal distributions, their reversibility, and the feasibility of interventions or changes of management to set the stage for recovery. Restoration measures must be prioritized based on the urgency of intervention, the likely consequences of failing to intervene or delaying the action, the availability and cost-effectiveness of suitable technology, and the risk of unintended side effects (Doppelt et al. 1993).

The most obvious action of direct benefit to salmon is in the curtailment of fishing or modification of fishing regimes to reduce cumulative mortality on fish destined to mature at older ages (Nicholas and Hankin 1990). This change directly affects population structure and within-population diversity and might also have long-range benefits by relaxing selection regimes against larger fish. In recent years, harvest on these chinook salmon stocks has been greatly curtailed. Although there has been little numerical response in spawning escapement, there is some evidence that the proportion of spawners in older age groups and larger body sizes has increased in recent runs. Such a qualitative response could be the first step toward improved reproductive success and future increases in spawning populations.

The reversibility of habitat damage is more problematic. The literature on post-logging recovery of habitat conditions is replete with examples of partial and arrested recovery patterns (Niemi et al. 1990, Frissell 1993b, Sear 1994). Although certain factors such as water temperature in small channels may recover rapidly with reestablishment of riparian vegetation, other facets of habitat condition (e.g., fine sediment load, channel stability, LWD abundance and stability, pool volumes and bar morphology, water temperature in larger streams) still retain the signature of disturbance 40 years to a century or more after logging and road construction (Hagans et al. 1986; Frissell 1991, 1992). Many aspects of channel morphology may not recover until a major flood occurs that is not accompanied by additional large influxes of sediment from hillslopes (i.e., the watershed is stabilized) and well-developed riparian forests are reestablished. This means that protection of presently productive habitats within relatively intact catchments is the most urgent and likely most cost-effective habitat conservation measure (Frissell 1993b, Doppelt et al. 1993). Given the high degree of uncertainty about recovery of altered aquatic ecosystems, watershed refugia (Sedell et al. 1990; Frissell 1991, 1992, 1993b), aquatic diversity areas (Henjum et al. 1994), or key watersheds (Reeves and Sedell 1992, Moyle and Yoshiyama 1994) provide the last best chance for these species to maintain a semblance of their historical distribution in the region. These streams, most draining roadless areas subject to little or no past human disturbance, can function as secure refugia if their drainage basins remain undisturbed. A network of such watersheds that is well distributed and redundant (Moyle and Sato 1991) is necessary to maintain the regional diversity of the species pool and provide sources for recolonization of adjacent habitats should they eventually be allowed to recover (Frissell 1993b). Protected streams offer fragments of historical, co-evolved habitat and biotic diversity within a landscape that has been otherwise severely degraded and simplified (Reeves and Sedell 1992, Reeves et al. 1993, Rhodes et al. 1994, Ebersole et al. 1996). Moreover, intact basins provide research and monitoring benchmarks that yield models for developing appropriate objectives and measures of restoration for comparable impacted basins (Ebersole et al. 1996).

Unfortunately, many of watersheds that remain largely undisturbed by human activities constitute the most naturally unstable and sensitive headwater portions of the landscape (Frissell 1992). Such habitats are subject to a range of natural disturbance events that render them only partly effective as long-term refugia. Although in many instances headwater refugia provide the best remaining habitat available for many species, they offer less than ideal habitat for fish such as fall chinook salmon, which were historically more prolific in downstream lowland areas. Protection of these refugia, while an absolutely necessary first step, must be closely followed by long-term restoration of critical habitats elsewhere in the river system (Doppelt et al. 1993, Frissell 1993b,).

In altered habitats, stabilization of chronic and potential sediment sources from roads and hillslopes will be required before site-specific, in-channel or riparian projects can be expected to succeed (Frissell and Nawa 1992, Frissell 1993b, Sear 1994). Treatments are available to prevent future erosion from roads, and such technology has been well developed in Redwood National Park and elsewhere in northern California (e.g., Weaver et al. 1987). Such projects will be necessary to enable future recovery of severely damaged watersheds, and these measures are urgently needed in basins that have been only recently disturbed by logging activities in headwater areas. In such drainages (primarily on federal lands), stream habitat and fish populations may still be healthy because road failures and other potential impact sources have not yet been exposed to a triggering storm event; however, these disturbances constitute "time bombs" that pose dire consequences for salmon and their habitats unless they are defused at the source (Hagans et al. 1986, Weaver et al. 1987, Frissell 1993b).

Conclusions and Recommendations

Exploitation-oriented approaches to the management of salmon resources, dominated by mechanistic thinking, have largely run their course in the Pacific Northwest. Widespread depletion and endangerment of populations and curtailment of salmon fisheries reflect the systematic impoverishment of Pacific salmon and their environment (Nehlsen et al. 1991, Frissell 1993a). Our examples illustrate how human-induced habitat change, and direct alteration of the life-history diversity of salmon populations through fishing and fish culture practices, can interact to debilitate salmon populations and their ecosystems in ways not well addressed in traditional models of management. Biological and cultural elements of ecosystems have suffered from this loss, and today we face the challenge of defining a major shift in the scientific and management approach for restoring these species and their ecological and cultural contexts.

In this paper, we frame a more contextualistic view of salmon management. Our perspective emphasizes that numerical performances of salmon populations and their habitats are simply indicators of the processes and relationships that give rise to the salmon resource. In our perspective, different empirical measures of the salmon resource assume importance, and parameters such as biomass and aggregate escapement or recruitment convey little meaning outside of a limited context. In the future, less mechanistic views of management must emphasize the natural capacity of living systems for self-regeneration and self-restoration. Management is redefined as identifying and alleviating environmental constraints (particularly artificial ones) that hinder these self-restorative modes. In many cases, restoration of the intrinsic self-sustaining capacity of ecosystems and salmon resources clearly will take a very long time. In some cases, it may no longer be possible. But experience has demonstrated the regrettable environmental and human consequences of previous approaches to the management of natural resources.

Empirical information useful to assess and focus this process includes measures of spatial and developmental diversity of populations, temporal continuity of populations, organization of colonization pools for reestablishing lost populations or species at a range of scales, and life-history variation and significance of adaptive strategies both among and within populations and species. We place similar emphasis on spatial and temporal diversity and pattern of habitats and the basin-wide processes that create them. Special attention to the spatial distribution and reversibility of habitat-altering processes (particularly those strongly influenced by human activities) is critical. Contrasts exist in the selection and interpretation of performance measures between more mechanistic and more contextualistic approaches to salmon management (Table 1). Contextualistic measures seem more complex than past ones, and they do span a much richer body of knowledge of natural resources and ecosystems. However, much of the apparent complexity of a contextual approach may simply be an artifact of its relative novelty to most of us.

Most past and many ongoing attempts to ameliorate declines in salmon resources have been by nature superficial, ineffective, and counterproductive. Professional managers and scientists squander vast human and fiscal resources daily on activities that fail to address the root causes of resource problems and merely serve to temporarily pacify or narcotize key segments of society (Meffe 1992). Science has served primarily to provide ecologically irrelevant or semi-relevant indicators, such as retrospective measures of aggregate catches and escapement generated by past generations of fish, that tell us little more than how far the biological system has already been reshaped and simplified to conform to our models. These data serve to demonstrate when we have finally failed (e.g., when a resource is depleted to virtual extinction) (Ludwig et al. 1993). If we are fortunate enough to have accurate and sufficient data of this kind, our science allows us to reconstruct, after the stock is lost, what the optimal harvest rate would have been, or at what point we critically damaged the productive capacity of the habitat beyond recovery.

We suspect the most effective measures to protect and restore salmon and other natural resources are often overlooked because they seem simple and relatively non-technological in concept, conservative in application, and inexpensive in terms of necessary capital investment. However, conservative protection and restoration actions can be contentious and locally costly

Table 1. Some comparative examples of the goals and performance measures of salmon management under more mechanistic (production- and exploitation-oriented) and contextualistic (restoration- and ecosystem-oriented) perspectives.

Conceptual perspective	Goals of management	Performance measures
HABITAT MANAGEMENT		
Production/exploitation view	Control environment to create optimal state for production Improve natural habitat structure Replace habitat structures lost by previous or proposed management Limit impact of human activities at the activity site when practicable Replace lost habitat off-site	Average pool-riffle ratio, pieces of woody debris per km Proportion of natural riparian structure or shade retained Acceptable implementation of Best Management Practices (BMPs) Dispersion of activities to less-disturbed areas to dilute impact Net area disturbed relative to threshold level assumed to limit impact Equivalent area treated with off-site mitigation Dollars invested per area
Restoration/Ecosystem View	Allow natural processes to maintain range of states required by species and life stages in variable environments Maintain processes that create and maintain habitats Alleviate constraints and restore processes that create habitats Reduce cumulative impacts of activities by preventing, deferring, or relocating local disturbances Maintain temporal continuity and connectivity of existing high-quality habitats	Habitat complexity and connectivity Spatial, historical, and biophysical context of impacted site (legacy of previous impacts) Effects of activity on future habitat-forming processes (proportion of natural capacity lost) Resistance, resilience of ecosystem to natural disturbances (e.g., flood, fire, drought) Ecological effectiveness of BMPs , on-site and off-site Risk that cumulative effects exceed known direct effects Deviation from natural disturbance regimes and habitat mosaics, all scales Risks to high-quality, biodiverse, or productive sites Reversibility of activities and effects Ability to monitor and detect response before irreversible impacts occur Cost efficiency and ecological effectiveness of restoration versus impact avoidance Self-sustaining maintenance of target species and other native biota

Table 1—cont.

Conceptual perspective	Goals of management	Performance measures
Harvest and production management		
Production/Exploitation View	Broaden production base and diversify fisheries through species introductions	Aggregate average catch or yield
	Maximize or supplement recruitment to increase harvestable stock	Harvest rate
	Maximize sustained yield (MSY) or catch, in aggregate	Stock or catch per unit effort response to changes in harvest rate
	Promote efficiency of harvest to ensure profit	Aggregate escapement near that required for MSY
	Compensate lost habitat with hatchery production	Number of juveniles released from hatcheries or presumed to be produced by habitat improvements
	Maximize access and number of fishers participating	Productivity as measured by dominance of ages and species deemed most productive and desirable under heavy fishing regime
Restoration/Ecosystem View	Maintain natural diversity of size, age, other life-history features	Trend in spatial distribution and diversity of spawning populations
	Maintain the diversity of nonconsumptive values and ecosystem services provided by target species	Trend in size, age, run-timing diversity within populations
	Protect spawning stock to ensure short- and long-term recruitment	Sustainability and equitability of catch
	Protect quality of fishing experience and economic and cultural security of fishers (compatible with other objectives)	Interannual variation in escapement and recruitment, likelihood of collapse
	Reduce dependency of fishers on unsustainable resources or methods	Correlation among subpopulations
	Prevent introduction or dispersal of nonnative species and artificial inflation of native populations	Fragmentation or connectivity among subpopulations
	Protect and restore diversity and capacity of habitats	Selective effect of fishing mortality and habitat change
		Status of other species and ecosystem functions dependent on the target species
		Community integrity, absence or scarcity of nonnative species
		Satisfaction and economic well-being, diversity of dependent fishers and communities

Table 1—cont.

Conceptual perspective	Goals of management	Performance measures
Mainstem reservoir/dam management		
Production/exploitation view	Introduce warm-water species to increase fishing opportunities in reservoirs	Fishing effort, licenses sold, and aggregate catch
	Alter or mitigate dam operations to improve survival of target species	Black-box survival rates through dam/reservoir reaches
	Control selected predators to reduce salmon mortality	Pounds of predators killed or dollars of bounties paid
	Add hatchery fish to compensate for inflated mortality rates, replace species or ecotypes unsuited to altered habitats	Numbers of hatchery fish released and caught
Restoration/Ecosystem View	Minimize proliferation of or reduce nonnative species	Long-term effect of reservoirs and introduced species complex on native biota
	Favor persistence and recovery of native species and life history types	Sources, timing, location, and trends in wild salmon mortality in reservoir and tributary food webs
	Manage reservoirs and their tributaries as integrated ecosystems	Presence of life history and behavioral strategies, habitat features that have allowed existing native populations to persist
	Ensure that hatchery salmon do not directly or indirectly compromise the survival of wild populations	Invasion of tributary habitats by nonnative fishes from reservoirs
	Eliminate dams or alter operations to restore less artificial flow patterns and habitat conditions where possible	Deviation from historical flow regime and habitat conditions

because they require forgoing some ongoing extractive activities and weaning communities from the short-term economic benefits and subsidies such activities sometimes have generated. For example, despite the efforts of scientists to identify refuge watersheds and prescribe at least interim protection of specific areas from future logging, grazing, and mining (e.g., Reeves and Sedell 1992, Henjum et al. 1994, Moyle and Yoshiyama 1994), many public and private land managers have continued to target such sensitive drainages for logging activity (Frissell et al. 1992). Continuing this old management strategy risks the viability of these refugia and further jeopardizes many native species, such as salmon, that have become extraordinarily dependent on such places for their long-term survival (Frissell et al. 1992, Frissell 1993b, Rhodes et al. 1994). However, despite ostensible changes in management goals and objectives in written plans, resource managers find that their budgets and staffing remain closely tied to performance measures reflecting the extractive exploitation of timber, fish, and other natural resources.

The models of management and the empirical measures that drove them over the past century have served well to rationalize and allocate the overexploitation of a once-thriving biological resource. Now that we inherit the task of restoration of that resource, a shift in our concepts of management and different measures of management success are in order.

Acknowledgments

This paper is dedicated to our friend and colleague C. Warren. K. Wildman assisted in preparation of figures. Field research in southwest Oregon was funded by the Federal Aid in Sport Fish Restoration program through a grant from the Oregon Department of Fish and Wildlife, with additional support from Oregon State University's Agricultural Experiment Station. We especially thank J. Lichatowich for his role in fostering that cooperative research project.

Literature Cited

Andrewartha, H.G. and L.C. Birch. 1984. The Ecological Web. The University of Chicago Press, Chicago.

Balon, E.K. 1993. Dynamics of biodiversity and mechanisms of change: a plea for balanced attention to form creation and extinction. Biological Conservation 66: 5-16.

Bella, D.A. 1987. Organizations and the systematic distortion of information. Journal of Professional Issues in Engineering 113(4): 360-370.

Bella, D.A. 1992. Ethics and the credibility of applied science, p. 19-32. *In* G.H. Reeves, D.L. Bottom, and M.H. Brookes (coords.), Ethical Questions for Resource Managers. USDA Forest Service General Technical Report PNW-GTR-288. Portland, Oregon.

Bisson, P.A., T.P. Quinn, G.H. Reeves, and S.V. Gregory. 1991. Best management practices, cumulative effects, and long-term trends in fish abundance in Pacific Northwest river systems, p. 189-232. *In* R.J. Naiman (ed.), Watershed Management: Balancing Sustainability and Environmental Change. Springer-Verlag, New York.

Cobb, J.N. 1930. Pacific salmon fisheries. US Department of Commerce, Bureau of Fisheries, Fisheries Document 1092. Washington, DC.

Cooney, C.X. and S.E. Jacobs. 1992. Oregon coastal salmon spawning surveys, 1990. Oregon Department of Fish and Wildlife, Fish Division, Information Report 92-2. Portland, Oregon.

den Boer, P.J. 1968. Spreading of risk and the stabilization of animal numbers. Acta Biotheoretica 18:165-194.

Dixon, J.A., and P.B. Sherman. 1990. Economics of Protected Areas: a New Look at Benefits and Costs. Island Press, Covelo, California.

Doppelt, B., M. Scurlock, C. Frissell, and J. Karr. 1993. Entering the Watershed: a New Approach to Save America's River Ecosystems. Island Press, Washington, DC.

Ebersole, J.L., W.J. Liss, and C.A. Frissell. 1996. Restoration of stream habitats in managed landscapes in the western USA: Restoration as re-expression of habitat capacity. Environmental Management, in press.

Eldridge, M.B. 1988. Life history strategies and tactics of striped bass (*Morone saxatilis* Walbaum). Ph.D. dissertation, Department of Fisheries and Wildlife, Oregon State University. Corvallis.

Frissell, C.A. 1991. Water quality, fisheries, and aquatic biodiversity under two alternative forest management scenarios for the west-side Federal lands of Washington, Oregon, and northern California. Consultant Report, The Wilderness Society, Seattle, Washington.

Frissell, C.A. 1992. Cumulative effects of land use on salmon habitat in southwest Oregon coastal streams. Ph.D. dissertation, Department of Fisheries and Wildlife, Oregon State University. Corvallis.

Frissell, C.A. 1993a. Topology of extinction and endangerment of fishes in the Pacific Northwest and California. Conservation Biology 7(2): 342-354.

Frissell, C.A. 1993b. A new strategy for watershed restoration and recovery of Pacific salmon in the Pacific Northwest. Pacific Rivers Council, Eugene, Oregon.

Frissell, C.A., W.J. Liss, C.E. Warren, and M.D. Hurley. 1986. A hierarchical framework for stream habitat classification. Environmental Management 10: 199-214.

Frissell, C.A. and R.K. Nawa. 1992. Incidence and causes of physical failure of artificial fish habitat structures in streams of western Oregon and Washington. North American Journal of Fisheries Management 12: 182-197.

Frissell, C.A., R.K. Nawa, and R. Noss. 1992. Is there any conservation biology in "New Perspectives?": a response to Salwasser. Conservation Biology 6: 461-464.

Gregor, M.E. 1982. A conceptual unification and application of biogeographic classification. MS thesis, Department of Fisheries and Wildlife, Oregon State University. Corvallis.

Gresswell, R E., and W.J. Liss. 1995. Values associated with management of Yellowstone cutthroat trout in Yellowstone National Park. Conservation Biology 9: 159-165.

Grumbine, R.E. 1994. What is ecosystem management? Conservation Biology 8: 27-38.

Hagans, D.K., W.E. Weaver, and M.A. Madej. 1986. Long-term on-site and off-site effects of logging and erosion in the Redwood Creek basin, northern California, p. 38-65. *In* Papers presented at the American Geophysical Union Meeting on Cumulative Effects. National Council for Air and Stream Improvement Technical Bulletin 490. New York.

Hall, C.A.S. 1990. Sanctioning resource depletion: economic development and neo-classical economics. The Ecologist 20(3): 99-104.

Hankin, D.G. and M.C. Healey. 1986. Dependence of exploitation rates for maximum sustained yield and stock collapse on age and sex structure of chinook salmon (*Oncorhynchus tshawytscha)* stocks. Canadian Journal of Fisheries and Aquatic Sciences 43: 1746-1759.

Healey, M.C. 1991. Life history of chinook salmon (*Oncorhynchus tshawytscha*), p. 313-393. *In* C. Groot and L. Margolis (eds.), Pacific Salmon Life Histories. University of British Columbia Press, Vancouver.

Healey, M.C. 1994. Variation in the life history of chinook salmon and its relevance to conservation of the Sacramento winter run chinook salmon. Conservation Biology 8: 876-877.

Henjum, M.G., J.R. Karr, D.L Bottom, D.A. Perry, J C. Bednarz, S.G. Wright, S.A. Beckwitt, and E. Beckwitt. 1994. Interim protection for late-successional forests, fisheries, and watersheds, national forests east of the Cascade Crest, Oregon and Washington. The Wildlife Society Technical Review 94-2. Washington, DC.

Hilborn, R. 1992. Hatcheries and the future of salmon in the Northwest. Fisheries 17(1): 5-8.

Hindar, K., N. Ryman, and F. Utter. 1991. Genetic effects of cultured fish on natural fish populations. Canadian Journal of Fisheries and Aquatic Sciences 48:945-957.

Holling, C.S. 1973. Resilience and stability of ecological systems. Annual Review of Ecology and Systematics 4:1-23.

Holtby, L.B. and M.C. Healey. 1986. Selection for adult size in female coho salmon (*Oncorhynchus kisutch*). Canadian Journal of Fisheries and Aquatic Sciences 43: 1946-1959.

Kuhn, T. S. 1970. The Structure of Scientific Revolutions. University of Chicago Press, Chicago.

Leary, R.F., F.W. Allendorf, and K.L. Knudsen. 1989. Genetic differences among rainbow trout spawned on different days within a single season. Progressive Fish-Culturist 51: 10-19.

Leopold, A. 1949. A Sand County Almanac. Oxford University Press, New York.

Leopold, L.B. 1990. Ethos, equity, and the water resource. Environment 32(2): 16-42.

Li, H.W., C.B. Schreck, C.E. Bond, and E. Rexstadt. 1987. Factors influencing changes in fish assemblages of Pacific Northwest streams, p. 193-202. *In* W.J. Matthews and D.C. Heins (eds.), Community and Evolutionary Ecology of North American Stream Fishes. University of Oklahoma Press, Norman.

Lichatowich, J., L. Mobrand, L. Lestelle, and T. Vogel. 1995. An approach to the diagnosis and treatment of depleted Pacific salmon populations in Pacific Northwest watersheds. Fisheries 20(1): 10-18.

Lisle, T.E. 1982. Effects of aggradation and degradation on pool-riffle morphology in natural gravel channels, northwestern California. Water Resources Research 18: 1643-1651.

Livingston, R.J. 1991. Historical relationships between research and resource management in the Apalachicola River estuary. Ecological Applications 1:361-382.

Ludwig, D., R. Hilborn, and C. Walters. 1993. Uncertainty, resource exploitation, and conservation: lessons from history. Science 260: 17, 36.

MacArthur, R.H. and E.O. Wilson 1967. The Theory of Island Biogeography. Monographs in Population Biology I. Princeton University Press, Princeton, New Jersey.

MacLean, J.A. and D.O. Evans. 1981. The stock concept, discreteness of fish stocks, and fisheries management. Canadian Journal of Fisheries and Aquatic Sciences 38: 1889-1898.

McEvoy, A.F. 1986. The Fisherman's Problem. Cambridge University Press, New York, New York.

McHugh, M.H. 1986. Landslide occurrence in the Elk and Sixes basin, southwest Oregon. MS thesis, Department of Geosciences, Oregon State University. Corvallis.

McSwain, M.D. 1987. Summer stream temperatures and channel characteristics of a southwestern Oregon coastal stream. MS thesis, Department of Forest Engineering, Oregon State University. Corvallis.

Meffe, G.K. 1992. Techno-arrogance and halfway technologies: salmon hatcheries on the Pacific Coast of North America. Conservation Biology 6: 350-354.

Moyle, P.B., and G.M. Sato. 1991. On the design of preserves to protect native fishes, p. 155-169. *In* W.L. Minckley and J.E. Deacon (eds.), Battle Against Extinction: Native Fish Management in the American West. University of Arizona Press, Tucson, Arizona.

Moyle, P.B. and R.M. Yoshiyama. 1994. Protection of aquatic biodiversity in California: a five-tiered approach. Fisheries 19(2): 6-18.

Nawa, R.K. and C.A. Frissell. 1993. Measuring scour and fill of gravel streambeds with scour chains and sliding-bead monitors. North American Journal of Fisheries Management 13: 634-639.

Nawa, R.K., C.A. Frissell, and W.J. Liss. 1988. Life history and persistence of anadromous fish stocks in relation to stream habitats and watershed classification. Annual progress report prepared for Oregon Department of Fish and Wildlife. Oak Creek Laboratory of Biology, Oregon State University Department of Fisheries and Wildlife. Corvallis, Oregon.

Nehlsen, W., J.E. Williams, and J.A. Lichatowich. 1991. Pacific salmon at the crossroads: stocks at risk from California, Oregon, Idaho, and Washington. Fisheries 16(2): 4-21.

Niemi, G.J., P. DeVore, N. Detenbeck, D. Taylor, A. Lima, J. Pastor, D. Yount, and R J. Naiman. 1990. Overview of case studies on recovery of aquatic ecosystems from disturbance. Environmental Management 14: 571-588.

Nicholas, J.W. and D.G. Hankin. 1988. Chinook salmon populations in Oregon coastal river basins: description of life histories and assessment of recent trends in run strength. Oregon Department of Fish and Wildlife Information Report 88-1. Portland, Oregon.

Nicholas, J.W. and D G. Hankin. 1990 Chinook salmon in Oregon coastal river basins: a review of contemporary status and the need for fundamental change in fishery management strategy. *In* Proceedings of the Wild Trout IV Conference, Yellowstone National Park, USA, 18-19 September, 1989. US Fish and Wildlife Service, US Government Printing Office 774-173/25037. Washington, DC.

Noss, R.F. and A.Y. Cooperrider. 1994. Saving Nature's Legacy: Protecting and Restoring Biodiversity. Island Press, Covelo, California.

Northwest Power Planning Council. 1993. Strategy for Salmon, Volume II. Portland, Oregon.

Opdam, P. 1991. Metapopulation theory and habitat fragmentation: a review of holarctic breeding bird studies. Landscape Ecology 5: 93-106.

Pepper, S.C. 1942. World Hypotheses. University of California Press, Berkeley.

Price, D.C. 1978. Development of settlement in Curry County, Oregon. MA thesis, Department of Geography, University of Oregon. Eugene, Oregon.

Pyke, C.B. 1972. Some meteorological aspects of the seasonal distribution of precipitation in the western United States and Baja California. University of California Water Resources Center Contribution No. 139. Los Angeles.

Reeves, G.H., F.H. Everest, and J.R. Sedell. 1993. Diversity of juvenile anadromous salmonid assemblages in coastal Oregon basins with different levels of timber harvest. Transactions of the American Fisheries Society 122: 309-317.

Reeves, G.H. and J.R. Sedell. 1992. An ecosystem approach to the conservation and management of freshwater habitat for anadromous salmonids in the Pacific Northwest. Transactions of the 57th North American Wildlife and Natural Resources Conference: 408-415.

Regier, H.A., R L. Welcomme, R.J. Steedman, and H.F. Henderson. 1989. Rehabilitation of degraded river ecosystems, p. 86-97. *In* D.P. Dodge (ed.), Proceedings of the International Large River Symposium. Canadian Special Publication of Fisheries and Aquatic Sciences 106.

Reimers, P.E. 1971. The length of residence of juvenile fall chinook salmon in Sixes River, Oregon. Ph.D. dissertation, Department of Fisheries and Wildlife, Oregon State University. Corvallis.

Ricker, W.E. 1981. Changes in average size and average age of Pacific salmon. Canadian Journal of Fisheries and Aquatic Sciences 38: 1636-1656.

Rhodes, J.J., D.A. McCullough, and F.A. Espinosa, Jr. 1994. A coarse screening process for evaluation of the effects of land management activities on salmon spawning and rearing habitat in ESA consultations. Columbia River Inter-Tribal Fish Commission, Technical Report Series 94-4. Portland, Oregon.

Ryan, S.E. and G.E. Grant. 1991 Downstream effects of timber harvesting on channel morphology in Elk River basin, Oregon. Journal of Environmental Quality 20: 60-72.

Salwasser, H. 1991. New perspectives for sustaining diversity in US forest ecosystems. Conservation Biology 5(4):567-569.

Schlosser, I.J. 1991. Stream fish ecology: a landscape perspective. BioScience 41: 704-712.

Sear, D.A. 1994. River restoration and geomorphology. Aquatic Conservation: Marine and Freshwater Ecosystems 4: 169-177.

Sedell, J.R., G.H. Reeves, F.R. Hauer, J.A. Stanford, and C.P. Hawkins. 1990. Role of refugia in recovery from disturbances: modern fragmented and disconnected river systems. Environmental Management 14: 711-724.

Smith, C.L. 1994. Connecting cultural and biological diversity in restoring northwest salmon. Fisheries 19(2): 20-26.

Snyder, J.O 1908. The fishes of the coastal streams of Oregon and northern California. US Bureau of Fisheries Bulletin 27: 153-189.

Southwood, T.R.E. 1977. Habitat, the template for ecological strategies? Journal of Animal Ecology 46: 337-367.

Spencer, C.N., R.B. McClelland, and J.A. Stanford. 1991. Shrimp stocking, salmon collapse, and eagle displacement: cascading interactions in the food web of a large aquatic ecosystem. BioScience 41: 14-21.

Stanford, J.A. and J.V. Ward. 1991 Management of aquatic resources in large catchments: recognizing interactions between ecosystem connectivity and environmental disturbance, p. 91-124. *In* R.J. Naiman (ed.), Watershed Management: Balancing Sustainability and Environmental Change. Springer-Verlag, New York.

Steward, C.R. and T.C. Bjornn. 1990. Supplementation of salmon and steelhead stocks with hatchery fish: a synthesis of published literature. Bonneville Power Administration and US Fish and Wildlife Service Technical Report 90-1. Portland, Oregon.

Swanston, D.N., G.W. Lienkaemper, R.C. Mersereau, and A.B. Levno. 1988. Timber harvest and progressive deformation of slopes in southwestern Oregon. Bulletin of the Association of Engineering Geologists 25: 371-381.

Suzuki, D. and P. Knudtson. 1992. Wisdom of the Elders: Honoring Sacred Native Visions of Nature. Bantam Books, New York.

Taylor, E.B. and J.D. McPhail. 1985. Variation in body morphology among British Columbia populations of coho salmon, *Oncorhynchus kisutch*. Canadian Journal of Fisheries and Aquatic Sciences 42: 2020-2028.

Verspoor, E. 1988. Reduced genetic variability in first-generation hatchery populations of Atlantic salmon (*Salmo salar*). Canadian Journal of Fisheries and Aquatic Sciences 45: 1686-1690.

Walters, C.J. 1986. Adaptive management of renewable resources. McGraw-Hill, New York, New York.

Walters, C.J. and P. Cahoon. 1985. Evidence of decreasing spatial diversity in British Columbia salmon stocks. Canadian Journal of Fisheries and Aquatic Sciences 42: 1033-1037.

Waples, R.S. 1991. Genetic interactions between hatchery and wild salmonids: lessons from the Pacific Northwest. Canadian Journal of Fisheries and Aquatic Sciences 48 (Supplement 1): 124-133.

Warren, C.E. 1979. Toward classification and rationale for watershed management and stream protection. US Environmental Protection Agency Ecological Research Series EPA-600/3-79-059. Corvallis, Oregon.

Warren, C.E. 1989. Resources, culture, and capitalism, p. 148-157. *In* C.L. Smith (ed.), Ocean Agenda 21: Passages to the Pacific Century. Oregon Sea Grant Publication ORESU-B-89-001. Oregon State University, Corvallis, Oregon.

Warren, C.E., M.W. Allen, and J.W. Haefner. 1979. Conceptual frameworks and the philosophical foundations of general living systems theory. Behavioral Science 24:296-310.

Warren, C.E. and W.J. Liss. 1980. Adaptation to aquatic environments, p. 15-40. *In* R.L. Lackey and L. Nielsen (eds.), Fisheries Management. Blackwell Scientific Publications, Cambridge, Massachusetts.

Warren, C.E., W.J. Liss, G.L. Beach, C.A. Frissell, M.E. Gregor, M.D. Hurley, D. McCullough, W.K. Seim, G.G. Thompson, and M.J. Wevers. 1983. Systems classification and modeling of watersheds and streams. Oak Creek Laboratory of Biology Report to the US Environmental Protection Agency and USDA Forest Service. Oregon State University, Corvallis.

Weaver, W.E., M.M. Hektner, D.K. Hagans, L.J. Reed, R.A. Sonnevil, and G.J. Bundros. 1987. An evaluation of experimental rehabilitation work, Redwood National Park. Redwood National Park Technical Report 19. Arcata, California.

Wilcox, B.A. 1980. Insular ecology and conservation, p. 95-117. *In* M.E. Soule´ and B.A. Wilcox (eds.), Conservation Biology: An Evolutionary-Ecological Perspective. Sinauer, Sunderland, Massachusetts.

Zeisenhenne, F.C. 1935. A stream improvement report and a preliminary survey of the Siskiyou National Forest, Oregon and California, 1935. Siskiyou National Forest. Grants Pass, Oregon.

Technological Solutions: Cost-Effective Restoration

Watershed Management and Pacific Salmon: Desired Future Conditions

Peter A. Bisson, Gordon H. Reeves, Robert E. Bilby, and Robert J. Naiman

Abstract

Natural disturbances are an important part of the ecology of Pacific Northwest watersheds and create a diversity of aquatic environments to which different stocks of salmon (*Oncorhynchus* spp.) and other native fishes have adapted over time. Objectives for managing habitat should be focused on maintaining the full range of aquatic and riparian conditions generated by natural disturbance events at landscape scales large enough to encompass the freshwater life cycles of salmon and other species. Because streams are dynamic, establishing fixed habitat standards for such parameters as temperature, fine sediment concentration, woody debris abundance, or pool frequency (especially when applied to limited stream reaches) is not likely to protect the overall capacity of watersheds to produce fish or to recover from natural or anthropogenic disturbances. Attempting to make streams conform to an idealized notion of optimum habitat through legal regulations or channel manipulations will not easily accommodate cycles of disturbance and recovery, and may lead to a long-term loss of habitat and biological diversity. Desired future conditions can be derived by examining how natural disturbances influence the distribution of aquatic habitats and development of riparian communities within relatively pristine watersheds and by using these patterns as target conditions for watersheds in which management activities are planned. Although it is not feasible to return watersheds to a pristine state in most cases, a complete or nearly complete range of aquatic habitats can be maintained if anthropogenic disturbances are compatible with natural disturbances to the extent possible. Protecting the interactions between streams and surrounding terrain during disturbances (e.g., by maintaining river-floodplain connections and inputs of coarse sediment and organic material during fires, windstorms, and periods of high streamflow) is fundamentally important to maintaining the productivity and biodiversity of river systems. Analysis of watershed condition and development of management prescriptions should include a consideration of the eventuality of large, infrequent natural disturbances to ensure that when these events do occur, important transfers of organic and inorganic materials from terrestrial to aquatic ecosystems are not significantly altered and riparian recovery processes are not impeded.

Introduction

Loss of habitat has played a significant role in the reduction or extinction of many stocks of anadromous salmonids (*Oncorhynchus* spp.) in the Pacific Northwest (Nehlsen et al. 1991). Environmental degradation has resulted from a variety of human activities involving water use and land management adjacent to rivers, lakes, and estuaries. Only a small fraction of the river basins along the Pacific Coast in which anadromous salmonids occur has remained relatively free from habitat loss. In the central and southern range of Pacific salmon in North America, virtually no large river basins remain in a completely pristine state; thus, no clear set of benchmark conditions exist against which habitat degradation can be measured or toward which restoration can be aimed. Increasing human populations in the Pacific Northwest will continue socioeconomic pressures on natural resources of the region and virtually ensure that recovery of entire watersheds to pristine conditions will not occur (National Research Council [NRC] 1996). The rehabilitation of salmon habitat becomes an issue of determining not only what is desirable but also what is realistic and feasible (Lee 1993). An important management question then is "What should the specific objectives of habitat restoration be?"

Invocation of the Endangered Species Act to protect salmon and other species at risk of extinction has compelled economic interests, fish and wildlife agencies, and a concerned public to acknowledge the widespread failure of previous attempts to maintain sustainable populations in the face of intense and often conflicting management activities (Volkman and Lee 1994). The inability of many enhancement projects directed at individual salmon stocks or other declining species to achieve conservation objectives (Meffe 1992, Hilborn and Winton 1993) has fueled the call for an alternative, less species-oriented approach involving ecosystem management at a broader landscape level (Franklin 1993). Although these terms remain operationally vague (Tracy and Brussard 1994, Stanley 1995), the notion of managing large land areas for the purpose of preserving patches of ecologically functional habitat and restoring degraded habitats at geographical scales that make biological sense for whole communities of organisms has become an important priority. It was a fundamental cornerstone of the Forest Ecosystem Management and Assessment Team (FEMAT) recommendations for federal forests in the United States (US) Pacific Northwest (FEMAT 1993, Franklin 1994). Following the federal example, ecosystem management has been embraced by state and private natural resource organizations (Salwasser 1994), but whether implementation of the new paradigm (an ecosystem-based approach guided by watershed analysis together with adaptive learning [Naiman et al. 1992]) will lead to recovery of Pacific salmon habitat is unclear. Success in the long term will likely depend on the willingness of land and water managers to clearly identify the changes they wish to make, engage in large-scale controlled experiments over extended periods, monitor the results, and learn from successes and failures.

Herein we review the notion of desired future conditions, a concept that has emerged as an important element of ecosystem management in the Pacific Northwest. Taken generally, desired future conditions are those that will ensure the maintenance of biological diversity and sustainability of harvestable natural resources (FEMAT 1993). Upon this general goal there has been little debate, but in specific terms, desired future conditions often mean different things to different people. Difficulties in identifying habitat goals for fish and wildlife, and salmon in particular, often derive from a failure to clearly address the following questions: What is desired, what constitutes the future, and what are the conditions we wish to manage? Current approaches to watershed management often stress attainment of habitat standards at spatial and temporal scales that may be geomorphically inappropriate, fail to consider the natural disturbance history of watersheds, and ignore the dynamics and locally adapted life-history requirements of salmon populations. We propose a view of desired future conditions that is less rigid at small scales but explicitly considers a mix of habitats generated by natural processes across larger landscape areas, thereby providing a broader ecological context within which environmental planning can take place. Identifying desired future conditions over geographic areas relevant to salmon life cycles should become an important component of integrated watershed management—the process by which resources within a drainage basin are extracted, nurtured, and conserved with a balance between environmental, social, and economic concerns and with a consideration of future generations.

Habitat Standards

Enactment of the Clean Water Act in the mid-1970s specified that surface waters of the US be maintained as fishable and swimmable. This legislation, more than any other, initiated a system of environmental requirements resulting in important and substantial improvements in the water quality of rivers and lakes. The Clean Water Act enabled the US Environmental Protection Agency (EPA) to identify water-quality standards that must be met by anyone introducing effluents into receiving waters through point-source discharges. It also mandated states to develop their own point-source effluent standards and to develop further plans for reducing water pollution from non-point sources (i.e., land uses in which water-quality impairment did not originate at a single location such as the end of a pipe). Since the original goal of the Clean Water Act was to reduce the discharge of pollutants, principally from industrial and municipal sources, many of the substances addressed in both point- and non-point water-quality regulations have been toxic chemicals.

As water-quality standards were refined with additional research (EPA 1986), regulatory organizations began to develop additional standards for fish habitat. Among the objectives of these standards were the designation of hazard thresholds beyond which significant degradation could be expected, and the definition of habitat states considered optimum for fish. Examples of hazard thresholds that have been applied to salmon habitat include the maximum percentage of fine sediment present in spawning gravels or the maximum allowable temperature for a stream during summer (Bjornn and Reiser 1991). For purposes of regulation, human activities could be allowed to alter fine sediment concentration or increase temperature as long as hazard thresholds were not exceeded. Examples of the second objective of defining optimum habitat include specified frequencies of pool and riffle habitat or the number of pieces of large woody debris (LWD) per unit of stream length believed necessary for pool formation (Washington Forest Practices Board 1993).

Despite improvements in water quality resulting from the implementation of standards (NRC 1992), many Pacific Northwest streams still exist in a highly altered state in which neither the range of natural conditions is present nor the full expression of ecological interactions between aquatic and terrestrial ecosystems is permitted (Gregory et al. 1991, FEMAT 1993, NRC 1996). In the two decades since enactment of the Clean Water Act, the general trend continues toward increasingly degraded aquatic habitat (Bisson et al. 1992, Karr 1994), and the consensus of professional fishery scientists in the region is that habitat loss continues to contribute to the decline of salmon (Nehlsen et al. 1991, Cederholm et al. 1993, Gregory and Bisson 1996).

The inability of water-quality and habitat standards to reverse the overall trend of habitat loss has stemmed from several problems. In some cases, water-quality and habitat standards have simply not been enforced, either because the standards were unrealistic, violations went unnoticed or unreported, or resources for adequate enforcement were insufficient (NRC 1992, Sauter 1994). Parameters selected as standards for salmonid habitat such as temperature, sediment, flows, dissolved oxygen, or pool-riffle ratios were often relatively easy and cheap to measure, or simple to model, but may not necessarily have been the factors exerting the greatest influence on salmonid production (Fausch et al. 1988, Shirvell 1989). Standards may have been based on habitat requirements of single life-cycle stages of individual species and thus were only partially effective (Bisson et al. 1992).

Perhaps most importantly, habitat standards have generally not accounted for the dynamic nature of aquatic ecosystems in which patterns of disturbance and recovery provide the local evolutionary template to which salmonid stocks have adapted. Rather, such standards have often described a set of conditions representing a compromise among the perceived needs of different species or life-history stages. When applied to restoration projects, habitat standards have potentially reduced habitat diversity by eliminating some conditions, even those that may occur naturally (Fig. 1).

We do not advocate abandoning water-quality and habitat standards. They may serve as useful signals of severe environmental degradation. Instead, we suggest that habitat standards not be taken sensu stricto as desired future conditions. If conservation of functional ecosystems supporting naturally occurring assemblages of plants and animals, including salmon, is the principal goal of watershed management (Franklin 1993), then environmental planning and regulation should preserve the dynamic changes that accompany disturbance-recovery cycles and protect essential energy and material transfers that take place between aquatic and terrestrial ecosystems during disturbance events (Reice 1994). Habitat standards should not be taken as surrogates for ecological function; restoration of productive watersheds will require management activities that allow the natural range of conditions to be expressed.

Dynamic Populations in Dynamic Watersheds

Pacific salmon exhibit characteristics of complex metapopulations in which local reproductive groups (demes) spawn in different areas and exchange genes with other groups through adult straying (Riddell 1993). The structure of salmon metapopulations may vary for different species, depending on the geoclimatic features of a particular area and the species' life-history requirements, but salmon spawning in multiple locations within a river basin usually have strong and weak demes at any given time (Scudder 1989). Whether a particular deme is large or small depends on many factors, some of which are related to freshwater conditions and some to oceanic conditions.

At large spatial and temporal scales, extirpation of local demes in marginal habitats may be relatively common (Harrison 1991), especially in salmon populations. Straying is an important adaptation for recolonizing suitable habitat as populations expand. The centripetal flow of genes from marginal demes to large central demes during periods of population contraction is an important means of enriching genetic diversity (Scudder 1989). Even in watersheds relatively free

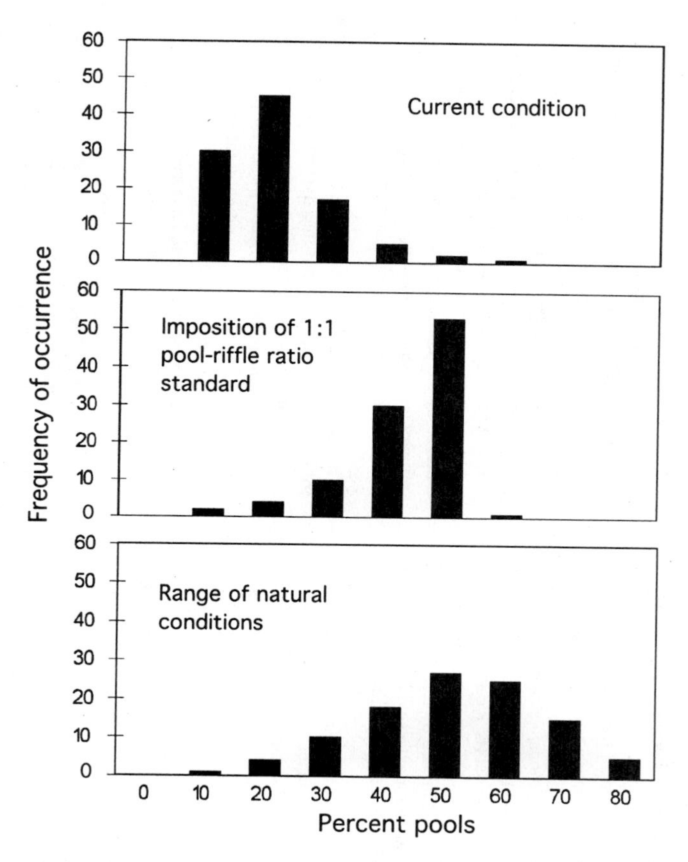

Figure 1. Three scenarios describing the frequency of pool habitat in streams within a hypothetical watershed. The graph at the top illustrates a situation in which pools have been lost due to widespread habitat alteration (e.g., Sedell and Everest 1991). The middle graph illustrates what might result from an attempt to rehabilitate altered streams to a pool-riffle ratio of 1:1 (some streams would likely fail to achieve this desired condition because of fundamental geomorphic constraints). The lower graph describes how pools might appear if the watershed were in a pristine state, where geomorphic controls and natural disturbance-recovery cycles have generated a wide range of pool frequencies.

of human influences, natural disturbances (Table 1) create a mosaic of habitat patches in various stages of post-disturbance recovery (Reice 1994). Particularly severe disturbances may extirpate local demes, while successional processes may generate exceptionally favorable conditions that promote survival and growth as habitat recovery occurs (Bisson et al. 1988, Minshall et al. 1989, Reice et al. 1990). Anadromous salmonids in the Pacific Northwest appear to be well adapted to reinvading areas of suitable habitat within their native drainage systems. Adults are strong swimmers with a relatively high fecundity for benthic spawning species, and juveniles often disperse over great distances from natal spawning sites in search of productive rearing habitat (Groot and Margolis 1991).

Table 1. Approximate occurrence rates of different types of natural and anthropogenic disturbances. Modified from National Research Council (1996), based on Swanston (1991).

Type of disturbance	Approximate recurrence interval (yr)		Physical and chemical factors influenced by the disturbance	Habitat effects
	Natural	Anthropogenic		
Daily and weekly precipitation and discharge patterns	0.01–0.1	0.001–0.1	Stream discharge, channel width and depth, storage and transport of fine particulate organic matter, fine sediment transport and deposition, nutrient concentrations, water current velocity	Minor alteration of particle sizes in spawning gravels, minor variations in rearing habitat, minor temperature change, altered turbidity, altered primary productivity
Seasonal precipitation and discharge, moderate storms, ice formation	0.1–1.0	0.01–1.0	Bank-full flows, moderate channel erosion, high base-flow erosion, increased mobility of sediment and woody debris, local damming and flooding, sediment transport by anchor ice, gouging of stream bed by ice movement, reduced winter flows with extensive freezing, seasonal changes in nutrient concentrations	Changes in frequencies of riffles and pools, changes in particle sizes in spawning gravels, increased channel width, flooding of side channels, removal (or sometimes addition) of cover, relocation of holding areas. In areas affected by ice: decreased water temperatures, lower primary and secondary productivity, egg dewatering or scour during anchor ice formation and breakup
Major floods, rain-on-snow events	10–100	1–50	Inputs of sediment, organic matter and woody debris from hillslopes, riparian zones and streambanks; localized scour and fill of streambeds; lateral channel movement; streambed mobilization resulting in redistribution of coarse sediment and flushing of fine sediment; redistribution of large woody debris (LWD); inundation of floodplains; transport of organic matter and LWD to estuaries	Changes in the frequencies of riffles and pools; formation of large log jams; burial of some spawning sites but creation of new areas suitable for spawning; increased amounts of fine particular organic matter for processing by the benthic community, resulting in increased secondary production; destruction or creation of side-channels along the floodplain; increased secondary production and cover habitat in estuaries
Debris flows and dam-break floods	100–1000	20–200	Large, short-term increases in sediment and LWD inputs; extensive channel scour; large-scale movement and redistribution of substrate, fine particulate organic matter, and LWD; damming and obstruction of channels at the terminus of the torrent track; accelerated streambank erosion, resulting in channel widening; destruction of riparian vegetation; very large short-term increase in suspended sediment; subsequent summer temperature increases from vegetative canopy removal	Extensive loss of pool habitat in the torrent track, loss of spawning gravels, loss of habitat complexity along edge of stream, destruction of side-channels and other overwintering areas, creation of new cover in the terminal debris dam, creation of new spawning areas in the sediment terrace upstream from the debris dam, short-term loss of aquatic invertebrates, possible damage to gills from heavy suspended sediment load, increased primary production

Table 1—cont.

Type of disturbance	Approximate recurrence interval (yr)		Physical and chemical factors influenced by the disturbance	Habitat effects
	Natural	Anthropogenic		
Beaver activity	5–100	0 (removal of beavers)	Channel damming, obstruction and redirection of channel flow, flooding of streambanks and side-channels, entrainment of trees from riparian zone, creation of large depositional areas for fine sediment, conditions that promote anaerobic decomposition and denitrification, resulting in nutrient enrichment downstream from the pond	Enhanced rearing and overwintering habitat, increased water volumes during low flows, refugia during floods, possible blockage to upstream migration by adults and juveniles, elevated summer temperatures and lower winter temperatures, local reductions in dissolved oxygen including areas under ice in winter, increased production of lentic invertebrates in pond, increased primary and secondary production downstream from pond
Major disturbances to vegetation				
1. Windthrow	100–500	50–150 (buffer strip blow-down)	Increased sediment delivery to channels, decreased litterfall, increased inputs of LWD, decreased riparian canopy, increased retention of sediment and fine organic matter in the channel	Increased pool habitat, localized sedimentation, increased in-channel cover, increased summer temperatures and decreased winter temperatures, creation of eddies and alcoves along channel margins, increased secondary production
2. Wildfire	100–750	40–200 (timber harvest rotation)	Increased sediment delivery to channels, inputs of large woody debris, loss of riparian canopy and vegetative cover, short-term increase in fine particulate organic matter and nutrients, decreased litterfall, increased peak discharge, short-term increase in summer flows from reduced evapotranspiration, short-term increase in biochemical oxygen demand in stream substrate	Increased sedimentation of spawning and rearing habitat, increased pool habitat and in-channel cover, increased water volume in summer, increased summer temperatures and decreased winter temperatures, increased secondary production, reduced dissolved oxygen in spawning gravels, scour of eggs and alevins in spawning gravels
3. Insects and disease	100–500	0 (chemical treatment)	Inputs of LWD, loss of riparian canopy and vegetative cover, decreased litterfall, short-term increase in summer flows from reduced evapotranspiration	Increased pool habitat and in-channel cover, increased summer temperatures and decreased winter temperatures, increased water volume in summer, increased primary and secondary production in treated areas, loss of secondary production due to toxicity of insecticides

Table 1—cont.

Type of disturbance	Approximate recurrence interval (yr)		Physical and chemical factors influenced by the disturbance	Habitat effects
	Natural	Anthropogenic		
Slumps and earthflows	100–1,000	50–200	Low-level, long-term contributions of sediment and LWDs to streams; partial blockage of channel; local base-level constriction below point of entry; shifts in channel configuration; long-term source of nutrients	Sedimentation of spawning gravels; scour of channels below point of entry; accumulation of gravels behind obstructions; possible blockage of fish migrations; increased pool habitat in coarse sediment and LWD depositional areas; destruction of side channels in some areas, creation of new side channels in others; long-term maintenance of primary productivity
Volcanism	100–1,000		Increased delivery of fine sediment and organic matter, scour of channels from mudflows, formation of mudflow terraces along rivers, destruction of riparian vegetation, damming of streams with creation of new lakes, increased nutrients	Sedimentation of spawning gravels, loss of pool habitat from mudflows but creation of pool habitat in areas with tree blowdown, creation of new overwintering habitat and side channels along mudflow terraces, short-term potentially lethal sediment and temperature levels during eruptions, long-term increases in primary and secondary production, formation of migration blockages, long-term benefits to lake-dwelling species
Climate change	1,000–100,000	10–100 (thermal discharges, riparian canopy removal, channelization)	Major changes in channel direction, gradient, and configuration; stream capture; long-term changes in temperature and precipitation regimes	Major changes in frequencies of dominant habitat types; shifts in species composition related to preferences for temperatures, substrates, and streamflows; faunal transfers during stream capture; reproductive isolation may lead to stock differentiation; founder effects

Salmon can exhibit multiple freshwater life histories, apparently a means of spreading the risk in uncertain environments (Schlosser and Angermeier 1995). Some juvenile coho salmon (*O. kisutch*) spend most of their time in freshwater close to natal spawning sites whereas others disperse downstream after emergence from spawning gravels to suitable but unoccupied habitats (Chapman 1962). In many river systems, juvenile coho occur throughout the drainage, including lowland sloughs and estuaries as well as headwater streams (Tschaplinsky and Hartman 1983). Some members of the population remain in headwater streams throughout the winter; others emigrate from headwaters to overwintering sites along riverine floodplains (Peterson 1982, Brown and Hartman 1988). Juvenile fall chinook salmon (*O. tshawytscha*) in Oregon's Sixes River have five distinctive rearing patterns with various periods of tributary, mainstem river, and estuarine residence (Reimers 1973).

Within a river basin, streams cycle between productive and unproductive conditions in response to disturbances and subsequent periods of physical and biological recovery (Minshall et al. 1989, Reice 1994), forming a dynamic setting within which anadromous salmonids exist through such adaptations as multiple freshwater rearing patterns, straying, and extended run timing. As streams move into a productive state, conditions favoring large demes develop; as streams become unproductive, carrying capacity declines (Sousa 1984). Productive streams with large populations provide colonists for recovering streams with underutilized habitat (Sheldon 1987). Emigrants from large demes can occupy marginal habitat in which survival is normally low but which may under unusual circumstances be superior to normally preferred conditions (Scudder 1989). River basins and their fish populations can thus be seen as a mosaic of habitat patches in various disturbance-recovery states that are interconnected physically by fluvial processes and interactions between aquatic and terrestrial ecosystems, and as population subunits of different sizes that are interconnected biologically by gene flow among locally reproducing demes.

This view of watershed processes suggests that both habitat and salmon populations exist in a state of dynamic equilibrium that is self-buffering over time, but this is often not the case (Reice 1994). The scale and impact of some types of disturbances (Table 1) may be so great that watershed boundaries are transcended, creating large landscape mosaics in which whole river basins undergo long-term recovery cycles. Examples of such disturbances in the Pacific Northwest include the coastal floods of 1962, 1964, and 1996 in Oregon, the 1980 eruption of Mount St. Helens, and 1994 wildfires in eastern Washington and Idaho. The spatial impact of these large disturbances may extend well beyond the boundaries of locally reproducing demes, resulting in depression of entire populations. Interdecadal climate changes now known to have significant effects on ocean productivity (Francis and Sibley 1991; Pearcy 1992, 1996; Beamish and Bouillon 1993) introduce even more long-term variability into salmon populations. Although salmon are well adapted to dynamic and unpredictable environments, their abundance is rarely if ever stable at either local or regional scales. Studies of juvenile and adult anadromous salmonids extending 5 or more consecutive years suggest that interannual variations of 40–70% are the general rule for coho salmon, steelhead (*O. mykiss*), and sea-run cutthroat trout (*O. clarki*) in Pacific Northwest streams (Fig. 2).

The point of the foregoing discussion is to emphasize that neither stream channels nor salmon populations are stable but vary in response to a variety of forces. Management policies often oppose variation, thinking that variability is the enemy of sustainability (Botkin 1990). Instead, policies should recognize that variability is an inherent property of aquatic ecosystems

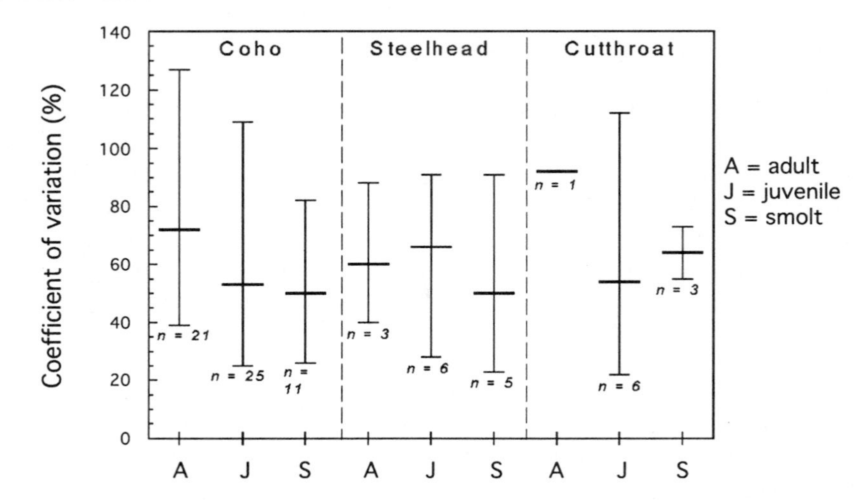

Figure 2. Comparison of the interannual variability (coefficient of variation [CV]) of coho salmon, steelhead, and cutthroat trout in western Oregon, Washington, and British Columbia streams, based on a compilation of published and unpublished studies in the Pacific Northwest where anadromous salmonid populations were censused for ≥5 consecutive years using consistent enumeration methods. Sample sizes refer to the number of studies examined. Horizontal lines represent averages and vertical lines represent the range of CVs for each species. Data sources: Johnson and Cooper (1986); Nickelson et al. (1986); Hall et al. (1987); Hartman and Scrivener (1990); Reeves et al. (1990); Ward and Slaney (1993); P.A. Bisson and R.E. Bilby, Weyerhaeuser Company, Tacoma, Washington, unpubl. data; S.V. Gregory, Oregon State University, Corvallis, unpubl. data; T.E. Nickelson, Oregon Department of Fish and Wildlife, Corvallis, unpubl. data; D. Seiler, Washington Department of Fish and Wildlife, Olympia, unpubl. data.

in the Pacific Northwest (Naiman et al. 1992, Stanford and Ward 1992), that habitat at any given location will change naturally from year to year, decade to decade, and century to century, and that the abundance of local breeding populations will similarly increase and decrease according to changes in freshwater and ocean conditions. Desired future conditions, then, must include variability as an integral and essential component of habitat and population objectives.

Disturbance-Recovery Cycles and Habitat Formation

The natural disturbance regime is the engine that drives habitat formation for salmon. Short-term impacts of natural disturbances on salmon populations are often negative. Death may result, habitat may be destroyed, access to spawning or rearing sites may be blocked, or food resources may be temporarily reduced or eliminated (Fig. 3). However, many types of natural disturbances introduce new materials into stream channels that are essential for maintaining productive habitat (Table 1). Mass soil movements such as earthflows and debris avalanches contribute coarse sediment and woody debris (Swanson et al. 1987). Wildfires and windstorms

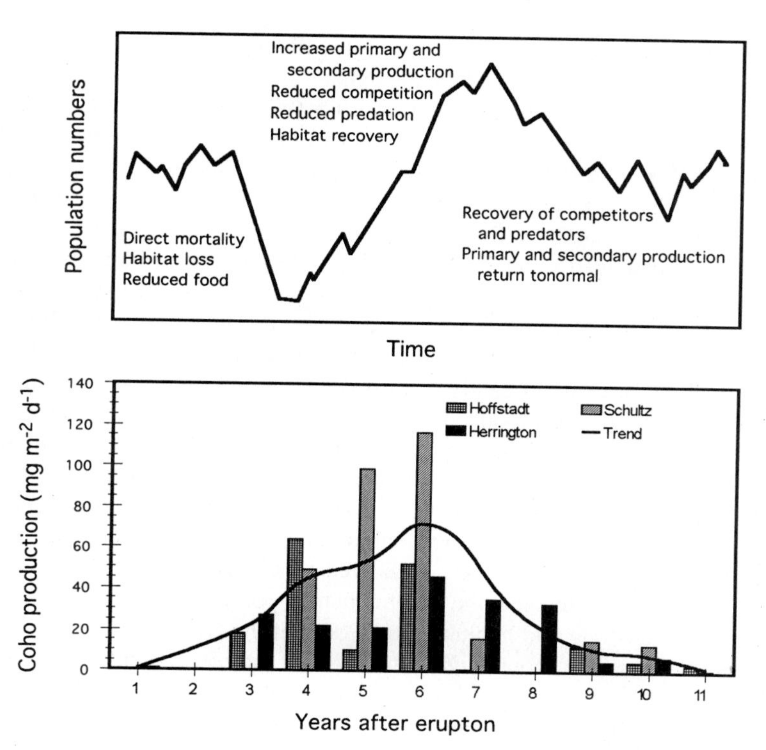

Figure 3. Top: factors potentially influencing the recovery of salmonid populations after large disturbances. Bottom: summer production of juvenile coho salmon in three 3rd-order streams impacted by the 1980 eruption of Mount St. Helens (Bisson et al. 1988; B.R. Fransen and P.A. Bisson, Weyerhaeuser Company, Tacoma, Washington, unpubl. data).

contribute both coarse and fine debris as well as nutrients (Minshall et al. 1989). Floods entrain nutrients, sediment, and particulate organic matter of all sizes (Bayley 1995). Volcanic eruptions create new soil, form new riparian terraces, and create new stream channels and lakes (Franklin et al. 1985).

Salmonid populations may rebound relatively quickly after large disturbances owing to their recolonizing abilities, temporarily abundant food, and a relative scarcity of competitors or predators (Fig. 3, top). Short-term recoveries of stream-dwelling salmonids have been observed after floods (Hanson and Waters 1974), volcanic eruptions (Fig. 3, bottom; Bisson et al. 1988), clearcut logging (Murphy and Hall 1981, Hawkins et al. 1983, Bisson and Sedell 1984, Holtby 1988, Bilby and Bisson 1992), and channelization (Chapman and Knudsen 1980). Quite often the recovery occurs unequally among species. One or two forms might be favored by altered habitat conditions and certain types of temporarily elevated foods while other species remain depressed (Waters 1983, Reeves et al. 1993).

Following a period of temporary superabundance, which seems to last ~3–15 years, salmonid populations typically decline to or below pre-disturbance levels as competitors and predators become reestablished and food levels return to normal (Fig. 3, top). Whether populations decline below pre-disturbance levels will be strongly influenced by the rate of habitat recovery. If the disturbance is such that stream channels remain depleted of important habitat elements such as LWD and coarse sediment, or if connections between the stream and its flood plain are disrupted, the carrying capacity of the watershed will probably be depressed until these elements are restored (Fig. 3, bottom; Bilby 1988, Gore and Shields 1995).

In many systems, periodic renewal of certain structural features of stream channels requires episodes of significant disturbance. Benda (1990, 1994) and Reeves et al. (1995) studied three streams in largely unmanaged watersheds of western Oregon with different time periods since their last major disturbance. They presented evidence that large quantities of coarse sediment and LWD eroded into streams after natural wildfires and accompanying landslides. Over periods of centuries, this material gradually washed out. A stream in an old-growth forest (330+ years old) that had not been significantly disturbed for a long time possessed a predominantly bedrock streambed with few deep pools or gravel riffles, a condition not particularly favorable for a high diversity of anadromous salmonids. Similarly, habitat diversity was relatively low in a stream of a watershed that had experienced a wildfire 90–100 years ago. This channel had deep sediment deposits from the post-fire erosion and was dominated by gravel riffles and deep but hydraulically simple pools. The greatest habitat diversity (and salmonid species diversity) was found in a third stream in which ~160–180 years had elapsed since the last major disturbance (a wildfire). In this watershed, the stream channel possessed the greatest array and most even distribution of substrate and habitat types. Some of the LWD and coarse sediment were still present from the erosion accompanying the wildfire and recruitment of new woody debris from the adjacent riparian zone was well underway (see also Grette 1985). Thus, diversity was greatest at an intermediate point in the long-term cycle of recovery after a major disturbance and was reduced at the beginning and end of the cycle.

The effects of debris flows differ somewhat from those of disturbances that simply add material to streams from adjacent riparian zones and hillslopes. In a debris flow, LWD and coarse sediment are scoured from the channel and concentrated at the terminus of the deposit, leaving in the wake of the flow a channel lacking in pool habitat (Sullivan et al. 1987, Swanson et al. 1987). Where mass wasting is a relatively common feature of the landscape, such as on the Queen Charlotte Islands of British Columbia, debris flow deposits constitute important nodes of complex habitat along the stream profile (Hogan 1985). In Oregon's Elk River, log jams created by debris flows trap gravels used extensively for both spawning and rearing by anadromous salmonids (G.H. Reeves, USDA Forest Service [Forest Service], Corvallis, Oregon, unpubl. data). As wood within a log jam decays and the matrix of sediment and smaller woody debris is washed out during subsequent storms, the debris deposit is breached, spreading woody debris and coarse sediment downstream to form new habitat. Very large log jams can be quite stable and can remain in place for centuries, depending on the size of the stream and its hydrologic regime. Although these jams may block upstream fish migrations, the material they contain will ultimately contribute to productive habitat when the log jam finally washes out.

The time required for new woody debris to be recruited to a stream channel or for a debris flow deposit to erode is far longer than the potential recovery time of a fish population after a disturbance, which may be a matter of only a few years (Fig. 3). Therefore, multiple cycles of population rebounding in response to small and intermediate-scale disturbances may be superimposed over much longer cycles of habitat change that accompany very large disturbances or long-term climate trends. Even if factors such as food availability are favorable for production, fish populations are limited to some extent by the amount of suitable habitat (Chapman 1966). Far greater production is possible if food resource abundance and habitat quality are high simultaneously than if one is elevated but the other depressed (Warren et al. 1964, Mason 1976). Response of salmon populations to smaller, more frequent disturbances (Table 1) is likely to be mediated by the overall condition of the stream and its watershed, which will itself be strongly influenced by large, infrequent natural disturbances (Frissell et al. 1986).

Land and water management actions often increase the frequency of small and intermediate disturbances, or they may increase the severity of impacts from natural disturbances (Table 2). In some cases, the magnitude of an anthropogenic disturbance (e.g., a dam) may be so great that irreversible changes to the aquatic community occur. The challenge facing natural resource managers is twofold: (1) where possible, to ensure that human activities do not increase the frequency or severity of disturbance events so greatly that the capacity of aquatic ecosystems to recover from either natural or anthropogenic disturbances is significantly impaired, and (2) to ensure that, when anthropogenic disturbances do occur, the essential linkages (e.g., coarse sediment and woody debris inputs, nutrient and fine organic matter transfers, floodplain connections) that promote habitat recovery are not disrupted.

In summary, we believe that the most important aspects of identifying desired future conditions are dealing with natural variability inherent in both habitat and fish populations, and accommodating the natural disturbance regime of a watershed. Spatial and temporal changes result from nested hierarchies of small-scale, short-term recovery cycles within very large-scale, long-term cycles. The dynamic nature of these processes suggests that prescribing desired habitat conditions at the scale of a stream segment or reach (the scale most often addressed in habitat restoration projects) may not be an appropriate or effective method of maintaining the ecological functions that govern productivity at the scale of entire watersheds or sub-basins. Attempting to engineer stream channels to conform to some idealized combination of pools, riffles, cover, substrate, depth, and water velocity is likely to be both prohibitively expensive and practically impossible (Sedell and Beschta 1991). In the following section, we propose an alternative approach that considers desired future conditions as ranges of states appropriate to the geomorphic setting and disturbance history of larger-order stream networks.

Identification of Desired Future Conditions

What Is Possible?

One of the first steps in defining desired future conditions is to identify geomorphic constraints imposed by the regional setting of a stream and its valley. At any given geographic scale, there exists a finite range of habitat conditions that can persist over time. These conditions may be affected by human activities but will nonetheless reflect the potential set of constraints dictated by the prevailing geoclimatic features of that particular river basin (Stanford and Ward 1992). Specific characteristics of a stream or lake and its associated riparian zone will be strongly influenced by a number of factors, including predominant rock type, substrate characteristics, and the degree of valley confinement; gradient; climate and flow regime; vegetative communities; natural disturbance history; and anthropogenic disturbance history.

Table 2. Spatial scales, recovery times, and some biological recovery mechanisms following natural and anthropogenic disturbances to aquatic ecosystems in the Pacific Northwest. Adapted from National Research Council (1996) based on Poff and Ward (1990).

Nature of disturbance	Spatial scale	Examples	Relative recovery time	Biological recovery mechanisms
Natural disturbance	Small	Flood of 1- to 3-year recurrence interval, local windstorm, minor landslide	Fast	Behavioral avoidance and refuge-seeking, increased growth among survivors, rapid recolonization of disturbed area
	Large	Major wildfire, dam-break flood (small streams), 50- to 100-year recurrence interval flood (large streams)	Slow to moderate	Adjustment of populations and community structure to new habitat conditions, migrations and establishment of new population units
Anthropogenic disturbance				
Acute				
Sublethal	Small	Minor landslide or streambank erosion, short-lived toxicant (local use), temporary water withdrawal	Fast	Behavioral and physiological avoidance, refuge-seeking, rapid recolonization
	Large	Short-lived toxicant (widespread use), various flood control practices	Slow	Physiological acclimation, selection for tolerant species, selection for species not dependent on floodplains, behavioral avoidance and refuge-seeking
Lethal	Small	Short-lived toxicant (e.g., spill), major debris torrent	Slow to moderate	Behavioral avoidance, recolonization; new species establishment
	Large	Introduction of pathogen to drainage system, channelization	Slow	Population and community adjustments, selection of tolerant species

Table 2—cont.

Nature of disturbance	Spatial scale	Examples	Relative recovery time	Biological recovery mechanisms
Anthropogenic disturbance—cont.				
Chronic				
Sublethal	Small	Gradual, small-scale sediment inputs, local thermal change; migration blockage in tributary	Slow to moderate	Local population and community adjustments, colonization by tolerant species, behavioral and physiological acclimation
	Large	Increased erosion at the watershed level, widespread loss of riparian vegetation, habitat simplification; multiple water withdrawals; dams	Slow	Population and community adjustments throughout system, selection of tolerant species, species migrations
Lethal	Small	Frequent discharges of long- or short-lived toxicants, chronic anoxia, temperature or flows beyond tolerance limits	Very slow	Selection of tolerant species
	Large	Frequent introductions of pathogens, frequent discharges of long- or short-lived toxicants, chronic anoxia, temperature or flows beyond tolerance limits	Very slow	Colonization by rare, resistant species

At the scale of a stream segment (reach), geomorphic classification systems (e.g., Rosgen 1985, Paustian 1992, Montgomery and Buffington 1993) assist in determining what is possible given the topographic constraints of a particular setting. Knowing the type and location of stream channels that are present in a watershed provides important clues to how different habitat types are likely to be distributed within the drainage (Frissell et al. 1986). In turn, the suitability of different channel segments for different species or their various life-cycle stages can be assessed if the frequencies of habitat types are known (Hankin and Reeves 1988).

The types of habitat available in individual streams and associated riparian zones change according to the watershed's history of natural and anthropogenic disturbances (Grant et al. 1990). Because watersheds exist in some phase of recovery from combinations of large and small disturbances (Reice 1994), it is likely that not all channels and riparian zones will look like those in old-growth forests, even in pristine watersheds. However, many watersheds in the Pacific Northwest have been altered by anthropogenic disturbances to such an extent that their channels possess the simplified appearance of streams in early post-disturbance recovery (Bisson et al. 1992), in which only one or two species have been favored and biodiversity has been significantly reduced (Reeves et al. 1993).

Determining what is possible is no easy task. There are few 4th- or 5th-order watersheds in pristine condition and no intact drainages of 6th-order and larger within the range of salmon south of Puget Sound that can serve as benchmarks. Much of the useful information on the range of natural conditions in the Pacific Northwest has come from surveys of small headwater streams (Naiman and Sedell 1979; Murphy et al. 1984; Bilby and Ward 1989, 1991; Hartman and Scrivener 1990). Descriptions of habitat in large rivers prior to human alteration are scarce (Sedell et al. 1984). Some of the best evidence has come from historical reconstruction of river channels and estuaries based on late 19th-century and early 20th-century maps, photographs, and stream surveys (Sedell and Luchessa 1982, Sedell and Froggatt 1984, Boulé and Bierly 1987, Sedell and Beschta 1991, Sedell and Everest 1991, Simenstad et al. 1992).

We believe it is imperative that a regional network of reference sites encompassing drainages of 5th- to 6th-order be established throughout the Pacific Northwest to determine the range of conditions over a variety of geoclimatic and disturbance regimes. These sites should contain substantial areas where stream channels have not been significantly altered by human activities or have recovered from such perturbations to the extent that they approximate conditions present in unmanaged watersheds (NRC 1996). Reference sites should be sufficiently large that the range of habitat conditions produced by natural disturbances is expressed and all or nearly all of the channel types found in the region are present.

Locating such sites will be difficult but not impossible. Some already exist in national and state parks. Large areas of the landscape have been designated as late-successional old-growth reserves, research natural areas, or key watersheds on federal forest lands (FEMAT 1993). Other examples of potential sites include large watershed areas designated as Watershed Administrative Units by the Washington Department of Natural Resources (Washington Forest Practices Board 1993), and the Class 1 Waters of the Aquatic Diversity Management Areas recently proposed for California by Moyle and Yoshiyama (1994) (i.e., those with the highest quality habitat and most intact native aquatic plant and animal communities).

Until a regional network of reference sites is established and systematic, long-term habitat monitoring programs are initiated, our knowledge of the range of conditions possible at a given location will be limited (Walters and Holling 1990). At present, we are forced to rely on the few existing studies of habitat in streams without significant anthropogenic influences and on the best professional judgment of watershed specialists. Information from a few small watersheds, while useful, may be so limited that the full range of potential states for all channel types throughout a region is poorly understood and cannot be applied to large streams. Without a more complete understanding of this range of possibilities, available evidence may lead us once again into embracing uniform sets of habitat standards that may be geomorphically inappropriate and that do not reflect naturally dynamic processes upon which aquatic productivity ultimately depends. We strongly urge land owners and resource management agencies to establish a regional network of reference sites and implement cooperative, long-term monitoring programs without delay. The information generated from such an effort will be essential to answering the following three questions.

What Is Desired?

A key goal of watershed management with respect to the protection of Pacific salmon and other members of aquatic ecosystems is to allow interactions between aquatic and terrestrial ecosystems to continue as unhindered as possible. Two of the most important steps to ensuring that ecological linkages will not be disrupted are (1) maintaining riparian zones of adequate width to mediate the full range of physical and biological exchanges of energy and materials between land and water, and (2) minimizing the occurrence of severe anthropogenic disturbances within these zones. These steps will enable natural disturbances to generate the changes necessary for habitat formation and long-term maintenance of aquatic productivity (Naiman et al. 1992). Although restoration of completely pristine conditions will not be possible in watersheds possessing a variety of human activities, management decisions should aim to maintain the range of conditions produced by natural disturbance regimes, including a frequency distribution of riparian forest successional stages that resembles those in unmanaged watersheds over long time periods.

The goal is not to maintain all streams in the same state over time but to allow disturbance and recovery processes to take place as normally as possible. Even streams with high-quality habitat (i.e., habitat that appears exceptionally favorable for a particular species of interest) will change over sufficiently long intervals. Disturbances that temporarily transform habitats favorable for salmon into unfavorable habitats need not necessarily be viewed as greatly harmful as long as other streams with productive conditions are readily available within the drainage system. The metapopulation structure of salmon stocks permits the strength of locally reproducing population units to expand and contract in response to habitat recovery cycles (Reeves et al. 1995). An appropriate goal might be to ensure that some biologically productive streams are always present in the drainage system and that their frequency approaches what might be expected in undeveloped but otherwise similar watersheds. Establishing a mix of streams in different stages of recovery from natural disturbances is, we believe, most appropriate at landscape scales encompassing intermediate to large watersheds (i.e., >1,000 ha). Drainages of at least 4th- to 6th-order may be the minimum size landscape units within which habitat goals can properly be matched to the life cycles of salmon populations (i.e., they may be the minimum size necessary to contain summer and winter habitats as well as migration corridors).

We caution habitat managers against basing restoration decisions solely on the needs of individual salmon species or life stages. Streams in the Pacific Northwest are often classified according to their potential use by adult or juvenile salmonids. Stream segments may be termed "steelhead streams" or "coho streams," or classified according to use by certain life-cycle stages such as "spawning areas" or "overwintering sites." While these descriptions may accurately reflect current utilization, the danger is that habitat improvement projects may be designed solely to enhance the characteristics of the channel that favor these particular species or life-history stages. Such projects may be successful in the short term for the limited purpose for which they were intended, but they often lead to unnatural channel characteristics that may impede successional changes ultimately favoring other species (Sedell and Beschta 1991). As a general rule, we recommend that objectives include a diversity of habitat types within a watershed sufficient to support complete assemblages of native species. These assemblages should include both salmonid and non-salmonid fishes as well as other aquatic vertebrates, invertebrates, and plants (Pister 1995).

What Conditions Should Be Managed?

Emphasis should be placed on protecting watershed processes that provide materials essential for maintaining the structure and functional properties of aquatic ecosystems. Key elements of ecologically healthy watersheds include fully functional riparian zones, unaltered streamflow regimes, floodplains connected with river channels, uninterrupted hyporheic zones, and natural input rates of sediment, organic matter, and nutrients (Naiman et al. 1992, Stanford and Ward 1992). Because most human activities change these elements to some extent, the objective should be to minimize adverse changes, restore biophysical connections where possible, and preserve remaining ecologically functional areas in which a range of habitats exist. By allowing natural disturbances to occur with a minimum of human intervention, and by providing the raw materials upon which these disturbances can act, it should be possible to maintain patches of productive habitat across the landscape that are spatially arrayed in a manner that provides for the needs of salmon populations.

Restoration of ecological functions in Pacific Northwest watersheds will require considerable time and patience on the part of natural resource managers (Bisson et al. 1992), but the current plight of many salmon stocks may call for short-term intervention to improve habitat. In theory, recovery can be accelerated by adding structures to streams to increase pool habitat (House and Boehne 1986), creating missing habitat types such as overwintering ponds (Cederholm et al. 1988), or actively managing riparian vegetation to speed development of late-successional coniferous forests (Bilby and Bisson 1991). If instream habitat improvement efforts are undertaken, care should be taken to utilize or mimic to the greatest extent possible the size and composition of material that would occur at the site naturally, the locations where the material is most likely to enter the channel and stabilize, and the ability of structures to interact dynamically with water flow, streambed, and streambanks. In practice, this will mean using native tree species and inorganic materials typical of those produced by local site characteristics, including whole trees with intact rootwads and branches, irregular spacing of structures similar to streams in pristine watersheds with similar channel types, limited use of exotic materials such as wire mesh and geotextile fabrics, and a minimum of anchoring to the streambed or streambanks.

Anchoring systems such as steel cables to prevent fluvial transport of introduced woody debris are often used to prevent damage to bridges, homes, and other capital structures downstream; however, anchored debris often does not function in the same way as do naturally produced materials, their failure rates may be high (Frissell and Nawa 1992), and their effectiveness in significantly increasing ecosystem productivity remains largely unproved.

What Constitutes the Future?

We suggest that the minimum planning horizon for watershed rehabilitation should be 100 years, a time period representing about 20–50 salmon generations. This interval should be long enough to allow natural disturbances and other ecological processes to provide genuine gains in overall watershed productivity. Planning efforts should recognize the likelihood that, over a century or more, very large disturbances will occur in most areas. When these events take place, the future capacity of watersheds to produce salmon will be strongly influenced by changes accompanying the disturbances. Taking steps to ensure that watersheds possess the full range of vegetative characteristics and hydrologic connections in appropriate locations upon which disturbances can eventually act is, we believe, more important for restoring the long-term capacity of drainage systems to produce salmon than modifying channels is to rehabilitating degraded habitat. Relying strictly on instream projects to increase salmon production, however well-intentioned, without a comprehensive plan for improving ecological functions and processes at the watershed scale will not achieve the diversity of channel characteristics upon which Pacific salmon depend and to which their population structure and life cycles are matched.

Tracking long-term changes in freshwater and marine environments facilitates a better understanding of interdecadal cycles of salmon abundance and the need to adjust management policies appropriately. The inverse relationship between marine survival of northern and southern salmon populations along the Pacific coast of North America (Beamish and Bouillon 1993, Francis 1996) illustrates the great importance of changing ocean conditions; yet, while ocean productivity may be low in some years and high in others the need for good land and water stewardship always remains (Pearcy 1992, Lawson 1993). For example, it seems prudent to ensure that salmon have access to all available productive freshwater habitat when ocean survival is relatively low (Lawson 1993). Success during spawning and juvenile rearing phases of the life cycle may help compensate for less than favorable marine conditions by maintaining a high rate of smolt production for each returning female salmon. High rates of freshwater survival and growth should help maintain relatively small demes during periods of weak marine production and support large, abundant demes during periods of strong ocean production.

Long-term vision is also needed in order to properly interpret the results of salmon enhancement programs in an adaptive learning context. Habitat managers should be careful not to proclaim habitat enhancement projects successful based on short-term increases in fish populations (Hilborn and Winton 1993), especially when those increases may have been caused by factors unrelated to freshwater habitat (e.g., reduced fishing rates or increased ocean survival). Recovery of ecosystem functions at large landscape scales will be gradual. The high interannual variability in salmon abundance (Fig. 2) will make it difficult to detect long-term, consistent improvement in the productivity of a watershed for various species. Even large increases in salmonid populations caused by habitat improvement projects are likely to require decades for statistical detection (Fig. 4).

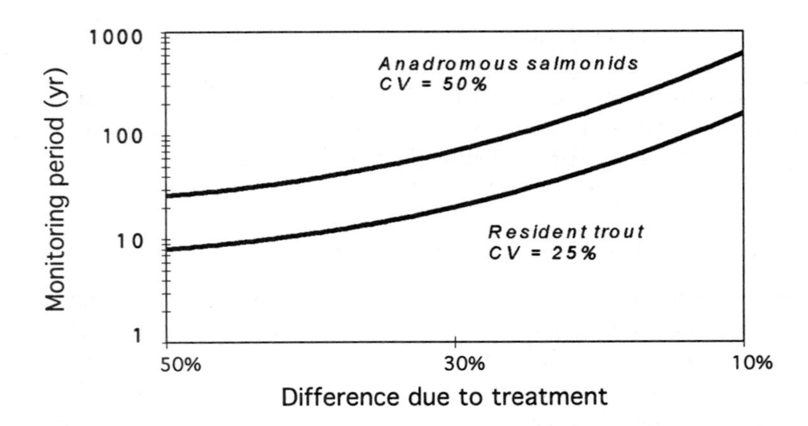

Figure 4. Years of monitoring required to be 80% certain of detecting true treatment differences at a Type 1 error level of $p \leq 0.05$ (Sokal and Rohlf 1981) for an anadromous salmonid population having an interannual coefficient of variation (CV) of 50% and a resident salmonid population having an interannual CV of 25% (based in part on population studies summarized in Fig. 2).

For these reasons we argue that planning horizons should extend well beyond the normal scope of long-term plans, which are more often on the order of decades rather than centuries. Large disturbances and long-term climate changes are usually impossible to forecast, but planning should proceed with the certainty that they will eventually occur and that they will create a mosaic of conditions on the landscape. We do not suggest that human activities must create no disturbances, but rather that the type and frequency of disturbances match to the extent possible those that exist naturally. With the advent of geographic information system technology and other new modeling tools, comprehensive views of current and proposed watershed characteristics can assist planners in designing management activities that better resemble natural disturbances and are more compatible with natural successional processes and hydrologic regimes. Such an approach was used by the FEMAT as an example of how timber harvest in the Augusta Creek (Oregon) watershed could simulate historical patterns of wildfire (FEMAT 1993).

Need for a Better Landscape Context

We have avoided specific recommendations for desired future conditions because in most cases our knowledge of the natural disturbance history of Pacific Northwest watersheds is far from complete. Although some habitat trends are abundantly clear and do not require additional verification before restoration can begin (e.g., loss of floodplain habitat, loss of late-successional and old-growth riparian forests, loss of lowland sloughs and estuaries, accelerated erosion, and highly altered streamflows), information on the distribution of aquatic and riparian habitat in various recovery states following natural disturbances is usually lacking. Without such information, our ability to provide a landscape context for establishing refugia or determining appropriate restoration strategies for extensively altered stream networks will lack the spatial and temporal dimensions needed to match salmon population dynamics (Gregory and Bisson 1996). The locations of favorable habitats within watersheds will change over time and management plans should be cognizant of these changes. For example, we should know the location of important overwintering sites for different species within a river system and be able to predict how proposed management activities will likely influence their condition and accessibility in the event of a large flood. As a better understanding of basic watershed processes is developed, our ability to identify desired future conditions appropriate to specific watersheds will improve. This understanding should begin with a careful examination of the local disturbance regime.

Watershed analysis, a procedure for assessing the current condition of streams with respect to biophysical watershed processes, aquatic resource values, and their vulnerability to disruption by forestry operations (Washington Forest Practices Board 1993, USDA Forest Service 1994) provides an opportunity to incorporate disturbance history into landscape planning. The NRC report on the protection and management of anadromous salmonids in the Pacific Northwest (NRC 1996) strongly endorsed the importance of recognizing natural disturbance patterns in the watershed analysis process. The NRC report identified four major categories of information to be addressed when describing current conditions and management objectives: (1) the spatial context of the watershed (i.e., the location and geomorphic setting of channel characteristics such as constrained versus unconstrained, sinuous versus braided, incised versus unincised, bedrock-controlled versus alluvial floodplains), (2) the temporal context and natural disturbance history of the watershed (i.e., its annual flow regime and sediment yield, long-term channel changes, climate trends, forest successional patterns, fire history, and other significant natural disturbances), (3) the range of riparian vegetation and availability of reference sites within the watershed (i.e., the variation in riparian community composition and the presence of unmanaged areas within the watershed or nearby that can be used as reference sites for identifying natural conditions), and (4) the history of human impacts within the watershed (i.e., the institutional, scientific, and social records of anthropogenic changes to the watershed's ecosystem).

A watershed analysis procedure that addresses the dynamic aspects of watershed processes in the context of spatial and temporal disturbance history will be very useful for determining the extent of habitat alteration caused by land management and for providing a rational framework for establishing target ranges of future conditions (Peterson et al. 1992). The protocols of Washington State (Washington Forest Practices Board 1993) and federal watershed analysis (USDA Forest Service 1994) are geomorphically based and recognize different types of natural and anthropogenic physical disturbances, particularly those related to erosional events. Analysis of physical watershed processes in both procedures explicitly acknowledges long-term changes in the sediment and woody debris properties of stream channels, yet a similarly dynamic perspective is often not applied in watershed analysis for the assessment of biological communities and their habitats (Karr 1994). Stream reach-specific habitat ratings (good, fair, or poor) rely on habitat standards relevant to certain salmonid species or life cycle phases (Table F-1, p. F-22 in Washington Forest Practices Board 1993). Stream segments are classified according to use by the predominant salmonid species with little consideration for long-term changes in population distribution (Figure F-2, p. F-13 in Washington Forest Practices Board 1993). Habitat parameters are considered individually without the possibility of synergism among elements (Table F-1, p. F-22 in Washington Forest Practices Board 1993; p. 2-112 in USDA Forest Service 1994), and fish population objectives do not adequately consider natural variability (p. 2-103 in USDA Forest Service 1994).

As watershed analysis protocols are further refined, we hope they will place less emphasis on enforcing reach-specific standards that benefit a few salmonid species, possibly for only a limited time, and more emphasis on restoring an appropriate range of natural aquatic-riparian ecosystem states throughout watersheds. For example, a large increase in fine sediment or large decrease in shade in a single stream segment might be less alarming than small to moderate increases in sediment or temperature throughout the entire watershed if other streams with favorable habitat are present in the drainage. Relatively severe but localized disturbances are part of a watershed's history, but widespread gradual changes may not be representative of the natural disturbance regime.

We conclude our discussion of desired future conditions with the observation that existing knowledge of natural disturbance regimes and aquatic-riparian habitat is often highly imperfect and, therefore, we must make many assumptions and inferences with regard to what target conditions should be. Because pristine watersheds are rare in the Pacific Northwest, because a regional network of reference sites has not been established, and because current knowledge of the response of aquatic and riparian ecosystems to disturbance is incomplete, we cannot define desired future conditions with certainty at regional or watershed scales. Natural resource managers should not be afraid to implement bold and innovative approaches to restoring functional aquatic and riparian ecosystems, provided the effects are carefully monitored and evaluated (Walters et al. 1988). Although it is impossible to describe in detail how desired future conditions should appear, we do not believe land-use planners should be deterred from attempting to restore functional ecological linkages between streams and their watersheds and, especially, from designing management actions that better simulate the effects of natural disturbances. The current status of many salmon stocks calls for conservative habitat protection, yet we argue that to ignore the historical disturbance context within which these and other aquatic organisms evolved is to ignore some of the most important processes upon which they depend.

Acknowledgments

While the views expressed in this paper are our own, we sincerely thank the USDA Forest Service, the Weyerhaeuser Company, and the University of Washington for their support. Many scientists freely shared their wisdom and insight with us as our ideas were taking shape and we thank them all, especially P. Angermeier, L. Benda, R. Beschta, A. Dolloff, F. Everest, K. Fausch, B. Fransen, K. Fresh, G. Grant, S. Gregory, G. Grossman, J. Hall, J. Karr, K. Lee, J. Lichatowich, J. Light, D. Montgomery, P. Peterson, T. Quinn, H. Regier, B. Riddell, I. Schlosser, J. Sedell, D. Stouder, K. Sullivan, F. Swanson, and J. Williams. Three anonymous referees provided helpful suggestions on an earlier draft.

Literature Cited

Bayley, P.B. 1995. Understanding large river-floodplain ecosystems. BioScience 45: 153-158.

Beamish, R.J. and D.R. Bouillon. 1993. Pacific salmon production trends in relation to climate. Canadian Journal of Fisheries and Aquatic Sciences 50: 1002-1016.

Benda, L.E. 1990. The influence of debris flows on channels and valley floors in the Oregon Coast Range, USA. Earth Surface Processes and Landforms 15: 457-466.

Benda, L.E. 1994. Stochastic geomorphology in a humid mountain landscape. PhD thesis, Department of Geological Sciences, University of Washington. Seattle.

Bilby, R.E. 1988. Interactions between aquatic and terrestrial systems, p. 13-29. *In* K.J. Raedeke (ed.), Streamside Management: Riparian Wildlife and Forestry Interactions. University of Washington, Institute of Forest Resources, Contribution No. 59. Seattle.

Bilby, R.E. and P.A. Bisson. 1991. Enhancing fisheries resources through active management of riparian areas, p. 201-209. *In* B. White and I. Guthrie (eds.), Proceedings of the 15th Northeast Pacific Pink and Chum Salmon Workshop. Pacific Salmon Commission, Vancouver, British Columbia.

Bilby, R.E. and P.A. Bisson. 1992. Allochthonous versus autochthonous organic matter contributions to the trophic support of fish populations in clear-cut and old-growth forested streams. Canadian Journal of Fisheries and Aquatic Sciences 49: 540-551.

Bilby, R.E. and J.W. Ward. 1989. Changes in characteristics and function of woody debris with increasing size of streams in western Washington. Transactions of the American Fisheries Society 118: 368-378.

Bilby, R.E. and J.W. Ward. 1991. Characteristics and function of large woody debris in streams draining old-growth, clear-cut, and second-growth forests in southwestern Washington. Canadian Journal of Fisheries and Aquatic Sciences 48: 2499-2508.

Bisson, P.A., J.L. Nielsen, and J.W. Ward. 1988. Summer production of coho salmon stocked in Mount St. Helens streams 3-6 years after the 1980 eruption. Transactions of the American Fisheries Society 117: 322-335.

Bisson, P.A., T.P. Quinn, G.H. Reeves, and S.V. Gregory. 1992. Best management practices, cumulative effects, and long-term trends in fish abundance in Pacific Northwest river systems, p. 189-232. *In* R.J. Naiman (ed.), Watershed Management: Balancing Sustainability and Environmental Change. Springer-Verlag, New York.

Bisson, P.A. and J.R. Sedell. 1984. Salmonid populations in streams in clearcut vs. old-growth forests of western Washington, p. 121-129. *In* W.R. Meehan, T.R. Merrell, Jr., and T.A. Hanley (eds.), Fish and Wildlife Relationships in Old-growth Forests. American Institute of Fisheries Research Biologists, Juneau, Alaska.

Bjornn, T.C. and D.W. Reiser. 1991. Habitat requirements of salmonids in streams. American Fisheries Society Special Publication 19: 83-138.

Boulé, M.E. and K.F. Bierly. 1987. History of estuarine wetland development and alteration: what have we wrought? Northwest Environmental Journal 3: 43-61.

Botkin, D.B. 1990. Discordant Harmonies: A New Ecology for the Twenty-first Century. Oxford University Press, New York.

Brown, T.G. and G.F. Hartman. 1988. Contribution of seasonally flooded lands and minor tributaries to the production of coho salmon in Carnation Creek, British Columbia. Transactions of the American Fisheries Society 117: 546-551.

Cederholm, C.J., B.A. Huether, and P. Wagner. 1993. What salmon biologists say about the status of wild coho salmon (*Oncorhynchus kisutch*) runs and their habitat conditions in large western Washington rivers: an opinion survey, p. 352-373. *In* L. Berg and P. Delaney (eds.), Proceedings of a Workshop on Coho Salmon, sponsored by the Association of Professional Biologists of British Columbia and the North Pacific International Chapter of the American Fisheries Society. Available from Habitat Management Division, Department of Fisheries and Oceans, Vancouver, British Columbia.

Cederholm, C.J., W.J. Scarlett, and N.P. Peterson. 1988. Low-cost enhancement technique for winter habitat of juvenile coho salmon. North American Journal of Fisheries Management 8: 438-441.

Chapman, D.W. 1962. Aggressive behavior in juvenile coho salmon as a cause of emigration. Journal of the Fisheries Research Board of Canada 19: 1047-1080.

Chapman, D.W. 1966. Food and space as regulators of salmonid populations in streams. American Naturalist 100: 345-357.

Chapman, D.W. and E. Knudsen. 1980. Channelization and livestock impacts on salmonid habitat and biomass in western Washington. Transactions of the American Fisheries Society 109: 357-363.

Fausch, K.D., C.L. Hawkes, and M.G. Parsons. 1988. Models that predict standing crop of stream fish from habitat variables: 1950-85. USDA Forest Service General Technical Report, PNW-213, Pacific Northwest Experiment Station. Portland, Oregon.

Forest Ecosystem Management Assessment Team. 1993. Forest ecosystem management: an ecological, economic, and social assessment. USDA Forest Service. Portland, Oregon.

Francis, R.C. 1996. Managing resources with incomplete information: making the best of a bad situation, p. 513-524. *In* D.J. Stouder, P.A. Bisson, and R.J. Naiman (eds.), Pacific Salmon and Their Ecosystems: Status and Future Options. Chapman and Hall, New York.

Francis, R.C. and T.H. Sibley. 1991. Climate change and fisheries: what are the real issues? Northwest Environmental Journal 7: 295-307.

Franklin, J.F. 1993. Preserving biodiversity: species, ecosystems, or landscapes. Ecological Applications 3: 202-205.

Frankliin, J.F. 1994. Ecological science: a conceptual basis for FEMAT. Journal of Forestry 92: 21-23.

Franklin, J.F., J.A. MacMahon, F.J. Swanson, and J.R. Sedell. 1985. Ecosystem responses to the eruption of Mount St. Helens. National Geographic Research 1: 198-216.

Frissell, C.A., W.J. Liss, C.E. Warren, and M.D. Hurley. 1986. A hierarchical framework for stream habitat classification: viewing streams in a watershed context. Environmental Management 10: 199-214.

Frissell. C.A. and R.K. Nawa. 1992. Incidence and causes of physical failure of artificial habitat structures in streams of western Oregon and Washington. North American Journal of Fisheries Management 12: 182-197.

Gore, J.A. and F.D. Shields. 1995. Can large rivers be restored? BioScience 45: 142-152.

Grant, G.E., F.J. Swanson, and M.G. Wolman. 1990. Pattern and origin of stepped-bed morphology in high-gradient streams, Western Cascades, Oregon. Geological Society of America Bulletin 102: 340-352.

Gregory, S.V. and P.A. Bisson. 1995. Degradation and loss of anadromous salmonid habitat in the Pacific Northwest, p. 277-314. *In* D.J. Stouder, P.A. Bisson, and R.J. Naiman (eds.), Pacific Salmon and Their Ecosystems: Status and Future Options. Chapman and Hall, New York.

Gregory, S.V., F.J. Swanson, and W.A. McKee. 1991. An ecosystem perspective of riparian zones. BioScience 40: 540-551.

Grette, G.B. 1985. The abundance and role of large organic debris in juvenile salmonid habitat in streams in second growth and unlogged forests. MS thesis, University of Washington. Seattle.

Groot, C. and L. Margolis (eds.). 1991. Pacific Salmon Life Histories. University of British Columbia Press, Vancouver, British Columbia., Canada.

Hall J.D., G.W. Brown, and R.L. Lantz. 1987. The Alsea watershed study: a retrospective, p. 399-416. *In* E.O. Salo and T.W. Cundy (eds.), Streamside Management: Forestry and Fishery Interactions. University of Washington, Institute of Forest Resources, Contribution No. 57. Seattle.

Hankin, D.G. and G.H. Reeves. 1988. Estimating total fish abundance and total habitat area in small streams based on visual estimation methods. Canadian Journal of Fisheries and Aquatic Sciences 45: 834-844.

Hanson, D.L. and T.F. Waters. 1974. Recovery of standing crop and production rate of a brook trout population in a flood-damaged stream. Transactions of the American Fisheries Society 103: 431-439.

Harrison, S. 1991. Local extinction in a metapopulation context: an empirical evaluation. Biological Journal of the Linnean Society 42: 73-88.

Hartman, G.F. and J.C. Scrivener. 1990. Impacts of forestry practices on a coastal stream ecosystem, Carnation Creek, British Columbia. Canadian Bulletin of Fisheries and Aquatic Sciences 223.

Hawkins, C.P., M.L. Murphy, N.H. Anderson, and M.A. Wilzbach. 1983. Density of fish and salamanders in relation to riparian canopy and physical habitat in streams of the northwestern United States. Canadian Journal of Fisheries and Aquatic Sciences 40: 1173-1185.

Hilborn, R. and J. Winton. 1993. Learning to enhance salmon production: lessons from the Salmonid Enhancement Program. Canadian Journal of Fisheries and Aquatic Sciences 50: 2043-2056.

Hogan, D. 1985. The influence of large organic debris on channel morphology in Queen Charlotte Island streams. Proceedings of the Annual Meeting of the Western Association of Fish and Wildlife Agencies 1984: 263-273.

Holtby, L.B. 1988. Effects of logging on stream temperatures in Carnation Creek, British Columbia, and associated impacts on the coho salmon (*Oncorhynchus kisutch*). Canadian Journal of Fisheries and Aquatic Sciences 45: 502-515.

House, R.A. and P.L. Boehne. 1986. Effects of instream structures on salmonid habitat and populations in Tobe Creek, Oregon. North American Journal of Fisheries Management 6: 38-46.

Johnson, T. H. and R. Cooper. 1986. Snow Creek anadromous fish research. Washington State Game Department, Fisheries Management Division, Report No. 86-18. Olympia.

Karr, J.R. 1994. Restoring wild salmon: we must do better. Illahee 10: 316-319.

Lawson, P.W. 1993. Cycles in ocean productivity, trends in habitat quality, and the restoration of salmon runs in Oregon. Fisheries 18: 6-10.

Lee, K.N. 1993. Compass and Gyroscope: Integrating Science and Politics for the Environment. Island Press, Washington, DC.

Mason, J.C. 1976. Response of underyearling coho salmon to supplemental feeding in a natural stream. Journal of Wildlife Management 40: 775-788.

Minshall, G.W., J.T. Brock, and J.D. Varley. 1989. Wildfire and Yellowstone's stream ecosystems. Bioscience 39: 707-715.

Meffe, G.K. 1992. Techno-arrogance and halfway technologies: salmon hatcheries on the Pacific coast of North America. Conservation Biology 6: 350-354.

Montgomery, D.R. and J.M. Buffington. 1993. Channel classification, prediction of channel response, and assessment of channel condition. Department of Natural Resources, Washington State Timber/Fish/Wildlife Agreement, Report TFW-SH10-93-002. Olympia.

Moyle, P.B. and R.M. Yoshiyama. 1994. Protection of aquatic biodiversity in California: a five-tiered approach. Fisheries 19: 6-18.

Murphy, M.L. and J.D. Hall. 1981. Varied effects of clear-cut logging on predators and their habitat in small streams of the Cascade Mountains, Oregon. Canadian Journal of Fisheries and Aquatic Sciences 38: 137-145.

Murphy, M.L., J.F. Thedinga, KV. Koski, and G.B. Grette. 1984. A stream ecosystem in an old-growth forest in southeast Alaska: Part V: seasonal changes in habitat utilization by juvenile salmonids, p. 89-98. *In* W.R. Meehan, T.R. Merrell, Jr., and T.A. Hanley (eds.), Fish and Wildlife Relationships in Old-growth Forests. American Institute of Fishery Ressearch Biologists, Juneau, Alaska.

Naiman, R.J., T.J. Beechie, L.E. Benda, D.R. Berg, P.A. Bisson, L.H. MacDonald, M.D. O'Connor, P.L. Olson, and E.A. Steel. 1992. Fundamental elements of ecologically healthy watersheds in the Pacific Northwest Coastal Ecoregion, p. 127-188. *In* R.J. Naiman (ed.), Watershed Management: Balancing Sustainability and Environmental Change. Springer-Verlag, New York.

Naiman, R.J. and J.R. Sedell. 1979. Relationships between metabolic parameters and stream order in Oregon. Canadian Journal of Fisheries and Aquatic Science 37: 834-847.

National Research Council. 1992. Restoration of Aquatic Ecosystems. National Academy Press, Washington, DC.

National Research Council. 1996. Upstream: Salmon and Society in the Pacific Northwest. National Academy Press, Washington, DC.

Nehlsen, W., J.E. Williams, and J.A. Lichatowich. 1991. Pacific salmon at the crossroads: stocks at risk from California, Oregon, Idaho and Washington. Fisheries 16: 4-21.

Nickelson, T.E., M.F. Solazzi, and S.L. Johnson. 1986. Use of hatchery coho salmon (*Oncorhynchus kisutch*) presmolts to rebuild wild populations in Oregon coastal streams. Canadian Journal of Fisheries and Aquatic Sciences 43: 2443-2449.

Paustian, S.J. 1992. A channel type user's guide for the Tongass National Forest, southeast Alaska. USDA Forest Service, Region 10, Technical Paper 26, Alaska Region. Juneau, Alaska.

Pearcy, W.G. 1992. Ocean Ecology of North Pacific Salmonids. Washington Sea Grant Program, University of Washington Press, Seattle.

Pearcy, W.G. 1996. Salmon production in changing ocean domains, p. 331-352. *In* D.J. Stouder, P.A. Bisson, and R.J. Naiman (eds.), Pacific Salmon and Their Ecosystems: Status and Future Options. Chapman and Hall, New York.

Peterson, N.P. 1982. Immigration of juvenile coho salmon (*Oncorhynchus kisutch*) into riverine ponds. Canadian Journal of Fisheries and Aquatic Sciences 39: 1308-1310.

Peterson, N.P., A. Hendry, and T.P. Quinn. 1992. Assessment of cumulative effects on salmon habitat: some suggested parameters and target conditions. University of Washington, Center for Streamside Studies, Technical Report. Seattle.

Pister, E.P. 1995. The rights of species and ecosystems. Fisheries 20: 28-29.

Poff, N.L. and J.V. Ward. 1990. Physical habitat template of lotic ecosystems: recovery in the context of historical pattern of spatial heterogeneity. Environmental Management 14: 629-645.

Reeves, G.H., L.E. Benda, K.M. Burnett, P.A. Bisson, and J.R. Sedell. 1995. A disturbance-based ecosystem approach to maintaining and restoring freshwater habitats of evolutionarily significant units of anadromous salmonids in the Pacific Northwest. American Fisheries Society Symposium 17: 334-349.

Reeves, G.H., K.M. Burnett, F.H. Everest, J.R. Sedell, D.B. Hohler, and T. Hickman. 1990. Responses of anadromous salmonid populations and physical habitat to stream restoration in Fish Creek, Oregon. USDA Forest Service, Pacific Northwest Forest and Range Experiment Station, Project Report 84-11. Portland, Oregon.

Reeves, G.H., F.H. Everest, and J.R. Sedell. 1993. Diversity of juvenile anadromous salmonid assemblages in coastal Oregon basins with different levels of timber harvest. Transactions of the American Fisheries Society 122: 309-317.

Reice, S.R. 1994. Nonequilibrium determinants of biological community structure. American Scientist 82: 424-435.

Reice, S.R., R.C. Wissmar, and R.J. Naiman. 1990. Disturbance regimes, resilience, and recovery of animal communities and habitats in lotic ecosystems. Environmental Management 14: 647-659.

Reimers, P.E. 1973. The length of residence of juvenile fall chinook salmon in Sixes River, Oregon. Research Report 4(2) of the Fish Commission of Oregon, Portland, Oregon.

Riddell, B.E. 1993. Spatial organization of Pacific salmon: what to conserve? p. 23-41. *In* J.G. Cloud and G.H. Thorgaard (eds.), Genetic Conservation of Salmonid Fishes. Plenum Press, N.Y.

Rosgen, D.L. 1985. A stream classification system, p. 91-95. *In* R.R. Johnson, C.D. Zeibell, D.R. Patton, P.F. Folliott, and R.H. Hamre (eds.), Riparian Ecosystems and their Management: Reconciling Conflicting Uses. USDA Forest Service General Technical Report RM-20, Rocky Mountain Research Station, Fort Collins, Colorado.

Salwasser, H. 1994. Ecosystem management: can it sustain diversity and productivity? Journal of Forestry 92: 6-10.

Sauter, K.F. 1994. Explaining variation in western Washington riparian zone management width on state and private lands. MS thesis, University of Washington. Seattle.

Schlosser, I.J. and P.L. Angermeier. 1995. Spatial variation in demographic processes of lotic fishes: conceptual models, empirical evidence, and implications for conservation. American Fisheries Society Symposium 17: 392-401.

Scudder, G.G.E. 1989. The adaptive significance of marginal populations: a general perspective, p. 180-185. *In* C.D. Levings, L.B. Holtby, and M.A. Henderson (eds.), Proceedings of the National Workshop on Effects of Habitat Alteration on Salmonid Stocks. Canadian Special Publication of Fisheries and Aquatic Sciences 105.

Sedell, J.R. and R.L. Beschta. 1991. Bringing back the "bio" in bioengineering. American Fisheries Society Symposium 10: 160-175.

Sedell, J.R. and F.H. Everest. 1991. Historic changes in pool habitat for Columbia River Basin salmon under study for TES listing. USDA Forest Service, Pacific Northwest Research Station, Draft General Technical Report. Portland, Oregon.

Sedell, J.R. and J.L. Froggatt. 1984. Importance of streamside forests to large rivers: the isolation of the Willamette River, Oregon, USA, from its floodplain by snagging and streamside forest removal. Internationale Vereinigung fur Theoretishe und Angewandte Limnologie Verhandlungen 22: 1828-1834.

Sedell, J.R. and K.J. Luchessa. 1982. Using the historical record as an aid to salmonid habitat enhancement, p. 210-223. *In* N.B. Armantrout (ed.), Acquisition and Utilization of Aquatic Habitat Inventory Information. Proceedings of a symposium held October 28-30, 1981, Portland, Oregon. The Hague Publishing, Billings, Montana.

Sedell, J.R., J.E. Yuska, and R.W. Speaker. 1984. Habitats and salmonid distribution in pristine, sediment-rich river valley systems: S. Fork Hoh and Queets River, Olympic National Park, p. 33-46. *In* W.R. Meehan, T.R. Merrell, Jr., and T.A. Hanley (eds.), Fish and Wildlife Relationships in Old-growth Forests. American Institute of Fishery Research Biologists, Juneau, Alaska.

Sheldon, A.L. 1987. Rarity: patterns and consequences for stream fishes, p. 203-209. *In* W.J. Matthews and D.C. Heins (eds.), Community and Evolutionary Ecology of North American Stream Fishes. University of Oklahoma Press, Norman.

Shirvell, C.S. 1989. Habitat models and their predictive capability to infer habitat effects on stock size, p. 173-179. *In* C.D. Levings, L.B. Holtby, and M.A. Anderson (eds.), Proceedings of the National Workshop on Effects of Habitat Alteration on Salmonid Stocks. Canadian Special Publication in Fisheries and Aquatic Sciences 105.

Simenstad, C.A., D.A. Jay, and C.R. Sherwood. 1992. Impacts of watershed management on land-margin ecosystems: the Columbia River estuary, p. 266-306. *In* R.J. Naiman (ed.), Watershed Management: Balancing Sustainability and Environmental Change. Springer-Verlag, New York.

Sokal, R.R. and F.J. Rohlf. 1981. Biometry. W.H. Freeman and Company, San Francisco, California.

Sousa, W.P. 1984. The role of disturbance in natural communities. Annual Reviews of Ecology and Systematics 15: 353-391.

Stanford, J.A. and J.V. Ward. 1992. Management of aquatic resources in large catchments: recognizing interactions between ecosystem connectivity and environmental disturbance, p. 91-124. *In* R.J. Naiman (ed.), Watershed Management: Balancing Sustainability and Environmental Change. Springer-Verlag, New York.

Stanley, T. R., Jr. 1995. Ecosystem management and the arrogance of humanism. Conservation Biology 9: 255-262.

Sullivan, K., T.E. Lisle, C.A. Dolloff, G.E. Grant, and L.M. Reid. 1987. Stream channels: the link between forests and fishes, p. 39-97. *In* E.O. Salo and T.W. Cundy (eds.), Streamside Management: Forestry and Fishery Interactions. University of Washington, Institute of Forest Resources, Contribution No. 57. Seattle.

Swanson, F.J., L.E. Benda, S.H. Duncan, G.E. Grant, W.F. Megahan, L.M. Reid, and R.R. Ziemer. 1987. Mass failures and other processes of sediment production in Pacific Northwest forest landscapes, p. 9-38. *In* E.O. Salo and T.W. Cundy (eds.), Streamside Management: Forestry and Fishery Interactions. University of Washington, Institute of Forest Resources, Contribution No. 57. Seattle.

Swanston, D.N. 1991. Natural processes. American Fisheries Society Special Publication 19: 139-179.

Tracy, C.R. and P.F. Brussard. 1994. Preserving biodiversity: species in landscapes. Ecological Applications 4: 207-208.

Tschaplinski, P.J. and G.F. Hartman. 1983. Winter distribution of juvenile coho salmon (*Oncorhynchus kisutch*) before and after logging in Carnation Creek, British Columbia, and some implications for overwinter survival. Canadian Journal of Fisheries and Aquatic Sciences 40: 452-461.

US Environmental Protection Agency. 1986. Quality criteria for water. Office of Water Regulations and Standards, EPA 440/5-86-001. Washington, DC.

USDA Forest Service. 1994. A federal agency guide for pilot watershed analysis, Version 1.2. Pacific Northwest Research Station. Portland, Oregon.

Volkman, J.M. and K.N. Lee. 1994. The owl and Minerva: ecosystem lessons from the Columbia. Journal of Forestry 92: 48-52.

Walters, C.J., J.S. Collie, and T. Webb. 1988. Experimental designs for estimating transient responses to management disturbances. Canadian Journal of Fisheries and Aquatic Sciences 5: 530-538.

Walters, C.J. and C.S. Holling. 1990. Large-scale management experiments and learning by doing. Ecology 71: 2060-2068.

Ward, B.R. and P.A. Slaney. 1993. Habitat manipulations for the rearing of fish in British Columbia, p.142-148. *In* G. Shooner and S. Asselin (eds.), Le Developpement du Saumon Atlantique au Quebec: Connaitre les Regles du Jeu pour Reussir, Colloque International de la Federation Quebecoise pour le Saumon Atlantique, Quebec, Decembre 1992. Collection *Salmo salar* no. 1. Montreal, Quebec.

Warren, C.E., J.H. Wales, G.E. Davis, and P. Doudoroff. 1964. Trout production in an experimental stream enriched with sucrose. Journal of Wildlife Management 28: 617-660.

Washington State Forest Practices Board. 1993. Standard methodology for conducting watershed analysis under Chapter 222-22 WAC, Version 2.0. Washington Department of Natural Resources, Olympia.

Waters, T.F. 1983. Replacement of brook trout by brown trout over 15 years in a Minnesota stream: production and abundance. Transactions of the American Fisheries Society 112: 137-146.

Restoration of Riparian and Aquatic Systems for Improved Aquatic Habitats in the Upper Columbia River Basin

Robert L. Beschta

Abstract

In this paper, I explore linkages among soils, channel morphology, riparian hydrology, and streamside vegetation to illustrate why, from an ecological perspective, instream structural manipulations to improve fish habitat may often be inappropriate. Examples are presented regarding the role of streambanks, flow regimes, and riparian vegetation in the development and maintenance of functional riparian zones and aquatic ecosystems. While many historical and current attempts to alter anadromous fish (i.e., *Oncorhynchus* spp.) habitat in the upper Columbia River Basin have been focused on the addition of structural elements, improving riparian and aquatic functions is a more important requirement for restoring degraded habitats. Furthermore, aquatic habitat restoration should be based upon an understanding of natural disturbance regimes, multiple roles of streamside vegetation, land-use history, and reference sites. Together, these factors provide an ecological context from which to address the restoration of riparian and aquatic habitats in the upper Columbia River Basin.

Introduction

Anthropogenic influences (e.g., roads, mining, forest harvesting, grazing, urbanization, flow alterations) have profoundly affected the habitats of fish and other aquatic organisms within the upper Columbia River Basin (i.e., east of the Cascade Mountains). Although natural disturbances such as fire, drought, and volcanic eruptions have always been a component of Pacific Northwest ecosystems, the widespread reductions of anadromous fish (i.e., *Oncorhynchus* spp.) stocks (Nehlsen et al. 1991) indicate that human-caused changes to riparian and aquatic systems represent a major departure from previous conditions. While there is an increasing emphasis on programs to restore, recover, or enhance the habitats of existing aquatic systems (Kusler and Kentula 1990, National Research Council [NRC] 1992, Pacific Rivers Council 1993, Forest Ecosystem Management Assessment Team 1993, Beschta 1994), an improved understanding of the interactions of hydrogeomorphic disturbance regimes, historical land uses, and streamside vegetation is increasingly needed. My objective in this paper is to provide an overview of historical stream system changes and a conceptual framework for undertaking restoration of aquatic ecosystems in the upper Columbia River Basin.

Prehistory

From the perspective of geologic time, landscapes in the North American West have experienced massive changes; mountains have come and gone, and valleys have been eroded and filled only to be eroded again. Within this geologic continuum of adjusting landscapes and climate, the flora and fauna of the Pacific Northwest evolved. Near the end of the most recent ice age, the late Pleistocene, at least some mountains were again sculpted by continental and valley glaciers that rafted soil and rock detritus to lower elevations. At lower elevations, the landscape indirectly experienced the effects of glacial activity in the form of outwash deposits and periglacial climates. The catastrophic release of water from Montana's glacial Lake Missoula cascaded and spilled across eastern Washington, eroded the loess deposits of the Palouse, and gouged deep channels and coulees; these floodwaters were ultimately collected into the mainstem Columbia River and funneled into the Pacific Ocean.

Approximately 10,000 years ago, the late Pleistocene was ending. The dramatic retreat of glaciers, both continental and valley, indicated that a major climatic adjustment was underway. Although general climatic warming and increased aridity followed during the Holocene era, the wide range in hillslope gradients, aspects, and elevations of the region's topography indicates that the consequences of these climatic changes on microclimates and geomorphic processes were not uniform across the Columbia River Basin. Nevertheless, it would appear that, with the start of the Holocene, many low-gradient valleys in the upper Columbia River system, and perhaps elsewhere, began a period of long-term aggradation (Peacock 1994, Sullivan 1994). The extent of this aggradation is sometimes represented by the long-term accretion of alluvial sediments above a layer of Mazama Ash ~6,800 years old (Welcher 1993, Peacock 1994).

The long-term accumulation of sediment as valley-fill deposits is a common feature along many low-gradient streams and river valleys in the upper Columbia River Basin and other areas of the intermountain west. Persistent accumulation of sediments over the last half of the Holocene (perhaps longer) indicates that riparian plants, channels, and aquatic habitats have been "stable" in terms of the general ecosystem for thousands of years. Stability occurred despite moderate shifts in climatic variables (long-term precipitation and temperature) and concurrent disturbance regimes, including droughts, floods, and fires, indicating the resiliency of many riparian and aquatic systems.

An important component of these valley-fill systems is the valuable hydrogeomorphic and ecological role of streamside vegetation. This vegetation influences patterns of sediment transport by adding complexity, hydraulic resistance, and stability to alluvial channels. It often affects the size, shape, and distribution of channel features such as pools, riffles, undercut banks, etc. (Sedell and Beschta 1991). Vegetation moderates temperature extremes because of its ability to reduce the amount of solar radiation that reaches a stream; it also moderates long-wave radiation losses (Beschta et al. 1987). The persistent reestablishment of vegetation along stream margins traps sediments along streambanks and narrow channels. Vegetation alters the relatively simple chemistry of nutrient production and transport into a complicated array of storage locations, transformations, and nutrient spirals (Gregory et al. 1987, Pinay et al. 1990). In addition, its presence helps to maintain the hydrologic connectivity between main and side channels and hyporheic zones (Stanford and Ward 1988, Gilbert et al. 1990). Thus, riparian vegetation is not only crucial to the long-term functioning of aquatic systems, but also has a fundamental role in the restoration of degraded aquatic systems (NRC 1992).

Over geologic time scales, riparian plant communities have evolved in response to changing regimes of climate, flow, and sediment transport. Such vegetation has continued as an important and persistent factor in moderating, initiating, offsetting, and assimilating a wide range of ecosystem functions. Thus, in combination with a range of physical, chemical, and biological processes, riparian vegetation is indispensable for sustaining high-quality and functional stream habitats upon which fish and other aquatic organisms depend for their survival.

The Development Period

With the immigration of Euro-American peoples into the Pacific Northwest over the last two centuries and the continuing economic development of the region, watersheds in the upper Columbia River Basin experienced a wide range of land-use and management regimes. Perhaps the first significant impact to streams and riparian systems began with the extensive trapping of beaver in the 1800s. Ranching, agriculture, water withdrawals, forest harvesting, stream damming, and other land-use practices soon followed. Although the anthropogenic impacts by Euro-Americans to the streams and riparian systems of the Pacific Northwest extended over many decades, perhaps the most significant changes occurred during the latter half of the 20th century. Some of the anthropogenic impacts were technologically driven while others were associated with an increasing population and economic base. For example, following World War II, a wide variety of heavy equipment (bulldozers and tractors) was increasingly available to land owners for roading, logging, draining, ditching, berming, and clearing land. United States (US) Census reports indicate that population levels in the Pacific Northwest doubled in <50 years. However, the increasing per capita economic growth caused an even greater impact on the region's natural resources than census levels indicate.

During the period of Euro-American development, the impacts on stream and riparian resources have not only been extensive but, in some instances, essentially permanent. It is difficult to reverse the effects of urban developments, highways, channelizations, dams, or other structural modifications. Similarly, the practice of appropriating water for other than instream purposes remains deeply ingrained in US western water law. A major consequence of the "use it or lose it" mandate is that natural flows can be seriously depleted. Even for streams with important fisheries habitat, it is common for consumptive demand in a given sub-basin to exceed available flows (Gauvin 1996).

By any standard, the alteration of watersheds, riparian systems, and streams in the upper Columbia River Basin has been extensive. Thus, substantial opportunities exist for restoring or improving many riparian and aquatic systems. Improvements are needed to provide not only adequate habitat for fish and other aquatic organisms, but to improve water quality and riparian wildlife habitat.

Valley-Fill Deposits and Channel Incision

Field observations of channel banks throughout the basin generally confirm that the long-term accumulation of fluvial sediments and development of extensive valley fills (e.g., floodplains, meadows) is a common feature of the recent geologic record (throughout much of the Holocene and perhaps earlier). However, with the influx of Euro-American immigrants and the land uses that accompanied them, stream incision, entrenchment, and channel downcutting replaced the continued accumulation of valley-fill sediments (Welcher 1993, Peacock 1994, Sullivan 1994). Inspection of the vertical sidewalls of incised channels indicates that streams often downcut through previous floodplain deposits, thus establishing contemporary terraces (Welcher 1993, Peacock 1994, Sullivan 1994). Furthermore, many profiles appear to have developed under very moist conditions. For example, soil layers may express indications of wetland characteristics: strong mottling features, oxidized rhizospheres, gleying, and accumulations of organic matter. The occurrence of paleosols interspersed vertically along sidewall profiles of incised channels suggests the persistent, long-term accumulation of sediments may have been punctuated by periods of relative watershed stability.

The widespread occurrence of stream incision in valley-fill deposits of the Columbia River Basin from the mid-1800s to present represents a major stream system response to land use (Elmore and Beschta 1987). Once incision starts, downcutting continues until a resistant layer (e.g., bedrock, hardpan, large gravels) or a regional water table is encountered. Gully erosion, consisting of predominantly up-valley migration of headwalls accompanied with channel downcutting, provide a simple mechanism for rapid development of an incised channel system (Brooks et al. 1991). After incision has occurred, channel widening generally follows as the stream begins to erode laterally, often undercutting steep gully sidewalls (Van Haveren and Jackson 1986).

A wide variety of land uses can contribute to stream incision: trapping of beaver; overgrazing of riparian areas; dewatering of channels by flow diversions; conversion of riparian vegetation to agricultural crops; extensive logging of riparian areas; construction of roads along streams, which straightens and steepens channels; construction of berms along floodplains, which reduces overbank flows and concentrates flood flows; riprapping of channels; and others. Although channel incision has multiple causes, it is often associated with factors that decrease or eliminate the role of riparian plants.

The potential for channel incision to affect floodplain and riparian functions depends upon several variables, such as the extent of valley fill deposits and incision depth. When channel incision is relatively small, perhaps ≤0.5 m lowering of a streambed, the extent of active floodplain surfaces and functions may still be reduced. For example, floodplain surfaces that formerly flooded every 2 years on average may now experience less frequent inundation; former 25-year floodplain surfaces might no longer flood, effectively becoming terrace landforms. Even so, some riparian-dependent plant species may continue to become locally established and survive on former floodplains.

When channel incision is >0.5 m, concurrent declines in subsurface saturated zones also occur, leading to devastating changes in riparian vegetation and stream functions. As local water tables decline following incision, plants previously established on former floodplain surfaces may satisfy their moisture demands with deep root systems; however, plants dependent on overbank flooding or moist subsoils for establishment are likely to become eliminated over time. Whereas sedge, willow, or cottonwood species previously dominated riparian plant communities, following incision they may be replaced by sagebrush, cheatgrass, juniper, or similar xeric species. Unfortunately, channel incision ≥0.5 m is common along many streams in the upper Columbia River Basin.

Restoring Instream Habitats

A report by the NRC (1992) defines restoration as the "reestablishment of pre-disturbance aquatic functions and related physical, chemical, and biological characteristics." This report further indicates that "restoration is different from habitat creation, reclamation, and rehabilitation—it is a holistic process not achieved through the isolated manipulation of individual elements." While such a definition seems relatively straightforward, applying this concept may entail considerable difficulty in degraded systems. For example, the extent of incised channels associated with many low-gradient streams in the upper Columbia River Basin may make restoration to pre-disturbance conditions ≥1 century ago impossible. However, in many of these situations the reestablishment of riparian plant communities approximating pre-disturbance functions is a reasonable restoration goal.

In the 1980s, federal and state fisheries managers undertook numerous instream projects in the Columbia River Basin in an attempt to enhance fish habitat; similar activities have continued into the 1990s (Kauffman et al. 1993). While corridor fencing sometimes has been used to reduce grazing impacts, most projects focused on the addition of various structures within the active channel: boulders, riprap, rock weirs, gabions, sill logs, deflector logs, and other materials in a variety of configurations. Certain types of habitat alterations (e.g., logs cabled to boulders, with their ends buried in streambanks, or stabilized with riprap and geotextile fabrics) have been widely used regardless of the geomorphic setting (e.g., valley constraint, stream gradient, sediment sizes), hydrologic disturbance regimes (e.g., peak flows derived from snowmelt versus peakflows from summer thunderstorms), or riparian vegetation (e.g., species composition, density, and age class distribution).

The addition of structural elements to channels for habitat management has been generally promoted in fisheries literature (Seehorn 1980, 1992; Wesche 1985; House et al. 1988; Hunter 1991; Reeves et al. 1991). While structural diversity may be an important need in some stream systems, little research demonstrates significant and long-term improvement in fisheries production from such practices (Hamilton 1989, Beschta et al. 1994). The mere addition of structure to a channel for specific habitat components is unlikely to satisfy the emerging goals and concepts associated with ecosystem management and, in some instances, such structural approaches may actually be counterproductive to restoring ecological functions.

The "restoration" of degraded aquatic systems in much of the Upper Columbia River Basin is unlikely to be accomplished simply by adding structural components in an increasingly complex array of configurations to stream channels. Restoration requires an understanding of not only fish biology and specific habitat needs, but also aquatic ecology, hydrology, soils and geomorphology, and plant ecology (Fig. 1). Attaining the intent of the NRC's (1992) restoration definition represents a major challenge to our understanding of complex systems and our ability to undertake improvement projects assisting in long-term habitat recovery and sustainability.

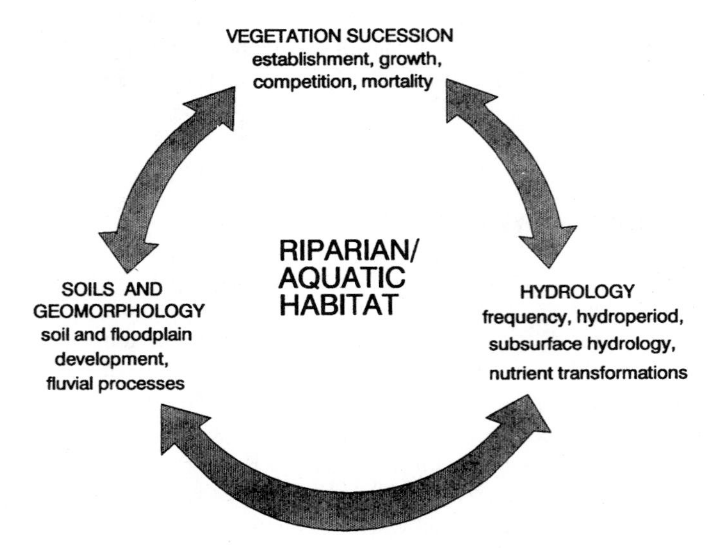

Figure 1. Processes and components of riparian and aquatic zones that influence habitat quality.

Restoration projects generally need information on hydrogeomorphic disturbance regimes, historical context, streamside vegetation, and reference sites. I discuss each of these needs in turn.

Hydrogeomorphic Disturbance Regimes

Understanding of hydrogeomorphic processes that occur over a range of spatial and temporal scales is prerequisite to developing aquatic habitat restoration plans. High and low flows, subsurface flow dynamics, sediment transport, channel adjustments, and other events combine in various ways to influence stream channel characteristics and adjacent riparian systems (Hill et al. 1991). For example, an important disturbance regime associated with streams in the upper Columbia River Basin is the occurrence of snowmelt peaks in late spring or early summer followed by recession to a baseflow condition that continues through the fall and winter months. These flow patterns are intrinsically intertwined with sediment movement, seed and plant propagule dispersal, riparian plant establishment, rewatering of floodplains and subsurface riparian environments, nutrient transfers, moisture availability for various riparian plant communities, and more. In some areas, midsummer thunderstorms provide another component of the hydrologic disturbance regime that can locally influence the general character of riparian systems, channels, and aquatic habitats.

Streamflow plays an essential role not only in forming and maintaining channels but also in establishing riparian vegetation; thus, it is crucial to understand both seasonal patterns and variability of flow to formulate strategies for restoring degraded riparian aquatic systems. For example, where reservoir construction has altered the frequency, magnitude, or timing of high flows, adjustments to the reservoir's operating plan may be required. Such adjustments should provide an improved downstream representation of a stream's natural peakflow hydrology to ensure the establishment of plant species such as cottonwoods and willows, which are dependent upon high flows, as well as to maintain the size and frequency of pools and other channel features (Hill et al. 1991). Similarly, if land-use practices within a watershed have significantly increased overland flow during large storm events, such practices may need to be altered to reduce or eliminate increased flows. Regardless of other land-use impacts on a watershed or stream system, the retention of a relatively natural flow regime (i.e., flows not significantly altered by diversions such as dams) is critical to the restoration of riparian and aquatic systems.

Channel characteristics such as the size and distribution of pools typically develop during periods of high flow and sediment transport. Thus, management practices that alter flows, sediment transport, or riparian vegetation are also likely to affect channel morphology. Alterations to pool size or frequency may have significant adverse effects on aquatic biota. The other end of the hydrologic spectrum, low flows, are also important to aquatic habitats. Low flows not only vary from year to year but also may encompass multiple years of drought. Additional reductions in summertime flows or dewatering of channels by instream diversions or other withdrawals can thus have major impacts on aquatic habitat quality. Therefore, productive aquatic habitats may not be sustainable without restoring instream flows to needed levels.

The movement of soil particles is often coupled with the movement of water across watershed surfaces. The sedimentation process (i.e., the detachment, transport, and deposition of soil particles by water) represents another important component of a hydrogeomorphic perspective. For many watersheds, accelerated soil erosion may affect aquatic habitat quality more profoundly than changes in flow regime. Where sedimentation rates have been increased significantly, a return to pre-anthropogenic levels likely represents a high priority for successful restoration of stream and aquatic systems.

Alterations of hydrogeomorphic disturbance regimes or riparian vegetation can significantly alter channel morphology, particularly the size and distribution pools and other channel units. Because pools represent a critical fish habitat feature, attempts to enhance pool habitat usually involve adding structural materials of various types, shapes, and sizes into a channel (Seehorn 1980, 1992; House et al. 1988; Hunter 1991). From an ecological perspective, such structures may be entirely out of context with the geomorphic setting, hydrologic disturbance regimes, or the needs of riparian plant communities (Beschta et al. 1994).

Naturally occurring pools in stream channels vary widely in configuration, size, and frequency depending upon the flow regime, sinuosity and gradient of the channel, composition of bed material, sediment sizes and transport rates, and streamside vegetation (Beschta and Platts 1986). Pools can form along well-vegetated channels by several mechanisms:

- scour of bed material along the outer bend of a meander owing to the velocities, secondary currents, and turbulence characteristic of these locations;
- locally accelerated flows because of channel constrictions;
- an interaction of high flows with root systems, particularly those of woody species that disrupt local patterns of flow and resist bank erosion, results in pool formation;
- banks may be narrowed and built (via deposition of fine sediments) among existing streamside vegetation during periods of bank overflow;
- flows deflected by individual pieces of large woody debris or by accumulations of both large and floatable woody debris create erosional and depositional areas; and
- rejoining of flows in braided channels.

Pools that develop as a result of interactions between high flows, sediment transport, channel adjustments, and flow resistance caused by the channel and its vegetation are typically complex and contain various microhabitats for fish and other aquatic organisms. Such pools are not simply depressions in a streambed profile that can be easily replicated by a hydraulic excavator or by the addition of large roughness elements to stream systems. Instead, they represent multiple factors and disturbance regimes operating over various temporal and spatial scales. Collectively, pools, riffles, and undercut banks, as well as other channel features and the functional attributes of streamside vegetation provide diverse aquatic habitats. These features within aquatic ecosystems have helped provide long-term sustainability to fish and aquatic organisms in Pacific Northwest streams.

Perhaps one of the most perplexing features of many instream habitat alterations is the extensive use of similar structural approaches across a wide range of biophysical regions. For example, many of the alluvial deposits associated with valley-fill systems in the upper Columbia River Basin are fine-textured, and the dominant hydrologic regime is driven by snowmelt hydrographs. Extensive floodplains and meadows are common. Large woody debris or boulders may never have been an important habitat feature in many of these systems. Yet, the addition of large wood and boulders, which may be more appropriate for steep-gradient forested streams, has been an almost universal means of attempting to improve fish habitat (Beschta et al. 1991, 1993; Kauffman et al. 1993). Regardless of the effectiveness in producing more fish, which is debatable (Hamilton 1989, Beschta et al. 1994), this structural approach to instream habitat management does not represent "restoration" (NRC 1992). While there has been increased awareness of the need to maintain native fish stocks in the various tributaries and sub-basins of the Columbia River system, there has been less appreciation and understanding of how local geomorphic settings and natural hydrologic disturbance regimes interact with native riparian plant communities to create sustainable aquatic habitat for fish and other aquatic organisms (Fig. 1).

Historical Context

An understanding of historical resource development and land-use patterns within specific watersheds and their effects on stream systems is essential before initiating restoration activities. Such information typically provides an important basis for interpreting the present status or conditions associated with riparian and aquatic habitats (Sedell and Luchessa 1982, Platts et al. 1987). Many of the widespread and diverse impacts on streams and riparian systems were initiated prior to the current generation of land-use managers. Therefore, we need to examine historical conditions and trends to develop insights regarding the characteristics and features of pre-development aquatic systems. For many upper Columbia River tributaries, the initial riparian and instream impacts associated with farming, grazing, logging, mining, roading, and other land uses occurred decades ago. Nevertheless, useful information can sometimes be gleaned from the journals of early trappers, traders, ranchers, and settlers. Historical maps, land survey notes, photographs, and anecdotal information can provide additional perspectives. The earliest available aerial photographs may be invaluable in identifying the historical context or setting of a stream and its riparian ecosystem. Several sets of aerial photographs through time can illustrate how rapidly and to what extent a stream and its riparian system may have changed. However, even aerial photographs dating back to the late 1930s or early 1940s may have recorded riparian and stream systems that had already been significantly altered by land use.

The basic objective of establishing a historical context is to develop an in-depth appreciation and understanding of the characteristics of riparian areas and stream systems that existed prior to human activities, the rapidity and extent that a watershed and its riparian system have changed over time as a result of land-use activities, and the land uses that have caused or contributed to these changes. Without knowing what the ecological status of a system has been and how it has been modified over time, it is inordinately difficult to develop an effective strategy for restoration.

Where stream incision has occurred, field inspection and interpretation of the alluvial deposits that compose the sidewalls of the incised channel system provide important information. The occurrence of buried paleosols, percent organic matter, textural composition, hydric soil characteristics, and other information often indicate how the riparian system functioned prior to human alterations. Such interpretations can be invaluable for establishing a prehistorical context from which to compare present conditions and determine the magnitude of changes that may have occurred.

Streamside Vegetation

The influence of streamside vegetation on the character of stream channels and aquatic habitats has been increasingly recognized (Johnson et al. 1985, Salo and Cundy 1987, Abell 1989). Vegetation, either as individual plants or collectively as plant communities, has a fundamental influence on local environmental conditions (stream microclimates), ecosystem processes, and an array of ecological functions. The aboveground importance of streamside vegetation has been widely recognized for shading (Beschta et al. 1987), allochthonous inputs of carbon and nutrients (Gregory et al. 1987), and large woody debris (LWD) recruitment (Sedell et al. 1988). Stream systems and their associated aquatic habitats are not simply combinations of physical processes (e.g., flow, sediment transport, energy exchanges) and physical features (e.g., pools, riffles, gravel bars). Instead, they represent a unique blend of both physical and biological components that interact with disturbance regimes.

The interaction of vegetation with varying flow regimes and sediment transport loads is an aspect of channel development and aquatic condition essential to fisheries and other aquatic organisms. In general, riparian areas with high vegetation densities are more conducive to narrow channels, the development of overhanging banks, long-term floodplain accretion, improved water quality, and food-web support. Streamside vegetation provides an array of ecological functions:

- Vegetation anchors streambank soils by fibrous and woody root systems that resist the erosive forces of flowing water;
- vegetation forms overhanging and undercut banks;
- plant establishment and sediment accumulation occur along stream margins, narrowing channels (decreasing width-to-depth ratios) and thus influencing spatial distributions and characteristics such as aquatic habitat units, flow patterns, temperature responses, effectiveness of streamside vegetation for shade, etc.;
- vegetation contributes to mid-channel bar stabilization, which in turn accentuates localized stream braiding and the formation of pools where side channels rejoin the main stream; and
- plants and their associated structures increase the interaction between instream and hyporheic flows.

Many of the above functions and processes are associated with the effects of root systems. The belowground components of plants add another layer of ecological complexity to a riparian system that is typically characterized, from both spatial and temporal perspectives, as complex, diverse, multifaceted, and interactive. Although root system development by riparian vegetation is an essential aspect of restoring aquatic ecosystems, the ecological role of such systems has had limited scientific study (Sedell and Beschta 1991).

From the mid-1970s to present, considerable research has been focused on the importance of LWD and its role in maintaining aquatic habitats of mountain streams that drain forested catchments (Harmon et al. 1986, Sedell et al. 1988). However, channel gradients, dominant sediment sizes, peakflow hydrographs, and streamside vegetation associated with forested systems are often quite different from those associated with upper Columbia River Basin streams (Kovalchik 1987). In many upper Columbia River Basin fish-bearing streams, the occurrence of deciduous shrubs and trees (particularly willows, alders and cottonwoods) and their accompanying understories (sedges, rushes, grasses, or forbs), and not LWD create the dominant vegetation that in turn influences channel morphology and ecological functions. For example, the extensive root systems of woody shrubs and sedges effectively bind sediments, thus assisting in the formation of undercut banks and irregularities in channel morphology. In addition, multiple aboveground stems provide increased resistance to flow and accentuate the deposition of fine sediments. Where cottonwoods or other deciduous trees are common along a stream, woody debris inputs may vary in size: entire trees, broken stems, or smaller, branch-sized material. Floatable pieces of woody debris may accumulate at mid-channel bars, along the sides of channels, or at other locations. In addition to providing a food base for aquatic biota and local cover for fish, such accumulations may affect microsites for seedling establishment and local channel morphology because of their hydraulic interaction with high flows.

Euro-American land uses throughout much of the Columbia River Basin have resulted in widespread and long-term reductions in riparian vegetation. Hence, the restoration of these plant communities may appear to be an overwhelming task. In instances where permanent changes have occurred (e.g., dams, urbanization, roads), it may not be possible to recover all stream functions associated with riparian vegetation. However, important restoration opportunities exist where destructive watershed or riparian land uses can be modified or eliminated. These restoration programs need to focus on factors that promote the establishment, growth, and succession of riparian-dependent plants.

Reference Sites

At their best, reference sites represent relatively large and intact aquatic ecosystems that continue to function without the influence of anthropogenic impacts. They not only provide important examples of how hydrogeomorphic processes, geomorphic setting, and vegetation interact (Fig. 2), but they also allow us an opportunity to understand and appreciate the interaction between disturbance regimes, plant communities, and aquatic habitats. As ecological benchmarks, reference sites provide an essential perspective of what degraded riparian and aquatic ecosystems might become if the appropriate restoration activities are undertaken. Thus, they provide a means of assessing whether proposed treatments or management activities will move a degraded system closer to a restored one. They also help predict the amount of time required for recovery.

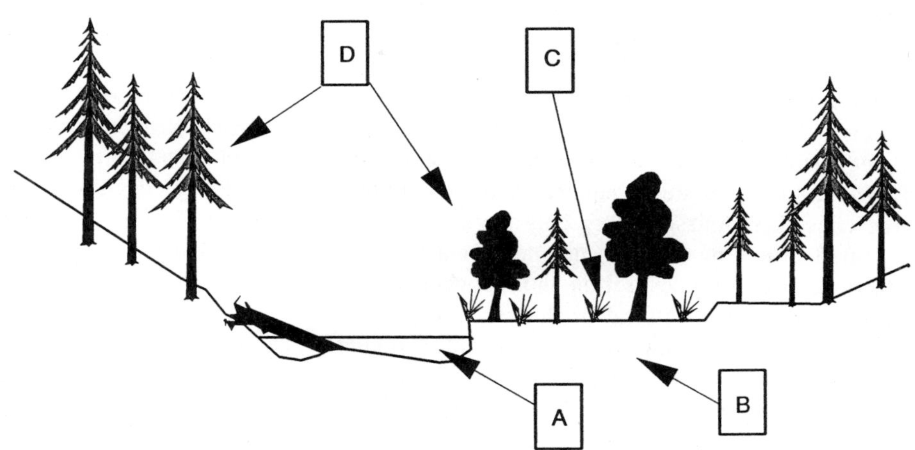

Figure 2. Functional attributes of riparian and aquatic ecosystems that support freshwater fisheries include the following: A. Aquatic zone—large woody debris (LWD) accumulations, short-term sediment storage (fines and gravels), primary productivity (autochthonous inputs), spawning/rearing habitat, anadromous fish migration route; B. Subsurface and hyporheic zone—long-term sediment storage (fines and gravels), seasonal soil moisture storage, nutrient transformations (oxidation/reduction), groundwater discharge moderation; C. Floodplain surface—short-term floodwater storage, fine sediments accumulations (sand/silt/clay), particulate organic matter storage, energy dissipation of overbank flows; and D. riparian vegetation—stream shading (thermal regulation), litterfall (allochthonous inputs), root system development (channel and bank stability), hydraulic roughness during overbank flows, nutrient cycling, LWD production.

Although reference sites can impart a sense of attainable future conditions, it may be unlikely that recovery can be quickly or simply achieved when riparian and aquatic degradation is severe.

Reference sites also provide important demonstration areas where scientists, managers, regulators, and interested citizens can interact on a common footing when trying to address restoration needs, priorities, and the potential for successfully attaining restoration goals. For example, assume that a wet-meadow reference site with fine-textured soils and riparian vegetation dominated by willows and sedges is available and that it typifies low- to moderate-elevation locations in a particular drainage. Assume also that nearby meadows have been sufficiently impacted by land uses such that the channels are wide and shallow and that shrub species have been nearly eradicated. Given this simple scenario, a proposal to load structural elements such as large wood and boulders into the degraded reaches is not likely to represent an appropriate restoration strategy. First, the fine-textured soils and geomorphic setting (floodplain) would indicate that boulders, and perhaps large woody debris, have never been a component of such systems. Second, large structural features are not part of the reference condition. Third, when structural approaches to habitat alteration are the primary focus, the underlying causes of habitat degradation may be ignored. Furthermore, the addition of structural elements to a riparian system significantly lacking in vegetative cover and associated root biomass may cause accelerated erosion or other impacts during periods of high flow, increasing the likelihood of additional degradation. A more appropriate restoration goal would be to reduce or eliminate those management practices that initially caused or have maintained the degraded condition. This "passive restoration approach" (Fig. 3) is recommended as a way to restore the appropriate plant communities as quickly as possible.

Within a given watershed or ecoregion, the search for reference sites in good ecological condition is an important task. Where such sites are abundant, they provide excellent models for developing restoration strategies associated with degraded stream reaches and watersheds. Conversely, where such sites are infrequent, small, or not present for commonly occurring types of stream reaches, their omission represents an important warning as to the extent and severity of habitat degradation. When limited amounts of intact and functional riparian and aquatic systems exist, they need to be protected from anthropogenic influences while restoration strategies are developed and implemented for degraded reaches.

Multiple reference sites, with vegetation that encompasses a range of serial stages and plant communities, provide an opportunity for assessing the relative importance of various functions and processes. Multiple reference sites can also provide insights regarding successional patterns

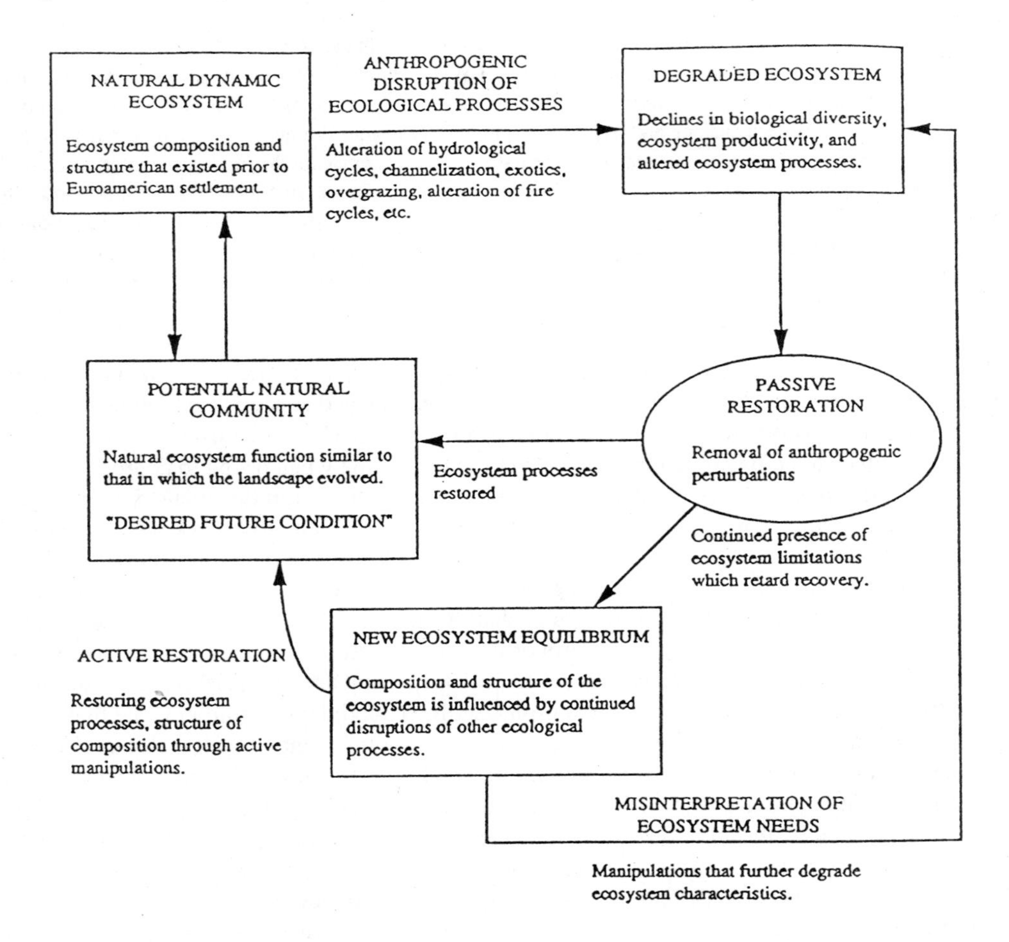

Figure 3. Conceptual pathways of ecological restoration. Source: Kauffman et al. (1993).

and their interaction with disturbance regimes. In the development of restoration prescriptions, reference sites help us understand how stream gradients, flow patterns, riparian plant communities, and other factors interact to create the resulting aquatic ecosystem. Where riparian and aquatic systems have been affected by anthropogenic impacts, reference sites can provide an indication of the time required to attain some degree of restoration for degraded reaches. For example, where historical grazing has prevented the establishment or height growth of woody plant seedlings, the restoration of mid- to late-seral shrub or cottonwood communities along a particular reach may require an altered grazing program, riparian fencing, or an extended period of grazing exclusion. A nearby reference site may provide an important indication as to the length of time required to attain these mid- to late-seral communities.

While passive restoration is urgently needed on many streams in the upper Columbia River Basin, the importance of allowing sufficient time to accomplish restoration goals needs to be acknowledged and emphasized by restoration practitioners. The restoration of complicated, diverse, interactive, and multifunctional stream and riparian ecosystems that will benefit fisheries and other aquatic organisms cannot occur overnight; nor can it be accomplished by simply adding structural elements to a channel (Elmore and Beschta 1989). If restoration activities are undertaken with little understanding of hydrologic disturbance regimes, geomorphic context, vegetation influences, or without evaluating the historical and ongoing anthropogenic impacts causing degradation, simply modifying the physical characteristics of aquatic habitats is likely to be counterproductive. Extensive channel modifications to currently degraded systems only increase the probability of additional impacts and adverse effects to many of these systems. For forested ecosystems, perhaps Bisson et al. (1992) said it best: "patience, patience, patience." Although ecological recovery of many degraded tributaries in the upper Columbia River Basin may proceed at a faster rate than required for old-growth forest ecosystems, patience is still an important ally of those working to recover these systems.

An appreciation of temporal recovery patterns has important social, institutional, and economic implications. Because full recovery of degraded systems seldom can be expected to occur quickly, the infusion of vast sums of money into local restoration programs may detract from ecological solutions and serve to emphasize only those activities that require large amounts of money (e.g., instream construction activities). Thus, a perception develops that large amounts of restoration funding can simply and quickly offset the impacts of historical land-use practices or of current land-use practices that continue to degrade riparian and aquatic systems. Unfortunately, there is little evidence that this is the case. Field evaluations of historical fish habitat enhancement projects in the upper Columbia River Basin indicate not only a potential for ecological degradation of aquatic systems, but also a lack of measurable fisheries success (Beschta et al. 1991, 1993, 1994; Kauffman et al. 1993). Even with significant investments for instream structural additions, these evaluations indicate that the potential for effective long-term habitat improvement often was not being achieved. Furthermore, because biological monitoring of these projects seldom occurs, there is no way to assess whether improvement has accrued to fisheries productivity as the result of various habitat manipulations.

Summary and Conclusions

On the basis of concepts proposed by the NRC (1992), many structural approaches to altering fish habitat cannot be construed as restoration. The extensive placement of structural additions to streams is surprising given the increasing application of ecosystem management principles to federal lands in the upper Columbia River Basin. The desire to protect native salmonid stocks on one hand while actively manipulating channels with a wide variety of mechanical approaches on the other represents an interesting paradox. If native stocks are to be protected and sustained, minimizing or modifying those anthropogenic influences and land uses that have historically caused habitat degradation would seem to be a higher priority. Resource managers and stakeholders need to familiarize themselves with the disturbance regimes and ecological functions of healthy aquatic ecosystems, thereby providing the greatest likelihood for sustaining productive instream habitats (Fig. 2).

An ecological approach to restoration involves increased understanding across stream reach, watershed, and ecoregion scales. Such an approach also involves an improved understanding of vegetation's multiple roles in riparian and aquatic habitats, the historical context of anthropogenic impacts, and an appreciation of how references sites function over a range of geomorphic settings and natural disturbance regimes. Thus, even when particular habitat manipulation techniques prove useful for one stream reach, they cannot simply be extrapolated as solutions to a wide variety of stream systems and ecoregions.

Land-use practices causing adverse impacts on both riparian and aquatic systems need to be modified, controlled, or eliminated to assist in recovering degraded stream systems. However, altering existing land-use practices, even along narrow riparian corridors, often requires an adjustment of cultural values and perspectives of local stakeholders. Whereas overcoming nature has been a central cultural theme of Euro-American immigrants intent on "winning the West," it would seem that society's current challenge is to learn how natural disturbance regimes and improved land-use practices can help restore riparian and aquatic habitats for sustaining healthy populations of fish and other aquatic organisms.

Because of the importance of the issues facing stakeholders interested in restoring aquatic ecosystems, there is a need for an accelerated, multidisciplinary research effort that helps establish an improved understanding of basic processes, functions, and interactions within these systems. As such information becomes available, active approaches to aquatic restoration may be desired (Fig. 3). However, until such information becomes available, perhaps the most effective and enlightened approach for restoring riparian and aquatic habitats is to simply focus on minimizing and eliminating those anthropogenic impacts that currently are known to cause degradation of these habitats.

Some stakeholders may want to delay restoration and protection efforts until the scientific community identifies fully the importance of all ecological functions and more completely understands the ramifications of past and current land-use practices. However, the widespread decline in anadromous fisheries in the upper Columbia River Basin indicates society has probably lost the opportunity for such an approach. An improved understanding of hydrogeomorphic disturbance regimes, historical context, streamside vegetation, and reference sites will enable us to make protecting healthy riparian and aquatic systems and restoring degraded ones a high regional priority.

Literature Cited

Abell, D.L. (Tech. Coordinator). 1989. Proceedings of the California Riparian Systems Conference: Protection, management, and restoration in the 1990's. US Department of Agriculture, Forest Service, General Technical Report PSW-110. Berkeley, California.

Beschta, R.L. 1994. Opportunities and challenges in the restoration of riverine/riparian wetlands, p. 18-27. *In* Partnerships & Opportunities in Wetland Restoration: Proceedings of a Workshop. US Environmental Protection Agency, Region 10, EPA 910-94-003. Seattle, Washington.

Beschta, R.L., R.E. Bilby, G.W. Brown, L.B. Holtby, and T.D. Hofstra. 1987. Stream temperature and aquatic habitat: Fisheries and forestry interactions, p. 191-232. *In* E.O. Salo and T.W. Cundy (eds.), Streamside Management: Forestry and Fisheries Interactions. Institute of Forest Resources, University of Washington, Contribution No. 57. Seattle, Washington.

Beschta, R.L., J. Griffith, and T.A. Wesche. 1993. Field review of fish habitat improvement projects in central Idaho. US Department of Energy, Bonneville Power Administration, Division of Fish and Wildlife, DOE/BP-61032-1. Portland, Oregon.

Beschta, R.L. and W.S. Platts. 1986. Significance and function of morphological features of small streams. Water Resources Bulletin 22(3): 369-379.

Beschta, R.L., W.S. Platts, and B. Kauffman. 1991. Field review of fish habitat improvement projects in the Grande Ronde and John Day River Basins of eastern Oregon. US Department of Energy, Bonneville Power Administration, Division of Fish and Wildlife, DOE/BP-21493-1. Portland, Oregon.

Beschta, R.L., W.S. Platts, J.B. Kauffman, and M.T. Hill. 1994. Artificial stream restoration-money well spent or an expensive failure? Universities Council on Water Resources Annual Conference, Big Sky, Montana. Southern Illinois University, Carbondale, Illinois.

Bisson, P.A., T.P. Quinn, G.H. Reeves, and S.V. Gregory. 1992. Best management practices, cumulative effects, and long-term trends in fish abundance in Pacific Northwest river systems, p. 189-232. *In* R.J. Naiman (ed.), Watershed Management: Balancing Sustainability and Environmental Change. Springer-Verlag, New York.

Brooks, K.N., P.F. Folliott, H.M. Gregersen, and J.L. Thames. 1991. Hydrology and the management of watersheds. Iowa State University Press, Ames, Iowa.

Elmore, W. and R.L. Beschta. 1987. Riparian areas: perceptions in management. Rangelands 9(5): 260-265.

Elmore, W. and R.L. Beschta. 1989. The fallacy of structures and the fortitude of vegetation, p. 116-119. *In* D.L. Abell (Tech. Coordinator), Proceedings of the California Riparian Systems Conference: Protection, Management, and Restoration in the 1990's. US Department of Agriculture, Forest Service, General Technical Report PSW-110. Berkeley, California.

Forest Ecosystem Management Assessment Team. 1993. Forest ecosystem management: an ecological, economic, and social assessment. Report of the Federal Ecosystem Management Assessment Team. US Department of Agriculture, Forest Service. Washington, DC.

Gauvin, C. 1996. Water-mangement and water-quality decision making in the range of Pacific salmon, p. 389-409. *In* D.J. Stouder, P.A. Bisson, and R.J. Naiman (eds.), Pacific Salmon and Their Ecosystems: Status and Future Options. Chapman and Hall, New York.

Gilbert, J., M.J. Dole-Oliver, P. Marmonier, and P. Vervier. 1990. Chapter 10, Surface water-groundwater ecotones, p. 199-225. *In* R.J. Naimen and H. Decamps (eds.), Volume 4, The Ecology and Management of Aquatic-Terrestrial Ecotones. Parthenon Publishing Group, Paris.

Gregory, S.V., G.A. Lamberti, D.C. Erman, KV. Koski, M.L. Murphy, and J.R. Sedell. 1987. Influence of forest practices on aquatic production, p. 233-255. *In* E.O. Salo and T.W. Cundy (eds.), Streamside Management: Forestry and Fisheries Interactions. University of Washington, Institute of Forest Resources, Contribution No. 57. Seattle.

Hamilton, J.B. 1989. Response of juvenile steelhead to instream deflectors in a high gradient stream, p. 149-158. *In* R.E. Gresswell, B.A. Barton, and J.L. Kershner (eds.), Practical Approaches to Riparian Resource Management, an Educational Workshop. US Department of Interior, Bureau of Land Management, Billings, Montana.

Harmon, M.E., J.F. Franklin, F.J. Swanson, P. Sollins, S.V. Gregory, J.D. Lattin, N.H. Anderson, S.P. Cline, N.G. Aumen, J.R. Sedell, G.W. Lienkaemper, K. Cromack Jr., and K.W. Cummins. 1986. Ecology of course woody debris in temperate ecosystems. Advances in Ecological Research 15: 133-302.

House, R., J. Anderson, P. Boehne, and J. Suther (eds.). 1988. Stream Rehabilitation Manual Emphasizing Project Design, Construction, and Evaluation. Proceedings of an American Fisheries Society Workshop, February 7 and 8, 1988, Bend, Oregon. Oregon Chapter, American Fisheries Society, Portland.

Hill, M.T., W.S. Platts, and R.L. Beschta. 1991. Ecological and geomorphological concepts for instream and out-of-channel flow requirements. Rivers 2(3): 198-210.

Hunter, C.J. 1991. Better Trout Habitat. Island Press, Washington, DC.

Johnson, R.R., C.D. Ziebell, D.R. Patton, P.F. Folliott, and R.H. Hamre (eds.). 1985. Riparian Systems and their Management: Reconciling Conflicting Uses. US Department of Agriculture, Forest Service GTR RM-120. Fort Collins, Colorado.

Kauffman, B., R.L. Beschta, and W.S. Platts. 1993. Fish habitat improvement projects in the Fifteen Mile and Trout Creek Basins of central Oregon. US Department of Energy, Bonneville Power Administration, Division of Fish and Wildlife, DOE/BP-18955-1. Portland, Oregon.

Kovalchik, B.L. 1987. Riparian zone associations: Deschutes, Ochoco, Fremont, and Winema National Forests. US Department of Agriculture, Forest Service, Pacific Northwest Region, Ecology Technical Paper 279-87. Portland, Oregon.

Kusler, J.A. and M.E. Kentula. 1990. Executive summary, p. xvii-xxv. *In* J.A. Kusler and M.E. Kentula (eds.), Wetland Creation and Restoration: The Status of the Science. Island Press, Washington, DC.

National Research Council. 1992. Restoration of Aquatic Ecosystems. Natural Research Council, National Academy of Sciences, Washington, D.C.

Nehlsen, W., J.E. Willaims, and J.A. Lichatowich. 1991. Pacific salmon at the crossroads: stocks at risk from California, Oregon, Idaho and Washington. Fisheries 16: 4-21.

Pacific Rivers Council. 1993. Entering the watershed: an action plan to protect and restore America's river ecosystems and biodiversity. A report to Congress by the Pacific Rivers Council. Eugene, Oregon.

Peacock, K.A. 1994. Valley fill and channel incision in Meyer's Canyon, northcentral Oregon. MS thesis, Department of Forest Engineering, Oregon State University. Corvallis, Oregon.

Pinay, G., H. Decamps, E. Chauvet, and E. Fustec. 1990. Chapter 8, Functions of ecotones in fluvial systems, p. 141-169. *In* R.J. Naimen and H. Decamps (eds.). Volume 4, The Ecology and Management of Aquatic-Terrestrial Ecotones. Parthenon Publishing Group, Paris.

Platts, W.S., C. Armour, G.D. Booth, M. Bryant, J.L. Bufford, P. Cuplin, S. Jensen, G.W. Lienkaemper, G.W. Minshall, S.B. Monsen, R.L. Nelson, J.R. Sedell, and J.S. Tuhy. 1987. Methods for evaluating riparian habitats with applications to management, p. 93-105. US Department of Agriculture, Forest Service, General Technical Report INT-221. Ogden, Utah.

Reeves, G.H., J.D. Hall, T.D. Roelofs, T.L. Hickman, and C.O. Baker. 1991. Chapter 15: Rehabilitating and modifying stream habitats, p. 519-557. *In* W.R. Meehan (ed.), Influences of Forest and Rangeland Management on Salmonid Fishes and Their Habitats. American Fisheries Society Special Publication 19, Bethesda, Maryland.

Salo, E.O., and T.W. Cundy (eds.). 1987. Streamside Management: Forestry and Fisheries Interactions. University of Washington, Institute of Forest Resources, Contribution No. 57. Seattle, Washington.

Sedell, J.R. and K.J. Luchessa. 1982. Using the historical record as an aid to salmonid habitat enhancement, p. 210-223. *In* N.B. Armantrout (ed.), Proceedings of a Symposium on Acquisition and Utilization of Aquatic Habitat Inventory Information. The Hague Publishing, Billings, Montana.

Sedell, J.R., P.A. Bisson, F.J. Swanson, and S.V. Gregory. 1988. What we know about large trees that fall into streams and rivers, p. 47-81. *In* C. Maser, R.F. Tarrant, J.M. Trappe, and J.F. Franklin (eds.), From the Forest to the Sea: A Story of Fallen Trees. US Department of Agriculture, Forest Service, General Technical Report PNW-GTR-229. Portland, Oregon.

Sedell, J.R., and R.L. Beschta. 1991. Bringing back the "Bio" in bioengineering. American Fisheries Society Symposium 10: 160-175.

Seehorn, M.E. (workshop coordinator). 1980. Proceedings of the Trout Stream Habitat Improvement Workshop. US Department of Agriculture Forest Service, November 3-7, 1980, Asheville, North Carolina. Atlanta, Georgia.

Seehorn, M.E. 1992. Stream habitat improvement handbook. US Department of Agriculture, Forest Service, Technical Publication R8-TP 16. Atlanta, Georgia.

Stanford, J.A. and J.V. Ward. 1988. The hyporheic habitat of river ecosystems. Nature 335: 64-66.

Sullivan, A.E. 1994. Selected small and ephemeral streams in arid central Washington State: a historical perspective with recommendations for salmon habitat enhancement. MS thesis, Department of Geography, Central Washington University. Ellensburg, Washington.

Van Haveren, B.P. and W.L. Jackson. 1986. Concepts in stream riparian rehabilitation, p. 280-289. Transactions 51st North American Wildlife and Natural Resources Conference, Wildlife Management Institute, Washington, DC.

Welcher, K.E. 1993. Holocene channel changes of Camp Creek; an arroyo in eastern Oregon. MA thesis, Department of Geography, University of Oregon. Eugene, Oregon.

Wesche, T.A. 1985. Stream channel modifications and reclamation structures to enhance fish habitat, p. 103-163. *In* J.A. Gore (ed.), The Restoration of Rivers and Streams. Butterworth Publishers, Boston, Massachusetts.

Rehabilitation of Pacific Salmon in Their Ecosystems: What Can Artificial Propagation Contribute?

Anne R. Kapuscinski

Abstract

Artificial propagation is only one of a number of tools whose application might assist rehabilitation of depleted Pacific salmon (*Oncorhynchus* spp.) populations in their ecosystems. The motivation for and implementation of hatchery programs have evolved as world-views of nature-human relations have changed, understanding of salmonid ecosystems has evolved, the diversity of stakeholders involved in salmon issues has increased, and adverse human impacts on aquatic ecosystems have accumulated. Building on lessons from past uses, this paper outlines changes needed to integrate future applications of artificial propagation into comprehensive, ecosystem-focused adaptive management. Changes cover all steps in the adaptive management cycle including setting goals and objectives, problem identification, actions, and evaluation. Hatchery programs should be compatible with management goals of retaining or rehabilitating all levels of biodiversity found in Pacific salmon ecosystems and the natural regenerative capacity of these ecosystems. Consistent with these goals, two guiding principles are as follows: (1) use hatcheries only as part of a comprehensive rehabilitation strategy, with rigorous adherence to adaptive management and (2) ensure that all hatchery programs maintain genetic diversity between and within salmon populations and avoid disruption of all other levels of biodiversity in salmonid watersheds. To be compatible with these goals and principles, hatchery programs need new tools in the following areas: planning that includes assessment of genetic and ecological risks; operational changes based on population inventories and principles of genetics, evolution, and ecology; and monitoring and evaluation for progress in achieving goals while minimizing risks.

Introduction

This paper calls for taking a humble approach to applying artificial propagation towards rehabilitation of depleted populations of Pacific salmon (*Oncorhynchus* spp.). Such an approach is predicated on our emerging appreciation of the dynamic variability of natural ecosystems. This variability places real constraints on the capability of any human intervention to rebuild Pacific salmon populations (e.g., Lawson 1993, National Research Council [NRC] 1996). It also demands a fundamental change in approach if we wish to succeed at conserving the remaining biodiversity found in Pacific salmon populations and species. The NRC (1996) reviewed problems posed by past approaches of many hatchery programs, including demographic risks; genetic and evolutionary risks; problems due to the behavior, health, or physiology of hatchery fish; and ecological problems. In light of that review, recommendations in this paper aim to reduce imposition of these problems.

Biodiversity in integrated terrestrial and aquatic systems manifests variability in biological composition, structure, and process (function) at numerous nested levels, starting from the building blocks of genetic diversity up through population and species diversity, then diversity in communities and ecosystems, and finally diversity in landscapes and regions (Noss 1990, Hughes and Noss 1992, Kapuscinski 1994). An example of a biological structure is the natural accumulation of genetic diversity between and within salmon populations. An example of a biological process is the evolution of salmon populations in terms of local adaptation. Thus, fisheries managers should aim to conserve "a shifting variability system, not a fixed and invariable entity" (Carson 1983).

Artificial Propagation Congruent with Biotic Variability and Within Adaptive Management

In light of contemporary appreciation of variability and nested levels of biodiversity, it is paramount to define the following two biological goals of salmon management:

1. to retain, or in places of extensive damage, rehabilitate variability at each of the nested and interacting levels of biodiversity; and
2. to retain or rehabilitate the natural regenerative capacity of ecosystems where capacity includes both biological and physical structures and processes.

These goals are consistent with the recently increasing focus on watershed rehabilitation (e.g., Doppelt et al. 1993). An important implication of these goals is that success in sustaining the remaining healthy populations and rebuilding the many depleted salmon populations throughout the Pacific Northwest will require management attention to the full range of biodiversity associated with salmon, not simply attention to Pacific salmon themselves (NRC 1996). Defining an appropriate future role of any technology in recovery programs for Pacific salmon populations involves determining whether the proposed application of the technology is consistent with these goals.

Artificial Propagation and the Conservation of Biodiversity

To assist the recovery of Pacific salmon populations, we must apply any technology in a manner that will conserve or rehabilitate, not disrupt, the variability at the different, nested levels of biodiversity. At the genetic and population levels, we need to conserve or rebuild genic diversity between and within salmon populations. These patterns of genetic diversity ensure evolution, including local adaptation, of salmon populations (NRC 1996, Reisenbichler 1996). Rethinking the role of artificial propagation under the goals presented above requires that we conserve or rehabilitate more than the genetic diversity component of biodiversity. It is clearly understood that Pacific salmon are interdependent with their surrounding natural systems; thus, salmon managers must also aim to retain or rebuild the broader, more complex variability systems of the Pacific salmon's surrounding communities, ecosystems, landscapes, and regions (see related comments by Gregory and Bisson 1996).

Therefore, future roles of artificial propagation in recovery must also accommodate important linkages between salmon and their surrounding ecosystems (Wilson and Halupka 1995, NRC 1996). In the freshwater landscape, artificial propagation programs must operate within constraints set by existing conditions of natural food webs, considering both prey and predators of salmon; nutrient cycles, such as important inputs from riparian vegetation and fish carcasses; and habitat structure, such as woody debris in streams. Artificial propagation protocols for collection of broodstock and release of progeny must also avoid direct damage to the ecosystem linkages they can most easily affect. Hatchery programs need to somehow accommodate at least three important linkages provided by salmon in the wild. First, spawned out carcasses supply marine-derived nutrients (carbon, nitrogen, and phosphorus) to riparian and lacustrine ecosystems, thus supporting many portions of the food web including juvenile salmon in the next generation (Donaldson 1967; Richey et al. 1975; Kline et al. 1990, 1993). Second, numerous terrestrial and aquatic vertebrates (including fish, mammal, and bird species) prey on salmon eggs, juveniles, live adults, or spawned out carcasses (Cederholm et al. 1989, Groot and Margolis 1991, Wilson and Halupka 1995). Third, the redd-digging activities of adult females affect physical features of stream ecosystems in ways that may benefit survival of progeny and affect other stream inhabitants (NRC 1996).

One specific example of how this ecosystemic perspective might revise artificial propagation practices is to reconsider what should be done with carcasses of adults after they have been artificially mated in the hatchery. Perhaps these carcasses should be returned to the stream from which the adults were taken or to which they would have returned if there had been no hatchery intake, in order to reestablish this natural source of nutrient inputs for stream productivity and food for myriad species of terrestrial and aquatic vertebrates. Salmon management entities might want to determine the pros and cons of returning spawned carcasses in a case-specific assessment of each existing and proposed hatchery program.

Adherence to the Adaptive Management Cycle

Rather than turning immediately to artificial propagation as a magic bullet for declining populations, we must base decisions to use this technology on rigorous adherence to the full cycle of adaptive management (Fig. 1; NRC 1996). Adaptive management refers to learning by planning, executing, and evaluating management actions as experiments (Krueger et al. 1986). Before the startup or expansion of an artificial propagation program can even be considered, a thorough discussion must occur—first, to establish salmon management goals and objectives for a particular watershed or group of watersheds, and second, to identify problems associated with those goals and objectives. As discussed in more detail in the following text, genetic conservation should be one goal of all salmon management programs. The quality of subsequent steps hinges heavily on the quality of questions identified and inferences drawn from existing information during these first two steps of adaptive management. Only after completing these steps does it make sense to consider alternative actions. Artificial propagation may be identified

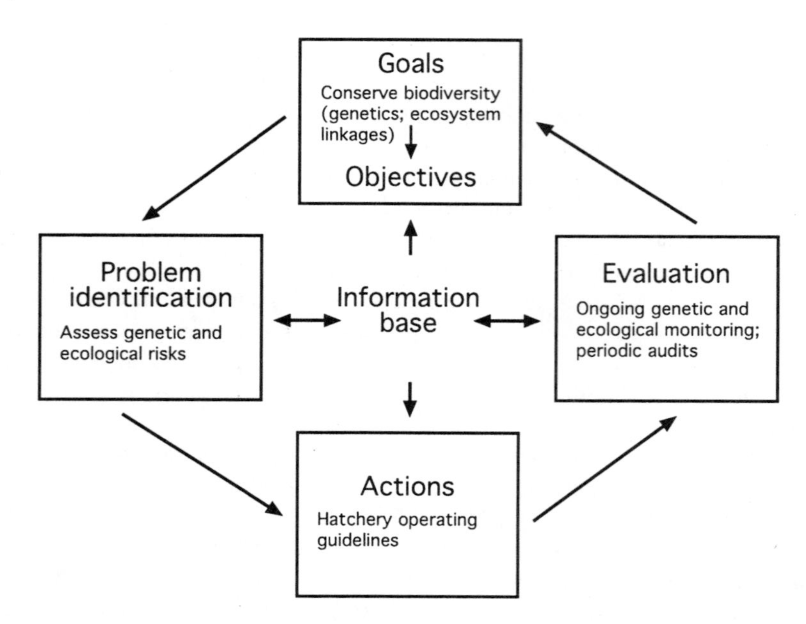

Figure 1. Schematic diagram of the four essential steps of adaptive management of hatcheries, all of which are guided by consulting the contemporary information base (after Krueger et al. 1986). Planning, risk assessment, operations, and monitoring of hatcheries should be integrated into the adaptive management cycle. Articulation of goals and objectives and identification of problems posed by goals and objectives should precede specific consideration of artificial propagation as one among numerous alternative actions. Whenever artificial propagation is chosen as an action, there needs to be rigorous and ongoing evaluation of its ability to meet stated goals and objectives.

as an appropriate management action in some cases but not in others. If an artificial propagation program is indeed implemented, it is critically important to complete the adaptive management cycle via rigorous and ongoing evaluation of the program's ability to achieve clearly defined objectives and to avoid adverse effects. As outlined later in this paper, future evaluations must be substantially improved compared with the generally inadequate or nonexistent evaluation of past hatchery programs (Kapuscinski 1991).

Ideally, fisheries managers should apply the entire adaptive management cycle to all existing artificial propagation programs, not only to new programs under proposal. They need to assess whether maintaining each hatchery program is still an appropriate management action after reexamining and perhaps revising goals, objectives, and problems for the geographical area in question.

Guiding Principles for Artificial Propagation under Adaptive Management

Consideration of artificial propagation within the context of ecosystem-focused adaptive management leads to two principles that should guide all future uses of artificial propagation in assisting recovery of Pacific salmon populations. The first principle is to use hatcheries only as part of a comprehensive rehabilitation strategy and with rigorous adherence to all steps of adaptive management. The second principle is that all hatchery programs must avoid disrupting any level of biodiversity in salmonid ecosystems, and must specifically maintain existing patterns of genetic diversity between and within salmon populations.

A hatchery program should be only one component of a comprehensive rebuilding strategy designed to remove or significantly reduce the human-induced causes of the population's decline (Northwest Power Planning Council [NPPC] 1992, NRC 1996). Different salmon populations in the Pacific Northwest have experienced different causes of decline, with several causes often acting in concert (Nehlsen et al. 1991, NRC 1996). Major problems have been habitat deterioration or loss due to barriers to passage (dams), reduced or altered stream flows, poor water quality, water diversions, destruction of spawning or nursery habitat, increased exposure to competition and predation from native species, or effects of introduced species (Li et al. 1987). Some natural populations that normally produce relatively low amounts of catchable surplus have been further depleted by overfishing. Sole focus on a hatchery, no matter how well designed and managed, cannot compensate for all root causes of decline and will not result in sustainable rehabilitation of naturally spawning populations.

Conservation of genetic diversity among and within salmon populations is critically important because it allows salmon populations to evolve in the face of changing local environments and thus persist far into the future. Sustainable restoration and maintenance of salmon populations can only be achieved if these genetic resources are maintained in perpetuity (Riddell 1993). This will require two interrelated improvements.

First, hatchery policies and practices of each management agency must be fully congruent with modern understanding of important genetic and evolutionary principles (Allendorf and Ryman 1987, NRC 1996). There are promising signs in this direction; for instance, the Washington Department of Fish and Wildlife is developing genetic elements of a wild salmon policy (C. Busack, Washington Dep. Fish and Wildlife, Olympia, pers. comm.). Independent assessment of the genetic policies of all salmon management agencies in the Pacific Northwest would be needed to evaluate overall progress on this issue. Other developments in assisting integration of genetic principles into hatchery programs are discussed in a later section on tools for planning, operations, and evaluation of artificial propagation.

Second, coordination between agencies responsible for management of the same or interacting Pacific salmon populations must greatly improve in order to achieve consistency in their hatchery genetic policies and practices. To date, serious attention to such coordination has occurred only in the Columbia River Basin although it is needed throughout the Pacific Northwest. I analyzed hatchery genetic policies and guidelines of the five agencies responsible for salmon management in the Columbia River Basin (Kapuscinski 1991). Tremendous variability existed across agencies in the comprehensiveness and biological quality of their treatment of five factors identified to be essential for genetic conservation. These factors were definition of key genetic terms and stock categories (i.e., wild, natural, naturalized, or hatchery), genetic goals of hatchery programs, hatchery broodstock and rearing guidelines, indirect genetic impacts of hatchery fish, and genetics research and monitoring. I concluded that there was a paramount need for greater consistency between agencies in the way that their policies addressed these five factors (Kapuscinski 1991).

Recent actions of the Integrated Hatchery Operations Team (IHOT) represent the first steps towards improved coordination and consistency of hatchery programs in the Columbia River Basin. The IHOT comprises representatives from 11 fisheries co-manager entities and 7 cooperating entities. Through persistent consensus building among its members, the IHOT developed five general policies (hatchery coordination, hatchery performance standards, fish health, ecological interactions, and genetics) and an implementation plan for all 90 anadromous hatchery facilities in the Columbia River Basin (Integrated Hatchery Operations Team [IHOT] 1994a, b, c). Laudable features of the implementation plan are that it calls for the following three steps: (1) ongoing independent hatchery audits to determine if each hatchery is complying with standards established in the five policies; (2) coordinated documentation of individual hatchery plans (including their updating in light of new information) and of data on hatchery compliance with regional policies; and (3) establishment of a forum for regular sharing of information, resolution of differences, and coordination.

However, shortcomings of the policies undermine the potential for IHOT to encourage all hatchery programs to follow a complete adaptive management cycle and to avoid disruption of the different levels of biodiversity in salmonid ecosystems. The primary obstacle to the former is that the policies are limited to "guidance for the technical operation of hatcheries" (IHOT 1994b). Guidance on the more fundamental decisions about whether to use artificial propagation and, when the decision is affirmative, about the appropriate level of artificial propagation is outside the jurisdiction of IHOT. Instead, fishery co-managers make these sorts of decisions about production priorities through plans developed during negotiations, primarily with the Columbia River Fish Management Plan's Production Advisory Committee. Whenever management entities in the Columbia River Basin make these fundamental hatchery decisions, they need a specific interagency policy ensuring that these decisions are consistent with the two principles presented above.

Although the IHOT policies lean in the direction of genetic and biodiversity conservation, the scope and clarity of the policies for ecological interactions and genetics are inadequate to ensure that implementation will conserve the different levels of biodiversity in Pacific salmon ecosystems, including genetic diversity between and within salmon populations. For instance, in the genetic policy the overarching statement of policy says that artificial propagation programs will "maintain adequate genetic variation and fitness in populations and protect the biological diversity of wild, natural, and cultured anadromous salmonid populations." This language effectively avoids a less ambiguous, more scientifically defensible, more concise policy statement that artificial propagation programs will conserve *existing* genetic diversity between and within populations of anadromous salmon.

Future Uses of Artificial Propagation

There are three possible scenarios for future uses of hatchery programs as part of a comprehensive effort to rehabilitate Pacific salmon and their ecosystems (Kapuscinski 1993). The first scenario is one in which a temporary hatchery program increases a greatly depleted population or reestablishes an extinct population to self-sustaining status while root causes are being reversed (e.g., while habitat is being restored). The second involves a long-term hatchery program to mitigate for irreversibly damaged or lost habitat, provided that adverse impacts of hatchery fish on other naturally spawning populations and other components of biodiversity are kept negligible. Last is the use of a long-term hatchery program to augment catches above levels supported by existing freshwater habitat, provided it is feasible to reliably separate hatchery fish from naturally spawning fish in freshwater habitats and in the fishery and to avoid adverse effects on other components of biodiversity. Implementation of any of these scenarios should include marking of all hatchery-released fish, via artificial tags (e.g., coded wire tags) or endogenous biological marks (e.g., DNA markers). This is needed to evaluate whether the hatchery-propagated fish are assisting rehabilitation without imposing adverse genetic and ecological effects on naturally reproducing populations.

Temporary Hatcheries

A temporary hatchery is the best possible scenario for artificial propagation for two reasons. First, hatchery expenses can be terminated once rates of natural reproduction and survival increase enough to achieve self-sustainability of the targeted population(s). Second, successful reestablishment of a self-sustaining population is a good indicator of improved health of the overall watershed ecosystem. A self-sustaining population is one that depends solely on natural reproduction to perpetuate itself, generally maintaining or increasing its abundance in offspring generations relative to parental generations. Relatively inexpensive and easy to dismantle rearing units, buildings, and water distribution systems may work for this scenario. Temporary hatcheries must have a written plan that describes how to determine when to terminate the program. This will require developing appropriate and measurable achievement indicators of self-sustainability by the rehabilitated population. The common past approach was to implement hatchery programs with no planned termination point and to build capital-intensive, permanent facilities. In contrast, temporary hatcheries would represent greatly reduced costs of control by human society because the various monetary outlays for this technical intervention would be relatively short-lived, after which natural regenerative processes would take over.

Captive broodstock propagation is an extreme case of a temporary hatchery. It involves collection of many or all remaining individuals of a greatly depleted population and captive propagation for at least one entire generation. This should be a last resort only when a population is dangerously close to extinction (Hard et al. 1992). Captive rearing of fish over the entire life cycle greatly increases the rate of their adaptive evolution to the hatchery environment and conversely reduces their adaptation to natural environments (Vincent 1960; Flick and Webster 1962, 1964; Mason et al. 1967; Moyle 1969; Johnson and Abrahams 1991). Also, catastrophic loss of the entire population is possible if there is a power outage, epizootic, or other hatchery accident.

Success of a temporary hatchery program in reestablishing a self-sustaining population is necessary but not always sufficient to achieve abundance levels that permit fishing a portion of returning adults. The rebuilt population might be hovering at or close to replacement, the point where the number of progeny surviving to a catchable size and returning to fishing waters (with the potential of escaping the fishery and returning to spawning grounds) is only equal to the number of successfully spawning parents in the former generation (Ricker 1954, see also NRC 1996). In such cases, the population may need more generations to rebuild to a higher abundance level at which there will be a catchable surplus. In such cases, socially desired fishing levels will have to be postponed. For fishing to proceed without driving the population to extinction, the spawners need to produce, on average, more progeny than are needed to replace the parents, and these surplus progeny must survive all sources of natural mortality from embryogenesis through adulthood. Fishers and other stakeholders must expect lag times in obtaining beneficial effects of the entire suite of management actions applied under comprehensive rehabilitation (e.g., Lawson 1993).

Long-Term Mitigation Hatchery

The second scenario, a long-term hatchery to mitigate for lost natural reproduction caused by irreversible habitat loss or damage, will be more difficult to implement than a temporary hatchery because it increases the possibility and exposure time of naturally reproducing populations to adverse genetic and ecological effects. Many existing hatcheries, particularly in the Columbia River Basin, were established as mitigation hatcheries to make up for lost habitat due to construction of dams. To be consistent with this second scenario as stated above, these existing hatcheries must successfully prevent adverse genetic and ecological effects from affecting other naturally spawning populations. Clearly, this necessitates developing, implementing, and evaluating hatchery policies and practices for genetic conservation and prevention of adverse ecological impacts (Kapuscinski 1991, Busack and Currens 1995, Currens and Busack 1995).

Long-Term Augmentation Hatchery

The third scenario involves indefinite releases of hatchery-propagated fish to augment catches in a particular fishery. This may not always be compatible with comprehensive efforts to rehabilitate Pacific salmon and their ecosystems. To avoid overfishing the less productive wild populations in a mixed-stock fishery, this scenario requires selective fishing of hatchery fish. To avoid other genetic and ecological risks, it requires separation of hatchery-propagated from naturally reproducing fish in freshwater habitats. These conditions will be hard to meet in many cases and would require changes in fishing methods and management. Meeting these conditions is particularly challenging in the face of the dynamic nature of Pacific salmon ecosystems and remaining social resistance to changes in existing fishing regimes. At least one sign of willingness to change fishing methods has occurred. The Strategy for Salmon, which was adopted by the NPPC after much input from different stakeholders in the Columbia River Basin, includes a recommendation in Section 5 to test alternative methods for selective harvest of hatchery fish (NPPC 1992).

Supplementation

Some recent proposals for artificial propagation in the Pacific Northwest have involved increasing the size of naturally reproducing populations by releasing hatchery fish with the expectation that they will interbreed with the depleted, naturally spawning population and that the mixed population will be managed as one unit. This approach is often called supplementation (Anonymous 1992, Cuenco et al. 1993). For example, this approach is planned for new hatcheries in the Yakima sub-basin of the Columbia River Basin. Because they encourage ecological interaction and interbreeding between hatchery-propagated and naturally spawning fish, supplementation programs make it difficult to avoid the range of genetic, demographic, behavioral, ecological, and other biological problems posed by traditional hatchery programs (NRC 1996). To maximize learning about risks, benefits, and desirable modifications, a strong commitment to monitoring and evaluating each supplementation program is imperative. This requires marking of all hatchery-released fish. It would be best to allow learning to proceed in a small number of carefully chosen cases, such as in the Yakima River basin, before considering widespread adoption of this approach.

The goal of supplementation programs would overlap with that of temporary hatcheries if they have an explicitly stated and definite ending point, such as terminating hatchery releases once natural reproduction is rebuilt to a given number of returning adults. Additionally, they need to be a part of a broader, integrated effort to remove root causes of decline to the fullest extent possible. Unfortunately, there remain many situations in which overall societal will is against removing root causes of decline. In such situations, the true goals of so-called supplementation propagation programs must be honestly examined and stated. In some cases, it might be more appropriate to openly plan for a long-term mitigation hatchery that includes the genetic and ecological safeguards mentioned above.

Tools for Adaptive Management of Artificial Propagation

Judicious use of artificial propagation as only one part of a comprehensive, ecosystem-focused rehabilitation effort carried out via adaptive management requires new or revised tools for case-specific planning, operation, and evaluation of each proposed or existing hatchery program (Fig. 1).

Planning Tools

Decision makers need to apply specific planning tools whenever they consider artificial propagation among several options for management action (steps two and three of the adaptive management cycle, Fig. 1). They need scientifically sound and practicable guidelines for genetic and ecological risk assessment. Conducting such assessments before deciding whether to institute or continue a hatchery program will allow managers to identify the most likely problems posed by the hatchery program. This will provide them the rationale for redesigning aspects of the program to avoid or, at least, reduce problems. If the assessments indicate particularly high risks, managers then have a rational basis for deciding against institution or continuation of the hatchery program.

Artificial propagation programs in the Pacific Northwest currently lack uniform and widely used guidelines for conducting such assessments although some recent steps in this direction are described in the following. A user-friendly format for risk assessment guidelines would involve the use of interactive, computerized decision support or expert systems (e.g., White 1995). Decision support systems can be designed so that users respond to prompted questions on a screen, have menu-driven access to explanatory text and annotated bibliographies, and obtain a computer-generated compilation or trace of their decision-making path. In light of new information, a computerized format also facilitates frequent revision of guidelines and widespread dissemination of updated versions.

Currens (1993, see also Currens and Busack 1995) developed a model for genetic risk assessment of Pacific salmon hatcheries. Currens (1993) applied it to the test case of a set of proposed hatcheries for supplementing chinook salmon (*O. tshawytscha*), steelhead trout (*O. mykiss*), and coho salmon (*O. kisutch*) in the Yakima River sub-basin, located upstream of four mainstem dams in the upper Columbia River Basin. The main elements of genetic risk assessment are quantitative or, when this is not possible, qualitative assessments of the vulnerability of the target and non-target Pacific salmon populations to four genetic hazards. Genetic hazards are consequences of artificial propagation of a target population that lead to losses of genetic structure or function. The four genetic hazards are (1) extinction of local breeding populations, (2) loss of genetic diversity within a given population, (3) loss of genetic diversity among different local breeding populations, and (4) domestication (or loss of fitness in the wild) of the artificially propagated population. To determine sources of these hazards, we need to search through proposed hatchery plans or existing hatchery operations at the following steps: (1) broodstock selection from different candidate populations in the wild, (2) collection of broodstock from adults returning to spawning grounds, (3) artificial mating of collected broodstock, (4) rearing of progeny, (5) release of progeny from the hatchery, and (6) patterns of downstream migration by released progeny. Once these sources of genetic hazard are identified, vulnerability of relevant target and non-target populations is estimated as the product of risk (probability that adverse consequences of the genetic hazard will occur) and hazards (e.g., the loss of within-population genetic diversity due to artificial mating of too small a number of collected broodstock).

Local, naturally breeding populations are the biological level at which the four genetic hazards might operate. Therefore, an essential precursor to conducting genetic risk assessment of a proposed hatchery program is to obtain an inventory of the genetic differentiation (i.e., genetic relationships) among local breeding populations (Busack and Currens 1995). Genetic inventories exist for a growing portion of the range of Pacific salmon, and continued compilation is a routine aspect of a number of Pacific salmon management programs.

Conducting genetic risk assessments should be an integral part of interagency coordinated hatchery genetic policies, as IHOT has initiated for hatcheries in the Columbia River Basin. Unfortunately, the IHOT genetics policy (IHOT 1994b) lacks an explicit commitment to conducting genetic risk assessments prior to initiation of new hatcheries and as part of periodic review of existing hatcheries (e.g., in independent audits). This omission weakens two essential linkages in the adaptive management cycle (Fig. 1). First, genetic risk assessments provide essential information for identifying genetic conservation problems in a proposed or existing hatchery program. Second, proper identification of problems is a necessary precursor to taking actions that will be consistent with the principle of conserving genetic diversity between and within salmon populations. Lack of rigorous and comprehensive genetic risk assessment at the outset will jeopardize the rigor and comprehensiveness of implemented hatchery operating practices and evaluation.

Compared with the recent advances in genetic risk assessment, ecological risk assessment of artificial propagation programs has received minimal attention. As discussed earlier, Pacific salmon are linked with their surrounding ecosystems in several important ways. The long-term sustainability of natural populations of salmon, and perhaps of other interdependent biota, depends heavily on protecting a range of important ecological structures and processes.

The ecological interactions policy for Columbia Basin anadromous salmonid hatcheries addresses only a narrow portion of potential ecological risks (IHOT 1994a). The policy focuses on how different release strategies affect interactions between hatchery fish and pre-existing anadromous and resident salmonids, with some minor reference to other fish species. These interactions include competition, predator-prey relationships, and social behavior. The preamble to the policy's performance standards explains that the state of the art precludes a "cookbook" approach and recommends that ecologists be directly involved in evaluating existing standards and developing new standards. This policy is a welcome step forward compared with the prior lack of any coordinated policy on ecological effects of Pacific salmon hatcheries in the Columbia River Basin.

However, all Pacific salmon hatchery programs need a comprehensive ecological risk assessment framework. Ecological risk assessment needs to address the full range of possible ecological effects associated with the full cycle of hatchery actions. These include management of returning adult passage upstream of hatchery weirs, collection of adults for hatchery propagation, preclusion of carcasses from natural streams, water quality of effluents, and smolt release strategies. To move beyond statements of broad principles laid out in the IHOT policy, a comprehensive ecological risk assessment framework should provide effective, step-by-step guidance to the user.

It is possible to design stepwise guidance through a risk assessment procedure and still retain sufficient flexibility for case-specific application. Two examples are the previously discussed genetic risk assessment guidelines (Currens and Busack 1995) and a suite of flowcharts developed by the US Department of Agriculture's Agricultural Biotechnology Research Advisory Committee (ABRAC) for assessing the genetic and ecological effects of accidental escapes of genetically modified fish and shellfish from research and development projects (Agricultural Biotechnology Research Advisory Committee 1995). The ABRAC flowcharts do not address impacts of traditional hatchery stocking programs but instead are aimed at effects of accidental escapes of transgenic, chromosome-manipulated, and interspecific hybrid organisms. Yet, the approach of guiding the user through interconnected decision trees or flowcharts, based on yes or no answers to carefully worded questions, is a good model for addressing ecological impacts of hatchery-propagated salmon.

Operational Tools

Completed genetic and ecological risk assessments will provide an environmentally comprehensive and scientifically sound basis for deciding whether to maintain an existing or initiate a new hatchery program. When the decision is made to proceed with the hatchery program, then each phase of hatchery operations needs to follow explicit guidelines for conservation of both genetic resources and ecological linkages associated with the propagated salmon population and other interacting biota. Although the IHOT genetic and ecological interactions policies are important first steps (IHOT 1994a), they are too narrow in scope and too sketchy in content to serve the role of comprehensive genetic and ecological guidelines for hatchery operations.

Kapuscinski and Miller (1993) presented step-by-step genetic guidelines for all phases of anadromous Pacific salmon hatchery operations. These guidelines are based on the integration of contemporary scientific principles of genetics and evolution, knowledge of Pacific salmon biology and practical constraints of hatchery operations, and the prior input of an interdisciplinary team of scientists and fisheries managers, which was convened in 1991 by the NPPC. Although they were designed with the proposed Yakima sub-basin hatcheries in mind, they are generally applicable to Pacific salmon hatcheries intended to rebuild depleted populations or reintroduce locally extirpated populations. These genetic operating guidelines are designed to prevent, or at least reduce, imposition of the four previously described genetic hazards. They also aim to prevent one type of ecological hazard, namely the overlapping of resource needs and uses (e.g., food, cover, spawning habitat) between hatchery-released and naturally reproduced salmon, which may trigger adverse genetic changes within the naturally reproduced salmon.

There are specific guidelines for every decision point in hatchery operations, including broodstock collection, broodstock spawning, rearing of progeny, and release of progeny (see outline in Appendix 1). For each guideline, explanatory text provides the genetic or evolutionary rationale and supporting literature and discusses practical issues of implementation. In order to minimize all four genetic hazards, the broodstock collection guidelines address three factors: (1) the choice of donor population, (2) collection of broodstock from returning adults, and (3) special considerations for captive broodstock. Spawning guidelines aim to minimize loss of within-population genetic diversity and domestication. On the basis of consideration of the numbers and sex ratio of available spawners and quality of male sperm, they guide the user to the mating scheme that is most appropriate for conserving the genetic diversity existing in the group of available spawners. Rearing guidelines, which focus on preventing domestication (loss of fitness in the wild), give specific recommendations for physical, chemical, and biological conditions for incubating embryos and rearing juveniles. Finally, release guidelines are designed to avoid extinction, loss of between- and within-population genetic diversity, and adverse ecological interactions between hatchery and naturally propagated fish. They address the number of fish released, life-stage at release, location of release, alternative release procedures, and special considerations for juveniles that neither smoltify nor voluntarily outmigrate from acclimation ponds.

Evaluation Tools

Once the operations of a given hatchery program are following genetic guidelines and, contingent on their development, ecological guidelines, appropriate evaluation tools are needed to complete the adaptive management cycle. Hatchery programs need to be explicitly evaluated for their contribution to rehabilitation of the targeted salmon populations and their ability to prevent or minimize genetic and ecological hazards. Both genetic monitoring and ecological monitoring protocols are needed. To reduce inherent problems with evaluations in natural environments, effective monitoring must involve (1) commitment to long-term collection of appropriate data in order to overcome difficulties in detecting a real trend against the normal background of interannual variation and (2) careful selection of appropriate control populations not affected by a hatchery program in order to reduce difficulties in properly identifying the cause of a detected trend (Eberhardt and Thomas 1991). Although these two requirements present major scientific, financial, and operational challenges, ignoring them could make a monitoring program completely useless and thus waste all resources expended on it.

Development of genetic monitoring protocols, to complement the genetic risk assessment guidelines and operating guidelines discussed previously, is underway (C. Busack, Washington Dep. Fish and Wildlife, Olympia, and J. Hard, National Marine Fisheries Service, Seattle, unpubl. data). The focus of such monitoring is the ability of hatchery operations to prevent or reduce imposition of the four genetic hazards. The intent is that the monitoring guidelines will explain specific techniques for detecting genetic changes over time in the hatchery broodstock and the targeted naturally reproducing population. In general, genetic monitoring involves the following types of analyses. Monitoring for increased exposure to extinction will require analysis of demographic trends via population viability analysis (Boyce 1992) or similar methodologies. This will require long-term collection of population dynamics data on separate, naturally reproducing populations. Monitoring for loss of between-population genetic diversity will require analysis of unnatural patterns of gene flow between hatchery and natural populations (due to straying and transplantations) using protein electrophoretic and DNA data. Monitoring for increased loss of within-population diversity will require analysis of random genetic drift (e.g., via estimation of effective population size). The two most tractable methods for estimating the effective size of naturally reproducing populations require either multilocus genotype data collected from a single random sample of the population (Hill 1981) or allele frequency data from two samples taken a generation or more apart (Pollak 1983, Waples 1989, Jorde and Ryman 1995). Finally, monitoring for domestication will require analysis of the amount of genetic variation available in life-history traits thought to be important to fitness in the wild. One approach would involve mating 10 or more parents that have returned to spawn naturally, recording parental phenotypes for chosen life-history traits, marking their offspring with family-specific marks prior to release, and then measuring the same life-history traits on returning offspring (Hard 1995).

Wahl et al. (1995) presented an ecological framework for evaluating the effects of stocked fishes. Although these authors focused on information about stocking warmwater fish species, their framework provides an excellent starting point for developing step-by-step ecological monitoring protocols for Pacific salmon hatcheries. Salmon ecologists can make an important contribution in this arena even if much of the monitoring has to rely on qualitative rather than quantitative measures. The most effective way to fill this gap would be to develop an integrated package of ecological guidelines for risk assessment, hatchery operations, and monitoring.

In the Columbia River Basin, the need for hatchery evaluations was recognized when the NPPC (1992) called for independent hatchery audits. In response to this call, a central element of the IHOT implementation plan is the recommendation to conduct hatchery audits of all Columbia Basin hatcheries (IHOT 1994a). Clearly, the consistency and scientific grounding of these audits would benefit greatly if comprehensive and step-by-step genetic and ecological monitoring guidelines were already in place.

The entire Pacific Northwest region, not just the Columbia River Basin, needs periodic audits of Pacific salmon hatcheries based on application of explicit genetic and ecological monitoring guidelines. Audits should be carried out by an independent party that has no stake in the outcome of the audit. A rigorous audit would involve a two-tiered evaluation based on information gathered from written agency documents, selected interviews with administrators and operators of hatcheries, review of hatchery records, and hatchery site visits.

In the first tier, the audit should determine the degree to which initiating new hatcheries and continuing long-established hatcheries is the outcome of watershed-scale planning. The planning should have involved consideration of all alternative actions for maintaining or rehabilitating target populations, drawing on state-of-the-art information about these populations and their ecosystems, and only then should the audit lead to the conclusion that the hatchery was an appropriate management action. The second tier of the audit should determine the following: whether the organization governing the hatchery has developed written genetic and ecological guidelines for risk assessment, hatchery operations, and monitoring; the adequacy of these written guidelines for conserving biodiversity of Pacific salmon ecosystems; and the degree to which the day-to-day functioning of the hatchery complies with these written guidelines.

Concluding Remarks on Public Involvement and Education

A growing diversity of local, regional, and even national or international groups of people, each with different sets of values and goals, have expressed and acted upon an interest in the fate of Pacific salmon. The term "stakeholder" refers to any such group, which has an economic, aesthetic, religious, political, or other cultural "stake" in the fate of salmon. Because the trend of increasing public involvement is likely to continue, there will be growing influence of diverse stakeholders on the controversial issue of the hatcheries' role in rebuilding salmon populations. Fisheries educators in academia, government agencies, and non-governmental settings need to increase the opportunities for cooperative learning by all parties about the strengths and weaknesses of emphasizing technologies, such as artificial propagation, versus emphasizing natural regenerative processes in efforts to rebuild salmon populations. There is a pressing need to develop more effective ways of communicating why technological applications can only be a small part of an overall strategy to rehabilitate salmon populations and why they must be used in an ecosystemic, adaptive management context. In support of rehabilitating Pacific salmon populations, more effective communication of these points is a prerequisite to achieving societal changes that are more widespread and comprehensive than those currently taking place in the Pacific Northwest.

Acknowledgments

Special thanks go to C. Busack, K. Currens, D. Campton, and L. Miller for extensive discussions, which influenced the text on new tools for planning, operations, and monitoring of hatchery programs. Support for manuscript preparation was provided in part by the Minnesota Agricultural Experiment Station and the Minnesota Sea Grant College Program supported by the NOAA Office of Sea Grant, Department of Commerce under Grant No. NOAA-NA46-RG0101, Projects No. R/F-26, J.R. 404. This is article 21,920 of the Minnesota Agricultural Experiment Station Scientific Journal Article Series.

Literature Cited

Agricultural Biotechnology Research Advisory Committee, Working Group on Aquatic Biotechnology and Environmental Safety. 1995. Performance standards for safely conducting research with genetically modified fish and shellfish. Parts I & II. United States Department of Agriculture, Office of Agricultural Biotechnology, Documents No. 95-04 and 95-05. Washington, DC.

Allendorf, F.W. and N. Ryman. 1987. Genetic management of hatchery stocks, p.141-159. *In* N. Ryman and F. Utter (eds.), Population Genetics and Fishery Management. University of Washington Press, Seattle.

Anonymous. 1992. Supplementation in the Columbia Basin Regional Assessment of Supplementation Project report series: part I. Background, description, performance measures, uncertainty, and theory. May 1, 1992. Bonneville Power Administration. Portland, Oregon.

Boyce, M.S. 1992. Population viability analysis. Annual Review of Ecology and Systematics 1992 (23): 481-506.

Busack, C. and K. Currens. 1995. Genetic risks and hazards in hatchery operations: fundamental concepts and issues. American Fisheries Society Symposium 15: 71-80.

Carson, H. L. 1983. The genetics of the founder effect, p. 189-200. *In* C. M. Schonewald-Cox, S. M. Chambers, B. MacBryde, and L. Thomas (eds.), Genetics and Conservation. The Benjamin Cummins Publishing Co., Inc., Menlo Park, California.

Cederholm, C.J., D.B. Houston, C.B. Cole, and W.J. Scarlett. 1989. Fate of coho salmon (*Oncorhynchus kisutch*) carcasses in spawning streams. Canadian Journal of Fisheries and Aquatic Sciences 46: 1347-1355.

Cuenco, M.L., T. W.H. Backman, and P.R. Mundy. 1993. The use of supplementation to aid in natural stock restoration, p. 269-293. *In* J.G. Cloud and G.H. Thorgaard (eds.), Genetic Conservation of Salmonid Fishes. Plenum Press, New York.

Currens, K. 1993. Genetic vulnerability of the Yakima fishery project: a risk assessment. Washington Department of Fish and Wildlife. Olympia.

Currens, K.P. and C. Busack. 1995. Genetic risk assessment. Fisheries 20(12): 24-31.

Donaldson, J.R. 1967. The phosphorous of Iliamna Lake, Alaska as related to the cyclic abundance of sockeye salmon. Ph.D. Dissertation, University of Washington. Seattle.

Doppelt, B., M. Scurlock, C. Frissell, and J. Karr. 1993. Entering the Watershed. A New Approach to Save America's River Ecosystems. Island Press, Washington, DC.

Eberhardt, L.L. and J. M. Thomas. 1991. Designing environmental field studies. Ecological Monographs 61(1): 53-73.

Flick, W.A. and D.A. Webster. 1962. Problems in sampling wild and domestic stocks of brook trout (*Salvelinus fontinalis*). Transactions of the American Fisheries Society 91: 140-144.

Flick, W.A. and D.A. Webster. 1964. Comparative first year survival and production in wild and domestic strains of brook trout, *Salvelinus fontinalis*. Transactions of the American Fisheries Society 93: 58-69.

Gregory, S. and P.A. Bisson. 1996. Degradation and loss of anadromous salmonid habitat in the Pacific Northwest, p. 277-314. *In* D.J. Stouder, P.A. Bisson, and R.J. Naiman (eds.), Pacific Salmon and Their Ecosystems: Status and Future Options. Chapman and Hall, New York.

Groot, C. and L. Margolis (eds.). 1991. Pacific Salmon Life Histories. University of British Columbia Press, Vancouver, British Columbia, Canada.

Hard, J.J. 1995. Genetic monitoring of life-history characters in salmon supplementation: problems and opportunities. American Fisheries Society Symposium 15: 212-225.

Hard, J.J., R.P. Jones, Jr., M.R. Delarm, and R.S. Waples. 1992. Pacific salmon and artificial propagation under the Endangered Species Act. National Marine Fisheries Service, Technical Memorandum NMFS-NWFSC-2. Seattle, Washington.

Hill, W.G. 1981. Estimation of effective population size from data on linkage disequilibrium. Genetical Research 38: 209-216.

Hughes, R. M., and R. F. Noss. 1992. Biological diversity and biological integrity: current concerns for lakes and streams. Fisheries 17(3): 11-19.

Integrated Hatchery Operations Team. 1994a. Policies and procedures for Columbia Basin anadromous salmonid hatcheries, October 1994. Bonneville Power Administration, Portland, Oregon.

Integrated Hatchery Operations Team. 1994b. Policies and procedures for Columbia Basin anadromous salmonid hatcheries. Executive Summary, October, 1994. Bonneville Power Administration, Portland, Oregon.

Integrated Hatchery Operations Team. 1994c. Implementation plan for integrating regional hatchery policies, October 1994. Bonneville Power Administration, Portland, Oregon.

Johnson, J.I. and M.V. Abrahams. 1991. Interbreeding with domestic strain increases foraging under threat of predation in juvenile steelhead trout (*Oncorhynchus mykiss*): an experimental study. Canadian Journal of Fisheries and Aquatic Sciences 48: 243-247.

Jorde, P.E. and N. Ryman. 1995. Temporal allele frequency change and estimation of effective size in populations with overlapping generations. Genetics 139:1077-1090.

Kapuscinski, A.R. 1991. Genetic analysis of policies and guidelines for salmon and steelhead hatchery production in the Columbia River Basin, p. 117-160 *in* Hearing before the Subcommittee on Environment and Natural Resources, Committee on Merchant Marine and Fisheries, US House of Representatives, One Hundred Third Congress, First Session on Watershed Management and Fish Hatchery Practices in the Pacific Northwest, March 9, 1993. Serial No. 103-3. US Government Printing Office, Washington, DC.

Kapuscinski, A.R. 1993. Testimony on role of hatcheries in the recovery of naturally-spawning salmon populations in the Pacific Northwest, p. 107-113. *In* Hearing before the Subcommittee on Environment and Natural Resources, Committee on Merchant Marine and Fisheries, US House of Representatives, One Hundred Third Congress, First Session on Watershed Management and Fish Hatchery Practices in the Pacific Northwest, March 9, 1993. Serial No. 103-3. US Government Printing Office, Washington, DC.

Kapuscinski, A.R. 1994. A common conservation ethic: communicating the way towards conservation of aquatic biodiversity, p. 36-50. *In* D. P. Philipp (ed.), The Protection of Aquatic Biotic Diversity: Proceedings of the World Fisheries Congress, Theme 3. Oxford and IBM Publishing Company, New Delhi, India.

Kapuscinski, A.R. and L.M. Miller. 1993. Genetic hatchery guidelines for the Yakima/Klickitat Fisheries Project. Washington Department of Fish and Wildlife. Olympia.

Kline, T.C., Jr., J.J. Goering, O.A. Mathisen, P.H. Poe, and P.L. Parker. 1990. Recycling of elements transported upstream by runs of Pacific salmon: I. $\delta^{15}N$ and $\delta^{13}C$ evidence in Sashin Creek, southeastern Alaska. Canadian Journal of Fisheries and Aquatic Sciences 47: 136-144.

Kline, T.C., Jr., J.J. Goering, O.A. Mathisen, P.H. Poe, and P.L. Parker. 1993. Recycling of elements transported upstream by runs of Pacific salmon: II. $\delta^{15}N$ and $\delta^{13}C$ evidence in the Kvichak River watershed, Bristol Bay, Southwestern Alaska. Canadian Journal of Fisheries and Aquatic Sciences 50: 2350-2365.

Krueger, C.C., D.J. Decker, and T.A. Gavin. 1986. A concept of natural resource management: an application to unicorns. Transactions of the Northeast Section of the Wildlife Society 43: 50-56.

Lawson, P.W. 1993. Cycles in ocean productivity, trends in habitat quality, and the restoration of salmon runs in Oregon. Fisheries 18(8): 6-10.

Li, H.W., C.B. Schreck, C.E. Bond, and E. Rextad. 1987. Factors influencing changes in fish assemblages in Pacific Northwest streams, p. 193-202. *In* W. J. Matthews and D. C. Hein (eds.), Community and Evolutionary Ecology of North American Stream Fishes. University of Oklahoma Press, Norman.

Mason, J.W., O.M. Brynildson, and P.E. Degurse. 1967. Comparative survival of wild and domestic strains of brook trout in streams. Transactions of the American Fisheries Society 96: 313-319.

Moyle, P.B. 1969. Comparative behavior of young brook trout of domestic and wild origin. The Progressive Fish Culturist 31: 51-56.

National Research Council. 1996. Upstream. Salmon and Society in the Pacific Northwest. National Academy Press, Washington, DC.

Nehlsen, W., J.E. Williams, and J. Lichatowich. 1991. Pacific salmon at the crossroads: stocks at risk from California, Oregon, Idaho, and Washington. Fisheries (Bethesda) 16(2): 4-21.

Northwest Power Planning Council. 1992. Strategy for salmon. Volume II. Northwest Power Planning Council. Portland, Oregon.

Noss, R.R. 1990. Indicators for monitoring biodiversity: a hierarchical approach. Conservation Biology 4: 355-364.

Pollak, E. 1983. A new method for estimating the effective population size from allele frequency changes. Genetics 104: 531-548.

Reisenbichler, R.R. 1996. Genetic factors contributing to declines of anadromous salmonids in the Pacific northwest, p. 223-244. *In* D.J. Stouder, P.A. Bisson, and R.J. Naiman (eds.), Pacific Salmon and Their Ecosystems: Status and Future Options. Chapman and Hall, New York.

Richey, J.E., M.A. Perkins, and C.R. Goldman. 1975. Effects of kokanee salmon (*Oncorhynchus nerka*) decomposition on the ecology of a subalpine stream. Journal of the Fisheries Research Board of Canada 32: 817-820.

Ricker, W.E. 1954. Stock and recruitment. Journal of the Fisheries Research Board of Canada 11: 559-623.

Riddell, B.E. 1993. Salmonid enhancement: lessons from the past and a role for the future, p. 338-355. *In* D. Mills (ed.), Salmon in the Sea and New Enhancement Strategies. Blackwell Scientific Publications, Ltd., Oxford, United Kingdom.

Vincent, R.E. 1960. Some influences of domestication upon three stocks of brook trout (*Salvelinus fontinalis* Mitchell). Transactions of the American Fisheries Society 89: 35-52.

Wahl, D.H., R.A. Stein, and D.R. DeVries. 1995. An ecological framework for evaluating the success and effects of stocked fishes. American Fisheries Society Symposium 15: 176-189.

Waples, R.S. 1989. A generalized approach for estimating effective population size from temporal changes in allele frequency. Genetics 121: 379-391.

White, G.M. 1995. Urban planning for the conservation of stream ecosystems: a case study from Indianapolis, Indiana. Ph.D. dissertation, University of Minnesota, St. Paul.

Wilson, M.F. and K.C. Halupka. 1995. Anadromous fish as keystone species in vertebrate communities. Conservation Biology 9(3): 489-497.

Appendix: Outline of Genetic Guidelines for Hatchery Operations Developed by Kapuscinski and Miller (1993)

Broodstock Collection

Priorities for Choice of Donor Population, Based on Three Similarity Criteria

Choose the donor stock that, compared with the stock targeted for supplementation, shows *greatest possible similarity* in (1) genetic lineage, (2) life-history patterns, and (3) ecology of originating environment.

Restoration [target population extirpated]

Best choice—a neighboring population from an environment best meeting the three similarity criteria.

Augmentation [target population at depressed level]

Best choice—the target population.

Collecting Broodstock: Years Before Hatchery Returns Are Present

Collect an appropriate number of fish in an appropriate manner to minimize genetic differences between the hatchery and wild spawning subpopulations and, ultimately, to minimize genetic alterations of the overall population.

Unless conducting a captive broodstock program for a critically small population (see Captive Broodstock Programs), collect ≤50% of the wild spawners for broodstock:

Collect <50% if minimum size requirements can be met easily.

Appropriate rate will be affected by genetic and demographic risks.

Important and generally unknown factors (research needs) are (1) relative return rates of hatchery and wild-spawned fish, and (2) reproductive success of matings involving hatchery descendants.

Reevaluate recommendation based on initial findings.

Collect adults for broodstock so that they are an unbiased sample of the naturally spawning donor population with respect to run timing, size, age, sex ratio, and any other traits identified as important for long-term fitness.

Collecting Broodstock: Years When Hatchery Returns Are Present

Generally exclude hatchery returns from broodstock collections BUT reevaluate this practice and the entire supplementation program if hatchery returns exhibit low fitness on spawning grounds.

Captive Broodstock Programs

Consider a captive broodstock program only if the target population is critically small and extinction is probable.

Maintain captive broodstock for only one generation.

Spawning Guidelines

Within each group of adults that are "ripe" for spawning on a given day, randomize matings with respect to size and other phenotypic traits.

For all fish collected as hatchery broodstock, choose the mating scheme that will maximize effective population size.

Mating Schemes

One-by-one and nested mating schemes

Within each group of males and females that are ripe for spawning on a given day, mate one female with one male. If one sex is in excess, divide gametes from individuals of the less numerous sex to allow matings with all of the excess sex (i.e., nest two or more of the excess sex under each individual of the less numerous sex).

Diallel mating scheme

For critically small populations, consider a diallel cross on each spawning date (i.e., cross all females with all males).

In cases where male fertility is highly variable (e.g., owing to poor sperm motility), pool sperm from overlapping pairs of males (1 and 2, 2 and 3, 3 and 4, etc.) and immediately use each pooled pair to fertilize one lot of eggs.

Rearing Guidelines

Produce fish that are qualitatively similar to wild fish of the targeted population in size, morphology, behavior, physiological status, health and other ecological attributes important for fitness in the wild.

Unless conducting experiments on effects of different rearing methods, rear all fish of a given population under the same conditions.

Mix families randomly as soon as possible, taking into consideration plans to mark families.

Incubation

Provide for volitional movement of swim-up fry to feeding areas in rearing troughs, raceways, or ponds.

Simulate natural intragravel environment in incubation for these attributes: incubation densities, substrate, light, temperature, and oxygen.

Juvenile Rearing

Simulate natural juvenile rearing conditions for these attributes: rearing densities, container hydraulics, habitat complexity, feeding conditions, feeding and predator avoidance behaviors, fish size and nutritional status at release.

Release Guidelines

Number of Hatchery Fish to Release

Sum of numbers released and naturally reproduced fish (with none displaced by hatchery fish) should be no more than the carrying capacity (K) of the freshwater environment for juvenile salmon.

Life Stage at Release

Release at earliest life stage possible considering the trade-off between adult return rate and genetic risks.

Release at a life stage consistent with that exhibited at the same time by naturally reproduced fish of the target population.

Location of Release

Release fish in locations where the target population's naturally produced fish of the same life stage occur.

Release Procedures

Acclimate smolts to release site by exposing them to water from the target stream.

Allow volitional movement of smolts to the target stream.

Release pre-smolts into the target stream well in advance of smoltification.

Managing Resources with Incomplete Information: Making The Best of a Bad Situation

Robert C. Francis

Abstract

The problems inherent in managing renewable resources with "incomplete information" imply that manager's problems could be resolved with "complete information." This thinking is based on concepts of constancy, stability, and reversibility of nature. They have formed the foundation of 20th-century scientific thinking about populations and ecosystems and have provided the scientific underpinnings for both renewable resource management and the environmental movement. Recently, scientists and managers have begun to view chance and change as fundamental aspects of life and death in the natural world, and as a result, they have begun to abandon the need for a pretense of scientific certainty and devise management plans that attempt to protect the resource, and the economy that depends on it, against a range of uncertain and, in many cases, unknowable and unpredictable possibilities. Salmonid fishery development and management has often filtered out or ignored uncertainty.

Essentially, there are three fundamental ways to deal with uncertainty: (1) treat it as noise in a deterministic system, (2) explicitly incorporate it into statistical decision models, or (3) acknowledge that unknowability is an inherent property of complex systems and restructure our science accordingly. Cases are drawn from salmonid ecosystems of Alaska and the Pacific Northwest to provide the following advice to resource managers in their attempts to deal with complex and large-scale salmonid management issues. First, controlling harvest is not the only way to manage salmonid ecosystems. Second, natural disasters (e.g., El Niño) provide convenient events upon which to focus blame. Third, we should always keep in mind the historical nature of salmonid ecosystem issues. And fourth, the only way to manage resources in this context is to use management as a method to experiment with system dynamics to achieve desired policy objectives.

Introduction

The idea that there are inherent technical problems associated with managing renewable resources with "incomplete information" implies that the problems would be solved if only we had "complete information." This idea stems from what Botkin (1990) calls the machine metaphor for order in nature being applied to the scientific understanding of the dynamics of the resources being managed. In fact, concepts of constancy and stability of nature form the foundation of 20th-century scientific thinking about populations and ecosystems and have provided the scientific underpinnings for both renewable resource management and the environmental movement. Recently, these paradigms have begun to break down. Chance and change are becoming regarded as fundamental aspects of life and death in the natural world. As a result of this shift in thinking, managers have begun to abandon the need for a pretense of scientific certainty and devise management plans that attempt to protect the resource, and the economy that depends on it, against a range of uncertain and, in many cases, unknowable and unpredictable possibilities. In this paper, I show how salmon fishery development and management has, in many cases, followed the old machine metaphor, essentially filtering out or ignoring uncertainty. I then discuss some new developments in explicitly recognizing and dealing with uncertainty in the area of assessment and management of salmonid ecosystems.

Setting the Stage

In order to set the stage, I briefly look at two salmonid fishery case histories. The first has to do with the pink salmon (*Oncorhynchus gorbuscha*) fishery in Prince William Sound, Alaska, and the second with the coho salmon (*O. kisutch*) fishery off the Oregon-Washington coast. In the mid-1970s, catches and runs of pink salmon in Prince William Sound (and throughout the Gulf of Alaska) had been at an all time low for a number of years (Fig. 1A). As a result, a group of fishers and processors formed a nonprofit hatchery corporation with the idea of enhancing depressed levels of wild production. Their hope was to create runs that would provide bountiful harvests even in years when wild runs were weak. By the mid-1980s, the consortium had created the largest man-made pink salmon run in North America. The run became so large that it outgrew its market in 1990, forcing the dumping of millions of fish which could not be sold, and then crashed in 1992 and 1993. The causes of both the rapid increase and the precipitous decline are hotly debated issues. Candidates to take credit for the increase are the new hatchery system and the marine environment, and for the subsequent decline include the 1989 *Exxon Valdez* oil spill, overfishing, over-production by the hatcheries, and changes in the marine environment.

The second case history involves coho salmon adult returns for a portion of the Pacific coastal water bounded by Leadbetter Point, Washington, on the north and Monterey Bay, California, on the south (Oregon Production Index [OPI]; Fig. 1B; Brodeur 1990, Emlen et al. 1990). As a result of increased hatchery output, there is a corresponding increase in smolt production (Fig. 2). However, during the mid-1970s there was an abrupt decrease in OPI coho run size, which persisted through the end of 1986. A number of factors have been implicated in both the run size increases of the 1960s and early 1970s and decreases of the late 1970s and 1980s. These factors include changes in the marine environment, compensatory and depensatory effects of increases in hatchery production, habitat loss, and overfishing.

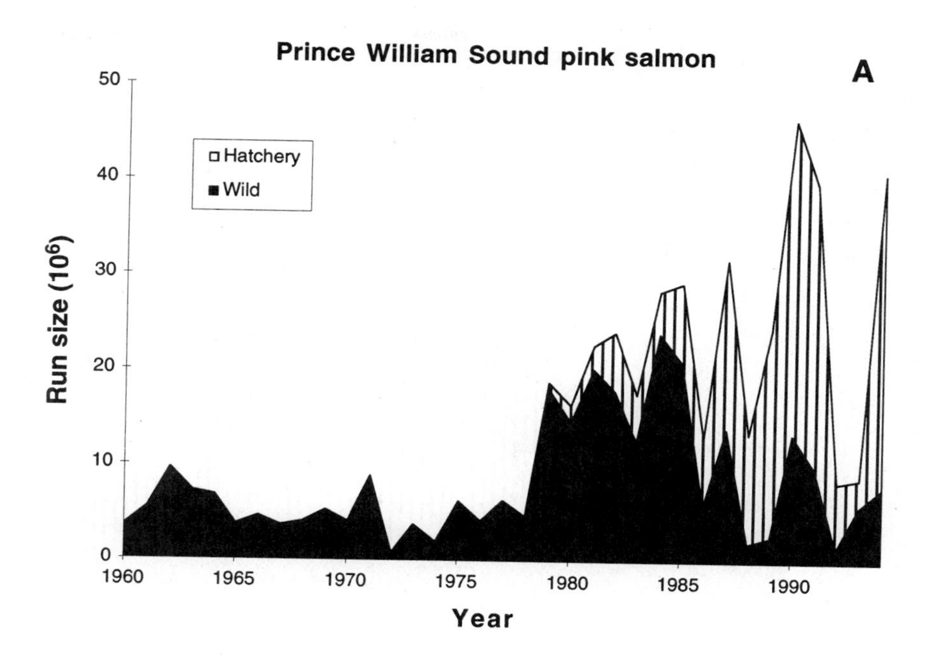

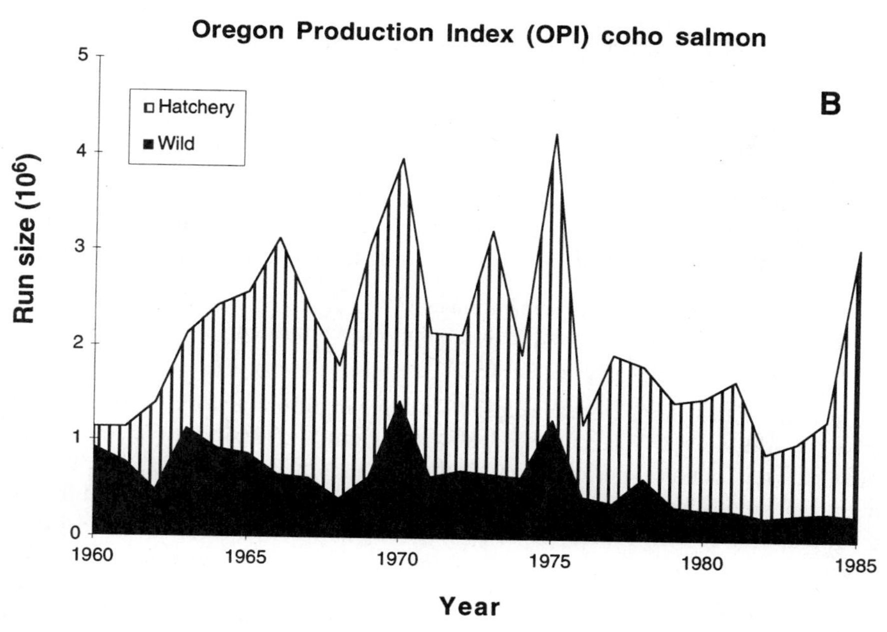

Figure 1. Prince William Sound pink salmon (A) and Oregon Production Index coho salmon (B) time series for estimated run size.

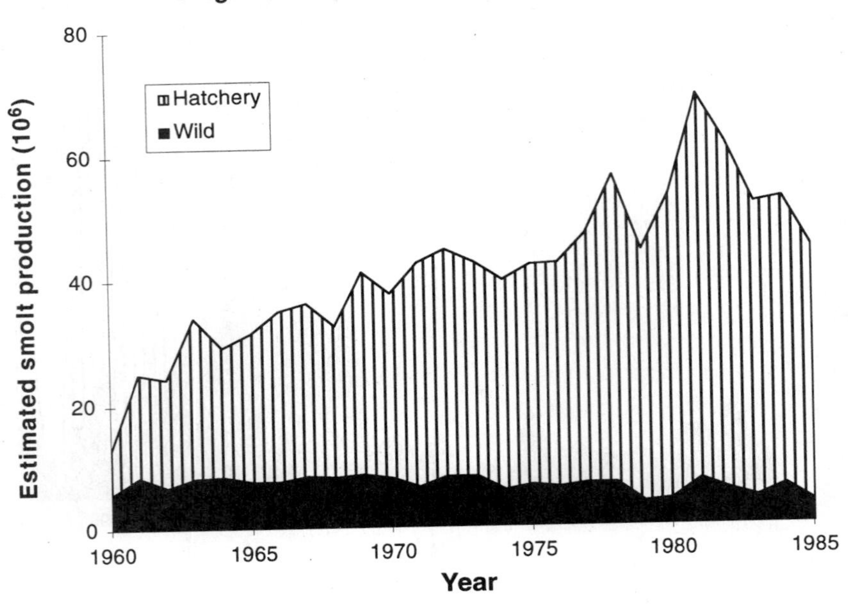

Figure 2. Oregon Production Index coho salmon time series for estimated smolt production.

The Problem

What happens when we manage resources with incomplete information? To address this question, I identify several key words and implications to provide more meaning and context to this question. I think that the real problem arises when one tries to answer the following questions:

> What do we mean by "incomplete information"?
> Would our problems be solved if we had "complete information"?
> Is this a "bad situation" or just reality?

A good deal of this stems from the clash between two different world views (Botkin 1990): the analytic world view, which tends to deal with nature by freezing it conceptually, and the rational world view, which experiences nature through motion, sound, and action. Botkin makes three important points concerning the former. First, concepts of constancy and stability of nature form the foundation of 20th-century scientific thinking about populations and ecosystems. This form of scientific thinking clearly has been around for a long time. The assumptions included in this point are that undisturbed (by man) nature remains constant and stable and that if nature is perturbed, then once the perturbation is removed nature will slowly return to its unperturbed state. Second, ecology depends on physical sciences and engineering for its theory, concepts, models, and metaphors. Therefore, the equations of physics and engineering have served to reduce the biological world to a mechanistic system. Because the analytic properties of many of these equations have used stable or equilibrium solutions to describe population and ecosystem properties and dynamics, these properties have been transformed from the equations to reality (nature) itself. Third, the machine metaphor for the balance (and order) of nature, expressed in formal mathematical models, has become the basis for renewable resource management (and the environmental movement).

This mechanical perception of nature, in which the complexity of the whole has been reduced to the sum of its independent (orthogonal) and replaceable parts, has tended to lead resource scientists and managers to focus on single causes and single effects (analogous to diagnosing the ailments of a machine) and to the belief that mechanistic nature can be re-engineered. Concepts familiar to fishery scientists and managers such as maximum sustained yield and equilibrium yield per recruit arise from the properties of analytic models used to address fishery resource issues. This analytic world view is, I believe, at the root of our scientific problems in dealing with salmon and their ecosystems. And it leads us to raise the question,

> Is there a silver bullet?

Certainly the machine metaphor encourages us to search for the silver bullet. If we can only find the one thing that controls what is wrong with the system, then we can either fix it or re-engineer the system to enable us to overcome the problem. And so, in Prince William Sound we focus on whether it is the marine environment, the oil spill, overfishing, or hatcheries that might provide us the silver bullet for pink salmon. Similar scientific treasure hunts take place with regard to OPI coho salmon. In fact, with regard to the effects of the marine environment on production, the debate has been raging for years as to whether predation on or starvation of juvenile salmonids might provide the silver bullet.

According to Holling (1993), this analytic world view has led us to do our science in a particular and constrained way. Holling (1993) calls this traditional or first-stream science and characterizes it as "disciplinary, reductionist, and detached from people, policies, and politics" (Table 1). One can see how the machine metaphor for nature pervades the first-stream science approach. For example, the methodology of sequentially testing independent hypotheses, which are either accepted or rejected, reflects this orthogonal view of nature. Management is oriented to smoothly changing and reversible conditions, and it operates under the view that one needs to know before taking action. The point is this: if only we study the problem a little harder and obtain better information, we will be able to develop the operational knowledge to effectively solve the problem. Much of this arises from the way that first-stream science has sold itself to the public. Certainty, predictability, and reducibility are words that are associated with this approach to resource science.

Reality

In reality, as Botkin (1990) eloquently points out, one of the main impediments to progress on environmental issues is that current beliefs about nature (represented by the machine metaphor and first-stream science) tend to blind us to the possibilities for constructive action, leading us to emphasize the benefits of doing nothing until we know more. In this context, Botkin (1990) makes four critical points. First, change is intrinsic and natural at most scales of time and space in the natural world. This is a difficult concept for society to accept. To abandon the belief in the constancy of undisturbed nature is psychologically uncomfortable. In addition, this concept tends

Table 1. A summary of Holling (1993) discussion of first-stream and second-stream science.

	First-stream science	Second-stream science
Assumptions	Ecosystems are inherently knowable and predictable	Ecosystems have inherently unknowable and unpredictable aspects
Approach	Science of parts Fixed scales Disciplinary Reductionist Detached from people	Science of integration of parts Multiple scales Transdisciplinary Holistic Considers humans as part of the system
Method	Experimental Independent hypothesis—accept, reject	Historical, comparative, experimental Integrates modes of inquiry Multiple sources of evidence
Goal	Explanation Prediction	Understanding
Management	Social or economic adaptation to smoothly changing, reversible conditions	Adaptive response to uncertainty and surprise
	Need to know before taking action	Yields learning

to create inherently moving targets for managers, something that they are unaccustomed to dealing with. Second, chance is a fundamental aspect of life and death. This view contradicts the idea that chance or randomness in nature arises simply from our ignorance (uncertainty), and moves on to the perception that randomness is inherent in nature. This being the case, then certain processes in nature are inherently unknowable and unpredictable, and must be dealt with accordingly. The idea of nature throwing dice becomes a reality. Third, the machine, with its internal controls, is not the right metaphor for nature. One of the real problems with applying the machine metaphor in trying to solve natural resource issues is that it tends to separate man from nature. Man is a user and manipulator rather than a member and citizen of the natural world (Table 1). We become an independent entity and, in fact, one not subject to study under the umbrella of natural science. Fourth, we must begin to accept nature for what we are able to observe it to be, not what we wish it to be.

Ecosystems are complex entities whose structure seems to be determined by the natural frequencies of their component parts (Allen and Starr 1982, O'Neill et al 1986). In other words, what seems to be important in determining order at the ecosystem level are rates and relationships rather than entities and their quantities. In addition, as is insightfully pointed out by Allen (1985), ecosystems, as thermodynamically open, dissipative systems, seem to demonstrate a capacity for self organization. In this regard, certain aspects of ecosystem structure form within the system. They obtain the energy for their formation from the dissipation of substances or energy flowing through the system. In many cases, what might appear to be system equilibrium is really a state of dynamic stability far removed from equilibrium. When this tension or disequilibrium reaches a certain intensity, "then many amazing and surprising things can happen" (Allen 1985). The system is in a state of self-organized criticality (Bak and Chen 1991, Waldrop 1992) where a mechanism that leads to minor events can also lead to major events. In this way, systems can shift from one state to another, causing quite drastic alterations in system structure even though these shifts may be triggered by relatively small changes in system fluxes.

Two Alaskan examples illustrate a system's ability to change drastically from one condition to another. Prince William Sound pink salmon run sizes (both wild and hatchery) show a radical reorganization in the early 1990s, with a sudden, precipitous decrease in the 1992 run (Fig. 1A). This continued through 1993 and then the runs seem to have increased equally suddenly in 1994. A similar sudden and unexplained increase in the production of the Egegik sockeye salmon (*O. nerka*) production system of the Bristol Bay region of western Alaska occurred in the early 1990s (Fig. 3). Both of these shifts were unexpected, very sudden and, seemingly, virtually impossible to explain. In addition, salmon production in vast regions of the Northeast Pacific Ocean seems to undergo rapid shifts in response to decadal-scale climate forcing (Francis and Sibley 1991. Francis and Hare 1994). This phenomenon is reflected in the sudden increase in Prince William Sound pink salmon production and decrease in OPI coho salmon production in the late 1970s (Fig. 1).

And so, it can be the tension within a system caused by relative fluxes of energy and material that creates what we call order. The greater the tension, the greater the capacity is for rapid reorganization. This implies that order is emergent and many important ecosystem processes are irreversible. Therefore, history plays a vital role in determining the nature of order in an ecosystem.

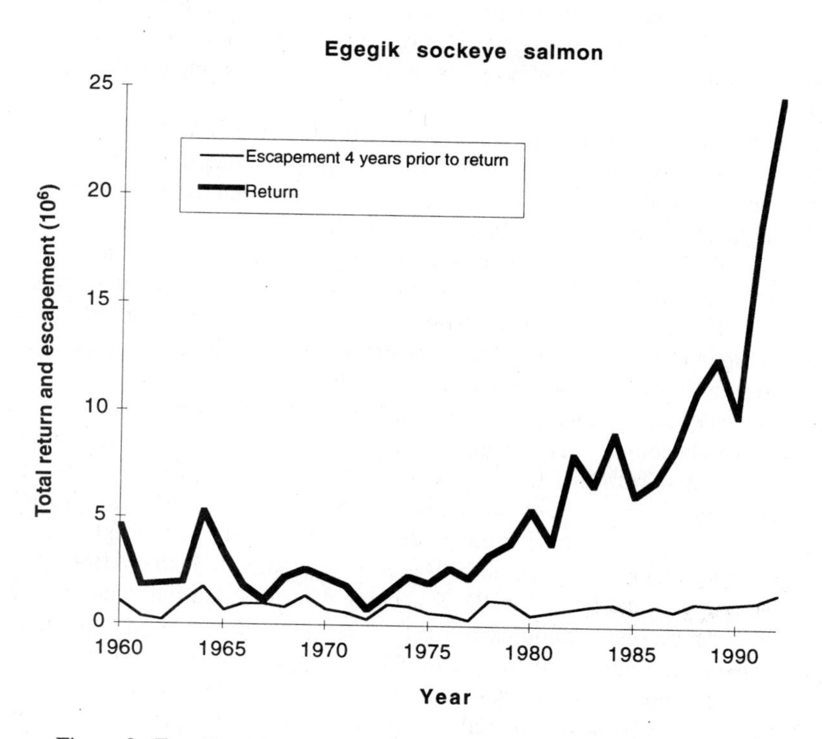

Figure 3. Egegik sockeye salmon total return and escapement 4 years earlier.

Reality thus leads us to answer our earlier question:

There is no silver bullet!

To put it simply, nature is not orthogonal. System properties do not necessarily result from the predictable interactions of independent components. In addition, historical contingency is an important aspect of natural processes. As Gould (1989) so clearly points out, a historical explanation of a natural process does not rest solely on direct deductions from the laws of nature; it also takes into account an unpredictable sequence of antecedent states where any major change in any step of the sequence would have altered the final result.

Returning once again to Holling (1993), there is another stream of science more appropriate for approaching some of these complex ecosystem issues represented in the salmonid world by Prince William Sound and OPI. The premise of this second-stream science is that knowledge will always be incomplete and, in fact, certain aspects are inherently unknowable and unpredictable (Table 1). In terms of science for management, uncertainty and surprise must become an integral part of a sequence of actions. Managers will incorporate historical information on how the system responded. In addition, learning becomes one of the goals of management.

Three Ways to Deal with Uncertainty

I propose three fundamental ways to deal with uncertainty. First, treat it as noise in a deterministic system. Second, explicitly incorporate it into statistical decision models. Third, acknowledge that unknowability is an inherent property of complex systems and restructure our science accordingly.

In the first case, uncertainty is treated as noise in a fundamentally deterministic system. One can either filter it out or systematically investigate it so that the uncertainty is removed and the processes accounting for it are explicitly incorporated into the deterministic representation of the system being studied or managed. In essence, we treat the ecosystem as a small number system (O'Neill et al. 1988)—a system with a small number of components, simple interactions, and a deterministic equation for the dynamics of each component and its interactions with the rest of the system. This is the realm of first-stream science. Everything is orderly, consistent in interaction, and fundamentally deterministic. Uncertainty is not an inherent property of the system. It simply results from a lack of information or inability to measure.

In the second case, uncertainty is explicitly incorporated into a framework for making decisions. In essence, this becomes a case for applying statistical mechanics. The ecosystem is treated as a large number system (O'Neill et al. 1988)—a system with a large number of independent components with random interactions (disorganized complexity). In this case, the only way to explicitly incorporate system behavior into a decision framework is through the analysis of average behavior. For example, the classical Bayesian approach (Sainsbury 1991, Hilborn and Walters 1992, McAllister and Peterman 1992) is being resurrected in fisheries science.

Finally, when dealing with complex ecosystem issues, there comes a point when we must acknowledge true uncertainty or indeterminacy. Certain aspects of system dynamics are inherently unknowable and must be regarded as such. The ecosystem is treated as a medium number system (O'Neill et al. 1988)—a system with too few parts to average their behavior reliably and too many parts to manage or represent each separately with its own equation. I think that Holling (1993) states the case beautifully:

> Not only is the science incomplete, the system itself is a moving target, evolving because of the impacts of management and the progressive expansion of the scale of human influences on the planet. . . . The essential point is that *evolving systems* require policies and actions that not only satisfy social objectives but, at the same time, also achieve continually modified understanding of the evolving conditions and provide flexibility for adaptation to surprises. Science, policy, and management then become inextricably linked. (Emphasis added.)

In this context, uncertainty and surprises are expected and, as best possible, planned for. In particular, system responses are fundamentally nonlinear. They demonstrate those properties of nonlinear systems that have come to be termed chaotic (e.g., multiple quasi-stable states and abrupt discontinuities). As such, policies that rely on slow and smooth change and reversible conditions simply will not be useful. The most effective way to operationally deal with uncertainties that fall into this category is to invoke a second-stream approach to science and management (Table 1). In particular, management needs to pay as much attention to learning as it does to optimal properties of whatever commodity is being managed. This is clearly the realm of active adaptive management (McAllister and Peterman 1992, Holling 1993,).

Clearing the Stage

In returning to the two salmon examples, can one determine where they fit in this quagmire of uncertainty and indeterminacy? Most of the major salmon issues of today involve complex interactions of physical, biological, and social components of ecosystems, all of which have evolutionary characteristics. If one examines the Prince William Sound pink and OPI coho salmon issues, one sees that they both display nonlinear characteristics. First, they are influenced by environmental variability occurring at multiple (seasonal, annual, decadal . . .) scales. As Francis and Sibley (1991) and Hare and Francis (1994) have clearly pointed out, salmon production in both the Alaska Current and California Current systems is strongly influenced by decadal-scale atmospheric and oceanic climate variability. Simultaneously, fisheries scientists and managers seem to have been studying annual scale variability in salmon production for hundreds of years. Clearly, managers are asked to confront the resultant effects of causes occurring at multiple time scales. Second, both of these issues have been strongly influenced by technological intervention (e.g., hatchery, harvest, and habitat impacts). All have had major impacts on salmon production in Pacific Northwest. In the case of OPI coho salmon, Francis and Brodeur (in press) propose new approaches to management that will interface environment and technology. Finally, uncertainty and indeterminacy have developed in areas that have been strongly influenced by natural and man-made disasters (e.g., the 1989 *Exxon Valdez* catastrophe in Prince William Sound and a series of strong El Niño events [in particular 1982–83] in the case of OPI coho).

Thus, the foregoing issues raise a number of (seemingly) unanswerable questions: What is the effect of density-dependence on the production of wild salmon populations and how has the introduction of large numbers of hatchery smolts influenced the abilities of the wild populations to sustain observed levels of production (see Figs. 1 and 2)? How does one compare and contrast the effects of the marine environment and artificial enhancement on the total production of salmonids of a particular region? As has been previously indicated, the marine environment caused major reorganizations of the salmonid production systems of the Northeast Pacific Ocean in the late 1970s, leading to large increases in Alaska salmon production and decreases in Pacific Northwest salmon production (Francis and Sibley 1991, Francis and Hare 1994, Hare and Francis 1994). For example, Alaska pink salmon production increased before large-scale hatchery production came on-line (Fig. 4A). However, the continuing increase in Alaska pink salmon fishery production in the late 1980s seems to track increases in hatchery output. Then there is the crash of 1992 (Fig. 4A). Pacific Northwest coho salmon show an equally complicated response. During the 1960s and early 1970s, increases in fishery production tracked the increases in hatchery output (Fig. 4B). Then, in the late 1970s, the marine environment experienced a strong El Niño Southern Oscillation event and a continuing and unexplainable decrease in coho salmon production into the late 1980s and early 1990s.

In summary, what insights might these examples provide to resource managers in their attempts to deal with complex and large-scale salmon management issues? First, controlling harvest is not the only way to manage salmonid ecosystems. It appears that, both in Prince William Sound and in the Pacific Northwest, hatchery releases remain virtually uncontrolled. Francis and Brodeur (in press) suggest that current levels of (OPI coho) hatchery releases far exceed those necessary to maximize total catch or realize increases in wild adult production. In fact, they further suggest that "the more one increases wild spawning potential, the more one must reduce hatchery production in order to realize that potential in terms of wild adults" (Francis and Brodeur in press).

Second, it seems that interested parties, be they fishery managers, harvesters, or conservationists, too often focus on man-made and natural disasters as convenient causes of fishery declines. There is little evidence that the 1989 *Exxon Valdez* oil spill or the 1982–83 El Niño had significant long-term effects on salmon production in Prince William Sound and the Pacific Northwest, respectively. Disasters provide convenient scapegoats upon which to place blame.

Third, we should always keep in mind the historical nature of salmonid ecosystem issues. Promises such as those made by the Northwest Power Planning Council to double the salmon runs in the Columbia River over a relatively short period of time ignore history and are doomed to failure. Salmon runs may double in the Columbia River Basin. However, it is more likely that this will result from some sudden shift in climate or changes in production, both of which have very little to do with the massive bioengineering effort currently being undertaken in the region.

Fourth, the only way to manage resources in this context is to use management to experiment with system dynamics to achieve desired policy objectives. We will only understand the effects of large-scale hatchery programs on the production of wild salmonids if we manage these programs in an actively adaptive manner. Management must be willing to operate at the appropriate temporal and spatial scales of the ecosystem being managed. If one is interested in evaluating the effects of water flow or spill over a dam on downstream survival and, ultimately, the production of returning adults, then experiments must be designed at the watershed level. Experiments at individual dams will simply disappear as noise within the overall system, particularly if they are performed over relatively short periods of time.

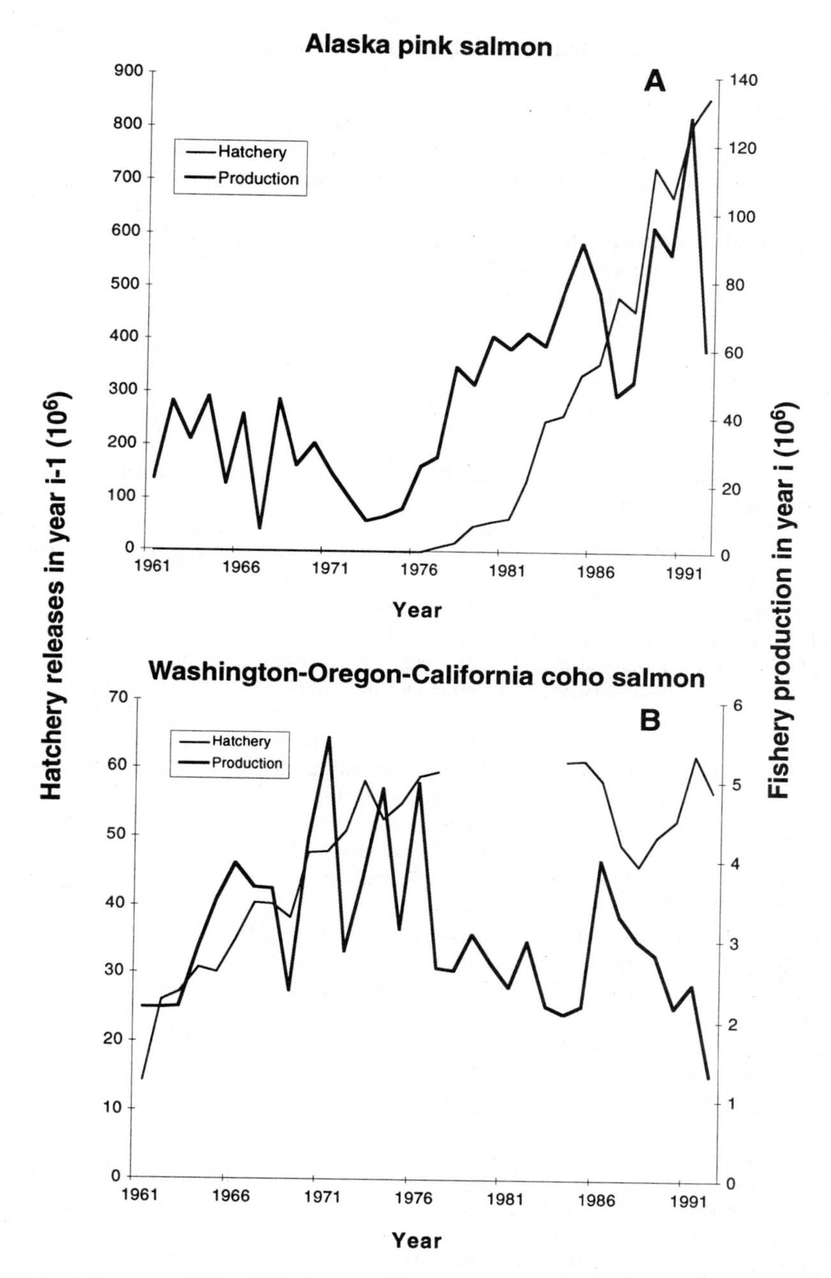

Figure 4. Alaska pink salmon (A) and Washington-Oregon-California coho salmon (B) fishery production and hatchery releases.

In closing, I return to the thoughts of Holling (1993). He says that we need to continue developing new science and management which openly acknowledge indeterminacy, unpredictability, and the historical nature of resource issues:

> The problems are therefore not amenable to solutions based on knowledge of small parts of the whole, nor on assumptions of constancy or stability of fundamental relationships, ecological, economic or social. . . . Therefore, the focus best suited for the natural science components is evolutionary, for economics and organizational theory is learning and innovation, and for policies is actively adaptive designs that yield understanding as much as they do product (Holling 1993).

Literature Cited

Allen, P.M. 1985. Ecology, thermodynamics, and self-organization: Towards a new understanding of complexity, p. 3-26. *In* R.E. Ulanowicz and T. Platt (eds.), Ecosystem Theory for Biological Oceanography. Canadian Bulletin of Fisheries and Aquatic Sciences 213.

Allen, T.F.H. and T.B. Starr. 1982. Hierarchy: Perspectives for Ecological Complexity. University of Chicago Press. Chicago.

Bak, P. and K. Chen 1991. Self-organized criticality. *Scientific American* (January): 46-53.

Botkin, D.B. 1990. Discordant Harmonies. Oxford University Press, New York.

Brodeur, R.D. 1990. Feeding ecology of and food consumption by juvenile salmon in coastal waters, with implications for early ocean survival. Ph.D. dissertation, University of Washington. Seattle.

Emlen, J.M., R.R. Reisenbichler, A.M. McGie and T.E. Nickelson. 1990. Density-dependence at sea for coho salmon (*Oncorhynchus kisutch*). Canadian Journal of Fisheries and Aquatic Sciences 47(9): 1765-1772.

Francis, R.C. and R.D. Brodeur. In press. Production and management of coho salmon: A simulation model incorporating environmental variability. Canadian Journal of Fisheries and Aquatic Sciences.

Francis, R.C. and S.R. Hare. 1994. Decadal-scale regime shifts in the large marine ecosystems of the Northeast Pacific: a case for historical science. Fisheries Oceanography 3(4): 279-291.

Francis, R.C. and T.H. Sibley. 1991. Climate change and fisheries: What are the real issues? Northwest Environmental Journal 7: 295-307.

Gould, S.J. 1989. Wonderful Life: The Burgess Shale and the Nature of History. W.W. Norton & Co., New York.

Hare, S.R. and R.C. Francis. 1995. Climate change and salmon production in the Northeast Pacific Ocean, p. 357-372. *In* R.J. Beamish (ed.), Climate Change and Northern Fish Populations. Canadian Special Publication of Fisheries and Aquatic Sciences 121.

Hilborn, R. and C.J. Walters. 1992. Quantitative fisheries stock assessment: Choice, dynamics and uncertainty. Chapman and Hall, New York.

Holling, C.S. 1993. Investing in research for sustainability. Ecological Applications 3(4): 552-555.

McAllister, M.K. and R.M. Peterman. 1992. Experimental design in the management of fisheries: A review. North American Journal of Fisheries Management 12: 1-18.

O'Neill, R.V., D.L. DeAngelis, J.B. Waide, and T.F.H. Allen. 1986. A Hierarchical Concept of Ecosystems. Princeton University Press, New Jersey.

Sainsbury, K.J. 1991. Application of an experimental approach to management of a tropical multispecies fishery with highly uncertain dynamics. International Council for the Exploration of the Sea Marine Science Symposia 193: 301-320.

Waldrop, M.M. 1992. Complexity: The Emerging Science at the Edge of Order and Chaos. Touchstone, New York.

Is Ecological Risk Assessment Useful for Resolving Complex Ecological Problems?

Robert T. Lackey

Abstract

Risk assessment has been suggested as a tool to help manage Pacific salmon (*Oncorhynchus* spp.) and solve other complex ecological problems. Ecological risk assessment is usually defined as the process that evaluates the likelihood that adverse ecological effects are occurring, or may occur, as a result of exposure to one or more stressors. The basic concept, while straightforward, is difficult to apply to all but the simplest ecological problems. Strong reactions, both positive and negative, are often evoked by proposals to use risk assessment. Ecological problems that might be addressed by risk assessment include how to accomplish the following: (1) estimate objectively the condition or health of ecological resources such as Pacific salmon; (2) reduce the cost of regulations and policies; (3) focus public and private expenditures on solving the most important priorities; (4) describe and incorporate uncertainty in decision analysis and public choice; (5) provide technical information in ways that help move beyond political gridlock; and (6) democratize the decision-making process. When applied to relatively simple ecological problems (chemical toxicity being the most common), risk assessment is popular. There are many vigorous supporters, particularly among scientists, administrators, and politicians; however, there are also critics. The intellectual history of the risk assessment paradigm does not follow a neat, linear evolution. A formidable problem in many risk assessments, and especially for complex questions such as managing salmon, is selecting the ecological component or system that is considered at risk; this selection is entirely social and political, but estimating the actual risk is technical and scientific. Defining what is at risk must be resolved within the political decision-making framework or the results of the risk assessment will be of limited utility. Use of risk assessment for the Pacific salmon problem would be difficult politically unless the boundaries of the assessment were extremely narrow. However, narrowly defining the salmon problem would make the results of the risk assessment of limited relevance in decision making. For Pacific salmon, ecological risk assessment will be of limited use except for policy problems defined by fairly narrow technical boundaries or constrained by limited geographic scope.

Introduction

Increasingly, there are calls for the use of risk assessment to help solve complex ecological policy problems. The Pacific salmon (*Oncorhynchus* spp.) situation—specifically, to reverse the decline of salmon in the Pacific Northwest—is a typical example of a complex ecological problem. The basic concept underlying risk assessment is relatively straightforward: Risk is something that can be estimated (i.e., risk assessment). In turn, that estimate can be used to manage the risk (i.e., risk management). Ecological risk assessment is usually defined as "the process that evaluates the likelihood that adverse ecological effects are occurring, or may occur, as a result of exposure to one or more stressors" (US Environmental Protection Agency [USEPA] 1992a, 1993). Extensive analyses of the options and procedures to conduct risk assessment are available for human health and ecological problems (National Research Council [NRC] 1983, 1993).

The basic concepts of risk assessment may be simple, but the jargon and details are not. Risk assessment (and similar analytical tools) is a concept that can evoke strong reactions whenever it has been used. At the extreme, some have even concluded that use of risk assessment in human health decision making is "premeditated murder" (Merrell and Van Strum 1990). A number of philosophical and moral reasons for such strong reactions are discussed in detail elsewhere (Newkirk 1992, Lackey 1994) but generally they are either concerns (1) that the analysis (risk assessment) and decisions (risk management) accept the premise that people will die to achieve the desired net benefits, or (2) that the process of risk assessment places too much power with technocrats.

Reaction to ecological risk assessment may be less harsh than reaction to risk assessment applied to human health problems, but even with ecological issues, strong positive and negative responses occur. Several legislative bills (e.g., Environmental Risk Reduction Act) have been introduced in the United States (US) Congress mandating that federal agencies use risk assessment to set priorities and budgets. Prestigious scientific panels have made similar recommendations (NRC 1993). Popular and influential publications argue for a risk assessment approach (Gore 1992). Alternatively, some conclude that risk assessment is a disastrous approach, one that is "scientifically indefensible, ethically repugnant, and practically inefficient" (M. O'Brien, Univ. Oregon, Eugene, unpubl. ms.).

Still, risk assessment has been widely used to link environmental stressors and their ecological consequences. The risks associated with chemical exposure are the typical concern. Quantifying the risk of various chemicals to human health is a logical outgrowth of risk assessment as applied in the insurance field. Over the past 20 years, a body of procedures and tools has become widely used for environmental risk assessment for human health (NRC 1983, 1993). Risk assessment applied to ecological problems is more recent but has also focused primarily on chemicals, with animals used as surrogates for "ecological health." The resulting literature is massive and is reviewed elsewhere (Cairns and Niederlehner 1992; Suter 1993; USEPA 1991, 1993). Most current and past applications of risk assessment to ecological problems deal with the toxicity of chemicals. Efforts to broaden the application of the concept to stresses beyond chemicals are summarized in several reports (USEPA 1990a, b; USEPA 1992b).

Adapting the risk paradigm from assessing insurance risks to assessing human health risks to assessing ecological risks has not been simple (Douglas and Wildavsky 1982, Lackey 1994). Some view ecological risk assessment merely as using new labels for old ideas. Whether ecological risk assessment will actually improve decision making and ultimately protect ecological resources is still unclear.

The Problem to Solve

Risk assessment is just beginning to be applied to the Pacific salmon problem and its utility has yet to be demonstrated. Massive scientific literature is available on Pacific salmon and the effects of fishing, hatcheries, logging, agriculture, predation, habitat alteration, etc. (see Stouder et al. 1996 and papers therein), but it is not particularly helpful in placing the "problem" in a risk assessment context. Nor is there any shortage of articles dealing with policy issues concerning salmon. In spite of this massive literature, defining risk for a complex ecological problem such as that for salmon is typically difficult. Eight general characteristics and problems must be considered. Because of very limited uses of risk assessment for Pacific salmon, experiences from elsewhere will have to be extrapolated for possible relevance.

Boundaries

First, defining the boundaries of ecological risk assessment for complex problems is never an easy task. This is the most important challenge that would face anyone using risk assessment on salmon problems. A problem can be defined along narrow, biological grounds similar to the approach taken by Nehlsen et al. (1991) or in more expansive terms such as ecosystem health (Rapport 1989). There is no consensus algorithm for deciding on the problem definition. In the successful uses of risk assessment, the problem has been defined in very narrow terms by legislation (i.e., registration of potentially toxic chemicals). If the risk problem is defined in narrow technical terms, the analysis can be accomplished relatively easily, but the result will often lack political or broader ecological relevance. For example, if the problem for salmon is limited to purely "fish" factors, such as the maintenance of all salmon stocks, then the other competing tradeoffs such as benefits from competing decisions and policy for agriculture, electricity generation, housing, forest use, and transport will not be included. If the risk problem is defined in more general, inclusive terms, the analysis may not be technically achievable. If the "entire" salmon problem is included, how do we describe the relationship between agricultural policy, for example, and salmon populations except in the most general terms? These two extremes require a tradeoff between technical tractability and political relevance (see Rubin 1989 for examples of this problem). It is important to determine precisely what problem needs to be solved and to clearly and formally reach consensus among the affected parties. Otherwise, additional technical analysis will be of limited value in resolving political choices.

Ecological Condition

Second, a basic ecological question that could theoretically be answered is to reach consensus on the condition of resources. After all, the effectiveness of policies, management decisions, and environmental protection programs must be measured against the benchmark of changes over time in resource conditions. If the measure of risk will be defined in terms of salmon, what is their condition? Often, depending on whom one believes, ecological resources are in terrible shape and becoming worse, or they are not in bad shape at all and improvements have been made. For example, look at the debate over biological diversity. Some contend that biological diversity is a measure of ecosystem health. A number of scientists contend that species are being lost at a record rate and the problem is at crisis levels (Ehrlich and Wilson 1991, Karr 1992, Angermeier and Williams 1993). Others challenge this view and point to the fairly recent (Pleistocene) major North American extinctions of the mastodon, mammoth, giant sloths, giant armadillos, giant beavers, American camel, American horse, and the big tigers and wolves, all lost to hunting, climate change, or both (Lewis 1992). Simon (1986) called into question much of what passes for rigorous analysis of species loss whereas Ehrlich and Wilson (1991) came to the opposite conclusion. The process of risk assessment, if done properly, forces the analyst and the participants to explicitly define ecological condition in terms of what is most desired.

Cost and Efficiency

A third problem that might be effectively addressed by risk assessment is the need to reduce the cost of environmental protection. Rules and regulations designed to protect the environment have cost the US$1 trillion since 1970, and cost ~$115 billion annually in the US according to the US General Accounting Office (1992). Is this money well spent? Could the same environmental benefits be achieved for less money? Some maintain that the cost of complying with environmental regulations, for example, could be reduced by a third to a half (Repetto and Dower 1992). There is no clear answer in this debate over how effectively money is being spent to protect the environment because there is no consensus on the condition of the environment.

Priorities

A fourth problem that might be solved by risk assessment is determining priorities for protecting the environment. Are public and private resources being used to solve the right ecological problems? What is the relative risk of various threats to the environment? For example, how are decisions made to save Redfish Lake sockeye salmon (*O. nerka*) by allowing the bull trout (*Salvelinus confluentus*) to disappear? Some argue that risk assessment should be used to totally set the regulatory and research priorities (Shifrin 1992). Others argue that scientists and the media have vastly overstated the risks of many chemicals to the environment (Whelan 1993). Whether your political rallying cry is "Back to the Pleistocene" or "Save people, not owls," there is little consensus on even the most critical risks to ecosystems.

Scientific Uncertainty

Fifth, uncertainty is difficult to address in all types of ecological analysis as well as to present to decision makers in an understandable way. As an illustration of analytical uncertainty, Hall (1988) evaluated several important and commonly used ecological models and the data that were offered to support their validity. While the models evaluated made intuitive sense, the data to support the models were weak. Others have looked at uncertainty and its effect on natural resource management problems (Ludwig et al. 1993) and concluded that classical deterministic models do not work in practice because uncertainty, incomplete information, and the dynamics of bureaucracies undermine management effectiveness. Use of classical population models to manage natural resources does not work well in practice because chaos, natural variation in ecosystems, and the unorderliness of natural processes and events make predictions unreliable for all but the most simplistic questions (Botkin 1990). For example, fish stock-recruitment models do not work well for management purposes in most situations.

Winners and Losers

Sixth, aside from the difficulties of uncertainty and the use of models, there is the issue of winners and losers in society. The policy questions surrounding the salmon issue immediately evolve to winners and losers. The question of who receives the benefits and who pays the costs is a major debate in all natural resource and environmental policies and regulations. As with the salmon problem, the benefits (and costs) of potential management and regulatory decisions are not soundly quantified. Risk assessment is often advocated as a tool that will force a clear statement of the benefits and costs. The decision of which option to adopt may not be easy, but at least (theoretically) the ecological consequences are clear within the bounds of uncertainty previously discussed.

Analytical Tools

Seventh, analytical tools are needed to help move beyond political gridlock. The salmon controversy is an excellent example. How much of that problem is analytical and how much is due to fundamental differences of opinion on what is most important to society? Major environmental change will almost certainly result in "violent conflict" within and between societies, and will likely become more severe as competition for natural resources intensifies (Homer-Dixon et al. 1993). Causes of gridlock are often exemplified by NIMBY (not in my backyard) or in the extreme case, BANANA (build absolutely nothing anywhere near anyone). Risk assessment, if executed properly, might at least create a framework for the debate.

Who Decides

Eighth, are environmental decisions driven by the will of the majority or by an elite (politicians and technocrats) that is out of touch with that majority? Control of the decision-making process may be exerted by controlling jargon (P. Loge, US Congressional staff, Tempe, Arizona, unpubl. ms.). Those who do not speak or understand the jargon are usually excluded. A formal, open analytical framework such as risk assessment might democratize the process, which, the argument goes, is now run by professionals with "after the fact" input from the public. There is a virtual cottage industry supported by attempts to "solve" the salmon problem. Each interest group, organization, and agency is represented by professionals who speak the same technocratic language.

Risk Assessment in Practice

In spite of the difficulties of defining problems in complex ecological policy questions, the use of risk assessment to help solve ecological problems is widely supported. The closest application to the salmon problem is the 1993 "Forest Summit" held in the Pacific Northwest to help resolve the timber versus spotted owl impasse; it was, in part, based on an implicit but highly simplified risk assessment framework. The approach to analyzing the spotted owl versus late successional forest issue is sometimes proposed as the model to follow for risk assessment with the salmon issue. Legislation being debated in the US Congress would mandate use of risk assessment by federal agencies for many problems. Clearly, many people think that risk assessment is a valuable tool and should be used extensively in solving ecological problems.

However, there is a vocal group of critics debating the use of risk assessment for ecological problems. They argue that risk assessment and risk management are essentially a triage approach: deciding which ecological components will be "saved" and which will be "destroyed." The theme of "biospheric egalitarianism" is a mindset that makes risk assessment a real anathema. Many risk assessment critics have a strong sense of technophobia and often view mainstream environmental organizations as co-opted by industrial or technocratic interests (Lewis 1992).

Risk assessment is also challenged from a different, more utilitarian perspective. The assertion is that, while the concept of risk assessment is sound, the process of risk assessment is often controlled by scientists and others who have political agendas different from the majority of people. Critics contend that "risk assessors" use science to support their position under the guise of formal, value-free risk analysis. Risk assessment thus has the trappings of impartiality, but is really nothing more than thinly disguised environmentalism (or utilitarianism). The apparent lack of credibility and impartiality of the science (and risk assessment) underlying the policy debates over acid rain, stratospheric ozone depletion, global climate change, and loss of biological diversity is often offered as an example of how science has allegedly been misused by scientists and others to advocate political positions (see Limbaugh 1992 for a sense of the concern).

Risk assessment has historically been separated from management (Ruckelshaus 1985). Such separation requires that scientists play clearly defined roles as technical experts, not policy advocates; these distinctions are blurred when scientists advocate political positions (Lewis 1992, Ray 1992). Further, some critics charge that scientists who use their position to advocate personal views are abusing their public trust. The counterargument is that scientists, and all individuals for that matter, have a right to argue for their views and, as technical experts, should not be excluded simply because of their expertise. Others conclude that the execution of the scientific enterprise is value-laden and therefore partially a political activity. Rather than attempting to be solely scientifically objective, a scientist should also be an advocate (O'Brien 1993). Either way, the role of the analyst must be clear to everyone using the results.

History of the Paradigm

Neither risk assessment nor any other commonly used management tool is completely new, but draws on earlier tools and shares some of the core principles. For example, both assessment and management are based on the fundamental premise that all benefits are accruable to man. This is a utilitarian approach and a necessary assumption in all the models or paradigms that follow. "All benefits are accruable to man" encompasses the fact that society might choose to protect wilderness areas that few visit, preserve species that have no known value to man, or preserve natural resources for their scenic beauty (Lackey 1979, Johns 1990, Foster 1992). Benefits may be either tangible (fish yield, tree harvest, camping days, etc.) or intangible (pristine ecosystems, species preservation, visual beauty, etc.). It is easy to jump past the fundamental premise of a utilitarian assumption, but much of the political debate revolves around the issue of whether a person operates with a utilitarian world view or an ecocentric (or other, usually religiously based) world view. It is not a trivial difference. However, in practice the split between those with utilitarian and ecocentric (or other) world views is not complete; most of us manifest features of both (Herzog 1993).

The multiple-use model of managing natural resources has been the basic paradigm in North America during this century. Popularized early in this century by Gifford Pinchot and others, it has been used extensively and widely in fisheries, forestry, and wildlife (Pinchot 1947). The idea is simple: there are many benefits that come from ecological resources (commodity yields, recreational fishing and hunting experiences, outdoor recreational activities, ecosystem services such as water purification, etc.) and the mix of outputs needs to be managed to produce the greatest good for the greatest number of people over a sustained period of time (Callicott 1990, 1991; Smith and Steel 1996). The concept is straightforward and works well if there is a high degree of shared values among the public.

A number of variations in the multiple-use model arose over the middle years of this century: maximum sustainable yield, maximum equilibrium yield, and optimum sustained yield. Widely used in teaching and management, these concepts have dominated mainstream professional thought and practice through current times (Baranov 1918, Graham 1935, Ricker 1954, Beverton and Holt 1957, Roedel 1975, Larkin 1977). As with all natural resource management paradigms and goals, none of these evolved in a linear manner (Walters 1986, Barber and Taylor 1990). The basic idea is that commodity yields (fish, trees, wildlife) could be produced annually from "surplus" production and could be continued in perpetuity with sound management. All these models suffered from the problem of a heterogeneous public with conflicting demands and an inability, in fisheries management at least, to control harvest pressure. Even today, there is still a struggle to control fishing pressure in politically acceptable and managerially efficient ways (e.g., individual transferable quotas).

Scientific management is a related management paradigm that includes operations research, management by objectives, optimization, linear programming, artificial intelligence, and other mathematical procedures (Lackey 1979, Sokoloski 1980). There are many outputs from ecosystems, both commodity and non-commodity, and these can (must) be measured and the integrated output optimized. The outputs are selected by experts, who then use mathematical tools to quantify and evaluate various combinations of outputs. Input from the public was not particularly important because there was a "correct" optimal set of decisions to maximize output. The natural resource professional is dominant in the process. The view that "if politicians and the public would just stay out of the process, we professionals would manage natural resources just fine" is characteristic of scientific management. There are many examples of the collapse of fisheries based on following this general management approach.

Ecosystem management, including permutations such as watershed management (Likens 1992, Naiman, 1992) and "new forestry" (Swanson and Franklin 1992, Slocombe 1993), has become popular in the past decade. Both ecosystem management and watershed management have ambiguous definitions such as the popular wall poster for ecosystem management: "Considering All Things." Usually other concepts, such as biological diversity, are embedded in ecosystem management, although biological diversity is an ill-defined concept in its own right (Cairns and Lackey 1992). For example, in our quest to restore salmon stocks, should we eradicate predators such as squawfish (*Ptycocheilus oregonensis*) and competitors such as walleye (*Stizostedion vitreum*) or do we restore ecosystems (habitat) to some desired state and let nature take its course? Does ecosystem management mean we optimize this mix of species? These and myriad others are policy questions and must be explicitly answered regardless of which management approach is used. They must be answered as policy questions, not scientific ones. Advocates of ecosystem management often see it as a fundamental shift in management and assessment thinking; skeptics see it as a "warmed-over" version of multiple-use management or, more pejoratively, as "policy by slogan."

A different approach is embodied in chaos theory and adaptive assessment and management, and is fundamentally a response to the high degree of uncertainty in ecosystems discussed previously. The basic idea is that ecosystems are unfathomably complex and that they react to unpredicted (chaotic) events; thus, it is pointless to develop sophisticated ecosystem models for decision making based on equilibrium conditions (Holling 1978; Walters 1986, 1990). There is also constant feedback between man's decisions and adjustments of the ecosystem to those decisions. Uncertainty is so great that it is not feasible to create useful predictive models. Also, for alternatives that preclude future options, adaptive environmental assessment and management will not work well (e.g., construction of dams on the mainstem of the Columbia River has had major ecological consequences and each major project was an irrevocable decision [Volkman and McConnaha 1992]). In general, the manager or analyst will make a series of small decisions, evaluate the results, and then make revised decisions. To make a big decision requires strong public support and acceptable ways to compensate the losers; there is little in the history of salmon management to show that this condition will be realized.

The concept of total quality management (TQM) is a concept that became popular in business and government in the 1980s and 1990s (Osborne and Gaebler 1992). The widespread efforts to "reinvent government" have their basis in TQM. The core idea is that the customer comes first and, in turn, management should be measured by what customers want. In natural resource management and environmental protection, the "customer" is often defined as the public. Hence, TQM presupposes that an agency can find out what the public wants in terms of ecosystem management and protection, and then deliver that product. There are difficulties in defining the public, but TQM has been successful in some business applications. However, its usefulness in managing and protecting public natural resources is open to question (Bormann et al. 1994). In a pluralistic society, a common public goal for salmon that will allow the principles of TQM to be used effectively is unlikely.

Risk assessment and management, the final management paradigm reviewed here, has been used as a tool in some of the previous paradigms, or as a stand-alone approach. Strongly advocated by some (USEPA 1990a), the approach has generally been used for assessing chemicals and their role in altering ecosystems or components of ecosystems. The basic idea is that there are a number of risks to the environment and ecosystems, and these risks ought to be identified, quantified, and managed. The application of risk assessment to ecological questions such as Pacific salmon management is a formidable challenge because there are no examples of it being used successfully for ecological problems of this complexity. For example, risk assessment has been used for addressing technical questions concerning hatchery and power plant operation but not for complex salmon policy problems.

Tools Used in Ecological Risk Assessment

There is a fairly standard set of tools and techniques used to generate the data for risk assessment. However, who assumes the "burden of proof" is another challenging element of conducting risk assessments (Brown 1987, Volkman and McConnaha 1992, Bella 1996). Do risk assessors assume that current ecological conditions are the norm and any proposed deviation from the status quo must be justified? Or do they assume some pristine ecological state as the norm? Or do they assume that the person or organization proposing the action must justify it? One of the reasons that the Endangered Species Act and Section 404 of the Clean Water Act are so potentially powerful is that they effectively shift the burden of proof to those who would change a defined condition (e.g., species must not go extinct, or wetlands must not be altered unless there is explicit government approval). The practitioners of ecological risk assessment often overlook values, ethics, and burden of proof in defining the problem, and operate instead on the purely technical level (Brown 1987). For example, why do we assume that the physical alterations of salmon rivers, such as the Columbia, are a given? Is it not an option to demand that the organizations responsible for dams either demonstrate that the dams are not adversely affecting salmon populations, or alter their operations (including removal) so as not to adversely affect salmon? Why should the burden of proof rest with those trying to protect or restore salmon?

Bioassays are the most commonly used tools in producing the basic data for ecological risk assessment dealing with exposure to chemicals. There are many permutations of the basic bioassay and the literature is massive (Jenkins et al. 1989, Cairns and Niederlehner 1992). Bioassays work well for certain types of ecological problems and especially for the "command and control" regulatory approach. Severe limitations occur in assessing multiple, concurrent stresses, assessing effects on ecosystems or regions, or assessing effects that are not chemically driven (e.g., land-use alterations). It is easy to lose sight of the fact that bioassays are simplifications of the ecosystems and regions with which risk assessors are concerned and are merely surrogates for the realistic tests or experiments that cannot be performed. On an administrative level, the use of bioassays has become institutionalized, and the public may now view such tests as more relevant to protecting the environment than is warranted.

Environmental impact analysis and monitoring are additional tools that are often used in risk assessment. Such analyses are relevant to real-world problems and are often targeted directly at the public choice issues. Because problems are relevant, they are often complex scientifically, and the resulting predictions lack the scientific rigor typically seen in peer-reviewed journals. As a result, users often lack confidence in the reliability of the predictions. Moreover, the process of developing an environmental impact statement may be more important than the actual document produced.

Modeling and computer simulation are tools that have proved to be very popular in ecological risk assessment. These tools have many desirable features, such as the ability to deal with complex problems, the ability to evaluate alternative hypotheses quickly, and the ability to organize data and relationships into a defined whole. However, modelers often fall into the trap of substituting analytical rigor for intellectual rigor. Very simplistic (and incorrect) ideas can be masked by mathematical complexity. Even some of the most widely accepted and applied models in ecology illustrate the problem with developing and applying models to actual management issues (Hall 1988). Further, the ease and beauty of tools such as computer-assisted geographic analysis can also cause the analyst to lose sight of intellectual rigor and common sense.

Because most ecological risk assessment problems are complex and do not lend themselves entirely to laboratory experiments, field experiments, or modeling, the use of expert judgment and opinion is desirable and necessary. Expert opinion is useful but is not without problems. For example, when experts have dramatically different opinions, how does a risk assessor handle this analytically? Experts can be wrong and history is filled with examples of experts being completely in error. Alternatively, risk assessors trust that experts are less wrong on topics of their expertise than are non-experts, and the use of experts and formal expert systems will continue to increase. When relying on the opinions of technical experts, there is the particularly insidious problem of separating clearly their personal and organizational values from their technical opinion (Philpot 1992).

Risk assessment, at least in the problem-formulation stage, must include an explicit determination of what the customer wants. The customer is usually the public or a subset of the public (or an institutional surrogate such as a law or a court determination). This is not as easy as it sounds. The typical type of information from the public is that people want to "protect the environment," "protect endangered species," or "maintain a sustainable environment." The same people may also say that they want to "protect family wage jobs," "maintain economic opportunities for our children," and "protect the sanctity of personal property." It is very difficult to move beyond such platitudes and obtain information that is really useful in risk assessment. On the other hand, individuals or elements of society with a direct and vested interest will have very specific preferences. Those less directly affected tend to have more general preferences. For example, Gale and Cordray (1994) found various elements of the public possessing at least nine different concepts of sustainability for forests, many of which are mutually exclusive. Some people may talk about democratization of the decision-making process (Bormann et al. 1994). This radical concept would fundamentally alter how business is done, but some argue that a radical change is needed.

Application to the Salmon Issue

The first step in conducting any analysis of ecological risk is to clearly define the problem. Unfortunately, this step is often overlooked or resolved simplistically (Stout and Streeter 1992). In many cases, agreeing on the problem is impossible because that is in itself the political impasse. There is also tension between analysts who want to simplify the problem so that it is technically tractable, and politicians who want to keep problem definition as realistic (i.e., technically complicated) as possible. Defining the problem is a political process, requiring technical input, but it is based on values and priorities.

In defining risk assessment for Pacific Northwest salmonids, an analyst explicitly resolves whether the focus is on preserving some or all stocks (or any other evolutionarily significant unit) from going extinct, or maintaining some or all stocks at fishable levels. These are largely mutually exclusive alternatives. They are also not scientific decisions. Further, defining which species, communities, or ecosystems are to be evaluated in risk assessment is a value-based, not a scientific, determination (Costanza 1992a, b). Does the analyst consider the baseline condition to be 10,000 years ago, 200 years ago, or for the Columbia Basin, pre-impoundment construction (basically before World War II)? Analysts should not decide these questions; society should. Depending on the baseline selected, the results of a risk assessment will differ.

Most practitioners argue that, to be more useful, risk assessment (estimating risk) must be separated from risk management (making choices) both in practice and in appearance (Ruckelshaus 1985, Lave 1990). There are counterarguments against separating assessment from management (Jasanoff 1993). Usually the arguments recognize that it is impossible to separate a person's values from his technical (risk assessment) activities, and therefore the separation is illusionary. Separating the two activities (management and assessment) is not as easy as it might appear. Many scientists have strong personal opinions on public choice issues that concern ecological resources. It is difficult for anyone to separate purely technical opinions from personal value judgments. Even more difficult is convincing all elements of the public (all stakeholders) that the assessment is being conducted without a bias on the part of scientists.

The best scientists and most credible scientific information must be used in risk assessment. Besides being independent, the assessors must not advocate their organization's political position or their personal agenda. If the risk assessment is not perceived to be independent, the results will be tainted (Botkin 1990). Further, the research and assessment function within an organization should be separated from the management and regulatory function (Risser and Lubchenco 1992). Credibility and impartiality are difficult to maintain, especially in the public eye (Smith and Steel 1996).

Risk analysis will result in a number of options to manage the risk. These may range from drastic, expensive options to those that maintain the status quo, which may also be expensive. Options must be presented as clear alternatives with statements of ecological benefits and costs, and measures of uncertainty, for each. There is not a lot of rationality in most decision making, but there should be in decision analysis (Douglas and Wildavsky 1982). For example, risk analysts (and scientists) deal with estimates of ecological change, and risk managers (and politicians) deal with ecological degradation and improvement. As one scientist concluded after listening to a discussion of the pro's and con's of salmon streams being filled with downed timber from trees in buffer strips blowing over, "Your ecological mess is my ecological nirvana." Such statements move the scientist out of the scientific realm and into the political, value-driven realm. It may well be true that such ecological conditions are better from the policy perspective of enhancing salmon, but they are not better from a scientific perspective.

For the salmon problem, my recommendation is to not conduct a risk assessment unless there is a high likelihood that it will be used in decision making. If expectations are raised, and if no decision is made, the public has a sense of government institutions not working. Further, risk analysis of any significant ecological problem will result in options that create both big winners and big losers. It serves no productive purpose to try to convince losers that they are really winners. If someone's property will be effectively expropriated for some larger societal good, that action should be clearly stated in the assessment. Conversely, if an owner is permitted to alter his property for short-term gain, but at huge expense to society at large or to future generations, that also should be clearly stated.

Future Research Needs

It is easy to create a long list of research needs for the salmon problem. An implied premise of creating such a list is that insufficient research is the limitation to better management. It is not. The salmon problem is complex. The evolutionary biology and zoogeography are complicated and not well understood. The genetics of the species remains unclear. Oceanic influences are ambiguous. Considerable information on salmon and their habitat exists, but much of it is not directly usable for risk assessment and management. The baseline condition for salmon populations is open to considerable debate. Genetics, habitat relationships, ocean productivity, interactions with other organisms, human harvest, water quality issues, the effects of stocking, and the general problem of uncertainty and random events are all areas of concern and interest to researchers. However, I propose that what we need most is to better link research to the way society makes decisions.

First, ecological risk assessment needs to be modified to create a paradigm of ecological consequence analysis. The concept of risk applied to natural resources (including salmon) will only work for a narrow set of problems where there is a clear public (and legal) consensus, and on issues where there is an agreed-upon timeframe of interest (e.g., are benefits and risks defined over 10 years or 10 centuries?). With all ecological risks, a probability (of cause and effect, or ecological change) is neither good nor bad, it is only a probability. The resolution of many ecological problems is not limited by lack of scientific information, or technical tools, but by conflict created by fundamentally different values and social priorities as to what is most important (e.g., cheap food via irrigation water use versus fishing, cheap power versus free-flowing rivers, personal freedom versus land-use zoning). If we are dealing with an ecological problem that is at an impasse because some of the stakeholders do not accept the utilitarian model, we should not be surprised when risk assessment and management will not resolve the issue. We need to do ecological consequence analysis and let the political process select the desired option.

Second, the concept of ecological health needs to be better defined and understood by politicians and the public. The fundamental problem is not lack of technical information but what is meant by health. Is a wilderness condition defined as the base, or preferred level, of ecological health? Is the degree of perturbation by human activity the measure of ecological health? The concept of ecological degradation is human value-driven; the concept of ecological alteration is scientific. If the consequences of chaotic events in ecosystems are considered, what is natural (Botkin 1990)? There are scientific answers for some of these questions, but political (social) answers to many others.

Third, risk assessors need better ways to use expert opinion. Most of the policy-relevant problems in ecology are too complex for easy scientific experimentation or analysis. An old rule in policy analysis is that if something can be measured, it is probably irrelevant to public choice. If problems are simplified to the point of making them scientifically tractable, then the result may lack policy relevance. Expert opinion must be used. Computer-generated maps and computer-assisted models may be elegant, but for really important decisions, the political process demands expert opinion.

Fourth, better ways to evaluate and measure public preference and priorities in framing ecological issues need to be developed. Public opinion polls always show that the public is very supportive of the "environment," as it is with "peace," "freedom," and "economic opportunity." The public is similarly supportive of preserving biological diversity, ecosystem management, and sustainable natural resource management (Smith and Steel 1996). Unfortunately, this type of information is of limited use in helping frame or make difficult environmental decisions. The public is not a monolith; it encompasses many divergent views, and individuals vary greatly in the intensity of their opinions. Individuals may argue forcefully for the industrial economic paradigm or for the natural economic paradigm, but practical political options are not framed in this context.

The fifth critical need is to develop better ways to present options and consequences to the public, to policy analysts, and to decision makers. Society is not well served by statements such as "it is a complicated problem and you need to have an advanced degree in ecology to understand it," or "you can select this option without significant cost to society" when there will be costs to some people. The main message in risk assessment must be that environmental protection is difficult and that decisions must be clearly framed as decision alternatives.

The Challenge to Biologists

Biological and social science must be linked if public decision making is to be improved. Too often fisheries, forestry, and wildlife problems are viewed as biological challenges. It is society that should define problems and set priorities, but the public speaks with not one, but many voices. As Gale and Cordray (1994) pointed out for forest management, many of the stated public demands are mutually exclusive. Ecological health, for example, is a social value defined in ecological terms (Costanza 1992a). But incorporating public input into risk assessment and management may be carried to the extreme (i.e., democratization). Such a radical concept is anathema to most scientists and politicians.

Scientists must maintain a real and perceived position of providing credible ecological information, information that is not slanted by personal value judgments. Those of us involved in risk assessment cannot become advocates for any political position or choice lest our credibility suffer (Lave 1990, Philpot 1992). Such a position may be painful at times because who among us can completely separate personal views from professional opinions?

Biologists must educate the public (and political officials) about what scientific and technical information can and cannot do in resolving public choice issues such as the salmon decline problem. We should never hear the question "When will the scientists be in a position to answer this policy question?"

Public choice for issues like the salmon problem is too important to be left completely in the hands of technocrats; we should seriously consider democratizing the decision-making process. Most current public decision making involves professionals controlling the process with public input requested as desired (Merrell and Van Strum 1990, P. Loge, US Congressional staff, Tempe, Arizona, unpubl. ms.). Values and priorities originate with what the public wants, but are we really ready for democratization? Do any professionals want to trust the public with defining goals? After all, biology, ecology, fisheries, wildlife, and forestry professions tend to be filled with people with a strong natural resource protection streak. Perhaps the public will not support such professional positions.

We should not become complacent that complex ecological problems, such as as illustrated by Pacific salmon, have only technological and rational solutions. Tools such as risk assessment might help at the margins of the political process but are not going to resolve the key policy questions. Non-rational ideas are extremely important in all crucial public choice issues. We must guard against technical hubris, the false sense of confidence in technology, technological solutions, and rational analysis . . . including risk assessment.

Acknowledgments

The author is Associate Director for Science at USEPA's Western Ecology Division. He is also Professor (courtesy) of Fisheries and Professor (adjunct) of Political Science at Oregon State University. The comments and views expressed in this paper do not necessarily represent policy positions of the Environmental Protection Agency or any other organization.

Literature Cited

Angermeier, P.L. and J.E. Williams. 1993. Conservation of imperiled species and reauthorization of the Endangered Species Act of 1973. Fisheries 18(7): 34-38.

Baranov, F.I. 1918. On the question of the biological basis of fisheries. Nauch. Issled. Ikhtiol. Inst. Izu. 1(1): 81-128.

Barber, W.E. and J.N. Taylor. 1990. The importance of goals, objectives, and values in the fisheries management process and organization: a review. North American Journal of Fisheries Management 10(4): 365-373.

Bella, D.A. 1996. Organizational systems and the burden of proof, p. 617-638. *In* D.J. Stouder, P.A. Bisson, and R.J. Naiman (eds.), Pacific Salmon and Their Ecosystems: Status and Future Options. Chapman and Hall, New York.

Beverton, R.J.H. and S.J. Holt. 1957. On the Dynamics of Exploited Fish Populations. Fisheries Investigations (Series 2) 19. United Kingdom Ministry of Agriculture and Fisheries.

Bormann, B.T., M.H. Brookes, E.D. Ford, A.R. Kiester, C.D. Oliver, and J.F. Weigand. 1994. Volume V: a framework for sustainable-ecosystem management. US Forest Service General Technical Report PNW-GTR-331. Washington, DC.

Botkin, D.B. 1990. Discordant harmonies: a new ecology for the twenty-first century. Oxford University Press, New York.

Brown, D.A. 1987. Ethics, science, and environmental regulation. Environmental Ethics 9: 331-349.

Cairns, J., Jr., and B.R. Niederlehner. 1992. Predicting ecosystem risk: genesis and future needs, p. 327-343. *In* J. Cairns, Jr., B.R. Niederlehner, and D.R. Orvos (eds.), Predicting Ecosystem Risk. Princeton Scientific Publishing Co., Princeton, New Jersey.

Cairns, M.A. and R.T. Lackey. 1992. Biodiversity and management of natural resources: the issues. Fisheries 17(3): 6-10.

Callicott, J.B. 1990. Standards of conservation: then and now. Conservation Biology 4(3): 229-232.

Callicott, J.B. 1991. Conservation ethics and fishery management. Fisheries 16(2): 22-28.

Costanza, R. (ed.). 1992a. Ecological Economics: The Science and Management of Sustainability. Columbia University Press, New York.

Costanza, R. 1992b. Toward an operational definition of ecosystem health, p. 239-256. *In* R. Costanza, B.G. Norton, and B.D. Haskell (eds.), Ecosystem Health. Island Press, Washington, DC.

Douglas, M. and A. Wildavsky. 1982. Risk and Culture. University of California Press, Berkeley.

Ehrlich, P.R. and E.O. Wilson. 1991. Biodiversity studies: science and policy. Science 253: 758-762.

Foster, J.W. 1992. Mountain gorilla conservation: a study in human values. Journal American Veterinary Medicine Association. 200(5): 629-633.

Gale, R.P. and S.M. Cordray. 1994. Making sense of sustainability: nine answers to 'what should be sustained?' Rural Sociology 59(2): 311-332.

Gore, A. 1992. Earth in the Balance: Ecology and the Human Spirit. Houghton Mifflin Company, Boston.

Graham, M. 1935. Modern theory of exploiting a fishery, and application to North Sea trawling. Journal du Conseil Conseil International pour l'Exploration de la Mer 10: 264-274.

Hall, C.A.S. 1988. An assessment of several of the historically most influential theoretical models used in ecology and of the data provided in their support. Ecological Modelling 43: 5-31.

Herzog, H. 1993. Human morality and animal research: confessions and quandaries. The American Scholar 62(3): 337-349.

Holling, C.S. (ed.). 1978. Adaptive Environmental Assessment and Management. John Wiley and Sons, Chichester, United Kingdom.

Homer-Dixon, T.F., J.H. Boutwell, and G.W. Rathjens. 1993. Environmental change and violent conflict. Scientific American 268(2): 38-45.

Jasanoff, S. 1993. Relating risk assessment and risk management: complete separation of the two processes is a misconception. EPA Journal 19(1): 35-37.

Jenkins, D.G., R.J. Layton, and A.L. Buikema. 1989. State of the art in aquatic ecological risk assessment. Entomological Society of America Miscellaneous Publications no. 75: 18-32.

Johns, D.M. 1990. The relevance of deep ecology to the third world: some preliminary comments. Environmental Ethics 12: 233-252.

Karr, J.R. 1992. Using biological criteria to insure a sustainable society, p. 61-70. *In* P.W. Adams and W.A. Atkinson (compilers.), Watershed Resources: Balancing Environmental, Social, Political, and Economic Factors in Large Basins. Oregon State University Press. Corvallis.

Lackey, R.T. 1979. Options and limitations in fisheries management. Environmental Management 3(2): 109-112.

Lackey, R.T. 1994. Ecological risk assessment. Fisheries 19(9): 14-18.

Larkin, P.A. 1977. An epitaph for the concept of maximum sustained yield. Transactions American Fisheries Society 106(1): 1-11.

Lave, L.B. 1990. Risk analysis and management. The Science of the Total Environment 99: 235-242.

Lewis, M.W. 1992. Green Delusions. Duke University Press, Durham, North Carolina..

Likens, G.E. 1992. The Ecosystem Approach: Its Use and Abuse. Ecology Institute, Oldendorf/Luhe, Germany.

Limbaugh, R. 1992. The Way Things Ought to Be. Pocket Books, New York.

Ludwig, D., R. Hilborn, and C.J. Walters. 1993. Uncertainty, resource exploitation, and conservation: lessons from history. Science 260: 17, 36.

Merrell, P. and C. Van Strum. 1990. Negligible risk: premeditated murder? Journal of Pesticide Reform 10(1): 20-22.

Naiman, R J. 1992. Integrated watershed management: science or myth, p. 5-20. *In* P.W. Adams and W.A. Atkinson (compilers.), Watershed Resources: Balancing Environmental, Social, Political, and Economic Factors in Large Basins. Oregon State University Press. Corvallis.

National Research Council. 1983. Risk Assessment in the Federal Government: Managing the Process. National Academy Press, Washington, DC.

National Research Council. 1993. Issues in Risk Assessment. National Academy Press, Washington, DC.

Nehlsen, W., J.E. Williams, and J.A. Lichatowich. 1991. Pacific salmon at the crossroads: stocks at risk from California, Oregon, Idaho, and Washington. Fisheries 16(2): 4-21.

Newkirk, I. 1992. Free the Animals! The Noble Press, Inc., Chicago.

O'Brien, M H. 1993. Being a scientist means taking sides. BioScience 43(10): 706-708.

Osborne, D.E. and T. Gaebler. 1992. Reinventing Government: How the Entrepreneurial Spirit Is Transforming the Public Sector. Addison-Wesley Publishing Company, Reading, Massachussetts.

Philpot, C.W. 1992. Institutional change: science, policy, and management, p. 127-130. *In* P.W. Adams and W.A. Atkinson (compilers.), Watershed Resources: Balancing Environmental, Social, Political, and Economic Factors in Large Basins. Oregon State University Press. Corvallis.

Pinchot, G. 1947. Breaking New Ground. Harcourt, Brace, and Company, New York.

Rapport, D.J. 1989. What constitutes ecosystem health? Perspectives in Biology and Medicine 33(1): 120-132.

Ray, D.L. 1992. Trashing the Planet: How Science Can Help Us Deal with Acid Rain, Depletion of Ozone, and Nuclear Waste (Among Other Things). Harper Perennial, New York.

Repetto, R. and R.C. Dower. 1992. Reconciling economic and environmental goals. Issues in Science and Technology Winter: 28-32.

Ricker, W.E. 1954. Stock and recruitment. Journal of the Fisheries Research Board of Canada 11: 559-623.

Risser, P.G. and J. Lubchenco. 1992. The role of science in management of large watersheds, p. 119-126. *In* P.W. Adams and W.A. Atkinson (compilers.), Watershed Resources: Balancing Environmental, Social, Political, and Economic Factors in Large Basins. Oregon State University Press. Corvallis.

Roedel, P.M. 1975. A summary and critique of the symposium on optimum sustainable yield, p. 79-89. *In* P.M. Roedel (ed.), Optimum Sustainable Yield as Concept in Fisheries Management. American Fisheries Society, Special Publication No. 9, Bethesda, Maryland..

Rubin, C.T. 1989. Environmental policy and environmental thought: Ruckelshaus and Commoner. Environmental Ethics 11: 27-51.

Ruckelshaus, W.D. 1985. Risk, science, and democracy. Issues in Science and Technology 1(3): 19-38.

Shifrin, R. 1992. Not by risk alone: reforming EPA research priorities. Yale Law Journal 102: 547-575.

Simon, J.L. 1986. Disappearing species, deforestation, and data. New Scientist May 15: 60-63.

Slocombe, D.S. 1993. Implementing ecosystem-based management. BioScience 43(9): 612-622.

Smith, C.L. and B.S. Steel. 1996. Values in the valuing of salmon, p. 599-616. *In* D.J. Stouder, P.A. Bisson, and R.J. Naiman (eds.), Pacific Salmon and Their Ecosystems: Status and Future Options. Chapman and Hall, New York.

Sokoloski, A.A. 1980. Planning and policy analysis, p. 185-195. *In* R.T. Lackey and L.A. Nielson (eds.), Fisheries Management. John Wiley and Sons, New York.

Stouder, D.J., P.A. Bisson, and R.J. Naiman (eds.). 1996. Pacific Salmon and Their Ecosystems: Status and Future Options. Chapman and Hall, New York.

Stout, D.J. and R.A. Streeter. 1992. Ecological risk assessment: its role in risk management. The Environmental Professional 14: 197-203.

Suter, G.W. (ed.). 1993. Ecological Risk Assessment. Lewis Publishers, Boca Raton, Louisiana.

Swanson, F.J. and J.F. Franklin. 1992. New forestry principles from ecosystem analysis of Pacific Northwest forests. Ecological Applications 2(3): 262-274.

US Environmental Protection Agency. 1990a. Reducing risk: setting priorities and strategies for environmental protection. Science Advisory Board SAB-EC-90-021. Washington, DC.

US Environmental Protection Agency. 1990b. The report of the ecology and welfare Subcommittee, relative risk reduction project: reducing risk. Appendix A. Science Advisory Board EPA-SAB-EC-90-021A. Washington, DC.

US Environmental Protection Agency. 1991. Summary report on issues in ecological risk assessment. Risk Assessment Forum EPA/625/3-91/018. Washington, DC.

US Environmental Protection Agency. 1992a. Framework for ecological risk assessment. Risk Assessment Forum EPA/630/R-92/001. Washington, DC.

US Environmental Protection Agency. 1992b. Safeguarding the future: credible science, credible decisions. EPA/600/9-91/050. Washington, DC.

US Environmental Protection Agency. 1993. A review of ecological assessment case studies from a risk assessment perspective. Risk Assessment Forum EPA/630/R-92/005. Washington, DC.

US General Accounting Office. 1992. Environmental protection issues. Transition Series, United States General Accounting Office GAO/OCG-93-16TR. Washington, DC.

Volkman, J. and W.E. McConnaha. 1992. A swiftly tilting basin: the Columbia River, endangered species, and adaptive management, p. 95-106. *In* P.W. Adams and W.A. Atkinson (compilers.), Watershed Resources: Balancing Environmental, Social, Political, and Economic Factors in Large Basins. Oregon State University Press. Corvallis.

Walters, C.J. 1986. Adaptive Management of Renewable Resources. Macmillan Publishing Company, New York.

Walters, C.J. 1990. Large-scale management experiments and learning by doing. Ecology 71(6): 2060-2068.

Whelan, E.M. 1993. Toxic Terror. Prometheus Books, Buffalo, New York.

An Ecosystem-Based Approach to Management of Salmon and Steelhead Habitat

Jack E. Williams and Cindy Deacon Williams

Abstract

The USDA Forest Service and USDI Bureau of Land Management have developed an ecosystem-based management strategy—known as PACFISH—to restore and maintain habitat for the natural production of anadromous salmonids (*Oncorhynchus* spp.) on public lands. The strategy provided the foundation for the aquatic and riparian elements of the Scientific Analysis Team report to US District Court Judge Dwyer, the Forest Ecosystem Management Assessment Team report to US President Clinton's Pacific Northwest Forest Conference, and PACFISH interim direction. Later, PACFISH will be considered as part of comprehensive, long-term strategies for eastern Oregon and Washington and the remainder of the Upper Columbia River Basin. The focus of PACFISH is on restoring and maintaining ecological processes and functions within broadly defined riparian zones rather than the more traditional approach of active instream manipulation. PACFISH consists of the following components: riparian goals; quantified riparian management objectives; standards and guidelines for all land management activities within broad riparian reserves; designation of riparian habitat conservation areas; networks of key watersheds that receive priority analysis, protection, and restoration; watershed analysis; watershed restoration; and monitoring. Individual components may be altered through compliance with the National Environmental Policy Act to meet regional needs.

Introduction

> The effort to control the health of land has not been very successful. It is now generally understood that when soil loses fertility, or washes away faster than it forms, and when water systems exhibit abnormal floods and shortages, the land is sick (Aldo Leopold 1941 in Flader and Callicott 1991).

The declining trend in condition of aquatic habitats in the United States (US) has become increasingly obvious during recent years. A nationwide rivers inventory reported that only 1.9% of 5.23 million km of streams in the contiguous US remained in "high natural quality" (Benke

1990). Also, in the lower 48 states, >50% of all wetlands have been destroyed during the past 100 years (Dahl 1990). In the western US (the West), where water and aquatic habitats always are in scarce supply, the picture often is more ominous. For example, Katibah (1984) reported that only 11% of the 373,100 ha of pre-settlement riparian forests remain in California's vast Central Valley.

The status of salmon (*Oncorhynchus* spp.), steelhead (*O. mykiss*), sea-run cutthroat trout (*O. clarki*), and other native aquatic species has overshadowed the decline of aquatic and riparian habitat conditions. A total of 214 stocks of anadromous salmonids from California, Idaho, Oregon, and Washington were considered "at risk" of extinction or "of special concern" by a committee of the American Fisheries Society (AFS) (Nehlsen et al. 1991). This same committee documented 106 additional stocks that already were extirpated from the four-state area. Using the AFS data, Williams et al. (1992) calculated that of the 192 stocks of salmon, steelhead, and sea-run cutthroat trout in the Columbia River Basin, 35% were extinct, 19% were at high risk of extinction, 7% at moderate risk, 13% were of special concern, and only 26% were secure. Subsequent surveys in California (Higgins et al. 1992) and by state agencies in Oregon (Nickelson et al. 1992) and Washington (Washington Department of Fisheries et al. 1993) have confirmed the broad scope of the decline of anadromous salmonid stocks and have added to the list provided by the AFS.

Federal lands provide habitat for a large proportion of the remaining anadromous salmonid stocks. The USDA Forest Service (Forest Service) and USDI Bureau of Land Management (BLM) estimate that of the 214 stocks included in the AFS report, 134 occur on Forest Service-administered lands and 109 occur on BLM-administered lands. An expanded list (Nehlsen et al. 1991, Higgins et al. 1992, Nickelson et al. 1992, Washington Department of Fisheries et al. 1993) indicates 259 rare stocks of anadromous salmonids occur on federal lands solely within the range of the northern spotted owl (Fig. 1; Forest Ecosystem Management Assessment Team [FEMAT] 1993).

The cumulative effects of timber harvest, road construction, agriculture, livestock grazing, hydroelectric development, mining, and other land-use activities have resulted in significant declines of anadromous salmonid habitat. In forested regions, one of the largest single factors affecting the loss of habitat quality on federal lands results from increased erosion and sediment yields from vast road networks (Swanson and Dyrness 1975, Furniss et al. 1991, FEMAT 1993). Federal lands within the range of the northern spotted owl are estimated to contain ~177,000 km of roads with 250,000 stream crossings through culverts, the majority of which cannot tolerate flows greater than a 25-year flood event without high likelihood of failure (FEMAT 1993). In rangelands, livestock grazing is the primary causal factor in declines of riparian condition and erosion of streambanks (Platts 1991).

During 1987–92, Forest Service personnel (McIntosh et al. 1994) resurveyed 116 stream systems in the Pacific Northwest and compared the number of large, deep pools (>50 m long and >2 m deep) per stream kilometer with those surveyed during 1935 to 1945. The comparison documented substantial decreases in the number of large, deep pools, which are a primary indicator of high-quality in-channel habitat conditions (FEMAT 1993). The changes in frequency of pools appear to be the result of land management activities. Within the Columbia River Basin, streams in managed watersheds lost an average of 31% of their pools, whereas streams in wilderness areas or other areas without timber harvest, road building, or livestock grazing, increased 200% in pool frequency during the past 50 years (McIntosh et al. 1994). For managed

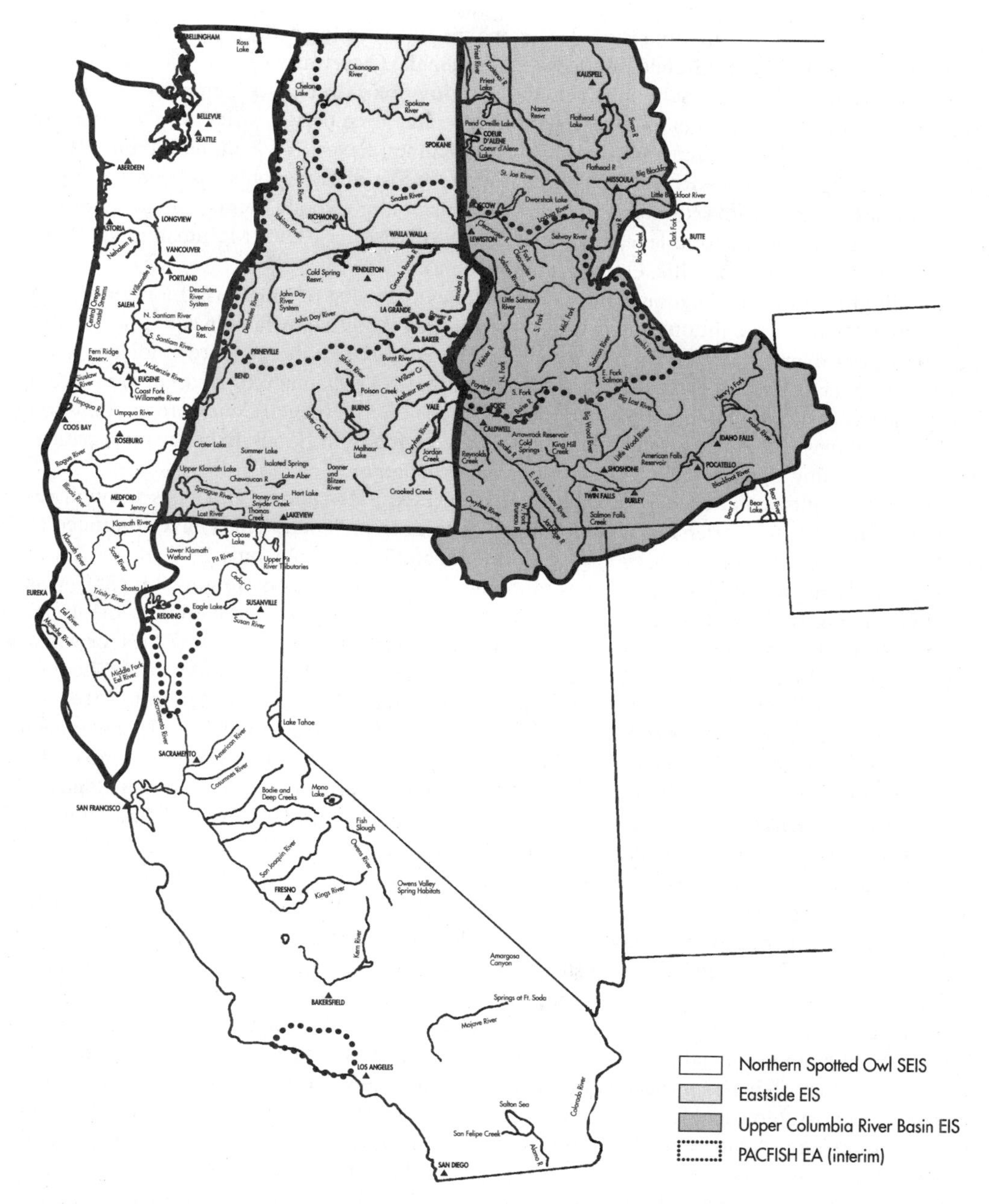

Figure 1. Map of PACFISH region comparing area of PACFISH Environmental Assessment interim direction (portions of Oregon, Washington, Idaho, and California) with the Supplemental Environmental Impact Statement (Supplemental EIS) for the northern spotted owl (west of the crest of the Cascades in Washington, Oregon, and California), the Eastside EIS (eastern Washington and Oregon), and Upper Columbia River Basin EIS (Idaho, western Montana, and portions of Nevada, Utah, and Wyoming).

portions of rivers in the Columbia River Basin, the frequency of large, deep pools from 1935 to 1992 decreased 66% in the Grande Ronde River, 65% in the Clearwater River, 57% in the Lewis and Clark River, 54% in the Salmon River, and 36% in Asotin Creek, yet increased 57% in the Wenatchee River (McIntosh et al. 1994). During the same time period, wilderness or roadless portions increased in pool frequency by 28% in the Salmon River, 144% in the Yakima River, and 156% in the Wenatchee River. Concomitant with the loss of large pools has been a regional trend towards finer stream sediments, reduced amounts of large woody debris, and decreases in pool-riffle ratios from historic levels of about 50:50 to 20:80 or 10:90 (McIntosh et al. 1994; USDA Forest Service, Corvallis, Oregon, unpubl. data).

While the above information clearly demonstrates the value of long-term monitoring and attention to historical conditions, many basins were highly modified by human activities prior to the 1930s. For example, in the Grande Ronde Basin, Oregon, historical records indicate problems from overgrazing as early as the 1880s (Skovlin 1991). Damage to riparian areas from sheep and livestock overgrazing can be severe. An Oregon Environmental Council report (Hanson 1987) found livestock grazing, rather than timber harvest, roads, or floods, to be the most critical factor determining riparian habitat condition in eastern Oregon.

Despite implementation of gradually improving Best Management Practices through revised land-use plans, riparian and aquatic habitat conditions on federal lands have continued to decline. In the BLM's Salem District of western Oregon, for example, only 23% of 339 km of streams providing habitat for anadromous salmonids was classified as optimum while 35% was rated fair and 42% as poor (House 1992). In eastern Oregon, 80% of fish habitat in the upper Grande Ronde River Basin failed to meet USDA Forest Plan standards and guidelines for water temperature, sediment, and riparian conditions; and in the Middle Fork Clearwater and Lochsa rivers on Idaho's Clearwater National Forest, 70% of stream habitats also failed to meet the same standards and guidelines (USDA Forest Service and USDI Bureau of Land Management 1994). Despite the degraded habitat conditions of many federal lands, they provide some of the best remaining habitats for anadromous salmonids. For comparison, non-federal rangelands in Oregon were classified as 40% poor condition, 38% percent as fair, 19% as good, and only 3% as excellent, with widespread declines in riparian and stream habitats (Wissmar et al. 1994).

The recent assessment documenting degradation of stream systems throughout the Pacific Northwest and the reports of region-wide stock declines prompted the Forest Service and BLM to prepare a comprehensive strategy designed to address and remedy these losses. The intent of the strategy is not to restore individual stocks but rather to restore ecological health and productivity to watersheds containing existing or potential habitat for anadromous salmonids. Such an ecosystem-based approach is necessary to halt declines in habitat condition, restore degraded systems, and maintain remaining high-quality habitat (FEMAT 1993). However, it is important to recognize that such an approach is focused solely on freshwater habitat conditions and associated riparian systems and fails to address the full variety of problems affecting sustainability of salmon resources. Alternatively, coldwater salmonids provide good indicators of biological integrity in stream ecosystems, and improvements in their status should indicate general improvements in habitat and water quality (Marcot et al. 1994). The purpose of this paper is to describe the ecosystem-based habitat strategy known as PACFISH and its development and implementation.

PACFISH Strategy Development

Both the Forest Service and BLM have pledged to take an ecosystem-based management approach to resolving the critical resource problems facing public lands. In April 1992, the Chief of the Forest Service directed his staff to undertake an assessment and develop a management strategy to address the habitat needs of all anadromous salmonid stocks on National Forests. The BLM officially joined the Forest Service strategy development in March 1993. To facilitate a strong linkage between management and research, the PACFISH effort was staffed with technical specialists from the Forest Service's National Forest System and BLM, and research scientists from the Forest Service's Pacific Northwest Research Station, Pacific Southwest Research Station, and the Intermountain Research Station. The organizational framework for PACFISH consisted of three components: (1) Washington Office Policy Group, (2) Washington Office Work Group, and (3) Field Team. Each component had co-leaders from both agencies.

Initially, eight alternative strategies were evaluated for PACFISH. Six were developed by the Field Team while another alternative was developed by the Scientific Panel on Late-Successional Forest Ecosystems (alternative 8A from Johnson et al. 1991), and the eighth was a draft riparian management strategy of the Pacific Southwest Region of the Forest Service. These eight alternatives included different combinations and applications of riparian goals, riparian management objectives, standards and guidelines, riparian area identification, key watershed identification, watershed assessment procedures, and watershed restoration. Many of the alternatives were deleted because they failed to adequately restore the integrity of aquatic habitats or restore depressed stocks (or both), despite improvements over existing conditions.

Many of the concepts explored within the PACFISH alternatives presented a management approach acceptable for late-successional forests of the Pacific Northwest (Appendix 5-K, Thomas et al. 1993). This process was eclipsed a few months later when US President Clinton held his Forest Conference to resolve forest management and endangered species controversies in the Pacific Northwest and established the Forest Ecosystem Management Assessment Team (FEMAT).The FEMAT adopted the PACFISH elements as the foundation for aquatic and riparian management in various options (e.g., Option 9 in FEMAT 1993). During April to May 1993, while PACFISH was being examined in the FEMAT strategy development process, it was being further refined by the BLM and the Forest Service.

The following section describes the elements of interim direction for PACFISH (USDA Forest Service and USDI Bureau of Land Management 1994, 1995) as modified by public comments and in compliance with Section 7 of the Endangered Species Act. These documents also include modifications produced in response to issues raised in the FEMAT process and during Range Reform efforts. As a result, elements of PACFISH and their application may be modified for regional or state conditions with the development of the Eastside Environmental Impact Statement (EIS) and Upper Columbia River Basin EIS for long-term strategy implementation (i.e., >18 months). Ultimately, the PACFISH approach may apply to Forest Service- and BLM-administered lands within all or part of an eight-state area (Fig. 1) plus Alaska.

The PACFISH Strategy

The PACFISH effort is an ecosystem-based management approach for maintaining and restoring the ecological health of watersheds containing habitat for anadromous fish on Forest Service and BLM lands in the West. The PACFISH strategy includes the following components: riparian goals, riparian management objectives, standards and guidelines, riparian habitat conservation areas (RHCAs), key watersheds, watershed analysis, watershed restoration, and monitoring. The PACFISH effort is designed to ensure the maintenance and restoration of ecosystem processes and functions, and the production and maintenance of complex habitats necessary for naturally reproducing native salmonid fishes. Components of the strategy are interconnected for concurrent implementation with the exception of restoration activities that are needed only where cumulative effects have substantially degraded watershed conditions. Fragmented federal lands, including "checkerboard" private/federal land ownership patterns, may complicate implementation of the strategy and impede the attainment of certain riparian management objectives. Alternatively, management of federal lands may require more protective measures to compensate for uses on non-federal lands to achieve a net positive effect at the watershed level. Effective coordination with adjacent landowners within watersheds is key to productive implementation of any management strategy, especially as scientific perspectives (e.g., emerging field of landscape ecology), policy issues (sustaining wide-ranging species), and new technologies (e.g., remote sensing) enable and require examination of cross-ownership ecosystem questions.

Riparian Goals

The goals of PACFISH establish an expectation of healthy, productive watersheds, riparian areas, and associated aquatic habitats consistent with legislative direction. The goals focus on elements needed for healthy, productive habitats by calling for the maintenance or restoration of the following:

- water quality necessary for stable (within range of natural dynamics) and productive riparian and aquatic ecosystems;
- stream channel integrity, natural channel processes, and the sediment regime under which the riparian and aquatic ecosystems developed;
- instream flows to support the effective function of stream channels and the ability to route sediment, woody debris, and flood discharges;
- natural timing, extent, and variability of water table elevation in meadows, floodplains, and wetlands;
- diversity and productivity of native and desired non-native plant communities in riparian zones;
- sufficient riparian vegetation for natural large woody debris characteristics, adequate thermal regulation, and rates of surface and bank erosion characteristic of those under which the communities developed;
- riparian and aquatic habitats necessary to foster the unique fish stocks that evolved locally; and
- habitat to support populations of well-distributed native and desired plant, vertebrate, and invertebrate populations that contribute to the viability of aquatic and riparian-dependent communities.

Riparian Management Objectives

The working hypothesis for PACFISH is that productive third- to seventh-order riverine habitats generally are characterized by excellent water quality, sufficient water quantity, complex channel characteristics, healthy riparian communities, and watersheds that are free from chronic and accelerated sedimentation. These habitats support mixed species assemblages of native salmonids. The PACFISH strategy includes quantified objectives for stream channel, riparian, and watershed conditions to measure whether attainment, or progress towards attainment, of the goals is being achieved. Interim riparian management objectives have been established based on data developed from resurveys of 116 stream systems in the Columbia River Basin conducted since 1988 and compared with surveys conducted by the Bureau of Fisheries between 1935 and 1945 (FEMAT 1993, McIntosh et al. 1994). These data have been supplemented by data collected from coastal stream systems in Alaska (USDA Forest Service, Pacific Northwest Research Station, Juneau, Alaska, unpubl. data). Certain habitat features were identified as surrogates of watershed health and used to develop the objectives. These features include pool frequency (forested and rangeland streams), water temperature (forested and rangeland streams), large woody debris frequency (forested streams only), streambank stability (rangeland streams only), lower bank angle (rangeland streams only), and width-to-depth ratio (forested and rangeland streams) (Table 1). Interim objectives still are under development for interior Alaska and are being validated for southeast Alaska. The interim riparian management objectives apply until they are tailored to local watershed conditions through watershed analysis. Each of these objectives must be met or exceeded before general habitat conditions would be

Table 1. Interim riparian management objectives for the PACFISH strategy. Interim objectives apply only until they are tailored to local watershed conditions through the watershed analysis process.

Habitat feature	Interim objective								
Pool frequency (all systems)	Varies by channel width as follows:								
wetted width (m):	3	6	8	15	23	30	38	46	61
number pools/km:	154	90	76	42	37	29	23	19	14
Water temperature (all systems)	Maximum water temperatures <18°C within migratory and rearing habitats, and <16°C within spawning habitats								
Large woody debris									
Coastal forested systems in California, Oregon, Washington, and Alaska	>129 pieces per km; >60 cm diameter; >15 m length								
Coastal forested systems east of Cascade Crest in Oregon, Washington, and Idaho	>32 pieces per km; >30 cm diameter; >11 m length								
Bank stability (rangeland systems)	>80% stable								
Lower bank angle (rangeland systems)	>75% of banks with <90° angle (i.e., undercut)								
Width/depth ratio (all systems)	<10								

considered favorable for anadromous salmonids. However, application of the interim riparian management objectives may require some latitude. For example, some headwater streams may have an abundance of boulders that contribute to pool formation, and though lacking large quantities of woody debris, still constitute good habitat. Watershed analysis will play an important role in establishing site-specific clarification of appropriate objectives.

STANDARDS AND GUIDELINES

Standards and guidelines were developed to establish consistent management within RHCAs to ensure that goals and objectives for stream channel, riparian, and watershed conditions are met. Standards and guidelines are designed to steadily improve habitat conditions, or if existing habitat is in good condition, ensure maintenance of such condition. Specific standards and guidelines have been developed to govern all practices related to land management (Table 2). A complete list of PACFISH standards and guidelines, which may vary somewhat depending upon regional differences, can be found in National Environmental Policy Act (NEPA) documents implementing PACFISH (USDA Forest Service and USDI Bureau of Land Management 1995).

Many of the standards and guidelines refer to attainment of riparian management objectives as the yardstick to determine whether activities should be allowed within RHCAs. For example,

Table 2. PACFISH standards and guidelines. These apply to lands within riparian habitat conservation areas (RHCAs) and will vary geographically as modified through regional National Environmental Policy Act compliance.

Category	General intent
Timber management	Prohibit harvest in RHCAs with limited exceptions to meet riparian objectives.
Roads management	Minimize sediment delivery; improve stream crossings; minimize disruption of hydrologic flow paths; require watershed analysis prior to new roads in RHCAs.
Grazing management	Modify or eliminate grazing practices that are inconsistent with riparian objectives; remove livestock facilities from RHCAs.
Recreation management	Modify or eliminate recreation practices that are inconsistent with riparian objectives; remove recreation facilities from RHCAs as needed.
Minerals management	Locate structures and support facilities outside RHCAs; prohibit surface occupancy in RHCAs for certain mining activities.
Fire/fuels management	Minimize disturbances during fire suppression activities; locate facilities outside RHCAs; design rehabilitation to meet riparian objectives.
Lands	Design and locate hydroelectric and other water developments to maintain/restore riparian resources and channel integrity.
General riparian area management	Control chemical applications within RHCAs; secure instream flows as appropriate.
Watershed and habitat restoration	Design restoration projects to meet long-term ecological objectives; coordinate with others on watershed-level activities.
Fisheries and wildlife management	Design fish and wildlife projects to meet riparian objectives.

the following standard and guideline would govern whether campgrounds would be allowed within RHCAs (USDA Forest Service and USDI Bureau of Land Management 1995):

> Design, construct, and operate recreation facilities, including trails and dispersed sites, in a manner that does not retard or prevent attainment of the Riparian Management Objectives and avoids adverse effects on listed anadromous fish. Complete Watershed Analysis prior to construction of new recreation facilities in RHCAs. For existing recreation facilities inside RHCAs, assure that the facilities or use of the facilities will not prevent attainment of Riparian Management Objectives or adversely affect listed anadromous fish. Relocate or close recreation facilities where Riparian Management Objectives cannot be met or adverse effects on listed anadromous fish avoided.

This reference of certain standards and guidelines to attainment of riparian management objectives provides a mechanism for applying management direction in a manner that is sensitive to local watershed conditions. Other standards and guidelines do not refer to attainment of objectives but simply require that certain conditions be met. For example, one of the standards and guidelines for road management requires new and existing culverts, bridges, and other stream crossings to be constructed or improved "to accommodate a 100-year flood, including associated bedload and debris" (USDA Forest Service and USDI Bureau of Land Management 1995). Standards and guidelines apply only to the RHCAs and to certain projects and activities outside RHCAs that have a high likelihood of degrading them. This focus on RHCAs arises because RHCAs require special management sensitivity. Standards and guidelines can be revised only if the following criteria are met: monitoring data clearly show that the ecological capability of the habitat is not being adequately addressed by the existing standards and guidelines, the proposed standards and guidelines are supported by analysis clearly establishing that the proposed modification better supports achievement of PACFISH goals and objectives, and the proposed revision will be reviewed by a team of watershed/fisheries managers and research scientists. Areas outside of RHCAs largely would be governed by standards and guidelines contained within existing land-use plans and appropriate regulations.

Riparian Habitat Conservation Areas

Riparian Habitat Conservation Areas (a.k.a. Riparian Reserves [FEMAT 1993]) are portions of watersheds where riparian and aquatic resources receive primary emphasis and where PACFISH standards and guidelines apply. Although receiving precise management direction, RHCAs are not "lock-out" zones where resource use is prohibited. RHCAs include areas required for maintaining hydrologic, geomorphic, and ecological processes that directly influence the quality of aquatic and riparian habitats. Of particular concern in the delineation of RHCA widths are the delivery and movement of water, sediment, and woody debris to the stream channel, which are defined as key processes in determining the ecological health of watersheds (Naiman et al. 1992). Unlike traditionally defined riparian areas, RHCAs include broader reaches of ecologically important lands and are defined in four categories: (1) permanent fish-bearing streams; (2) permanently flowing, nonfish-bearing streams; (3) ponds, lakes, reservoirs, and wetlands >0.4 ha (1 acre); and (4) intermittent streams, associated wetlands <0.4 ha (1 acre), landslides, and landslide-prone areas. The PACFISH strategy provides interim widths for RHCAs that apply until watershed and site analyses are completed (Table 3). These analyses redefine RHCA widths based on local watershed conditions.

KEY WATERSHEDS

The PACFISH strategy includes delineation of key watersheds to provide a network of protected areas across the landscape and to identify top-priority watersheds for conducting watershed analysis and restoration and maintenance activities. Criteria for designating key watersheds consist of the following: watersheds containing one or more anadromous fish stocks at risk of extinction, watersheds that contain high-quality remaining habitat, or degraded watersheds with a high restoration potential (or any combination of the three). Not all watersheds meeting these criteria would be selected, but rather, a representative selection would be made of those containing large tracts of lands administered by the Forest Service or BLM or both. Determination of key watersheds would be coordinated with existing reserve areas, such as research natural areas, wilderness areas, areas of critical environmental concern, national parks, etc. Within the range of the northern spotted owl, a network of 162 key watersheds totaling 3.5 million ha was selected (FEMAT 1993). Key watersheds receive additional and priority protection through use of expanded RHCAs in intermittent streams in order to maintain high-quality habitat and to accelerate restoration of degraded areas.

Table 3. Interim widths of Riparian Habitat Conservation Areas from the edge of the stream channel, water body, wetland, landslide and landslide-prone area.

Habitat area	Forested ecosystems	Non-forested rangelands
Fish-bearing streams	Top of inner gorge; extent of 100-year floodplain; distance equal to height of 2 site potential trees or 91 m, whichever is greatest	Extent of 100-year floodplain
Permanently flowing nonfish-bearing streams	Top of inner gorge; extent of 100-year floodplain; distance equal to height of 1 site potential tree or 46 m, whichever is greatest	Extent of 100-year floodplain
Ponds, lakes, reservoirs, and wetlands >.4 ha	Outer edges of riparian same as forested vegetation, extent of seasonally saturated soils; distance equal to height of 1 site potential tree or 46 m, whichever is greatest	Same as forested ecosystems
Intermittent streams, wetlands <.4 ha, landslide areas in key watersheds	Distance equal to height of 1 site potential tree or 30 m, whichever is greatest	Same as forested ecosystems
Intermittent streams, wetlands <.4 ha, landslide areas outside key watersheds	Distance equal to 1/2 height of site potential tree or 15 m, whichever is greatest	Same as forested ecosystems

Watershed Analysis

Watershed analysis compiles, develops, and integrates physical, biological and social information at a catchment scale ranging from 50 to 500 km^2. Goals of the watershed analysis process include the following.

1. Determine the type, areal extent, frequency, and intensity of watershed processes, including landslides, fire, flood, drought, surface erosion, and other processes affecting the flow of water, sediment, organic material, or disturbance within a watershed.
2. Using the above information, interpret the natural disturbance regime and compare with those under managed conditions.
3. Identify parts of the watershed that are either sensitive to disturbance processes or critical for at-risk resources.
4. Determine the distribution, abundance, life histories, habitat requirements, and limiting factors of critical species.
5. Identify human uses of the watershed, societal concerns, issues, and public perceptions.
6. Integrate the information generated to describe physical, biological, and social conditions into a set of management options, opportunities, and constraints.
7. Use ecologically and geomorphologically appropriate criteria to define the proper extent of RHCA boundaries and other areas of special protection and to define watershed-specific riparian management objectives.
8. Develop monitoring procedures to determine ecological condition and the effectiveness of management actions.
9. Identify restoration opportunities and priorities.

Watershed analysis will be conducted by an interdisciplinary team of four to six specialists. Areas of expertise of team members will vary depending upon local needs, but typically will include aquatic ecology, rangelands/forestry/botany, geomorphology/hydrology, terrestrial ecology, and planning/economics. The Regional Interagency Executive Committee (1995), which includes federal agencies in the Departments of Interior, Agriculture, and Commerce as well as the Environmental Protection Agency, has produced a Federal Guide for Watershed Analysis that establishes an accepted process for conducting watershed analysis. The six-step process established by the Executive Committee consists of the following:

1. characterization of the dominant processes affecting ecosystem function,
2. identification of key elements of the ecosystem that are relevant to management questions,
3. description of current watershed condition,
4. description of historical and reference condition,
5. synthesis of historical versus current conditions, and
6. resulting management recommendations.

The watershed analysis process should be coordinated with other landowners within watersheds. The US Environmental Protection Agency and USDA Soil Conservation Service landowner-assistance programs can assist coordination efforts. Watershed analysis requires that one of the following four spatial scales for analysis be considered in ecosystem management planning: (1) broad regional assessments that may span a multi-state area; (2) river basin or physiographic province assessments that may span many 1000s of km^2; (3) watershed analysis, which covers 50–500 km^2; and (4) site analysis (Fig. 2).

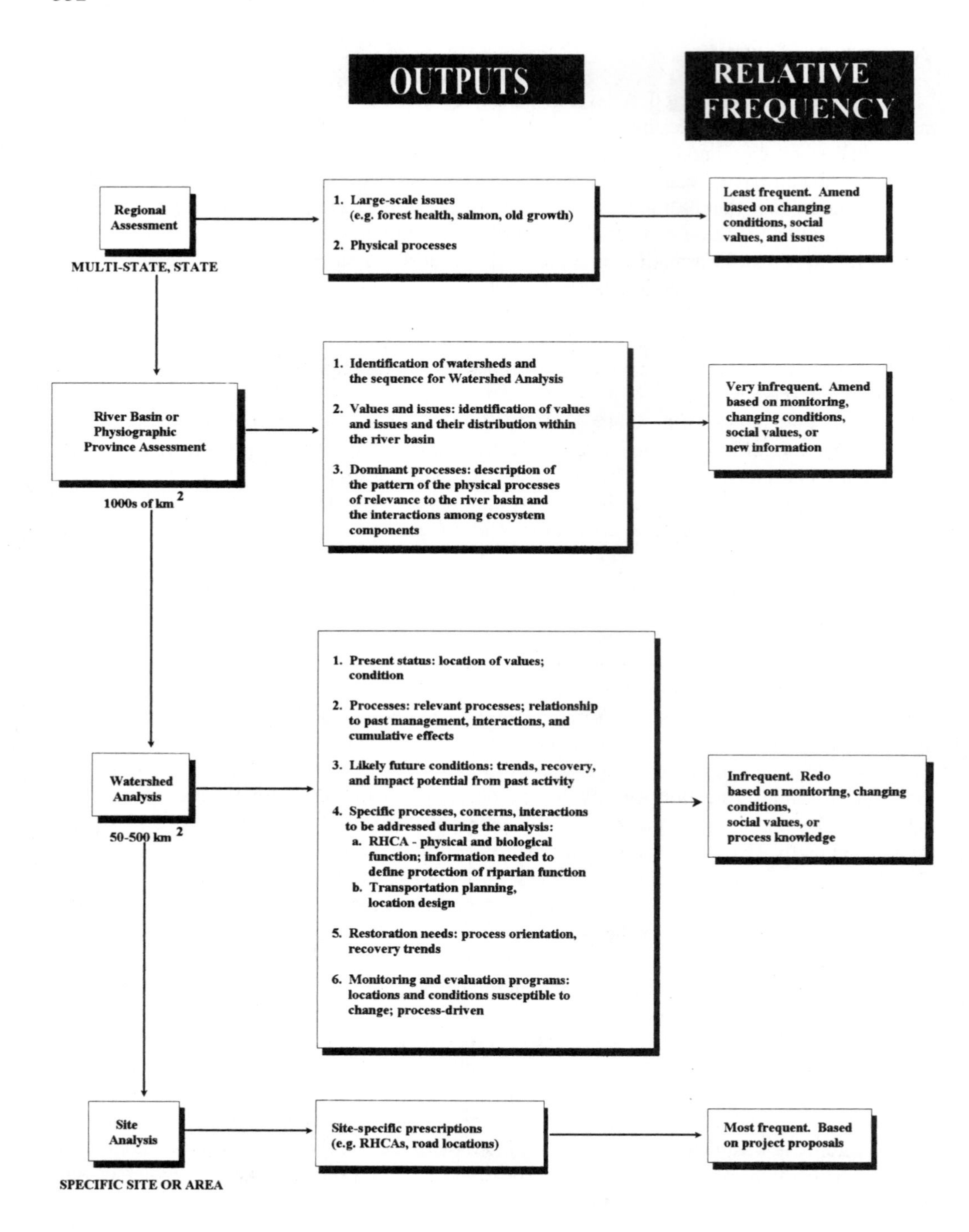

Figure 2. Scales of analysis for ecosystem management showing context for watershed analysis. Modified from Forest Ecosystem Management Assessment Team (1993).

Watershed Restoration

The PACFISH strategy proposes broad changes in land management activities that, if successful, will reverse the processes that have degraded watersheds. Natural restoration, however, may be too slow to save critically endangered, anadromous fish stocks and to restore highly degraded ecosystems. Short-term restoration work may be necessary in such cases to prevent loss of critical resources. Natural recovery of streams in heavily logged riparian areas can be accelerated, for example, by placing large woody debris along stream banks.

Restoration should be designed for individual watersheds through watershed analysis. Repair, removal, or replacement of inadequately designed bridges, culverts, or other stream crossings will be an important element of most restoration programs (Reeves and Sedell 1992), as will riparian silviculture, road rehabilitation, and control of exotic species. Successful stream restoration strategies typically begin in headwater areas utilizing existing ecological processes rather than relying primarily on in-channel structures to mitigate problems in riparian and upland areas (Frissell and Nawa 1992, Doppelt et al. 1993). Such long-term restoration is designed to restore ecological processes and functions and to reconnect rivers to floodplains.

Adaptive Management

An iterative process of goal seeking, management implementation, monitoring of results, and management reformulation is essential for successful ecosystem management (Lee and Lawrence 1986, Walters 1986, Lee et al. 1992). This iterative process of analyzing past actions and refining management strategies often is referred to as adaptive management. A well-designed and implemented monitoring program is a key element of adaptive management for the PACFISH strategy. Successful aquatic and riparian monitoring programs, especially those sufficiently long to differentiate between effects of natural stochastic events and human disturbance, are invaluable but exceedingly rare (Bisson et al. 1992).

Monitoring Programs

In general, monitoring will determine the extent to which the PACFISH strategy elements have been appropriately implemented (implementation monitoring); standards and guidelines, RHCAs, and restoration activities have been successful in achieving riparian goals and management objectives (effectiveness monitoring); and modifications to the strategy assumptions and/or elements are needed (validation monitoring). An inter-agency team of scientists (Chen et al. 1994) has developed an interim protocol for monitoring the effectiveness of the PACFISH strategy. Keys to successful monitoring efforts include establishing clear objectives and quantitative descriptions of management success, which for PACFISH consist largely of the quantified riparian management objectives. The interim parameters selected by the team for monitoring consist of pool frequency, large woody debris, width-to-depth ratio, water temperature, bank stability, and lower bank angle. Additional parameters are likely to be added to the protocol. Monitoring protocols then can be tailored to local watershed conditions through watershed analysis. Further descriptions of monitoring concepts applicable to PACFISH can be found in Chen et al. (1994)

or in Platts et al. (1987), and MacDonald et al. (1991), from which much of the conceptual framework and methods for the PACFISH monitoring protocol were derived.

Monitoring also determines the effectiveness of PACFISH strategy implementation at multiple scales, from individual sites to local watersheds to larger basins. Long-term monitoring will be conducted in reference watersheds. Coordination of federal, state, and tribal resources ensures the necessary consistency across watersheds and ensures availability of adequate skills and other resources.

Benefits of Implementation

The PACFISH strategy is an ecosystem-based management plan for restoring the integrity of stream systems on multiple-use lands and the species dependent upon them. It is not a comprehensive "protection" strategy but, rather, an active management strategy that encourages protection, restoration, and utilization although the latter may be significantly curtailed in some areas. Implementation of PACFISH also will assist in attaining Clean Water Act goals for biological integrity, turbidity, and water temperature. In addition, PACFISH provides an entirely new way of managing land use based on habitat conditions critical to the health of aquatic systems. Successful implementation of PACFISH, FEMAT, and similar ecosystem-based strategies will not necessarily save all at-risk salmon stocks because additional causal factors beyond habitat degradation may be operating. However, their implementation should restore ecological health to watersheds that all salmon depend upon. The "sick lands" observed by Aldo Leopold have spread too far since the 1930s and 1940s and now cover too broad an expanse of the West for the diverse salmon resources as we have known them to survive without a quantum leap in improving habitat condition. Even if all at-risk salmon, steelhead, and sea-run cutthroat stocks become extirpated, implementation of PACFISH would be warranted because it focuses on restoring riparian and aquatic ecosystem processes and functions. Redband trout (*O. mykiss* sspp.), shorthead sculpin (*Cottus confusus*), and other species sensitive to changes in land management practices would benefit substantially from implementation of PACFISH. A slightly modified version of PACFISH has been implemented on Forest Service-administered lands throughout the range of the bull trout (*Salvelinus confluentus*) to help conserve inland native fishes (USDA Forest Service 1995). The salmon are but one beneficiary of healthy riparian and aquatic systems.

Acknowledgments

Development of the PACFISH strategy involved the dedicated efforts of dozens of USDA Forest Service and BLM scientists and managers. The authors are pleased to acknowledge D. Unger, E. Ross, and K. Conn (co-leaders in the Washington Office Policy Group); P. Janik, J. Sedell and G. Secrist (co-leaders of the Washington Office Work Group); and R. Joslin, F. Everest, R. Bastin, and E. Zielinski (co-leaders of the Field Team). L. Brown, K. Burnett, M. Crouse, H. Forsgren, M. Furniss, G. Grant, D. Harr, G. Haugen, R. House, G. Reeves, F. Swanson, R. Wiley, C. Wood, and R. Ziemer provided invaluable support to the various teams in developing

PACFISH. The authors appreciated the comments of P. Janik, F. Swanson, and two anonymous reviewers on earlier versions of this manuscript.

Literature Cited

Benke, A.C. 1990. A perspective on America's vanishing streams. Journal North American Benthological Society 9: 77-88.

Bisson, P.A., T.P. Quinn, G.H. Reeves, and S.V. Gregory. 1992. Best management practices, cumulative effects, and long-term trends in fish abundance in Pacific Northwest river systems, p. 189-232. *In* R.J. Naiman (ed.), Watershed Management: Balancing Sustainability and Environmental Change. Springer-Verlag, New York.

Chen, G., D. Konnoff, G. Reeves, B. House, A. Thomas, R. Wiley, and G. Haugen. 1994. Section 7 fish habitat monitoring protocol for the Upper Columbia River Basin. Sixth version, June 1994. USDA Forest Service. Portland, Oregon.

Dahl, T.E. 1990. Wetlands losses in the United States 1780's to 1980's. US Department of Interior, Fish and Wildlife Service. Washington, DC.

Doppelt, B., M. Scurlock, C. Frissell, and J. Karr. 1993. Entering the Watershed: A New Approach to Save America's River Ecosystems. Island Press, Covelo, California.

Flader, S.L. and J.B. Callicott (eds.). 1991. The River of the Mother of God and Other Essays by Aldo Leopold. University of Wisconsin Press, Madison.

Forest Ecosystem Management Assessment Team. 1993. Forest ecosystem assessment: an ecological, economic, and social assessment. USDA Forest Service. Portland, Oregon.

Frissell, C.A. and R.K. Nawa. 1992. Incidence and causes of physical failure of artificial habitat structures in streams of western Oregon and Washington. North American Journal of Fisheries Management 12: 182-197.

Furniss, M.J., T.D. Roelofs, and C.S. Yee. 1991. Road construction and maintenance, p. 297-323. *In* W.R. Meehan (ed.), Influences of Forest and Rangeland Management on Salmonid Fishes and Their Habitats. American Fisheries Society Special Publication 19. Bethesda, Maryland.

Hanson, M.L. 1987. Riparian zones in eastern Oregon. Oregon Environmental Council. Portland, Oregon.

Higgins, P., S. Dobush, and D. Fuller. 1992. Factors in northern California threatening stocks with extinction. Humboldt Chapter, American Fisheries Society, Arcata, California.

House, R.A. 1992. Management of anadromous salmon and trout habitat and their status in the Salem District. US Bureau of Land Management. Salem, Oregon.

Johnson, K.N., J.F. Franklin, J.W. Thomas, and J. Gordon. 1991. Alternatives for management of late-successional forests of the Pacific Northwest. Agriculture Committee and the Merchant Marine Committee of the US House of Representatives. Washington, DC.

Katibah, E.F. 1984. A brief history of riparian forests in the Central Valley of California, p. 23-29. *In* R.E. Warner and K.M. Hendrix (eds.), California Riparian Systems: Ecology, Conservation, and Productive Management. University of California Press, Berkeley, California.

Lee, K.N. and J. Lawrence. 1986. Adaptive management: learning from the Columbia River Basin fish and wildlife program. Environmental Law 16: 431-460.

Lee, R.G., R. Flamm, M.G. Turner, C. Bledsoe, P. Chandler, C. DeFerrari, R. Gottfried, R.J. Naiman, N. Schumaker, and D. Wear. 1992. Integrating sustainable development and environmental vitality: a landscape ecology approach, p. 499-521. *In* R.J. Naiman (ed.), Watershed Management: Balancing Sustainability and Environmental Change. Springer-Verlag, New York.

MacDonald, L.H., A.W. Smart, and R.C. Wissmar. 1991. Monitoring guidelines to evaluate effects of forestry activities on streams in the Pacific Northwest and Alaska. EPA Technical Report EPA 910/9-91-001. Region X, Seattle

Marcot, B.G., M.J. Wisdon, H.W. Li, and G.C. Castillo. 1994. Managing for featured, threatened, endangered, and sensitive species and unique habitats for ecosystem sustainability. USDA Forest Service Technical Report PNW-GTR-329. Portland, Oregon.

McIntosh, B.A., J.R. Sedell, J.E. Smith, R.C. Wissmar, S.E. Clarke, G.H. Reeves, and L.A. Brown. 1994. Management history of Eastside ecosystems: changes in fish habitat over 50 years, 1935 to 1992. USDA Forest Service Technical Report PNW-GTR-321. Portland, Oregon.

Naiman, R.J., T.J. Beechie, L.E. Benda, D.R. Berg, P.A. Bisson, L.H. MacDonald, M.D. O'Connor, P.L. Olson, and E.A. Steel. 1992. Fundamental elements of ecologically healthy watersheds in the Pacific Northwest coastal ecoregion, p. 127-188. *In* R.J. Naiman (ed.), Watershed Management: Balancing Sustainability and Environmental Change. Springer-Verlag, New York.

Nehlsen, W., J.E. Williams, and J.A. Lichatowich. 1991. Pacific salmon at the crossroads: stocks at risk from California, Oregon, Idaho, and Washington. Fisheries 16(2): 4-21.

Nickelson, T.E., J.W. Nicholas, A.M. McGie, R.B. Lindsay, D.L. Bottom, R.J. Kaiser, and S.E. Jacobs. 1992. Status of anadromous salmonids in Oregon coastal basins. Oregon Department of Fish and Wildlife. Portland.

Platts, W.S. 1991. Livestock grazing, p. 389-423. *In* W.R. Meehan (ed.), Influences of Forest and Rangeland Management on Salmonid Fishes and Their Habitats. American Fisheries Society Special Publication 19. Bethesda, Maryland.

Platts, W.S., C. Armour, G.D. Booth, M. Bryant, J.L. Bufford, P. Cuplin, S. Jensen, G.W. Liekaemper, G.W. Minshall, S.B. Monsen, R.L. Nelson, J.R. Sedell, and J.S. Tuhy. 1987. Methods for evaluating riparian habitat with applications to management. USDA Forest Service Technical Report INT-GTR-221. Ogden, Utah.

Regional Interagency Executive Committee. 1995. Ecosystem analysis at the watershed scale: federal guide for watershed analysis (Vers. 2.2). Regional Ecosystem Office. Portland, Oregon.

Reeves, G.H. and J.R. Sedell. 1992. An ecosystem approach to the conservation and management of freshwater habitat for anadromous salmonids in the Pacific Northwest. Transactions of the 57th North American Wildlife and Natural Resources Conference 57: 408-415.

Skovlin, J.M. 1991. Fifty years of research progress: a historical document on the Starkey Experimental Forest and Range. USDA Forest Service Technical Report PNW-GTR-266. Portland, Oregon.

Swanson, F.J. and C.T. Dyrness. 1975. Impact of clear-cutting and road construction on soil erosion by landslides in the western Cascade Range, Oregon. Geology 3: 393-396.

Thomas, J.W., M.G. Raphael, R.G. Anthony, E.D. Forsman, A.G. Gunderson, R.S. Holthausen, B.G. Marcot, G.H. Reeves, J.R. Sedell, and D.M. Solis. 1993. Viability assessments and management considerations for species associated with late-successional and old-growth forests of the Pacific Northwest. USDA Forest Service. Portland, Oregon.

USDA Forest Service. 1995. Inland native fish strategy environmental assessment. USDA Forest Service, Intermountain, Northern and Pacific Northwest regions. Missoula, Montana.

USDA Forest Service and USDI Bureau of Land Management. 1994. Environmental assessment for the implementation of interim strategies for managing anadromous fish-producing watersheds on federal lands in eastern Oregon and Washington, Idaho, and portions of California. Washington, DC.

USDA Forest Service and USDI Bureau of Land Management. 1995. Decision notice/decision record for the interim strategies for managing anadromous fish-producing watersheds in eastern Oregon and Washington, Idaho, and portions of California. Washington, DC.

Walters, C. 1986. Adaptive management of renewable resources. Macmillan, New York.

Washington Department of Fisheries, Washington Department of Wildlife, and Western Washington Treaty Indian Tribes. 1993. 1992 Washington state salmon and steelhead stock inventory. Washington Department of Fisheries. Olympia, Washington.

Williams, J.E., J.A. Lichatowich, and W. Nehlsen. 1992. Declining salmon and steelhead populations: new endangered species concerns for the West. Endangered Species UPDATE 9(4): 1-8.

Wissmar, R.C., J.E. Smith, B.A. McIntosh, H.W. Li, G.H. Reeves, and J.R. Sedell. 1994. A history of resource use and disturbance in riverine basins of eastern Oregon and Washington (early 1800s–1990s). Northwest Science 68(Special issue): 1-35.

Institutional Solutions: Effective Long-Term Planning and Management

Do We Need Institutional Change?

Robert M. Hughes

Abstract

Despite their cultural and economic values, many populations of Pacific salmonids (*Oncorhynchus* spp.) continue to decline in abundance. The anthropogenic and natural stressors limiting salmonids have been described, and technological or management strategies have been designed to accommodate these stressors. However, these attempts will fail if the fundamental cultural and institutional stressors limiting salmonids are not remedied. The root causes of depleted salmonids lie in a cultural tendency to overpopulate and overconsume, to use reductionist scientific approaches, and to employ outdated institutions and ethics. Although these causes are deeply rooted in civilization, the adaptability of human cultures, particularly western democratic societies, suggests considerable promise for their resolution.

Introduction

> A stationary condition of capital and population implies no stationary state of human improvement (Mill 1857).

> There are a thousand hacking at the branches of evil to one who is striking at the root (Thoreau 1893).

This section of the book is focused on some of the institutional foundations that must be considered for effectively managing the ecosystems that support Pacific salmonids (*Oncorhynchus* spp.). First, I briefly outline how prevailing scientific and management procedures, utilitarian ethics, and population and economic growth influence salmonids. Next, Bottom (1996) provides a historical perspective that illustrates fishery managers' long attraction to fish domestication and culture, and their links with the prevailing social values of the times. Smith and Steel (1996) then compare contrasting social, political, and economic paradigms governing the habitats of Pacific salmon and summarize the differences in attitudes of Oregon and national survey respondents. Bella (1996) describes how the organizations to which most of us belong reinforce some activities while suppressing others and he discusses the need to produce frequent disorder in those organizations. Using a model to describe how organizations function, Bella (1996) demonstrates why the burden of proof must be shifted to protect ecosystems. The final paper in this section is by Lee (1996), who explains why differences in temporal and spatial scales in human values, policies, and salmon hinder problem resolution; he then suggests a process for overcoming these obstacles.

Why do problems that are serious and obvious to ecologists, such as the depletion or extirpation of many Pacific salmon populations, appear of little concern to many people? In answering, I discuss how reductionism impedes scientific progress and natural resource management. Next, the origins of dominant western ethical perspectives and their effects on natural resource management are described. This is followed by a short depiction of human overpopulation and overconsumption. The paper ends with a discussion on why there is considerable hope that such obstacles can be overcome and what aquatic ecologists can do about them.

Scientific Reductionism

The demise of many salmon populations occurs at least partly because ecologists rarely communicate outside their own scientific culture. Moreover, many ecologists tend to ignore the effects of human economic and ethical systems (Fig. 1) on the environment even though humans use or eliminate about 40% of the potential terrestrial net primary production (Vitousek et al. 1986). Instead, natural resource scientists often focus on immediate stressors rather than on the socio-economic policies that stimulate unsustainable resource uses. For example, our restoration efforts typically examine the effects of logging or fish bypass practices instead of addressing how to conserve wood, paper, and water or the need to develop alternative sources of fiber and power. Perhaps this is a useful way to study simple systems as championed by Descartes (Ponting 1991), but it ignores the reality that hatcheries, hydrological developments, harvest, and other types of habitat disruption typically co-occur and are driven by the same utilitarian cultural paradigm (i.e., make it useful for human consumption).

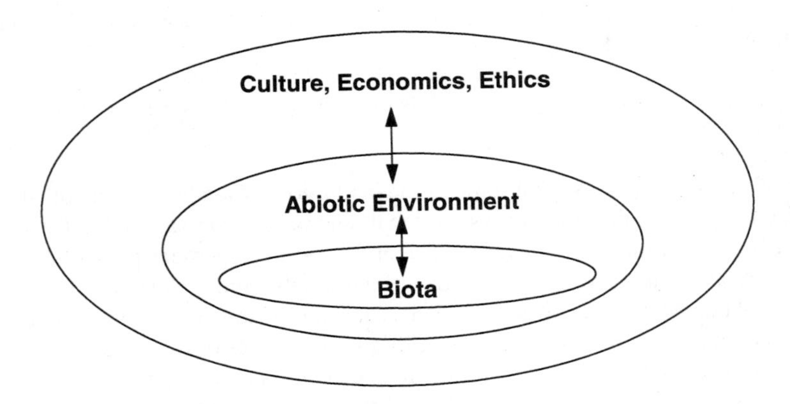

Figure 1. Natural and cultural environments of biota. Note the two-way influence of biota and natural environment and of natural environment and cultural environment. For ecologists and fishery scientists to successfully manage natural resources, the biota and their natural environment cannot be viewed as separate from the cultural environment.

We also tend to be reductionist in the spatial and temporal scope of our research (Hinch 1991). For example, few ecologists examine regional stressors, despite the value of regional and national faunal assessments (Williams et al. 1989, 1993; Nehlsen et al. 1991; Huntington et al. 1996). In a similar temporal vein, we retain the myth of a harvestable surplus, ignoring that large numbers of dead salmon and dead trees are important for supporting other species, cycling nutrients, and buffering against natural fluctuations in abundance (Maser and Sedell 1994, Willson and Halupka 1995, Bilby et al. 1996). This lack of a sufficiently long temporal perspective is of particular concern given the depleted and fragmented state of so many species and populations, and the great amount of variance and disturbance common to nature. We have simply left salmon and many other species with little flexibility to adjust to a series of severe years (Lawson 1993).

Usually biologists direct their efforts at the lower level components such as biological systems and their immediate habitats (Fig. 2). However, if we are to preserve salmon, we must link human values (such as biological diversity and wild salmon) with quantitative indicators (such as native species richness and wild salmon abundance). We must also link the species and their habitat conditions with the socioeconomic policies that generate the stressors. Reductionism in our research organizations, what Daly and Cobb (1989) call disciplinolatry, hinders forming the research teams necessary to demonstrate such links. Recent papers have demonstrated that landscape-level stressors such as land use are associated with degradation of aquatic habitats and organisms (Omernik 1976, Smart et al. 1985, Osborne and Wiley 1988, Hughes et al. 1993, Richards and Host 1994). Continued developments in quantitative geographic analysis, particularly through use of remote sensing and geographic information systems (GIS), should further clarify these connections.

The schism between leaders and ecologists is increased by reductionism in scientific exchanges, problem definition, management strategies, and spatial and temporal perspectives, which hinders meaningful communication between scientists and leaders. Decreasing the reductionism by establishing the links among socioeconomic and environmental values and policies, and

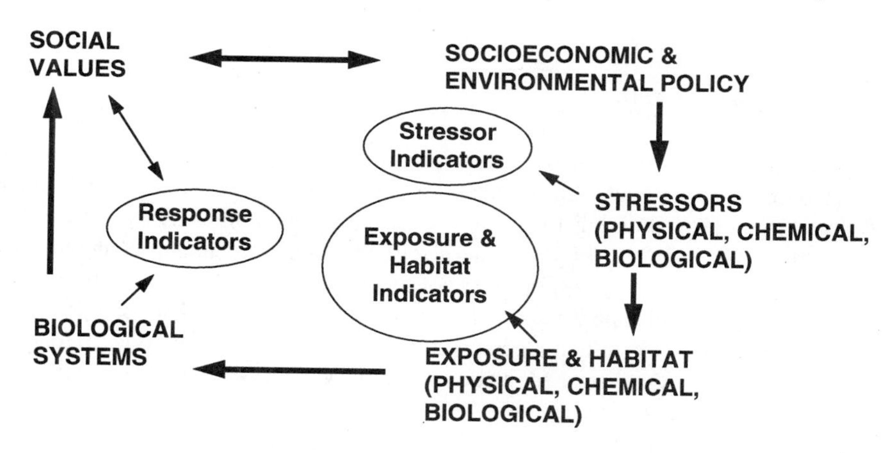

Figure 2. Linkages among social values, policy, and indicators of stressors, habitat, and biological condition. Note that useful biological condition indicators are linked to social values, that biological condition assessments influence policy only through our social values, and that socioeconomic and environmental policies generate stressors or resource use.

habitat, biota, and resource use or stressors would facilitate communication and the policy changes needed to reduce stressor levels (Alverson 1995, Byerly and Pielke 1995).

Management Reductionism

In most resource management agencies, there is a clear focus on the continued production of commodities, such as board feet or cans of salmon. This perspective occurs because of the dominant utilitarian paradigm that stimulated formation of resource management agencies at the turn of the 20th century (Callicot 1991). Such agencies remain poorly prepared to operate within the more holistic ecological or partnership paradigms needed for ecosystem management. It is also questionable whether such terms as resource management and ecosystem management are appropriate when it is people that require the managing (Ludwig et al. 1993). Resources and ecosystems can do quite well without us despite our desire to modify them for our own uses.

Because of a commodity orientation, agency staff also tend to identify more with the individuals that consume or harvest the resources than with people desiring to preserve them. Their major clients are the consumers rather than the non-consumptive users of resources or the ecosystems themselves. Fishery agencies are often concerned with increasing the supply of fish through use of lake fertilization, hatcheries, and stocking (Bottom 1996) instead of reducing the rapidly growing demand. Therefore, utilitarian managers have difficulty understanding and implementing ecological management philosophies.

Even when operating within a utilitarian paradigm, state or provincial and federal agencies often work at cross purposes and rarely cooperate closely (Yoder 1994). Within a single agency, the research, regulatory, monitoring, restoration, and planning components are often poorly coordinated. New ideas, positive criticism, and serious concerns rarely influence senior administrators until situations approach a critical state (Bella 1996).

Environmental Ethics

Ethics are one of the most difficult aspects of any culture to change. Our current environmental ethics have their roots in the Jewish and Greek religions (Ponting 1991, Bowler 1993), medieval views of nature (Sale 1990, Ponting 1991), and 18th-century economics (Daly and Cobb 1989, Hargrove 1989, Kassiola 1990). Major western religions began as human-centered codes of ethics and they have remained so despite clear concerns with nature (Hargrove 1986, Daly and Cobb 1989). Medieval Europeans viewed wild nature as distasteful and fearsome, or simply as a source of resources to be exploited, a perspective that is not uncommon today (Hoage 1989, Sale 1990, Ponting 1991). The rise of economics as both a value system and a science has given it substantial influence over current cultures.

These homocentric utilitarian and technological ethics were combined with an opposition toward indigenous peoples that Europeans brought wherever they immigrated. Rather than learn the indigenous residents' technologies and religions, which had persisted for thousands of years, the immigrants and their offspring imposed their own (Sale 1990). Although the western European perspective produced technological superiority and apparent dominance over nature, its successes required ever-expanding markets and resources. Aboriginal cultures have been the victims of both (Sale 1990, Grove 1995).

The collapse of some Pacific salmonid fisheries (as well as other industries based on natural resources) is a direct consequence of believing that nature is only a source of resources for humans. There are numerous historical examples of this relationship between homocentrism and the collapse of civilizations (Marsh 1885, Ponting 1991). Presumably, if we do not cede some rights to nature, we will continue to destroy it. When combined with advanced technology and greater demand from a quickly growing human population, such destruction can occur with great rapidity. Stewart (1987) has compared the pre-Copernican viewpoint of a flat world and a geocentric universe with the current homocentric perspective, which ignores the evolutionary and ecological revolutions that cast humans as simply another species. A more humble view of ourselves and our role on earth is needed if people are to effectively limit their population sizes and consumption levels and avoid the global collapse predicted by Meadows et al. (1972).

Human Overpopulation and Overconsumption

In addition to the need to change our scientific and ethical paradigms, we must also reduce the growth rate of the human population and our material consumption rate. In simple mathematical terms:

$$I = PC$$

where I = environmental impact,
P = population size, and
C = per capita impact or material consumption rate per capita (Ehrlich and Holdren 1971).

In the United States (US), with a population growth rate of 1.1% yr^{-1} and an economic growth (largely material consumption) rate of ~2%, the environmental impact can be expected to double in 32 years. On a global scale, if population and economic growth rates remain at 2% and 4% respectively, impact would double in 9 years. Clearly these rates or the type of economic growth cannot continue for long. A doubling of impact could only raise Vitousek et al.'s (1986) estimate of human consumption of global net primary production to 80%; a second doubling would be impossible. This means even more massive environmental disruption, substantial changes in population or economic growth, or all three. It is also little wonder that we have been experiencing atmospheric changes, species extinctions, and losses of open space in our lifetimes (Gore 1992).

The preceding equation has three correlates. First, if either population or consumption is held constant and the other continues to increase, environmental impact will also increase. Second, when both components are high, a slight increase in either results in a large impact because of the multiplicative nature of the equation. Third, when both factors are high, a large decrease in either can produce improvement only if species have not already been lost or if habitat has not been irretrievably degraded.

Regardless of the improvements we make in harvest levels, hatchery practices, habitat restoration, or hydrological manipulations, aquatic resources will continue to deteriorate as long as human population and material consumption continue to increase. The cited improvements are largely aimed at the aquatic systems themselves, while most anthropogenic disturbances are directed towards land use. Because aquatic systems are inseparable from the lands they drain, more intensive land use produces degraded aquatic ecosystems.

We must broaden our spatial and temporal perspective as ecologists, managers, and citizens. At least as ecologists, we should realize that prevailing attitudes have produced chronic ecological disasters. Pacific salmon are not the only species in trouble in the Pacific Northwest. For example, Oregon and California rank second and fifth among US states in the number of fishes of concern (Williams et al. 1989). In addition, a focus on nature and technological solutions without incorporating human cultures clearly is bound to fail in protecting ecosystems. The same cultural forces that eliminated or depleted Atlantic salmon (*Salmo salar*) and Great Lakes native salmonids (*Salvelinus* spp., *Coregonus* spp., *Prosopium* spp.) are still at work in the Pacific Northwest (Regier 1996). They will have the same result here if we do not implement fundamental cultural and institutional change, and begin treating the diseases instead of the symptoms. As Hardin (1968, 1972, 1993) wrote, we cannot solve such deeply rooted problems without revolutionary changes in our life styles; the easy, pleasant choices would have been found long ago.

The collapse of many Pacific salmon populations and widespread environmental deterioration are indicators of equally serious social problems. The classical Greeks viewed an insatiable hunger for consumption as a pathological addiction, while neoclassical economics defines it as the engine for economic growth and progress (Kassiola 1990; C. Cobb, Redefining Progress, San Francisco, California, pers. comm.). It is little wonder that a value system, such as modern economics, which promotes selfishness and greed as rational approaches to human action, creates environmental and sociological problems (Hargrove 1989, Kassiola 1990). The same materialism and overpopulation that decimate aquatic and terrestrial ecosystems are also associated with increased crime, lower physical and mental health, and a deteriorated sense of community and freedom (Schumacher 1973, Kassiola 1990).

Both socialist and capitalist economies share a homocentric utilitarian perspective based on continued economic growth (Daly and Cobb 1989, Caldwell 1990, Ponting 1991). Such nations differ mostly in which elites control most of the wealth, and in the degree to which governments and individuals are involved in environmental protection. The desires of capitalists and socialists to stimulate population growth to provide cannon fodder, a source of cheap labor, and a market for cheap products cannot continue to improve the quality of life for workers (Packard 1960, Daly 1973); thus laborers are generally better off when populations are small. Also, the relative benefits of economic growth go largely to the rich or the politically powerful while the costs go largely to the poor (Daly 1973, England and Bluestone 1973, Ehrlich et al. 1977), whether the economy is socialist, capitalist, or more commonly a hybrid of the two.

Sources of Optimism

Social change may seem like a frighteningly difficult prospect, but there is considerable cause for hope. As a group, people of the Pacific region of the US are ecologically progressive (Hoage 1989), and the large amount of public lands in the region often receive much less damaging management than occurs on private lands (Henjum et al. 1994). However, as Smith and Steel (1996) show, people who are most affected by cultural change resist it most strongly. Nonetheless, western and US cultures have periodically granted increased natural rights to both humans and nonhumans (Table 1). Continued ethical evolution of humans can be expected to further extend natural rights to ecosystems. Once our society becomes concerned with ecosystem integrity and protection, the need to reduce our population and material consumption will also become more evident (Gore 1992).

As cultures evolve, so do our sciences and management strategies. Perhaps it is the salmon that have forced us to think more broadly and integrate the waters and lands through which they pass. Whatever the reason, a growing number of aquatic ecologists in the Pacific Northwest have begun to resist the reductionistic nature of their disciplines and have sought out experts in other fields with whom to collaborate. This has greatly benefited our thinking and our sciences, and it must occur more often.

As increasing numbers of scientists collaborate, the institutions themselves have begun to cooperate more closely, reducing the prevailing interagency reductionism. Although the Forest Ecosystem Management Assessment Team (FEMAT) (1993) process is perhaps the best-known current example of interagency cooperation, several others come to mind. Monitoring staff of the water quality agencies of Idaho, Oregon, and Washington, together with the US Environmental Protection Agency (EPA), now meet annually. They have produced a bioassessment protocol (Hayslip 1993) that has been tested in a cooperative stream monitoring program from 1994 to 1996. The EPA, US Fish and Wildlife Service, and National Marine Fisheries Service are developing a common approach for protecting and restoring aquatic habitat on nonfederal lands (B. Spence et al., Umpqua Land Exchange Project, Corvallis, Oregon, unpubl. ms.). The Oregon Departments of Agriculture, Energy, Environmental Quality, Fish and Wildlife, Forestry, State Lands, and Water Resources have begun to develop a watershed assessment program (Daggett 1994). A similar program has arisen with the Washington Departments of Natural Resources and Fish and Wildlife (Carey 1994). These are all recent developments; if they continue to evolve, human and environmental management will become more rational and effective.

In closing, I return to the question with which I began: Why does the depletion of Pacific salmon appear to be of so little concern to managers and the public? First, scientists clearly have

Table 1. Development of natural rights in American and western culture (from Nash 1989).

Type	Development
Endangered species	Endangered Species Act (1973)
	Marine Mammals Protection Act (1972)
Natural ecosystems	Wilderness Act (1964)
African Americans	Civil Rights Act (1957)
Native Americans	Indian Citizenship Act (1924)
European American women	Nineteenth Amendment (1920)
Natural ecosystems	National Park Service (1916)
African Americans	Emancipation Proclamation (1863)
Livestock	Martin's Act (1822)
European American men	Declaration of Independence (1776)
English lords	Magna Carta (1215)
Israelite tribesmen	Ten Commandments (2500 BC)

not communicated the issues definitively to these parties. To effect change, we must do a much better job of getting our message out to the public. Second, most of us tend to watch from the sidelines while environmentalists, agency managers, and resource-based industry representatives argue over harvest levels, hatchery releases, dams, and aquatic habitat restoration.

We must enter these debates ourselves for the following reasons: (1) doing nothing supports the status quo (continued environmental degradation), (2) asking and answering questions about environmental problems is inherently political because human policies generate the stressors, (3) scientific theories and observations are value-laden because they are human constructions, and (4) we and our families live here. When we choose to add our scientific knowledge and professional judgment to these discussions, it is also useful do so through our professional associations. Acting as individuals, we are easily ignored. However, clear statements from professional fishery and ecological organizations on the need for social change are much more likely to be taken seriously by the public and leaders.

Acknowledgments

I thank D. Stouder for inviting me to make this presentation, for supporting my expenses to do so, and for having the patience to wait for me to complete the requested revisions. An earlier draft was improved by suggestions from P. Pister, S. Bryce, and three anonymous reviewers. The viewpoints described in this paper are my own and do not represent policies of the institutions with which I am affiliated. The manuscript was produced outside of normal work hours and it was not subjected to any of the institutions' review or clearance processes.

Literature Cited

Alverson, D.L. 1995. Fisheries management: a perspective over time. Fisheries 20(8): 6-7.

Bella, D.A. 1996. Organizational systems and the burden of proof, p. 617-638. *In* D.J. Stouder, P.A. Bisson, and R.J. Naiman (eds.), Pacific Salmon and Their Ecosystems: Status and Future Options. Chapman and Hall, New York,

Bilby, R.E., B.R. Fransen, and P.A. Bisson. 1996. Incorporation of nitrogen and phosphorus from spawning coho salmon into the trophic system of small streams: evidence from stable isotopes. Canadian Journal of Fisheries and Aquatic Sciences 53: 164-173.

Bottom, D.L. 1996. To till the water: a history of ideas in fisheries conservation, p. 569-597. *In* D.J. Stouder, P.A. Bisson, and R.J. Naiman (eds.), Pacific Salmon and Their Ecosystems: Status and Future Options. Chapman and Hall, New York.

Bowler, P.J. 1993. The Norton History of the Environmental Sciences. W.W. Norton, New York.

Byerly, R., Jr. and R.A. Pielke, Jr. 1995. The changing ecology of United States science. Science 269: 1531-1532.

Caldwell, L.K. 1990. Between Two Worlds: Science, the Environmental Movement, and Policy Choice. Cambridge University Press, New York.

Callicott, J.B. 1991. Conservation ethics and fishery management. Fisheries 16(2): 22-28.

Carey, A.B. 1994. Washington forest landscape management project-progress report. Washington Department of Natural Resources. Olympia.

Daggett, S. 1994. Stage 1 watershed assessment. Oregon Division of State Lands. Salem, Oregon.

Daly, H.E. 1973. The steady-state economy: toward a political economy of biophysical equilibrium and moral growth, p.149-174. *In* H.E. Daly (ed.), Toward a Steady-State Economy. W.H. Freeman, San Francisco.

Daly, H.E. and J.B. Cobb. 1989. For the Common Good. Beacon Press, Boston.

Ehrlich, P.R., A.H. Ehrlich, and J.P. Holdren. 1977. Ecoscience: Population, Resources, Environment. W.H. Freeman, San Francisco.

Ehrlich, P.R. and J.P. Holdren. 1971. Impact of population growth. Science 171: 1212-1217.

England, R. and B. Bluestone. 1973. Ecology and social conflict, p. 190-214. In H.E. Daly (ed.), Toward a Steady-State economy. W.H. Freeman, San Francisco.

Forest Ecosystem Management Assessment Team. 1993. Forest ecosystem assessment: an ecological, economic, and social assessment. USDA Forest Service. Portland, Oregon.

Gore, A. 1992. Earth in the Balance: Ecology and the Human Spirit. Houghton Mifflin, New York.

Grove, R.H. 1995. Green Imperialism: Colonial Expansion, Tropical Island Edens and the Origins of Environmentalism, 1600-1860. Cambridge University Press, Cambridge, England.

Hardin, G. 1968. The tragedy of the commons. Science 162: 1243-1248.

Hardin, G. 1972. Exploring New Ethics for Survival. Viking Press, New York.

Hardin, G. 1993. Living within Limits: Ecology, Economics, and Population Taboos. Oxford University Press, New York.

Hargrove, E.C. 1986. Religion and Environmental Crisis. University of Georgia Press, Athens.

Hargrove, E.C. 1989. Foundations of Environmental Ethics. Prentice Hall, Englewood Cliffs, New Jersey.

Hayslip, G.A. 1993. EPA Region 10 instream biological monitoring handbook. US Environmental Protection Agency, EPA 910/9-92-013. Seattle.

Henjum, M.G., J.R. Karr, D.L. Bottom, D.A. Perry, J.C. Bednarz, S.G. Wright, S.A. Beckwitt, and E. Beckwitt. 1994. Interim protection for late-successional forests, fisheries, and watersheds: the national forests east of the Cascade crest, Oregon and Washington. The Wildlife Society, Bethesda, Maryland.

Hinch, S.G. 1991. Small- and large-scale studies in fisheries ecology: the need for cooperation among researchers. Fisheries 16(3): 22-27.

Hoage, R.J. 1989. Perceptions of Animals in American Culture. Smithsonian Institute Press, Washington, DC.

Hughes, R.M., C. Burch Johnson, S.S. Dixit, A.T. Herlihy, P.R. Kaufmann, W.L. Kinney, D.P. Larsen, P.A. Lewis, D.M. McMullen, A.K. Moors, R.J. O'Connor, S.G. Paulsen, R.S. Stemberger, S.A. Thiele, T.R. Whittier, and D.L. Kugler. 1993. Development of lake condition indicators for EMAP-1991 pilot, p.7-90. *In* D.P. Larsen and S.J. Christie (eds.), EMAP-Surface Waters 1991 Pilot Report. US Environmental Protection Agency EPA/620/R-93/003. Corvallis, Oregon.

Huntington, C., W. Nehlsen, and J. Bowers. 1996. A survey of healthy native stocks of anadromous salmonids in the Pacific Northwest and California. Fisheries 21(3): 6-14

Kassiola, J.J. 1990. The Death of Industrial Civilization: The Limits to Economic Growth and the Repoliticization of Advanced Industrial Society. State University of New York Press, Albany.

Lawson, P.W. 1993. Cycles in ocean productivity, trends in habitat quality, and the restoration of salmon runs. Fisheries 18(8): 6-10.

Lee, R.G. 1996. Salmon, stewardship, and human values: the challenge of integration p. 639-654. *In* D.J. Stouder, P.A. Bisson, and R.J. Naiman (eds.), Pacific Salmon and Their Ecosystems: Status and Future Options. Chapman and Hall, New York.

Ludwig, D., R. Hilborn, and C.J. Walters. 1993. Uncertainty, resource exploitation, and conservation: lessons from history. Science 260: 17, 36.

Marsh, G.P. 1885. The Earth as Modified by Human Action. Charles Scribner's Sons, New York.

Maser, C. and J.R. Sedell. 1994. From the Forest to the Sea: The Ecology of Wood in Streams, Rivers, Estuaries, and Oceans. St. Lucie Press, Delray Beach, Florida.

Meadows, D.H., D.L. Meadows, J. Randers, and W.W. Behrens III. 1972. The Limits to Growth. Signet, New York.

Mill, J.S. 1857. Principles of Political Economy. Vol. 2. J.W. Parker & Son, London.

Nash, R.F. 1989. The Rights of Nature: A History of Environmental Ethics. University of Wisconsin Press, Madison, Wisconsin.

Nehlsen, W., J.E. Williams, and J.A. Lichatowich. 1991. Pacific salmon at the crossroads: stocks at risk from California, Oregon, Idaho, and Washington. Fisheries 16(2): 4-21.

Omernik, J.M. 1976. The influence of land use on stream nutrient levels. US Environmental Protection Agency, EPA-600/3-76-014. Corvallis, Oregon.

Osborne, L.L. and M.J. Wiley. 1988. Empirical relationships between land use/cover and stream water quality in an agricultural landscape. Journal of Environmental Management 26: 9-27.

Packard, V. 1960. The Waste Makers. Penguin Books, Middlesex, England.

Ponting, C. 1991. A Green History of the World. St. Martin's Press, New York.

Regier, H.A. 1996. Old traditions that led to abuses of salmon and their ecosystems, p. 17-28. *In* D.J. Stouder, P.A. Bisson, and R.J. Naiman (eds.), Pacific Salmon and Their Ecosystems: Status and Future Options. Chapman and Hall, New York.

Richards, C. and G. Host. 1994. Examining land use influences on stream habitats and macroinvertebrates: a GIS approach. Water Resources Bulletin 34: 729-738.

Sale, K. 1990. The Conquest of Paradise. Penguin Books, New York.

Schumacher, E.F. 1973. Small is Beautiful: Economics as if People Mattered. Harper & Row, New York.

Smart, M.M., J.R. Jones, and J.L. Sebaugh. 1985. Stream-watershed relations in the Missouri Ozark Plateau Province. Journal of Environmental Quality 14: 77-82.

Smith, C.L. and B.S. Steel. 1996. Values in the valuing of salmon, p. 599-616. *In* D.J. Stouder, P.A. Bisson, and R.J. Naiman (eds.), Pacific Salmon and Their Ecosystems: Status and Future Options. Chapman and Hall, New York.

Stewart, G.W. 1987. The leading question: a paper related to a project on economics and environment. Hydrobiologia 149: 141-157.

Thoreau, H.D. 1893. Walden. F.H. Allen (ed.), Houghton Mifflin, Cambridge, Massachussetts.

Vitousek, P.M., P.R. Ehrlich, A.H. Ehrlich, and P.A. Matson. 1986. Human appropriation of the products of photosynthesis. BioScience 36: 368-373.

Williams, J.D., M.L. Warren Jr., K.S. Cummings, J.L. Harris, and R.J. Neves. 1993. Conservation status of freshwater mussels of the United States and Canada. Fisheries 18(9): 6-22.

Williams, J.E., J.E. Johnson, D.A. Hendrickson, S. Contreras-Balderas, J.D. Williams, M. Navarro-Mendoza, D.E. McAllister, and J.E. Deacon. 1989. Fishes of North America endangered, threatened, or of special concern: 1989. Fisheries 14(6): 2-20.

Willson, M.F. and K.C. Halupka. 1995. Anadromous fish as keystone species in vertebrate communities. Conservation Biology 9: 489-497.

Yoder, C.O. 1994. Toward improved collaboration among local, state, and federal agencies engaged in monitoring and assessment. Journal of the North American Benthological Society 13: 391-398.

To Till the Water—A History of Ideas in Fisheries Conservation

Daniel L. Bottom

Abstract

Ideas about fisheries conservation in the United States (US) have been shaped by social and ecological changes that accompanied rapid economic development and westward expansion in the 19th century. The first major conservation movement in the country occurred after the Civil War as a result of severe fishery depletion in New England and the development of new technology for propagating fish in hatcheries. Fish culture offered a means to restore eastern US fisheries, to provide an income for farmers, and to increase the food supply of an expanding nation. The creation of state fish commissions and the passage of many new protective laws in the 19th century were largely a response to the economic potential of new hatchery technology. The agricultural goals of the US fish culture movement dictated the kinds of scientific questions that were relevant and may explain why fisheries science developed its own ideas and theories distinct from those of community or systems ecology. These ideas emphasized the improvement of fish through hatchery selection, the introduction and acclimatization of species in new environments, and the efficient production and yield of economically viable populations. The utilitarian ideas of fishery conservation presumed a natural world created to satisfy human needs and sought evidence and developed theories to confirm this design. New understanding about the adaptations of fish stocks to their local environments and the recent collapse of salmonid production in the Pacific Northwest have undermined the old agricultural model of applied fisheries science. Yet the choice of ecosystem management as a new metaphor for conservation implies a continuing search for an analytical solution to a value-based problem. A more important role for fisheries than ecosystem management will be to foster a better understanding and appreciation of human ecosystem dependence.

Introduction

The American West, as defined by historian Walter Prescott Webb (1957), is a region containing all or part of 17 large states westward of a line that runs from the southernmost tip of Texas to central North Dakota. With the exception of the Pacific Northwest coast, a distinguishing characteristic of the region is its lack of rain. Early explorers described a treeless, desert plain east of the Rocky Mountains. By 1835 its location was printed prominently on school maps. But

"After the Civil War," Webb (1957) writes, "the Great American Desert was abolished." It was replaced with a garden—first, an imaginary garden supported by ideas that planting trees on the edge of the plains or plowing the soil would somehow increase rainfall (Smith 1947). Such notions may have been bolstered by a temporary wet cycle in the region (Smith 1947, Webb 1957), but as A.M. Simmons described: "Following times of occasional rainy season, the line of social advance rose and fell with rain and drought, like a mighty tide beating against the wall of the Rockies. And every such wave left behind it a mass of human wreckage in the shape of broken fortunes, deserted farms, and ruined homes" (cited in Webb 1957). Ultimately, ideas about the rain gave way to the technological means—dryland farming techniques and large-scale irrigation projects—to create the equivalent of rain. Indeed science and technology did abolish the desert, at least in places, to produce, as Webb (1957) called it, "an oasis civilization" in the West.

The myth of an agrarian utopia in a virgin land has long been a powerful symbol in US literature and history (e.g., Marx 1964). It captured the image of a new beginning for Western society. It affirmed the doctrine of progress and merged with technology and industry to expand civilization across a continent and to create a productive "garden of the world" (Smith 1950). Americans forged a partnership with science and government to shape that vision into reality. Among the active participants in this transformation were the state and federal fisheries agencies that were established (not coincidentally) just as the Great American Desert was being erased from the map. Much more than a symbol, the garden myth had a concrete and literal meaning for fisheries science, which, from the beginning, was defined as a branch of agriculture to help create the garden empire.

The ideals of early fishery management agencies were shaped by the same social forces that promoted westward expansion and economic progress across North America. From this social context, the planting of civilization and its economy was viewed as a struggle of humans against nature. An imaginary wall was built around the garden to insulate it from the hostile forces that lay immediately outside. The wall segregated civilized from uncivilized, harmony from chaos, plenty from scarcity. Like the parallel effort to abolish the desert, fisheries science first built the image of a garden and then fortified it with technology.

Today an entire profession of scientists and resource managers are debating the causes and solutions to severe fishery depletion throughout the Pacific Northwest. This and similar environmental crises worldwide have inspired a creation of new terminology and metaphors to redefine the goals of resource conservation—sustainability, biodiversity, biotic integrity, ecosystem health, and, most recently, ecosystem management. If, as Kuhn (1970) suggests, such change is often a prelude to a sudden shift in the paradigm or world view of a scientific discipline, then this indeed is a critical time for us (natural resource professionals) to understand our scientific heritage. My purpose is to trace the early history of ideas in fisheries management for what they may teach us about ecosystem management. This history shows that scientific beliefs about nature are not separable from society's aspirations but often serve to legitimize them. The same ideas and tools that the fisheries profession developed to produce an agricultural commodity may not be very useful if we now wish to promote the health and integrity of complex natural-cultural systems.

The Myth of the Garden

Following the Civil War, a new technology for propagating fish became widely popular throughout the US and inspired the first significant conservation movement in this country. Between 1870 and 1900 the nation's population more than doubled, and many immigrants settled in eastern cities. At the same time, large expanses of grazing land were converted to corn and wheat production, raising fears of a possible shortage of affordable meat to support an increasingly urban society (Pisani 1984). Fish culture offered a means to restore severely depleted fisheries in the eastern US, to expand the nation's food supply, to provide a source of income for farmers, and to make widely available to all classes of US citizens the most desirable food and game fishes of the world. Fish culture transformed the anxiety of resource scarcity into an engineering opportunity of unlimited potential.

In the 1860s many articles began to appear in the popular press to describe the decline of fisheries in the northeastern states. These often expressed regret and nostalgia over the loss of once seemingly inexhaustible supplies of Atlantic salmon (*Salmo salar*), brook trout (*Salvelinus fontinalis*), and American shad (*Alosa sapidissima*). Articles fondly reported the same tale of fish so abundant that they once were treated with contempt: an 18th-century law in Connecticut reportedly "prohibited masters from forcing trout on their apprentices oftener than three times a week" (Anonymous 1868; see also New York Times, October 24, 1868 and United States Commission of Fish and Fisheries [USCFF] 1880). Various versions of the story give it a mythical character. Goode's (1887) version suggested that apprentices in colonial Connecticut had refused to eat salmon more than twice a week. He also noted that in the Middle Ages, frequent consumption of salmon was believed to contribute to leprosy. For this reason a law in Scotland at the time of Cromwell reportedly required "all masters, and others not to force or compel any servant or apprentice to feed upon Salmon more than once a week" (Goode 1887).

Remorse about fishery depletion was more than compensated by an ebullient enthusiasm for the restoration potential of fish culture. Artificial propagation, it was believed, might restore and even increase abundance of fish as shown by the tremendous survival of their eggs when reared in a controlled environment. One reviewer estimated that with a 90% survival rate for artificially reared eggs, 100,000 salmon released into the upper Merrimack River might produce enough offspring for the next generation to yield 450 million adult salmon or about 2.25 billion lb (~1.03 million mt) of fish (Anonymous 1868)! Fish culturist Stephen Ainsworth predicted that the New England fish commissioners "will soon improve and restore all the rivers of the Eastern States with salmon, brook trout, and shad . . . worth hundreds of millions of dollars to the country (Anonymous 1868)." Forest and Stream, an official journal of the American Fish Culturists' Association, pronounced "our depleted and heretofore barren waters will soon team with unnumbered millions of edible fish, providing greater blessing to the masses than any other discovery of the last quarter of a century" (cited in Pisani 1984). And so the desert was abolished.

Fish propagation offered a new source of income for farmers. Pastoral scenes of fish hatcheries and ponds were depicted on the pages of Harper's New Monthly Magazine (Fig. 1). Like the ancient myth of a Golden Age when the land was so fertile that food was abundant without human toil (Glacken 1990), proponents of fish culture suggested that the water could be sown to produce great quantities of fish with hardly any effort. Numerous articles, often written by fish culturists themselves, offered different versions of the superior productivity of the water compared with the

A

THE TROUTDALE SPRING.

B

TURNING SALMON INTO THE POND.

Figure 1. Views of the garden: (A) The Troutdale Spring, site of the trout hatchery and rearing ponds of Dr. J. H. Slack, Bloomsbury, New Jersey. (Source: Harper's New Monthly Magazine, Anonymous 1868.) (B) Pond near Bucksport, Maine, used to hold adult Atlantic salmon from the Penobscot River prior to hatchery spawning. (Source: Harper's New Monthly Magazine, Anonymous 1874.)

land. These articles variously proclaimed that an acre of water could produce food (or profit) that was (Atkins 1867) 2 times (Anonymous 1868; New York Times, October 24, 1868), 5 times (Green and Roosevelt 1879), or even 10 times (Koss 1883) as much as an equal amount of land. Hume (1893), a canneryman and fish culturist on the Rogue River, boasted that an acre (.4 ha) of any salmon stream in Oregon was worth more in food potential "than forty acres [16 ha] of the best land in the state." New York fish culturist Seth Green, who was among the most eloquent proponents of artificial propagation, described the utility of fish culture in the image of the garden:

> It has been said that an acre of water would produce as much as five acres of land, if it were tilled with equal intelligence. In making such a comparison, it must be borne in mind that the crop of one needs no manure, requires no care during its period of growth, and after it has once been planted, . . . it is harvested by simply taking it from the water in which it dwells. It is almost wholly profit. The other must not merely be planted but must be fertilized at great expense, and worked and cultivated with assiduous labor of man and beast, and . . . it yields but a meagre advance upon the cost of time and trouble . . . . The land we value dearly, because to till it costs us dear in sweat and thought, and the water we despise because it yields its free will offering without an effort on our part. We have tilled the ground four thousand years, we have just begun to till the water (Green and Roosevelt 1879).

Economic Perspectives on Fish Protection

From the outset support for government action on behalf of fish was closely tied to the economic possibilities of artificial propagation. Expectation about hatchery production helped to stimulate the rapid spread of state fish commissions. Maine, Massachusetts, New Hampshire, Connecticut, and Vermont all appointed commissions in 1867. A decade later more than 30 state fish commissions were organized (Goode 1886). The early efforts of the New England states to restore and enhance their fisheries established the basic principles for the other commissions to follow: fishways for dams, artificial propagation of fish, and restrictions on harvest (Allard 1978).

In 1871, Spencer Fullerton Baird became the first commissioner of the USCFF, which was originally established to investigate the causes of fishery decline along the northeast coast. Through Baird's skillful efforts and the lobbying of the American Fish Culturists' Association (the progenitor of the American Fisheries Society), that mission was quickly expanded to include the production and introduction of foodfishes throughout the United States (Allard 1978). From 1871 to 1883, 75–85% of the USCFF budget was dedicated solely to its fish culture program (Goode 1886).

Despite the broad-based support for fish propagation, efforts to conserve fish through restrictive laws generated controversy from all sides of an unsupportive public. Fish culture immediately became the focal point in a debate that boiled down to a simple choice between production or protection. The specific arguments varied with each economic interest.

A growing number of wealthy sportsmen, including many prominent fish culturists who were also among the era's most avid anglers, were the most vocal proponents of habitat protection and restrictive fishing laws (Reiger 1986). However, their motives were not purely altruistic or nonutilitarian. Angling clubs introduced exotic game fishes into fishing streams and lakes, controlled potential predatory species, and constructed artificial ponds for fishing and profit. In the 1860s, for example, many trout ponds were constructed on New York's Long Island by draining wetlands and excavating natural streams. The ponds reportedly were valued anywhere between $15,000 and $100,000 (New York Times, October 24, 1868). On July 7, 1873, the New York Times wrote that "the trout-breeding business is becoming very extensive on the south side of the island, and is proving very profitable." Robert B. Roosevelt—angler, conservationist, state fish commissioner, Ex-Congressman, uncle of Theodore Roosevelt, and member of the Southside Sportsmen's Club—was listed among those who planned to develop a trout pond on Long Island "for going into the business on an extensive scale."

Before the 1860s, hunting and fishing had not been considered a socially acceptable activity unless its purpose was to provide food or income (Reiger 1986). Following the Civil War, the new financial incentives created by fish culture and the establishment of fishing preserves for elite sportsmen may help to explain the greater popular acceptance of angling. If, as the editor of the American Sportsman claimed, fishing was no longer considered "a pursuit for loafers only" (Reiger 1986), it may have been because wealthy individuals were now willing to pay for it.

From the economic perspective of commercial fishermen, harvest restrictions to protect fish were simply unwarranted. The purpose of commercial fishing was to provide food for the public, and artificial propagation promised to ensure a plentiful supply. Therefore, fishermen argued, laws regulating the types of gear or the time periods open to fishing needlessly interfered with their moral purpose and threatened their livelihood. This viewpoint was in obvious conflict with the wealthy angler who advocated restrictions to protect their sport. An anonymous net fisherman from New York described the conflict succinctly:

> The object of stocking our streams is not alone to provide for the amusement of a few wealthy sportsmen . . . but to provide cheap fish food for the masses. It matters not how the fishes are captured, and fish food placed within the reach of all; therefore, nets should be of no objection (New York Times, January 17, 1874).

The commercial fishermen of the American West adopted a similar position. Salmon fishermen and industry representatives on the Columbia River, for example, argued for fish culture in lieu of regulation. More than 170 fishermen, canneries, and industry representatives signed a public letter of protest in the Portland Oregonian (January 22, 1877), flatly stating that the establishment of a hatchery

> is the only protection we want for the future prosperity of this important business. . . . We would respectfully remonstrate against . . . any law preventing [us] from taking fish from the Columbia River at any particular time, or in any particular manner. . . .

Well before the Civil War, the eastern states had struggled with fish protection laws with little success. However, when fish became a form of property through their artificial propagation, this established a new role for government to promote and protect an economic investment. One means the eastern states sought to provide this support was to grant exclusive property rights to lakes or streams dedicated to fish propagation. In 1869, for example, the Massachusetts legislature changed a colonial law that had reserved all ponds >4 ha (10 acres) as a common property for all citizens. The new legislation increased to 8 ha (20 acres) the size of ponds available for exclusive control by riparian owners. Furthermore, ponds >8 ha could be leased to any private citizens for use in fish culture (Steinberg 1991). Similar laws to support artificial propagation or the leasing of private fishing rights were established in other northeast states, including New York (New York Times, July 19, 1868), Maine (New York Times , September 7, 1873), and Connecticut (Milner 1874). The conversion of waters from common property to private real estate helped to stimulate investment in a new economic enterprise.

The development of fish culture also provided new justification for cooperatively managing those fish that migrated across state boundaries. The appointment of state fish commissions began at about the time that fish culturists developed the techniques to propagate commercially valuable anadromous species. Little incentive had existed to restore passage for fish at dams or to seek cooperative legislation between adjoining New England states when fish culturists were experimenting with resident brook trout. Successful propagation of salmon and shad, however, raised the possibility of producing vast quantities of economically valuable species and thereby increased interest in fish restoration. Yet these efforts would be fruitless, it was argued, if the fish were not allowed to reestablish themselves or if all the fish produced in one state only benefited the fishermen downstream in another. Artificial propagation inspired the New England states to push harder for laws requiring fishways at mill dams and to establish temporary fishing moratoria to aid reestablishment of anadromous populations (Steinberg 1991). Thus, the apparent justification for protection became clearer when fish production came under human ownership and control.

Despite such regulatory efforts, fishways generally were ineffective, and the lack of restraint by fishermen continued to erode anadromous populations (Steinberg 1991). The USCFF tried to encourage better fish protection and even threatened to request federal legislation (Allard 1978) or to withhold shipments of hatchery fish unless the states enforced adequate protective measures (e.g., Oregon Fish and Game Protector 1894). However, such threats were rarely if ever carried out. The USCFF had no direct regulatory authority and preferred to stress the politically positive side of its mission—production. Conflicts among net and line fishermen or between sporting clubs and commercial interests were left safely in the hands of the individual states.

The USCFF's emphasis on the production of food for the masses easily undermined any arguments it had made for restricting fisheries or protecting habitat. If artificial production was as successful as the Commissioner himself claimed, then protection of the natural rearing capacity of rivers and lakes was difficult to defend on utilitarian grounds. Spencer Baird could only argue protection as a means to establish self-sustaining runs and admitted "that by continuing indefinitely the practice of artificial impregnation of the eggs and introduction of the young into the water, the supply of fish can be maintained (USCFF 1874a)." On the Columbia River, for example, Baird urged constructing a salmon hatchery to maintain or increase fish production instead of harvest regulations "which cannot be enforced except at very great expense and with much ill feeling" (Oregonian, March 3, 1875). By 1884, Baird's successor, G. Brown Goode, openly acknowledged that, with the exception of some closures on fisheries, the various protective statutes of the states are "generally worse than useless." Given the reality of poor legislation and lack of enforcement, Goode proclaimed:

> The Policy of the United States Commissioner has been to carry out the idea *that it is better to expend a small amount of public money in making fish so abundant that they can be caught without restriction, and serve as a cheap food for the public at large,*

> *rather than to expend a much larger amount in preventing the people from catching the few that still remain after generations of improvidence* (Goode 1886, emphasis in original).

Fish production thus had become the justification both for and against fish protection. From the viewpoint of commercial fishermen, the federal fish culture program removed any need for restrictive laws. From the perspective of many anglers, fish farmers, and state fish commissioners, the new hatchery technology was the principal reason that such laws were now required. While the commercial fishermen who caught wide-ranging anadromous species fought to maintain their unrestricted access to the commons, state commissions and fish culturists sought to protect a private right of property that resulted from their investment in fish propagation. Neither argument, however, had much at all to do with the protection of fish outside the hatchery garden.

Scientific Perspectives on Fish Production

While governments struggled with managing the garden, it was science that provided the means to establish it. The techniques and ideas associated with the early fish culture movement formed the foundation upon which applied fisheries science was built. The agricultural goals of management defined the kind of science that was needed and may help to explain why fishery science developed its own ideas and theories distinct from those of community or systems ecology (McIntosh 1985). Implicit to these roots are basic assumptions about the relationship between humans and the rest of nature, the behavior of natural systems, and the relevant questions for scientific study.

DOMESTICATION

Fish propagation represented a form of animal husbandry. A major theme of the early fish culture movement was the "taming" of wild fish to serve as domestic animals. Animal domestication has had an important influence on ideas about nature since ancient times. The utilitarian ethic was rooted in domestication. Utilitarian ideas were prevalent in the writings of many ancient Greeks, for example, who believed that the inherent order of the earth was the result of divine creation to serve human needs. The very existence of useful plants and animals was offered as evidence of an earth that was designed for human use and improvement (Glacken 1990). Similar utilitarian ideas inherent to modern fisheries conservation are another legacy of the early fish culture movement.

Before the turn of the century, fish culturists frequently compared domestication of fish with the raising of livestock. Kellogg (1857) stated, "I can see no reason why fish may not be stall-fed, so to speak, and fattened as well as other animals." In an address to the New England Fish Commissioners, naturalist Louis Agassiz concluded that fish culture "stands in the same relation as the first attempts stood to domesticate bulls and raise cattle" (New York Times, March 9, 1868). Domestication of fish involved human control of breeding and production to meet certain standards of quality and number. It also implied an exclusive right to the products as personal property. Milner (1874) described the culture of brook trout as providing the means for individual land owners to "have [trout] as a possession almost as much under their control as their cattle or their horses."

Because trout were considered relatively easy to raise and hold in a confined space, they were a preferred species for land owners with ponds or small streams. Some fish culturists described the domestication of trout as they would the taming of a wild animal to become a faithful servant or pet. Koss (1883) noted that trout "come to know man, approach him without fear, and even jump out of the water to get some food which is held out to them." Stone (1901) asked whether artificially bred trout might become domesticated in the same way as "dogs and fowls" so that "they will *prefer* to seek the shelter and food which they find around the homes of men to the precarious chances of a wild and roaming life."

The goal of artificial propagation was to improve fish by accentuating the qualities most valued by fish culturists, anglers, and consumers. Popular reviews of fish culture techniques in the 19th century often referred to the fish preserves of ancient Romans, who were thought to have developed the means to improve the growth and flavor of food fish (e.g., Anonymous 1862). Early fish culture efforts included hybridization experiments to combine the most desirable qualities of paired species. As early as 1873, breeding experiments had been conducted among a variety of European salmon and trout and, in the US, between whitefish (*Coregonus clupeaformis*) and lake trout (*Salvelinus namaycush*) and between alewife (*Alosa pseudoharengus*) and shad (Milner 1874). Whereas many fisheries scientists today ask questions about the effects of hatchery selection on the gene pools of native fish populations, for most of its history, the primary purpose of fish culture was selection. Domestication meant tame and improved.

The most dramatic improvement made possible by artificial propagation was the increased efficiency of fertilization and survival of fish eggs in a hatchery setting. Some early fish culturists were concerned, however, that the rules for selecting breeding pairs in a hatchery were not understood. A.S. Collins and Stephen Ainsworth, for example, experimented with spawning screens to collect eggs that were fertilized naturally. Collins favored natural spawning because he thought it was safer to assume that "fish can do this part of the business best" and would likely choose partners that are "best adapted to each other" (Collins 1870). Eventually, however, such concerns were set aside in favor of the more efficient dry method of fecundation (Fig. 2). As early as the 1850s, Swiss professor Karl Vogt had recognized the advantage of spawning fish manually "in a shallow vessel containing barely water enough to cover the eggs" (Vogt 1857). The technique minimized dilution of the milt and greatly increased the percentage of eggs successfully fertilized. If the method of dry fecundation indeed marked the "first great advancement in fish culture (Bowen 1970)," it also represented the most complete control over the reproductive process. More than any subsequent refinements in fish culture, dry fecundation in a shallow vessel symbolized the complete abstraction of fish reproduction from its aquatic and evolutionary context.

Acclimatization

Fish culture enabled the easy transport of large numbers of fish for introduction into new waters throughout the world. The term acclimatization was applied to fish introductions in the same way it was used to describe the transfer of new varieties of domesticated plants from distant lands. Among the many private and government organizations that became actively involved in fish introductions were various acclimatization societies—the Société d'Acclimatation in Paris

Figure 2. "Dry fecundation in a shallow vessel" as demonstrated by New York fish culturist, Seth Green. (Source: Green and Roosevelt 1879.)

(e.g., USCFF 1879), the Ballarat Acclimatization Society of Australia (USCFF 1887), the California Acclimatizing Society (Milner 1874). Many fish culturists believed that the quality of fish would be improved by simply placing them in a new environment. In a letter to promote a fish culture station for New York's Central Park, Fish Commissioner R.B. Roosevelt again used agricultural crops and domestic animals as examples of the possible benefits of fish introductions:

> Acclimatization as applied to fish culture is yet an almost unexplored field of scientific research. We import plants, vegetables, trees and domestic animals from all parts of the world, and find great profit in so doing. . . . The crossing of our horses and cattle with foreign breeds has been found to improve them greatly . . . while the improvement of plants, under change of climate and domestication, has been in certain cases remarkable. By analogy of reasoning there seems no impediment to our having, in the rivers and lakes of America, the best varieties of fish that exist anywhere throughout the world (New York Times, September 10, 1870).

The USCFF actively experimented with fish introductions to provide a source of protein where it seemed most needed. The Commission tried salt-tolerant species (salmon, shad, alewives) in the Great Salt Lake, Utah (New York Times, January 23, 1873), and shad in the Mississippi River where high-quality food species were absent (Allard 1978). The effects of temperature on fish became an area of particular scientific interest in the search for species that could be adapted to new locations. Fish culturists searched for temperature-tolerant species that could adapt to small ponds in rural areas of the southern US to provide a much-needed protein source (Bowen 1970). It was believed that Pacific salmon and trout found in the upper Sacramento River basin of California—"one of the hottest places in the United States"—could tolerate warmer temperatures than their eastern counterparts (USCFF 1874b). Thus, millions of chinook salmon (*Oncorhynchus tshawytscha*) from the upper Sacramento River were introduced, albeit unsuccessfully, into tributaries of the Mississippi River, the Gulf of Mexico, and the Atlantic Ocean (USCFF 1880).

Among the most famous acclimatization experiments was the introduction of carp (*Cyprinus carpio*), which Spencer Baird emphasized as a solution to his "often-expressed want of a fish for the South." Baird predicted in 1877 that "within ten years to come this fish will become, through the agency of the United States Fish Commission, widely known throughout the country and esteemed in proportion" (USCFF 1879). Carp indeed became widely known.

When the USCFF began its efforts to introduce carp in 1875, the species was already established in private waters in California and New York. In May 1877, after one unsuccessful attempt, the USCFF transported from Germany 227 leather and mirror carp and 118 scaled carp. The fish were temporarily held at a state hatchery near Baltimore, but Baird was able to gain a Congressional appropriation of $5,000 for development of a more suitable facility. In keeping with its important status, the carp were transferred to quarters that Baird had constructed on the grounds of the Washington Monument (Fig. 3; USCFF 1879, Smith 1896, Bowen 1970). Although they were not produced in as great a quantity as salmon and shad, carp were planted throughout the country, and introductions were immediately successful. Smith (1896) reviewed the USCFF's acclimatization experiments in the western US and boasted that carp had become one of the most widely distributed species in the Pacific and Rocky Mountain states.

The ideas of domestication and acclimatization dictated the relevant scientific point of view for early fishery conservation. Acclimatization asked the question: What is the effect of this new environment on this fish? This same basic frame of reference—environment on fish—would continue to direct the priorities of applied fisheries science and management in the 20th century (e.g., see Northcote 1988). The effects of fish on their respective environments generally were not relevant concerns of the agricultural model of fisheries science except when obvious conflicts occasionally arose with other economic interests. For example, when the USCFF made plans to introduce catfish in Oregon in 1908, a newspaper editorial responded with a reminder of the effects of fish on environment:

> These flabby, coarse-meated representatives of the fish family are a shade better than the carp for food purposes, but lack some of the merits of the carp as a fertilizer, and for that reason will not be much more welcome than were the cultus "rooters" which have dug up all of the wapato on the Columbia River bottoms, and have even destroyed the hay on the overflowed meadows along the stream (Portland Oregonian, Oct. 22, 1908).

Widespread introduction of fish across America was made possible by a rapidly expanding transportation network. In 1869, just 2 years after Seth Green had developed the techniques for culturing shad and 2 years before the establishment of the USCFF, a transcontinental rail line was completed, opening a vast corridor for the USCFF's acclimatization experiments. Cross-country transportation of fish began almost immediately. Railroad companies provided support by allowing cans of fish to be carried in baggage cars at no additional cost above the passenger

A

B

Figure 3. A monument to carp: (A) A load of carp fry in fish cans en route to the control station at the carp ponds on the grounds of the partially constructed Washington Monument (in the background); (B) sorting fry at the carp ponds with Tiber Creek (now Constitution Avenue) in the background. Source: Rivinus and Youssef 1992. Photographs courtesy of the United States National Archives.

fare for the messenger. Ten years after its establishment, the USCFF built the first of two specialized rail cars each capable of carrying 5 million fry or 119 kg of fingerling-sized fish (Bowen 1970). By 1900, the USCFF was shipping annually >1 billion eggs and fry of 34 species of fish and invertebrates across a rail system with >400,000 km of track (Smith 1896, Kagan 1966).

Nature's Perfect Ledger

The agricultural goals of early fishery management required a relevant body of scientific theory to support them. Most importantly the utilitarian ideals of fishery conservation were upheld by an 18th- and 19th-century idea of a stable balance of nature (Egerton 1973, Botkin 1990, Callicott 1991). According to this view, nature is in a constant equilibrium and recovers to its former condition after a disturbance. These ideas were consistent with an analytical and mechanistic brand of science. Nature, like a machine, could be reduced to its critical (economic) parts and manipulated to achieve the most efficient production and yield of natural resources. Stability and utility of nature thus were mutually reinforcing ideas: the presumed balance of nature not only permitted cultivation of America's garden, it seemed to provide evidence of a world designed expressly for that purpose.

Early fish culturists were impressed with the tremendous fecundity of fish and the terrible "waste" of that potential in nature (e.g., Atkins 1867). This fecundity was seen as part of a natural balance sheet that precisely took into account the annual losses to disease or predators. Humans, not being a part of nature, were not factored into the ledger and could disrupt the balance. Karl Vogt argued that although large numbers of predators destroy the eggs and fry of fish, "all [nature's] economy is calculated for this proportion of loss; the populations of the waters, if left undisturbed by man, would still remain at the same average level. . ." (Vogt 1857). Thirty years later, Livingston Stone proclaimed that "the quantity of salmon in any specified river before they were molested by man, unquestionably remained constant from year to year" (Stone 1884).

These were not arguments against human intervention in fish populations. To the contrary, the stable and predictable equilibrium enabled certain conscious "improvements" to be made by controlling the sources of natural mortality to claim for humans what was wasted on "useless" predators. Like the utilitarian ideas of ancient agricultural writers, it all seemed part of a purposeful design. Stone (1884) described the large number of fish eggs as "a vast store that has been provided by nature, to hold in reserve against the time when the increased population of the earth should need it and the sagacity of man should utilize it." Early fishery science largely associated fish yield with reproductive survival. These ideas were consistent with an agricultural view that equated production of a food crop with the sowing and survival of the seeds. The success of artificial propagation seemed self-evident since anywhere from 70% to 90% of the eggs survived in the hatchery compared with only a few percent or less in nature.

By the 20th century, fishery conservation sought continued improvements of the technology needed to fulfill its agricultural purpose. Efforts to further minimize "waste" of fish production were expanded to include aggressive predator control programs for certain birds and mammals, and chemical "rehabilitation" projects to remove undesirable "trash fish" from lakes and streams. These promised to further readjust the natural ledger in favor of human use by removing sources of apparent mortality or competition during later stages of fish life. The most intensive management effort, however, involved incremental expansion of the hatchery system and improvements in hatchery technology, nutrition, and disease resistance to increase the numbers, pounds, and finally, the sizes of fish produced and distributed.

A major step in the progressive expansion of hatchery production in the Pacific Northwest came in 1909 with the construction of Oregon's Central Hatchery at Bonneville on the Columbia River. With a rearing capacity of 60 million eggs, the new hatchery sought to maximize production efficiency through economies of scale and centralization of the rearing process. Oregon fish warden H.C. McAllister argued against the construction of numerous small hatcheries in remote locations since, for approximately half the cost, eggs could now be delivered by rail from distant collection sites for rearing at a single facility (Oregon Department of Fisheries 1911). In this way, Central Hatchery epitomized the 20th-century shift toward an industrial model of fish propagation based on the mass production of a factory product (Fig. 4).

As the production capacity of modern hatcheries increased, the distribution system was expanded to accommodate a growing fish supply. More flexible modes of transportation capable of reaching even the most remote locations—truck, pack horse, and finally, airplane—enabled 20th-century acclimatization experiments in an ever greater number and variety of aquatic ecosystems (Fig. 5).

By the first several decades of the 20th century, increasing concern about the rapid development of fishery technology and examples of overharvest of marine species worldwide gave birth to new government research on exploited fish populations (Thompson 1919, McEvoy 1986). Biological investigations by the US Bureau of Fisheries (successor to the USCFF in 1903) during this period represented a shift from earlier management policies that were based on the assumption that hatcheries alone could maintain salmon production. Biologists now also argued for improved detection and prevention of overharvest by fisheries, which required a detailed understanding of salmon life histories, ocean and spawning migrations, and fluctuations in abundance. Early salmon tagging and scale studies began to expand the role of science in order to establish a more rational basis for managing fisheries.

By the late 1930s and 1940s, growing population pressures and resource demands required even greater management vigilance to ensure efficient use and equitable distribution of fishery resources. Fish and game regulations greatly expanded, data collection increased, and the principle of scientific management became firmly established in fishery conservation (Larkin 1977). Scientific management emphasized the importance of systematic measurements by technical experts to determine the most appropriate stocking and harvest levels for fish. The first issue of the Oregon State Game Commission Bulletin (Oregon State Game Commission 1946), for example, was dedicated to "the desirability of a management program based on facts alone." The relevant facts for scientific fishery management involved economic measures of production and yield expressed in numbers and pounds of fish per acre (or kg ha^{-1}), catch per hour, and angler-days of use.

Scientific management operated on the principle of supply and demand. This is perhaps best illustrated by the "catchable trout" programs that were initiated in the 1940s to provide a cost-effective and ready-to-harvest hatchery product in areas that were heavily fished or readily accessible to anglers. Annual reports of the Oregon State Game Commission evaluated trout stocking activities by comparing the total pounds of trout released each year to the total number of licensed sportsmen as an index of "the need of increased production . . . for the growing army of anglers" (Koski 1963). Lost in this index, however, was the fact that hatcheries were helping

A

B

Figure 4. Views of the factory: (A) pond system and (B) troughs at Oregon's Central Hatchery at Bonneville. (Source: Finley 1912.)

to stimulate the demand they hoped to satisfy by expanding the market for their products. Stocking tended to inflate densities of fish and concentrate or increase angling pressure "as evidenced by the number of calls received at the Portland [Oregon Game Commission] office asking where fish have been planted" (Oregon State Game Commission 1952). Many trout stocking programs also were designed to create new fisheries by releasing fish at time periods or in locations—barren lakes in the Oregon's Cascade Mountains, for example—where they did not occur naturally. As

A

B

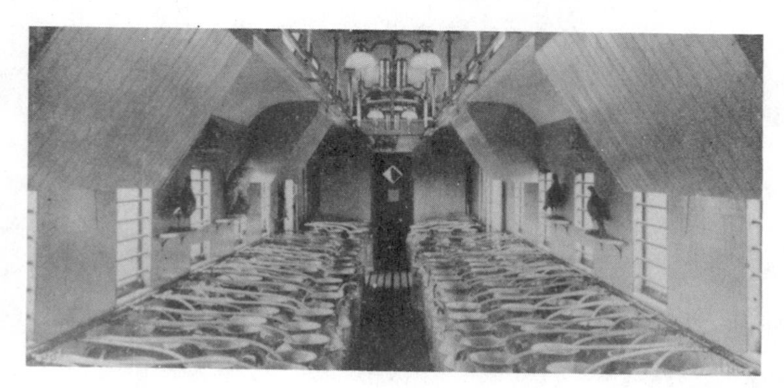

C

Figure 5. Progress in fish transportation: this page—(A) the "Rainbow," Oregon's fish distribution car used from 1913 to 1922, (B) interior of Rainbow with 180 ten-gallon milk cans and air tubes attached capable of carrying up to 180,000 fry, (C) liberation trucks for transporting cans of fish; facing page—(D) pack horses for transporting fish to remote locations, (E) airplane equipped for aerial stocking of Oregon high lakes, and (F) 1969 refrigerated and aerated tanker truck for hauling up to 10,000 legal-sized fish. (Sources: (A) Oregon State Game Commission 1958, (B) Finley 1912, (C) undated, from Oregon Wildlife Commission 1974, (D) Johnson 1914, (E) Oregon State Game Commission 1962, and (F) Koski 1969.)

D

E

F

Figure 5—cont.

suggested by the Game Commission's procedures for allocating hatchery fish, scientific trout management had more to do with the economics of supply and demand than with the ecology of fish and their aquatic environs:

> In general, yearling fish are planted on a put-and-take basis with little expectation of carry-over into the succeeding year. . . . With this in mind, it is unnecessary or illogical to allocate fish to streams on the basis of their "carrying capacity" and it is difficult indeed to determine how many yearling fish should be planted in a stream using the physical and biological aspects of the stream as an index. Certain obvious factors can be considered and for the purposes of allocating fish in an equitable manner, should be sufficient. . . . Aside from the question of suitable trout habitat, the two factors given the most weight in determining the number of fish to be allocated to a particular stream from the available supply are its size and the angling pressure on it (Oregon State Game Commission 1952).

A principal tool of scientific management was the concept of maximum sustainable yield (MSY), which offered a predictive model for managing harvest of the renewable fish crop (Ricker 1948, Larkin 1977). While the details were not fully developed, biologists had become quite familiar with the sustained yield concept as early as the 1930s (McEvoy 1986). By 1950, the Oregon State Game Commission (1950), for example, defined its management goal as "the maintenance of the game resources, while providing the maximum degree of orderly utilization on a sustained yield basis." Maximum sustainable yield, another product of the balance of nature concept, was based on a logistic growth curve developed from animal populations held under constant food supply and environmental conditions (Barber 1988, Botkin 1990). Like the early fish culturists' notion of a vast store of eggs beneficently created for human use, scientific management now defined in precise mathematical terms the harvestable surplus ordained by nature for exploitation by fisheries. Larkin (1977) interpreted the scientific dogma of MSY as follows: "Any species each year produces a harvestable surplus, and if you take that much, and no more, you can go on getting it forever and ever (Amen). You only need as much effort as is necessary to catch this magic amount, so to use more is wasteful of effort; to use less is wasteful of food."

Whereas fisheries focused on the production and growth of populations, other branches of aquatic and academic science—ecology, limnology, and oceanography, for example—developed alternative points of view from other historical roots (e.g., Worster 1977, McIntosh 1985, O'Neill et al. 1986, Northcote 1988). These included ideas about the ecological role of fish within aquatic communities, trophic interactions among species, and the physical and biological processes of stream, lake, and marine ecosystems. However, the effects of fish on or in their native environments had little relevance for a discipline in which ideas were more closely linked to Adam Smith than to Charles Darwin (see Worster 1977). Fisheries conservation viewed nature as a vast agricultural warehouse to supply human needs and sought evidence and developed theories to confirm this purposeful design (e.g., the vast store of eggs just waiting to be propagated, a surplus of fish just waiting to be caught). Non-economically valuable species and ecological interactions were either irrelevant ("useless") to this design or were "limiting factors" to be stricken from nature's ledger so that the warehouse could be fully stocked. Unless the rest of nature could prove itself necessary for fish production, subtle ecological relationships had little significance for the utilitarian ideals of fisheries conservation.

Beyond the Garden

One problem with the agrarian myth is that it did not account for the possibility of failure of the garden. The idea of a stable and benign balance of nature did not account for unpredictable fluctuations in the productive capacities of aquatic ecosystems. Like the infrequent wet periods that some early settlers encountered on the desert plains, large runs of Pacific salmon seemed to confirm both the scientific assumptions and the noble aspirations of fish culture. The presumption of hatchery success, an idea that was consistent with the unrestricted flow of resources in a free market, also was sufficiently appealing to withstand the occasional recession in nature's economy. In years of declining salmon runs, fish managers cited the need for better protective laws while they refined and expanded technology in order to "do hatcheries better." Yet the possibilities that salmon production might spontaneously increase independently of hatchery programs or actually decrease because of them simply were not relevant hypotheses of traditional fisheries conservation. The mere suggestion of these ideas involved rejection of the garden myth.

For example, increased salmon runs in the late 1870s, soon after the establishment of a federal hatchery on the Sacramento River, provided apparent proof that fish culture had succeeded. The USCFF reported that "the increase in yield to the canneries for ten years has been almost exactly proportionate to the increase in the deposition of fry" (Smiley 1884). Although the federal hatchery closed for a few years following a fishery collapse, it reopened in 1888. Increased runs of Sacramento salmon in 1892 were again offered as evidence of hatchery success although, as they had been during the 1870s, such effects were likely the result of a change in climate and a reduction in mining activity in the basin (McEvoy 1986). By 1906, California fish managers reported that the salmon runs of the Sacramento had been completely restored. In retrospect, however, McEvoy (1986) concluded, "it is not clear that the hatchery program ever had any favorable impact on the fisheries with which it was concerned" but rather, "egg-taking activities on the McCloud [in the Sacramento River basin] and other California rivers may actually have contributed to the fishery's decline."

In Oregon, trends in salmon production also were interpreted primarily in their hatchery context. A record pack of cannery salmon for the Columbia River in 1895 was attributed to an increased number of fry released 4 years earlier from the first hatchery in the basin (Oregon Fish and Game Protector 1896). At first, when salmon landings fell from these high levels, the Oregon Department of Fisheries concluded that hatchery production had managed to maintain a sizable harvest and, given the lack of adequate protective laws, had been responsible for preventing an even more serious decline. Fish Warden H.G. Van Dusen claimed a moral and an agricultural victory: "[A]rtificial propagation is the one thing that is preserving the great salmon industry. It is to the fisheries what the sowing of seed is to the farmer, excepting that it is more far-reaching in its effect, as it is laying the foundation for a gradually increasing harvest year by year" (Oregon Department of Fisheries 1905).

Salmon returns in the Columbia River remained low for another decade despite continued hatchery effort. Unconditional faith began to wear thin. Dry weather, overfishing, and predation on salmon fry released from hatcheries were cited among the possible causes of poor returns (Portland Oregonian, October 13, 1906; Oregon Department of Fisheries 1909). However, in 1914, when the pack of salmon from Columbia River canneries again jumped above 500,000 cases, the development of an experimental rearing and feeding program at Oregon's Central Hatchery was credited with the result (Johnson 1984, Lichatowich and Nicholas in press). The

Oregon Fish and Game Commission surmised that "it is practically a foregone conclusion that if hatcheries are run to their full capacity and the laws are permitted to remain as they now exist a steady average increase in the output of the river can be looked for" (Oregon Fish and Game Commission 1917).

Some individuals sounded early warnings that hatcheries were creating an unjustifiable complacency about the need to protect fish or their native habitats. Fish Warden H.G. Van Dusen, for example, cautioned against an exclusive reliance on fish culture and recommended that it "should be invoked as an aid and not entirely as a substitute for reproduction under natural conditions" (Oregon Department of Fisheries 1903). Elmer Higgins of the US Bureau of Fisheries noted that the early popularity of fish culture had "induced a complacent confidence" that had been proven unjustified by a widespread decline in US fisheries, including those for Pacific salmon, American shad, and whitefish (Higgins 1927). Cobb (1930) decried the "almost idolatrous faith in the efficacy of artificial culture of fish" without conclusive evidence to support this belief. Cobb (1930) argued that the fishing industry could not be maintained without also enforcing adequate restrictions on harvest. Despite such warnings and Canada's decision in 1936 to permanently close all of its salmon hatcheries because of poor performance (Columbia Basin Fish and Wildlife Authority 1989), improved fish production through artificial propagation remained a primary objective of Oregon fishery management (Lichatowich and Nicholas in press).

Although Oregon had been propagating fish for 75 years, the first substantive evidence that its hatcheries might directly influence salmon production did not come until the 1960s (Johnson 1984). When hatcheries developed the means to rear vast quantities of large yearling coho (*O. kisutch*) salmon (smolts), both the survival rate of hatchery fish and the total return of adults measurably increased. Like the 1870s return of Sacramento salmon that was "almost exactly proportionate to the . . . deposition of fry," production of Oregon coho salmon in the 1960s apparently increased in proportion to the number of hatchery smolts released (Fig. 6). After 1970, however, adult returns began to fluctuate independently of releases and finally declined despite a steady increase in hatchery output. Subsequent analyses have suggested that the apparent relationship between hatcheries and fish production was fortuitous—the expanding smolt release program in the 1960s coincided with a cyclic increase in regional oceanic productivity (Bottom et al. 1986, Nickelson 1986, Ware and Thompson 1991, Lichatowich 1993). Today, incremental fishery decline and the threat of extinction of numerous Pacific Northwest stocks of salmon (Nehlsen et al. 1991) have undermined traditional assumptions about the effectiveness of hatcheries and about the insignificance of the ocean environment to salmon production.

Long before the collapse of Pacific Northwest salmon runs, new information about the life history of salmonids had already begun changing ideas in fishery conservation. Government-sponsored tagging and scale studies began to provide new insight about the complex life cycles and wide-ranging migrations of salmon species. By 1917 the Oregon Fish and Game Commission acknowledged the "parent stream theory," which held that salmon populations return to their native streams to spawn. At the time, this idea was considered noteworthy because it suggested that it may be "necessary to establish a hatchery on each individual stream in order to maintain an adequate supply [of salmon]" (Oregon Fish and Game Commission 1917). Broader implications of homing behavior, however, gradually became apparent. Rich and Holmes (1928) summarized more than a decade of tagging experiments in the Columbia River and concluded that "the fish belonging to any given tributary enter the river from the ocean at a definite and characteristic time. This is an important point, as it gives additional evidence of the existence of local races

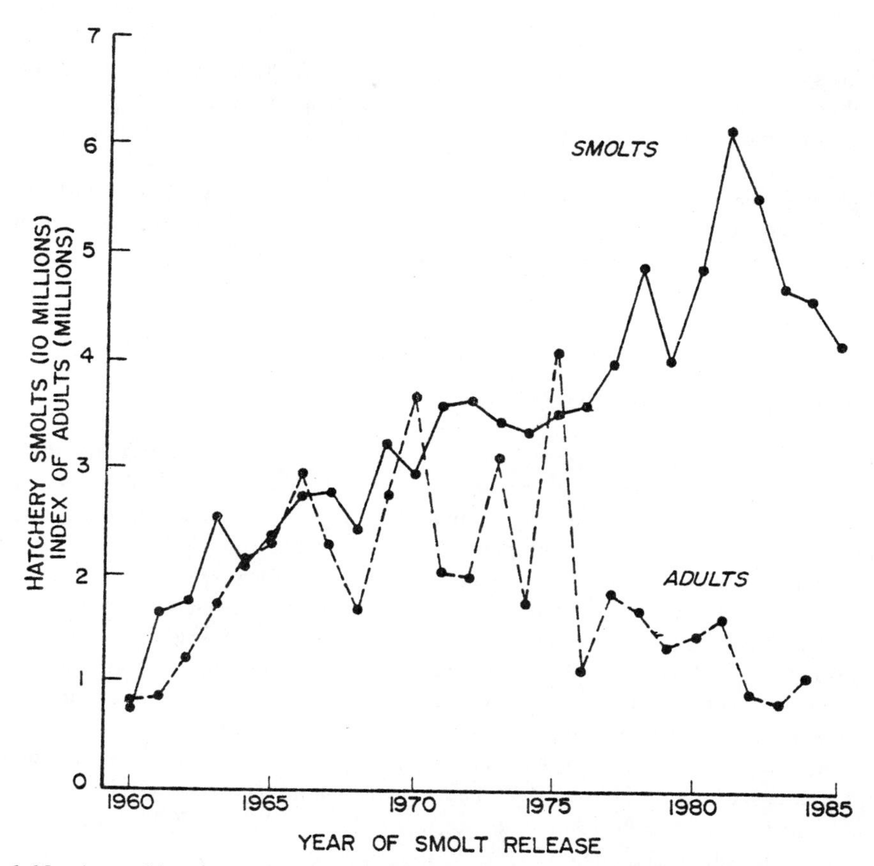

Figure 6. Numbers of hatchery coho salmon smolts released and estimated abundance of adult salmon produced the following year in the Oregon Production Area. (Source: Nickelson 1986.)

in the tributary streams and shows that each race is present in the main river only a comparatively short time. Knowing, further, that each race is self-propagating, it becomes perfectly apparent that all parts of the salmon run in the Columbia River must be given adequate protection if the run as a whole is to be maintained." Rich (1939) later described in detail the importance of local breeding populations of salmon as the fundamental units of conservation for entire species. This stock concept gained wider scientific acceptance as biologists identified unique, heritable characteristics of individual populations—resistance to disease, migration behavior, period of spawning, for example—that were associated with increased fitness in local environments (Ricker 1972, Nehlsen et al. 1991).

The stock concept placed the conservation of fish in a spatial and historical context based on principles of natural selection. In stark contrast to the dry method of fecundation, the stock concept offered evidence that nature indeed may be quite necessary. The adaptations of local stocks implied that fish populations and their native ecosystems are inseparable parts of a co-evolutionary whole. Thus, salmon populations could not be conserved apart from the ecological processes and disturbance histories that created them. Conversely, certain ecological and human cultural systems may not persist in the absence of their local stocks of naturally reproducing salmon. Contrary to agricultural goals based on fish domestication, acclimatization, and production, the stock concept argued for ecological goals based on fish life history, adaptation, and reproduction.

The gradual change in fisheries conservation that began with a steady accumulation of knowledge about locally adapted stocks is today driven by a biodiversity crisis. If resource conservation now seeks to sustain the reproductive capacities of fish within their native habitats, then the agricultural model of fishery management is not very helpful.

Ecosystem Citizenship

More than a century later, the reenactment in the Pacific Northwest of a biological collapse (Nehlsen et al. 1991) similar to the one in New England that led to the establishment of fishery-management agencies in the first place is causing many professionals to reexamine the traditional goals and assumptions of applied fisheries science. Consensus seems to be building in favor of ecosystem management as a new conservation metaphor. In 1994, for example, annual meetings of the Oregon Chapter of The Wildlife Society, Oregon Chapter of the American Fisheries Society, and Western Division of the American Fisheries Society all were dedicated to variations on an ecosystem management theme. State and federal governments and their conservation agencies have established watershed restoration initiatives in an effort to develop and demonstrate principles of ecosystem management. Such changes acknowledge the failure of many single-species protection and recovery efforts (e.g., Williams et al. 1989, Tear et al. 1993) and the increasing scale and interactive effects of human disturbance on habitat linkages, ecosystem processes, and entire landscapes (e.g., Sheldon 1988; Sedell et al. 1990; Frissell 1992, 1993; Forest Ecosystem Management Assessment Team [FEMAT] 1993).

The worthy intent of whole-system conservation would be as difficult to dispute today as the goal of unlimited fish abundance was in the 19th century. Ecosystem management, however, sounds like a contradiction of terms; it suggests that the old reductionist notion of "management" is directly transferable to the conservation of complex systems.

Ecosystems are so complicated that we are unable to precisely control or manage them. Their components are tightly coupled, their behavior depends on an entire history of unknown events, and they do not respond to disturbance in a simple linear way. Many possible causes, therefore, might reasonably explain the same ecological result; present conditions are not necessarily a useful indicator of future performance (e.g., Ehrenfeld 1991, Ludwig et al. 1993). As the history of fish culture attests, such uncertainty often reinforces our preconceived notions because science is unable to absolutely disprove the most socially palatable (e.g., economically satisfying) interpretations of how nature ought to work. If scientific management could not predict or maintain the production of individual species, it seems a bit presumptuous to expect that we could now anticipate and direct all the interacting components of an entire ecosystem. A mechanistic notion of management is not compatible with the concept of an ecosystem.

The idea of ecosystem management sounds like the old idea of scientific management at a grander scale. It implies an analytical brand of science that segregates observer from object (e.g., manager from ecosystem) in the same way that the garden myth separated humans from the rest of nature. If ecosystem management is intended to find better ways for us to live within our ecological means, then the model of the detached observer directing the system from outside may be counterproductive. The term ecosystem management creates a vision of technical experts poring over data, programming computers, and pulling the appropriate ecological levers; the goal to conserve natural-cultural systems suggests an interdependent community with a common history, shared values, and a strong sense of place. Conservation of ecosystems—that is, maintenance of the qualitative context and adaptive processes of life—implies an intimate understanding and firsthand experience unavailable to the disconnected observer. In short, the term ecosystem management seems to offer society an analytical-technical solution to what is principally a normative, value-based problem (e.g., Sagoff 1985, 1991; Kassiola 1990).

Perhaps the more honest role for applied science in matters that never were purely analytic, is to serve as an advisor, like a physician or counselor, to help society identify alternatives and to promote our common interest in something we might imprecisely call ecosystem or community health. Rather than the control *of* nature, this role is intended to help society make sufficient accommodation *for* nature. From this viewpoint, conservation of ecosystems becomes an ongoing process for society to define its cultural-ecological goals rather than simply a means for achieving a particular quantity or predictable endpoint (Sagoff 1985). Unlike the analytic-reductionist methods of scientific management, ecosystem health demands an integrative science to foster a better understanding and appreciation of our own ecological dependence. This role involves fundamental changes from traditional scientific management (see Lichatowich 1996) to lend support for responsible ecosystem citizenship. The following examples illustrate these types of changes:

Reconstructing Nature (Synthesis). The simple balance of nature idea or equilibrium models of populations do not provide a sufficient theoretical or ethical foundation to support the conservation of natural and cultural systems. This new goal requires a framework to help us synthesize rather than further fragment our knowledge (Warren et al. 1979). Recent ideas about self-organization and adaptation in nonlinear systems (Kauffman 1993, Bella 1996) and about the hierarchical structure of biological systems (Frissell et al. 1986, O'Neil et al. 1986) may offer a useful context for applied fishery science. These ideas provide a unified view of both change and stability in complex systems. They logically lead to conservation strategies based on ideas of resilience (Holling 1973, 1986) or system capacities (Warren et al. 1979, Frissell et al. 1986) as alternatives to static equilibrium models of production and yield.

Redirecting Human Development (Planning). Human economies are built on changes in the quality of finite resources. Ecological and economic health require that the patterns and intensities of such change do not exceed the regenerative capacities of living systems. Local actions, therefore, must be placed in a regional context to evaluate whole-system risks and effects. For example, landscape and watershed plans are being developed in the Pacific Northwest to avoid continued degradation of forested ecosystems by logging activities (FEMAT 1993). Yet the spatial patterns of other development actions superimposed on these as well as other (nonforested) landscapes have received little attention in resource conservation. Intensive fish stocking programs, agricultural activities, livestock grazing, and urban developments, for example, extend uniformly across large expanses of the western US landscape without consideration of the cumulative patterns of disturbance or the risk of fragmentation of native ecosystems (e.g., Henjum et al. 1994). Humans now use >40% of the world's net terrestrial primary productivity (Vitousek et al. 1986). Ecological health simply cannot be maintained if human populations irreversibly appropriate ever greater proportions of a finite supply of land, water, and energy.

Restoring Degraded Landscapes (Redevelopment). Throughout the western US, entire aquatic faunas of major drainage basins, not just individual species or populations, are threatened with extinction (Sheldon 1988, Minckley and Douglas 1991, Hughes and Noss 1992, Frissell 1993). A common thread in the decline of native fishes—those that are and are not harvested in fisheries, those that do and do not migrate past dams, and those that do and do not go to sea—is the degradation and simplification of habitat. Even with no further habitat loss, entire landscapes of the Pacific Northwest have been degraded until they have become highly vulnerable to delayed responses from "triggering" events such as floods, droughts, or mass soil movement (Henjum et al. 1994). Restorative or preventative actions will be needed to help protect these systems from future disturbance. Effective restoration must begin with the protection of those few remaining healthy ecosystems that have representative native populations capable of recolonizing adjacent areas. Classification of the diverse types of native ecosystems and assessment of their relative condition are necessary for restoration planning. In this regard, recent efforts to define key watersheds in Oregon and Washington (e.g., FEMAT 1993, Henjum et al. 1994) provide a useful first step toward redevelopment of the western landscape (see Regier and Baskerville 1986).

Restoring Public Choice (Democracy). Public choice in conservation decisions was abolished by the myth of unlimited production, of a benign and stable balance of nature, and of scientific management by experts. If nature is unlimited or can be scientifically controlled to meet society's every desire, then no choices are necessary. Restoring public choice will relieve management agencies of having to defend their inability to provide unlimited quantities of natural resources. This will require greater public understanding of resource issues, greater humility by resource scientists and agencies, and more direct public participation in resource decisions. Conservation of ecosystems requires a society that knows enough to care for them. Applied science has an important role in helping society develop this knowledge—what Orr (1992) has called "ecological literacy." A knowledgeable public is fundamental to open debate and value selection, which are also basic requirements of a functional democracy.

Donald Worster (1977) notes that "every generation writes its own description of the natural order . . . which generally reveals as much about human society and its changing concerns as it does about nature." In the 19th century, to a society settling in a strange land and trying to wrest a living from a harsh environment, the wilderness was a frightening place. The desire was to create a hospitable refuge. America's early fish culture movement helped to replace a chaotic wilderness with a well-ordered garden tended by virtuous yeoman farmers. The task for government was to establish the laws to encourage and protect investment in the garden. The task for science was to provide the theories and techniques necessary to cultivate it. By the 20th century the garden myth had merged with industrial technology to create an economic empire. The model of fisheries science became that of the agro-industrial machine efficiently and precisely managed by technical experts to redesign nature to fit an economic ideal.

Like the 19th-century settlers who faced unaware the reality of the Great American Desert, Northwest fishermen, loggers, and others today risk "broken fortunes, deserted farms, and ruined homes" as a result of unexpected natural-resource collapse. The goals of conservation are rapidly changing because the myth of the garden has been exposed. The science and resource institutions that were built around this myth will become irrelevant unless they can adapt to the reality of life in the desert. If we now wish to redefine conservation in a way that expresses reverence for the complexity of life and an appreciation for ways of knowing that are not purely mechanistic, then we may need to find new metaphors that lend support to ideas of interdependence and community. We, as citizens of the ecosystem, might begin by rethinking the term ecosystem management.

Acknowledgments

I appreciate the many helpful comments on drafts of this paper provided by D. Buchanan, J. Hall, B. Hughes, J. Lichatowich, P. Pister, and three anonymous reviewers. I am very grateful to Jim Lichatowich for the many wide-ranging conversations, ideas, and information sources that have helped to stimulate my curiosity about the history of our profession.

Literature Cited

Allard, D.C., Jr. 1978. Spencer Fullerton Baird and the US Fish Commission: A study in the history of American science. Arno Press, New York, New York.

Anonymous. 1862. Fish culture. Harper's New Monthly Magazine 24: 541-545.

Anonymous. 1868. Fish-culture in America. Harper's New Monthly Magazine 37: 721-739.

Anonymous. 1874. Collecting salmon spawn in Maine. Harper's New Monthly Magazine 49: 29-38.

Atkins, C.G. 1867. Fish culture. The American Naturalist 1(6): 296-304.

Barber, W.E. 1988. Maximum sustainable yield lives on. North American Journal of Fisheries Management 8: 153-157.

Bella, D.A. 1996. Organizational systems and the burden of proof, p. 000-000. *In* D.J. Stouder, P.A. Bisson, and R.J. Naiman (eds.), Pacific Salmon and Their Ecosystems: Status and Future Options. Chapman Hall, New York.

Botkin, D.B. 1990. Discordant harmonies: A new ecology for the twenty-first century. Oxford University Press, New York, New York.

Bottom, D.L., T.E. Nickelson, and S.L. Johnson. 1986. Research and development of Oregon's coastal salmon stocks. Oregon Department of Fish and Wildlife, Fish Research Project NA-85-ABD-00111, Annual Progress Report. Portland, Oregon.

Bowen, J.T. 1970. A history of fish culture as related to the development of fishery programs, p. 71-93. *In* N.G. Benson (ed.), A Century of Fisheries in North America. American Fisheries Society Special Publication no. 7, Washington, DC.

Callicott, J.B. 1991. Conservation ethics and fishery management. Fisheries 16: 22-28.

Cobb, J.N. 1930. Pacific salmon fisheries. US Department of Commerce, Bureau of Fisheries. US Government Printing Office, Washington, D. C.

Collins, A. S. 1870. The impregnation of eggs in trout breeding. The American Naturalist IV(10): 601-610.

Columbia Basin Fish and Wildlife Authority. 1989. Review of the history, development, and management of anadromous fish production facilities in the Columbia River Basin. US Fish and Wildlife Service.

Egerton, F.N. 1973. Changing concepts of the balance of nature. The Quarterly Review of Biology 48: 322-350.
Ehrenfeld, D. 1991. The management of diversity: a conservation paradox, p. 26-39. *In* F.H. Bormann and S.R. Kellert (eds.), Ecology, Economics, Ethics: The Broken Circle. Yale University Press, New Haven, Connecticut.
Forest Ecosystem Management Assessment Team. 1993. Forest ecosystem management: An ecological, economic, and social assessment. USDA Forest Service. Portland, Oregon.
Finley, W.L. 1912. Game and fish protection and propagation in Oregon, 1911-12. Boyer Printing Company, Portland, Oregon.
Frissell, C.A., W.J. Liss, C.E. Warren, and M.D. Hurley. 1986. A hierarchical framework for stream habitat classification: viewing streams in a watershed context. Environmental Management 10: 199-214.
Frissell, C.A. 1992. Cumulative effects of land use on salmonid habitat in southwest Oregon coastal streams. PhD dissertation, Oregon State University, Corvallis, Oregon.
Frissell, C.A. 1993. Topology of extinction and endangerment of native fishes in the Pacific Northwest and California (USA). Conservation Biology 7: 342-354.
Glacken, C.J. 1990. Traces on the rhodian shore: Nature and culture in western thought from ancient times to the end of the eighteenth century. Fifth printing. University of California Press, Berkeley, California.
Goode, G.B. 1886. The status of the US Fish Commission in 1884, p. 1139-1180. *In* Report of the Commissioner for 1884, Part XII. United States Commission of Fish and Fisheries. Government Printing Office, Washington, DC.
Goode, G.B. 1887. American fishes: A popular treatise upon the game and food fishes of North America with especial reference to habits and methods of capture. Estes and Lauriat, Boston, Massachusetts.
Green, S. and R. B. Roosevelt. 1879. Fish hatching and fish catching. Union and Advertiser Co.'s Book and Job Print, Rochester, New York.
Henjum, M.G., J.R. Karr, D.L. Bottom, D.A. Perry, J.C. Bednarz, S.G. Wright, S.A. Beckwitt, and E. Beckwitt. 1994. Interim protection of late-successional forests, fisheries, and watersheds: National forests east of the Cascade Crest, Oregon and Washington. Eastside Forests Scientific Society Panel. Institute for Environmental Studies, University of Washington, Seattle, Washington.
Higgins, E. 1927. The policy of the Bureau of Fisheries with regard to biological investigations, p. 588-598. *In* E. Higgins (ed.), Progress in Biological Inquiries 1926, Appendix VII, Annual Report of the Commissioner of Fisheries for the Fiscal Year ended June 30, 1927, Bureau of Fisheries Document No. 1029. US Government Printing Office, Washington, DC.
Holling, C.S. 1973. Resilience and stability of ecological systems. Annual Review of Ecology and Systematics 4: 1-23.
Holling, C.S. 1986. The resilience of terrestrial ecosystems: local surprise and global change, p. 292-317. *In* W.C. Clark and R.E. Munn (eds.), Sustainable Development of the Biosphere. Cambridge University Press, New York, New York.
Hughes, R.M. and R.F. Noss. 1992. Biological diversity and biological integrity: current concerns for lakes and streams. Fisheries 17(3): 11-18.
Hume, R.D. 1893. Salmon of the Pacific coast. Schmidt Label and Lithographic Co., San Francisco, California.
Johnson, G. 1914. Stocking Cascade mountain lakes, Part II. The Oregon Sportsman, II(9): 14-20.
Johnson, S.L. 1984. Freshwater environmental problems and coho production in Oregon. Oregon Department of Fish and Wildlife, Information Reports Number 84-11, Portland, Oregon.
Kagan, H.H. (ed.). 1966. The American Heritage Pictorial Atlas of United States History. American Heritage Publishing Company, Inc., New York, New York.
Kauffman. S.A. 1993. The Origins of Order: Self-Organization and Selection in Evolution. Oxford University Press, New York, New York.
Kassiola, J.J. 1990. The death of industrial civilization: The limits to economic growth and the repoliticization of advanced industrial society. State University of New York Press, Albany, New York.
Kellog, E.C. 1857. Experiments in artificial fish-breeding. Appendix pages 41-62 *in* G.P. Marsh, Report Made Under Authority of the Legislature of Vermont on the Artificial Propagation of Fish. Free Press Print, Burlington, Vermont.

Koski, R.O. 1963. Fish distribution, p. 390-396. *In* C.J. Campbell and F.E. Locke (eds.), Annual Report, Fishery Division, 1963. Oregon State Game Commission, Portland, Oregon.

Koski, R. 1969. All about fish stocking. Oregon State Game Commission Bulletin 24(6): 3-6.

Koss, R.A. 1883. Trout and trout culture. Pages 473-476 *in* Bulletin of the United States Fish Commission, Volume III for 1883. Government Printing Office, Washington, DC.

Kuhn, T.S. 1970. The structure of scientific revolutions. Second Edition. The University of Chicago Press, Chicago, Illinois.

Larkin, P.A. 1977. An epitaph for the concept of maximum sustained yield. Transactions of the American Fisheries Society 106: 1-11.

Lichatowich, J.A. 1993. Ocean carrying capacity. Mobrand Biometrics, Inc., Vashon Island, Washington.

Lichatowich, J.A. 1996. Evaluating salmon management institutions: the importance of performance measures, temporal scales, and production cycles, p. 69-87. *In* D.J. Stouder, P.A. Bisson, and R.J. Naiman (eds.), Pacific Salmon and Their Ecosystems: Status and Future Options. Chapman Hall, New York.

Lichatowich, J.A. and J.W. Nicholas. In press. Oregon's first century of hatchery intervention in salmon production: Evolution of the hatchery program, legacy of a utilitarian philosophy, and management recommendations. *In* Proceedings of the Symposium of Biological Interactions of Enhanced Wild Salmonids, June 17-20, 1991. Department of Fisheries and Oceans, Pacific Region, Nanaimo, British Columbia, Canada.

Ludwig, D., R. Hilborn, and C. Walters. 1993. Uncertainty, resource exploitation, and conservation: Lessons from history. Science 260: 17 (cont. 36).

Marx, L. 1964. The machine in the garden: technology and the pastoral ideal in America. Oxford University Press, New York, New York.

McEvoy, A.F. 1986. The fisherman's problem: ecology and law in the California fisheries. Cambridge University Press, New York, New York.

McIntosh, R.P. 1985. The background of ecology: Concept and theory. Cambridge University Press, New York, New York.

Milner, J.W. 1874. The progress of fish-culture in the United States, p. 523-558. *In* Report of the Commissioner for 1872 and 1873, Part II. United States Commission of Fish and Fisheries. Government Printing Office, Washington, DC.

Minckley, W.L. and M.E. Douglas 1991. Discovery and extinction of western fishes: A blink of the eye in geologic time, p. 7-17. *In* W.L. Minckley and J.E. Deacon (eds.), Battle Against Extinction: Native Fish Management in the American West. University of Arizona Press, Tucson, Arizona.

Nehlsen, W., J.E. Williams, and J.A. Lichatowich. 1991. Pacific salmon at the crossroads: stocks at risk from California, Oregon, Idaho, and Washington. Fisheries 16(2): 4-21.

Nickelson, T. 1986. Influences of upwelling, ocean temperature, and smolt abundance on marine survival of coho salmon (*Oncorhynchus kisutch*) in the Oregon Production Area. Canadian Journal of Fisheries and Aquatic Sciences 43: 527-535.

Northcote, T.G. 1988. Fish in the structure of freshwater ecosystems: a "top-down" view. Canadian Journal of Fisheries and Aquatic Sciences 45: 361-379.

O'Neill, R.V., D.L. DeAngelis, J.B. Waide, and T.F.H. Allen. 1986. A hierarchical concept of ecosystems. Princeton University Press, Princeton, New Jersey.

Oregon Department of Fisheries. 1903. Annual reports of the Department of Fisheries of the State of Oregon to the Legislative Assembly, twenty-second regular session (for years 1901-1902). Salem, Oregon.

Oregon Department of Fisheries. 1905. Annual reports of the Department of Fisheries of the State of Oregon for the years 1903 and 1904 to the twenty-third Legislative Assembly. Salem, Oregon.

Oregon Department of Fisheries. 1909. Annual reports of the Department of Fisheries of the State of Oregon to the Legislative Assembly, twenty-fifth regular session (for years 1907-1908). Salem, Oregon.

Oregon Department of Fisheries. 1911. Biennial report of the Department of Fisheries of the state of Oregon to the twenty-sixth Legislative Assembly (for years 1909 and 1910). Salem, Oregon.

Oregon Fish and Game Commission. 1917. Biennial report of the Fish and Game Commission of the State of Oregon to the Governor and the Twenty-ninth Legislative Assembly. Portland, Oregon.

Oregon Fish and Game Protector. 1894. First and second annual reports of the Fish and Game Protector to the Governor, 1893-1894. Salem, Oregon.

Oregon Fish and Game Protector. 1896. Third and fourth annual reports of the state Fish and Game Protector of the State of Oregon, 1895-1896. Salem, Oregon.

Oregon State Game Commission. 1946. Dedication. Oregon State Game Commission Bulletin 1(1): 1. Portland, Oregon.

Oregon State Game Commission. 1950. Biennial report of the Oregon State Game Commission, 1949-1950. Portland, Oregon.

Oregon State Game Commission. 1952. Annual report, Fishery Division, 1951. Portland, Oregon.

Oregon State Game Commission. 1958. Photos then and now, p. 24-25. *In* Oregon State Game Commission Biennial report, 1957-1958. Portland, Oregon.

Oregon State Game Commission. 1962. Fish production and stocking, p. 14-16. *In* Oregon State Game Commission Biennial report, 1961-62. Portland, Oregon.

Oregon Wildlife Commission. 1974. Historical highlights. Oregon Wildlife 29(1): 6-7.

Orr, D.W. 1992. Ecological literacy: Education and the transition to a postmodern world. State University of New York Press, Albany, New York.

Pisani, D.J. 1984. Fish culture and the dawn of concern over water pollution. Environmental Review 8: 117-131.

Regier, H.A. and G.L. Baskerville. 1986. Sustainable redevelopment of regional ecosystems degraded by exploitive development, p. 75-103. *In* W.C. Clark and R.E. Munn (eds.), Sustainable Development of the Biosphere. Cambridge University Press, New York, New York.

Reiger, J.F. 1986. American sportsmen and origins of conservation. Revised Edition. University of Oklahoma Press, Norman, Oklahoma.

Rich, W.H. and H.B. Holmes. 1928. Experiments in marking young chinook salmon on the Columbia River, 1916 to 1927. Bull. US Bureau of Fisheries 44: 215-264.

Rich, W.M. 1939. Local populations and migration in relation to the conservation of Pacific salmon in the western states and Alaska. Contribution No. 1. Fish Commission of Oregon, Portland, Oregon.

Ricker, W.E. 1948. Methods of estimating vital statistics of fish populations. Indiana University Publications Science Series No. 15. Bloomington, Indiana.

Ricker, W.E. 1972. Hereditary and environmental factors affecting certain salmonid populations, p. 19-160. *In* R.C. Simon and P.A. Larkin (eds.), The Stock Concept in Pacific Salmon. University of British Columbia, Vancouver, British Columbia, Canada.

Rivinus, E.F. and E.M. Youssef. 1992. Spencer Baird of the Smithsonian. Smithsonian Institution Press, Washington, DC.

Sagoff, M. 1985. Fact and value in ecological science. Environmental Ethics 7: 99-116.

Sagoff, M. 1991. On making nature safe for biotechnology, p. 341-365. *In* L. Ginzburg (ed.), Assessing Ecological Risks of Biotechnology. Butterworth-Heinemann, Boston, Massachusetts.

Sedell, J.R., G.H. Reeves, F.R. Hauer, J.A. Stanford, and C.P. Hawkins. 1990. Role of refugia in recovery from disturbances: modern fragmented and disconnected river systems. Environmental Management 14: 711-724.

Sheldon, A.L. 1988. Conservation of stream fishes: Patterns of diversity, rarity, and risk. Conservation Biology 2: 149-156.

Smiley, C.W. 1884. What fish culture has first to accomplish, p. 65-68. *In* Bulletin of the United States Fish Commission for 1884, Volume IV. Government Printing Office, Washington, DC.

Smith, H.M. 1896. A review of the history and results of the attempts to acclimatize fish and other water animals in the Pacific states, p. 379-472. *In* Bulletin of the United States Fish Commission for 1895, Volume XV. Government Printing Office, Washington, DC.

Smith, H.N. 1947. Rain follows the plow: The notion of increased rainfall for the Great Plains, 1844-1880. Huntington Library Quarterly X: 169-193.

Smith, H.N. 1950. Virgin land: the American West as symbol and myth. Vintage Books, New York, New York.

Steinberg, T. 1991. Nature incorporated: Industrialization and the waters of New England. Cambridge University Press, New York, New York.

Stone, L. 1884. The artificial propagation of salmon in the Columbia River basin. Transactions of the American Fish-Cultural Association. Thirteenth Annual Meeting: 21-31.

Stone, L. 1901. Domesticated trout: How to breed and grow them. Sixth Edition. Cape Vincent, New York.

Tear, T.H., J.M. Scott, P.H. Hayward, and B. Griffith. 1993. Status and prospects for success of the Endangered Species Act: A look at recovery plans. Science 262: 976-977.

Thompson, W.F. 1919. The scientific investigation of marine fisheries, as related to the work of the Fish and Game Commission in Southern California. US Fishery Bulletin No. 2. State of California Fish and Game Commission, Sacramento.

United States Commission of Fish and Fisheries. 1874a. Concluding remarks, p. lxxvii-lxxxiii. *In* Report of the Commissioner for 1872 and 1873, Part II. United States Commission of Fish and Fisheries. Government Printing Office, Washington, DC.

United States Commission of Fish and Fisheries. 1874b. Reports of special conferences with the American Fish-Culturists' Association and State Commissioners of Fisheries, p. 757-773. *In* Report of the Commissioner for 1872 and 1873, Part II. United States Commission of Fish and Fisheries. Government Printing Office, Washington, DC.

United States Commission of Fish and Fisheries. 1879. Propagation of food fishes, p. 18-45. *In* Report of the Commissioner for 1877, Part V. United States Commission of Fish and Fisheries. Government Printing Office, Washington, DC.

United States Commission of Fish and Fisheries. 1880. Human agencies as affecting the fish supply, and the relation of fish culture to the American fisheries, p. xlv-lvii. *In* Report of the Commissioner for 1878, Part VI. United States Commission of Fish and Fisheries. Government Printing Office, Washington, DC.

United States Commission of Fish and Fisheries. 1887. Courtesies to foreign countries, p. xxxviii-xlii. *In* Report of the Commissioner for 1885, Part XIII. United States Commission of Fish and Fisheries. Government Printing Office, Washington, DC.

Vitousek, P.M., P.R. Ehrlich, A.H. Ehrlich, and P.A. Matson. 1986. Human appropriation of the products of photosynthesis. Bioscience 36:368-373.

Vogt, K. 1857. Artificial fish breeding, p. 22-52 (appendix). *In* G.P. Marsh, Report Made Under Authority of the Legislature of Vermont on the Artificial Propagation of Fish. Free Press Print, Burlington, Vermont.

Ware, D.M. and R.E. Thomson. 1991. Link between long-term variability in upwelling and fish production in the Northeast Pacific Ocean. Canadian Journal of Fisheries and Aquatic Sciences 48: 2296-2306.

Warren, C.E., M. Allen, and J.W. Haefner. 1979. Conceptual frameworks and the philosophical foundations of general living systems theory. Behavioral Science 24: 296-309.

Webb, W.P. 1957. The American West, perpetual mirage. Harper's Magazine 214: 25-31.

Williams, J.E., J.E. Johnson, D.A. Hendrickson, S. Contreras-Balderas, J.D. Williams, M. Navarro-Mendoza, D.E. McAllister, and J.E. Deacon. 1989. Fishes of North America endangered, threatened, or of special concern: 1989. Fisheries 14: 2-20.

Worster, D. 1977. Nature's economy: A history of ecological ideas. Cambridge University Press, New York, New York.

Values in the Valuing of Salmon

Courtland L. Smith and Brent S. Steel

Abstract

Ecosystems are one context in which to assess options for restoring Pacific Northwest salmon (*Oncorhynchus* spp.). The future of salmon, however, rests on broader and more fundamental ethical issues that relate to economy and ecology. Concern for the relative weight given to economy and ecology is not new, and it is not limited to the Pacific Northwest. For over a hundred years, the push and pull of ecology and economy has been debated. We review the background of several paradigms that contrast ecological and economic values. We present evidence on Pacific Northwest and national public thinking about the questions of ecology and economy with respect to forest, fish, and water resources. Survey data show strong support for protection and restoration of salmon populations. We find that the people who exhibit greatest support for protection and restoration are urban, younger, female, and have resource use experience. We think that the politics of salmon that characterizes the fishery management process is indispensable for setting the socio-political priorities for Pacific Northwest salmon and other natural resources. To move forward, we must engage public debate on how salmon and ecosystems fit into the our future economic and ecological vision.

Introduction

> Examine each question in terms of what is ethically and esthetically right, as well as what is economically expedient (Leopold 1949: 262).

Maintaining and restoring salmon (*Oncorhynchus* spp.) in the Pacific Northwest depend, in part, on the values people have for salmon and a quality environment. People's values and their valuing of salmon determine the support for recommended actions. Values indicate what people think is right ethically and esthetically. Valuing is what they think the economic costs and benefits might be.

Politicians expend considerable effort and expense to determine public attitudes for major political decisions. Unfortunately, for many environmental policy decisions, there is little systematic study of people's attitudes. This is the case with the values people hold about salmon. What is the level of support for restoration and environmental programs? What groups are more likely to support these programs? Has the support for environmental values been changing over the last generation? Are people comfortable with the recommendations of scientists?

One way to assess people's values is to observe the actions they take and decisions they make. Often managers want to know how people might react to policy choices ahead of time. A tool for doing this is the attitude survey (Dillman 1991). Federal privacy policy and limited financial resources severely restrict the opportunities of federal agencies to assess people's attitudes using a survey. Recent surveys done by university social scientists working on forest problems offer insights into people's values with respect to fish, wildlife, and especially salmon (Steel et al. 1992, 1994; Steel and Brunson 1993).

Values in the Ecology–Economy Debate

Many scholars studying public attitudes note the tensions between ecological and economic values (Olsen et al. 1992, and see below). From simple dichotomies to distinguishing several intermediate positions, ecology and economy seem rooted in different value positions about what is right. The persistence of each position suggests that perhaps each perspective offers a set of values that touch people materially, ethically, and esthetically. In the development of the American west, the ecology–economy value dichotomy is seen in the arguments between John Muir and Gifford Pinchot, the 1960s revival of interest in Henry Thoreau, and emphasis on Aldo Leopold in the 1990s.

The Pinchot–Muir debate in the early 1900s over a proposal to construct the Hetch Hetchy water and power project near the Yosemite Valley, California, shows the contrast between the more ecologically oriented preservation and economically oriented conservation value positions (Jones 1965). In 1901 San Francisco filed a claim 32 km north of the Yosemite Valley for a new source of water and electric energy. Muir and the newly founded Sierra Club fought to establish Yosemite as a national park and to protect it from the depredation of developers. "Pinchot's philosophy of conservation for use collided with Muir's conviction that the best parts of the woodlands should be preserved" (Udall 1963). The Hetch Hetchy project was approved in 1913 despite Muir's bitter charges that "these temple destroyers, devotees of ravaging commercialism seem to have a perfect contempt for Nature, and instead of lifting their eyes to God of the Mountains, lift them to the Almighty Dollar" (Udall 1963).

Conservation values have guided resource management for much of the 20th century. Conservation promised that human manipulation of nature could maintain sustainability and actually increase the quantity of resources available. The conservation philosophy promoted the wise use of resources. Conservationists opposed "locking up" resources that could be used for the benefit of society. They saw themselves as the moderates between those supporting human-centered development and those wanting to withdraw areas and resources from human use.

The conservation–preservation debate of the 1960s was influenced by the writings of Thoreau, who had strongly influenced Muir. The writings of Thoreau provided the value base for people concerned about too much consumption and too high a priority given to economic growth. In 1858 Thoreau advocated living with less. Walden Pond, Massachusetts, was Thoreau's experiment in simple living and having less impact on the environment. Thoreau said, "The necessaries of life . . . may, accurately enough, be distributed under the several heads of Food, Shelter, Clothing, and Fuel." When someone has enough of these, that person's desires are "surely not more warmth of the same kind, as more and richer food, larger and more splendid houses, finer and more abundant clothing, more numerous incessant and hotter fires, and the like." Those bent on material acquisitions Thoreau criticized as being "not simply kept comfortably warm, but unnaturally hot." "I went to the woods," he said, "because I wished to live deliberately, to front only the essential facts of life, and to see if I could not learn what it had to teach, and not, when I came to die, discover that I had not lived" (Krutch 1962).

Two oil price shocks, a severe recession, limited job growth, and stagnant wage growth in the 1970s and 1980s found Thoreau's values wanting. With the rise of the 1990s environmentalism, Aldo Leopold's broader, ecosystem approach served as a model. In the A–B cleavage (Table 1), Leopold contrasts commodity production (A) with natural production (B). Leopold (1949) says, "In each field one group (A) regards land as soil, and its function as commodity–production; another group (B) regards the land as biota and its function as something broader." Leopold's views serve as the basis from which to build ecosystem management.

Through the 1970s and 1980s, many scholars tried to capture the values that distinguish the economy–ecology contrast and name the different value positions. The phrasing includes developmentalists versus preservationists, shallow versus deep ecology, and humans above and apart from nature versus humans one with and a part of nature. A number of bipolar conceptualizations have been proposed. For example, Hirshleifer (1978) contrasts the "natural" versus "political" economy. Schurr et al. (1979) propose "two diametrically opposed world views," the "expansionist" and "limited" views. Sale (1985) differentiates the "industrio–scientific" and "bioregional" paradigms. Devall (1985) identifies the "dominant worldview" and "deep ecology" positions. On the economic side of each contrast, emphasis is on exploitation, domination, materialism, consumerism, centralization, competition, hierarchy, and technology as dimensions of current practice as opposed to the limits to growth, the need to live in harmony with nature, to recycle and to simplify material needs, to cooperate, and the desire for egalitarianism and appropriate technology that should characterize human interaction with the environment. Olsen et al. (1992) document a shift in public values away from the "old Technological Social Paradigm" toward accepting "a new Ecological Social Paradigm." This trend is reflected in the ecology–economy positions of Lutz (1992) and Callenbach et al. (1993). They see an evolution from industrial to superindustrial to postindustrial. Some of the dimensions of this evolution are as follows:

- growth euphoria to growth limits to principle of sustainability,
- environmental pollution to environmental laws to environmental restoration,
- material basic attitude to saturation and stagnation to postmaterial orientation, and
- mechanistic models to cybernetic models to systemic models.

Adapting the ecology–economy views to issues of salmon in the Pacific Northwest, Lichatowich (1992) depicts an "industrial" and "natural" economy (Table 2). These two economies "reflect the difference between conservation as defined by Aldo Leopold (1966) and

Table 1. The characteristics of Leopold's (A) commodity–production (B) and natural–production cleavages. Source: Leopold (1949).

A. Commodity–production	B. Natural–production
Agronomic production	Biotic
Scientific agriculture	Biotic farming
Conqueror	Biotic citizen
Land as slave and servant	Land as the collective organism

Table 2. Differences between the industrial and natural economies. Source: Lichatowich (1992).

Industrial economy	Natural economy
Fossil fuel	Solar energy
Centralized production	Dispersed production
Monocultural	Diversity
Linear, extractive	Circular, renewable
Emphasizes production	Emphasizes reproduction
Waste & poisonous by products	No waste, recycle

immediate economic gain; between thinking of the future and only of the present . . ." (Lichatowich 1992). In addition to characterizing the two positions, Lichatowich (1992) identifies the measures appropriate to each type of economy. The industrial economy uses productivity measures such as angler days, catch per unit of effort, number of juveniles released, and escapement. The natural economy measures system qualities such as diversity, system productivity, and fitness.

Some try to identify intermediate positions in the economy–ecology dichotomy as if searching for a middle ground. Norton (1989) discusses "exploitationism, conservationism, naturalist–preservationism, and extensionist–preservationism" strategies. Exploitationism is viewing resources as unproductive until brought under human control. The naturalist–preservationism rests on Leopold's view (Norton 1989) about "human obligations toward the land community." For the extensionist–preservationism value system, using the term "resources" is not appropriate because of the anthropocentric values this shows toward nature. Using Rawls's rational decision model, Norton (1989) argues that the rational chooser operating under the veil of ignorance would select the naturalist–preservationist strategy, "which imposes constraints on the pervasive exploitation of ecological systems and thereby protects biological diversity in the long run. . . ." The naturalist–preservationist strategy, according to Norton, best protects future generations from the changes to the environment that can be made by the current generation.

Since attempts in the 1970s to measure public attitudes on economic versus ecological issues, a shift has occurred toward greater support for understanding ecology (the interrelationships between organisms and their environment) (Olsen et al. 1992). Where initially the exploitationism end of Norton's four positions, also referred to as developmentalist, had a substantial number of adherents, by the end of the 1980s the conservationist–preservationist contrast was more commonly used.

The Pinchot–Muir, Thoreau, Leopold, and Norton discussions show a continuing concern for the relative importance given economic and ecological values. From these dimensions political scientists and sociologists have developed survey instruments to assess public attitudes. While there has been a trend toward the rightness of ecology, many Pacific Northwest environmental issues are phrased in an economy versus ecology way. In the forests it is owls versus jobs; old growth versus timber supply. For salmon it is fish versus hydropower; buffer strips versus access to forest resources.

Survey Evidence About Fish and Environmental Quality Values

Scarnecchia (1988) says that in order to make management more effective, "More surveys are needed to assess the values of the public toward salmon resources." Our data come from several surveys conducted over the last 3 years. Several questions on these surveys ask people to evaluate preferences for fish and wildlife protection and use. The first, a mail survey (the Oregon Forestry Survey), was completed in the fall of 1991 by faculty at Oregon State University (Steel et al. 1992, Shindler et al. 1993). This survey contrasts a representative sample of national and Oregon publics concerning forestry and natural resource issues. The second survey used is the Western Rangelands Survey, conducted in the spring of 1993 (Steel and Brunson 1993, Brunson and Steel 1994). This was a telephone survey of national and regional publics. Additional public opinion data come from urban and rural mail surveys conducted in Oregon, Washington, and British Columbia. For each survey utilized, approximately two-thirds of households sampled returned surveys (Table 3).

Mail and telephone surveys carry some bias because of the distribution of housing and phone communication in society. Respondents can be expected to be economically prosperous and better educated than the population as a whole. Surveys find that wealth, income, and education correlate positively with favoring environmental values and being active in the environmental movement (Inglehart 1991, Dunlap and Mertig 1992, Olsen et al. 1992). Political scientists also observe that individuals who are better educated and prosperous participate more in activities and organizations that influence governmental policy (Erikson et al. 1991).

Surveys do not necessarily predict behavior, but they give an idea of public perception at a point in time. Public attitudes change with a variety of factors, and they are highly subject to change. The questions used in these surveys build upon observations discussed above about the economy-ecology debate and changes occurring in society.

In the decades following the World War II, a number of fundamental changes transpired in the industrial nations, especially those usually identified as the "Western Democracies" (Dalton 1988). In contrast to the pre-war period, economic growth in the 1950s and 1960s was so rapid that fundamental structures of society were altered, and social commentators began to note a new stage of development (Fiala 1992). This new stage of socioeconomic development in advanced industrial society has been labelled "postindustrial."

A number of studies are available that examine the social, economic, and political implications of postindustrialism (Touraine 1971, Bell 1973, Heisler 1974, Huntington 1974, Gibbins

Table 3. The sample sizes and response rates for the two main surveys used.

Survey	Sample size	Surveys completed	Response rate	Survey type
OSU National and Oregon Forestry	1603	1094	68.4%	mail
Western Rangelands	2000	1360	68.0%	phone

1989). While some definitional disagreement is present, a few central features of this new type of society can be identified. Postindustrial societies are characterized by the following:

- economic dominance of the service sector over that of manufacturing and resource extraction,
- complex nationwide and international communication networks,
- a high degree of economic activity based upon an educated work force employing scientific knowledge and technology in their work,
- a high level of public mobilization in society (including the rise of new social movements such as the environmental movement),
- increasing population growth and employment in urban areas and subsequent depopulation of rural areas, and
- historically unprecedented affluence associated with suburban living (Bell 1973, Inglehart 1977, Tsurutani 1977, Galston 1992).

Correlated with the advent of postindustrial society were individual value structures, particularly among younger individuals, that placed greater emphasis on the higher order needs such as meeting ego needs and personal self-fulfillment. These supplanted more fundamental subsistence needs as motivators for much societal behavior (Inglehart 1977, 1991; Yankelovich 1981; Flanagan 1982). Value changes entailing greater attention to "postmaterialist" needs are thought to have brought about changes in many types of personal attitudes, including those related to natural resources and the environment (Catton and Dunlap 1980, Steger et al. 1989). Some observers suggest that the development of the environmental movement was, in great measure, a product of the vast socioeconomic changes evident in postwar advanced industrial societies (Van Liere and Dunlap 1980, Milbrath 1984).

Along with development of environmental consciousness and the environmental movement in the late 1960s, many traditional political and economic institutions were questioned by people (Habermas 1981, Offe 1985). These changing value orientations among the public have led to changing expectations concerning the management of natural resources (Mather 1990). Some scholars even suggest that beliefs supportive of environmental protection form the basis for a new intellectual "paradigm" in postindustrial societies (Catton and Dunlap 1980, Milbraith 1984). The beliefs of this "New Environmental Paradigm" challenge traditional approaches to natural resource management (Brown and Harris 1992). Brown and Harris (1992) convert this into a "New Resource Management Paradigm" in contrast with the "Dominant Resource Management Paradigm" (Table 4). The new resource management paradigm is more biocentric than anthropocentric and advocates the protection and preservation of both game and nongame wildlife and fish habitats. Statements as represented in the two paradigms became the basis for the design of the forestry and rangelands surveys to assess where the public stands on these issues.

Results from a 1989 Gallup survey indicate strong support for environmental protection and the environmental movement: 41% of respondents considered themselves "strong environmentalists" and 35% called themselves (not strong) "environmentalists" (Gallup Report 1989). Only 20% of survey respondents stated that they were "not an environmentalist." In addition, 85% of respondents indicated that they worry about the loss of natural habitat a "fair amount" to a "great deal" (58% said a "great deal"). Perhaps one of the most important findings of the survey was that 49% of respondents "contributed money to an environmental, conservation or wildlife preservation group." Given these findings, it is not surprising to see that declining populations of certain animals and fish have generated much public concern and activism. Comparative surveys

Table 4. Contrasting elements of the new resource and dominant resource paradigms. Source: Modified from Brown and Harris (1992).

New resource management paradigm	Dominant resource management paradigm
Amenity outputs have primary importance	Amenities are coincident to commodity production
Nature for its own sake (biocentric perspective)	Nature to produce goods and services (anthropocentric perspective)
Environmental protection over commodity outputs	Commodity outputs over environmental protection
Generalized compassion toward other generations (long-term)	Concern for this generation (short-term)
Less intensive forest management, such as "new forestry" and selective cutting	Intensive forest management—clearcutting, herbicides, and slash burning
Less intensive rangeland management, stream protection, less grazing, etc.	Intensive rangeland management—maintain traditional grazing patterns, etc.
Limits to resource growth, conservation—earth has a limited "carrying capacity"	No resource shortages, production, and consumption—science and technology will resolve shortages
New politics—consultive and participatory	Old politics—determination by experts
Decentralized/devolved decision making	Centralized/hierarchical decision making

of United States (US), Japanese, and German publics by Kellert (1993) show strong support for particular animals species, but the people interviewed were less aware and concerned about the ecological contexts that threatened these animals.

The 1993 rangelands survey was conducted nationally and regionally via telephone by researchers at Washington State University and Utah State University (in conjunction with the Sustainable Forestry Program, Oregon State University). Randomly selected individuals were asked their level of agreement or disagreement with a variety of statements concerning natural resources and the environment (Steel and Brunson 1993).

National and Oregon respondents, respectively, had 55% and 54% disagreeing with the statement that "plants and animals exist primarily for human use." In addition, a majority of national respondents (51%) and near majority of Oregon respondents (48%) disagreed with the anthropocentric statement "humankind was created to rule over the rest of nature." Most striking is the strong support by both national and Oregon respondents for the statements "humans have an ethical obligation to protect plant and animal species" and "wildlife, plants and humans have equal rights to live and develop on the earth." Of the national respondents 89% agreed with this statement, and for Oregonians 86% agreed. While one may take issue with the wording of these statements or the commitment of responding individuals, they do indicate very strong support for protection of nature in both the national and Oregon context.

National and Regional Values Toward Salmon and Salmon Habitat

The 1993 rangelands and 1991 forestry surveys indicate the level of national and Oregon public support for salmon and salmon habitat protection (Table 5). While few of the survey questions deal specifically with the salmon issue, they do provide useful information on public support for protecting endangered species (including salmon) and salmon habitat. These data indicate regional support and even stronger national support for protecting salmon and associated critical habitat. Little support is evident for setting aside endangered species laws to protect ranching and timber jobs.

Random samples of national and Oregon publics were surveyed. Some observers suggest important regional differences in value orientations toward the environment and natural resources (interestingly, Gallup Polls have found some of the strongest environmental orientations in the western US; see Gallup Report 1989). National and Oregon differences were summarized from over 1,300 respondents to the rangelands survey and over 2,000 to the forestry survey (Table 5). The results show significantly stronger national preferences for protecting fish and not changing endangered species laws, as well as concern for riparian habitats.

Urban and Rural Values Toward Salmon Protection

Values differed between urban and rural responses to various opinion surveys sent to random samples of the public in metropolitan Portland, Oregon; metropolitan Vancouver, Washington; rural Washington counties contiguous to the Gifford Pinchot National Forest; metropolitan Spokane and Seattle, Washington; and Prince George and metropolitan Vancouver, British Columbia. All surveys were conducted between 1991 and 1993. While the purpose for each survey varied (e.g., forestry, environmental and contemporary policy issues), the same question concerning protection of salmon was included in each survey.

Similar to the national and Oregon data presented previously (Table 5), the results suggest agreement among all samples for giving "Greater protection . . . to fish such as salmon on federal/provincial forest lands" (Table 6). A majority of respondents from each sample stated that greater efforts should be made to protect salmon, ranging from over 58% of the Prince George respondents to 81% of Vancouver, British Columbia, respondents. Clearly a strong mandate exists among the urban and rural publics examined here for protecting salmon and their habitat in the Pacific Northwest, with support being stronger among urban than rural respondents.

To further identify public orientations toward the environment, Steel et al. (1992) examined responses from a national and regional attitude survey concerning federal forestry issues (Table 7). Respondents were asked to self-select their position on a scale regarding the importance of managing for environmental and economic considerations. The largest single response for each sample is at the midpoint; near majorities favored a balance between environmental and economic components in natural resource management (Table 7).

These data show strong general support for protecting wildlife and fish, such as salmon, on federal lands, with urban publics again giving greater support than rural ones. Grouped by gender, age cohort, rural/urban residence, economic interest, and outdoor recreation activities, all of the various national groups presented still showed strong support for protection of habitat and salmon on federal lands.

Table 5. National and Oregon concern for wildlife, salmon, and their habitat: results from rangeland and forestry surveys.

	Agree (%)	Neutral (%)	Disagree (%)	
1993 WSU/USU Rangelands Survey				
A. Greater protection should be given to fish, such as salmon, in rangelands.				
National	78	10	12	$\chi^2 = 120$
Oregon	55	21	25	$p < 0.0001$
				$n = 1929$
B. Endangered species laws should be set aside to preserve ranching jobs.				
National	17	18	65	$\chi^2 = 101$
Oregon	37	15	48	$p < 0.0001$
				$n = 1929$
C. The loss of streamside vegetation is a serious problem on federal rangelands.				
National	82	10	8	$\chi^2 = 20$
Oregon	80	15	5	$p < 0.001$
				$n = 1933$
1991 OSU Forestry Survey				
D. Greater protection should be given to fish and wildlife habitats on federal forest lands.				
National	78	10	12	$\chi^2 = 130$
Oregon	55	21	25	$p < 0.0001$
				$n = 1936$
E. Endangered species laws should be set aside to preserve timber jobs.				
National	17	18	65	$\chi^2 = 104$
Oregon	37	15	48	$p < 0.0001$
				$n = 1919$
F. More "wild and scenic rivers" should be designated.				
National	66	21	13	$\chi^2 = 92$
Oregon	49	21	30	$p < 0.0001$
				$n = 1937$

Table 6. Local community support for protecting salmon.[a]

		Agree (%)	Neutral (%)	Disagree (%)
Greater protection should be given to fish, such as salmon, on federal/ provincial forest lands.	Vancouver, British Columbia	81	13	7
	Seattle, Washington	79	14	7
	Portland, Oregon	71	21	8
	Vancouver, Washington	69	19	13
	Spokane, Washington	62	20	18
	Rural SW Washington	59	22	19
	Prince George, British Columbia	58	19	23

[a] Funding for the various surveys was provided by a Canadian Embassy Grant (Spokane, Seattle, Washington; Prince George and Vancouver, British Columbia), Oregon State University Sustainable Forestry Project, Washington State University at Vancouver, and the Consortium for the Social Values of Natural Resources, USDA Forest Service (Portland, Vancouver, and rural SW Washington). Sample sizes are Portland, OR = 622; Vancouver, WA = 609; Rural SW Washington = 304; Spokane, WA = 399; Seattle, WA = 417; Prince George, BC = 555; Vancouver, BC = 376.

Table 7. Economic versus environmental trade-offs. Comparison of national and Oregon samples. Source: Steel et al. (1992) OSU Survey of Natural Resource and Forestry Issues.

Many natural resource management issues involve difficult trade-offs between natural environmental conditions (wildlife, old growth forests) and economic considerations (employment, tax revenues).

Where would you locate yourself on the following scale concerning these issues?

	National (%)	Oregon (%)
The highest priority should be given to maintaining natural environmental conditions even if there are negative economic consequences.	42	37
Both environmental and economic factors should be given equal priority in forest management policy.	47	44
The highest priority should be given to economic considerations even if there are negative environmental consequences.	11	19

$\chi^2 = 24$, $p < 0.001$, $n = 1844$

Demographic Characteristics of Those Supporting Salmon Protection

Previously we suggested that there are growing generational differences in value orientations toward the environment and natural resources in postindustrial society. These differing orientations were argued to be the product of "postmaterialist" values among the younger post-World War II baby-boom generation. Some evidence supports this hypothesis based on data from the Oregon respondents (Table 8), and this same pattern holds true for the national sample.

In general those most strongly in favor of protection include women, younger people, environmental group members, urban residents, and those respondents who report they go fishing frequently. For Oregon residents, most support greater efforts to protect wildlife habitat and salmon, but at levels slightly lower than the nation as a whole. There are a few notable exceptions where less than a majority of respondents in a category are supportive. These exceptions include the oldest cohorts, men, and those respondents dependent on the ranching and timber industry for their economic livelihoods. These categories of respondents tend to have values consistent with the dominant resource management paradigm. It must be noted, however, that less than 40% of these individuals oppose efforts to protect wildlife and salmon, indicating no clear mandate for the status quo.

Kellert (1993) also observed differences by age. One explanation is a shift toward greater ecological concern with younger age groups. A second is that, as people age, they become less environmentally conscious, which might be explained by the effects of assuming greater financial obligations. Longitudinal studies, however, suggest that the age differences reflect a shift in the direction of a postindustrial value system. Olsen et al. (1992) differentiate a Technological Social Paradigm (material accomplishments, economic efficiency) and an Ecological Social Paradigm (future-oriented, human scale and development, effectiveness-oriented), and claim that the Technological Social Paradigm is gradually being replaced by the Ecological Social Paradigm in modern industrial societies, especially in the US.

Data show that humans have different valuations of nature according to their age (Table 9). Younger people are more concerned about the environment than are older people. The most striking finding is for the first indicator, where 80% of the 18–29 age group disagreed with the statement "plants and animals exist primarily for human use." This compares with 37% of the 61 plus age group disagreeing with the statement (and 41% in agreement). If this pattern continues into the future, we can expect continued and perhaps increasing concern about environmental issues in general and about declining runs of salmon and other fishes.

The question about "plants and animals existing for human use" shows a clear gradation across generations, with those 46 and over being most similar. This raises the question of where the age break occurs. The response to looking at whether "humankind was created to rule over the rest of nature" shows a clear split between the baby boomers and their elders. The support is strong across all generations for the "ethical obligation to protect plant and animal species." The issue of "equal rights to live and develop on the earth" again shows the split with the baby-boomer generation, especially when looking at the disagreement side.

Table 8. Support for protection of salmon and wildlife on federal timber and rangelands controlling for various demographic and interest variables

	National % Agree		Oregon % Agree	
1993 WSU/USU Rangelands Survey				
"Greater protection should be given to fish such as salmon in rangelands."				
GENDER:				
Women	84	$\chi^2 = 35$	65	$\chi^2 = 10$
Men	69	$p < 0.001$	53	$p = 0.01$
AGE COHORT:				
18–29 years	87	$\chi^2 = 50$	60	$\chi^2 = 31$
30–45 years	85	$p < 0.001$	59	$p < 0.001$
46–60 years	72		57	
61 plus years	63		47	
RESIDENCE:				
Urban	78	$\chi^2 = 12$	63	$\chi^2 = 9$
Rural	69	$p < 0.003$	51	$p = 0.011$
INTEREST:				
Environmental group member	84	$\chi^2 = 35$	77	$\chi^2 = 22$
Timber/ranching dependent	65	$p < 0.001$	43	$p < 0.001$
OUTDOOR RECREATION:				
Fishing	80	$\chi^2 = 22$	84	$\chi^2 = 42$
No fishing	74	$p < 0.001$	51	$p < 0.001$
1991 OSU Forestry Survey				
"Greater protection should be given to fish and wildlife habitats on federal forest lands."				
GENDER:				
Women	86	$\chi^2 = 34$	64	$\chi^2 = 16$
Men	71	$p < 0.001$	44	$p < 0.001$
AGE COHORT:				
18–29 years	88	$\chi^2 = 55$	55	$\chi^2 = 31$
30–45 years	83	$p < 0.001$	61	$p < 0.001$
46–60 years	68		47	
61 plus years	64		47	
RESIDENCE:				
Urban	79	$\chi^2 = 10$	63	$\chi^2 = 19$
Rural	73	$p < 0.05$	52	$p < 0.001$
INTEREST:				
Environmental group member	84	$\chi^2 = 25$	76	$\chi^2 = 22$
Timber/ranching dependent	60	$p < 0.001$	48	$p < 0.001$
OUTDOOR RECREATION:				
Fishing	80	$\chi^2 = 28$	57	$\chi^2 = 15$
No fishing	76	$p < 0.001$	50	$p < 0.001$

Table 9. Generational differences in environmental values. Source: Steel and Brunson (1993) Western Rangelands Survey.

	Agree (%)	Neutral (%)	Disagree (%)	
A. Plants and animals exist primarily for human use.				
18–29 years	8	13	80	$\chi^2 = 115$
30–45 years	24	11	65	$p < 0.0001$
46–60 years	39	8	53	$n = 1067$
61 plus years	41	22	37	
B. Humankind was created to rule over the rest of nature.				
18–29 years	21	15	64	$\chi^2 = 51$
30–45 years	28	11	62	$p < 0.0001$
46–60 years	45	10	45	$n = 1069$
61 plus years	43	16	41	
C. Humans have an ethical obligation to protect plant and animal species.				
18–29 years	86	4	10	$\chi^2 = 28$
30–45 years	93	4	3	$p < 0.0001$
46–60 years	85	2	13	$n = 1072$
61 plus years	88	5	7	
D. Wildlife, plants & humans have equal rights to live and develop on the earth.				
18–29 years	91	2	7	$\chi^2 = 29$
30–45 years	68	23	9	$p < 0.0001$
46–60 years	63	11	26	$n = 1064$
61 plus years	72	7	21	

The Public and Remediation Strategies for Salmon Recovery

While the survey data presented in this paper show widespread public support for protection of wildlife and salmon habitat, the public also is cynical about the ability of scientific and technological experts to solve natural resource and environmental problems. The rangelands survey (Steel and Brunson 1993) shows public ambivalence about science and technology in general, and very little confidence in government bureaucrats, resource professionals, and industry to solve natural resource problems (Table 10). Apparently, salmon recovery plans are not just a "science and technology problem" for the public. The public most clearly wants to be involved in the process, which is illustrated by over 77% of national and Oregon respondents agreeing with the statement: "Citizen participation is of great value even if it adds to the cost of government" (Steel et al. 1992).

National and regional data from the rangelands and forestry surveys suggest widespread concern for the plight of wildlife and salmon on federal lands in the Pacific Northwest. Findings also indicate that most respondents are willing to make sacrifices to protect the wildlife and

Table 10. National and Oregon public attitudes toward science and technology.

	Agree (%)	Neutral (%)	Disagree (%)	
Technology will find a way of solving the problem of shortages of natural resources.				
National	40	22	38	$\chi^2 = 117$
Oregon	39	25	36	$p < 0.0001$
				$n = 1064$
People would be better off if they lived without so much technology.				
National	36	22	42	$\chi^2 = 36$
Oregon	33	26	41	$p < 0.001$
				$n = 1064$
Technical and scientific experts are usually biased.				
National	42	36	22	$\chi^2 = 11$
Oregon	46	30	23	$p < 0.09$
				$n = 1052$

Source: Steel and Brunson (1993).

salmon. Assuming that young people will retain the philosophic base of values through their lifetime, the age distribution of responses suggests that public environmental consciousness and biocentric orientations will not subside and may even grow in the future with generational replacement.

Socio-Political Adjustment of Ethical and Esthetic Values

A survey of public attitudes does not mean that people's preferences get carried out. This merely reflects public attitudes at one point in time. Interest group organization, institutional dynamics, and ability to communicate ideas affect which parts of the public have the greatest impact.

These survey data show that people outside the region support preserving and restoring forests and fisheries more strongly than those in the region. Urban people are more supportive than rural people. Younger people and women show stronger agreement with environmental programs. Those who use recreation opportunities are much more supportive than those who lack recreational experience. This does not mean that public attitudes direct policy. Actions are taken based on the attitudes and basic values people are willing to support. However, it is people with power, not attitudes, that direct policy. With salmon management coming under the Endangered Species Act, decisions will be more influenced by the national, urban, and user constituencies.

The surveys show pluralities, not consensus. Since consensus is lacking on what to do, the policy issue is one of choosing direction—giving greater priority to ecological or economic values. The data suggest that people with a more environmental ethical position perceive that protecting resources leads to a better future. These people feel that the weight of economy is damaging to resources. In the minds of those favoring protection, the market does not value items according to what is ethically and esthetically right. Different value priorities and ways to finance them have to be considered.

Extractive use of resources emphasizes economic means of valuing. In extractive approaches to forestry and fisheries, harvest revenues have been the primary measures used in valuing these resources. The valuing of watersheds for flood control and water supply, of wetlands for productivity and water storage, and of recreation, esthetics, and other resource qualities to which people attach importance is much more difficult.

If people's tastes cannot be reflected in the market, then other mechanisms for valuing resources come into play for adjusting costs, prices, and the distribution of benefits. This is the nature of the conflict over endangered species in the Pacific Northwest. Extractive and economic growth practices are not adequate in valuing the fish, wildlife, flood control, water storage, and nutrient cycling functions of riparian habitats (Bessey 1963, Castle 1993).

For many extractive activities, people question whether the extractors pay their fair share of the social costs for these activities. Mining, grazing, and water users are accused of being free-riders on the public lands and resources. Recreational users of Northwest streams and rivers, too, have not borne the full share of the costs of their activities, at least from the perspective of those with more of an environmental value orientation.

From an environmental perspective, an expanding service economy in the postindustrial society produces more economic activity that meets people's egos and self-fulfillment needs. The problem is that markets have not developed to provide the revenues for using resources in other than extractive ways. Coupled with preferences for improved salmon stocks, there has to be some source of revenues that can accomplish the preferred goals. Creating a market for goods and services is one way to help people express their preferences. For items like leaving areas intact without human intrusion, a market is difficult to create. Alternatives to market mechanisms are the political process.

If the current ways of valuing resources are inadequate, an alternative for those with an environmental concern is to raise the costs of extraction. This is done by making administrative processes more costly and burdensome. It is also done by highlighting violations of existing law. Sometimes change requires new laws and institutions to incorporate changed values for forest and fishery resources.

Postindustrial preferences about environmental quality demand less extractive uses of ecosystems. Those taking an environmental position provide political support to develop the financial resources to restore ecosystems. Many people expressing preferences for restoration feel past actions have been wrong and a debt is owed to society for the benefits received at great environmental cost. The ethical stance that a debt is owed does not lead environmentalists to offer to pay for the new environmental services being championed. Funding more holistic and less extractive resource management from the perceived debt owed is difficult. First, resource extractors do not perceive themselves to owe a debt to society. Second, even if they did, many of the corporate organizations that received the benefit no longer exist. Thus, one of the critical issues to be worked out in fostering new resource uses is who pays and how much.

Another way to look at valuing salmon restoration expenditures is to say that they are an investment in greater future returns from forests and fish stocks. Giving up one's job now for some uncertain benefit in the future, however, is not a very compelling prospect for those losing their jobs. From the very beginning, Pinchot made the argument to US Congress that management by the USDA Forest Service could be thought of as a business venture. He predicted that wise management would yield a profit. His experience was not the promise. Pinchot then shifted to arguing that management of the National Forests was in the public welfare (Steen 1976). This meant the public had to be willing to support these efforts with new or redirected tax monies.

Reduced levels of extraction mean that the people most hurt are current workers who depend on extraction of natural resources for their livelihood. These people take the position that economic means of valuing should be given more weight. They complain that environmental concerns are given greater weight than human economic concerns. As a society, we have not dealt very effectively with the problems of occupational transition and change.

The question of who pays gets worked out in a political process. Ethical and esthetic values are too messy for market forces to value monetarily. Political valuing makes past activities that went on with less scrutiny more costly. Political action typically runs behind the actions of development because economic change occurs incrementally, and it takes time to build evidence of undesirable change. The socio-political problem for the future is whether people's ability to do the ethical and aesthetic calculation that raises concern and fosters change in the process for valuing resources will evolve as fast as the ways people can think of to modify the environment.

A solution is to let people into the decision-making process early and often. Too often decisions are left to scientific analysis. This may work when the goals of science and society are well matched. When the management paradigms diverge from the preferences of large segments of the public, new ways of valuing need to be found. New ways of managing are suggested in the concepts of adaptive, bioregional, and cooperative management (Berkes 1989, McCay and Acheson 1989, Pinkerton 1989, Lee 1993, Duncan 1994, Smith 1994).

Can science decide the values that should be assigned to salmon, owls, jobs, justice, and ecosystem management? How do people want to weigh economy and ecology? Science can inform people about the projected impacts of various actions or inactions. The Northwest salmon problem is fundamentally a social and political one in which the relative valuations given to economy and ecology are being worked out. Science can inform, but it cannot determine relative weightings. To move forward, we need to be willing to engage the public in debate on how salmon and ecosystems fit into our future economic and ecological vision.

Acknowledgments

Support for the surveys providing data for this paper came from the Sustainable Forestry Program at Oregon State University; Washington State University at Vancouver; US Department of State, Man in the Biosphere Program; the Consortium for the Social Values of Natural Resources of the USDA, Forest Service and Forest Research Laboratory; and the Canadian Embassy. We acknowledge the thoughtful and helpful suggestions of R. Lackey, C. Younger, and several anonymous reviewers.

Literature Cited

Bell, D. 1973. The Coming of Postindustrial Society. Basic Books, New York.

Berkes, F. (ed.). 1989. Common Property Resources: Ecology and Community-Based Sustainable Development. Belhaven Press, New York.

Bessey, R.A. 1963. Pacific Northwest Regional Planning—A Review. Bulletin 6 of the Washington Department of Conservation, Division of Power Resources. Olympia, Washington.

Brown, G. and C.C. Harris. 1992. The US Forest Service: Toward the New Resource Management Paradigm? Society and Natural Resources 5: 231–245.

Brunson, M.W. and B.S. Steel. 1994. National public attitudes toward federal rangeland management. Rangelands 16: 77–81.

Callenbach, E, F. Capra, L. Goldman, R. Lutz, and S. Marburg. 1993. EcoManagement. Berrett-Koehler Publications, Inc., San Francisco, California.

Castle, E.N. 1993. A pluralistic, pragmatic and evolutionary approach to natural resource management. Forest Ecology and Management 56: 279–295.

Catton, W. and R.E. Dunlap. 1980. A new ecological paradigm for post-exuberant sociology. The American Behavioral Scientist 24: 15–47.

Dalton, R. 1988. Citizen Politics in Western Democracies: Public Opinion and Political Parties in the United States, Great Britain, West Germany and France. Chatham House, Chatham, New Jersey.

Devall, B. 1985. Deep Ecology. Peregrine Smith Book, Layton, Utah.

Dillman, D.A. 1991. Mail Surveys: A Comprehensive Bibliography, 1974–1989. Council of Planning Librarians, Chicago, Illinois.

Duncan, A. 1994. A proposal for a Columbia Basin watershed planning council. Illahee 10: 287-303.

Dunlap, R.E. and A. Mertig. 1992. The evolution of the US environmental movement from 1970 to 1990: an overview, p. 1–10. *In* R.E. Dunlap and A. Mertig (eds.), American Environmentalism. Taylor & Francis, Philadelphia, Pennsylvania.

Erikson, R.S., N.R. Luttbery, and K.L. Tedin. 1991. American Public Opinion. Macmillan Publishing Company, New York.

Fiala, R. 1992. Postindustrial society, p. 1512–1522. *In* E.F. Borgatta and M.L. Borgatta (eds.), Encyclopedia of Sociology. Macmillan, New York.

Flanagan, S. 1982. Changing values in advanced industrial society. Comparative Political Studies 14: 99–128.

Gallup Report. 1989. The Environment. Gallup Report No. 285. June, 1989.

Galston, W. 1992. Rural America in the 1990s: trends and choices. Policy Studies Journal 20: 202–211.

Gibbins, J. (ed.). 1989. Contemporary Political Culture: Politics in a Postmodern Age. Sage Publications, London, England.

Habermas, J. 1981. New social movements. Telos 49: 33–37.

Heisler, M. 1974. Politics in Europe: Structures and Processes in Some Postindustrial Democracies. McKay, New York.

Hirshleifer, J. 1978. Competition, cooperation, and conflict in economics and biology. American Economic Review 68: 238–243.

Huntington, S. 1974. Postindustrial politics: how benign will it be? Comparative Politics 6: 147–177.

Inglehart, R. 1977. The Silent Revolution: Changing Values and Political Styles Among Western Publics. Princeton University Press, Princeton, New Jersey.

Inglehart, R. 1991. Culture Shift in Advanced Industrial Society. Princeton University Press, Princeton, New Jersey.

Jones, H.R. 1965. John Muir and The Sierra Club, The Battle for Yosemite. Sierra Club, San Francisco, California.

Kellert, S.R. 1993. Attitudes, knowledge, and behavior toward wildlife among the industrial superpowers: United States, Japan, and Germany. Journal of Social Issues 49: 53–69.

Krutch, J. W. 1962. Thoreau: Walden and Other Essays. Bantam Books, New York.

Lee, K.N. 1993. Compass and Gyroscope: Integrating Science and Politics for the Environment. The Island Press, Washington, DC.

Leopold, A. 1966. A Sand County Almanac with Essay on Conservation from Round River. Ballantine Books, New York.

Lichatowich, J. 1992. Management for sustainable fisheries: some social, economic, and ethical questions, p. 11–18. *In* G.H. Reeves, D.L. Bottom, and M.H. Brookes (eds.), Ethical Questions for Resource Managers. Pacific Northwest Research Station, US Forest Service, General Technical Report PNW-GTR-288. Corvallis, Oregon.

Lutz, R. 1992. Innovationsokologie. Baun Aktwell, Munich, Germany.

Mather, A. 1990. Global Forest Resources. Timber Press, Portland, Oregon.

McCay, B.J. and J.M. Acheson. 1989. A Question of the Commons. University of Arizona Press, Tucson, Arizona.

Milbrath, L. 1984. Environmentalists: Vanguard for a New Society. State University of New York Press, Albany, New York.

Norton, B.G. 1989. Intergenerational equity and environmental decisions: a model using Rawls' veil of ignorance. Ecological Economics 1: 137–159.

Offe, C. 1985. New social movements: challenging the boundaries of institutional politics. Social Research 52: 817–868.

Olsen, M.E., D.G. Lodwick, and R.E. Dunlap. 1992. Viewing the World Ecologically. Westview Press, Boulder, Colorado.

Pinkerton, E. 1989. Co-operative Management of Local Fisheries. University of British Columbia Press, Vancouver, British Columbia.

Sale, K. 1985. Dwellers in the Land. The Bioregional Vision. Sierra Club Books. San Francisco, California.

Scarnecchia, D.L. 1988. Salmon management and the search for values. Canadian Journal of Fisheries and Aquatic Sciences 45: 2042–2050.

Schurr, S. H., J. Darmstadter, H. Perry, W. Ramsey, and M. Russell. 1979. Energy in America's Future: The Choices Before Us. Johns Hopkins University Press, Baltimore, Maryland.

Shindler, B., P.C. List, and B.S. Steel. 1993. Managing federal forests. Journal of Forestry 91(7): 36–42.

Smith, C.L. 1994. Connecting cultural and biological diversity in restoring Northwest salmon. Fisheries 19: 20–26.

Steel, B.S. and M. Brunson. 1993. Western Rangelands Survey: Comparing the Responses of the 1993 National and Oregon Public Surveys. Washington State University at Vancouver, Vancouver, Washington.

Steel, B.S., P.C. List, and B. Shindler. 1992. Oregon State University Survey of Natural Resource and Forestry Issues: Comparing the Responses of the 1991 National and Oregon Public Surveys. Department of Political Science, Oregon State University, Corvallis, Oregon.

Steel, B.S. P.C. List, and B. Shindler. 1994. Conflicting values about federal forests: a comparison of national and Oregon publics. Society and Natural Resources 7: 137–153.

Steen, H.K. 1976. The US Forest Service: A History. University of Washington Press, Seattle, Washington.

Steger, M.A., J.C. Pierce, B.S. Steel and N.P. Lovrich. 1989. Political culture, postmaterial values, and the new environmental paradigm: a comparative analysis of Canada and the United States. Political Behavior 11: 233–254.

Touraine, A. 1971. The Post-Industrial Society: Tomorrow's Social History. Random House, New York.

Tsurutani, T. 1977. Political Change in Japan. McKay, New York.

Udall, S.L. 1963. The Quiet Crisis. Holt, Rinehart, and Winston, New York.

Van Liere, K. and R.E. Dunlap. 1980. The social bases of environmental concern: a review of hypotheses, explanations, and empirical evidence. Public Opinion Quarterly 44: 43–59.

Yankelovich, D. 1981. New Rules: Searching for Self-Fulfillment in a World Turned Upside Down. Bantam Books, New York.

Organizational Systems and the Burden of Proof

David A. Bella

Abstract

To understand the salmon crises, we must come to a radically new understanding of human systems. The vast organizational systems of technological society should be seen as "CANL" systems—they are complex, adapting, and nonlinear. They display emergent behaviors that cannot be reduced to the intentions and values of individuals. We depend upon the emergent behaviors of these CANL systems to provide the technological services of modern society: "Benefits do emerge." However, destructive tendencies also arise as emergent phenomena. We misperceive such phenomena when we reduce them to the intentions and values of individuals. This paper departs from such reductionism. It develops a model to identify, explain, and even predict emergent tendencies that are destructive within these CANL systems of the human kind. The space shuttle explosion, the decimation of old growth forest ecosystems, and the decline of salmon (*Oncorhynchus* spp.) stocks are shown as outcomes of the same emergent phenomenon, "systemic imbalance." In this paper, I warn that we are unlikely to protect the salmon unless we recognize how our own perceptions are transformed to conform with systemic imbalances. This paper is radically new: it departs from conventional science, drawing upon recent notions from nonlinear dynamics—self-organization, the interplay of order and disorder, and the pull of attractors. Alternatively, this paper comes back to old notions such as independent checks and balances, responsible citizenship, and the democratic and prophetic traditions.

Introduction

In 1892, Livingston Stone warned that the salmon (*Oncorhynchus* spp.) of the Pacific Northwest faced "the slow but inexorable march of those destroying agencies of human progress, before which the salmon must surely disappear as did the buffalo of the plains and the Indian of California." Today these agencies can be described as the vast organizational systems upon which our modern technological world depends. Such systems provide many benefits; in Stone's words, they contribute to human progress. Without denying these benefits, my objective here is to ask: What is it about such vast systems that allows them to shape the activities of ordinary people so that we ourselves contribute to "the slow but inexorable march of those destroying agencies of human progress. . ."? Conventional answers, I claim, are insufficient. This paper develops unconventional answers.

A Paradigmatic Case Study

On January 28, 1986, the space shuttle Challenger exploded during liftoff, killing all seven passengers. The technical cause of the accident was a failed O-ring in one of the solid boosters that allowed hot gas to leak and ignite the adjacent liquid fuel. This accident, I claim, can be seen as arising from a general phenomenon known as systemic imbalance. Imbalance implies a tendency to reinforce some activities (type A) while suppressing other activities (type B). Systemic implies that imbalance arises from the normal behaviors of organizational systems. Such systems are not fixed structures. Instead, they are complex webs of information and resource transfers that continually adapt, displaying nonlinear behaviors such as self-organization. I call them CANL systems (complex, adaptive, and nonlinear) and I describe the accident from this perspective.

In the late 1960s, the Apollo Moon Program came to an end. The loss of program funding was a major disruption to the National Aeronautics and Space Administration (NASA) system. The system adaptively shifted to draw in funding needed to survive. In more general terms, the organizational system adapted to dampen disorders to nondisruptive levels. Thus, the space shuttle program emerged.

In the early and formative stage of this shift, disorders that threatened the emergent system included criticisms that the shuttle program would be too expensive and unable to meet its own schedules. From the perspective of individuals, such threats resulted in no small amount of wheeling and dealing and rearranging in ways that often seemed chaotic. From a broader perspective, however, one saw the emergence of complex patterns and networks that tended to reinforce disorder-damping activities (type A) while suppressing disorder-producing activities (type B).

In the earlier stages of development, type A activities included favorable assessments concerning cost and schedules while type B activities included unfavorable assessments. Systemic imbalance (A reinforced and B suppressed) emerged in the form of distorted assessments that justified the program and dampened experienced disorders (i.e., objections by critics). Funding was secured. However, the imbalanced assessments had a cumulative effect; they influenced expectations and created a systemic need to reinforce those activities that served to meet the demanding schedules and cost requirements that had been employed to justify the program. Given this adaptive history, we can identify two broad types of activities: (A) activities that served schedule demands and held costs down and (B) activities that promoted safety and avoided accidents. Good reasons were given for both types of activities. But type A (meeting schedules, cutting costs) tended to dampen disorders while type B (promoting safety) tended to promote disorders within the system. Adaptive change shifted toward A and away from B, not because of any deliberate plan but rather because of the adaptive tendency of CANL systems to gravitate toward those arrangements that dampen disorders to nondisruptive levels. Systemic imbalance emerged.

This CANL view helps us to explain why the United States (US) Commission (1986) found that the accident was "rooted in history." The report states "heavy emphasis was placed on the schedule"; at the same time, "the safety, reliability and quality assurance work force was decreasing which adversely affected mission safety" (US Commission 1986). As the number and severity of O-ring problems grew, "NASA minimized them in management meetings." In a review of organizational activities, Vaughan (1990) found the following:

> NASA's incentive/award fee contract itself prioritized cost savings and meeting deadlines—the fee system reinforced speed and economy rather than caution . . . when rewards were great for cost savings and meeting deadlines and punishment was not forthcoming for safety infractions, contractors would tend to alter their priorities accordingly.

The eventual outcome of this systemic imbalance was dramatic—the tragic explosion.

This accident reveals a general phenomenon at work, that of systemic imbalance within organizational systems. If unchecked, such imbalance leads to the inexorable march of destroying agencies. Because the shuttle explosion was sudden and dramatic, attention was drawn to the destructive consequences. In the case of salmon, however, the destructive outcomes are gradual and more easily overlooked. Nevertheless, the same systemic tendencies are involved.

Such systemic tendencies emerge through the interplay of order and disorder within CANL systems. To illustrate, consider for a moment succession in natural ecosystems (Odum 1966). On the one hand, succession involves the emergence over time of complex ecosystems, networks of interrelationships, as if an attractor drew out levels of organized complexity. On the other hand, disorders caused by forest fires, floods, and storms shape the character of ecosystems in important ways. In other words, ecosystems are not static structures. Instead, they are dynamic CANL systems, shaped and formed through the continuing interplay of order and disorder over time. I examine organizational systems from a similar perspective.

Order and Disorder

Within organizational systems, we can observe humans attempting to sustain some form of order in their own affairs. They rush to meet schedules, work to complete assignments, fight to secure funding, and struggle to resolve and contain disorders of all kinds. Their activities and personalities are extremely diverse and change over time. Confusions, misunderstandings, conflicts, and other forms of disarray are not uncommon. Evidence of disorder is not hard to find. At the same time, however, not all these busy activities are random and disconnected. Amid a sea of apparent disorder, one can notice relational patterns, vast webs of information and resource transfers that exert nonarbitrary influences upon behaviors far beyond the individual viewpoint. These patterns shift, adjust, and emerge in new forms. Amid constant shuffling and apparent disarray, activities tend to settle into mutually reinforcing patterns. Activities coalesce into arrangements of codependency linked by complex networks of information and resource exchange that tell individuals to do an activity and receive a reward in return. These networks, webs, and patterns constitute systemic order, a dynamic state of affairs that draws together activities to form CANL systems with behaviors and capabilities far beyond those of individuals.

The nature and character of systemic order are radically different from the deliberate order implied in organizational charts. Such charts imply that order, the coordination of activities to form integrated wholes, is the result of deliberate human design and control. Those in higher position direct the activities of those below. Order is presumed to be deliberate, arising from the intentions, designs, and control of particular humans and groups. Such deliberate order presumes that people know what they are doing and do it. Systemic order implies no such thing.

Clearly, evidence for deliberate order (intentional design) can be found in many human undertakings. But the scale of modern technology and the order it depends upon have grown far beyond the capacity of anyone to grasp, much less design and direct. Deliberate order becomes a misleading model for describing the vast scale of modern technology and its consequences.

To appreciate the vast scale of organized complexity in modern technological systems, consider the impenetrable mazes of codes, regulations, specifications, laws, contracts, and software. Consider the bewildering assortments of specialists, each with their own techniques, jargon, and paradigms incomprehensible to all but a few. Consider the networks and hierarchies—overt and covert—through which status, position, and authority are gained and denied. Consider the endless procession of instruments, machines, devices, and computer codes. The world that you and I depend upon, even for a drink of water, depends upon such things continuously coming together in coherent ways that are ever changing. All this is far beyond the capacity of deliberate order. Thus, people must limit their notions of deliberate order to local activities of limited complexity contained within and shaped by vast systems of organized complexity far beyond the capabilities of intentional design. Of course, intention, design, and deliberate acts by powerful individuals and groups can often influence and exploit such vast systems (Jackall 1988), but this is quite different than designing, directing, and controlling organized complexity itself. In exploring the emergence of organized complexity, I suggest we do not presume intentional design on vast scales.

To grasp the dynamic character of systemic order, imagine an order-disorder scale (Fig. 1). Systems located to the far right are characterized by organized complexity; that is, they contain vast interconnected webs of resource and information exchange that provide coherence to many diverse but widely separated activities. Toward the left end of the scale one finds disorganized commotion, diverse and disconnected activities with little coherence. The emergence of reinforcing patterns of order can be seen as a shift from left to right (much like a maturing ecosystem) as if drawn by an R attractor. A disruption can be seen as a flip back to the left (much like a forest fire or, in organizational systems, a cutback in funding as NASA faced).

Human activity can be seen as constant shuffling. Interrelationships and codependencies form patterns. Some patterns are more persistent, renewing, and enduring than others, and tend to be self-reinforcing. When such patterns emerge, the behaviors within them tend to be enduring. Organized complexity emerges in the form of multiple webs of self-referencing loops of influence that mutually reinforce their own internal behaviors. These CANL systems display the properties of self-organization. They are self-renewing, drawing in behaviors from a sea of disorder toward coherent patterns on vast scales.

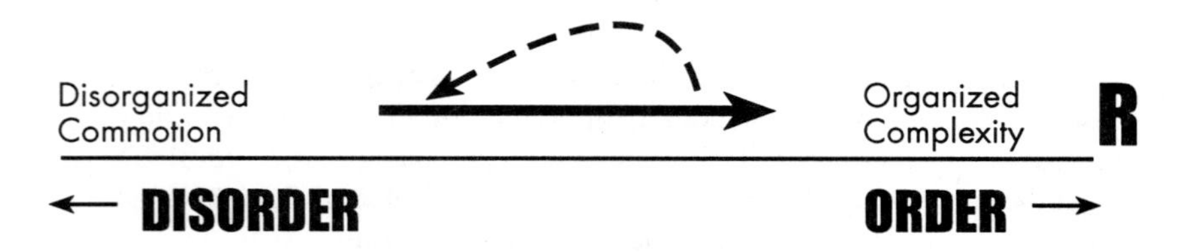

Figure 1. The order-disorder scale. The R attractor (R) pulls toward order (solid arrow); disruptions flip the system toward disorder (dashed arrow).

From this CANL perspective, systemic order emerges as behaviors coalesce into enduring forms of mutual reinforcement. The emergence of organized complexity is a systemic outcome having to do with the tendency of human activities to adaptively gravitate toward patterns of self-reinforcement. Amid endemic shuffling around, indeed because of it, behaviors tend to settle into those self-supporting patterns able to endure. Thus, from disorganized commotion, behaviors are attracted toward relational loops, webs, and networks that provide mutual reinforcement. This emergence of systemic order results in an adaptive shift (e.g., to the right in Fig. 1).

The R Attractor, Disorder, and Time

Systemic order in human affairs requires that systems develop good reasons for people to accommodate to systemic demands. While good reasons of many kinds can and do emerge, a common reason is that people get paid for some activities and not others. Behaviors that accommodate systemic demands are rewarded with financial support (salary, contract payment, etc.). Since too many required behaviors are not sufficiently enjoyable on their own, payment is the common incentive to get them done. The payment of money (along with position, status, and security) and the threat of its loss provide good reasons for acting out the behaviors organizational systems require. But to provide such good reasons, these systems must reinforce those patterns of behavior sufficient to secure funding. Contrary behaviors promote disorder. All this should sound familiar to managers who must continually deal with funding requirements. The difference, however, is that I suggest that people view these requirements as systemic. Systemic order requires mutually accommodating behaviors that allow a system to secure the funding needed to promote mutually accommodating behaviors—a circular pattern. Reinforcing behavioral loops forms the basis for self-generating order.

Systemic order in organizational systems can be sketched as a characteristic pattern (Fig. 2) that emerges under many different conditions. Note that each statement in this pattern supports (provides good reasons for) another statement, which in turn loops back to support the first statement. This pattern allows those behaviors that support organizational survival to become mutually supporting. Arrangements of human activity that do not conform to such a pattern are likely to be replaced by arrangements that conform more closely. Amid different conditions of continual change and shifting arrangements, organizational systems are likely to gravitate toward this overall form. Thus, this sketch (Fig. 2) describes what I define as the attractor for organizational systems. The term R attractor is used to describe a characteristic or property of CANL processes, the tendency of behaviors to be drawn into (attracted toward) mutually reinforcing patterns. The R attractor acts like a gravitational field, pulling activities within its grip toward organized complexity on the order-disorder scale (Fig. 1). While particular forms of systemic order are not predetermined or predictable, one can say that, amid constant change, adjustment, and turmoil, a tendency toward those forms more consistent with the R attractor is likely to emerge.

The pull of the R attractor itself, however, is insufficient to explain the emergence of organized complexity on vast scales (far beyond the capabilities of intentional human design and the boundaries of individual firms and agencies). The R attractor should be viewed in a broader context involving the dynamic tension between order and disorder over time (Fig. 3). The trajectory toward organized complexity lies on the plane formed by these two dimensions.

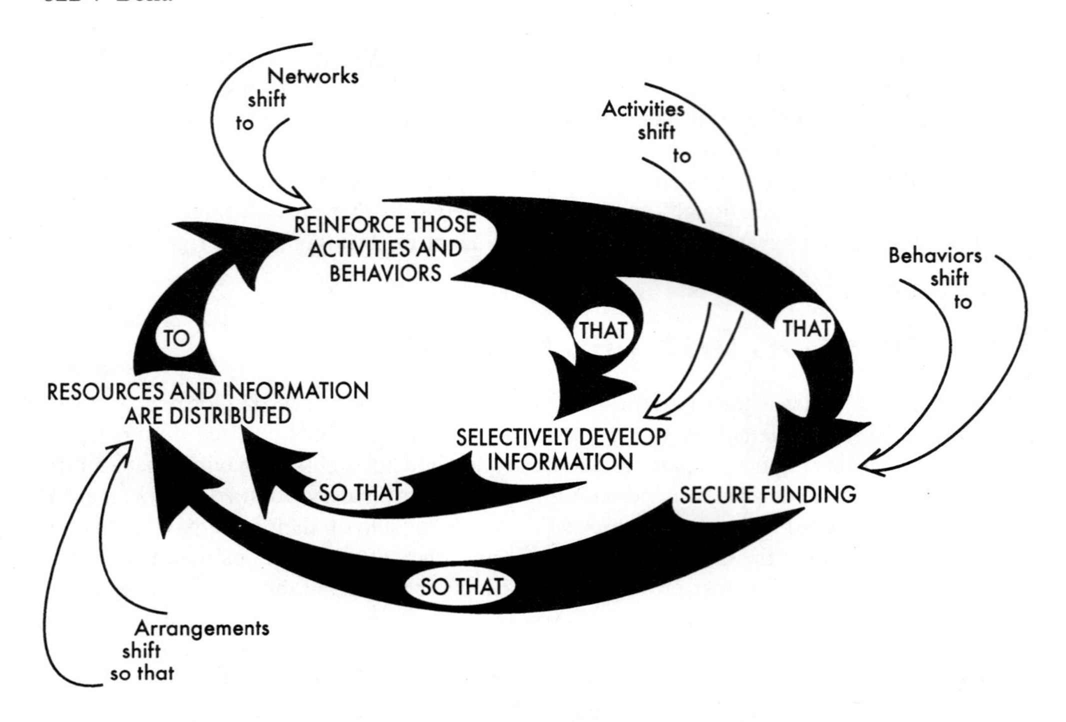

Figure 2. The R attractor: sketched as self-reinforcing loops that draw in (attract) activities, behaviors, arrangements, and networks.

A significant feature of this trajectory is its jagged shape, illustrating the dynamic interplay of order and disorder over time. Behaviors tend to coalesce into self-sustaining patterns that constitute systemic order (solid lines, Fig. 3). This adaptive process involves the emergence of self-regulating systems that draw behaviors into mutually reinforcing patterns while dampening disorders that would disrupt such patterns. This tendency is described as the pull of an R attractor (toward the right). If only this tendency were involved, systemic order would likely stabilize at some limited but static level. Systemic development would cease. But, as experience shows us, such stable states are not likely to persist in either natural or human systems. Disorders do arise to disrupt reinforcing patterns (Fig. 3).

Disruptions are characterized by shifts toward disorder. Such shifts also open up possibilities for order to re-emerge in different and more complex ways. Given the continual pull of the R attractor, disruptions exert a creative influence that allows order to emerge at higher and more complex levels. Engineers recognize this as a trial and error approach; much of our technological progress has arisen from failures and disasters that disrupted old ways of doing things, thereby clearing the way to construct things in new and more advanced ways. Albert Speer (1970), the Minister of Armaments under Hitler, claimed that the earlier bombing of German industries in World War II opened up new possibilities for higher levels of industrial productivity. These raids were regarded as helpful (Speer 1970). The development of new scientific paradigms has been seen as arising from revolutionary crises that disrupt established ways of doing things (Kuhn

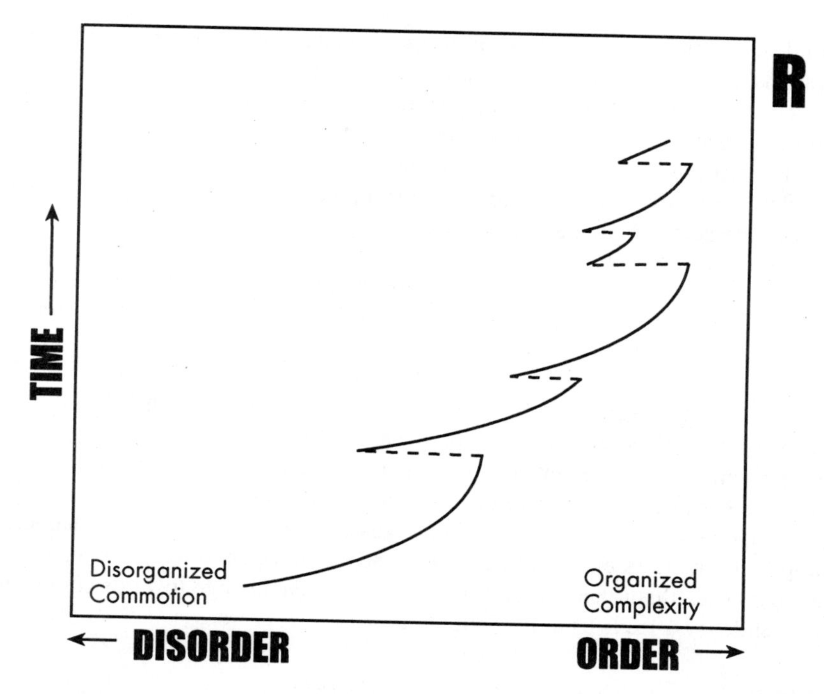

Figure 3. The interplay of order and disorder over time. The R attractor (R) draws activities and behaviors toward order (solid line); disruptions shift system toward disorder (dashed line).

1962). Innovations disrupt established organizational systems, allowing new technological development to emerge as competition promotes a scramble to secure favorable positions within the R attractor. In all such examples, we see the interplay of order and disorder over time (Fig. 3).

The tension between the pull of the R attractor and the experienced disorders serves as a systemic process seeking new forms of order. But the new patterns of order may also amplify small disturbances to disruptive levels. Moreover, disorder can be self-generated, allowing new forms of order to emerge. The R attractor can be viewed as pulling behaviors toward the continual interplay of emergent order and disorder on vast scales. Nothing is final in such a CANL world. Discoveries, innovations, entrepreneurial initiatives, mistakes, and even disasters (e.g., the shuttle explosion) promote disorders, thus renewing the adaptive search for order. We should not look for fixed structures of order; instead we should explore the dynamic interplay of order and disorder over time and the systemic tendencies that emerge from it. Such dynamic interplay occurs at different temporal and spatial scales and is similar at different scales (much like fractals).

Evidence of this developmental interplay of order and disorder abounds but is often missed because we see order in mechanistic rather than CANL ways. A mechanism, such as a lawn mower, is not likely to develop into a higher level machine if we kick it or drop it from the garage roof. But a mechanism is not an adaptive system; thus, it is meaningless to describe its systemic behavior in terms of an attractor that shapes components into coherent wholes through a history of disorder. Deliberate order is indeed required for the design and repair of lawn mowers, but it is insufficient for organized complexity within CANL systems.

Beyond mechanisms such as lawn mowers, the interplay of order and disorder in many CANL systems is evident, including the vast organizational systems of modern technology. We begin to recognize the limitations of organizational charts; they misleadingly show order in human affairs in a manner similar to order in a lawn mower. Instead of the intentions, designs, and plans of decision makers, the interplay of order and disorder over time—the pull of the R attractor and the formative role of disorders—must be examined. We can summarize this CANL process with statements of general tendencies.

> Tendency 1: Adaptive change leading to organized complexity involves the emergence of reinforcing relational patterns that tend to dampen disorders to nondisruptive levels.

This tendency can be described as the pull of an R attractor (Fig. 2) that draws activities and behaviors away from disordered commotion toward organized complexity (Figs. 1 and 3).

Disorders can be seen as events that disturb and disrupt reinforcing patterns, thus allowing and forcing the system to respond to the R attractor in different ways. The adaptive process adjusts to dampen experienced disorders below disruptive levels. Thus, the character of organized complexity, with its emergent behaviors and capabilities, reflects its history of experienced disorders (i.e., disturbance regime), forcing adaptive change towards some patterns and not others. Tendency 1 leads us to the following.

> Tendency 2: The emergent behaviors of organized complexity reflect the history of experienced disorders that the adaptive process was forced to accommodate.

By considering disorders common to organizational systems (i.e., the common disturbance regime), one can infer generic behaviors that tend to emerge through the interplay of order and disorder. Common disorders involve events and threats such as budget crises, neglect of assignments, and schedule failures. Given this common disturbance regime, we should expect organized complexity to emerge in ways that tend to dampen such disorders. That is, we should expect CANL systems to reinforce some kinds of behavior while suppressing others (Table 1). These systemic tendencies can be observed in the behaviors of millions of people caught up within organizational systems throughout the world. That such behaviors come as no surprise reflects the pervasive and persistent pull of the R attractor upon human behavior.

Systemic Imbalance

The capacity of a CANL system to selectively reinforce some activities while suppressing others is a necessary property of systemic order. People experience such order when they are rewarded for carrying out some tasks and not rewarded for others. The requirements are defined for workers in assignments, contracts, and instructions; they are implied in evaluations, job descriptions, and opportunities for advancement or profit. Rewards and reprimands, overt and covert, serve to reinforce some activities while suppressing others. All of this is common experience within organizational systems. People don't get to do whatever they want to do. If

Table 1. Generic behaviors reflecting the common disturbance regime of organizational systems.

R (reinforced) behaviors tending to sustain order	S (suppressed) behaviors tending to promote disorder
Securing and distributing funds to support revenue-producing activities	Undertaking activities not promoting and possibly threatening the funding and support of activities
Accommodating to established arrangements, schedules, assignments, objectives, information channels, and authority	Departing from established arrangements, contrary to schedules, beyond assignments, going outside of information channels, and around authority
Gaining approval for activities; shaping behavior to performance evaluations	Acting without, and possibly contrary to, prior approval; sustaining behaviors not favored by established performance evaluations

everyone did, disorganized commotion would result. Thus, if organized complexity is to endure, human behaviors must be shaped in nonarbitrary ways (some activities reinforced, others suppressed) so that coherence and coordination emerge despite individual inclinations.

Consider such behavior from the perspective of the system rather than the individuals within it. Again, think of systems not as the outcomes of deliberate order, but rather as adaptive outcomes—networks of resource and information transfer that continually emerge as self-reinforcing patterns. While detailed prediction is impossible, one can describe general characteristics of activities likely to be reinforced or suppressed within such systems. In other words, the R attractor tends to pull together (coordinate, promote, organize) some activities and not others. Such selectivity reflects systemic requirements. Stated simply, one should expect the adaptive process to reinforce those activities that allow reinforcing patterns to continue and expand while suppressing those activities that tend to disrupt these patterns. Only when such selective reinforcement emerges over time can organized complexity grow to levels beyond the limited capabilities of intentional design.

Consider an imagination experiment. Assume that two types of activities, A and B, are sustained at equal levels within a CANL system. Both activities involve the coordinated efforts of many people, including a wide range of specialists. Thus, each activity can be considered as a systemic outcome. Assume that both systemic activities are found to be equally worthy on the basis of independent and credible assessments. With respect to activities A and B, the system is initially balanced, showing no preference of one activity over the other. Now, consider how adaptive change proceeds from this initial balanced condition.

The CANL system is not static. Mistakes, budget crises, personnel changes, conflicts, political pressures, entrepreneurial initiatives, and competition promote a continual shuffling around, forcing people to often rush about solving a seemingly endless variety of local problems. Amid such disorders, shifting arrangements tend toward self-reinforcing patterns as though drawn by an R attractor. Systemic order emerges in ways that tend to dampen rather than amplify disorders (Tendency 1). Imagine activities A and B within this dynamic process.

Activities A and B will fluctuate and shift in response to the dynamic nature of the system and the disorders that continually arise within it. Assume that variations of A and B produce different responses within the adapting system such that when A is increased, disorders tend to be dampened and, in contrast, when B is expanded, disorders experienced within the system tend to be magnified. As activity A decreases, greater and more widespread disorders tend to arise within the system. The reverse tends to occur within B. In other words, disorders experienced within the system tend to be inversely related to A and directly related to B (Table 2).

Given that CANL systems are dynamic, arrangements tend to shift around. When reforming tends to reinforce activity A, disorders tend to be reduced. In comparison, when these shifts tend toward activity B, disorders tend to grow. Thus, in the shifting around, arrangements that reinforce A while suppressing B tend to be more enduring (less disrupted) than arrangements that reinforce B while suppressing A. Given the normal shuffling around, reinforcing arrangements will adaptively shift toward A rather than B. Any adaptive shift contrary to this tendency would increase disorder, pressing more people to take more steps to contain disorders, thus forcing rearrangements. The tendency of shifting arrangements (the pull of the R attractor) will be to settle into less disorderly and hence more enduring arrangements that reinforce A while suppressing or neglecting B.

Through such adaptive change, the networks and reinforcing loops of the system shift away from the balanced initial condition. Activity A is reinforced while, in comparison, activity B is suppressed. Systemic imbalance emerges. All that is required from individuals are their countless local efforts to address the immediate disorders they experience within the system. Through the countless activities of many people, each striving to solve local and immediate problems in some coherent manner, the system adapts in ways that contain the growth and spread of experienced disorders (Tendency 1). We should expect adaptive systems to do this. Given the continuing fluid tension between order and disorder, a systemic tendency emerges: activity A is accommodated and reinforced while activity B is neglected and suppressed. No conspiracy, plan, or deliberate decision is needed. The system adapts to attract A and not B. This is a normal systemic outcome (Tendency 2).

To express a general model, envision a behavioral phase plane (Fig. 4). Using the order-disorder scale (Fig. 1) and rotating this scale 90° counterclockwise results in the R attractor acting much like gravity pulling systems downward toward "order" while disruptions shift systems upward toward "disorder." Consider the horizontal (x) axis to represent the relative emphasis upon type A and type B activities (as defined in Table 2). Toward one end (left), A activities dominate. Toward the other end (right), B activities dominate. A balance between A and B would be indicated by a "state" position within the central region of this dimension. Together, these two dimensions constitute the behavior phase plane (Fig. 4).

Nonarbitrary behavioral shifts emerging from CANL systems are shown as trajectories toward the R attractor located in the lower left region of the plane. Such trajectories lead to the emergence of systemic order (y-axis; previously described in Figs. 1 and 3). These same trajectories involve a systemic shift toward type A activities and away from type B activities (x-axis).

Table 2. General characterization of activities based upon disorder response.

Activity variation	Disorder response
A — Increase, expand	Decrease, decline
A — Decrease, decline	Increase, expand
B — Increase, expand	Increase, expand
B — Decrease, decline	Decrease, decline

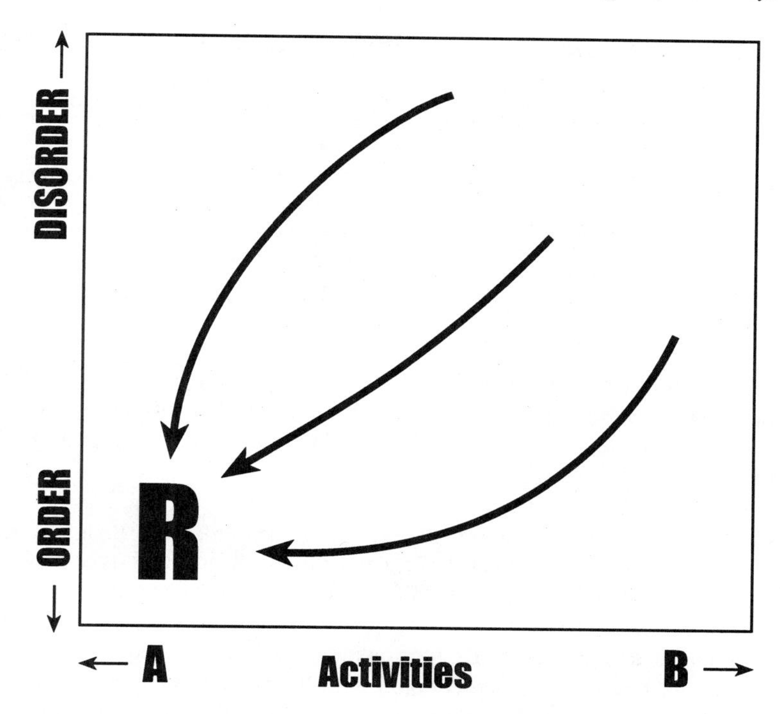

Figure 4. The behavioral phase plane showing trajectories toward the R attractor (see Table 2 for a characterization of A and B activities)

Imbalance (A over B) emerges through the normal adaptive process that gravitates toward self-reinforcing patterns.

Given the character of the R attractor (Fig. 1), we can infer general characteristics of activities likely to be reinforced or suppressed (Table 3). Type A activities are those consistent with the systemic need to sustain reinforcing patterns of order. They tend to dampen disorders, allowing patterns to endure and expand. Such activities allow schedules to be met and assignments completed in a coordinated way. Type A activities tend to enhance funding, rather than undermine it. The R attractor should be expected to draw out type A activities. In contrast, type B activities would be more disruptive, less accommodating, and more troublesome to schedules and coordination. Type B activities would pose funding difficulties, threatening existing and future funding. If, in a relative sense, a particular activity fits the generic characteristics of A more than B, expect the attractor to reinforce it. If an activity conforms more to the generic qualities of B, expect systemic neglect and even suppression. Thus, some guidelines to identify and even predict systemic imbalance exist.

Again consider the shuttle explosion. The vast NASA system had reconfigured into a new form of order. Organized complexity emerged involving the coordinated activities of thousands of specialists. An adaptive transformation occurred. But this same adaptive shift involved a systemic reinforcement of type A activities (meeting cost and schedule requirements) and a relative suppression of type B (meeting safety requirements). Systemic imbalance emerged.

Table 3. Generic characteristics of activities with different systemic responses (see Table 2).

Type A (disorder damping) activities tend to	Type B (disorder promoting) activities tend to
• secure funding	• threaten funding
• support systemic activities	• undermine support for systemic activities
• avoid and resolve budget crises	• promote budget crises
• meet schedules, quotas, or deadlines	• disrupt schedules, quotas, and deadlines
• satisfy those in higher position	• upset those in higher position
• reassure the public	• expose shortcomings, failures, and risks

The fate of the shuttle passengers and the fate of the salmon, I claim, both display the some general phenomena (illustrated in Fig. 4). Systemic imbalance is a normal outcome that should be expected as CANL systems of the human kind adaptively shift to dampen disorders (Tendency 1). Thus, the third tendency emerges:

Tendency 3: As organizational systems shift to dampen disorders over time, they tend toward systemic imbalance, reinforcing some activities (type A) while suppressing others (type B), sometimes to destructive extremes.

Conspiracies, schemes, and intentional designs by individuals are not required. What is required of individuals is that they become preoccupied with the normal behaviors reinforced by organizational systems (Tables 1 and 3), rushing to meet schedules, seeking funding, defending budgets, completing assignments, and having little time to stand back and question what is going on.

Applying the Model to Ecosystem Protection

Stone (1892) described two types of activities relevant to the salmon of the Pacific Northwest. Type A included mining, road building, and timber harvest. Type B included the protection of natural habitats and ecosystems. Stone (1892) proposed a Salmon National Park to protect the salmon as Yellowstone Park had protected the buffalo. Ross L. Leffler (1959), Assistant Secretary of the Interior in the Eisenhower administration, expanded the type A activities to include the construction, operation, and use of dams. Type B activities again included the protection of natural habitats and ecosystems. Leffler (1959) recommended that "we can and should forthrightly declare that certain river basins of the Pacific Northwest should be dedicated to the conservation and development of fish and wildlife resources as their highest and best use." Nehlsen et al. (1991) lead one to expand type A activities, including the operation of hatcheries. Echoing Stone's (1892) concern from a century before, they too call for the protection of habitat and ecosystems, again type B activities. Good reasons can be given for both types of activities but the model (Fig. 1) suggests that the normal adaptive tendencies of organizational systems is toward imbalance, A over B (Tendency 3).

Consider US federal forests. Two kinds of activities can be identified: (A) harvesting timber ("getting out the cut") and (B) protecting diverse forest ecosystems (Overton and Hunt 1974).

Good reasons can be given for both activities. Laws were established to promote a balance between A and B. Given that older forest ecosystems contained dense stocks of harvestable timber, activities A and B were often mutually exclusive to a significant degree. The funding of the US Forest Service was heavily dependent upon timber harvest (activity A). Under such conditions, shifts away from A toward B tended to amplify disorders (including political pressures) to disruptive levels. In contrast, a shift away from B toward A served to dampen immediate disorders and their disruptive threats. Given the dynamic interplay of order and disorder over time, the system shifted (as the NASA system did) toward A and away from B (Fig. 4). In both cases we see an adaptive shift, damping experienced disorders. Systemic imbalance emerged.

The salmon crisis in the Pacific Northwest provides evidence that systemic imbalance is a persistent and pervasive phenomenon. There are many salmon stocks. Their life cycles carry them across wide geographic areas. The decisions that contribute to their decline span more than a century, involving countless individuals, agencies, and organizations. Many voices have called for their protection. Laws, regulations, and technological mitigation have been employed, each claiming to offer protection, and yet, the decline has continued. Many stocks are threatened with extinction. The model developed herein offers some explanation for such an outcome.

Consider Stone's (1892) call for a Salmon National Park and Leffler's (1959) recommendation that certain river basins be dedicated to the conservation and development of fish and wildlife resources. As part of a comprehensive approach, couldn't we now develop a distributed Salmon National Park to protect a wide range of habitats and ecosystems? Given the popularity of existing national parks and their overcrowding, one could hardly dismiss such a proposal as unpopular or wasteful. People seem to be more tolerant of type A financial disasters than type B proposals that would protect natural ecosystems for this and future generations. Why? People are caught up within CANL systems that shape their activities. Their perceptions of what is reasonable, possible, and proper gravitate toward a vast R attractor, and the world is being transformed in its image (Fig. 2). People will not come to understand this transformation by merely observing the salmon. They must also observe the influence of systemic imbalance upon our own behaviors and perceptions.

Systemic Imbalance and the Burden of Proof

Inquiry, discourse, and decisions are shaped by a background of premises (givens) that form the basis of activity (Simon 1976). Individuals work within environments of givens. Techniques, attitudes, and language reflect implicit premises that are seldom critically examined. As an example, Overton and Hunt (1974) describe how an exploitive forest policy is implicit in the techniques and language commonly employed by foresters. The premises behind inquiry, discourse, and decision, implicit within techniques and language, cannot be explained solely in terms of the reasoning and concerns of individuals. Something else is occurring!

Imagine two broad classes of possible premises, A and B as defined in Table 2. Type A premises influence inquiry, discourse, and decisions in ways that tend to dampen and contain disorders. In contrast, type B premises tend to promote and amplify disorders to disruptive levels. The proper balance between A and B should itself be a topic of ongoing concern and reasoned inquiry. But the relative persistence of A and B within organizational systems, hence their relative influence, is also an emergent outcome that cannot be reduced to individual reasoning and concern. Systemic imbalance should be expected. The normal adaptive behavior of organizational systems, the pull of the R attractor, involves a persistent and pervasive shift in premises toward A and away from B (Fig. 4). This tendency can be stated as follows:

> Tendency 4: As organizational systems adaptively shift in their normal manner, they tend toward the reinforcement of some premises (type A) and the suppression of others (type B).

Consider two classes of activities, A and B. Upon reflection, individuals might agree that a particular activity (type B) is important, but then they rush off to meet more pressing (type A) demands. Clearly, the premises that people employ to address how they devote their time influence their activities. For each individual, the shift of such premises, toward A and away from B, may appear as an innocent and practical necessity. But, if this shift is seen as a nonarbitrary tendency common to vast organizational systems (evidence of an R attractor of grand scale), a powerful force in human affairs becomes evident. There exist subtle, widespread, and nonarbitrary shifts of premises (Tendency 4) that shape human responses to the following kinds of questions. To what should people devote their efforts? How should resources be allocated? What problems deserve attention? What must an individual do to be successful? What are an individual's responsibilities? The R attractor draws people to answers consistent with its own character.

Accommodation to systemic imbalance (Tendency 4) can lead to emergent outcomes including distortion of information (Bella 1987, 1992) and de facto decisions contrary to human ideals, values, and intent. Consider again the shuttle explosion. Prior to the fatal launch, engineer Rodger Boisjoly strongly objected to the O-ring design that became the technical cause of the accident. Pre-launch meetings were held to examine safety issues. Boisjoly testified:

> This was a meeting where the determination was to launch and it was up to us to prove beyond a shadow of a doubt that it was not safe to do so. This was a total reverse to what the position usually is in a preflight conversation or a flight readiness review. It is usually exactly opposite that. (US Commission 1986)

Robert Lund, an engineer and manager who played a key role in this meeting, also testified:

> We had to prove to them that we weren't ready, and so we got ourselves into the thought process that we were trying to find some way to prove to them it wouldn't work, and we were unable to do that. We couldn't prove absolutely that the motor wouldn't work. (US Commission 1986)

Given the pressures of type A activities (Table 3, e.g., meeting schedules, sustaining order, etc.), the burden of proof shifted so that those raising type B objections (e.g., R. Boisjoly) were unable to marshal evidence sufficient to disrupt the launch schedule. Once this shift in the burden of proof occurred, the decision to proceed with the launch had in effect been made. Those caught up within the system came to accept type A premises in ways that they themselves did not recognize. Within their discourse and deliberation, the burden of proof shifted (Tendency 4) in a way compatible to the pull of the R attractor (Fig. 4). The burden of proof shifted to dampen disorders that might disrupt the ongoing process (toward A, away from B). Given this premise shift, the decision to launch had essentially been made; the individuals merely acted it out under the givens. If a systemic shift of premises can occur under these conditions (where everyone agreed that a catastrophic possibility did exist, responsibilities were clearly defined, and the policy had been clearly stated), then we should expect similar shifts to occur for environmental concerns where catastrophic possibilities are often subtle, responsibilities are diffused, and policy is not so clearly stated.

The systemic shift of burden of proof (consistent with Tendency 4) is expressed in three steps clearly seen in the Challenger episode. First, resources devoted to the collection of information tend to promote A and neglect B. Second, when evidence supporting B and challenging A does arise, it tends to be dismissed as insufficient, allowing A to continue; often those presenting such evidence face criticism. Third, activity A continues until some serious and irreversible consequences become so apparent that they cannot be denied.

In the case of salmon decline, these same three steps recur in less dramatic but more persistent ways. From a broad historical perspective, a recurring theme emerges. Attention to type B concerns tends to follow rather than avoid irreversible losses. At each step in this process, we find that evidence is dismissed as incomplete and inconclusive. Activity A continues until consequences become too apparent to deny. With time, evidence does accumulate to place some constraints upon A. But, too often, evidence only becomes convincing after irreversible losses occur. The pattern repeats itself as other type A activities arise. This approach has worked reasonably well for acute and reversible environmental impacts that can be directly related to specific activities (e.g., dissolved oxygen reductions and turbidity increases). For irreversible losses resulting from many activities (e.g., habitat loss), however, this pattern has led to cumulative impacts not adequately addressed. Any plan to prevent the further decline of salmon stock must break this pattern by protecting habitats and ecosystems before degradation occurs and shifting the burden of proof onto those who would alter these systems.

Systemic imbalance expresses itself in the form of an ongoing tendency for protection actions to arise after irreversible losses become apparent. Given an expanding array of type A activities, losses accumulate. Something must almost be lost before actions are taken to protect it. Protection of riparian zones and alteration of road construction techniques, as examples, have occurred only after considerable evidence accumulated and irreversible losses occurred. The Endangered Species Act (1973) attempts to reverse the burden of proof, but this occurs only after significant and potentially irreversible losses have occurred. Moreover, at present, it remains to be seen whether organizational systems will adapt to correct the conditions that led to the threatened extinction of species or, instead, force alterations of the Endangered Species Act to accommodate the needs of the organizational systems themselves.

Credible Disorders

How can systemic imbalance be corrected? To address this question, consider again the interplay of order and disorder that forms the basis of the model developed herein. From this perspective the character of organizational systems reflects the history of disorders they have experienced (Tendency 2). The explosion of the Challenger served as a credible disorder that disrupted the NASA system, forcing it to reform in less imbalanced ways. By a credible disorder I mean an event that is both compelling and contrary to established arrangements (patterns, networks) and hence disruptive. Clearly, there are more desirable ways to produce credible disorders so that systemic corrections occur before catastrophic disorders emerge. Testing, monitoring, and assessment are efforts to produce credible (reasonable and compelling) disorders (departures from established order) before catastrophes occur. The intent is to produce credible disorders earlier in the history of a system rather than allowing systemic imbalance to continue until larger disorders emerge. However, there is a systemic problem. Testing, monitoring, and assessment must themselves be sustained and accommodated by organizational systems. They depend upon sustained resource transfers to support potentially disruptive inquiries. They depend upon widespread information networks that will distribute disruptive information when it arises. The problem is this: systemic imbalance that emerges in CANL human systems (Tendency 3) tends to suppress such arrangements.

Independent checks and balances are experienced as type B activities; they promote disorders (Table 2). A future catastrophic possibility is not yet experienced as a disorder because it hasn't happened yet. Organizational systems adapt to dampen experienced disorders (Tendency 1); this can occur through the suppression of type B activities. This happened in NASA and the USDA Forest Service. It is expected that CANL systems require no deliberate conspiracy or plan. It only requires that people accept type A activities, striving to meet schedules, secure budgets, complete assignments, and sustain some sense of order (Table 3). How can people avoid such destructive tendencies? They must strive to sustain those independent social conditions from which credible disorders arise before catastrophic disorders emerge.

In the case of salmon, the Native American tribes and professional societies have served as independent checks upon organizational systems. They have been a source of disorders that forced adaptive change toward the protection of salmon. The Endangered Species Act is a source of similar disorders. Organizational systems with a history of disorders of this kind are more likely to accommodate the survival of salmon and the ecosystems upon which they depend (Tendency 2).

Unfortunately, in the case of US federal forests, the notion of independent checks and balances has been misperceived. On organizational charts, federal agencies were separated from the private firms. Nevertheless, the funding of federal agencies was strongly dependent upon activity A (harvesting timber) and weakly dependent upon activity B (protecting diverse ecosystems). Thus, the R attractor for both federal agencies and private firms pulled in similar directions. In effect, a single system emerged, a powerful timber-industrial complex. The R attractor of this system became strong, pulling the entire system toward activity A and away from B (Fig. 4).

Some will object, claiming that I have overstated the problem of imbalance. They will point to organizational systems that do serve environmental protection, even at substantial cost. I am aware of such outcomes. But to understand these outcomes, the history of credible disorders experienced by such organizational systems should be considered (Tendency 2). Such credible disorders require human activity not defined by organizational systems, behaviors not yielding and often contrary to the pull of the R attractor. How should people describe such activities and behaviors?

In his farewell address, US President Eisenhower (1961a) warned of the power of vast organizational systems, in particular, the "military industrial complex." He also warned that the nation's scholars could become dominated by "the power of money" and "that public policy could itself become the captive of a scientific technological elite." He described the dangers in terms of "imbalance," warning that we "must avoid the impulse to live only for today, plundering for our own ease and convenience the precious resources of tomorrow." "We should take nothing for granted," Eisenhower warned, saying "Only an alert and knowledgeable citizenry can compel the proper meshing" of the vast organizational systems upon which that society depends.

If activities and premises employed to assess their consequences are openly examined and found to be deficient (imbalanced in significant ways), potential disorders within existing organizational systems arise. If reasoned inquiry and discourse expose such deficiencies, then these disorders gain credibility. If civil discourse exposes these deficiencies and if people take them seriously, these credible disorders persist and grow as long as the deficiency continues. Then, given the interplay of order and disorder, CANL systems adapt to accommodate such disorders.

Of course, suppression of the discourse can be the adaptive response. But if such suppression is not tolerated, if suppression of discourse expands rather than dampens disorders, then the adaptive response is more likely to be credible. That is, in adapting to dampen disorder, the system shifts, adjusts, and rearranges to correct the deficiency that was the source of the disorder. In such a manner, the imbalanced activities and premises of organizational systems become sources of credible disorders that force adaptive change in less imbalanced directions. But for this to occur, the background of credible discourse and inquiry must not be shaped and molded by organizational systems themselves. To sustain such inquiry and discourse constitutes what Eisenhower (1961b) called "the duties of responsible citizenship," the essential role of "an alert and knowledgeable citizenry" (Bella 1992). Credible disorders arising from civil discourse broadly practiced serve to ensure that systemic coalitions can seldom persist "on any other principles than those of justice and the general good" (Madison 1778 in Hamilton et al. 1982).

Responsible citizenship requires inquiry and discourse that is both credible and independent of organizational influences, rewards, threats, and demands. Ideally, the inquiry and discourse of a responsible citizenry provide the fertile ground from which credible disorders arise and grow to disruptive levels. Organizational systems are shaped by the history of experienced disorders (Tendency 2). If this history contains credible disorders, then, as adaptive change shifts to dampen disorders (Tendency 1), less imbalanced arrangements are more likely to emerge. If, however, responsible citizenship is not practiced, if inquiry and discourse are superficial and uncivil, if citizens see themselves merely as customers, and if people act out the roles of functionaries merely completing assignments, then systemic imbalance is allowed to continue (Tendency 3), transforming not only our physical and social world but the premises (Tendency 4) upon which our inquiry and discourse are based. From this model, the character of organizational systems reflects the history of inquiry and discourse sustained by the citizenry. For the model developed herein, one can state the following:

> Tendency 5: The degree of systemic imbalance that emerges within modern society and its consequences are inversely proportional to the credible disorders that organizational systems experience through the continuing activities of independent checks, which ultimately depend upon an alert and responsible citizenry.

In Tendency 5 one finds the good news of the model. One can point to a wide range of laws and institutional actions including the protection of a range of habitats and ecosystems. Stone's (1892) example of Yellowstone Park has been followed by actions that continue to this day. This gives hope that wild salmon will not be completely driven to extinction. But such hope is likely to be misplaced if people fail to appreciate the essential role of independent checks and balances and an alert and knowledgeable citizenry. Of course, such notions are not new but they are not well accommodated by current modes of thinking.

In current discourse productivity is the ideal; people hear little of checks and balances appropriate to the challenges they face. In today's world people speak of customers rather than citizens. People defer to "policy makers" and "decision makers"; the notion of an alert and knowledgeable citizenry is too often cynically dismissed. Despite such views, I suggest that the history of environmental protection provides strong evidence supporting Tendency 5 (Paehlke and Torgerson 1990). Where environmental protection has occurred, one is likely to find a history of questions, challenges, and credible disorders sustained by people acting independently and often contrary to the assignments and interests of organizational systems. Without their actions, I doubt people would even recognize that a salmon crisis exists.

Moral and Spiritual Traditions

The model developed herein has drawn upon concepts from a broad field of scientific study loosely described as nonlinear dynamics (Prigogine and Stengers 1984, Waldrop 1992). Chaos, self-organization, organized complexity, and emergent behaviors are among the topics arising from such study. Briggs and Peat (1989) warn us that to see the world from such perspectives is to enter a "twilight zone," to encounter an "alternative reality" where "exact prediction is both practically and theoretically impossible." In describing such views, the Wall Street Journal (Farney 1994) depicts "one of history's great intellectual upheavals unfolding. It amounts to a reshaping of the thought-world that Western man inhabits like a turtle in its shell." The belief in progress through rational control has been challenged. Instead, one finds a world far less predictable and controllable, a world where history and disorder play crucial roles. "Clearly," Farney (1994) writes, "something has eroded away over time: The supreme confidence—or hubris—that characterized earlier eras of science."

The model (described in Fig. 4) and five tendency statements do challenge established views—in particular, those that assume that decision makers and policy makers can direct CANL systems if given the proper information from specialists. At the same time, however, the model can allow us to appreciate traditions of human responsibility that predate the reductionist (analytical, technological) traditions that have dominated modern technological societies. Such traditions do not limit responsibility to assignments, roles, jobs, and positions but rather broaden responsibilities in ways that cannot be delegated or turned over to experts and authorities.

In "Habits of the Heart," Bellah et al. (1985) examine the involvements of citizens within the United States. They describe the notion of citizenship in terms of the Biblical and Republican traditions. The prophetic tradition described in Biblical tales calls people to speak out against unjust systems despite the often high personal costs and remote promise of success. It is a tradition of criticism and empowerment that acts in history, promoting credible disorders over time (Brueggemann 1978). The Republican tradition calls for civil discourse, practiced as virtue. The Biblical tradition calls for this discourse to be inclusive, not defined by position and authority but rather directed toward the creation of a community where a genuine ethical and spiritual life could be lived (Bellah et al. 1985). The tradition of democratic deliberation (Reich 1990) emerges from these earlier traditions. It cannot be reduced to occasional voting. Instead, the democratic tradition calls for an independent, alert, and knowledgeable citizenry that, like the Biblical prophets, sustains that kind of discourse from which credible disorders arise despite the suppression by established systems.

The tradition of democratic deliberation calls for civil discourse and credible inquiry often suppressed within organizational systems. This tradition refuses to submit human responsibility to the pull of the R attractor. Responsibilities of a different moral kind are called for. Such responsibilities often lead people to Type B activities that produce disruptions within organizational systems (see Table 2).

From its prophetic roots, the democratic tradition calls for behaviors that disrupt systems of power in morally compelling ways. This calling can be described in terms of an S attractor that emerges whenever the R attractor and systemic imbalance become strong (Fig. 5). The existence and character of both attractors is inferred by nonarbitrary trajectories in behavior space, much as gravity is inferred by trajectories in physical space. The R attractor pulls toward nonarbitrary systemic requirements (Fig. 2). The S attractor pulls toward moral requirements, also nonarbitrary, sketched in those traditional stories (e.g., the Good Samaritan parable) that people find compelling even when sacrifice is required. An essential tension exists between the two, sometimes strong, sometimes not. The S attractor is moral and nonsystemic, drawing out morally compelling disorders. The R Attractor is amoral and systemic, drawing behaviors and activities into

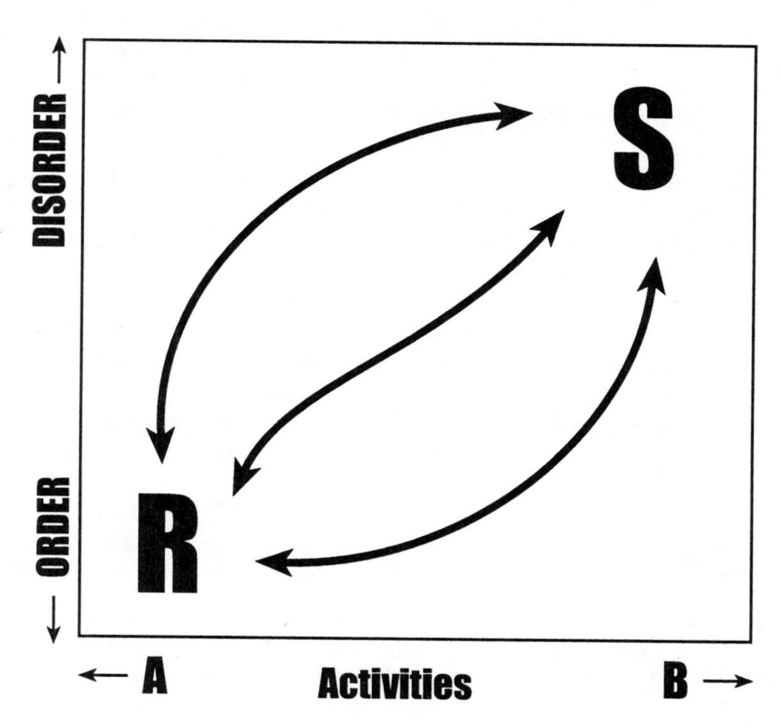

Figure 5. The essential tension between the R attractor and S attractor. Note: Each attractor is inferred by trajectories in behavioral phase space.

coherent patterns of order, which may or may not be morally compelling. The essential tension between the two is played out through a history of credible disorders (drawn out by the S attractor) that force amoral systems of order (drawn out by the R attractor) to continually adapt (Tendency 2) in ways that are not morally arbitrary. Remove the R attractor and disordered commotion results. Remove the S attractor and amoral (nihilistic) order takes over.

The democratic tradition can be described as an effort to sustain this dynamic tension between attractors (Fig. 5) so that the interplay of order and disorder over time (Fig. 3) tends toward systems of order that have moral worth, not because of order itself or the goods it provides, but because of the morally compelling behaviors people act out over time. The stories of such people point toward what I have defined as an S attractor, which draws people to those behaviors and activities from which credible disorders emerge. The degree of systemic imbalance that emerges is inversely proportional to the history of such credible disorders (Tendency 5).

The essential tension between R and S attractors describes the frustrations of actual people caught up within organizational systems, pulled in by the demands of the R attractor and yet drawn toward something else. Consider the salmon as a symbol, one of many, of the S attractor. This book provides ample evidence of people who had been drawn to type B activities by the condition of the salmon rather than the conditioning of their jobs. Over time, their activities constitute a compelling disturbance regime, a history of credible disorders, that can serve to correct systemic imbalance in non-trivial ways. The notion of contrary attractors (Fig. 5) will not resolve their frustrations nor will it tell them what to do, but it does provide a way of saying that their struggles are not meaningless. In other words, there are good reasons for being maladjusted to systemic order and "it may well be that the salvation of our world lies in the hands of the maladjusted" (King 1961).

I realize that this model (Fig. 5) cannot fully account for the implications of moral and spiritual traditions, but it can help people to better appreciate their relevance within a world of vast organizational systems. Of course, a purpose of such traditions is not disorder but rather a more meaningful balance (some traditions speak of harmony) than the organizational systems of modern technology will provide independently. Credible disorder, as described herein, is like tuning in music. It is disruptive, but without it, disharmony (or imbalance) emerges over time. Credible disorders arise from responsibilities that transcend the assignments, incentives, and roles defined for us by established organizational systems. This model accommodates such notions. It may lead us to appreciate that different moral and spiritual traditions have more in common with each other than with the hubris that leads us to transform the world into the image of the R attractor (Fig. 2), a world that has no place for wild salmon or anything else that does not serve the self-reinforcing needs of ever more powerful organizational systems.

Conclusion

All models have limitations that must be recognized if they are to be properly applied. Because the model developed in this paper has a broad perspective, it will not provide the sharp focus needed to evaluate specific decisions. The specific decisions of individuals and groups can have significant consequences, and these decisions must be evaluated under the particular conditions in which they are made. Individuals must be held accountable. But decisions are shaped by pressures, premises, and available choices that constitute the contexts within which decision makers act. A narrow (high resolution, reductionist) perspective focuses upon particular decision makers within specific contexts. From a broader (wide scope) perspective, however, this paper examines systemic tendencies that influence the contexts of many decisions over extended periods of time. Such tendencies can influence outcomes in ways not apparent when people focus upon specific decisions. Thus, both narrow and broad perspectives are necessary to understand the outcomes of human activities and the responsibilities they demand. The model developed herein takes a broad perspective and serves to complement rather than replace the narrow perspective models more familiar to us.

From this broad perspective, tendencies are identified that must be described as emergent behaviors of organizational systems, and are not reducible to the deliberate decisions of individuals. Such tendencies are observable and their consequences are significant. From this perspective, one sees the decline, decimation, and extinction of salmon stocks in the Pacific Northwest as an outcome of a general and widespread phenomenon, systemic imbalance within the vast organizational systems of modern technological societies. The notion of human responsibility is expanded to include responsibilities of citizenship essential to prevent systemic imbalance from developing to catastrophic proportions.

To some, these notions will seem esoteric, remote from the pressing issues we now face. This is a serious mistake! To illustrate, try an experiment. For several days, watch the television and play the video games that young people actually watch and play (activity A). Then withdraw from this world; spend time in a natural ecosystem watching and listening to salmon, birds, trees, and other participants on their own terms (activity B). Then ask the following question. What credible values and beliefs would lead a sane, concerned, and reasonable person to conclude that human efforts should be devoted to providing our children—the future—with an ever-expanding abundance of A while reducing and eliminating B? I cannot imagine values that would rationally support such a choice. Nevertheless, the evidence of such an outcome (A expanded and B eliminated) is overwhelming. How can we explain such outcomes? What is required so our own actions do not contribute to them? What is needed to direct human activity away from such imbalanced outcomes? What is required to hold accountable the vast organizational systems of modern technological society?

This paper addresses such questions. It leads us to see such outcomes (e.g., A expanded and B eliminated) as the normal and predictable tendencies of the organizational systems humans have grown dependent upon. It warns us that our reductionist habits—the dominance of narrow perspectives—fail to grasp such tendencies and serve to divert attention elsewhere. Reductionism in human affairs serves to protect organizational systems by leading people to dismiss emergent properties as esoteric. This paper challenges such views and claims that as a society we cannot address the salmon crisis and countless other consequences of systemic imbalance without taking a broader view that allows us to better grasp the nature of modern organizational systems and hold them accountable.

Literature Cited

Bella, D.A. 1987. Organizations and systemic distortion of information. Journal of Professional Issues in Engineering 113: 360-370.

Bella, D.A. 1992. The system and the citizenry: Eisenhower's warning, p. 57-97. *In* G.B. Walker, D.A. Bella, and S.J. Sprecher (eds.), The Military Industrial Complex. Peter Lang, New York.

Bellah, R.W., R. Madsen, W.M. Sullivan, A. Swidler, and S.M. Tipton. 1985. Habits of the Heart. Harper and Row, New York.

Briggs, J. and F.D. Peat. 1989. Turbulent Mirror. Harper and Row, New York.

Brueggemann, W. 1978. The Prophetic Imagination. Fortress Press, Philadelphia.

Eisenhower, D.D. 1961a. Farewell radio and television address to the American people, p. 1035-1040. *In* Public Papers of the Presidents. US Government Printing Office, Washington, DC.

Eisenhower, D.D. 1961b. The president's news conference of January 18, 1961, p. 1045. *In* Public Papers of the Presidents. US Government Printing Office, Washington, DC.

Farney, D. 1994. Chaos theory seeps into ecology debate, stirring up a tempest. The Wall Street Journal, July 11. CXXXI 6:1.

Jackall, R. 1988. Moral Mazes, The World of Corporate Managers. Oxford University Press, New York.

King, M.L. 1961. The American Dream, p. 208-216. *In* J.M. Washington (ed.), A Testament of Hope. Harper, San Francisco.

Kuhn, T.S. 1962. The Structure of Scientific Revolutions. University of Chicago Press, Chicago.

Leffler, R.L. 1959. The program of the U.S. Fish and Wildlife Service for the anadromous fishes of the Columbia River. Oregon State Game Commission Bulletin, October: 3-7.

Hamilton, A., J. Madison, J. Jay, and G. Wills (Introduction). 1982. The Federalist Papers. Bantam Books, New York. Reprint from the Federalist No. 51.

Nehlsen, W., J.E. Williams, and J.A. Lichatowich. 1991. Pacific salmon at the crossroads: stocks at risk from California, Oregon, Idaho, and Washington. Fisheries 16: 4-19.

Odum, E.P. 1969. The strategy of ecosystem development. Science 164: 262-270.

Overton, W.S. and L.M. Hunt. 1974. A view of current forest policy, with questions regarding the future state of forest and criteria of management, p. 334-353. *In* Transactions of the Thirty-Ninth North American Wildlife and Natural Resources Conference. Wildlife Management Institute, Washington, DC.

Paehlke, R. and D. Torgerson (eds.). 1990. Managing Leviathan: Environmental Politics and the Administrative State. Broadview Press, Peterborough, Ontario, Canada.

Prigogine, I. and I. Stengers. 1984. Order Out of Chaos. Bantam Books, Toronto, Canada.

Reich, R.B. 1990. Introduction, p. 1-12. *In* R. Reich (ed.), The Power of Public Ideas. Harvard University Press, Cambridge.

Simon, H.A. 1976. Administrative Behavior. The Free Press, New York.

Speer, A. 1970. Inside the Third Reich. Avon Books, New York.

Stone, L. 1892. A national salmon park. Transactions of the American Fisheries Society. Twenty-fifth Annual Meeting: 149-162.

US Commission. 1986. Report of the Presidential Commission on the Space Shuttle Challenger Accident. US Government Printing Office, Washington, DC.

Vaughn, D. 1990. Autonomy, independence, and social control: NASA and the space shuttle Challenger. Administrative Quarterly 35/2: 225-257.

Waldrop, M.M. 1992. Complexity. Simon and Schuster, New York.

Salmon, Stewardship, and Human Values: The Challenge of Integration

Robert G. Lee

Abstract

The challenge of conserving Pacific salmon (*Oncorhynchus* spp.) involves integrating an unusual complexity of human activities ranging across wide spatial and temporal scales. The life cycle of salmon covers a vast area ranging from mountain tops to migration routes in the ocean. Although life cycles average 3 to 5 years, genetic adaptations to environmental stressors require over hundreds to thousands of years. Since salmon are a wide-ranging common-pool resource, stewardship cannot become the responsibility of particular owners, regulators, or fishers. Moreover, centralized planning and top-down coordination is unlikely to be successful in conserving salmon. Successful stewardship instead requires the initiative of myriad small-scale institutions and land and water users, as well as coordination among users, institutions, and governmental authorities at all levels. A local-initiated salmon habitat conservation plan is described and evaluated to illustrate these general principles. Stemming from the evaluation of the case study, salmon stewardship boards are recommended as an ecologically appropriate institutional arrangement for coordinating the aggregation of independent initiatives by small-scale institutions and users.

Introduction

Social scientists have studied how humans perpetuate preferred states of their environment by regulating both ecological processes and the human activities that affect these processes. Cooley (1963) drew on the work of Philip Drucker to describe how the Pacific salmon (*Oncorhynchus* spp.) harvest by the native Tlingit and Haida peoples of coastal Alaska and Canada was regulated by myth:

> They believed that salmon were a race of supernatural beings who went about in human form feasting and dancing beneath the sea. When the time came for annual runs, these "Salmon People" assumed the form of fish and ascended the streams to sacrifice themselves for the benefit of mankind. The salmon migration was considered to be a voluntary act and it was thought that if human beings were careful not to offend their

> benefactors, the spirit of each fish would return to the sea, resume its original life and humanlike form, and prepare for the trip next season (Cooley 1963).

Annual salmon harvests were regulated by a norm of human reciprocity requiring people to limit harvesting so that most of the salmon could return to the ocean to dance again.

All enduring civilizations have found ways of integrating human activities with natural ecological processes to perpetuate valued species or environmental conditions. Firey (1963) described the basis of what we now call sustainability when he said "there are many kinds of human activities which, by their very nature, require some kind of orientation on the part of human agents to a remote future."

Americans have for decades valued the protection of soil, water, and air and sought to perpetuate forests, unique natural areas, and highly valued species such as salmon. Yet these environmental values have been expressed as ideals, and moral exhortation has been used to urge actions that promise attainment of these ideals. The same idealism and moral exhortation has characterized salmon conservation. As a result, there is a huge gap between the idealistic commitment to salmon conservation and the actions that are necessary to successfully perpetuate natural stocks of salmon. After years of moral exhortation, North Americans are only now learning how to institutionalize the conservation of salmon and other resource and environmental values.

In this paper, I briefly review the life cycle of salmon and examine the cumulative impacts of human activities on this life cycle. My objective is to suggest an approach for more effectively fostering the human stewardship of dwindling stocks. Integration of human activities with the life cycle of salmon must successfully address the challenge of coordinating a large number of independent human activities that take place in river basins and ocean migration routes. Each of these independent activities is guided by attempts to capture or celebrate the values of salmon values inevitably conflict with one another, leading to laws and regulations designed to reduce the effects of one human activity on the values sought by another.

Hence, the attempt to integrate human activity with the life cycle of salmon is sought through an aggregation of relatively independent laws and regulations, each of which is targeted on particular conditions of the biophysical system or the activities affecting these conditions, or both. Until recently, there have been no comprehensive attempts to coordinate these laws and regulations or to otherwise integrate the wide range of human activities that affect the entire life cycle of salmon.

Small-scale institutions seem to offer the most promising vehicle for salmon conservation, but they require the development of means for coordinating the independent activities to which they give rise. I illustrate general principles with a case study of a locally authored salmon habitat management plan for a small county in the United States (US) Pacific northwest. Since this plan has not been effectively integrated with federal salmon conservation efforts, I suggest the creation of salmon stewardship boards to provide an institutional vehicle for coordinating independent laws, regulations, and human initiatives across the entire geographic range of selected salmon stocks.

Cumulative Effects on Salmon Survival

Human actions affect the survival of salmon at all stages of their life cycle. Effects on survival accumulate from one stage in the life cycle to the next. The Pacific salmon life cycle is relatively well known (Groot and Margolis 1991). It begins in the spawning grounds of rivers and creeks where adults return to lay and fertilize the eggs, and continues in the stream environment with a developmental progression of embryos, alevins, fry, parr, and smolts.

Migration back to spawning grounds by returning adults requires adequate water flow, stream temperature, and water quality, and the absence of predators and physical impediments to movement upstream. Adequate stream flows, water temperature, and water quality, and a suitable substrate also are required for successful spawning. Survival of embryos, alevins, and emergent fry is achieved with adequate flow of well-oxygenated water, absence of a buildup of fine sediments on the stream bed, and protection from deep scouring of the stream channel during storms.

The length of time young salmon remain in the stream environment varies considerably within and between species. Pink (*O. gorbuscha*) and chum (*O. keta*) salmon fry soon start migrating downstream and reach estuaries or the open the ocean in ~2 months, while coho (*O. kisutch*) salmon may spend ≥1 years developing in freshwater before migrating to ocean feeding grounds (Groot and Margolis 1991). Groot and Margolis (1991) also report studies showing that sockeye salmon (*O. nerka*) may take from 1 to 2 years to reach the ocean, while chinook (*O. tshawytscha*) salmon start migrating soon after hatching and may remain in rivers until reaching the ocean as yearlings.

We know more about salmon life cycles than we do about how humans affect survival at each progressive stage in the salmon's life cycle, and we know even less about practical approaches for increasing survival throughout the cycle. Salmon conservation is made exceptionally complex by the wide range of temporal and spatial scales, the diversity of human values that affect salmon survival, and the lack of any effective means for integrating the life-cycle requirements of salmon with a plurality of human actions that affect survival. Salmon migrate over thousands of kilometers of ocean and hundreds of kilometers of rivers as smolts and adults, and yet spend the first months or years of life within a few kilometers of where they were hatched.

The life cycle of salmon averages from 3 to 5 years, and genotypes evolve over hundreds to thousands of years. Although salmon populations have developed genetic and behavioral characteristics that enable them to cope with stress induced by variations in climate, salinity, and many short-term environmental fluctuations, they are not adapted to many of the stresses imposed by heavy fishing, chemical pollution, widespread siltation, extreme water temperature changes, physical barriers such as dams, changes in stream structure, and other historically unprecedented effects of humans on the physical environment of rivers and streams (Salo and Cundy 1987). The mass production of hatchery fish has simplified the gene pool and reduced the behavioral and genetic flexibility essential for adapting to stressors ranging from natural predators and disease to changes in water temperature and stream conditions (Meffe 1992).

The cumulative effects of human actions on salmon survival are not well understood. Survival of eggs, fry, and parr can be affected by a wide variety of human activities including, but not limited to, road building, flood control, logging, urban development, agriculture, mining, hydropower development, irrigation, and chemical pollution. Some of the most important causes for degradation of spawning and rearing habitats include flood control construction, which removes riparian vegetation and alters stream structure; road building, which accelerates rates of natural erosion and deposition of organic and inorganic debris in streams; logging and agricultural development, which removes trees and other vegetative cover in the riparian zone; and irrigation, which reduces stream flows during critical stages of salmon development or spawning (Salo and Cundy 1987). Impacts on survival vary widely but have been estimated to reduce coho egg and parr survival from 5% to 3–4%, and reduce the survival of fry from ~20% to 12–17% (Cederholm et al. 1982; C.J. Cederholm, Washington Dep. Natural Resource, Olympia, unpubl. data and pers. comm.).

The survival of coho smolts as they mature to adults in the ocean environment ranges from <1% to >20% (Groot and Margolis 1991). Most of the natural mortality occurs within the first 6 months of life in the ocean or in estuaries at the mouth of rivers. Dredging or filling of these estuaries removes critical rearing habitat. The overall average survival rate for coho salmon smolts has been estimated at ~7% (C.J. Cederholm, Washington Dep. Natural Resource, Olympia, pers. comm.). Fish surviving to adulthood face other challenges.

Direct human effects have multiple impacts on the survival of salmon returning to spawn. These effects include fishing harvest in oceans, rivers, and lakes, as well as the physical obstructions caused by dams. Predation by marine mammals and other freshwater fish species are enhanced by the presence of restricted migration corridors. Additionally, inadequate stream flows (due to alterations of historical hydrologic patterns or drawdowns for irrigation and hydropower impoundments), high water temperatures, chemical pollution, and other physical obstructions (e.g., log jams) can reduce spawner survival. As a result, roughly half of the females survivors reach the spawning grounds and deposit their eggs.

There have been few attempts to estimate the cumulative effects of these activities across the life cycle of salmon. Hypothetical estimates of how reduced survival of coho salmon accumulates across the life cycle have been made (C.J. Cederholm, Washington Dep. Natural Resource, Olympia, unpubl. data, pers. comm.). Cederholm's estimates suggest that the most important causes of reduced survival are fishing and disruption of spawning and rearing habitats.

No studies have compared salmon survival in the contemporary, human-modified environment with baseline conditions subject to natural disturbances alone. This comparison is needed to show how the natural disturbance regimes such as fire, debris flows, population cycles of predators, climate change, and weather cycles may be mimicked. Also needed is information on how adverse human disturbances (e.g., dams) may be mitigated by undertaking practices that compensate for reduction of natural conditions by improving survival at other stages of the life cycle (e.g., construction of artificial spawning beds and rearing areas to increase smolt production).

The Problem: Complexity and Scale of a Common-Pool Resource

Salmon are a fugitive resource. They are highly mobile, do not remain in any one place very long, and use a wide variety of habitats over their life cycle. These characteristics have made it impossible to deny benefits to people who do not invest in protecting or producing salmon (Lee 1992). Hence, salmon are described as a common-pool resource (CPR)—a resource to which private property rights are not assigned (Ostrom 1990). They belong instead to public bodies such as Native American, states, or federal governments. As such, CPRs are especially difficult to regulate.

Myriad relatively independent federal, state, and local laws and regulations has been designed to regulate fishing, dam construction and associated fish ladders, agricultural land-use practices, use of water for irrigation, urban and residential development, wetlands management, and forest harvesting, in addition to mining and other land uses. Authority for regulating the quality of water and protection of wetlands originates at the federal level and is delegated to states for implementation. Salmon fishing in the open ocean outside national waters is regulated by international agreements while fishing in coastal waters and in rivers and streams is regulated by federal or state authorities or by federal/interstate commissions established to protect and replenish this fugitive resource as it moves across state boundaries or other jurisdictions (K. Lee 1991).

The role of the federal government in flood control works, hydroelectric projects, and irrigation projects has given it responsibility for mitigating impacts on salmon fisheries. Except as affected by federal water-quality and wetland standards, urban and residential development falls under the authorities of counties and cities, often with state requirements for land-use planning and urban growth management. Regulation of forest harvesting is similarly affected by state implementation of federal water-quality and wetland standards, with authority for controlling harvesting residing with the states.

The use of chemicals in silvicultural and agricultural practices is regulated first by the licensing of these chemicals by the federal government, and then by federal and state governments which set standards for application and disposal (Arbuckle et al. 1993). Agricultural land-use practices are not as tightly regulated as forest practices, but both, along with urban land uses, are affected by state laws regulating the use of land along the shores of salt water, freshwater lakes, rivers, and streams.

What is immediately evident from this incomplete and cursory survey of legal regulations affecting salmon and their environment is the enormous complexity of multiple, independent, and overlapping regulatory jurisdictions. Throughout their life cycle, salmon cross international, state, county, and city boundaries and fall under the jurisdiction of multiple, often conflicting, authorities that implement actions in accordance with a multitude of uncoordinated responsibilities designed to be responsive to concerns that may have very little to do with salmon conservation.

The intuitive response to a problem of this scale and complexity is to advocate a rational and comprehensive solution. Ecosystem management, involving attempts to establish centralized regulation of ecological processes that cross ownership boundaries and political jurisdictions, is an example of how intuition leads many to advocate a centralized holistic approach to a problem of enormous complexity (Forest Ecosystem Management Assessment Team 1993). Realignment and centralization of laws and regulations and resource practices to bring them into conformity with ecological processes such as the salmon life cycle satisfies a felt need for replacing overwhelming complexity and fear of chaos with a sense of order and simplicity directed by a central purpose. But a centralized approach to ecosystem management may not be the most promising means for integrating diverse values, laws, regulations, activities, impact mitigation, habitat restoration, and salmon biology (Lee 1993).

Promising Solutions and Criteria for Success

Studies of ecological regulation suggest that the coordination of decentralized management by small-scale institutions may be far more effective than centralized regulation in addressing complex problems involving large-scale salmon conservation efforts. Two simple principles summarize much of what has been learned about ecological regulation: (1) large-scale ecological systems are most effectively regulated by self-governing small-scale social institutions, and (2) the building, maintenance, protection, and integration of small-scale institutions are necessary to facilitate ecological regulation (Drysek 1987, Paehlke and Torgerson 1990, Wheatley 1993). These principles summarize a paradox of ecological learning: effectiveness in learning how to solve macro-level ecological problems is best attained by promoting and coordinating ecological effectiveness of micro-level actions (Dunn 1971, Friedman and Hudson 1974, Korten 1982, Lee 1992, Lee and Stankey 1993).

Practical applications of planning theory and natural resource management principles in developing countries have demonstrated the ineffectiveness of centralized control guided by rational and comprehensive blueprints. Such top-down planning has been rejected as impractical and has been replaced by far more successful experiments with bottom-up planning or combinations of bottom-up planning with centralized coordination (Korten 1982; IUCN/UNEP/WWF 1991). Both resource development and ecological regulation are more successful when the people most directly affected by change are intimately involved in discovering goals and formulating means for implementing them.

Self-organizing and self-governing institutions tend to be associated with greater success in learning about and maintaining the integrity of ecological functions (Ostrom 1990, Wheatley 1993). Ostrom's (1990) studies show that institutions governing CPRs are often less ecologically disruptive; she used several design principles to characterize ecologically successful and long-enduring common-property institutions. The following eight design principles are adopted from Ostrom (1990) to provide criteria for evaluating the salmon habitat management case study:

1. Clearly defined boundaries. In the absence of private property rights, individuals or households have clearly defined rights to withdraw resources from the CPR, and the boundaries of the CPR itself are clearly defined.
2. Congruence between appropriation and provision rules and local conditions. Rules restricting the use of resources to time, place, technology, and number of objects are related to local conditions and to rules regulating the use of labor, material, and money for taking of resources.
3. Collective-choice arrangements. Most individuals affected by the operational rules can participate in modifying the operational rules.
4. Monitoring. Monitors, who actively audit CPR conditions and appropriator behavior, are accountable to the resource users or are themselves resource users.
5. Graduated sanctions. Resource users who violate operational rules are likely to be assessed graduated sanctions (depending on the seriousness and context of the offense) by other resource users, by officials accountable to these resource users, or by both.
6. Conflict resolution mechanisms. Resource users and their officials have rapid access to low-cost arenas to resolve conflicts among resource users or between resource users and officials.

7. Minimal recognition of rights to organize. The rights of resource users to devise their own institutions are not challenged by external governmental authorities.
8. Nested enterprises. For CPRs that are parts of larger systems, resource use, provision, monitoring, enforcement, conflict resolution, and governance activities are organized in multiple layers of nested enterprises.

The principle of monitoring, along with the ecological learning to which it gives rise, is most fundamental to effective regulation of complex ecological systems. For this reason, several additional criteria are summarized for purposes of evaluating the potential of case study organization for promoting ecological learning. Human ecologists have discussed ecological learning as a problem involving how people acquire information about their environment.

McGovern et al. (1988) described transactions between humans and their environment in terms of information flow, by assuming that humans react "not to the real world in real time, but to a cognized environment filtered through expectations and a world view which may or may not value close tracking of environmental indicators." They noted the difficulty humans have in tracking information on the state of ecological systems, including the challenges of maintaining current, accurate, and properly scaled (localized) information. This is especially a problem for advanced industrial societies in which social control systems rely increasingly on standardization and centralization of ecological information. Chandler (1990) has shown how simpler agrarian societies learned ecological information much more slowly but were far more adept in accommodating ecological variability and scale. He drew on McGovern et al. (1988) to identify six factors that contribute to maladaptive information flow, or "information flow pathologies." Lee (1992) added a seventh factor.

1. False analogy. The managers' cognitive model of ecological characteristics (potential productivity, resilience, and stress signals) is derived from another ecological process or species, whose surface similarities mask critical threshold differences from the ecological processes or species of concern to the manager.
2. Insufficient detail. The managers' cognitive model is over-generalized and focused on species abundance, and does not adequately allow for the range of spatial or temporal variability in an ecological process where patchiness and sudden fluctuations may assure resilience.
3. Short observation series. Management lacks a sufficiently long memory of events to accurately track or predict spatial and temporal variability in key environmental factors over a multigenerational period and suffers from a chronic inability to separate short-term and long-term ecological processes.
4. Managerial detachment. The managers are socially or spatially detached (or both) from people who implement managerial decisions on the ground and who are normally in closest contact with microscale environmental variability feedbacks.
5. Reactions out of phase. Partly as a result of the last three factors, the managers' attempts to avert unfavorable impacts are too little, too much, or too late, or they prescribe the wrong remedy.
6. SEP (Someone else's problem). Managers act in accordance with their personal interests and may perceive a potential environmental problem but fail to take corrective action because their own short-term interests either are not immediately threatened or would be threatened by taking appropriate action.

7. Ideological beliefs. Managers may conform to ideological beliefs shared by attentive publics (e.g., equilibrium concepts of carrying capacity, natural regulation, government regulation of the economy, and free markets) and overlook ecological details, ignore environmental feedbacks or trends, and impose politically correct solutions (Schiff 1962, Botkin 1990, R. Lee 1991).

McGovern et al. (1988) emphasize that the first three causes for maladaptive information flows are most likely to be encountered when people first colonize a region and establish highly decentralized organizations, but the effects of these causes diminish as the new ecological system is learned. The fourth, fifth, and sixth factors become important after the development of complex private and public institutions for subjecting ecological systems to centralized management. Large public land management bureaucracies in advanced industrial societies are most susceptible to the seventh factor although ideological distortions of information flow can be found in societies at all stages of development.

Managers are more responsive to localized ecological conditions in societies that have acquired permanent communities with strong local institutions. They are less likely to suffer from the distortion of information flows caused by managerial detachment, out-of-phase reactions, or attributing responsibility for the problem to somebody else. Ideologically induced distortions of information can be reduced by making decision-making processes more accountable to the scientific method, especially methodologies that rely on the falsification of hypotheses (Bella 1987, R. Lee 1991). Managers can overcome the first three causes of maladaptive information flow by relying on established institutions, using scientific discipline, and paying greater attention to spatial and temporal variation.

Wallowa County Salmon Habitat Enhancement Plan

Background

I selected a case of salmon habitat protection and enhancement to illustrate how decentralized, self-organizing initiatives have the potential for promoting salmon conservation. I describe and then evaluate this case study by discussing how well it incorporates Ostrom's (1990) eight institutional design principles for governing CPRs. Also, I discuss the study's potential for overcoming information flow pathologies that would inhibit effective monitoring and ecological learning.

This grassroots planning effort addressed the need for coordinated action to help protect and enhance the mountain habitat for the Snake River chinook salmon within the boundaries of Wallowa County in the northeast corner of Oregon State. In both the spring and the fall, chinook salmon travel ~800 km up rivers to spawn in the mountain habitats of tributaries found in Wallowa County. As has occurred throughout the Columbia River Basin, by the early 1990s the number of anadromous fish returning to the county had dropped to 10–15% of historical averages. Degradation of mountain habitat was only a partial cause for the decline of the species; commercial and recreational fishing in the Pacific Ocean and at the mouth of the Columbia River, as well as river predators and dams along the Columbia and Snake rivers, are by far the most important causes (K. Lee 1991).

The Nez Perce Tribe and all major stakeholders residing in Wallowa County joined in developing a plan to assess existing habitat conditions and recommend remedial projects on every river and stream in the county, including upstream water impoundments, commercial thinning of forests, exclusion fencing around riparian zones, control of weeds, removal or addition of woody material, rotation of grazing, surfacing of roads, and relocation of campgrounds. Salmon habitat planning required voluntary self-organization, resulting in the coordination of activity among multiple local, state, and federal government land management and regulatory agencies, private land owners, local businesses, and private interest groups.

SETTING

Wallowa County has an area of 8,168 km^2 and a 1990 population of 6,950 residents. The county contains the upper reaches of two major river systems (Grand Ronde and Imnaha). These rivers run into the Snake River, which in turn enters the Columbia River. Sixty-five percent of the land in Wallowa County is a CPR owned by the federal government, and about 50% of this land is leased to stockgrowers for grazing cattle and sheep. The remaining 50% of the land is heavily forested or consists of high elevation lands covered by rock, seasonal snow packs, and brush. Agriculture is an important industry along the rivers in the fertile valley bottoms.

Over half the population of Wallowa County is concentrated in four rural communities, with the remainder dispersed in rural residences. These communities are nestled in picturesque high mountain valleys, which slope up to mountain ranges that define the county boundaries. The majority of residents share cultural values associated with an economy based on natural resource production, and they earn a living by ranching, farming, logging, manufacturing, and trading the commodities produced from these natural resources. The natural beauty and small-town way of life has attracted a few urban residents seeking remote places to practice arts and crafts and to enjoy unspoiled nature.

Local ranchers were given long-term leases to graze their cattle on the federal lands. Both the USDA Forest Service (Forest Service) and USDI Bureau of Land Management (BLM) recognized early in the 20th century that land and resource stewardship would be encouraged by extending long-term usufructory rights. Security of grazing land stimulated investments in fences, water development, and adoption of rangeland conservation practices, much as the modern literature would predict (Firey 1963, IUCN/UNEP/WWF 1991).

Federal lands also supplied wood to local sawmills and contract loggers. Sustained-yield plans for the federal forests assured local industries a wood supply that would attract long-term investments in improved utilization and manufacturing technology. Wood supplies have accumulated far more rapidly than they have been harvested and, in the absence of the natural thinning effect of fire, forests have grown crowded and unhealthy. Catastrophic wildfires have increased in frequency and intensity, and threaten the upstream mountain habitat of chinook salmon by killing riparian zone vegetation and altering the stream chemistry.

In 1885, the Nez Perce Tribe was granted treaty rights to salmon in the rivers and to gather, hunt, fish, and graze livestock on federal lands. Many private owners have also granted tribal members access for gathering, hunting, and fishing on lands traditionally used by the Nez Perce (see Note at end of this paper).

As a result of these resource-use activities, local residents and members of the Nez Perce Tribe (most of whom reside elsewhere) are sentimentally attached to the land, its resources, and the traditional uses of land and water-based resources. They feel they have inherent rights to maintain their culture and traditions and that these rights are protected by federal statutes (especially the National Environmental Policy Act, National Forest Management Act, and Hells Canyon Preservation Act), since these statutes require federal administrative agencies to respect the custom and culture of rural communities and Native American nations. They also feel that these rights are an expression of the "pursuit of happiness" guaranteed by the US Constitution.

Formation of the Planning Committee

A shared commitment to protect and restore the chinook salmon population was the unifying motivation for forming a committee to develop a cooperative plan. An additional motivation was the concern of land owners and resource users that an externally imposed salmon recovery plan might unacceptably reduce grazing, timber harvesting, and sand and gravel extraction.

Also contributing to a willingness to cooperate was growing insecurity of longstanding rights to use federal lands for grazing and wood production. Several years prior to this planning effort, the federal agencies began to reduce the term of grazing leases, culminating in revised regulations that threatened to remove many of the incentives for land and resource stewardship and replace these incentives with penalties.

The Wallowa County-Nez Perce Tribe Salmon Recovery Strategy Committee (the Committee) was formally established in 1992 following the listing of the Snake River chinook salmon as a threatened species under the Endangered Species Act (ESA). Planning started when a small group of concerned citizens in local government and the livestock and wood products industries united in asking the Nez Perce to join them in forming a leadership group that could organize a broadly based cooperative planning effort.

An inclusive list of stakeholders was developed, and individuals representing each stakeholder group were invited to an organizational meeting. Twenty persons representing 11 stakeholder groups covering all relevant aspects of salmon habitat issue were ultimately involved as members or alternates (Wallowa County-Nez Perce Tribe 1993). The committee consisted of local representatives from Oregon State and US government regulatory and land management agencies, the Nez Perce Tribe, and local citizen groups spanning livestock grazing and logging to environmental protection and recreation and tourism interests.

As the Committee coalesced into a working group, the original leadership group blended into the Committee and a local citizen respected for his knowledge and fair-mindedness was asked to serve as facilitator. The facilitator provided leadership for the process by keeping discussion focused, identifying issues, bringing out silent partners, and providing feedback to clarify positions. Every discussion emphasized the need for each participant to honor the legitimacy of all other participants and their interests. Decisions were made by talking until a consensus emerged. A spirit of interpersonal trust and confidence emerged among all committee members.

The Committee issued the following mission statement (Wallowa County-Nez Perce Tribe 1993):

> To develop a management plan to assure that watershed conditions in Wallowa County provide the spawning, rearing, and migration habitat required to assist in the recovery

> of Snake River salmonids by protecting and enhancing conditions as needed. The plan will provide the best watershed conditions available consistent with the needs of the people of Wallowa County, the Nez Perce Tribe, and the rest of the United States, and will be submitted to the National Marine Fisheries Service for inclusion in the Snake River Salmon Recovery Plan.

The National Marine Fisheries Service (NMFS) was mentioned because it is the US government agency responsible for preparing a plan for recovering the Snake River chinook salmon under the authority of the ESA.

Plans for restoring the mountain habitat of the chinook salmon were carefully coordinated with five federal agencies. In addition to the land management responsibilities of the Forest Service and BLM, the Committee coordinated its plans with the hydroelectric and irrigation responsibilities of the Bonneville Power Administration (BPA), the water resource management of the Bureau of Reclamation (BuRec), and the anadromous fish protection responsibilities of NMFS (Wallowa County-Nez Perce Tribe 1993). The Committee expected and was informally assured by local representatives of federal agencies that its plan would be incorporated into the comprehensive Snake River chinook salmon recovery plan.

Implementation

Initial financial support for planning came from participating agencies (especially the BPA) and in-kind contributions of Committee members. Leadership in seeking federal assistance with habitat restoration and impact mitigation came from the BuRec, since, unlike other federal agencies, its charter allows it to work on problems where public and private jurisdictions abut or overlap.

The total costs of implementing recommendations of the plan will be ~$19 million (Wallowa County-Nez Perce 1993). The plan identified 35 potential sources of external funds for implementation projects, including 5 federal agency programs, 8 Oregon State programs, 2 regional integrated resource assistance programs, and 6 private sources for wildlife and ecological restoration.

Three independent but interconnected implementation committees were formed soon after the plan was published: the Grand Ronde River Watershed Council, the Forest Enhancement Committee, and the Watershed Element Committee. Whenever opportunities and funding arise, these committees coordinate their activities with federal, state and local officials to implement recommended mitigation and restoration measures on both private and federal lands. Informal communication networks established during the preparation of the salmon recovery plan were expanded and became the basis for continuous communication among the three implementation committees.

Two major obstacles have arisen in implementing the plan. First, soon after the Committee reached consensus and issued a plan embraced by all participants, the Hells Canyon Preservation Council (a locally based Snake River environmental preservation group) broke from the Committee. It joined the Pacific Rivers Council and Sierra Club Legal Defense Fund in a lawsuit that sought to protect the chinook salmon (and restore the ecosystem to its pre-settlement wildness) by terminating all timber harvesting and grazing on watersheds of the Wallowa-Whitman National Forest. This left no environmental interests involved in implementation of the plan.

Second, the Forest Service, BLM, and NMFS have been slow to embrace the plan as an element in the federal planning efforts to manage regional ecosystems and recover the Snake River Chinook salmon. The Forest Service is leading a multi-agency ecosystem management planning effort initiated by US President Clinton's administration late in 1993—an effort that began after the Wallowa County-Nez Perce plan had been completed. This ecosystem management planning effort encompasses a region covering over 75 counties in four states with tributaries draining into the Snake and Columbia rivers. Federal agencies are waiting until the larger plan is complete to see how it articulates with the Wallowa County-Nez Perce Tribe plan. Comprehensive federal ecosystem management planning will not be completed until the end of 1996.

Case Study Evaluation

Ostrom's (1990) design principles for institutions governing use of CPRs provide appropriate criteria for evaluating the Wallowa County-Nez Perce salmon habitat conservation plan (Table 1). Both the Snake River chinook salmon and two-thirds of the land base in Wallowa County influencing salmon habitat are CPRs. Hence, Ostrom's (1990) criteria and the resource base addressed by the plan are similar enough to justify a direct comparison.

The collaboration that went into developing the plan laid an important foundation for institutionalizing salmon habitat conservation by clarifying who has what rights to use various kinds of resources that affect the quality and quantity of salmon habitat. Stakeholders, such as cattle grazers and sand and gravel extractors, conceded to substantial modifications to their usufructory

Table 1. Summary evaluation of Wallowa County salmon management plan using design principles of Ostrom (1990).

Design principles	Wallowa County Plan results
1. Boundaries	Stakeholders accepted modification to usufructory or private property rights
2. Congruence of rules regulating use and production	Plan initiated process of developing new rules to fit local context
3. Collective choice arrangements	Plan involved all affected stakeholders
4. Monitoring	Plan was effective in avoiding false analogies, insufficient detail, blaming problems on somebody else, ideological beliefs, and managerial detachment; partially successful at avoiding short observation series and reactions out of phase
5. Graduated sanctions	Plan did not contain provisions for sanctions
6. Conflict resolution mechanisms	Plan based on consensus processes worked for all but environmental advocacy groups
7. Rights to organize	Rights to plan, but limited rights to implement remedies for common pool resources
8. Nested enterprises	Undeveloped linkages of local, state, regional, national remedies, essential to success

rights to use the river beds or riparian zones. Moreover, stakeholder interaction clarified the nature of salmon as a fugitive resource that migrates across property boundaries, necessitating a more flexible view of private and public property rights in order to define the boundaries of salmon habitat as a CPR.

The collective assessment of salmon habitat protection and restoration began a process for defining, creating, and clarifying locally appropriate rules governing how resources could be produced and used. While the planning process only initiated the institutionalization of rules governing salmon, successful implementation will require a more explicit collective definition of rules.

The most important aspect for successful institutionalization of salmon habitat conservation was that the grassroots process of planning involved all affected citizens in the development and application of rules. At this early stage in the development of new institutions, graduated sanctions for rule violators were not established. However, full participation in the development and application of sanctions will need to occur before regulation can assume a more effective self-governing status.

The implementation process has already begun to involve and train local citizens as monitors of salmon habitat conditions. The three implementation committees are actively engaged in developing monitoring standards and remedial measures for forest, riparian, and streambed conditions. More will be said later in this paper about how the planning process may affect ecological learning.

The consensus process developed by the Committee was an especially effective and low-cost means for resolving conflicts between resource users choosing to participate. However, the process failed to resolve conflict between participants and non-participants, especially when stakeholders instead chose to use the federal courts or centralized federal planning processes to resolve conflicts. Hence, it appears that consensus can only be effective in resolving conflicts when stakeholders are committed to participate.

The formation of the Committee and development of a cooperative, multi-stakeholder plan have not been challenged by external government authorities. Resource users have been recognized as having the right to devise their own institutions. However, institutions arising from resource users have no legal authority with respect to the use of federal lands or the protection and enhancement of salmon habitat. As a result, federal land management agencies and the NMFS have chosen to ignore the considerable moral authority embodied in the development of the salmon habitat management plan. Thus, rights to organize may prove ineffective unless they are coupled with a sharing of authority to implement action plans.

A disjunction between federal legal authorities and local conservation initiatives and moral authorities is the primary obstacle to institutionalizing effective salmon habitat management in Wallowa County. Ostrom's (1990) eighth principle, nested enterprises, may provide valuable guidance for removing this obstacle by suggesting the need for nesting local initiatives and authorities in the larger context of salmon management. Rather than local initiatives being viewed as impediments to federal planning, they could be viewed as vehicles for institutionalizing salmon conservation. State and federal authorities play an important role in defining the requirements that need to be met by local salmon habitat protection and enhancement plans, and can significantly multiply the effectiveness of remedial actions by coupling the extension of shared authority to delegated responsibility (also see Selznick 1949). Salmon conservation could be best institutionalized by encouraging and rewarding the exercise of freedom at the local level through granting shared authorities along with coordination of hierarchically nested plans (also see Schieve and Allen 1982, Wheatley 1993).

Local institutions will be most effective as vehicles for salmon conservation if they facilitate citizen involvement in learning how best to conserve and enhance salmon habitat. Small-scale institutions can provide a means for systematically tracking information on the state of salmon habitat because they are congruent with the scale of ecological systems in which salmon spawn and young salmon emerge, grow, and begin their downstream migration.

Locally oriented institutions are generally more effective than centralized institutions in avoiding maladaptive information flows that originate in false analogies, insufficient detail, short observation series, and managerial detachment (Chandler 1990, Lee 1992). Wallowa County citizens currently share strong residential attachments and unusually positive sentiments toward restoring salmon, thereby providing conditions far more favorable for overcoming these four sources of maladaptive information flow than were found a century ago when local inhabitants were forming settled communities out of the instability and resource exploitation of the frontier era.

A shared history of over a century of experience working with land and resources also enables local citizens to understand long-term cycles and avoid reactions that are out of phase. And the type of mutual responsibility demonstrated in reaching consensus on salmon habitat conservation revealed a willingness on the part of all stakeholders to take responsibility for their facets of the issue and avoid attributing the problem to somebody else. The multiple-stakeholder committees involved in implementing the salmon habitat plan are sufficiently pluralistic to make it unlikely that one set of ideological beliefs will dominate thinking. Openness to new ideas and practical solutions gives local institutions a distinct advantage in developing ways for monitoring, evaluating, and remedying salmon habitat.

Conclusion: Salmon Stewardship Boards

Coordination of activities involving small-scale institutions and larger-scale land-management and regulatory authorities remains by far the most difficult challenge facing salmon conservation efforts. As a partial answer to this difficulty, I close by recommending the creation of salmon stewardship boards. These boards could best reflect the complexity of salmon biology by a nested, hierarchical structure, beginning with local institutions and culminating in international coordinating boards. They would include representatives of all the institutional and citizen stakeholders that value salmon or affect the life cycle of salmon. Membership would range from the various sorts of fishing interests (both foreign and domestic) to the users of water, and forest, agricultural, and urbanized land. Environmental and recreational interests would also be included, along with representatives of governmental authorities at all levels.

The Northwest Power Planning Council (NPPC), established to coordinate independent activities in the Columbia River Basin, serves as a trial prototype for a salmon stewardship board at the scale of a river basin since the NPPC brought together diverse interests across all aquatic and land-use activities in the Columbia Basin (K. Lee 1991, 1993). Salmon stewardship boards would have to be even more inclusive and international in scope, as well as more responsive to local institutions. Complexity could be reduced by designing a hierarchical, nested governance structure in which groups of representatives could be brought together on functionally defined panels to address particular ecological processes or geographic areas. The main board, or its technical subcommittees, would have responsibility for integrating input from these functional panels and would obtain feedback on proposed policies prior to deciding on how integration should proceed.

For example, stream restoration, habitat enhancement, riparian vegetation management, and erosion and deposition processes affecting spawning and rearing grounds could be addressed by panels comprising representatives of forest and agricultural landowners, urban developers, flood control authorities, related state and federal agencies with jurisdiction over these processes, and others with relevant knowledge and authority. Similar panels could be formed for wetland conservation and restoration, chemical pollution, commercial and sport fishing, hydropower development, regulation of water flow, and other discrete ecological functions.

To succeed as coordinating institutions, salmon stewardship boards would require the participation of small-scale institutions ranging from irrigation districts and fishing associations to neighborhood associations and restoration clubs. Above all else, salmon conservation is a human enterprise, and human-scale institutions are most likely to successfully facilitate the integration of independent human activities and diverse values with the biological requirements of the salmon's life cycle.

Note

Much of the information on the Wallowa County-Nez Perce Salmon Recovery Plan was gathered through personal interviews by the author. Following sociological convention, individual interviews are not noted.

Literature Cited

Arbuckle, J.G., F.W. Brownell, D.R. Case, W.T. Halbleib, L.J. Jensen, S.W. Landfair, R.T. Lee, M.L. Miller, K.J. Nardi, A.P. Olney, D.G. Sarvadi, J.W. Spensley, D.M. Steinway, and T.F.P. Sullivan (eds.) 1993. Environmental Law Handbook, 12th ed.. Government Institutes, Inc., Rockville, Maryland.

Bella, D.A. 1987. Organizations and the systematic distortion of information. Journal of Professional Issues in Engineering 113: 360-370.

Botkin, D.B. 1990. Discordant Harmonies: A New Ecology for the Twenty-First Century. Oxford University Press, New York, New York, USA.

Cederholm, C.J., L.M. Reid, B.G. Edie, and E.O. Salo. 1982. Effects of forest road erosion on salmonid spawning gravel composition and populations of the Clearwater River, Washington, p. 1-17. *In* K.A. Hashagen (ed.), Habitat Disturbance and Recovery. Proceedings of a symposium. California Trout, Inc., San Francisco.

Chandler, P. 1990. Ecological knowledge in a traditional agroforest management system among peasants in China. Dissertation, University of Washington, Seattle, Washington, USA.

Cooley, R.A. 1963. Politics and Conservation: The Decline of the Alaska Salmon. Harper and Row, New York.

Drysek, J. 1987. Rational Ecology: Environment and Political Economy. Oxford University Press, Oxford, England.

Dunn, E.S., Jr. 1971. Economic and Social Development: A Process of Social Learning. The Johns Hopkins Press, Baltimore, Maryland.

Forest Ecosystem Management Team. 1993. Forest ecosystem management: an ecological, economic, and social assessment. USDA Forest Service. Portland, Oregon.

Firey, W. 1963. Conditions for the realization of values remote in time. Pages 147-159 *in* E.A. Tiryakian, editor. Sociological Theory, Values, and Sociocultural Change: Essays in Honor of Pitirim A. Sorokin. Free Press, Glencoe, Illinois, USA.

Friedman, J. and B. Hudson. 1974. Knowledge and action: a guide to planning theory. Journal of the Institute of Planners 40:2-16.

Groot, C. and L. Margolis. 1991. Pacific Salmon Life Histories. University of British Columbia Press, Vancouver, British Columbia, Canada.

IUCN/UNEP/WWF (The World Conservation Union, United Nations Environmental Program, and World Wide Fund for Nature). 1991. Caring for the Earth: A Strategy for Sustainable Living. Gland, Switzerland.

Korten, D.C. 1982. Community organization and rural development: a learning process approach. Public Administration Review 40: 480-511.

Lee, K.N. 1991. Rebuilding confidence: salmon, science, and law in the Columbia Basin. Environmental Law 21: 745-805.

Lee, K.N. 1993. Compass and Gyroscope: Integrating Science and Politics for the Environment. Island Press, Washington, DC, and Covelo, California.

Lee, R.G. 1991. Scholarship versus technical legitimation: avoiding politicization of forest science. *In* R.A. Leary (ed.). Proceedings of Philosophy and Methods of Forest Research, XIX Congress of the International Union of Forestry Research Organizations, Montreal, Canada, August 7, 1990.

Lee, R.G. 1992. Ecologically effective social organization as a requirement for sustaining watershed ecosystems, p. 73-90. *In* R.J. Naiman (ed.), Watershed Management: Balancing Sustainability and Environmental Change. Springer-Verlag, New York.

Lee, R.G. and G.S. Stankey. 1992. Major issues associated with managing watershed resources. *In* P.W. Adams and W.A. Atkinson (eds.), Balancing Environmental, Social, Political, and Economic Factors in Managing Watershed Resources. Oregon State University, Corvallis.

McGovern, T.H., G. Bigelow, T. Amorosi, and D. Russell. 1988. Northern Islands, human error, and environmental degradation: a view of social and ecological change in the medieval North Atlantic. Human Ecology 16: 225-270.

Meffe, G. K. 1992. Techno-arrogance and halfway technologies: salmon hatcheries on the Pacific coast of North America. Conservation Biology 6: 350-354.

Ostrom, E. 1990. Governing the Commons: The Evolution of Institutions for Collective Action. Cambridge University Press, New York, New York, USA.

Paehlke, R. and D. Torgerson (eds.). 1990. Managing Leviathan: Environmental Politics and the Administrative State. Belhaven Press, London.

Salo, E.O. and T.W. Cundy (eds.). 1987. Streamside Management: Forestry and Fishery Interactions. University of Washington Institute of Forest Resources Contribution 57. Seattle.

Schiff, A.L. 1962. Fire and Water: Scientific Heresy in the US Forest Service. Harvard University Press, Cambridge, Massachusetts.

Schieve, W.C. and P.M. Allen (eds.). 1982. Self-Organization and Dissipative Structures. University of Texas Press, Austin.

Selznick, P. 1949. TVA and the Grass Roots. University of California Press, Berkeley.

Wallowa County–Nez Perce Tribe. 1993. Salmon recovery plan. Wallowa County, Oregon.

Wheatley, M. 1993. Leadership and the New Science: Learning About Organization from an Orderly Universe. Berrett- Koehler Publishers, San Francisco.

Where Do We Go from Here?

Where Do We Go From Here? An Outsider's View

James F. Kitchell

Abstract

This brief paper offers an overview of the book. A synthesis of the ideas advanced serves as the basis for identifying goals worthy of sustained effort, essential new directions, non-productive pursuits, and intractable problems. Main conclusions of the review include three recommendations:

1. Accept humans as real and continuing components of these systems. The "balance of nature" and "leave it alone and it will recover" arguments are dangerous myths.
2. Carefully select allies from the ranks of those engaged in policy making. Legislation and legislators have shorter life spans than treaties.
3. Design and conduct a small set of restoration experiments. A demonstrable success offers evidence to scientists and encouragement to the public.

Introduction

Readers of this book and the recent one by Lee (1993) have an excellent overview of a complex set of environmental and resource-related issues set in an equally complex institutional context. The central conclusion is that there is a major problem. I agree, of course, and use the following pages in an attempt to offer the views of an observer from outside the political and scientific centers of the problem of Pacific salmon (*Oncorhynchus* spp.) and their ecosystems. In addition to the problems, I see a major opportunity.

Implicit in the title of this book is a holistic view of the Pacific salmon problem. People often move toward a larger view when faced with issues that cannot be remedied from their current perspective. The quest for a larger view offers hope. In the case of salmon, that view includes all of the habitats and each of the life-history stages. In other words, it includes virtually all types of surface waters, fresh or marine, many of the species they contain, and the land uses that influence them. Thus, the holistic view is a formidable challenge. So, too, is the call for insight and advice when the issues of concern range across all of those habitats and the effects imposed upon them by human use of their resources.

I guarantee that what follows herein offers neither new insights nor sage advice about immediate reparations. Those who have read this far in the book will know that to be unlikely if not impossible. The only way to fully resolve the salmon problem in the Pacific Northwest is to simply declare the problem solved and move on to the next set of issues. For those who deem that unacceptable, the alternatives include lesser objectives and some compromise. We cannot put things back the way they were. Humans and their effects are a real and continuing part of the future of salmon. Coexistence will require concessions and affirmative action.

In the largest sense, there is not a salmon problem. At the global scale, the supply of salmon is probably at or near its highest (Francis 1996). For the producer, the problem is a glutted market and the modest price that follows. Record catches in Alaskan waters, marine ranching by the Japanese, and the successes of the pen-rearing ventures in Norway, Scotland, Chile, and even Minnesota have created a remarkable supply of salmon. While this is good news in most parts of the world, it is but cold comfort for those in the regions (e.g., Washington, Oregon, and California) where many salmon stocks are declining, imperiled, or extirpated.

Although there are reasons to doubt that the current yields will be sustained (Beamish and Bouillon 1993), we can predict neither the timing nor the magnitude of declines in native stocks. If large-scale and long-period oscillations in oceanic conditions account for the current high and future low of salmon, there is nothing we can do about that except expect it (Pearcy 1996). That turns out to be a major advantage for those interested in restoring the salmon to systems either where they are currently in low numbers or from which they have been excluded, or both. Acting on the potential of this advantage is the main theme of my brief paper.

The Problem

The logical first step in effecting reparations is to identify the cause of the problem. If asked what caused the decline of any or some set of salmon stocks, the answers would include both the disciplinary bias of each person questioned and a mix of causes that increases in proportion to the number of persons queried. Geneticists would point to dilution and depletion of the gene pool. Ecologists would focus on habitat losses and exotic species. Fisheries biologists would argue that overexploitation was a major contributor, and marine scientists would deduce that the estuarine nurseries are compromised. Environmentalists would point to the impediments caused by dams and irrigation systems that alter supply of the basic medium while engineers might offer the need for more structural improvements. Foresters would reason that effects move downhill (thence downstream) and oceanographers would argue that the problem is far from land. Social scientists would point out that the structure of our institutions and our expectations are a source of fundamental, continuing conflict (Lee 1993). There is room in this set of issues for many opinions and justification for more facts (Walters 1996). As detailed in the pages of this book, each view offers key elements of a larger truth and, unfortunately, none offers a complete explanation. In other words, the problem demonstrates effects of multiple causality. Thus, the solutions must include a polythetic approach. I believe that choosing to restore a subset of salmon populations is the most effective path toward more solutions.

The Solutions

I propose a three-step approach toward effecting some reversals in the current vector, which is dominated by declining salmon stocks, acrimonious debate about cause and effect, and legal actions that distance the issues from constructive alternatives. The first step is to change our conceptual model. We must reject the unrealistic view that a harmonious balance of nature can be restored. That is wrong and it impedes understanding. As a second step, we must recognize our allies and entrain their participation in a common goal. One way to do this is to identify salmon as an icon or symbol of things set back on the right course. The third step is to effect a set of discrete victories that may be used to demonstrate the prospects of restoration. Having accomplished all three, we can have greater confidence in continuing the call for public support.

Changing Expectations

Changing our conceptual model is the most difficult. We must give up the illusion that things can be put back the way they were when some harmonious balance of nature kept the salmon at high and sustained levels in all their native environments. Every evaluation of large-scale ecological issues reveals that unperturbed nature does not behave as we might hope (Shrader-Frechette and McCoy 1993). Populations vary hugely and respond to a miscellany of causes ranging from local landslides that block streams to weather effects and global climate changes. Humans contribute to those through deforestation, overexploitation, eutrophication, hydroelectric developments, dredging, agriculturalization, urbanization, selective breeding in hatcheries, etc. Salmon are probably evolving in direct response to our effects on their environment (Walters and Juanes 1993). While we can argue about the merits, debits, and limits of the balance of nature assumption, we cannot disregard the reality that human effects are a major cause of the current state and will continue to bring changes in the ecosystems occupied by salmon.

Humans and their effects are now a reality in the ecological future for salmon. Thus, a common dream of environmentalists—leave it alone and it will return to equilibrium—is a noble, seductive goal and a dangerous fallacy (Ludwig et al. 1993). Similarly, waiting to act until the research and management experts come to consensus about what caused the problem is a guarantee that we will run out of time to solve it. Instead, we must recognize that the future of salmon will be substantially dictated by choices we make. We must make those choices on the basis of admittedly incomplete understanding (Francis 1996). In my opinion, the role of the scientist is to elucidate the choices that seem most reasonable and their likely outcomes. Going back to the way things used to be is not one of those choices. Humans are here to stay. The rate of change in our environment outstrips our ability to fully understand it.

Salmon—The Icon

Salmon are a symbol of "the good life" in the Pacific Northwest. Their anadromous habit makes them visible to the public and brings an integrated message about environmental conditions that range from high mountain natal streams to the middle of the ocean. Seeing a wild salmon—especially in downtown Seattle, Washington, or Portland, Oregon—confirms that nature is functioning in spite of all the other immediacies that seem mostly negative. Buying a salmon in a Midwestern supermarket is less potent but does take the message to a larger public. As John Muir argued, salmon bring the solitude and restorative powers of wilderness to a public that cannot otherwise get there (Lee 1993). In the terminology of resource economists, salmon have a very large existence value. Their disappearance evokes a strong public call to action (Lee 1996, Smith and Steel 1996). Logically, fishery scientists are the first to hear that call. As the remedies become increasingly complex, other kinds of natural and social scientists are entrained. In my view, science alone will not be sufficient to the task.

Scientists need allies if we are to effect major improvements in the future of salmon. While there is a huge diversity of fishery issues confronting policy makers, salmon problems continue to be among the most widely represented in the present suite of electronic mail newsletters and briefings to legislative staffs. Those briefings arise because of current and pending legal action. Their domain of effect will occur in courtrooms, on the floors of state and federal legislatures, and in the committee meetings that yield authorization and appropriation. These are the domains of lawyers. By extension, among our most valuable allies should be a set of lawyers. Lawyers who like to fish might be even more motivated.

The Endangered Species Act (ESA) is widely viewed as among the most powerful pieces of legislative action available in support of protecting salmon and their habitat. It is the cornerstone of current restoration efforts (Nehlsen 1996). However, recognize that the ESA is an act of legislative will in a democratic process that periodically reevaluates its priorities. The ESA can be diminished or dissolved by the same process that created it. If the environmentalist's demands become overly burdensome to the larger public and become the tyranny of a minority, the ESA may itself become threatened, endangered, and extinct. The ESA is a powerful ally but one that might suffer from overexposure if employed in too many battles.

An alternative legal leverage exists in the several international agreements between federal governments. Reauthorization of the Magnuson Fishery Conservation Act will likely strengthen its role in regulating the losses to exploitation and will serve as a better focus for international negotiations over resource use. If, as recommended by a recent National Research Council panel, the Magnuson Act is modified to provide for interests of the public at large, the role of salmon as a symbol or icon will likely be strengthened (National Research Council 1994).

International treaties are even more powerful. Those with the Native American nations are particularly germane because they focus on a guarantee of sustained supply to purposes greater than brutal economics or the vagaries of elections and public opinion. While the disagreements over priority, access, and rights will continue to be difficult, the fact remains that all parties have a common goal—maintenance of Pacific salmon. Thus, the rigorous pursuit of treaty rights will likely be among the key components of a "bottom line" for legal and environmental actions in support of salmon conservation and restoration. In this case, salmon are the essential icon.

Staging for Victory

In the preceding text, I have argued that the future of salmon stocks must be viewed with some pragmatism and that the keys to recovery will involve strong support from public institutions. I believe we need a set of success stories that will demonstrate to legislators, the courts, and the public what can be done if resources are channeled to bringing about recovery of selected salmon stocks. Scientists can contribute the expertise required to demonstrate that restoration can be successful and, therefore, that restoration is among the policy options. Knowing the limitations, costs, and benefits allows policy makers to choose which stocks might be next for restoration.

Restoring aquatic ecosystems is a significant challenge; its history is mixed (National Research Council 1992). More specifically, the history of the Salmon Enhancement Program (Hilborn and Winton 1993) offers examples of successes and failures that can help identify the most likely prospects for a series of effective future efforts. Although I know too little about the specific prospects, I can offer some general advice derived from experiences in the Laurentian Great Lakes region of North America. Foremost is the need for historical perspective, replicates, and reference systems. Too many idiosyncrasies can enter the ecological arithmetic of individual sites. A set of discrete, but similar sites must be selected if we are to distinguish pattern and process from accidents and events. In other words, a case history approach must be pursued but with a sufficient number of individual cases assembled to allow some generality (Shrader-Frechette and McCoy 1993). The reference systems provide a measure of changes owing to something other than restorative manipulations. The elements I have outlined are familiar to those who know the literature of adaptive management (Walters 1986, Lee 1993).

The best of scientific advice and strong commitment from regulatory agencies will be called upon to repair selected sites and restore their salmon populations. Those efforts could range from improving riparian corridors on small streams to removing some hydroelectric facilities from major rivers. They could include constraints on local fisheries and might extend to major changes in regional exploitation practices. Obviously, site selection is crucial because the prospect for success must be maximized.

Among the first choices should be a series of spawning streams (current or former) proximate to urban centers. Effective distance should be estimated as something less than the average daily range of remote camera crews from major television stations. The development, progress, and result of salmon restoration efforts need to be regularly reinforced for the public. Successes must be advertised beyond the realm of reports and professional journals. In other words, we need to cause some good news, know of its likelihood in the next application, and be sure that it is well known to the taxpayers who supported it.

At the other end of the scale is the challenge of large river systems that have felt the effect of hydroelectric dams as well as the set of habitat and exploitation problems common to many salmon stocks. Issues surrounding salmon stocks in the Columbia, Snake, and Sacramento rivers of the western United States are probably too deeply confounded for any of those systems to serve as appropriate sites. (However, note that court decisions create resources for remediation and large-scale manipulations.) More likely choices would include places like the Elwah River on Washington State's Olympic Peninsula or some of the major rivers on the southern Oregon coast. Removing, rebuilding, or making major changes in the operation schedule of a dam or two (i.e., engineering to fully maximize salmon survival rather than power production) will be both a challenge and an opportunity.

To some extent, the courts are currently imposing analogous expectations on operations in the Columbia River system but are doing so to avoid further damage to severely diminished stocks. Instead, I propose that the goals of a new effort should focus on commitment to restoration. A set of successful examples of restoration has enormous potential benefits for participating scientists, policy makers, and the confidence of public opinion. A less-than-desired result is equally informative to the scientist, lets the public and the courts know that some pursuits may be futile and, therefore, allows resources to be channeled toward more effective alternatives. Having accomplished some to several restoration examples, we must recognize that they will need continued attention. As Lee (1993) points out, sustainability is not an endpoint, it is a continuous process.

Summary

My goal in the preceding text may be summarized by simply acknowledging that the public, our policy makers, and the scientific community are reasonably well informed about the miscellany of problems visited upon Pacific salmon. We (scientists) do not fully understand the cause and effect relationships and we cannot ask for time to do that because the rate of change in salmon ecosystems exceeds our rate of learning about the causes of those changes. In addition, some of the key agents of change (e.g., large-scale oceanic changes and increases in human population) are not within the realm of control by current management practices. We must accept the current reality and choose to act on those things we can change.

I believe that we need to emphasize solutions and our capacity to effect them. Doing so requires a realistic recognition of the possible, commitment to sound scientific methods, a willingness to learn, and an obligation to convey results from test cases in a form that allows choices among the accomplishable alternatives. Salmon are a sustainable resource, an integrating indicator of environmental conditions, and a symbol of societal values. That makes our charge both an important scientific opportunity and an essential obligation.

Acknowledgments

I thank D. Stouder and the symposium organizing committee for inviting my participation. My familiarity with salmon issues derives largely from research in the Laurentian Great Lakes, which has been supported by the University of Wisconsin Sea Grant Institute. Discussions with S. Carpenter, J. Magnuson, and D. Schindler improved the manuscript.

Literature Cited

Beamish, R.J. and D.R. Bouillon. Pacific salmon production trends in relation to climate. Canadian Journal of Fisheries and Aquatic Sciences 50: 1002-1016.

Francis, R.C. 1996. Managing resources with incomplete information: making the best of a bad situation, p. 513-524. *In* D.J. Stouder, P.A. Bisson, and R.J. Naiman (eds.), Pacific Salmon and Their Ecosystems: Status and Future Options. Chapman and Hall, New York.

Hilborn, R.H. and J. Winton. 1993. Learning to enhance salmon production: lessons from the Salmonid Enhancement Program. Canadian Journal of Fisheries and Aquatic Sciences 50: 2043-2056.

Lee, K.N. 1993. Compass and Gyroscope. Island Press, Washington, DC.

Lee, K.N. 1996. Sustaining salmon: three principles, p. 665-675. *In* D.J. Stouder, P.A. Bisson, and R.J. Naiman (eds.), Pacific Salmon and Their Ecosystems: Status and Future Options. Chapman and Hall, New York.

Ludwig, D., R. Hilborn, and C.J. Walters. 1993. Uncertainty, resource exploitation, and conservation: lessons from history. Science 260: 547-549.

National Research Council. 1992. Restoration of Aquatic Ecosystems: Science, Technology, and Public Policy. National Academy Press, Washington, DC.

National Research Council. 1994. Improving the Management of US Marine Fisheries. National Academy Press, Washington, DC.

Nehlsen, W. 1996. Pacific salmon status and trends—a coastwide perspective, p. 41-50. *In* D.J. Stouder, P.A. Bisson, and R.J. Naiman (eds.), Pacific Salmon and Their Ecosystems: Status and Future Options. Chapman and Hall, New York.

Pearcy, W.G. 1996. Salmon production in changing ocean domains, p. 331-352. *In* D.J. Stouder, P.A. Bisson, and R.J. Naiman (eds.), Pacific Salmon and Their Ecosystems: Status and Future Options. Chapman and Hall, New York.

Shrader-Frechette, K.S. and E.D. McCoy. 1993. Method in Ecology. Cambridge University Press, Cambridge, England.

Smith, C.L. and B.S. Steel. 1996. Values in the valuing of salmon, p. 599-616. *In* D.J. Stouder, P.A. Bisson, and R.J. Naiman (eds.), Pacific Salmon and Their Ecosystems: Status and Future Options. Chapman and Hall, New York.

Walters, C.J. 1986. Adaptive Management of Renewable Resources. MacMillan, New York.

Walters, C.J. 1996. Information requirements for salmon management, p. 61-68. *In* D.J. Stouder, P.A. Bisson, and R.J. Naiman (eds.), Pacific Salmon and Their Ecosystems: Status and Future Options. Chapman and Hall, New York.

Walters, C.J. and F. Juanes. 1993. Recruitment limitations as a consequence of natural selection for use of restricted feeding habitats and predation risk taking by juvenile fishes. Canadian Journal of Fisheries and Aquatic Sciences 50: 2058-2070.

Sustaining Salmon: Three Principles

Kai N. Lee

Abstract

Three ideas, simple to state but difficult to implement, would lead to the rehabilitation of many of the troubled salmon (*Oncorhynchus* spp.) runs of North America. These principles—to cooperate in the face of conflict, act on biologically appropriate scales of space and time, and learn by deliberate experimentation—are defined and the barriers to their implementation described. Salvaging salmon populations pressed by the forces that transform their habitat and expose them to unsustainable levels of harvest is a historical rather than rational process, a process requiring institutional change in which conflict among human interests lies at the heart of decision making. Considered from this vantage, a lot of good has been done over the past generation and seems likely to continue, although probably not fast enough to save all the salmon populations at risk.

Three Principles

Three simple rules would lead to the rehabilitation of many of the dwindling salmon (*Oncorhynchus* spp.) runs of western North America:

1. Decide in light of the practical limitations of one's powers, collaborating when necessary (cooperation principle).
2. Plan and act at biologically appropriate scales of space, time, and function (bioregional principle).
3. Act and learn so as to expand practical knowledge of salmon and their ecosystems (adaptive principle).

These rules face daunting barriers, however. Not one is easy to follow, particularly at the outset. They are interdependent and are likely to work only in combination. Moreover, these principles fly in the face of the way we humans have organized our activities with regard to salmon and other valuable aspects of the natural environment, both before and during the 3 decades of reform that followed the passage of the Wilderness Act of 1964. Following these rules even calls into question lessons that environmentalists have inferred from a generation of success.

Faced with the legal authorities of the United States (US) Endangered Species Act (ESA), the political economy of the US Pacific Northwest is confronting its greatest crisis since the 1970s, when a major reorganization of public lands and resources was touched off by another federal law, the National Environmental Policy Act. Today, the forest products industry, long an important economic presence in the region, has yet to find an accommodation with the USDA Forest Service and suburban environmentalists, in the controversy known by its connection to the northern spotted owl. Electric utilities and irrigated agriculture, together with allied industries nurtured for more than half a century by the Bonneville Power Administration, USDI Bureau of Reclamation, and US Army Corps of Engineers (the Corps), are harried over Pacific salmon and face unprecedented challenges to the operations of the Columbia River and other domesticated rivers. These challenges may force economically significant changes in the cost and availability of hydropower, irrigation water, navigation, and recreation in Northwest waterways, undermining ways of life that emerged from the existing arrangements. The forces are so large and disruptive that observers are unsure who will win or lose when the dust settles, and the forces are so complex that thoughtful participants recognize that the contestants may no longer understand their own interests.

Embroiled in urgent confusion and conflict on all fronts, both the powerful and the committed pause to take stock. This is a moment in which ideas are uncommonly important, one of the rare times when nothing is as practical as a good theory. The conference on Pacific Salmon and Their Ecosystems (10-12 January 1994, Seattle, Washington) came at a propitious moment when frustration spurs analysis. Notice that both Kitchell's (1996) conclusions and the ones offered here are pragmatic rather than theoretical; as both of us acknowledge, there is no theory that is good enough. Yet. Finding one in real time—that is, in time to salvage a good portion of what is being lost—is a practical task, one for which the three rules discussed here may serve as helpful guidance.

Cooperation

The salmon problem is mired in human conflict. Over their life histories, salmon rely upon large, varied habitats that span freshwater and marine environments. Humans exploit different parts of these habitats for different purposes (e.g., logging, urban development, and irrigated agriculture). Salmon need all the parts of their habitat, including those denied or polluted by human activities. Yet there is rarely a "smoking gun"; salmon face many hazards, and our ability to sense most of them is sparse. Thus, when fish decline, blame is hard to assign and cures are hard to create. Salmon have a single set of coherent biological needs but face a divided, inconsistent, and shifting array of human interests.

Finding ways to align human interests with salmon need is an obvious goal, but progress in cooperative management is hard-won. As Ostrom's (1990) review suggests, we have as yet little clear theory about how to foster cooperation although a body of empirical generalizations is beginning to emerge (see Pinkerton 1992). In the salmon ecosystem, the harvesters have made the most progress in cooperation (Rutter 1996), forging working relationships among state and tribal managers in the fires of legal combat as Native American tribes' fishing rights were litigated and then translated into genuine cooperation. However, harvest is only one life stage, and allocation of a dwindling population frays the fragile weave of joint decision making. The international treaty that should govern the sharing of harvests among Canada, the continental US, and Alaska (where circumstances and interests differ from the rest of the nation) falters for this reason.

The frailty of cooperative arrangements does not make them less valuable, however. Salmon stocks decline in the absence of affirmative action; the human interests that drive that decline are often firmly entrenched. Working with the powers-that-be is a moral imperative when fish populations are dangerously depleted.

Bioregional

The title of this book illustrates the seductions of conventional wisdom. We all agree, nowadays, that salmon and their ecosystems form a seamless whole, that ecosystem management is what resource policy should be, that to save the salmon we must conserve their ecosystems. The idea is right: we need to design human governance along biological lines, so that human actions better match the spatial, temporal, and functional needs of salmon (Lee 1993a). In this way, humans can better recognize when their actions affect fish and can better adjust their interests to the salmon's needs. But realizing this bioregional principle presently lies beyond our social grasp. We do not know how to do it.

Consider two maps from the justly famed report of the Forest Ecosystems Management and Assessment Team (FEMAT 1993). The biogeographical range of the northern spotted owl (Fig. 1) can be organized around key watersheds and the biological provinces that delineate the habitats and species upon which the endangered owl depends. But among human institutions, the more influential organizing principle is county government (Fig. 2). Ecology follows drainages drawn by topography and flowing water. Human property lines are drawn by ruler or occasionally by riverbed; even that choice of natural feature is wrongheaded, since rivers are the centers of ecosystems, not their edges.

Humans interact with salmon at several separate life stages: (1) at birth, in habitats and hatcheries; (2) during migration, down and up waterways modified by human impacts including impoundments; and (3) in fisheries, both marine and freshwater. The wealth of research represented in the papers in this book and in the posters that festooned the conference hall obscures the reality that the human capacity to grasp the state of the salmon's ecosystems is limited, fragmentary, and usually untimely. The abundance and health of the fish themselves remain in most cases the best integrated measure of the ecosystems that salmon traverse and inhabit. Salmon are often likened to the canary in the coal mine; in the Pacific Northwest, these canaries are the coal, too—a circumstance that confuses the judgment of those who would protect and exploit salmon.

The broad problem here is basic to environmental issues: natural resources develop along characteristic biological and geophysical scales. However, in industrial market economies, humans have organized the exploitation of natural resources in ways that take little account of natural scales, and this leads to trouble (Lee 1993a). The migration of salmon puts high- and low-productivity stocks in the same place at the same time, resulting in mixed-stock fisheries. Fishing the high-productivity stocks to reap their harvestable surplus puts unsustainable pressure on low-productivity populations. Water law in the western US declares river water to be the property of the earliest historical human user, foreshortening a protean resource of an arid landscape into a single-use commodity prized for its unnatural utility in irrigation or stock watering. As a consequence, some waterways dry up in the summer, blocking salmon migration and reproduction. Both of these scale mismatches, like the multitude that can be found throughout the

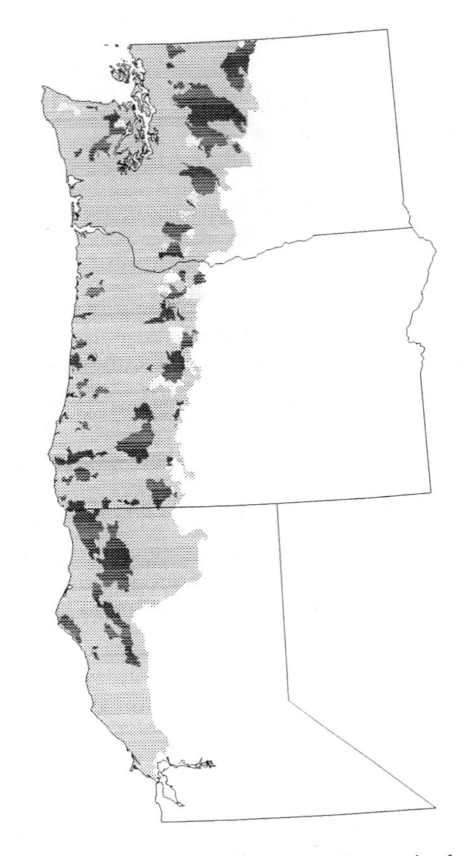

Figure 1. Key watersheds for protection and rehabilitation of salmon and bull trout in the range of the northern spotted owl. Dark shading = areas where watersheds likely contributed directly to fisheries conservation; light shading = areas where watersheds supply clean water for fish migration and rearing. Source: Forest Ecosystem Management Assessment Team (1993).

salmon's geographic range (Wilkinson and Conner 1983), foster decline and help to trigger the ESA.

Adaptive Management

Policies, practices, mandates, and organizations have also been designed around the assumption that we know how to exploit ecosystems. However, to exploit is not to conserve, but we barely recognize the implications of so basic a difference.

Ecosystems are complex, dynamic, and surprising. Human interventions in ecosystems may transform them outright, as when forest is converted to farmland. But even when we harvest valuable species, we move ecosystems away, often far away, from the paths they would have

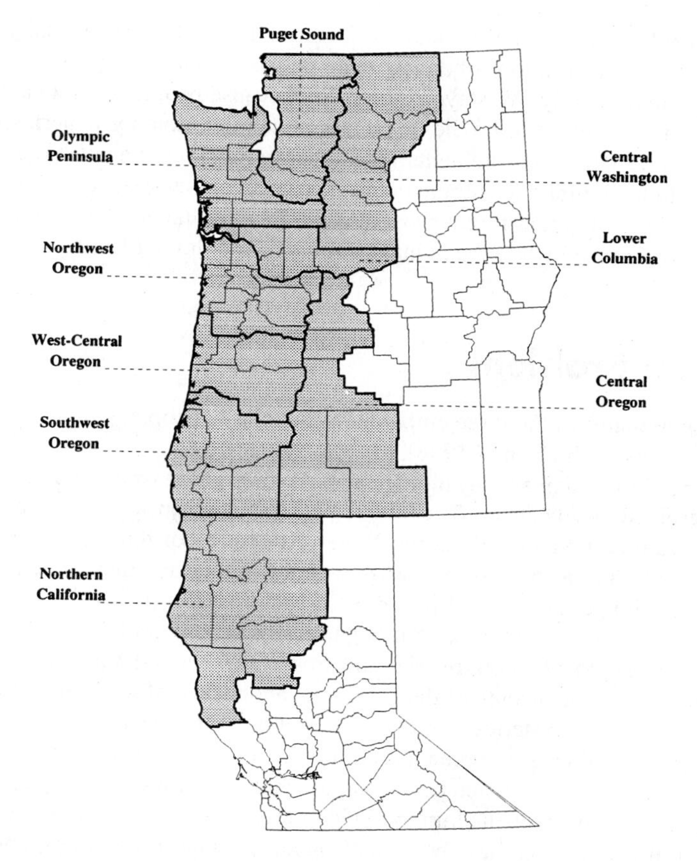

Figure 2. Counties affected by the management of federal forests in the range of the northern spotted owl. Source: Forest Ecosystem Management Assessment Team (1993).

followed. Theories aimed at understanding equilibria and small departures from equilibria are not obviously conceived in the appropriate fashion to describe exploited ecosystems. Yet notions like maximum sustainable yield, built upon theories of just this kind, have been the standard framework within which management is imagined and formulated (Ludwig et al. 1993).

An alternative approach is adaptive management: to recognize that human interventions are in effect experimental disturbances. They are probes of the behavior of dynamic, complex ecosystems (Walters 1986, Lee 1993b). Adaptive management studies these probes rigorously; it does not substitute study for action. Today, however, the experiments we run are not recognized to be tests of our limited understanding. So they have not been designed to yield reliable knowledge: they are poorly instrumented, information is collected (when it is) for short-term reasons such as economic accounting, and controls, replication, and statistical reliability are all ignored. We stay dumb.

Since the 1970s, conflict has escalated over the biologically appropriate way to reprogram the flow of the Columbia River and its major tributary, the Snake River, to benefit migrating juvenile salmon (Volkman and McConnaha 1993). After 1978, few scientifically usable measurements were done on the millions of downstream migrants each spring even though many of them were marked individually. We stayed dumb. Not because people did not care but because those who believed themselves to be the defenders of salmon, notably fisheries management agencies, feared that reliable measurements would weaken their case against the Corps, which clung tenaciously to the belief that salmon could be effectively transported by barges. We still do not know which set of humans might be right about this and under what conditions. And the salmon have mostly declined while the humans struggled and did not learn.

The Salmon Problem

The common denominator of the three embattled principles is people. People are the salmon's problem (National Research Council 1996).

Salmonids are highly adaptable by biological reckoning, able to recolonize habitats scoured by glaciers or sterilized by volcanism. The time scales, spatial reach, and biological productivity of these species were grand enough that the Native Americans of the Pacific Northwest were among the few peoples of our species to flourish without agriculture; their annual harvest swam in from the sea, worshipped but not cultivated.

Industrial-scale fishing of the kind that began in the second half of the 19th century, together with transformation of the terrestrial and freshwater ecosystems that support salmon, took a massive toll. Less than a tenth of the aboriginal abundance of wild salmon in the Columbia River remains, and the mysterious, implacable decline of hatchery production continues.

Over the past generation, environmental reforms grounded in science and law have altered in important ways the fashion in which people approach the natural world (Vig and Kraft 1994). Decision processes have been opened up to a wide range of voices; assumptions that once set agendas and determined decisions have been challenged; information that was ignored is weighed and sometimes drives outcomes.

But these changes have bypassed the three principles listed at the outset. Because control has been the goal of all the disputing parties, cooperation has been seen as a fallback or a compromise, not a necessary correlate of acting and learning at the relevant bioregional scale. Because the federal government is the major landowner in the US west, and its choices are most open to external influence, federal lands and resources have been the primary battleground even when such an approach was incommensurate with the bioregional logic of the habitats in dispute. Because science was used to challenge the once-confident assertion of exploitative use, uncertainties in science have often been ignored or used only to raise questions rather than to frame experimentation.

Ecosystem management, the newer fashion, draws environmentalists, managers, and resource users away from the battle lines of the past and moves them toward cooperation, bioregional scales, and systematic learning. Yet ecosystem-scale thinking and action pose a challenge that is less clearcut, in which past advantages may turn into liabilities or vulnerabilities. The alternative is deadlock, a frustrating and undesirable result. But deadlock leaves one with familiar enemies, at least, and the possibility of marshaling familiar allies and resources. All too often, the salmon's problem is again people.

An Institutional Perspective

The way that people become the salmon's problem is through institutional arrangements. Wilkinson (1992) coined the helpful term "lords of yesterday" to describe some of these arrangements—laws, physical infrastructure such as dams, economic interests, and habits. In general, these factors foster and reinforce an exploitative bias toward the riches of America. That bias is now outmoded and embattled but it retains a substantial grip on the way things get done.

Institutions matter. Institutions are to human societies what trees are to a forest. Trees are large, long-lived parts of an ecosystem. Trees pattern the fluxes of energy and materials over regions of time and space with sufficient stability that those patterns become microhabitats of evolutionary significance (Wilson 1993). Institutions are elements of societies that endure and channel resources to such an extent that social phenomena like crime or individual choices and life histories are affected by institutions in historically significant ways. Institutions are social, as trees are biotic; institutions are built, maintained, and contested by people. But institutions are larger than individual persons. Individuals can be influential, and in many institutional forms one or a few individuals are vested with many of the powers of an entire institution. Yet institutions modify individuals more than individuals change institutions. This is obviously the case for decentralized institutions like economic markets, but it is also a valid generalization applied to kingship or Christianity or nuclear weapons policy—to take three instances in which individual actions are pivotal, but in quite different senses. (Of course, the parallels between institutions and ecosystems are incomplete. That which is historically significant is not usually significant to evolution in a biological sense. Institutions matter even when, as is normal, they act over much shorter times—measured in years or in lifetimes—than tree species in a forest.)

The word "institution" includes more than formal organizations, such as government agencies, corporations, or interest groups. Formal organizations clearly count as tree-like entities in the ordering of human society. So do formalized patterns such as property rights, the exercise of which does not occur through any single organization (see Wilkinson 1992). So do many nonformalized patterns such as the political beliefs we associate with democracy.

As social creations, institutions often embody purposes. One consequence of this fact is that people judge institutions by the standards of rationality, testing whether an institution meets criteria of efficiency or substantive coherence. Yet as social entities, institutions exist through the agency of more than one person, usually many people. The activities of institutions are accordingly enacted by the actions and choices of many individuals, each of them pursuing particular, individual purposes. Institutions, as a result, embody both the teamwork of coordinated purposes and the cacophony of divergent ones. Institutions frequently fail tests of rationality.

Institutions remain somewhat mysterious to natural scientists with interests in public policy. Like others with only an instrumental interest in institutions, natural scientists are strongly tempted by the rationalist approximation: an institution's purposes should provide a sufficient basis for predicting its behavior. This is often wrong. Yet there is scant comfort available from social scientists, whose intellectual interests and language differ from natural scientists. Social scientists usually cannot offer transparent or helpful ways of generating more accurate predictions. This is partly due to the low power of social science as science. The theories we have are often wrong. But the larger problem is the complexity of social life, a world in which the number of parameters and boundary conditions that affect outcomes is typically larger than the number of reliable observations.

Nonetheless, managing ecosystems is an inherently institutional task because the ecosystems that matter are already exploited by human institutional arrangements. Changing the way that humans interact with ecosystems requires changing institutions. Today, institutional arrangements splinter landscapes; one cannot manage two-thirds of an ecosystem. Reassembling the human mosaic into a coherent vision for salmon is institutional work that goes well beyond science and law, which have been the primary components of environmental policy until now.

Increments of Change

Each of the three principles cited previously suggests institutional reforms. Indeed, much has already been done, often in their name:

Cooperative principle. Beginning with the public participation that was a hallmark of contemporary environmentalism, the decision processes of large organizations, both public and private, have become markedly more transparent over the past 25 years. The salmon problem has also been the setting for genuine power-sharing, as Native American treaty rights and governmental authority have become the crossbeams of a durable cooperation, particularly in harvest management.

Bioregional principle. Watersheds are increasingly recognized as logical management units. Indeed, the sudden prominence of ecosystem management as a policy idea marks the emergence of the bioregional principle as a practicable template for governmental organization.

Adaptive principle. Led by the Northwest Power Planning Council, adaptive management has made halting but influential gains. Petitions to protect salmon under the ESA disrupted the Council's attempt to organize salmon enhancement in the Columbia River Basin around bioregions called sub-basins (Volkman and McConnaha 1993). But while the ESA taketh away, it giveth as well. Rigorous studies of juvenile migrant salmon survivals in the Snake and Columbia rivers have started again after 15 years of inconclusive wrangling.

These developments give reason for hope, but signs that they are biologically effective are few as yet. Salmon population declines continue and the sense of foreboding persists.

The institutional reforms, with two exceptions, have been incremental. One exception is the reorganization of harvest by the legitimization of Native American treaty rights, the Magnuson Fishery Conservation and Management Act of 1976, and the US-Canada Pacific Salmon Treaty of 1985. The other is the commitment of a substantial fraction of the Columbia River's federal hydropower revenues to salmon protection and enhancement under the Northwest Power Act of 1980. In institutional terms, the current crisis, forced by the ESA petition process, is a battle over how far to carry the mandate of the Northwest Power Act, which provided no clear guidance on how much the river operations needed to be altered to benefit fish migration, particularly in the absence of clear evidence.

Human institutions have made major adjustments in harvest and in the terrestrial-freshwater habitat. Ocean harvest was completely closed in US waters in 1993-94. Economically impressive reapportionments of high-value river waters have been made in favor of fish, amounting to roughly $100 per adult salmon in the Columbia River's populations, excluding harvest costs. All of this has not clearly reversed the decline, and that is heart-rending and frustrating.

One view of the three principles, in this light, is that we need to be patient. We should persevere over times of biological significance: the ideas are right but they need time to work.

This notion is probably correct in a practical sense. Yet it is plainly inadequate in a setting of social crisis to call for the patient, persistent pursuit of cooperative, bioregional learning of how people and salmon can coexist. The tendency, instead, is to call for drastic change, whether by lowering the levels of reservoirs behind dams or by turning to a conservative US Congress to undermine the ESA. We do not know if a radical nostrum will work; we do know that the radical changes of the past 30 years have yet to halt the decline of salmon populations. However, talk of crisis misses a crucial point, which is how profound the changes wrought by the three principles will be as they take practical effect.

Adaptive management will produce more than knowledge. Learning can shift expectations of what is feasible. The discrediting of hatcheries, once the cure of choice, should be seen as a demonstration of how quickly humans can learn from experience in a controlled environment. We should anticipate equally dramatic shifts in expectations from an earnest attempt to manage at the ecosystem level although the shift may be toward greater pessimism rather than optimism.

A *bioregional* approach will foster changes in ideas about property and jurisdiction. Consider pollution prevention. Manufacturing industries in the US and Europe are only now modifying their production processes to eliminate or minimize the creation of pollutants, toxic wastes, worker exposures, and other targets of regulatory concern. Pollution prevention is a redefinition of property duties; what was formerly waste to be discarded or worker risk of no importance to managers is now a cost to be minimized in the pursuit of profit. Shifting decision making to watersheds is likely to affect political jurisdictions as towns or counties across rivers find they are in the same watershed. Conversely, a watershed approach should further exacerbate upstream-downstream tensions as the vulnerability of salmon and other migratory species becomes more visibly a problem not addressed by bioregionalism at this scale.

Cooperative management typically begins from hostility as parties with opposed interests find it instrumentally useful to work together in pursuit of limited objectives. Yet as state fisheries agencies and Native American tribes have found, cooperation breeds wider change by bringing issues to the fore in parties' agendas. An alliance in opposition to electric utilities and irrigators would have seemed a remote possibility when treaty-rights litigation made Native American tribes and state agencies bitter antagonists. But it has happened.

Now, as electric power becomes an increasingly important commodity in trade along western North America, the economics of Northwest river flows are changing. As net-pen rearing of salmon alters the world market for ocean-caught Pacific salmon, the economics of the salmon fishery is changing. Cooperative management is likely to channel the way attention is directed and, with it, the dynamics of conflict. Sabatier and Jenkins-Smith (1993) argue that, on a time scale of decades, the dynamics of policy evolution are controlled by "advocacy coalitions," political alliances whose members may also cooperate in managerial activities. This has happened in the Pacific Northwest and will continue to do so.

The institutional changes that grow out of incremental reforms can be large and surprising. The prominence of Native American tribes in contemporary salmon policy would have been predicted by few observers in the 1960s. Not only do the actors and means change, so also do the ends, as illustrated by the penumbra of hope that surrounds the idea of managing ecosystems, as compared with the promise, a generation ago, of cheap power and hatchery-bred fish.

In this respect, the most important aspect of the three principles in the long term may be their cultural significance for the institutions they reshape.

Coevolution

Coevolution of humans and their habitats began to break down ~10,000 years ago when agriculture and trade began in earnest (McNeill 1980). Agriculture produced surpluses in most years, fueling trade, and trade linked places that had not been connected by biogeography (Cronon 1983). For the last 350 years or so, since the European Enlightenment that gave the world both industrialization and democracy, the break between humans and their native landscapes has become an open breach. We have sought to replace with technology, markets, and deliberative social choice the adaptations of what we scornfully called "primitive" societies.

Like the wider concern for the environment, the protection of ecosystems endangered by what used to be called "progress" reminds us that coevolution may be the way of the future as well as the past. The tempo and scale of life on the planet is biological at its base. To the extent our technologies and organizations can heed and harness those rhythms, our species may prosper. But we ignore them at our peril.

As trees grow in a forest, they alter the ecosystem they shape. The push and pull of environmental politics is a social analog to second growth. Having cleared much of the old growth of traditional society in the fires of industrialism, we are groping toward some sort of coevolution with the rest of nature.

The shibboleths of environmental policy (including "sustainable development," environmental "balance," "wise use," and now "ecosystem management") all have the uneasy humor of oxymorons. The names of the activities we pursue sound either impossible or foolish. But that may be instead the accent of coevolution at the cultural level: the struggle to fashion new meanings for things we already do, putting them within a different frame of reference. Over time, that new frame of reference will change the mix of what we do on the landscape in the same way that recycling trash has gone from student enthusiasm to suburban routine.

A notion like social coevolution sounds implausibly grand when set against the fight over the spotted owl and the fears elicited by the ESA. But the Wilderness Act is only 32 years old, and from the vantage of natural resources policy, the pace of change is breathtaking.

The brooding crisis of the Pacific salmon has been more than a century in the making. The salmon problem cannot be solved in a much shorter time, in part because the life cycle of a semelparous species like salmon sets terms on how quickly population regrowth could occur. But to an important degree, the humans of the Pacific Northwest are also seeking a better understanding of what it means to sustain salmon runs—an understanding that includes economic costs, cultural change, propensities toward risk, and shifts of spiritual significance.

We do not know whether principles such as those I have called cooperative, bioregional, and adaptive will suffice to halt the loss of salmon. We can project that the pragmatic changes inspired by that continuing loss will have surprising and significant effects upon humans, who have lived with salmon since the glaciers retreated to reveal the land and waters that are our joint heritage.

Acknowledgments

I acknowledge helpful discussions with S. Beebe, R.L. Beschta, P.A. Bisson, D. Chapman, W.K. Jaeger, J. Magnuson, B. McCay, D. Policansky, C.L. Smith, and J.M. Volkman. A. Holden, USDA Forest Service, provided copies of Figures 1 and 2.

Literature Cited

Cronon, W. 1983. Changes in the Land. Hill and Wang, New York.

Forest Ecosystem Management and Assessment Team 1993. Forest ecosystem management: an ecological, economic, and social assessment. USDA Forest Service. Portland, Oregon.

Kitchell, J.F. 1996. Where do we go from here? An outsider's view, p. 657-663. *In* D.J. Stouder, P.A. Bisson, and R.J. Naiman (eds.), Pacific Salmon and Their Ecosystems: Status and Future Options. Chapman and Hall, New York.

Lee, K.N. 1993a. Greed, scale and mismatch. Ecological Applications 3: 560-564.

Lee, K.N. 1993b. Compass and Gyroscope. Island Press, Washington, DC.

Ludwig, D., R. Hilborn, C. Walters. 1993. Uncertainty, resource exploitation, and conservation: lessons from history. Science 260: 17, 36.

McNeill, W. 1980. The Human Condition. Princeton University Press, Princeton, New Jersey.

National Research Council. 1996. Upstream: Salmon and Society in the Pacific Northwest. National Academy Press, Washington, DC.

Ostrom, E. 1990. Governing the Commons. The Evolution of Institutions for Collective Action. Cambridge University Press, Cambridge, United Kingdom.

Pinkerton, E.W. 1992. Translating legal rights into management practice: overcoming barriers to the exercise of co-management. Human Organization 51: 330-341.

Rutter, L.G. 1996. Salmon fisheries in the Pacific Northwest: How are harvest management decisions made? p. 355-374. *In* D.J. Stouder, P.A. Bisson, and R.J. Naiman (eds.), Pacific Salmon and Their Ecosystems: Status and Future Options. Chapman and Hall, New York.

Sabatier, P. and H. Jenkins-Smith. 1993. Policy Change and Learning. An Advocacy Coalition Approach. Westview Press, Boulder, Colorado.

Vig, N.J. and M.E. Kraft (eds.). 1994. Environmental Policy in the 1990s, 2nd ed. CQ Press, Washington.

Volkman, J.M. and W.E. McConnaha. 1993. Through a glass darkly: Columbia River salmon, the Endangered Species Act, and adaptive management. Environmental Law 23: 1249-1272.

Walters, C. 1986. Adaptive Management of Renewable Resources. MacMillan, New York.

Wilkinson, C.F. 1992. Crossing the Next Meridian. Island Press, Washington.

Wilkinson, C.F. and D.K. Conner. 1983. The law of the Pacific salmon fishery. University of Kansas Law Review 32: 17-109.

Wilson, E.O. 1993. The Diversity of Life. Harvard University Press, Cambridge.

Index

D

G

H

I

O

P

Q

R

S

T

U